# Transfusion Medicine and Hemostasis

## Clinical and Laboratory Aspects

# Transfusion Medicine and Hemostasis
## Clinical and Laboratory Aspects

*Edited by*

Christopher D. Hillyer, MD
Beth H. Shaz, MD
James C. Zimring, MD, PhD
Thomas C. Abshire, MD

AMSTERDAM • BOSTON • HEIDELBERG • LONDON •
NEW YORK • OXFORD • PARIS • SAN DIEGO •
SAN FRANCISCO • SINGAPORE • SYDNEY • TOKYO

ELSEVIER

Elsevier
30 Corporate Drive, Suite 400, Burlington, MA 01803, USA
32 Jamestown Road, London NW1 7BY, UK
525 B Street, Suite 1900, San Diego, California 92101-4495, USA
360 Park Avenue South, New York, NY 10010-1710, USA

First edition 2009

Notice
No responsibility is assumed by the publisher for any injury and/or damage to persons
or property as a matter of products liability, negligence or otherwise, or from any use or
operation of any methods, products, instructions or ideas contained in the material herein.

Medicine is an ever-changing field. Standard safety precautions must be followed, but as new
research and clinical experience broaden our knowledge, changes in treatment and drug therapy
may become necessary or appropriate. Readers are advised to check the most current product
information provided by the manufacturer of each drug to be administered to verify the
recommended dose, the method and duration of administrations, and contraindications. It is the
responsibility of the treating physician, relying on experience and knowledge of the patient, to
determine dosages and the best treatment for each individual patient. Neither the publisher nor
the authors assume any liability for any injury and/or damage to persons or property arising
from this publication.

**Library of Congress Cataloging in Publication Data**
A catalog record for this book is available from the Library of Congress

**British Library Cataloguing in Publication Data**
A catalogue record for this book is available from the British Library

ISBN: 978-0-12-374432-6

For all information on all Elsevier publications
visit our Web site at www.elsevierdirect.com

Typeset by Macmillan Publishing Solutions
(www.macmillansolutions.com)

Printed and bound in China

# Working together to grow
## libraries in developing countries

www.elsevier.com | www.bookaid.org | www.sabre.org

ELSEVIER     BOOK AID
             International     Sabre Foundation

# Contents

## Laboratory Testing of Coagulation
### Introduction

### Screening Tests

### Laboratory Assessment of Platelet Disorders

# About the Editors

**Christopher D. Hillyer, MD,** is the Emory Distinguished Service Professor of Pathology, and a tenured professor in the Department of Pathology and Laboratory Medicine, Emory University School of Medicine. In these capacities, he serves as director of the Emory Center for Transfusion and Cellular Therapies with responsibility for all aspects of clinical and academic transfusion medicine at Emory's seven principle hospitals. Dr. Hillyer is chief editor of three textbooks in transfusion medicine, author of over 120 articles pertaining to transfusion, human immunodeficiency virus (HIV), cytokines, and herpes viruses, most notably Cytomegalovirus (CMV), as well as more than 25 book chapters. Nationally recognized as an expert in hematology blood banking, and transfusion medicine, Dr. Hillyer also is a past-president, board of directors of AABB (formerly known as the American Association of Blood Banks) and a former trustee of the National Blood Foundation (NBF). Dr. Hillyer has been awarded many million dollars in research funding from the National Institutes of Health (NIH), the Centers for Disease Control and Prevention (CDC), NBF and other agencies. He is an associate editor of the journal *Transfusion* and serves on several other editorial boards including the journal *Blood*. Dr. Hillyer was recently recognized as a Healthcare Hero for his work in Africa and is a recipient of the Tiffany Award from the American Red Cross where he also serves as a medical director. Dr. Hillyer is board certified in four specialty areas including transfusion medicine, hematology, medical oncology and internal medicine. He received his BS from Trinity College (1980), and his MD from the University of Rochester School of Medicine (1984), with postgraduate training and fellowships in hematology-oncology, transfusion medicine and bone marrow transplantation at Tufts-New England Medical Center in Boston.

**Beth H. Shaz, MD,** is an associate professor in the Department of Pathology and Laboratory Medicine, Emory University School of Medicine, and serves as director of the blood bank at Grady Memorial Hospital and co-director of clinical research at the Center for Transfusion and Cellular Therapy. Prior to moving to Emory she was the assistant medical director of the Transfusion Services and medical director of the Transfusion and Apheresis Unit at Beth Israel Deaconess Medical Center. She has written many articles and book chapters pertaining to transfusion, cellular therapy, and therapeutic apheresis. Nationally, she is chair of the Clinical Applications Committee of the American Society for Apheresis, which has recently published new and evidence-based guidelines for the use of therapeutic apheresis. In addition, she is the chair of the Transfusion Services Accreditation Program Unit and the Blood Banking/ Transfusion Medicine Fellowship Program Directors Subsection of the AABB. Dr. Shaz is board certified in transfusion medicine, clinical and anatomic pathology. She received her BS from Cornell University and her MD from the University of Michigan School of Medicine. Her postgraduate training includes a general surgery internship at Georgetown University, and residency in clinical and anatomic pathology and fellowship in transfusion medicine at Beth Israel Deaconess Medical Center, a major teaching hospital of Harvard Medical School in Boston.

**James C. Zimring, MD, PhD,** is a tenured associate professor in the Department of Pathology and Laboratory Medicine, Emory University School of Medicine, and serves as the assistant director of the special hemostasis laboratory at Emory University Hospital. Dr. Zimring oversees the laboratory diagnosis of patients with disorders in thrombosis and hemostasis. In addition, Dr. Zimring has an active NIH funded research program studying the immunology of transfusion and bone marrow transplantation. Dr. Zimring is a recent recipient of the David B. Pall award for advances in transfusion medicine, and frequently presents original research in the field of transfusion at national and international meetings. He is on the editorial board of the journal Transfusion and serves as a reviewer for additional journals in the fields of transfusion and coagulation. Dr. Zimring has authored several book chapters in the fields of transfusion, coagulation and molecular pathology. He is board certified in Clinical Pathology and received a BS in Chemistry, PhD in Immunology, and an MD all from Emory University.

**Thomas C. Abshire, MD,** is a professor in the Department of Pediatrics Emory University School of Medicine, and the director of the Comprehensive Hemostasis Program, Emory University and Children's Healthcare of Atlanta. He also serves as the Pediatric Hematology/Oncology Fellowship Program Director at Emory and Children's. Prior to joining the faculty at Emory University, Dr. Abshire completed a career in the United States Air Force with an emphasis as a clinician-educator, where he was the recipient of several medical student and housestaff teaching awards. Dr. Abshire's research interest is in evaluation of mild bleeding disorders and therapeutic intervention in hemophilia, von Willebrand disease and thrombosis. He has published over 60 peer-reviewed manuscripts, review articles or editorials and is a frequent invited speaker at national and international meetings. He also serves on the Steering Committee of the International Immune Tolerance Study and recently was a member on the CDC Uniform Data Collection Committee. He has received research funding from the NIH, CDC, and several non profit organizations and pharmaceutical companies. Dr. Abshire is a past-president of the Hemophilia and Thrombosis Research Society. Dr. Abshire is a distinguished graduate of the US Air Force Academy. He attended medical school at Tulane University, fulfilled his pediatric residency training at David Grant USAF Medical Center and completed pediatric hematology/oncology fellowship training at the University of Colorado Health Science Center.

# Contributors

**THOMAS C. ABSHIRE, MD**
Professor, Department of Pediatrics, Emory University School of Medicine
Director's Chair in Hemostasis, Hemophilia of Georgia, Inc.
Director, Comprehensive Hemostasis Program, Emory University and Children's
Healthcare of Atlanta, Atlanta, Georgia, USA

**CAROLYN M. BENNETT, MD**
Assistant Professor, Department of Pediatrics, Emory University School of Medicine
Attending Physician, Aflac Cancer Center and Blood Disorders Service, Children's
Healthcare of Atlanta, Atlanta, Georgia, USA

**MICHAEL A. BRIONES, DO**
Assistant Professor, Department of Pediatrics, Emory University School of Medicine
Attending Physician, Aflac Cancer Center and Blood Disorders Service, Children's
Healthcare of Atlanta, Atlanta, Georgia, USA

**MARY D. DARROW, MD**
Resident Physician, Department of Pathology and Laboratory Medicine, Emory
University School of Medicine, Atlanta, Georgia, USA

**AMY L. DUNN, MD**
Assistant Professor, Department of Pediatrics, Emory University School of Medicine
Attending Physician, Aflac Cancer Center and Blood Disorders Service, Children's
Healthcare of Atlanta, Atlanta, Georgia, USA

**JIMMIE L. EVANS, MT (ASCP)**
Manager, Transfusion Medicine Laboratory Services, Emory University Hospital,
Atlanta, Georgia, USA

**LAWRENCE B. FIALKOW, DO**
Fellow, Center for Transfusion Medicine and Cellular Therapies, Department of
Pathology and Laboratory Medicine, Emory University School of Medicine, Atlanta,
Georgia, USA

**COURTNEY E. GREENE, MD**
Resident Physician, Department of Pathology and Laboratory Medicine, Emory
University School of Medicine, Atlanta, Georgia, USA

**ALFRED J. GRINDON, MD**
Professor Emeritus, Department of Pathology and Laboratory Medicine, Emory
University School of Medicine, Atlanta, Georgia, USA

**ELEANOR S. HAMILTON, MT (ASCP)**
Medical Technologist, Cellular Therapies Laboratory, Emory University Hospital, Atlanta, Georgia, USA

**JEANNE E. HENDRICKSON, MD**
Assistant Professor, Department of Pediatrics, Emory University School of Medicine
Assistant Director, Blood Banks and Transfusion Services, Children's Healthcare of Atlanta, Atlanta, Georgia, USA

**CHRISTOPHER D. HILLYER, MD**
Distinguished Service Professor of Pathology
Professor, Department of Pathology and Laboratory Medicine, Emory University School of Medicine
Professor, Department of Hematology and Medical Oncology, Emory University School of Medicine
Director, Center for Transfusion and Cellular Therapies, Atlanta, Georgia, USA

**KRISTA L. HILLYER, MD**
Assistant Professor, Department of Pathology and Laboratory Medicine, Emory University School of Medicine
Chief Medical Officer, Southeast Division, American Red Cross Blood Services, Douglasville, Georgia, USA

**SHAWN M. JOBE, MD, PhD**
Assistant Professor, Department of Pediatrics, Emory University School of Medicine
Attending Physician, Aflac Cancer Center and Blood Disorders Service, Children's Healthcare of Atlanta, Atlanta, Georgia, USA

**CASSSANDRA D. JOSEPHSON, MD**
Associate Professor, Department of Pathology and Laboratory Medicine, Emory University School of Medicine
Associate Director, Blood Banks and Transfusion Services, Children's Healthcare of Atlanta, Atlanta, Georgia, USA

**CHRISTINE L. KEMPTON, MD**
Assistant Professor, Department of Pediatrics
Assistant Professor, Department of Hematology and Medical Oncology, Emory University School of Medicine
Attending Physician, Aflac Cancer Center and Blood Disorders Service, Children's Healthcare of Atlanta, Atlanta, Georgia, USA

**SHANNON L. MEEKS, MD**
Instructor, Department of Pediatrics, Emory University School of Medicine
Attending Physician, Aflac Cancer Center and Blood Disorders Service, Children's Healthcare of Atlanta, Atlanta, Georgia, USA

**CONNIE H. MILLER, PhD**
Director, Clinical Hemostasis Laboratory,
Division of Blood Disorders, National Center on Birth Defects and Developmental Disabilities, Centers for Disease Control and Prevention, Atlanta, Georgia, USA

**JOHN D. ROBACK, MD, PhD**
Associate Professor, Department of Pathology and Laboratory Medicine, Emory
University School of Medicine
Co-Director, Center for Transfusion and Cellular Therapies, Atlanta, Georgia, USA

**BETH H. SHAZ, MD**
Associate Professor, Department of Pathology and Laboratory Medicine, Emory
University School of Medicine
Director, Blood Bank, Grady Memorial Hospital, Atlanta, Georgia, USA

**CHELSEA A. SHEPPARD, MD**
Assistant Professor, Department of Pathology and Laboratory Medicine, Emory
University School of Medicine
Associate Medical Director, Southern Region, American Red Cross, Blood Services,
Douglasville, Georgia, USA

**ANNIE M. WINKLER, MD**
Resident Physician, Department of Pathology and Laboratory Medicine, Emory
University School of Medicine, Atlanta, Georgia, USA

**JAMES C. ZIMRING, MD, PhD**
Associate Professor, Department of Pathology and Laboratory Medicine, Emory
University School of Medicine
Assistant Director, Special Hemostasis Laboratory, Emory University Hospital,
Atlanta, Georgia, USA

# Preface

The editors present this inaugural edition of *Transfusion Medicine and Hemostasis – Clinical and Laboratory Aspects* with great excitement and anticipation.

As the title indicates, we view, and have done so for some time, that both transfusion medicine and hemostasis, particularly laboratory methods and diagnostic tests employed in routine and specialized coagulation laboratories, are linked. More specifically, coagulation tests, and indeed other laboratory tests and results, are often used to guide and monitor transfusion, blood component and coagulation factor therapy. Also, transfusion strategies are used to treat patients with disorders of hemostasis which have been previously diagnosed using specialized and vital hemostasis and coagulation laboratory tests and methods. In addition, pathology, pediatric and internal medicine residents, and transfusion medicine and hematology fellows, often receive training (to varying extents) in blood banking and transfusion medicine, and in disorders and diseases of congenital and acquired coagulation defects, though a combined and appropriately focused book does not appear to exist. Moreover, books that focus on critical and technical elements of laboratory testing methodologies, in the context of transfusion and hemostasis, are few – if any. Indeed, in our institution both transfusion medicine and the coagulation/hemostasis laboratories are housed in the same clinical and administrative place – the Emory Center for Transfusion and Cellular Therapies – and joint consultation services exist. It is thus that we envisioned a manual-style book combining transfusion medicine and hemostasis/coagulation medicine. Indeed, the book before you is divided into these two main parts, as the title obviously suggests.

Not surprisingly, others have noted the interdigitation of these two fields – such as those who conceived the Transfusion Medicine/Hemostasis Clinical Trials Network, now funded by the National Institutes of Health. Also, board certification in blood banking and transfusion medicine is currently accomplished via the American Board of Pathology, and the examination has a significant emphasis on clinical hemostasis and laboratory aspects of coagulation as well as blood banking and transfusion medicine. Interestingly, board eligibility in transfusion medicine, following an accredited transfusion medicine fellowship, can be achieved directly via internal medicine, pediatrics and anesthesia residencies, in addition to pathology and adult and pediatric hematology. Thus, it is our intention that this offering be of value to trainees in many specialties, as well as fellows in transfusion medicine and hematology. Moreover, as transfusion is the most common therapeutic modality in use in the US, it is anticipated that this book will provide valuable information for practitioners in a variety of fields, including anesthesiology, surgery and obstetrics.

The format of this book is intentional, with shorter, concise chapters being the norm. In each chapter, the intent is to have clear paragraphs, bullet points, and tables and figures, with a hierarchy of headings so that presentation of key points will be most efficient. Where possible and appropriate, a common hierarchical format has been employed. Instead of an exhaustive list of references, the editors have chosen to offer a small number of key articles from which the reader may gain additional information. Even without the usual citation-based technical writing style typically employed in books, we have tried to incorporate evidence-based and data-driven prose, and have referred to specific authors, studies, and publications in a way that will allow the reader to find a specific reference even when it is not cited at the end of a chapter. We fervently hope this style will be efficient, informative and enjoyable.

As stated above, the book is divided into two main parts – transfusion medicine and hemostasis. That on transfusion medicine is further subdivided into blood banking (that is, blood collection, processing, testing and storage) and transfusion medicine (that is, components for transfusion, pretransfusion immunohematology testing, product modifications, approaches to transfusion therapy in specific clinical settings, and transfusion reactions and complications). Also included are apheresis, cellular therapy, and an introduction to tissue banking in the hospital setting. These chapters and sub-parts follow introductory chapters with common elements defining the field of blood banking and transfusion medicine, a brief history, and quality and regulatory principles. We have also included chapters focused on the role of the physician in both the blood collection facility and the transfusion service, as well as descriptions of careers in the field.

In the second main part of the book, dedicated to hemostasis, the chapters are further divided into sub-parts – clinical coagulation, laboratory-based coagulation testing, and coagulation factor products. While these three sub-parts could have been compressed into a disease-based chapter system, we felt that focus on the clinical issues, the laboratory and testing methods and details to diagnose and guide and monitor therapy, and finally the products available to treat the conditions, would be helpful and ease assimilation of the material. We hope you will agree.

Finally, in many places throughout the book, we have sections or paragraphs dedicated to international considerations. These are not exhaustive and were not meant to be, especially as regulatory requirements and practice variations vary widely throughout the developed and the developing, world. Nonetheless, the editors felt it was important to broaden the horizon of this book to have somewhat of a "world view." Again, we hope you will agree.

In closing, we are deeply dedicated to the fields of transfusion medicine and hemostasis, the ambition of striving towards optimal diagnosis, laboratory testing and therapies for patients worldwide, and the education of trainees, practitioners and future leaders in these inter-related disciplines. We hope this book will fill a void, achieve a new vision, and perhaps be seen as unique and inspired in its presentation. We ask sincerely for feedback, general or specific criticism, and suggestions as to how to improve this book, and thank you for the confidence and trust you have placed in the authors and editors.

*Christopher D. Hillyer, MD*
*Beth H. Shaz, MD*
*Thomas C. Abshire, MD*
*James C. Zimring, MD, PhD*

# Acknowledgments

We, the editors, would like to acknowledge the outstanding technical and professional support of Sue Rollins, and the expertise and guidance of Megan Wickline and Mara Conner and other team members at Elsevier. Each of these individuals played an instrumental role in the creation of this textbook, and we sincerely thank them. We would like to thank our friends and families for their unconditional love and support, without which this project could not have come to fruition. We thank especially Krista, Whitney, Peter, Margot, Jackson and James Hillyer; David, Samara, Jacob and Andrew Shaz; Kim Jallow and Alexandra Zimring; and Diane, Jonathan, Jennifer, Erin and Matthew Abshire.

# PART I

## Blood Banking and Transfusion Medicine

# CHAPTER 1

# Blood Banking and Transfusion Medicine – the Field, the Discipline and the Industry

Christopher D. Hillyer, MD

A safe, reliable, and available blood supply is critical to the function of complex healthcare systems worldwide. Over the past more than 100 years, blood transfusion has grown from the transfusion of small amounts of fresh whole blood, to one of the most common therapeutic medical practices (a brief history of blood transfusion can be found in Chapter 2). Indeed, since approximately 1980, blood banking and transfusion medicine has been a subspecialty in its own right, credentialed by the American Board of Pathology.

**Blood Banking and Transfusion Medicine as a Discipline:** Over the past more than 25 years, blood banking and transfusion medicine has expanded to include a number of related laboratory disciplines, services and therapeutics, depending in part on the degree of complexity of the medical institution. These may include therapeutic apheresis, coagulation or specialized laboratories (hence the inclusion of clinical and laboratory coagulation in this text), hospital tissue banking, and cellular therapy – which itself may include the collection, processing, storage and distribution of human hematopoietic progenitor cells (HPC), pancreatic islet cells, and related minimally and highly manipulated cells. Some facilities classically designated as blood banks also offer HLA typing, ABO and other histocompatibility antigen/antibody and cross-matching for bone marrow, peripheral blood, and umbilical cord HPC and solid organ transplantation. While credentialed as a single entity, specifically *blood banking and transfusion medicine*, the terms have increasingly come to have different meanings.

**Blood Banking Defined:** *Blood banking* now typically refers to the collection, processing, storage and distribution of whole blood and apheresis-derived blood and blood components at a blood collection facility or blood center, defined by the Food and Drug Administration (FDA) during registration or licensure as a community blood bank, although a small percentage of units are collected in the hospital setting, defined by the FDA as a hospital blood bank. Chapters 4–16 are dedicated to "blood banking."

**Transfusion Medicine Defined:** *Transfusion medicine* most often connotes pretransfusion and compatibility testing; post-manufacture processing, including irradiation, washing and volume reduction; and administration of appropriate products to the appropriate patients. Transfusion medicine thus occurs predominately in hospitals usually under the FDA designation at registration as a hospital transfusion service. The AABB (formerly the American Association of Blood Banks) *Standards* define these activates as simply a "transfusion service," since a number of blood centers will also perform these functions or offer these services. The hospital transfusion service is also typically responsible for consultation to clinicians regarding complex transfusion and coagulation issues, and the choice of specialized products including recombinant and human-derived coagulation-related concentrates, intravenous gammaglobulin and albumin. Chapters 17–77 are dedicated to "transfusion medicine."

**The Blood Pipeline:** The provision of blood is best seen as a continuum or pipeline "from vein to vein" – that is, from the vein of the donor to the vein of the recipient. From a safe practices standpoint and from the view of federal regulators, including the FDA, blood banking is not just "from vein to vein" but rather "from vein to vein and back." This allows for tracking of products transfused to a particular recipient "back" to a given donor, and thus determination of infectious disease transmission and other donor-specific complications such as transfusion-related lung injury (TRALI). The blood pipeline is shown diagrammatically in Figure 1.1.

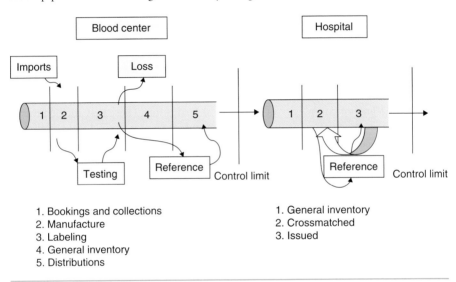

FIGURE 1.1  The blood pipeline.

It is valuable to consider that while blood centers typically provide blood and blood components to hospitals, and while hospital transfusion services typically provide blood components to medical and surgical services throughout the clinics, wards and operating theaters within the hospital confines, the ultimate destination of the products is the patient – and thus it is the patient as the recipient of blood that is ultimately served by the blood industry.

**Additional Services:** Both blood banks and transfusion services most commonly offer therapeutic (in contradistinction to donor) apheresis services, though in some locations dialysis services will provide this therapeutic modality. Preoperative autologous whole blood collection, intraoperative cell salvage and acute normovolemic hemodilution can also be performed under the auspices of the hospital transfusion service, though blood centers and many anesthesiology and surgery-run programs are in operation throughout the US.

**Structure of Blood Banks and Transfusion Services:** Worldwide, the predominate structure of blood banks and transfusion services is as a national system, often with organizational reporting to the Ministry of Health, likely via a national countrywide medical system. One example is the UK National Blood Service. In other countries the national blood service predominates but is not exclusive, and some fragmentation or sharing of responsibilities exists – such as within the South African National Blood Service and the National Blood and Transfusion Service of Tanzania. National blood and transfusion services have traditionally been well positioned to formulate nationally adopted clinical guidelines for transfusion and establish hemovigilance programs.

**Structure of Blood Banks and Transfusion Services within the US:** A national blood and transfusion service does not exist in the US, nor do any blood centers or hospital transfusion services (exclusive of the US Department of Defense) report organizational or legally to the federal government. Instead, a nationwide network of blood centers has evolved including some blood systems, the largest of which are the American Red Cross Biomedical Services (ARC; Washington, DC) and Blood Systems, Inc. (BSI; Tucson, AZ). In addition to the ARC and BSI, many other blood centers exist as community blood centers and are highly regarded within their locale. These blood establishments account for approximately 95% of the blood collected and distributed in the US, and are often members (in a variety of different configurations) of several trade organizations, including America's Blood Centers (ABC), Blood Centers of America (BCA) and the Council of Community Blood Centers (CCBC). Internationally, the US blood establishments interrelate through organizations such as the Alliance of Blood Operators (ABO) and the European Blood Alliance. About 5% of blood collected in the US is collected in hospitals, and these too must be registered with the FDA as hospital blood banks.

In the US, the formulation of clinical guidelines for transfusion has in many cases been accomplished by medical services that use large amounts of blood products, such as the American Society of Clinical Oncology regarding platelet transfusion guidelines, and the Practice Guidelines for blood component therapy by the American Society of Anesthesiologists – though recently the AABB (see below) has begun a broad-reaching initiative in earnest. Furthermore, in 2006 the AABB and a number of public and private organizations began the development of a hemovigilance system with broadened hegemony to include tissues, cells and organs, termed *biovigilance*.

*Centralized Transfusion Services:* In some locations, perhaps most notably Pittsburg and Seattle, a single corporate entity staffs and serves as both the blood collection facility and the hospital transfusion service. This model, referred to as the centralized transfusion service, has a number of real or potential advantages, including centralized tracking of blood components and physician consultation, and the ability rapidly and easily to move blood components in order to service areas of increased need or shortage.

**The Blood Industry:** The collective term *the blood industry* includes manufacturers (also called vendors) of reagents, appliances and devices used by blood establishments, as well as the blood establishments themselves, which include blood banks (centers) and hospital transfusion services. The vendors have also established an organization called the Advanced Medical Technology Association (AdvaMed). Indeed, AdvaMed has taken a leadership role in a number of areas, including defining appropriate corporate–customer relationships to be in compliance with federal trade, financial and tax regulations, and advancing the understanding of the challenges facing the industry as regards government and third-party insurer reimbursement for blood and blood components.

**Blood Banking and Transfusion Medicine as a Medical Specialty:** Blood banking and transfusion medicine is presently credentialed in the US by the American Board of Pathology, and as such represents a subspecialty of clinical pathology. In order to be eligible for board certification, a 1-year ACGME-accredited transfusion medicine fellowship is required. One logical pathway to fellowship in transfusion medicine is via a residency in clinical pathology or a combined anatomic pathology/clinical pathology program. However, a broad range of training can allow a physician to be eligible for transfusion medicine fellowship, including internal medicine, pediatrics and anesthesiology, and adult and pediatric hematology. More information can be obtained at www.abpath.org. Most transfusion medicine fellowships reside in university and related medical centers, though some blood centers have particularly excellent programs.

**Subspecialties within Blood Banking and Transfusion Medicine:** There are no specifically defined subspecialties within the field of blood banking and transfusion medicine, though board-certified transfusion medicine specialists will often become focused in either blood center or hospital transfusion service operations. Recent recognition of the specialized needs, processes and technologies required for optimal transfusion of the pediatric-aged patient has led some to ask whether there should be specialized centers of excellence and/or training in pediatric transfusion medicine. Additionally, a number of individuals have concentrated their clinical and academic efforts in a variety of more specialized areas, including the transfusion of the patient with sickle cell anemia, and in ensuring that non-transfusion alternatives are used appropriately, safely and effectively, the latter under the terms blood conservation or blood management.

**The Role of the Physician in Blood Centers and Hospital Transfusion Services:** The Code of Federal Regulations (CFR) and AABB *Standards for Blood Banks and Transfusion Services* both designate that blood collection and other functions must be under the supervision of a licensed physician, and that the physician must be readily available to the blood establishment. Chapters 3 and 18 focus specifically on the role of the physician in blood centers and hospital transfusion services, respectively.

**Oversight and Regulation of the Blood Industry:**

**The Federal Food and Drug Administration (FDA):** Regulation is the government's oversight and control of a non-governmental entity's operations and practices through the enactment and enforcement of laws called regulations. Compliance with these regulations is not voluntary, and the potential consequences for failure to comply with regulations include civil and criminal penalties.

In the US, the authority for federal regulatory oversight and enforcement is delegated by law to the FDA, an organizational arm of the Department of Health and Human Services (HHS) often referred to as the Agency. The FDA has regulatory authority over drugs, biologics and devices as mandated by the Federal Food, Drug and Cosmetic (FDC) Act of 1938 and the Public Health Services (PHS) Act of 1944. Additional information can be found at www.fda.gov/cber/blood.htm.

In particular, blood establishments in the US are therefore: (1) regulated by the FDA, (2) required to follow federal regulations provided by the FDA in the composite form of the CFR, and (3) inspected at regular intervals by agents of the FDA. In this context, blood establishments that ship products across state lines must be *licensed* by the FDA, while blood banks and hospital transfusion services that operate within state boundaries must be *registered* with the FDA. Hospital transfusion services that do not perform any manufacturing processes as defined by the FDA are not required to be licensed or registered, though they must be in compliance with applicable portions of the CFR and state regulations.

**Other Applicable Regulatory Agencies:** Additional oversight in specific circumstances is under the authority of the Nuclear Regulatory Commission (NRC), and for clinical laboratory testing elements is under the jurisdiction of the Centers of Medicare and Medicaid Services (CMS). Also, federal regulations from the Department of Transportation (DOT) and the Occupational Safety and Health Administration (OSHA) have bearing on important areas of blood center and hospital transfusion service operations.

**Current Good Manufacturing Practices:** As above, the regulations of the FDA are contained in the CFR. With respect to blood establishments, the CFR focuses primarily on current good manufacturing practices, also known as cGMP. In general, cGMP requirements are set forth to ensure the safety, quality, identity, potency and purity of drugs and biologics. These are collectively referred to by the acronym *SQUIPP*. In the broadest sense, the CFR, cGMP and SQUIPP are intended to provide the highest quality of blood collection practices for the protection of the blood donor, and of blood components for the protection of the transfusion recipient. In addition to blood establishments, manufacturers of reagents, devices and computer software are regulated by the FDA, often through licensing and/or approval processes required for these items. They must also follow cGMP.

**Oversight by Non-governmental Organizations:** In advanced systems worldwide, non-governmental organizations (NGO) serve a vital role in ensuring optimal function of blood establishments and, by extension, optimal blood donor care and blood component quality. Participation in these programs is, on a strict basis, voluntary, though the NGOs listed below have become accepted as *de facto* community standards.

**AABB:** In the US and worldwide, the AABB has served in this capacity via the formulation of its *Standards* and through widespread acceptance of its accreditation programs. The AABB, a professional organization, also serves a number of other key functions for the blood industry, including education, supporting science and research and its dissemination, and public advocacy. Indeed, the National Blood Foundation

(NBF), a subsidiary of AABB, was created with the sole mission of providing seed funding and financial support for scientific inquiry related to transfusion medicine and biology and cellular therapies. More information can be found at www.aabb.org.

**The Joint Commission:** The Joint Commission (TJC; formerly the JCAHO), an NGO, also provides oversight as it has a defined set of standards for hospitals, including hospital laboratories and hospital blood banks and transfusion services. More information can be found at www.jointcommission.org.

**College of American Pathologists:** The College of American Pathologists (CAP) also participates in laboratory accreditation and proficiency testing. The CAP program uses a system of peer review, and participating institutions must provide inspectors for future accreditation visits. More information can be found at www.CAP.org.

## Recommended Reading

AuBuchon JP, Whitaker BI. (2007). America finds hemovigilance! *Transfusion* **47**, 1937–1942.

Hillyer CD, Mondoro TH, Josephson CJ *et al.* (2009). Pediatric Transfusion Medicine (PTM): development of a critical mass. *Transfusion* **49**, 596–601.

MacPherson, J (ed.). (2007). The role of blood centers in transfusion recipient care. Second Joint Conference of America's Blood Centers and the European Blood Alliance, 13–14 November 2006, London, England. *Transfusion* **47**, S101–S204.

McCullough J, Benson K, Roseff S *et al.* (2008). Career activities of physicians taking the subspecialty board examination in blood banking/transfusion medicine. *Transfusion* **48**, 762–767.

# CHAPTER 2
# Brief History of Blood Transfusion
Alfred J. Grindon, MD

The history of blood transfusion is part of the fabric of the history of humankind: it has included religion and superstition as well as science, and has ranged from circulating humors to modern medicine. Few other substances cause the same emotions, have the same associations, lead to the same fears, or have found as many ways into our common parlance. Indeed, blood transfusion and blood letting (now called therapeutic phlebotomy and apheresis) are some of the oldest and most common medical practices. In the US, blood transfusion is the most common discharge code in surveyed hospitals. It is likely that one out of every three Americans will require transfusion of a blood product at some time in his or her life.

**Early Transfusion:** The record of man's attempt to treat suffering and disease by blood transfusion extends back at least to 1667, when Jean Denis published in the *Philosophical Transactions* his experience in Paris with transfusing blood from a lamb (because of its presumed soothing qualities) to an agitated man. Therein, Denis also recorded the first hemolytic transfusion reaction in, and subsequent death of, a patient and transfusion recipient. In 1818, James Blundell was the first successfully to transfuse human blood into a patient with post-partum hemorrhage. Blundell recognized that he was replacing lost blood volume, not providing a "vital force". Advances in a number of major areas have allowed the development of modern blood banking and transfusion medicine. These are described below.

**Blood Groups:** In 1900, Karl Landsteiner published the first of a series of papers demonstrating the presence of the ABO blood group system, stating that "the serum of healthy men will agglutinate not only the red cells of animals, but also often those of other individual humans." Although the use of this information to improve the safety of blood transfusion began within a few years, it was not until about 1920 that ABO testing was regularly used.

The Rh blood group system was discovered in 1939–1940 by Landsteiner, Weiner, Levine and Stetson, explaining the cause of many unexpected transfusion reactions. In 1945, Coombs, Mourant and Race described the use of anti-human globulin sera to detect IgG antibodies in compatibility testing (unaware that Moreschi had described the use of such sera in 1908), thus providing the still-used Coombs test.

**Blood Storage:** Direct transfusion (donor artery anastamosed to recipient vein) was performed by Alexis Carrel in 1908, and direct transfusion using a three-way stopcock was used until World War II. While sodium citrate as an anticoagulant was considered as early as 1914 and was employed (with glucose, by Rous and Turner) on a small scale during World War I by setting up blood depots before a battle, the blood could be stored for only a few days. In 1943, Loutit and Mollison developed acid citrate dextrose (ACD) solution, allowing storage of blood for weeks instead of days, and thus facilitating the "banking" of blood. In addition, the acidification of the anticoagulant-preservative solution allowed it to be autoclaved, and reduced the occurrence of bacterial contamination in storage solutions.

**Blood Derivatives:** The cold ethanol fractionation process (*Cohn fractionation*), allowing plasma to be broken down into albumin, gamma globulin and fibrinogen, among other constituents, was developed by Edwin Cohn in 1940. This allowed the use of albumin as a volume expander, and of the fibrinogen fraction (containing Factor VIII) to treat hemophilia A. Fibrinogen, as a therapeutic, fell into disuse because of the risk of hepatitis B, and treatment of hemophilia remained limited to fresh frozen plasma. When Pool and Shannon recognized in 1961 that the precipitate that formed when plasma was thawed in the cold (cryoprecipitate) contained Factor VIII, they revolutionized the treatment of hemophilia A. In 1985, dry-heated, lyophilized Factor VIII and IX concentrates became available. Genetically engineered (recombinant) Factor VIII became available in 1993, and Factor IX in 1998. Most recently, factor products are engineered without any human components, including Factor VIIa.

**Blood Component and Derivative Therapy:** In 1950, Walter and Murphy introduced plastic bags as a replacement for glass bottles for the collection and storage of blood. This allowed the development of component therapy, with the use of refrigerated centrifuges to separate components by density, and pre-collection attached satellite bags to store the prepared components. Concentrated blood platelets, prepared from whole blood, were recognized as useful for the treatment of thrombocytopenia by 1961, and platelets for transfusion were collected by apheresis by 1972.

In 1967, a concentrated Rh immune globulin was introduced commercially, beginning the gradual reduction and ultimately the near-elimination of Rh hemolytic disease of the fetus and newborn.

**Apheresis:** In the 1950s, Cohn designed a centrifuge to separate cellular components from plasma. As a result of work from many fields the advanced instrumentation for apheresis (both therapeutic and donor) became available. The development of donor apheresis allowed collection of therapeutic doses of platelet and granulocyte components from a single donor, and the collection of sufficient volumes of plasma for further manufacturing into factor concentrates, albumin, immunoglobulin and other derivatives. Full automation of therapeutic apheresis devices has expanded and simplified the use of this modality, which is vital to the treatment of many diseases (e.g. thrombotic thrombocytopenic purpura, sickle cell disease). Starting in approximately 2000, automated apheresis machine-based collection of red blood cells as double red cell products, or as a combination red blood cell unit with platelet or concurrent plasma components, began in earnest and continues to grow today.

**Adverse Effects of Transfusion:** In the 1960s blood banks became increasingly aware that paid donors were associated with higher rates of hepatitis transmission, and by 1970 the slow and difficult transition to an all-volunteer blood supply had begun. In 1971, commercial testing for hepatitis B surface antigen began, and further reduced the rate of post-transfusion hepatitis. A decade later, in 1985, and only slightly more than 2 years after transfusion-transmitted AIDS was described, a test for HIV antigen was introduced. By 1990, testing for hepatitis C had become routine. By 2000, most blood collection facilities in the developed world had adopted testing for viral nucleic acids and further reduced the residual risk of transfusion-transmitted HIV and HCV to less than 1 : 1,000,000 screened units. Since that time there has been growing recognition of the transmissibility of infectious agents, such as prions (the agent of new variant Creutzfeldt-Jakob disease), West Nile virus and *Trypanosoma cruzi* (the agent of Chagas' disease), and approaches to protect the safety of the blood supply from each of these have been implemented. In some cases, new tests have been developed; in others, the blood or components have been treated to minimize the risk of infection from a given agent using pathogen reduction or inactivation technologies, though none of these technologies is currently approved for use in the US. In still other disease conditions, those donors at risk for transmission of a given agent have been deferred.

Over the past 50 years, non-infectious complications of transfusion have also become apparent, such as febrile non-hemolytic transfusion reactions and transfusion-associated graft versus host disease (TA-GVHD). In 1970 Graw and colleagues used irradiation to prevent TA-GVHD, and in 1962 Greenwalt and colleagues demonstrated that leukocyte reduction filters prevented febrile reactions. Widespread use of leukoreduction started in the late 1980s, and became a customary blood center pre-storage processing step in the 1990s in most of the developed world.

**Current Transfusion Medicine and Blood Banking:** The field of transfusion medicine and blood banking continues to evolve as new infectious agents emerge and as there is increased recognition of non-infectious hazards of transfusion. Increased risk associated with blood transfusion in certain patient populations, such as those critically ill or undergoing cardiac surgery, has brought a new desire to understand better the beneficial and adverse effects of blood transfusion and of the blood storage lesion, with the continued goal to improve transfusion therapy.

Patient safety has expanded in hospitals, where patient identification and appropriate and directed therapy are critical. New technologies to prevent ABO incompatible transfusions have emerged. In addition, RBCs are being engineered in research environments with removal of the group A and group B antigens to form universal group O RBC products to minimize the risk of ABO incompatible transfusions.

Work continues on the development of blood substitutes, to eliminate the risks of human transfusion completely; however, to date these products have not been approved for human use, since they carry additional unique risks and in clinical trials have shown increased risk of mortality.

Blood transfusion is a critical element of medical and surgical therapies. The field will continue to expand, and there will be fine-tuning of the collection, processing, dispensing, and transfusing of blood components; automation of blood component preparation and processing; and automation of immunohematology and related testing, including donor and recipient genotyping and cellular therapies and regenerative medicine.

## Recommended Reading

*Blood: An Epic History of Medicine and Commerce in Red Gold*, available at www.pbs.org/wnet/redgold (accessed 26 June 2008).

Kilduff RA, DeBakey M. (1942). *The Blood Bank and the Technique and Therapeutics of Transfusion*. Mosby, St. Louis, MO.

Starr D. (2002). *Blood: An Epic History of Medicine and Commerce*. New York, NY: Alfred A Knopf.

# CHAPTER 3

# Introduction to Quality Systems and Quality Management

Jimmie L. Evans, MT (ASCP) and Christopher D. Hillyer, MD

As described in Chapter 1, blood banking is highly regulated, with government agencies writing the governing laws or regulations and performing inspections, and non-governmental organizations writing standards and providing accreditation. Quality management systems form a process that guides and tracks all elements of the facility's operations.

In the blood center, quality management helps to ensure: (1) that products have the highest level of safety, quality, identity, potency, and purity (SQUIPP); (2) that manufacturing is always in compliance with current Good Manufacturing Practices (cGMP) and always under review; and (3) that variation in both process and products is minimized.

In the hospital transfusion service, quality systems provide a process and procedure structure for optimal pre-transfusion testing, post-manufacture processing, crossmatching, issuing and administration of products, and validation, preventative maintenance and quality control (QC) of technologies, to name a few of the elements incorporated into the quality plan. Quality management systems are always ongoing, and should be constructed in a way that allows a facility to be in compliance with both regulations and standards.

**Compliance versus Quality:** In this setting, the term *compliance* does not equal the term *quality*. Compliance can be considered as meeting the minimal acceptable performance that should satisfy regulations. Quality can, and should, be at a higher level that should be set by the facility to ensure *optimal* performance, and cyclic and ongoing evaluation, review, and improvement. While most blood establishments are (or have been deemed by inspection to be) in compliance, the pursuit of the highest level of quality, especially in the most complex of blood centers and transfusion services, is paramount and difficult to achieve. It requires commitment, support and agreement at all levels of employees regarding the principles and practice of quality.

**Background:** In the late 1980s, Food and Drug Administration (FDA) investigators were using an inspection checklist to inspect blood establishments and thus determine compliance with FDA regulations. In addition, senior leadership within the FDA noted a significant difference between the level of quality performance pharmaceutical manufacturers were being judged against and the level at which blood manufacturers were being adjudicated as in compliance. At a presentation in 1995 the Commissioner of FDA reported that, in a 12-month period starting in 1993, the agency had issued more than 60 warning letters, more than 5 letters of intent to revoke license, approximately 5 actual suspensions of license to practice, 3 injunctions, and near to 500 recalls to blood centers and hospital transfusion services. This led to the formulation and publication of the 1995 Quality Assurance in Blood Establishments Act, which required that a blood facility's quality program must include processes to prevent, detect and correct deficiencies that could compromise blood product quality. At that time, the AABB was in the process of shifting its accreditation program from a checklist to a Quality System Essentials and Quality Program based approach. Europe was also moving to further define ISO-based system(s). Thus, there was, in essence, a worldwide movement towards requiring the blood industry to have both an increased focus on quality and an increase in quality itself. In addition to the FDA and AABB, the Centers of Medicare and Medicaid Services (CCMS), Department of Health and Human Services (HHS), College of American Pathologists (CAP), and The Joint Commission (TJC) all have verbiage relating to quality, quality control and quality assurance. Finally, laboratories must be in compliance with the Clinical Laboratory Improvement Amendments of 1988 (CLIA '88), which also contain quality requirements.

**Quality System Essentials:** Ten quality system essentials (QSEs) were and are used by the AABB in laying out the approach to building a functional quality plan. These QSEs are based on the 20 clauses of ISO 9000 and, to some extent, on cGMP. The QSEs are listed in Table 3.1 and described below.

| TABLE 3.1 Quality System Essentials | |
|---|---|
| 1. Organization | 6. Documents and records |
| 2. Human Resources | 7. Non-conformances and deviations |
| 3. Equipment | 8. Assessments: internal and external |
| 4. Supplier and customer issues | 9. Process improvement through corrective and preventive action |
| 5. Process control | 10. Facilities and safety |

Under each QSE, three or more critical control points (CCPs) are to be identified that, if not performed correctly, could adversely affect SQUIPP. Below each CCP are key elements (KEs) which identify key steps in particular processes that require written standard operating procedures (SOPs). Training employees (followed by documentation of training and competency testing) in the SOPs reduces errors and person-to-person variation, thus improving quality. One key element of a quality system or program is also the identification by the facility of facility-specific key performance indicators (KPIs).

*Organization*: Organization entails a table of organization and a reporting structure designating who is in charge of the facility and in charge of quality, followed by reporting relationships of employees. This clearly delineates the facility's structure, and how inspection reports, corrective actions and other determinations are communicated. In this QSE, the quality goals are presented.

*Human Resources*: Here, a summary of each position, including job description, orientation, documentation of training, performance standards, competency assessments and other related documents, is provided and/or described. Training plans can also be included.

*Equipment*: This QSE typically covers validation, calibration, preventative maintenance and proficiency testing. It includes the computer system, as well as equipment ranging from water baths to fully automated immunohematology testing platforms. Under each section there should be specific SOPS with methods, requirements and schedules, minimum and standard expectations for outcomes, descriptions of QC procedures, documentation of findings, actions to be taken, and how data will be communicated, along with follow-up including improvement plans. Documentation of repairs, maintenance, calibration and the acceptance of daily QC must be traceable to individual instruments and readily available. New equipment and equipment returning from repair must be assessed for functional ability before it can be placed in service. In general, daily controls with acceptable results must be achieved before patient results can be reported. Recommendations for instrument calibrations, repairs and scheduled maintenance from instrument manufactures should be followed.

*Supplier and customer issues*: Critical in this QSE are the SOPs for supplier qualification, especially as relates to reagents and all other supplies. A tracking system and associated log of documentation should be included, as well as appropriate conditions for storage and inventory control.

*Process control*: Process control includes procedures to minimize variation and errors. Acceptance criteria should be defined, as well as the process employed if products, results or particular endpoints are deemed to be either non-standard or non-conforming. All process control data should be monitored and evaluated, and communicated to relevant authorities with an appropriate plan of action to be taken in the event of values found outside of the defined range.

*Documents and records*: A complete list of documents and records must be included, as well as details of how they are to be reviewed and stored, including the duration of storage, and how they will be retrieved when needed. A plan for outdating or archiving documents should be presented, as well as a plan for version control on in-date documents. If appropriate, specific SOPs should be included in this QSE that regard labeling.

*Non-conformances and deviations*: Terms such as non-conformances and deviations change often, and have included such labels as "error and accidents" and "problems and problem management." There should be a mechanism in place to evaluate critical non-conformances and deviations, and pathways such as "potential system problems" and "urgent health hazard." Incidents, errors, accidents and other adverse occurrences should be described and a system to detect, report and categorize them should be included, as well as there being processes in place to ensure timely reporting of specific events to appropriate authorities including the FDA and state regulatory agencies. Donor- and transfusion-related fatalities, mislabeling, and release of unsuitable blood are included in this QSE. A process describing who is to be notified in the event of serious non-conformances and deviations must be developed, and records of all such events maintained in accordance with federal, state and institutional regulations and policies. Methods must be in place for determination of adverse events resulting from transfusion, including transfusion-transmitted diseases and immediate and delayed reactions.

*Assessments internal and external*: Internal assessments are highly recommended, and there should be a procedure for how such assessments are conducted, what systems will be used, and how results will be recorded and reports prepared. Once deficiencies or potential deficiencies have been identified, corrective actions can be taken and further monitoring strategies determined.

*Process improvement*: This QSE allows the facility to demonstrate and even showcase improvements that have decreased rework; lessened the incidence of deviations, errors and accidents; increased user or customer satisfaction; improved resource utilization; decreased outdating; and/or increased compliance with regulations. Within this QSE are SOPs describing how the results of internal assessments are to be used to process improvement. The wording of this QSE suggests a significant goal in quality management – that is, the shift from corrective action to prevention.

*Facilities and safety*: The inclusion of facilities and safety as a QSE is to ensure a safe and adequate workplace environment and that continuous evaluation and improvement occur.

In 2007, the AABB started to change the 10 QSEs to 12 QSEs, which are planned to be universal throughout all AABB Standards for all areas; release of this new categorization scheme has not occurred at the time of writing but is expected in 2010. The 12 new QSEs, as currently drafted, are now grouped into three hierarchical categories as follows: *Management*, which will include QSEs 1, organization and leadership, 2, facilities, work environment, and safety, 3, human resources, 4, customer focus, 5, suppliers and supply management and 6, equipment and management; *Work*, which will include QSEs 7, process management, 8, documents and records and 9, information management; and *Measurement*, which will include QSEs 10, non-conforming events, 11, monitoring and assessment and 12, process improvement.

## Other Systems:

NCCLS GP26-A: The NCCLS (formerly the National Committee for Clinical Laboratory Standards) has published a quality system model known as NCCLS GP26-A. This model comprises 10 QSEs that are based on AABB QSEs with minor changes, and include organization, personnel, equipment, purchasing and inventory, process control, documents and records, occurrence management, internal assessment, process improvement, and service and satisfaction. In some ways these terms may have more direct applicability in the hospital transfusion service, but the intent of both sets of QSEs is identical.

World Health Organization: World Health Organization (WHO) *Aide-Memoire for National Blood Programmes* (WHO, 2002), under the title of Quality Systems for Blood Safety, approached the development of quality and quality systems somewhat differently, and in a way that allows ease of use for undeveloped nations. The memoire states a single goal, four strategies to achieve this goal, and five key elements that make up the quality systems. The goal is "to ensure quality blood and blood products for clinical use." The four strategies to achieve this aim are as follows: (1) a well-organized, nationally-coordinated blood transfusion service, (2) blood collected from regular, voluntary non-remunerated blood donors from low-risk populations, (3) testing of all donated blood, including screening for transfusion-transmissible infections, blood grouping and compatibility testing, and (4) appropriate clinical use of blood. The key elements are listed in Table 3.2.

| TABLE 3.2 Key Elements of Quality Systems |
|---|
| Organizational management |
| Standards |
| Documentation |
| Training |
| Assessment |

**Quality as a Philosophy and Culture:** The pursuit of quality must be embraced at all levels of the facility, from top management to the most basic employee. Reading of the quality manual at orientation and training is important, but the demonstration of a smooth, broad and capable process to "live and breathe" quality is critical.

## Additional and Related Terms:

*Quality assurance*: Quality assurance is a systematic and planned review of the entire patient experience, providing a high level of confidence that quality goals are met. From the time the patient presents for care until he or she is discharged, every service process is assessed. Process review includes preanalytic, analytic and postanalytic assessments.

*Continuous quality improvement*: Continuous improvement in the blood banking industry is an ongoing, continuous process, with the goal of identifying ways to reduce waste and improve patient care.

*Quality control*: Quality control is a process of assessing the management of operational techniques, reproducibility and accuracy of results. The assessment provides verification that the operating systems are performing correctly.

*Quality system*: Quality system is the collection of policies, procedures, and processes that are organizationally structured to guide the daily quality efforts of an organization.

**Facility Inspections:** Blood establishments are subject to unannounced FDA inspections every two years. The objective of the inspections is to evaluate the blood establishment's compliance with all regulations. When an establishment is founded to be non-compliant, the FDA issues observations on Form 483. Depending on the type and severity of the violations, the blood establishment may be subject to other forms of enforcement activities (Table 3.3).

| TABLE 3.3 Other Forms of Enforcement | |
|---|---|
| Citation | Product recall or destruction |
| Suspension | Criminal prosecution |
| Civil penalty | License revocation |
| Seizure | Consent decree |

After a blood establishment has developed a history of non-compliance, the FDA may choose to impose enforcement by filing in court a "Consent Decree of Permanent Injunction." A consent decree defines how a blood establishment must operate to ensure compliance with cGMPs. The establishment is given a timeframe for making the required compliance adjustments for continued operations.

## Recommended Reading

Berte LM. (1995). A quality system for transfusion medicine ... and beyond. *Med Lab Obs* **27**, 61–64.

Berte LM. (2000). New quality guidelines for laboratories. *Med Lab Obs* **32**, 46–50.

Kim DU. (2002). The quest for quality blood banking program in the new millennium the American way. *Intl J Hematol* **76**(Suppl 2), 258–262.

McCullough J. (2006). The role of physicians in blood centers. *Transfusion* **46**, 854–861.

Smith DS. (1999). Performance improvement in a hospital transfusion service: the American Association of Blood Banks' quality system approach. *Arch Pathol Lab Med* **123**, 585–591.

World Health Organization, Department of Blood Safety and Clinical Technology. (2002). *Aide-Memoire for National Blood Programmes*. Geneva: World Health Organization.

# CHAPTER 4

# The Role of the Physician in the Blood Center

Christopher D. Hillyer, MD

Physicians play a central and vital role in virtually all blood center elements and operations typically with direct and indirect responsibilities. As blood centers are the collection, manufacturing, storage and distribution facilities for biologics (blood and blood components) and therapeutics (apheresis and hematopoietic progenitor cells [HPCs]), no element in the manufacturing chain is without a medical and scientific underpinning and perhaps no other person in the center better understands the ramifications to the donor and to the patient/recipient of non-conformances and deviations that can and do occur at every step in the process. In this regard, the physician's role has been sometimes described as the "conscience of the organization," and indeed it is the broad knowledge base of the blood center physician and his or her overarching responsibility that makes the role both challenging and rewarding.

In a blood center, the physicians have significant and often unparalleled responsibility, though their true lines of authority are often determined by the chief executive officer of the center/corporation and, at times, the degree of medical responsibility and the specifically designated lines of authority are incongruous. Diagrammatically, the physician's role in the blood center can be depicted as in Figure 4.1.

**Blood Center Table of Organization:** Physicians in blood centers are often given the title of Chief Medical Officer (CMO) or Medical Director, with a reporting line to the Chief Executive Officer (CEO). A typical blood center table of organization is presented in Figure 4.2. Furthermore, it is not unusual for the CMO to act as CEO when needed and appropriate.

## Specific Roles in Blood Center Core Operations:

### Collections: Donor Selection and Complications:
The blood center physician is required to oversee all aspects of blood collection, including donor selection and complications arising from the blood donation process. Indeed, blood collection is a medical procedure, and as such the physician's role constitutes the practice of medicine. Thus, the physician must be licensed in the states in which he or she has responsibility for this element of blood center operations.

Specific tasks of the blood center physician may include: 1) determination of volunteer allogeneic donor suitability specific to a donor with comorbid conditions or who is taking medication(s) both for the safety of the donor him or herself, but also for the safety, quality, identity, potency and purity (SQUIPP) of the product and thus for the ultimate safety of the patient, 2) determination of autologous, directed and dedicated donor suitability, as most of these donors do not satisfy all of the collection

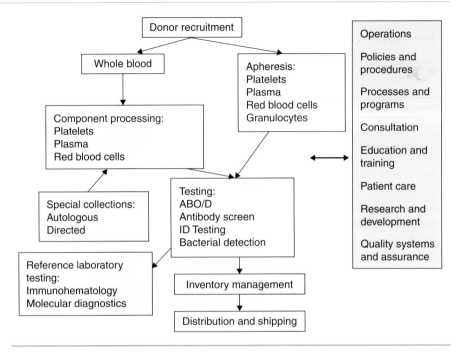

FIGURE 4.1  Physician oversight in the blood center.

criteria for allogeneic donation, and 3) determination of specialized donor suitability, including planned collection of granulocytes and maternal platelets.

The physician should also play a central role in the determination of technology acquisition, including double red blood cell collection by apheresis, and in determining the best course of action if errors and accidents occur during the collection procedure itself. The physician is typically the central figure in evaluating donor complications and injuries both for the medical safety of a given donor and in order to determine if a potential system problem or urgent health hazard exists.

Manufacturing and Processing: The physician typically has only a limited role in well-functioning and standard manufacturing operations, including component manufacturing, and preparation and processing (including leukoreduction), though he or she should be significantly involved with the *selection* of technologies related to these areas (e.g. leukoreduction filters, automated component preparation devices) and *during changes in operating procedures* that could adversely affect the SQUIPP characteristics of the manufactured components or products. This latter event is often governed by a change control process for which the physician should be a key signatory. Any non-conformances, errors or deviations that occur in manufacturing and processing require, at least, consultation with the blood center physician, and more often include his or her determination of product acceptability.

Quality Systems and Laboratories: The blood center physician is directly involved in the quality systems function of the institution, and again he or she is well

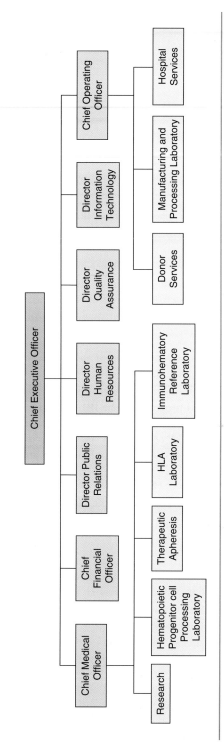

**FIGURE 4.2** Organizational chart of a blood center.

suited to understand the impact of procedural changes as well as evaluation of product acceptability following deviations from procedures or errors. This last responsibility is often accomplished by a material review board (MRB) process. The individual make-up and focus of MRB process is included in Table 4.1.

---

**TABLE 4.1  Material Review Board Process: Key Individuals and Functions**

1. Administrative leadership: typically the CEO, the administrative element of the MRB will focus on fiscal issues regarding the loss of unit or units.
2. Medical office leadership: typically the CMO, the medical office element of the MRB will focus on medical, scientific, SQUIPP and specific donor or patient issues as affects donor, unit and patient safety.
3. Regulatory/quality program leadership: typically, personnel within the regulatory or quality departments of a blood center will be responsible for the MRB process overall and focus on regulatory, FDA-related and procedural aspects of deviations, errors and accidents that occur.

---

The physician is also directly involved with inspections and accreditation, internal audits and problem management, and the plan(s) for resolution of any noted deficiencies. In addition to quality systems and programs, the medical officer will often have direct authority over and responsibility for quality control laboratories and the interpretation of proficiency testing data.

**Clinical Services:** In the blood center, *clinical services* is a collective term that usually includes the immunohematology reference and/or the histocompatibility laboratories (IRL and HLA laboratories, respectively), therapeutic apheresis (TA) services, and HPC collection and processing. The IRL and HLA laboratories are regulated under current Good Manufacturing Practices (cGMP) and Good Laboratory Practices (cGLP), and require CLIA certification. HPC collection and processing is now regulated under cGTP (current Good Tissue Practices), and as such is controlled by the FDA. TA is considered the practice of medicine, and as such is not under the regulatory authority of the FDA. The medical officer usually also serves as CLIA laboratory director and holds states' licensure. If the medical officer serves as the HLA laboratory director, additional training and/or certification by the American Society of Histocompatibility and Immunology (ASHI) is either required or desirable.

**Therapeutic Apheresis:** In many blood centers the physicians make all determinations about the TA service, including procedure medical appropriateness and technical aspects (replacement fluids, volume processed, etc.), and provide consultation to ordering physicians. In many locations TA is performed by registered nurses or specially qualified medical technologists, and the procedures occur outside the blood center confines. It is important to define the exact role of the blood center physician and where his or her hegemony stops relative to the hospital ordering and/or attending physician responsible for the ongoing care of the patient. In some cases the blood center physician will accompany TA staff to the patient's bedside. It is important for blood center physicians to know if senior management also expects the physician to be directly responsible for TA staff and, if indicated, to be given direct authority regarding this aspect of the center's operations.

**Immunohematology Reference, HLA and HPC Laboratories:** In the IRL, the blood center physician plays a highly medical and technical role, providing input, interpretation and decision-making regarding immunohematology testing, including blood typing (both phenotyping and genotyping), complex Rh determinations, testing for unexpected antibodies in donor samples and in patient samples sent to the blood center IRL, and panel and eluate data interpretation. As with TA above, it is important for blood center physicians to know if senior management expects them to be directly responsible for IRL staff and to have this authority so designated if required. In addition, blood center physicians provide clinical consultation to hospital transfusion medicine and transfusing clinicians regarding complicated donor–recipient matching, as often relates to the transfusion of patients with sickle cell anemia, and adverse outcomes posttransfusion. Finally, the blood center physician is intimately involved with policy, procedural, technical, regulatory and medical automation, device, computer and technology aspects of the IRL. These duties parallel the functional activities of the physician who serves as the HLA or HPC laboratory director.

**Financial Issues and Conflict of Interest:** Blood centers serve a humanitarian need, and mission statements of blood centers often include elements regarding donors and patients. A model blood center mission statement is included in Table 4.2. Still, blood centers are corporate entities and have financial performance measures and expectations. Thus, blood center physicians may be in situations of perceived and/ or real conflict of interest between the financial interests and the viability of the blood center for which they work, and the best medical interests of blood donors and transfusion recipients for which they have taken an oath.

| TABLE 4.2  Model Blood Center Mission Statement |
| --- |
| To provide the highest level of transfusion care and therapy to patients via optimal collection, processing and distribution of safe, effective and available blood components, with respect, appreciation and care for the donor, and excellent customer service. |

**Research and Advancement of the Field:** Blood centers offer rich environments for research and development. In fact, the largest blood centers have research and development laboratories, including the Jerome H. Holland Laboratory of the American Red Cross, the Blood Systems Research Institute of Blood Systems, Inc., and the Lindsey F. Kimball Research Institute of the New York Blood Center. Indeed, major NIH-funded research programs could not function without significant blood center and blood center physician/scientist involvement, including the Retrovirus Epidemiology Donor Study-II (REDS-II) consortium which involves domestic and international efforts dedicated to blood donor safety and blood availability issues, and the Transfusion Medicine/Hemostasis Clinical Trials Network (TMH CTN) charged with conducting clinical trials in transfusion medicine and hemostasis.

## Recommended Reading

McCullough J. (2006). The role of physicians in blood centers. *Transfusion* **46**, 854–861.

# CHAPTER 5

# The Blood Donor, Donation Process and Technical Aspects of Blood Collection

Krista L. Hillyer, MD

At present, there are no Food and Drug Administration (FDA)-approved, or clinically equivalent, substitutes for the whole blood and blood components donated by voluntary non-remunerated donor (VNRDs) in the US, or elsewhere. Indeed, without blood donations of red blood cells (RBCs), platelets, and plasma, modern transfusion therapy, which allows complex surgery, damage control resuscitation for trauma, transplantation of major organs, care of the complicated postpartum patient, and chemotherapy treatment for cancer patients, would not be possible. Thus, the blood donor is a vital element of blood banking and transfusion medicine, which in turn is critical to the function of advanced healthcare systems.

In the US, approximately 15 million units of whole blood are collected each year (~70 million worldwide) from VNRDs and are processed into blood components (RBCs, platelets, plasma, and cryoprecipitate) used in the treatment of surgical, obstetric, medical and other patients. Blood components can also be collected from VNRDs by automated apheresis (see Chapter 6). The blood donation process is: 1) overseen by a licensed, qualified physician, 2) regulated by the FDA, and 3) designed to provide protection for both the blood donor and the transfusion recipient by ensuring appropriate donor eligibility, donation procedures, and donor and product testing.

In this chapter, general considerations regarding whole blood donors are presented. Apheresis donation is discussed in Chapter 6, and autologous, directed, HLA-matched, crossmatched, paid and hemochromatosis donors, and other "special" donors and donations will be covered in Chapter 7.

**Blood Donor:** In the US, approximately 15 million whole blood units are collected each year from approximately 10 million donors. This blood donor pool is approximately 7% of 18 years of age or older individuals and represents both first time and repeat blood donors. The eligible donor pool was recently estimated by Riley in 2007 to be 38% of all individuals within the US. Blood donation is considered an act of selflessness or altruism and has no tangible benefit to the donor. Because of this, blood donors are afforded special protection under federal and state regulations and standards, and via careful and confidential practices of blood collection facilities.

The traditional US blood donor is a white, college-educated, middle-aged male. Recent changes in population demographics have led to increased participation in community blood donation programs by all racial and ethnic groups. Understanding the motivation and barriers to blood donation is essential for an adequate blood supply. The motivators to blood donation include altruism, humanitarianism, awareness of the need, sense of social obligation, social pressure, need to replace blood used, and increased self-esteem. Barriers to blood donation include fear, inconvenience, being too busy, perceived (or true) medical disqualification, not knowing there is a need for blood, distrust and apathy. Blood donation recruitment focuses on enhancing the motivators while minimizing the barriers to blood donation. Few studies, if any, have addressed the social and/or motivating factors that lead first-time donors to become repeat blood donors. Repeat blood donors are advantageous to the blood industry as they have fewer adverse reactions to donation and are a population with less positive infectious diseases markers, as donors with infections (in existence at the time of first donation) will have been previously deferred.

**Blood Donation Process:** The blood donation process consists of sequential steps prior to phlebotomy in order to ensure both donor and recipient safety. These steps include recruitment materials, educational materials, a donor health history questionnaire (DHQ), physical examination, and informed consent.

**Recruitment Materials:** Donor recruitment materials usually highlight the reasons to give blood by using phrases such as "give the gift of life". In addition, they provide some of the basic criteria for whole blood donation such as weight and age, thus allowing for self-deferral.

**Educational Materials:** Educational materials provide an overview of the donation process, donor eligibility, especially focusing on high-risk behaviors, signs/symptoms of AIDS, infectious disease transmission, and blood donation risks. Potential donors are informed of the need to provide accurate information and the importance of withdrawing themselves from the donation if appropriate in order to maximize recipient safety. The AABB has created AABB Blood Donation Education Materials (http://www.fda.gov/cber/dhq/dhq.htm).

**Registration:** The potential donor must provide the following information: name and contact information (to contact donor for future donations or to provide test results or other information), age or date of birth, and the date of last donation (to ensure adequate time has elapsed between donations). At this time, a donor deferral database is searched in order to prevent an individual who was previously determined to be ineligible from donating. Correct donor identification is also key to ensuring the donor can be informed of any positive infectious disease markers found following his or her donation, and for follow up if certain recipient reactions are reported.

**Donor History Questionnaire:** The DHQ contains a series of questions asked of donors prior to the start of the donation process. Whatever the format (verbal, written, or computerized) the questions on the DHQ are based on regulations and guidance from the FDA and on AABB *Standards for Blood Banking Banks and Transfusion Services (Standards),* covering a wide range of topics intended to protect the health of the donor during the blood donation process and to protect a future recipient from being harmed by the donor's blood.

Many questions on the DHQ are designed to qualify a donor and include determination of weight, date of most recent donation, etc., and are considered to be part of the donor physical exam as well (Table 5.1).

Specific questions on the DHQ are designed to protect recipients from transfusion-transmitted diseases for which there are no tests available or applied, including

| TABLE 5.1  Acceptable Allogeneic Blood Donation Criteria | |
|---|---|
| **Test** | **Acceptable Value(s)** |
| Age | ≥17 years or conforming to applicable state law |
| Weight | 110 lb; If weight <110 lb, then a maximum of 10.5 ml/kg, including samples, can be collected |
| Donation Interval | ≥56 days after whole blood donation |
|  | ≥112 days after 2-unit RBC donation |
|  | ≥28 days after infrequent plasmapheresis |
|  | ≥2 days after plasma-, platelet-, or leukapheresis |
| Hemoglobin; Hematocrit | ≥12.5 g/dl/; 38% (not by earlobe puncture) |
| Blood pressure | ≤180 mm Hg systolic |
|  | ≤100 mm Hg diastolic |
| Heart rate | 50–100 bpm without pathologic irregularities |
|  | <50 bpm if otherwise healthy athlete |
| Temperature | ≤37.5°C (99.5°F) measured orally |
| Antecubetal fossa | Free of lesions, "track marks", scars (i.e. signs of IV drug use) |

Modified from AABB *Standards for Blood Banks and Transfusion Services*, 25th ed.

malaria, babesia, vCJD, and others. Some of the elements, criteria and specific deferral periods for certain responses to DHQ responses are listed below. To be effective, questions asked of the donor must be truthfully answered and lead to deferral of a potential donor with high-risk behavior, those at risk for carrying certain transfusion-transmitted diseases, and those taking medications which may harm a recipient.

Drug Therapy:  The donor is asked questions about certain medications he or she is taking or has taken recently prior to giving blood.

*Indefinite deferral* occurs if the donor is taking Etretinate (Tegison) or bovine insulin manufactured in UK;

*3-year deferral* occurs if the donor is taking Acitretin (Soriatane);

*6-month deferral* occurs if the donor is taking Dutasteride (Avodart);

*1-month deferral* occurs if the donor is taking Finasteride (Proscar, Propecia) or Isotretinoin (Accutane); and

*48-hour deferral* occurs if the donor is the sole source of platelets for a recipient and is taking any medication that irreversibly inhibits platelet function (e.g. aspirin).

*Other medication deferrals* are defined by the blood collection facility medical director.

Medical History:

*General health:* Appearance of good health, free of major organ disease, free of cancer, and free of bleeding disorder. No skin lesions at the venipuncture site. Indefinite deferral with a family history of Creutzfeld-Jakob disease.

*Pregnancy:* Deferral if pregnant in last 6 weeks or currently pregnant.

*Receipt of human blood, blood component, or other human tissue:* Indefinite deferral for receipt of dura mater or pituitary growth hormones of human origin; 12-month deferral for receipt of blood, components, human tissue, or plasma-derived clotting factor concentrates.

*Immunizations and vaccinations:* See Table 5.2.

### TABLE 5.2 Donor Deferral Duration* for History of Recent Vaccination

| | |
|---|---|
| **No Deferral** | Receipt of toxoids, synthetic, or killed viral, bacterial, or rickettsial vaccines, if donor is symptom-free and afebrile. |
| | These vaccines include: anthrax, cholera, diphtheria, hepatitis A, hepatitis B, influenza, Lyme disease, paratyphoid, pertussis, plague, pneumococcal polysaccharide, polio (Salk/injection), Rocky Mountain spotted fever, tetanus, typhoid (by injection). |
| **2-Week Deferral** | Receipt of live attenuated viral and bacterial vaccines. |
| | These vaccines include: measles (rubeola), mumps, polio (Sabin/oral), typhoid (oral), yellow fever. |
| **4-Week Deferral** | Receipt of live attenuated viral and bacterial vaccines. |
| | These vaccines include: German measles (rubella), chickenpox (Varicella zoster). |
| **12-Month Deferral** | Receipt of other vaccines, including unlicensed vaccines. |

*For smallpox vaccine, the reader is referred to the current FDA Guidance.

*Infectious Diseases:*

Indefinite Deferral:
- History of viral hepatitis after 11th birthday;
- Confirmed positive test for hepatitis B surface antigen;
- Repeat reactive test for hepatitis B core antibody on more than one occasion;
- Present or past clinical or laboratory evidence of HCV, HTLV, or HIV infection, or as excluded by current FDA regulations and recommendations for the prevention of HIV transmission by blood and components;
- Donated the only component implicated in an apparent transfusion-transmission of hepatitis, HIV, or HTLV;
- History of babesiosis or Chagas' disease;
- Evidence or obvious "stigmata" of parenteral (IV) drug use;
- Using a needle to inject non-prescribed drugs; and
- Donors that are recommended for deferral for risk of variant Creutzfeld-Jakob disease (vCJD), as defined in most recent FDA Guidance.

12-month deferral from the time of any of the following:
- Mucous membrane exposure to blood;

- Non-sterile needles or other instruments penetrating donor skin, contaminated with blood or body fluids other than the donor's; includes tattoos or permanent make-up unless applied in a sterile manner;
- Sexual contact with anyone with a confirmed test for hepatitis B surface antigen;
- Sexual contact with anyone who is symptomatic with any viral hepatitis;
- Sexual contact with anyone with hepatitis C who has had symptomatic hepatitis in the last 12 months;
- Sexual contact with anyone who has HIV or is at high risk for HIV;
- Incarceration for more than 72 hours;
- Syphilis or gonorrhea: completion of therapy, or a reactive screening test for syphilis in the absence of a negative confirmatory test; and
- History of syphilis or gonorrhea.

West Nile virus (WNV) and malaria require specialized consideration.

- *WNV:* Defer according to FDA Guidelines.
- *Malaria:* Go to http://www.cdc.gov/malaria/risk_map/ for current endemic malarial risk areas on the Centers for Disease Control and Prevention's website. Then,

   **3-year deferral**: Individuals who have been diagnosed with malaria, or those who have traveled to or lived in a malaria-endemic area and have had symptoms suggestive of malaria (must be asymptomatic for 3 years); Anyone who has lived at least 5 consecutive years in an area where malaria is considered endemic by the CDC (3 years from departure from that area).

   **12-month deferral:** Anyone who has traveled to an area where malaria is endemic but does not have symptoms of malaria.
- *Travel:* Anyone who has traveled outside the country in which they are donating blood should be evaluated for potential travel risks. Go to http://www.cdc.gov/travel for more details.

   A Uniform Donor History Questionnaire (UDHQ) contains 48 questions, was developed by an interorganizational task force, is recommended by AABB, and endorsed by FDA (http://www.fda.gov/cber/dhq/dhq.htm). It is not required that the UDHQ is used by blood centers, but center-specific DHQs must comply with FDA regulations and AABB *Standards*.

   Some blood centers have sought and received FDA approval for an abbreviated DHQ (aDHQ) for repeat donors. Kamel *et al.* demonstrated the use of an aDHQ with 34 questions, compared to the full-length questionnaire, which is 55 questions, for frequent donors who had donated at least 3 previous times, once within the last six months. In this setting, the aDHQ resulted in increased satisfaction with the donation process and increased intention to donate in the future.

**Donor Informed Consent:**  Donor informed consent is required and must contain, according to AABB *Standards*, the risks of the procedure, the tests performed, that the donor had the ability to ask questions and had them answered, and the ability to withdraw from the donation process. The informed consent has come under scrutiny recently as it appears that many potential donors do not fully comprehend what they are consenting to and the risks involved. Alaishuski *et al.* demonstrated this in whole blood donors, where <3/4 of the information was found to be "understood".

Some states require that minor donors (i.e. those under the age of 18) have a parent or guardian give written consent for his or her minor to donate blood. Some authorities have suggested that separate and more comprehensive informed consent forms be required for blood collection from minor age donors.

**Technical Aspects of Blood Donation:** Collection of blood from donors is a highly regulated and technically-specific process. As the starting point for all blood manufacturing, the selection and preparation of healthy blood donors and a safe collection process is vital to the eventual preparation of high-quality blood components for transfusion and further manufacture. Facilities that collect blood for transfusion or further manufacture must have appropriate and adequate standard operating procedures, trained staff, and properly validated and functioning equipment.

**Identification:** Identification of the donor, and thus the donor's blood product(s), from collection to final disposition, is critical to the safety and protection of both donor and recipient. A numeric or alphanumeric system must be used to identify both the donor and the donor's blood unit(s), and it must identify and relate to the source donor, the donor record, the blood samples/tubes used for testing, the collection container, and all components that were prepared from the initial donation. Care must be taken to avoid mix-ups or duplication of these identifiers. If an error is found, an investigation must be performed to ascertain the underlying cause for the error. Any numbers that are voided must be kept in a record that can be accessed in the future for any type of investigation that may take place.

**Preparation of the Venipuncture Site:** The antecubital area is the best site for locating a firm vein from which to collect blood. This area must be free of skin lesions. An inspection of both antecubital fossae must be done to identify any skin disease, scarring or evidence of intravenous drug use. A tourniquet or a blood pressure cuff inflated to 40–60 mmHg (but no tighter, for donor comfort) can enhance the size of the veins in the antecubital fossae being considered. The donor may be asked to open and close his or her hand, also to enhance the prominence of the veins being inspected. The blood pressure cuff or tourniquet should be removed prior to preparing the skin for phlebotomy. Adequate skin cleansing and diversion pouches (when collecting platelet products) protect against bacterial contamination.

**Phlebotomy:** Blood must be collected by a trained phlebotomist using aseptic methods, and a sterile closed system. The blood unit should be collected from a suitable vein in a single venipuncture, within the antecubital fossa of the donor's arm, after the pressure device (blood pressure cuff or tourniquet) has been deflated and re-inflated. Whole blood is collected into a closed blood collection set consisting of a sterile collection bag containing anticoagulant, integrally attached tubing, and a large bore needle. A balance system or scale is used to measure the amount of blood being collected.

The total amount collected from the donor, including segments and specimen tubes, should not exceed 10.5 ml/kg of donor weight (whole blood collections); typically 70 ml of anticoagulant is in a container to collect $500 \pm 50$ ml of whole blood. During collection, the blood should be mixed well with the anticoagulant in the bag, to prevent clotting. Some blood centers use staff to agitate the bag from side to side to mix the fresh blood with anticoagulant; others utilize a specialized "shaking" device to mix the blood and anticoagulant mechanically and hands-free.

**Post-donation Care:** A pressure dressing is applied to the venipuncture site for hemostasis. The donor should proceed to the refreshment area to drink liquids and rest for 15 minutes. In addition, the donor should receive post-donation instructions, which include care of the venipuncture site, drinking liquids, and appropriate post-donation activity level. The contact information for the donor center should be provided for the donor to report any adverse reactions, to modify his or her donor history information, or to request his or her blood not be used for transfusion.

**Donor Adverse Reactions:** Blood donation is not without risk. The rate of adverse events increases based on age, weight, sex, and first-time versus repeat donations. In a study of blood donors which performed 1000 post donation interviews, 36.1% of whole blood donors had adverse events, including arm injuries (22.7% bruise, 10.0% sore arm, 1.7% hematoma, and 0.9% sensory changes), fatigue 7.8%, vasovagal findings 5.3%, and nausea and vomiting 1.1%. These adverse reactions are further discussed in Chapter 6.

## Recommended Reading

Alaishuski LA, Grim RD, Domen RE. (2008). The Informed Consent Process in Whole Blood Donation. *Arch Pathol Lab Med* **132**, 947–951.

Gillespie TW, Hillyer CD. (2002). Blood donors and factors impacting the blood donation decision. *Transfus Med Rev* **16**, 115–130.

Kamel HT, Bassett MB, Custer B, Paden CJ, Strollo AM, McEvoy P, Busch MP, Tomasulo PA. (2006). Safety and donor acceptance of an abbreviated donor history questionnaire. *Transfusion* **46**, 1745–1753.

Newman BH, Satz SL, Janowicz NM, Siegfried BA. (2006). Donor reactions in high-school donors: the effects of sex, weight, and collection volume. *Transfusion* **46**, 284–288.

Riley W, Schwei M, McCullough J. (2007). The United States' potential blood donor pool: Estimating the prevalence of donor-exclusion factors on the pool of potential donors. *Transfusion* **47**, 1180–1188.

Stowell C, Sazama K (eds). (2007). *Informed consent in blood transfusion and cellular therapies: Patients, donors, and research subjects*. Bethesda, MD: AABB Press.

Zou S, Eder AF, Musavi F, Notari Iv EP, Fang CT, Dodd RY. (2007). ARCNET Study Group. Implementation of the Uniform Donor History Questionnaire across the American Red Cross Blood Services: increased deferral among repeat presenters but no measurable impact on blood safety. *Transfusion* **47**, 1990–1998.

# CHAPTER 6
# Apheresis Blood Component Collections

Krista L. Hillyer, MD

Blood and blood components can be collected either by whole blood (WB) collection or by automated apheresis procedures from non-renumerated volunteer donors. Most red blood cell (RBC) and transfused plasma products in the US are collected by WB collection. In contrast, the majority of platelet products (75% in the US) are collected through automated apheresis. The manufacturing of WB into various components is time- and labor-intensive, and therefore apheresis collection is frequently preferable as there are fewer steps in component processing, especially the elimination of the need for post-donation centrifugation and leukoreduction. Additional benefits of apheresis collection over WB collection include the ability to control product volume and dose, the ability to use donors more effectively and to collect the products that are most needed, the ability to collect multiple components from the same donor, and the decreased risk of mislabeling or misidentifying a component because of the minimum processing required after collection. Disadvantages of automated apheresis collection include increased expense (which may be offset by an overall increase in efficiency), an increase in phlebotomist training, and the inability to collect a large number of products quickly.

Most of the regulations and standards are the same for blood donors undergoing automated apheresis collection as for those undergoing WB collection (see Chapter 5). However, a number of donor qualifications are unique to automated apheresis collections, which are the focus of this chapter. Of note, granulocyte apheresis collections are discussed in Chapter 32.

**Apheresis RBC Collections:** RBC apheresis products can be collected as two products (termed a double RBC procedure) or as multicomponent donations with one RBC component and one plasma and/or platelet component (termed *double* or *triple component procedure*).

In contrast to WB, the minimum hematocrit for double RBC collections is higher, at 40% instead of 38%. The deferral period after a double RBC collection is 16 weeks, in contrast to WB or a single RBC collection, where it is 8 weeks. In addition, there are gender-specific minimum height and weight collection requirements: men must weigh more than 130 pounds and be taller than 5'1" to donate double RBCs, while women must weigh more than 150 pounds and be taller than 5'5".

RBC apheresis donations are well tolerated, with lower adverse event rates, especially vasovagal reactions, than WB donations. Reasons for lower vasovagal rates in RBC apheresis compared to WB donation include the larger weight requirement (the highest risk of vasovagal reactions is in WB donors with lower weights, i.e. ≤130 pounds), saline replacement given during apheresis, reinfusion of plasma, and a longer time of procedure (30–40 minutes for apheresis versus 8–10 minutes for WB donation). These factors decrease hypovolemia and thus the risk of vasovagal reactions.

**Platelet Apheresis (Plateletpheresis) Collections:** Apheresis platelet collections allow for single or multiple platelet products to be collected from a single donor. Like RBCs, a platelet component can be collected along with plasma and/or RBC components in double or triple component procedures. Apheresis platelet products are commonly used in lieu of WB-derived platelet products in the US because plateletpheresis products eliminate the need for pooling multiple products, decrease donor exposure and ease bacterial testing. Outside the US there are automated systems which pre-pool and store leukoreduced WB-derived platelets, making their use easier and, therefore, more common.

Apheresis products are preferred in specific patient circumstances. These situations include HLA matched or crossmatched platelet products for a patient who is refractory to platelet transfusions secondary to alloantibodies, and maternal platelets or platelet antigen negative platelets for a newborn with neonatal alloimmune thrombocytopenia.

Plateletpheresis donors must meet the same criteria as WB donors, except they may donate plateletpheresis products more frequently: up to 24 times per year and twice per week, with intervals between donations of 2 or more days. If RBC loss exceeds 100 ml, then 8 weeks must lapse prior to the next donation. These criteria can be waived if approved by the medical director, as in the special situations cited above.

Additional deferral criteria include deferral for anti-platelet medications. The specific deferral criteria are defined by the collection facility's medical director, but at minimum include 48 hours after last dose of aspirin and Feldene, and 14 days after last dose of Plavix (clopidogrel) and Ticlid (ticlopidine).

A platelet count is not required before the first plateletpheresis collection, or if 4 or more weeks have elapsed since the last plateletpheresis procedure. However, if less than 4 weeks have elapsed since the last plateletpheresis procedure, the apheresis platelet donor must have a platelet count $\geq 150,000/\mu L$. Platelet count testing of the donor may be performed either before the current apheresis procedure, or before or after the preceding plateletpheresis procedure. Acceptance of a donor who does not meet all criteria can be made in the specific situations described above, with medical director approval.

The maximum total plasma volume removal requirements for each plateletpheresis collection are 500 ml for donors weighing less than 175 pounds, and 600 ml for donors weighing more than 175 pounds.

Unlike WB collections or double RBC collections, detailed records must be kept for each plateletpheresis procedure performed, including adverse events and the platelet count of each product collected. All platelet apheresis laboratory values and collection records must be reviewed at least once every 4 months (per FDA guidelines) by the blood center medical director.

**Plasmapheresis Collections:** Plasmapheresis products are collected either as *transfusable plasma* or as *source plasma* for further manufacturing into human-based plasma derivatives (such as intravenous immunoglobulin, albumin, and factor concentrates). Source plasma is usually collected by large commercial plasmapheresis collection centers from paid plasma donors. Source plasma is subject to different requirements per the FDA and is also overseen by the PPTA (Plasma Protein

Therapeutics Association), which has developed an International Quality Plasma Program (IQPP) certification.

Plasmapheresis programs can either be *infrequent plasmapheresis* (plasma donations more than 4 weeks apart) or *frequent plasmapheresis* (donations less than 4 weeks apart). Infrequent plasmapheresis donors have the same donor eligibility requirements as WB donors. Frequent plasmapheresis donors have other requirements, including the following:

- At the initial plasmapheresis, and at 4-month intervals, donor serum or plasma must be tested for total protein, and for serum protein electrophoresis or for quantitative immunoglobulins. These results must be normal for plasmapheresis procedures to continue.
- Donors can donate up to every 2 days, or twice per week.

## Recommended Reading

Burgstaler EA. (2006). Blood component collection by apheresis. *J Clin Apher* **21**, 142–151.

Department of Health and Human Services. (2006). *2005 Nationwide Blood Collection and Utilization Survey Report.* Washington, DC: DHHS.

Food and Drug Administration. (2007). *Guidance for Industry and FDA review staff: Collection of platelets by automated methods.* Rockville, MD: CBER Office of Communication, Training, and Manufacturers Assistance.

Food and Drug Administration. (2001). *Guidance for Industry: Recommendations for collecting red blood cells by automated apheresis methods.* Rockville, MD: CBER Office of Communication, Training, and Manufacturers Assistance.

Smith JW, Gilcher RO. (2006). The future of automated red blood cell collection. *Transfus Apher Sci* **34**, 219–226.

# CHAPTER 7
# Recipient-specific Blood Donations

Krista L. Hillyer, MD

Recipient-specific blood components are donated in three circumstances: exceptional medical need, directed donation, and autologous donation. *Exceptional medical need* reflects blood collected for a specific patient who requires a rare product. Examples of this include RBC products for a patient with multiple RBC antibodies or antibodies to high frequency RBC antigens, or a neonate with neonatal alloimmune thrombocytopenia who requires an antigen negative platelet product. *Directed donors* donate to a specific patient, who is usually a family member. *Autologous donors* donate for themselves, usually prior to a specific procedure which will require the use of blood (typically RBCs). Further discussion about the use of autologous donations can be found in Chapter 51; the current chapter will focus on donor requirements particular to recipient-specific donation.

**Exceptional Medical Need:** There are patient-specific situations where the blood center medical director may accept a blood donor donating at a shorter time interval than normally allowed in order for the donor to supply a product which is medically special and indicated for a particular patient. For RBC donation this can occur as frequently as every three days. The donor typically must fulfill all other allogeneic blood donation criteria. Circumstances where blood is collected for this reason include the following:

- Provision of RBC products for a patient with multiple alloantibodies or alloantibodies to high incidence RBC antigens, where rare RBC products are needed.
- Provision of platelet products for a newborn with neonatal alloimmune thrombocytopenia who requires platelet antigen negative products or maternal platelets.
- Provision of IgA deficient plasma or platelet products for a patient with anaphylactic transfusion reactions secondary to IgA deficiency and anti-IgA.

**Directed Donation:** Directed blood donors donate blood to be used for a friend, family member or another specific individual in need of blood products. Directed donors must meet all the criteria for allogeneic blood donation, and therefore these products can be released into the general inventory by the consignee.

There is a perceived notion by the lay public that directed donor units have a lower risk of viral infection, especially HIV, because the directed donor is "known" by the patient or a requesting family member. In fact, directed donors have a higher rate of viral marker positivity overall than volunteer donors, because they are often first-time donors. There is also the additional concern that a person who is requested to be a directed donor may not be willing to divulge high-risk behaviors on the donor history questionnaire. In general, the transfusion community considers directed donor units to be, at best, no safer than the general inventory of allogeneic blood products. In addition, directed donors should be aware that their ABO blood type may not be compatible with the recipient and therefore may not be available for the specified recipient's use. Finally, directed donor units have an increased cost compared to routine volunteer units, secondary to the need to label them for a specific patient, and to ensure they are sent to the correct transfusion service and reserved for the specific patient.

**Autologous Donation:** Autologous donors donate their own blood units for their own use, usually prior to scheduled surgeries. The use of autologous blood was highest in the late 1980s and the early 1990s when the risk of transfusion-transmitted diseases was high, and therefore the risk of transfusion-transmitted disease could be decreased with its use. Its use has declined and stabilized, such that currently 0.5 million units of autologous RBC products are collected annually. Only 50% of these units are transfused. Autologous units are not placed into the general inventory except in exceptional circumstances where a given blood component has an extremely rare phenotype. As with directed units, autologous units are more costly to produce secondary to the need to label and store them for a specific patient. In order to defray some of the additional costs, some blood centers will keep the unit as whole blood instead of manufacturing it into packed RBCs.

In order to prevent the collection and subsequent discarding of autologous units, autologous donation should only be allowed for planned procedures that have a high likelihood of RBC use. These include: major orthopedic procedures, vascular surgery, and cardiothoracic surgery. Procedures which are not indicted for autologous collection include uncomplicated obstetric delivery, cholecystectomy, and hysterectomy.

As well as mitigating the patient's risk of contracting transfusion-transmitted diseases, the benefits of autologous blood may include minimizing exposure to foreign RBC or white blood cell antigens that may stimulate alloimmunization and thus perhaps avoiding transfusion-associated immunomodulation (see Chapter 64).

In order to collect autologous blood, the patient's physician must write an order/prescription which includes the date of the anticipated transfusion (usually date of the surgery) and the number of blood products needed (typically RBC products are needed, but some complicated surgeries may also require the use of plasma or platelet products). The criteria for acceptance of autologous donors are less stringent than those for allogeneic donors, because the goal of the questionnaire is to protect the individual from harm as a result of donation. Table 7.1 compares the criteria for allogeneic and autologous blood donation.

The blood center medical director has the ability to modify or set criteria for autologous donors, based on discussions with the donor/recipient's ordering physician and weighing the risk/benefit analysis of the safety of donation for the donor versus having or not having the blood products available for the planned surgery (or other reasons for the need for blood products by the recipient). In some cases, the health risk to the donor of donating blood may outweigh the benefit of having his or her blood available for transfusion.

## TABLE 7.1  Comparison of Criteria for Allogeneic and Autologous Blood Donors

| Criterion | Allogeneic Donors | Autologous Donors |
|---|---|---|
| Age | Conform to applicable state law or ≥16 years old | As defined by medical director |
| WB volume collected | ≤10 ml/kg, including samples | ≤10 ml/kg, including samples; collection of small volume units requires adjustment of the anticoagulant volume in order to obtain the appropriate ratio |
| Hemoglobin | 12.5 g/dl | 11.0 g/dl |
| Hematocrit | 38% | 33% |
| Intervals between donations of RBCs | 56 days after each WB donation | Typically 72 hours after each WB donation* |
| Blood pressure | ≤180/100 mmHg | As defined by medical director |
| Donor history | Full set of DHQ questions | Abbreviated, as defined by medical director: Heart and lung diseases Blood disorders Bleeding conditions Pregnancy |
| Medication and vaccination deferrals | Finasteride, dutasteride, isotretinoin, acitretin, etretinate, growth hormone from human pituitary glands, bovine or beef insulin, hepatitis B immune globulin, unlicensed vaccines | As defined by medical director |
| Heart rate (pulse) | 50–100 bpm, or <50 bpm if otherwise healthy athlete | As defined by medical director |
| Temperature | ≤37.5°C (99.5°F) orally | As defined by medical director; defer for conditions that present risk of bacteremia |
| Venipuncture site | Free from skin lesions and punctures/scars indicative of self-injectable drug (narcotic, etc.) use | As defined by medical director |
| Platelet count (for platelet donors) | ≥150,000/ul before procedure, if performed more frequently than once every 4 weeks Not required before first procedure, or if procedure is not performed more often than every 4 weeks | As defined by medical director |
| Contraindications | As per DHQ and blood center criteria | As defined by medical director, likely including: Unstable angina Recent myocardial infarction or cerebral vascular event Symptomatic heart or lung disease Untreated aortic stenosis |

*Last donation no less than 72 hours prior to date and time of need
Modified from Eder A (2008). Allogeneic and autologous blood donor selection. In: Roback JR, Combs MR, Grossman BJ, Hillyer CD (eds), *Technical Manual*, 16th edition. Bethesda, MD: AABB Press.

# CHAPTER 8

# Adverse Donor Reactions

Krista L. Hillyer, MD

A variety of adverse reactions and injuries can and do occur in blood donors as a result of the blood collection process. Approximately 4% of donors will experience some form of reaction, though most are mild vasovagal reactions and hematomas. The American Red Cross classifies complications into defined categories, and reports syncopal-type (vasovagal) and phlebotomy-related complications to occur at rates of ~300/10,000 and ~75/10,000 whole blood donations, respectively (rates are modestly lower in donors of apheresis-derived red blood cells). First-time donors, women, and donors less than 20 years of age are at higher risk for adverse events (see Figure 8.1). The most common reaction is classified as a vasovagal event, which can lead to loss of consciousness, seizure-like activity, and severe injury if precautions are not taken to avoid falls. Occasionally, adverse reactions such as nerve injury and injury from falling can lead to permanent disabilities, which have been estimated to occur with a frequency of <1:250,000 donations. Death from blood donation has been investigated by the FDA, including 22 events reported to the agency from October 2004 to September 2006. In their report the FDA concluded that no deaths were a direct result

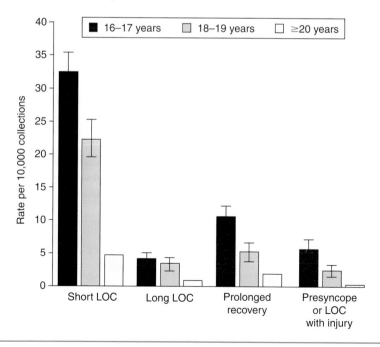

FIGURE 8.1 Adverse reactions in donors by age. LOC: loss of consciousness. From Eder AF, Hillyer CD, Dy BA *et al.* (2008). Adverse reactions to allogeneic whole blood donation by 16- and 17-year-olds. *JAMA* **299**, 2279–2286.

of the blood donation, though one could not be ruled out (http://www.fda.gov/Cber/blood/fatal0506.htm#g).

## Complications of Whole Blood Donation:

**Minor Local Tissue Injury:** Most adverse donor reactions are minor, and include bruising, soreness and hematoma. Small hematomas occur in ~75/10,000 donations. These complications can be treated with warm compresses and mild analgesics, and generally resolve completely within 7–14 days.

**Nerve Injury:** Local nerve injury occurs with an incidence reported in one study of 1 in 6,300 donations. Donor complaints include sensory changes in the forearm, wrist, hand or shoulder, as well as radiating pain. About 15% of affected donors will also report decreased arm or hand strength. Nerve injury can be due to nerve compression as a result of a large hematoma, which can be due to unrecognized arterial puncture. Direct nerve damage from the phlebotomy needle is an unusual event. Nerve injury is almost always transient, however, and 70% usually disappear within a month; almost all resolve within 1 year. Rare cases of complex regional pain syndrome (reflex sympathetic dystrophy) have been reported.

**Vasovagal Reactions:** The most common systemic donor reaction is the vasovagal reaction, occurring in ~250/10,000 donations. Predisposing factors include: (1) first-time donors; (2) donors with low weight; and (3) a history of previous donor adverse reaction. Complication rates after allogeneic whole blood donation are known to be higher in young and first-time donors, though the reasons for this are not known. Proposed mechanisms include changes in central thalamic pathways, vascular baroreceptor sensitivity, and age-dependent responses to physical and emotional stress. Anxiety related to blood donation, blood or needles in general may play a large part in vasovagal reactions as they can precipitate the reaction even before donation, and clusters of fainting ("epidemic fainting") can occur when prospective donors witness an untoward event at a blood drive. Symptoms of vasovagal reactions include lightheadedness, nausea, pallor, sweating, weakness, hyperventilation, and declaration of nervousness or complaints of any of these symptoms on the part of the donor. Vasovagal reactions can progress to loss of consciousness and seizure activity (tonic–clonic convulsive movements, tetany, or tonic–clonic seizures with loss of bladder or bowel control).

Interventions for symptoms of vasovagal reactions, including fainting, include:
- having the donor breathe into a paper bag,
- calmness and reassurance from the phlebotomist,
- elevation of the donor's legs above his or her heart,
- having the donor breathe in ammonium salts,
- cold compresses to the neck or forehead,
- loosening of tight clothing,
- ensuring that the donor has an adequate airway,
- monitoring blood pressure, pulse and respirations (vital signs) periodically until recovery, and
- reassurance from the collections staff.

**Prolonged Recovery:** Prolonged recovery is defined as presyncopal symptoms, with or without loss of consciousness, that do not resolve within 30 minutes. If the donor has not recovered from a vasovagal reaction or has unstable vital signs during the typical 5–30 minutes expected for complete recovery from the reaction, most blood collection centers not located within a hospital (in the US) should call 911 for emergency medical services (EMS) response. The staff members should also notify their facility medical director either prior to or following the 911 call, depending on the severity of the symptoms and/or resulting injury that occurs. The EMS staff will travel to the blood center or mobile collection site and proceed to evaluate whether or not the donor should be transported to an emergency room for further medical evaluation and treatment.

**Arterial Puncture:** Arterial puncture occurs at a rate of ~1 per 10,000 donations, and is more common among inexperienced phlebotomists than those with experience. The most common indicator is rapid filling of the collection bag, a total phlebotomy time of less than 4 minutes and a bright red color of the blood. This event can occur with the initial puncture, or after needle adjustment. Proper treatment requires immediate removal of the needle and application of pressure at the site of the injury. Rarely, pseudoaneurysms will complicate arterial puncture.

## Complications of Apheresis Donations:

**Apheresis-specific Reactions:** Apheresis (the withdrawal of blood with a portion being separated and the remainder re-transfused into the donor) removes a limited amount of blood from the donor, and with some systems the volume removed is replaced, so vasovagal reactions are less frequent (~1%) than for whole blood donation. In addition, apheresis donors are required to have higher blood volumes than whole blood donors. Hematomas and venepuncture complications can also occur in apheresis donors, as described above for whole blood donation. There are donor reactions which are specific to apheresis, including hypocalcemia, allergic reactions, and machine malfunction leading to thrombi, hemolysis, air emboli and leakages.

**Hypocalcemia:** Hypocalcemia is due to citrate anticoagulant, which binds calcium, used to keep the blood from clotting in the apheresis machine tubing. Usually hypocalcemia results in perioral or peripheral parathesias, nausea and shivering. Severe hypocalcemia can result in loss of consciousness, convulsions or cardiac arryhthmia. Hypocalcemia occurs in up to 10% of collections, but is severe in less than 0.03% of collections. Treatment of hypocalcemia includes reducing citrate infusion rate, administering calcium tablets and, for severe cases, terminating the collection and administering intravenous calcium.

**Machine Malfunction:** Machine malfunction leading to thrombi, hemolysis, air emobolism or leakage occurs in less than 0.5% of collections, but the effects can be severe. It is usually secondary to improper manufacturing of the disposable set or improper mounting of the set. Newer apheresis devices have improved safety features to protect the donor and mitigate these complications.

**Allergic Reactions:** Donors may be hypersensitive to sterilizers, especially ethylene oxide, or to components of the apheresis set.

**Platepheresis/Granulocytapheresis:** There are other considerations that must be taken into account for processes that extract platelets or granulocytes. In some instances frequent plateletpheresis can cause a drop in platelet levels that, in some cases, will cause collections to be discontinued altogether. Granulocytapheresis can also cause certain side-effects. Donors who take small doses of corticosteroids may experience insomnia. Furthermore, a combined dose of G-CSF and dexamethasone may cause insomnia, mild bone pain, and headaches (see Chapter 32).

**Approach to the Donor and Donation Process:** For all adverse reactions, the staff should terminate the donation. The donor experiencing an adverse reaction should then be taken to a private area away from others. It is important to reassure the donor, contact the blood bank physician and perhaps call for emergency medical services, depending on the severity of the reaction and the length of the recovery period after the reaction occurs. Another important consideration is to provide the donor with personal attention and reassure him or her throughout the phlebotomy procedure. It is of equal importance to maintain an adequate number of staff members. Interventions to prevent injury from falling should also be implemented when possible. Newman has studied ingestion of water within 30 minutes of the start of donation, and reported a decrease in vasovagal reactions.

Recommended Reading

Eder AF, Dy BA, Kennedy JM *et al.* (2008). The American Red Cross donor hemovigilance program: complications of blood donation reported in 2006. *Transfusion* **48**, 1809–1819.

Eder AF, Hillyer CD, Dy BA *et al.* (2008). Adverse reactions to allogeneic whole blood donation by 16- and 17-year-olds. *JAMA* **299**, 2279–2286.

Fatalities reported to the FDA following blood collection and transfusion (available at http://www.fda.gov/Cber/blood/fatal0506.htm#g).

Newman BH. (2004). Blood donor complications after whole-blood donation. *Curr Opin Hematol* **11**, 339–345.

Newman BH, Waxman DA. (1996). Blood donation-related neurologic needle injury: evaluation of 2 years' worth of data from a large blood center. *Transfusion* **36**, 213–215.

Newman BH, Pichette S, Pichette D, Dzaka E. (2003). Adverse effects in blood donors after whole-blood donation: a study of 1000 blood donors interviewed 3 weeks after whole-blood donation. *Transfusion* **43**, 598–603.

# CHAPTER 9

# Component Preparation and Manufacturing

Courtney E. Greene, MD and Christopher D. Hillyer, MD

Blood components, including red blood cells (RBCs), plasma, platelets and granulocytes, are prepared either from whole blood donation or by automated apheresis. Blood component manufacturing allows each component to be stored under optimal conditions, and component therapy has largely replaced the use of whole blood due to the ability to choose components and component constituents that target specific patient needs. Component therapy is also considered to be medically and fiscally efficient.

**Whole Blood:** Whole blood is the most common starting product for component preparation and manufacturing. Whole blood is generally not stocked in most blood banks today because component therapy is more appropriate to target a patient's specific indications for transfusion (for example, RBC products for symptomatic anemia, plasma products for coagulation factor deficiencies, and platelet products for thrombocytopenia). In addition, platelet function is lost when whole blood is stored in the refrigerator, and the coagulation factor activity, especially for labile factors (Factors V and VIII), deteriorates over time. In general, 500 ml of whole blood is collected into a bag with 70 ml of anticoagulant-preservative solution (see below), creating a product with a final hematocrit ≥38%. Whole blood is stored at 1–6°C. Depending on the manufacturer's system, different hold times are allowed for before the preparation of blood components from whole blood. If platelet products are to be manufactured from whole blood, the whole blood product must be stored at room temperature until the platelets are removed.

Fresh whole blood is occasionally being used by the US military via "walking blood donors" (donors who are ABO/D typed, infectious disease tested and available to donate in times of need). This product is transfused as soon as possible and has a shelf-life of 24 hours when stored at room temperature, allowing maintenance of platelet function while still minimizing the risk of bacterial overgrowth.

**Component Manufacturing:** When whole blood is manufactured into components, it is collected into a primary bag containing an anticoagulant-preservative solution. The primary bag has up to three satellite bags attached, for RBC, platelet, plasma and/or cryoprecipitate component manufacturing. Due to the different specific gravities of RBCs (1.08–1.09), plasma (1.03–1.04) and platelets (1.023), differential centrifugation of the whole blood product is used to prepare blood components. Optimal component separation requires specific centrifugation variables, such as rotor size, speed, and duration of spin (see Figure 9.1).

**Anticoagulant-preservative Solutions:** Anticoagulant-preservative solutions allow blood components to be stored for extended periods of time, generally weird spacing without a significant detrimental effect on the quality of the RBC. The shelf-life of a

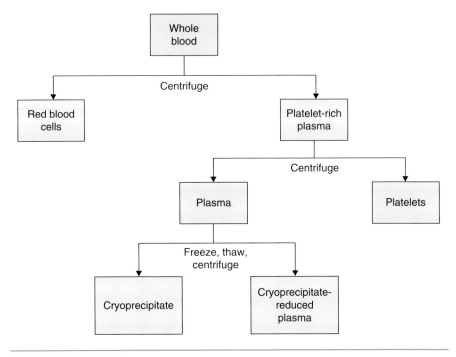

FIGURE 9.1  Manufacturing of blood components from whole blood.

given product is determined at the time of licensing of the product, and is chosen so that 75% of the transfused RBCs remain viable 24 hours after transfusion. The commonly used anticoagulant-preservative solutions are acid-citrate-dextrose (ACD), citrate-phosphate-dextrose (CPD), citrate-phosphate-dextrose-dextrose (CP2D) and citrate-phosphate-dextrose-adenine (CPDA-1). The citrate (sodium citrate and citric acid) acts as an anticoagulant, and the phosphate (monobasic sodium phosphate and trisodium phosphate), adenine and dextrose are substrates for cellular metabolism. ACD, CPD and CP2D allow for RBC storage of 21 days, whereas RBCs stored in CPDA-1 have a shelf-life of 35 days. Additive solutions with a number of differing constituents included to minimize the RBC storage lesion include Adsol 1, 3 and 5, and these solutions allow an outdate of 42 days (see below).

**RBC Components:** RBC products are used primarily for the treatment of symptomatic anemia or hemorrhage, to increase tissue oxygenation. RBC products are prepared either by centrifugation of whole blood followed by removal of the platelet-rich plasma layer, or by automated apheresis collection. RBCs stored in CPD or CP2D have a shelf-life of 21 days and in CPDA-1 of 35 days. RBCs stored in CPD, CP2D or CPDA-1 have a hematocrit of ~80% with a final volume of 225–350 ml and up to 110 ml of plasma. RBCs stored in additive solutions (see below) have a hematocrit of ~60% and can be stored for up to 42 days.

**Additive Solutions:** As above, additive solutions, such as AS-1, AS-3 and AS-5, can be added to the primary anticoagulant-preservative to increase the shelf-life of RBCs to 42 days. All of these solutions contain dextrose, adenine and sodium chloride; AS-1 and AS-5 contain mannitol, and AS-3 contains monobasic sodium phosphate, sodium citrate and citric acid. Additive solution products have a final volume of 350–400 ml and hematocrit of 55–65%, containing 100–110 ml of additive solution and 10–40 ml of plasma. In general, additive solutions must be added to the product within 24 hours of collection, depending on manufacturer's specifications.

RBC Modification: RBC products can be further modified for specific patients' needs. Multiple modifications can be performed on a single product. Examples of product modifications and their use include the following:

1. *Freezing*: Frozen RBC products, which are cryopreserved with glycerol, have an extended storage life of up to 10 years when stored at −65°C or below. The primary indications for frozen RBCs are for rare RBC products (i.e. products from donors who lack high frequency antigens) or autologous donations, or to maintain product inventory in disaster management. Frozen RBC products must be deglycerolized prior to transfusion. Deglycerolized products have a 24-hour outdate.

2. *Volume reduction*: RBC products are volume-reduced to remove the supernatant. The primary indications are to remove the accumulated potassium, which leaks from the RBCs during storage, in order to prevent hyperkalemia in at-risk patients such as neonates; to reduce the Adsol concentration, though this is rarely required; or to reduce the volume infused in volume-sensitive pediatric patients.

3. *Washing*: RBC products can be washed to remove the plasma and, with it, plasma proteins. This process can be used to minimize the risk of recurrence of severe allergic/anaphylactic reactions.

4. *Rejuvenation*: Rejuvenated RBCs have 2,3-DPG and ATP levels near to those of a freshly drawn unit. During storage, 2,3-DPG levels fall linearly towards zero during the first 2 weeks. ATP levels fall to 50–70% of initial levels during storage, depending on the anticoagulant-preservative solution used. Rejuvenation is usually performed prior to freezing RBCs. The RBCs must be washed prior to use to remove the cryoprotectant solution.

5. *Leukoreduction*: Leukoreduced RBCs are used to mitigate the risk of febrile non-hemolytic transfusion reactions, HLA alloimmunization, CMV infection and transfusion-related immunomodulation, each in appropriate and clinically indicated populations. Some centers and clinicians favor the availability of universally leukoreduced products.

6. *Irradiation*: Irradiated RBCs are used to prevent transfusion-associated graft versus host disease.

7. *Aliquots*: RBC products can be divided into smaller volumes for neonatal transfusions.

Plasma Components: The administration of plasma products is primarily to treat or prevent coagulopathy secondary to multiple plasma factor deficiencies. Plasma contains all of the coagulation factors in addition to albumin and fibrinogen. Plasma products are manufactured either from whole blood or by automated apheresis. The plasma products that are used in the transfusion service include fresh frozen plasma (FFP), plasma frozen within 24 hours of phlebotomy (FP24), and thawed plasma (TP). FFP must be frozen within 6–8 hours of collection, while FP24 is frozen between 8 and 24 hours of collection. FP24 has lower levels of Factors V and VIII than FFP, but this reduction is rarely of clinical significance. FFP and FP24 are stored at −18°C or colder for up to 1 year (with approval from the FDA, FFP can be stored at −65°C or colder for up to 7 years). Prior to transfusion, plasma must be thawed at 30−37°C, which requires ~20 minutes; it is then stored at 1−6°C if not transfused immediately. Once thawed, FFP and FP24 must be transfused within 24 hours, otherwise it becomes classified as TP, which is stored at 1–6°C for up to 5 days. During the 5 days of storage, Factor VIII activity levels decrease by approximately 30%. FFP and FP24 are used interchangeably in most situations, and TP is often acceptable in many clinical circumstances; its use is driven primarily by the human and fiscal cost of discarding previously thawed products, and by the necessity of having plasma available in emergent situations.

Plasma products may be further manufactured into cryoprecipitate and cryoprecipitate-reduced plasma. In addition, outside of the US, large pools of plasma are manufactured into solvent/detergent-treated plasma, which has minimal (if any) risk of transmitting lipid-enveloped viruses (HIV, HCV, and HBV).

Plasma (stored at −18°C or below) and liquid plasma (stored at 1−6°C) are manufactured from whole blood no later than 5 days after the whole blood expiration date, and are converted into an unlicensed product termed "recovered plasma" that is shipped for fractionation into albumin, antithrombin III, Factor VIII or Factor IX concentrates, or immunoglobulin preparations.

Cryoprecipitate: The primary indication for the use of cryoprecipitate (the FDA term is *cryoprecipitated AHF*) is for replacement of fibrinogen. Cryoprecipitate is prepared by slowly thawing FFP to 1–6°C, thereby leading to formation of a precipitate, which is collected and refrozen to make cryoprecipitate. Each unit of cryoprecipitate, which is approximately 15 ml, must contain fibrinogen (≥150 mg) and Factor VIII (≥80 IU). In addition, each unit contains von Willebrand's Factor (VWF) (80–120 IU), Factor XIII (40–60 IU) and fibronectin. Cryoprecipitate is usually administered to adults in doses of 8–10 donor units, which are pooled prior to transfusion (known as a cryopool). Some blood centers manufacture pre-pooled cryoprecipitate, where five units are pooled together prior to storage. This product is easier for the transfusion service to use in cases of emergency than pooling prior to issue (post-storage). Cryoprecipitate units can be stored for up to 1 year at −18°C, but must be transfused within 4 hours of pooling using an open system, or within 6 hours of thawing.

Cryoprecipitate-reduced Plasma: Cryoprecipitate-reduced plasma is used exclusively for plasma exchange or transfusion in patients with thrombotic thrombocytopenic purpura. Cryoprecipitate-reduced plasma is FFP from which the cryoprecipitate has been removed, and therefore contains decreased amounts of VWF, Factor VII, Factor XIII and fibrinogen. Cryoprecipitate-reduced plasma is stored at −18°C with an expiration date of 1 year, and once thawed at 30–37°C is stored at 1–6°C for up to 5 days.

Plasma Derivatives: Plasma derivatives are prepared by cold ethanol fractionation of large pools of plasma, and include albumin, immune globulin (Rh immune globulin and intravenous immunoglobulin), and coagulation factor concentrates (Factor VIII, Factor IX and antithrombin III).

**Platelet Components:** Platelet products are used for the treatment or prophylaxis of bleeding secondary to thrombocytopenia or dysfunctional platelets. Platelets are prepared from whole blood (also known as whole blood derived, platelet concentrates, or random donor platelets) or via automated apheresis (also known as apheresis derived platelets, "platelets,pheresis" (said platelets comma pheresis), plateletpheresis, or single donor platelets). Whole blood derived platelets have a volume of 40–70 ml and must contain $\geq 5.5 \times 10^{10}$ platelets. The usual adult dose of whole blood derived platelets is a pool of four to six units. Apheresis platelets have a volume of ~300 ml and must contain $\geq 3.0 \times 10^{11}$ platelets in 75% of the products tested. Most apheresis systems use ACD-A as the anticoagulant-preservative solution. The majority of apheresis devices provide products which are leukoreduced to a residual WBC count of $<5 \times 10^6$ white blood cells (termed process leukoreduction) and have minimal RBC contamination.

Platelets are stored at room temperature (20–24°C) with gentle agitation for a maximum of 5 days. Once the system has been opened or the platelets pooled, the product must be transfused within 4 hours.

**Buffy Coat Platelets:** Outside of the US, platelets can be prepared from whole blood using a buffy coat method, which differs from the US method (typically referred to as the platelet-rich plasma (PRP) method). Instead of removing the platelet-rich plasma after softly centrifuging whole blood followed by separation of platelets from plasma using a hard spin (the PRP method), the whole blood first goes through a hard spin, after which the RBCs and plasma are removed leaving the buffy coat. The buffy coat is then softly spun, after which the residual white blood cells are removed, leaving the platelet concentrate. Four to six buffy coat platelet concentrates are then pooled to create a single pooled platelet product. Studies have shown no difference in quality between these two methods of preparing platelet concentrates; however, the buffy coat method has several advantages, including a decrease in residual WBC, possible automation of platelet preparation from whole blood, and the ability to prepare platelet pools up to 24 hours after collection rather than within 4 hours of transfusion, as is the case for PRP platelets in the US.

**Platelet Modification:** Platelet products can be further modified for specific patients' needs. Multiple modifications may be performed on a single product. Examples of product modifications and their use include the following:

1. *Volume-reduction*: Platelet products are volume-reduced to remove the supernatant. The primary indications are to prevent hemolytic reactions by removing the plasma in ABO-incompatible products, or to prevent volume overload in at-risk patients.
2. *Washing*: Platelet products are washed to remove plasma proteins, in order to prevent recurrence of severe allergic/anaphylactic reactions.
3. *Leukoreduction*: Leukoreduced platelets are used to mitigate febrile non-hemolytic transfusion reactions, HLA alloimmunization, CMV infection, and transfusion-related immunomodulation.

4. *Irradiation*: Irradiated platelets are used to prevent transfusion-associated graft versus host disease.

5. *Aliquots*: Apheresis platelet products can be divided into smaller volumes for neonatal transfusions.

**Granulocytes:** Granulocytes, administered for treatment of patients with neutropenia and life threatening infection, have a 24-hour shelf-life. The use of granulocytes varies due to issues with rendering timely viral testing and in transporting of the product from collection to transfusion service during that timeframe, and the presumption of the collection of inadequate doses of granulocytes using currently accepted methods of donor preparation and collection. Granulocytes are usually collected using apheresis devices, but they may also be prepared from the buffy coat of centrifuged whole blood for neonatal transfusions. Apheresis-derived granulocytes have a volume of about 200 ml and contain RBCs (20–50 ml), platelets ($3 \times 10^{11}$) and plasma in addition to granulocytes ($\geq 1 \times 10^{10}$). Modalities used to improve the granulocyte dose are to stimulate the donor with granulocyte colony stimulating factor (G-CSF) and/or corticosteroids, though this method and its resultant clinical efficacy have not been proven as yet in randomized clinical trials, and/or to use hydroxyethyl starch (HES) to improve sedimentation and collection during apheresis. The use of all three modalities should increase the product yield to $>1 \times 10^{11}$ granulocytes. These products should be transfused as soon as possible, but they may be stored for up to 24 hours after collection at 20–24°C without agitation. All granulocyte products should be irradiated to prevent transfusion-associated graft versus host disease, but they cannot be leukoreduced and, therefore, CMV-seronegative products may be indicated for at-risk patients.

**Component Labeling:** Blood components must be labeled in accordance with FDA regulations and AABB *Standards for Blood Banks and Transfusion Services* (Standard 5.1.6.3). The following requirements apply:

- The labeling must conform with the most recent version of the United States Industry Consensus Standard for the Uniform Labeling of Blood and Blood Components using ISBT 128 (as of May 1, 2008).
- The original label and added portions must be attached to the container, including the ABO/Rh type, donation identification number, product code, and facility identification.
- All modifications must be specified.
- If a new label is applied, the process must be accurate.
- The labeling process must include a second check to ensure accuracy.

**Recommended Reading**

Kauvar DS, Holcomb JB, Norris GC, Hess JR. (2006). Fresh whole blood: a controversial military practice. *J Trauma* **61**, 181–184.

Slichter SJ. (2007). Platelet transfusion therapy. *Hematol Oncol Clin North Am* **21**, 697–729.

# CHAPTER 10
# Serologic Testing of Donor Products

Krista L. Hillyer, MD

Each component must have its ABO and D-antigen status tested as a primary mechanism of preventing ABO-incompatible blood transfusions and D-antigen sensitization. Additionally, components must be tested for the presence of unexpected, clinically significant antibodies to prevent potential recipient hemolysis. RBC products often undergo further RBC antigen characterization in order to supply blood products for patients with alloantibodies or patients with chronic anemia (e.g. sickle cell disease) who require "phenotypically-similar" RBC components.

Most blood centers use serologic methods to perform these tests, but the use of molecular testing is growing. Serologic methods include tube testing (also known as the "wet" method), the gel method (also known as "column agglutination"), or solid phase testing. These tests, and the various methods, are discussed in more detail in Chapter 19.

ABO Group Typing: The ABO blood group phenotypes include A, B, AB and O. The reciprocal antibodies are consistently present in the sera of a majority of individuals without previous red blood cell (RBC) exposure (e.g. anti-B antibodies in blood group A patients), and these antibodies may result in severe intravascular hemolysis after transfusion of ABO-incompatible blood components. The prevention of ABO-incompatible transfusion is the primary objective of pretransfusion testing.

ABO type is determined by testing donor RBCs with anti-A and anti-B (also known as a "forward" or "front" type) and the donor plasma with group $A_1$ and group B RBCs (also known as a "reverse" or "back" type). Discrepancies between the front and back type must be resolved prior to labeling a blood component (see Chapters 19 and 22). RBC genotyping can be used to resolve typing discrepancies.

D-antigen Phenotype: The D antigen (also termed the Rh or Rh(D) type) is the most immunogenic of the RBC antigens, and therefore the D antigen is second to the ABO blood group in importance in transfusion medicine. The donor center must ensure that D-negative products are appropriately labeled, such that a recipient of a D-negative product does not form anti-D in response to transfused RBCs. Currently donor centers use typing reagents to detect the presence of the D antigen, and these reagents must be sensitive enough to detect the presence of weak D. Weak-D testing can be performed either by including the antihuman globulin phase, which takes additional time, reagents and controls, or through the use of certain automated techniques which are sensitive enough for "weak-D" detection such that a separate "weak-D" test is not required. D-negative donors in Germany and Austria were *RHD* tested by molecular methods, and about 1 in 1000 was found to have the *RHD* gene and express the D antigen, but at levels which are not detected by routine serologic tests (e.g. $D_{el}$). In addition, transfusion of these products with low levels of D antigen resulted in sensitization of the D-negative recipient in some instances. Therefore, these donors were removed from the D-negative donor pool (see Chapters 19 and 23).

**Antibody Screening for the Presence of Unexpected RBC Antibodies:**
Alloantibodies other than anti-A and anti-B (which are expected RBC antibodies) are found in <1% of the US donor population. Immunization to these RBC antigens typically requires RBC exposure, usually as a result of prior pregnancy or transfusion. Most blood centers test all donated blood for unexpected antibodies. While the AABB *Standards* state that blood from donors with histories of transfusion or pregnancy must be antibody-tested, it is often not practical to separate donor specimens by these criteria. In addition, the use of automated platforms eases testing by incorporating ABO and D testing and antibody screening on a single instrument.

In contrast to the patient setting, antibody testing of a donor may use donor plasma or serum tested against either pooled or individual reagent RBCs of known phenotypes (because of the decreased sensitivity of using pooled RBCs, patient pre-transfusion testing must be performed on at least two individual screening RBCs). Antibody screening procedures are discussed in Chapter 19, and antibody identification, including determining whether the antibody is clinically significant, is discussed in Chapter 20. Blood components that contain RBC antibodies can be used for transfusion, given the transfusion service has appropriate policies in place, this is usually the form of providing the products to patients who are antigen negative for the corresponding antibody.

**Phenotyping RBC Products:** In the US, the current practice is to match only for the ABO and D antigens, but there are two common situations when multiple antigen-negative RBC products are requested: (1) when recipients have corresponding alloantibodies; and (2) for special chronically transfused patient populations, such as patients with sickle cell disease (SCD), thalassemia and autoantibodies. The transfusion management of sickle cell disease and thalassemia patients is discussed in Chapter 45 and the transfusion management of patients with autoantibodies is discussed in Chapter 44.

Alloimmunization occurs in approximately 2–6% of patients who receive RBC transfusions, but the rate of alloimmunization may be as high as 36% in patients with SCD. Recipients with multiple alloantibodies challenge a blood bank's ability to provide antigen-negative, compatible RBCs for transfusion, because blood banks must phenotype many times the number of units necessary to find an appropriate product. The number of products screened multiplies, depending on the RBC antigens' prevalence in the donor population and the number of antigens to which the recipient has made antibodies. This requires the donor center to utilize more laboratory technologists' time, maintain a larger RBC inventory, and have available all appropriate reagents. A major goal of the mass-scale genotyping process is to allow for the expansion of phenotype/genotype matching for a greater number of patients, thereby improving unit availability.

There are patient populations that benefit from receiving phenotype-matched products, especially those who are chronically transfused or at increased risk for alloantibody formation. Phenotype-matched products can be limited to the C, E and K antigens or extended to include $Fy^a$, $Jk^a$, $Jk^b$, S and other antigens. The fundamental reason to provide phenotypically-similar products is to prevent alloantibody formation and the subsequent negative consequences of hemolytic transfusion reactions.

The downside of this precise matching practice is that it makes routine transfusion more difficult for both the donor center and the transfusion service. Therefore, currently, phenotype matching is only applied to specific patient populations as described above.

Donor centers currently screen and stock RBC products to keep a pool of frequently needed antigen-negative products. Usually donor centers screen products from repeat group O donors and family members of patients who have formed alloantibodies to high-prevalence antigens for rare blood types. Batch serologic screening is technologist time-intensive to perform, typing reagents are expensive, and appropriate controls are required. As a result, donor centers have algorithms to help determine the likelihood of an antigen-negative product, based on limited phenotyping, and donor race and ethnicity.

The American Rare Donor Program (ARDP) has a list of over 30,000 individuals, compiled by the AABB and the American Red Cross (ARC), who are active blood donors with a blood type that occurs in less than 1 in 10,000 people. This Program supplies these rare RBCs all over the world, but mostly within the United States (US). It relies on the above methods to find these products, and the continued goodwill of these donors to maintain an adequate supply. In addition, the ISBT maintains a rare blood donor program which is compiled and maintained by the International Blood Group Reference Laboratory (IBGRL). The use of large-scale genotyping methods would enable increased identification of these rare RBC products and corresponding donors.

High-throughput molecular technologies in the field of transfusion medicine make it possible to determine multiple RBC antigen single nucleotide polymorphisms (SNPs) simultaneously. There are multiple mass-scale genotyping platforms in use and in development in North America and Europe. The above concept was demonstrated when 2355 donors were genotyped to predict for the presence of K, k, $Jk^a$, $Jk^b$, $Fy^a$, $Fy^b$, M, N, S, s, $Lu^a$, $Lu^b$, $Di^a$, $Di^b$, $Co^a$, $Co^b$, $Do^a$, $Do^b$, $Jo^a$, Hy, $LW^a$, $LW^b$, $Sc^1$, $Sc^2$ and HgbS, and 21 rare donors were identified (Co(a − b+), Jo(a−), S − s−, and K + k−).

**Antibody Titer Anti-A, Anti-B:** Blood products, especially platelet products from group O donors, may be tested for the presence of high-titer anti-A (more commonly) and/or anti-B in order to prevent hemolytic transfusion reactions secondary to ABO incompatibility. Currently in the US this is inconsistently practiced, with a variety of methods and cut-off values used to determine which products are "high titer" (and therefore at higher risk of causing a hemolytic reaction). The Biomedical Excellence for Safer Transfusion (BEST) collaborative is currently investigating this practice throughout the world.

**International:** Outside the US, more extensive RBC phenotype matching and other serological testing may occur. For example, in the UK it is routine to test RBC components for the K antigen, and K-negative RBC products are recommended for the transfusion of females with child-bearing potential (i.e. under the age of 55 years). In addition, the UK screens blood components for the presence of high-titer anti-A and/or anti-B using a 1/100 dilution with group AB RBCs on an automated platform. Components that are negative are labeled appropriately. Components with high-titer anti-A and/or anti-B are reserved for ABO group-specific transfusions.

## Recommended Reading

Hillyer CD, Shaz BH, Winkler AM, Reid M. (2008). Integrating molecular technologies for RBC typing and compatibility testing into blood centers and transfusion services. *Transfus Med Rev* **22**, 117–132.

MacLennan S. (2002). High titre anti-A/B testing of donors within the National Blood Service (NBS) (monograph on the Internet; available at: http://hospital.blood.co.uk/library/pdf/hightit.pdf, last accessed June 30, 2008).

# CHAPTER 11

# Overview of Infectious Disease Testing

Christopher D. Hillyer, MD

Safe and available blood products are required for optimal function of any advanced medical system. Indeed, ensuring blood product safety relies on a number of important measures, including appropriate selection of donors from safe populations; careful and consistent donor history screening by questionnaire; laboratory testing for infectious disease markers by a variety of methods; and the use of good manufacturing practices, quality systems, and accreditation and inspection systems. The focus of this chapter is laboratory-based infectious disease testing of samples collected from, or concomitantly with, a whole blood or apheresis-derived allogeneic donation from non-remunerated volunteer donors in the US. Outside the US, practices are very similar in developed nations, though national requirements and regulatory bodies obviously differ. In developing nations, restricted fiscal resources have typically required decreased use of some tests, though this is changing dramatically.

**Background:** A substantial number of infectious diseases can be transmitted by transfusion, and the agents responsible for these infectious diseases have four elements in common, as listed in Table 11.1.

| TABLE 11.1  Criteria for Transfusion Transmission of Infectious Agents |
| --- |
| • There must be an asymptomatic, infectious phase in the blood donor. |
| • Agent viability must be maintained during storage. |
| • There must be a seronegative recipient population. |
| • The agent must be capable of inducing disease following transfusion. |

Starting in the 1940s, with the recognition that syphilis could be transmitted by transfusion, routine testing was implemented. Later, viral hepatitis and human immunodeficiency virus (HIV) would become the biggest concerns, and advances in testing methodologies and strategies, immunology and molecular biology led both to improved understanding of the biology of these infectious agents – including hepatitis B virus (HBV), hepatitis C virus (HCV) and HIV – and the development and implementation of sensitive and specific tests to mitigate transmission by transfusion.

**Approach to Testing:** In the US, both blood and blood components, and the pre-release tests used in component manufacturing are classified as "biologics" by the US Food and Drug Administration (FDA). As such, infectious disease tests used in the blood industry undergo extensive clinical trials followed, typically, by FDA licensure. A list of current FDA-licensed tests can be found at www.FDA.gov/cber/products/testkits.htm.

In general, the following scheme is utilized:

1. Each and every blood donation is tested, testing is performed on a sample drawn at the time of a donation, and donated units are placed in quarantine until test results are completed, deemed negative, and reviewed.

2. With the exception of most viral nucleic acid testing (NAT), each sample is tested individually. NAT is an expensive yet sensitive technology that reduces the window period. In order to decrease the cost without significantly decreasing the sensitivity, NAT is performed in most circumstances using a pooled aliquot from 16–24 donor samples.

3. Non-reactive tests are considered to be negative and the related blood unit and its components may be released from quarantine and issued as appropriate and needed.

4. Samples with initially positive tests are retested in duplicate. If both of the repeat test results are non-reactive, the sample is classified as negative and the corresponding components may be released from quarantine. If one or both of the repeat tests is reactive ("repeatedly reactive"), the unit cannot be released for transfusion and is destroyed.

5. Samples that are repeatedly reactive undergo confirmatory or supplementary testing, typically with different sensitivity and specificity characteristics than the initial screening tests described above. The results of the confirmatory and/or supplementary tests improve the accuracy and predictive value of the results and aid significantly in donor notification and counseling.

**Determination of Need and Requirement for Testing:** While a large number of infectious agents can be transmitted by transfusion, testing is performed for only a relatively small number of these (Table 11.2). Many factors are considered in determining the requirement for testing. These include prevalence of the organism in the blood donor population, the severity of the disease the organism can transmit, the availability and efficacy of treatment in the recipient population, and the existence and availability of testing technologies, reagents, methodologies and test kits.

In the United States, there is no agreed upon formula for the implementation of new tests or the removal of tests that no longer substantially improve the safety of the blood supply. Some tests are mandated in the FDA Code of Federal Regulations (CFR), others have been adopted as a "community standard" without being mandated, and some are used only in specialized patient populations (Table 11.2). At the time of this writing, tests for Dengue and transmissible spongiform encephalopathies are being considered for development, testing or adoption.

**Selection of a Testing Methodology:** In general, test methodologies employed as screening tests for blood products have higher sensitivity (and correspondingly reduced specificity) as compared to diagnostic tests that are used in populations with a high infectious agent prevalence. Analytic and epidemiologic sensitivity are important in order to detect the presence of an infectious agent in the blood of a donor.

**The Biology of an Infection, the Window Period and Testing Strategies:** Reflecting on the biology of any given infection allows for an understanding of the

**TABLE 11.2 Agents that can be Transfusion-transmitted and Related Initial Testing Practices**

| Class | Name | Method | Requirement |
|---|---|---|---|
| *Viruses* | HIV | Ab, Ag*, NAT | CFR (Ab), CS (NAT) |
| | HBV | Ag,[1] Ab,[2] NAT** | CFR (Ag) |
| | HCV | Ab, NAT | CFR (Ab), CS (NAT) |
| | WNV | NAT | CS |
| | HTLV I/II | Ab | CFR |
| | CMV | Ab, NAT | SPP (Ab), RO (NAT) |
| | EBV | Ab | SPP (Ab) |
| *Bacteria* | *T. pallidum* | Ab | CFR |
| *Parasites* | *T. cruzii* | Ab | CS |

*HIV Ag is not typically used when HIV NAT is employed; **HBV NAT is not routinely used in the United States. [1]HBV surface antigen; [2]HBV core antibody. Ab, antibody; Ag, antigen; CFR, Code of Feceral Regulations; CS, community standard; HBV, hepatitis B virus; HCV, hepatitis C virus; HIV, human immunodeficiency virus; HTLV I/II, human T-lymphotrophic virus I and II; NAT, nucleic acid amplification test; SPP, specialized patient populations; RO, research only; WNV, West Nile virus.

clinical significance of any given infectious disease test, and directs the testing strategy. Figure 11.1 shows the clinical course of disease and related test markers following infection with HBV, which also closely mirrors that seen with HIV, HCV and West Nile Virus (WNV).

Immediately following infection (time 0) is the so-called *window period*. Initially described for HIV antibody, the window period was the time during which a blood product was infectious but antibody could not be detected. With the institution of HIV antigen and, later, HIV NAT, the window period "closed" significantly. Indeed, a window period exists for any viral infection and includes both the tissue or eclipse phase, in which there is early infection in tissues but no viremia, and the viremic phase, in which viral particles are present in the bloodstream but are below the limits of detection by current screening test methodologies.

Figure 11.1 shows that antigen and NAT positivity occur early in the course of infection, while the generation of antibody occurs sometime later following an immune response. Thus, testing strategies could conceivably employ HBV antibody, antigen and DNA detection. Specific testing strategies for each of the infectious agents are described in detail in the following chapters, and each of these strategies is based on understanding a diagram similar to that in Figure 11.1. During the course of infection there is an initial period where the patient is viremic, but current testing methodologies are unable to detect the virus (i.e. the window period), next the virus can only be detected through nucleic acid testing (presented in the figure as HBV DNA), then the virus can be detected through the serological testing of the antigen (HBsAg), and lastly the individual will have serologically detectable antibody (anti-HBc).

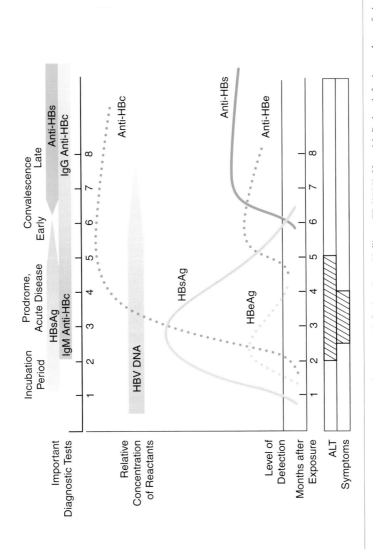

FIGURE 11.1 Serologic and clinical patterns during the stages of acute HBV infection. From Hollinger FB (2008). Hepatitis B virus infection and transfusion medicine: science and the occult. *Transfusion* **48**, 1001–1016.

## Recommended Reading

Coste J, Reesink HW, Engelfriet CP *et al.* (2005). Implementation of donor screening for infectious agents transmitted by blood by nucleic acid technology: update to 2003. *Vox Sang* **88**, 289–303.

Hollinger FB. (2008). Hepatitis B virus infection and transfusion medicine: science and the occult. *Transfusion* **48**, 1001–1016.

Stramer SL, Glynn SA, Kleinman SH *et al.* (2004). Detection of HIV-1 and HCV infections among antibody-negative blood donors by nucleic acid-amplification testing. *N Engl J Med* **351**, 760–768.

Stramer SL, Custer B, Busch MP, Dodd RY. (2006). Strategies for testing blood donors for West Nile virus. *Transfusion* **46**, 2036–2037.

Zou S, Notari EP IV, Stramer SL *et al.* (2004). Patterns of age- and sex-specific prevalence of major blood-borne infections in United States blood donors, 1995 to 2002: American Red Cross blood donor study. *Transfusion* **44**, 1640–1647.

# CHAPTER 12
# HIV Screening of Donor Products

Krista L. Hillyer, MD

Human immunodeficiency virus (HIV) transmission by intravenous administration of infected blood products is highly efficient. Accordingly, shortly after the discovery of HIV-1 in 1983 donor deferral measures were initiated, and the first test was licensed for donor screening in 1985. The effectiveness of donor deferral and testing procedures has decreased the rate of infection from blood transfusion from a peak incidence of HIV of ~1:100 transfused products in 1982 in the San Francisco area to current estimates of <1:2 million products.

**Description:** HIV is a lentivirus, which is a subgroup of the retrovirus family, and is the causative agent of Acquired Immune Deficiency Syndrome (AIDS). Retroviruses are RNA viruses with the presence of viral particle-associated reverse transcriptase and a unique replication cycle. The virus particles attach to the cell membrane (in the case of HIV, CD4+ lymphocytes) and subsequently enter the host cell, then the reverse transcriptase enzyme copies viral RNA into cDNA (complementary double-stranded DNA), and the cDNA is then integrated into the host cell's genome. Subsequent transcription, processing and translation of viral genes are mediated by the host cell enzymes. Particles then bud from the plasma membrane and infect other cells. In addition, the virus can spread by fusion of infected and uninfected cells, or by replication of the integrated viral DNA during mitosis or meiosis.

**Infection:** HIV infection can be transmitted through sexual contact, childbirth, breast-feeding, and parenteral exposure to blood. The CDC reported that, in 2006 in the US, the highest incidence of new disease was in African Americans, men, aged 25–44, who have male-to-male sex. Of new infections, 50% were from male-to-male sex, 33% from high-risk heterosexual contact, and 13% from intravenous drug use.

Approximately 60% of acute HIV infections result in a non-specific flu-like illness with an incubation period of 2–4 weeks. The acute infection resolves in weeks to a few months, resulting in an asymptomatic period that may last for years. During this period the HIV viremia persists, and the number of CD4+ lymphocytes, which are the primary target of HIV, gradually declines. This loss of CD4+ lymphocytes results in opportunistic infections and the virus has direct viral effects on multiple organs; these together result in death on average after 8–10 years. The course of disease has dramatically changed with the advent of potent anti-retroviral therapy, which has greatly prolonged survival. These medications do not eradicate HIV, though, and have multiple side-effects. Moreover, resistant viral strains have developed, which adds to the difficulty in treating HIV-infected patients.

**HIV Types:** HIV types 1 and 2 (HIV-1 and HIV-2) infections both cause AIDS. The HIV-1 family is divided into main (M), outlier (O), and non-M non-O (N) groups. Group M has 11 distinct subtypes or clades (A–K). In the US, clade B is almost exclusively prevalent. The greatest genetic diversity is in central Africa. Group O is most common in Cameroon and the surrounding West African countries, where it represents 1–2% of HIV infections. Group O infection is very rare in the US, and usually occurs in immigrants. Previous generations of HIV antibody assays did not reliably detect group O, but current assays have increased sensitivity to group O and other unusual variants. Still, the FDA continues to recommend permanent deferral of blood donors who have been born, resided or traveled in West Africa since 1977, or have had sexual contact with someone meeting these criteria. HIV-2 is also rare in the US (one infected donor identified out of 7.2 million donations), with no reported cases of HIV-2 transfusion transmission in the US.

## Methods and Interventions to Ensure the Safety of the Blood Supply:

**Donor Deferral:** Prior to the discovery and characterization of HIV, the risk of transfusion-transmitted HIV (TT-HIV) was estimated to be as high as 1:100 units in some cities. As the growing epidemic was thought at that time to be transmitted only by male-to-male sexual contact, individuals with any history of male-to-male sexual contact since 1977 were asked to self-defer (i.e. not donate blood). Donor deferral measures initiated in 1983 significantly decreased TT-HIV risk before tests for HIV-1 became available in 1985.

**Early HIV Testing – Anti-HBc and Anti-HIV:** For the period of time prior to the discovery of HIV but during which the signs and symptoms of AIDS were present in individuals in the US, hepatitis B core antibody testing (anti-HBc) was utilized as a surrogate marker for screening donor blood for HIV. Once the pathogen causing AIDS was identified as HIV, screening tests for antibodies to HIV-1 (anti-HIV-1) were quickly developed and became available in 1985.

In 1992, the FDA required that donor blood be tested for HIV-2 as well as HIV-1; combination tests for HIV-1/2 were available by that time and thus were quickly implemented.

Anti-HIV testing was, and is, accomplished by ELISA, and repeat-reactive tests are confirmed by western blot testing for the presence of antibodies directed against the HIV proteins of the *gag, pol* and *env* genes. Any blood units with a positive test of any type are destroyed, and the donors are indefinitely deferred from future blood donation. Introduction of anti-HIV testing decreased the risk of TT-HIV from an estimated 1:1000 products to 1:63,000 products transfused; later, with improvements in antibody test sensitivity, the estimated risk further decreased to 1:110,000.

Testing for HIV antibodies alone was not sufficient for the blood donor population, as "window period" infections accounted for a significant number of cases of transfusion-transmitted HIV. Immediately following infection (time 0) is the so-called "window period" (Figure 12.1). Initially described for HIV antibody, the window period is the time during which a blood product is infectious but antibody cannot be detected. The initial anti-HIV tests had a window period of approximately 3 months. With the institution of improved HIV antibody tests (and subsequent HIV antigen and, later, HIV NAT testing), the window period "closed" significantly.

**HIV p24 Antigen:** HIV antigen (p24 antigen) testing was approved for use for testing blood donors in 1996. As per Figure 12.1, finding the presence of an identifiable

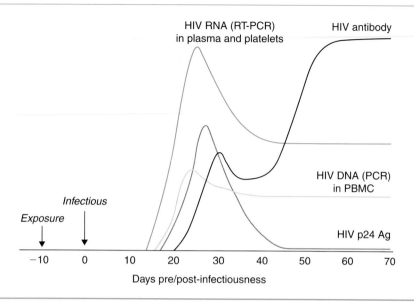

**FIGURE 12.1** Virologic events during primary HIV infection. After initial infection and propagation of HIV in lymph nodes, a blood donor becomes infectious (defined as day 0), with HIV RNA being detectable in plasma on days 14–15. HIV DNA is detectable in leukocytes at days 17–20, and HIV antibodies are detectable between days 20 and 25. Anti-HIV persists indefinitely but may be lost in the pre-terminal stage of the disease, in parallel with a surge in viral burden, indicating collapse of the immune system. HIV, human immunodeficiency virus; RT-PCR, reverse transcriptase polymerase chain reaction; PBMC, peripheral blood mononuclear cells. From Brecher ME (ed.). (2005). *AABB Technical Manual*, 15th edition, Bethesda: AABB Press.

HIV antigen was a significant advance, as p24 antigen could be identified in the blood of infected donors earlier than anti-HIV. By decreasing the window period to ~3–4 weeks, it was estimated that p24 antigen screening prevented 25% of cases of TT-HIV, which translated to the elimination of approximately five to ten cases of TT-HIV per year and an approximate risk of 1 : 200,000 to 1 : 450,000 transfused products. Although significantly improved, even greater safety from TT-HIV was demanded by the public as well as the FDA. Thus, a new method was applied to the testing of blood donors for HIV; a test that identified the presence of the viral nucleic acid.

**Nucleic Acid Testing:** With the advent of the polymerase chain reaction (PCR), which amplifies DNA, and reverse transcriptase-PCR, which reverse-transcribes RNA to DNA and then amplifies the DNA, it was recognized that nucleic acid amplification technology had the potential to identify HIV-positive blood donors very early following initial HIV infection (Figure 12.1). Thus, nucleic acid testing (NAT) for HIV (as well as for hepatitis C virus) was developed and tested in pooled donor samples under an FDA investigational new drug (IND) application.

By the year 2000, virtually all blood centers in the US were testing donated blood for HIV RNA by NAT. After the introduction of NAT, p24 antigen testing no longer provided added benefit in reducing the risk of TT-HIV in the US (or the EU), and the FDA allowed discontinuation of antigen testing, in the US, as long as HIV NAT was being used. Introduction of HIV NAT allowed the risk of TT-HIV to decrease from

~1:650,000 products to ~1:1,000,000 products transfused. Based on further refinements of the NAT tests, current-day estimates of TT-HIV in the US now range from ~1:2,000,000 to ~1:5,000,000 blood products transfused.

**Automation:** Currently, much of donor testing is automated, including HIV testing. One example is a multiplex NAT that is available for HIV/HCV combination ("combo") testing. Moreover, some large blood centers have centralized and consolidated their testing to regional centers across the US, while some continue to perform their testing in-house.

**Determination of Need and Requirement for Testing:** HIV meets all the criteria, as outlined in Chapter 11, for determining the importance of screening for a transfusion-transmitted infectious agent within the blood-donor population. Specifically, for HIV, these tenets are as follows:

- There may be asymptomatic, unidentified HIV-infected potential blood donors.
- HIV viability is maintained in all blood product storage conditions (RT 4°C −18°C).
- The potential recipient population is largely seronegative (in US likely >95% seronegative).
- HIV transmitted by transfusion efficiently causes HIV infection and AIDS.

Thus, HIV testing is of vital importance to all potential recipients of volunteer blood and/or components.

In the US, some HIV tests are mandated in the FDA Code of Federal Regulations (CFR). Others have been adopted as a "community standard" without being mandated. AABB *Standards* indicate that donated blood must be tested for anti-HIV-1/2 and HIV RNA by NAT; cited below are rules for how to notify and counsel donors who are positive for HIV at the time of blood donation.

> *Standard 5.8.4: Tests Intended to Prevent Disease Transmission by Allogeneic Donations.* A sample of blood from each allogeneic donation shall be tested for anti-HIV-1/2 and HIV RNA...Whole blood and components shall not be distributed or issued for transfusion unless the results of these tests are negative.
>
> *Standard 5.2.3: Donor Notification of Abnormal Findings and Test Results.* The medical director shall establish the means to notify all donors (including autologous donors) with any medically significant abnormality detected during the predonation evaluation or as a result of laboratory testing or recipient follow up...Appropriate education, counseling and referral should be offered. 21 CFR 630.6.

**Approach to Testing:** Blood, blood components and pre-release tests used in component manufacturing are classified as "biologics" by the US FDA. As such, infectious disease tests used in the blood industry undergo extensive clinical trials, followed in many cases by FDA licensure. FDA-licensed tests can be found at www.FDA.gov/cber/products/testkits.htm.

**Current US Methods and Approach:**

- Every blood donation is tested for HIV antibody (anti-HIV) and HIV RNA nucleic acid (HIV NAT). Testing is performed on a sample drawn from the donor at the time of blood donation, and each donated unit is placed in quarantine until test results have been completed, reviewed and found to be negative.

- HIV-1/HIV-2 antibody tests utilize a capture approach with reagents manufactured from specific viral peptides. The sensitivity of HIV antibody tests has increased over time, with studies demonstrating greater than 99% sensitivity with ELISA testing confirmed by western blot or immunofluorescense assay (IFA). Consequently, the current HIV *seronegative window period* has been reduced to ~20 days in the US. According to data from the CDC, the ELISA has a false-positive rate of 1 to 5 per 100,000 assays. Samples with initially positive results are retested in duplicate. Units with any repeat positive test results are destroyed and the donor indefinitely deferred from future donation. The confirmatory and/or supplemental tests for HIV are the western blot and an indirect immunofluorescence procedure. These allow determination of infectious state for donor counseling.
- HIV NAT testing can be done by two methods: (1) branched DNA (bDNA) signal amplification, or (2) transcription mediated amplification (TMA). The bDNA assay utilizes signal amplification from annealed DNA probes rather than direct amplification of the target sequence. In the TMA assay, viral RNA is captured on magnetic particles, copied by reverse transcriptase to produce transcription complexes, and amplified over one-billion-fold. Studies indicate that the TMA methodology is associated with higher sensitivity and specificity, while the bDNA technology is able to recognize a broad spectrum of HIV sequences and variants.
- HIV NAT testing is performed on pools of 16–24 donor samples (so called "mini-pool NAT"). NAT used on single donor samples (i.e. not pooled) is called individual donor NAT (ID NAT). For initial NAT screening, mini-pool NAT is used rather than ID NAT, as it significantly reduces expense without significantly decreasing safety risk, and results in less false positives to investigate. Any pool that is positive is "resolved" by ID NAT on all samples from the pool until it is clear which donor sample caused the positive result.
- HIV non-reactive tests are considered to be negative, and the related blood unit and its components may be released from quarantine and issued as appropriate.

**International Considerations:** The US testing methodologies and approach serve well as a template for that used in the rest of the developed world. Individual national experiences have allowed determination of a similar residual TT-HIV risk. Hourfar and colleagues published a recent review of the German Red Cross' experience with HIV-1, HCV and HBV NAT from 1997–2005; the 31,524,571 donations they studied represented approximately 80% of the blood collected in Germany during that time period. Of those >30 million donations, only 7 were positive for HIV-1. Thus, the residual risk in Germany of contracting HIV from a blood product was estimated at 1:4.3 million blood products.

In developing nations TT-HIV remains a significant issue, as population prevalence is substantially higher, quality control and monitoring of tracking and testing systems are less well developed and constructed, and fiscal resources for HIV testing are more limited. As the HIV epidemic in these areas, such as sub-Saharan Africa, is largely transmitted by male-to-female sexual contact, many currently accepted donor deferral criteria are unable to identify high-risk donors. In some locales, rapid test strips allow for pre-testing of donors.

## Recommended Reading

Alter HJ, Stramer SL, Dodd RY. (2007). Emerging infectious diseases that threaten the blood supply. *Semin Hematol.* **44**, 32–41.

Dodd RY, Notari EP, Stramer SL. (2002). Current prevalence and incidence of infectious disease markers and the estimated window-period risk in the American Red Cross blood donor population. *Transfusion* **42**, 975–979.

Evatt B. (2006). Infectious disease in the blood supply and the public health response. *Semin Hematol* **43**, S4–9.

Field SP, Allain JP. (2007). Transfusion in sub-Saharan Africa: does a Western model fit? *J Clin Pathol* **60**, 1073–1075.

Food and Drug Administration. (1996). FDA approves first test for HIV antigen screening of blood donors. Available at: http://www.fda.gov/bbs/topics/NEWS/NEW00529.html (last updated March 14, 1996).

Goodnough LT. (2005). Risks of blood transfusion. *Anesthesiol Clin North Am* **23**, 241–252.

Hourfar MK, Jork C, Schottstedt V. (2008). Experience of German Red Cross blood donor services with nucleic acid testing: results of screening more than 30 million blood donations for human immunodeficiency virus-1, hepatitis C virus, and hepatitis B virus. *Transfusion* **48**, 1558–1566.

Leiss W, Tyshenko M, Krewski D. (2008). Men having sex with men donor deferral risk assessment: an analysis using risk management principles. *Transfus Med Rev* **22**, 35–57.

Stramer SL, Chambers L, Page PL *et al.* (2003). Third reported US case of breakthrough HIV transmission from NAT screened blood (Abstract). *Transfusion* **43**, 40A.

Yee TT, Lee CA. (2005). Transfusion-transmitted infection in hemophilia in developing countries. *Semin Thromb Hemost* **31**, 527–537.

# CHAPTER 13

# Hepatitis B Screening of Donor Products

Krista L. Hillyer, MD

Post-transfusion hepatitis was a common consequence as a result of blood transfusion from 1940 to 1970. The initial step to decrease post-transfusion hepatitis was to eliminate high-risk donors (prisoners and paid donors); in addition, liver function screening tests were widely adopted (e.g. alanine aminotransferase [ALT]). Prior to the adoption of testing for hepatitis B surface antigen (HBsAg), the incidence of transfusion-transmitted hepatitis B virus (HBV; TT-HBV) infection was 6% in multi-transfused individuals. The continued decrease in incidence of TT-HBV infection has resulted from the increased sensitivity of HBsAg testing. Some countries outside the US have also adopted HBV nucleic acid testing (NAT).

In the US, the current risk of contracting HBV from a transfusion is 1 in 205,000 to 488,000 (first-time versus repeat blood donor). Of note, clinical cases of TT-HBV are infrequently reported: of the 7381 cases of acute hepatitis B reported to the Centers for Disease Control and Prevention in 2003, only 10 were confirmed to be transfusion associated; of these 10, only 1 case could be associated with an infected donor. Therefore, the calculated residual risk may be overestimated. This may be secondary to overestimates of HBV incidence, overestimates of the HBsAg-negative window period, underreporting (possibly due to asymptomatic infections) or incomplete investigations of cases, or widespread use of the HBV vaccine.

**Determination of Need and Requirement for Testing:** FDA requires testing allogeneic donors for evidence of infection with HBV (21 CFR 610.40), and AABB *Standards* specifically requires the use of HBsAg and anti-HBc to test donors for HBV infection (Standard 5.8.4). In addition, the donor must be notified and counseled if he or she has a positive test for HBV.

**Approach to Donor Testing:** In addition to the donor health questionnaire, which is reviewed in Chapter 5, to eliminate high-risk donors, each donor is tested for HBsAg and anti-HBc.

**HBsAg:** HBsAg, a HBV viral coat antigen produced in infected-cell cytoplasm in large quantities and that continues to be produced in patients with chronic, active HBV infection (Figure 13.1), is the primary screening test for HBV infection in the blood donor. HBsAg can be identified in an infected donor's serum or plasma by enzyme immunoassays (EIA) using animal antibodies (anti-HBs) as the solid-phase capture reagent and conjugated anti-HBs as the probe. Chemiluminescent labels are effectively replacing enzyme conjugates and chromogenic detection methods in automated testing strategies.

Although the epidemiologic sensitivities and specificities of current testing methods for HBsAg are high, the prevalence of detectable HBsAg in blood donors is low; therefore, the positive predictive value (PPV) of these tests is low as well. Thus, if a

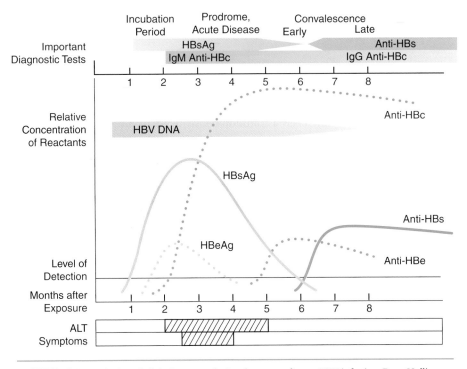

FIGURE 13.1 Serologic and clinical patterns during the stages of acute HBV infection. From Hollinger FB (2008). Hepatitis B virus infection and transfusion medicine: science and the occult. *Transfusion* **48**, 1001–1016.

donor tests repeatedly reactive for HBsAg, the positive test result must be confirmed by a neutralization test. Confirmation follows this general scheme:

- A donor is confirmed HBsAg positive when anti-HBsAg is added to the donor serum, and the HBsAg signal decreases by 50% or more from the control signal (where the antibody is not added), such that the Ag is neutralized. Confirmed donors are permanently deferred.
- If the donor serum is not neutralized but the donor is repeat reactive for anti-HBc, then the donor must be permanently deferred.
- If the donor's HBsAg reactivity is unconfirmed and the anti-HBc is non-reactive, then the donor may be retested after 8 weeks and reinstated if HBsAg is subsequently non-reactive.

**Anti-HBc:** Donors are tested for the presence of antibody to the core antigen of HBV (anti-HBc). This test was implemented in the mid-1980s as a surrogate marker to reduce the transmission of other hepatidities, especially non-A, non-B hepatitis (hepatitis C virus). Anti-HBc is present in either the chronic HBV carrier state or at the end of an acute resolving HBV infection. In the US, 1 in 37,000 to 54,000 donors is anti-HBc positive, HBV DNA positive and HBsAg negative.

The two available test methods for anti-HBc are: (1) a solid-phase inhibition immunoassay, with recombinant HBc as the capture reagent and a labeled, partially purified anti-HBc antibody as the probe (the presence of antibody in the donor serum inhibits the signal); and (2) a direct antiglobulin assay for anti-HBc, which has better specificity than the inhibition immunoassay method. No confirmatory test method for anti-HBc exists, but utilizing two different test methods increases the probability that any individual repeat reactive anti-HBc test result is either positive or negative. For this reason, most blood centers in the US allow a donor to continue to donate blood after one "confirmed" repeat reactive anti-HBc test, but donors are deferred following a second repeat reactive anti-HBc test or repeat reactivity using different test methods.

**HBV NAT:** TT-HBV occurs when blood donors who are negative for HBsAg are infected with HBV (i.e. during the window period). In an effort to decrease the window period and therefore reduce the risk of TT-HBV, the use of nucleic acid testing (NAT) has been investigated in the US, and implemented in some countries outside of the US. Figure 13.2 demonstrates the window periods using NAT versus HBsAg testing (minipool (MP)-NAT reduces the window period by 2–7 days compared to sensitive HBsAg screening tests). Therefore, MP-NAT has little effect on reducing the risk of TT-HBV compared with the most sensitive HBsAg screening methods in the US. HBV NAT testing is performed on large pools of samples for plasma sent for further manufacturing.

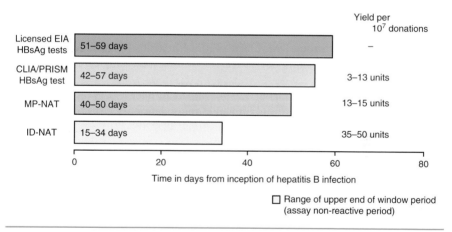

FIGURE 13.2  Detection differences and expected yields of hepatitis B surface antigen (HBsAg) and hepatitis B virus (HBV) nucleic acid amplification tests (NAT) assuming a NAT test sensitivity of 1000 copies/ml for a given sample in a pool of 20. CLIA, chemiluminescent assay; EIA, enzyme immunoassay; ID, individual donor; MP, minipool; NAT, nucleic acid amplification technology for HBV. From Comanor L, Holland P (2006). Hepatitis B virus blood screening: unfinished agendas. *Vox Sang* **9**, 1–12.

**International Considerations:**  Countries vary in their prevalences of HBV infection, such that the risk of TT-HBV increases in countries with a higher prevalence of HBV. Therefore, high-prevalence countries (such as South Africa and Indonesia) have the most to gain from implementing HBV NAT. In these countries, the yield of HBsAg-negative, HBV NAT-positive units ranges from 0.06% in Singapore to 0.5% in Indonesia and Ghana, in contrast to the yield in the US (which also uses anti-HBc) of 0.0003%. Yet some of these countries lack the resources for large-scale implementation of HBV NAT for blood donors.

Japan, a country with moderate prevalence of HBV infection, modifies anti-HBc screening to accept donors with high amounts of anti-HBs, if also HBV NAT negative, because presence of anti-HBs in the absence of detectable HBV usually signifies a resolved infection and absence of circulating virus.

Some HBV low-prevalence countries have implemented HBV NAT testing. Italy currently performs HBV MP-NAT, ALT and HBsAg testing on all units and recently reported that 57.8 in 1,000,000 donors are HBV MP NAT positive/HBsAg negative/ALT normal; 2.3 in 1,000,000 donors were in the window period; 55.5 in 1,000,000 had occult infection (all donors with occult infection were anti-HBc positive). Germany recently published their residual risk for TT-HBV, using MP-NAT testing, as 1 in 360,000 transfusions.

## Recommended Reading

Comanor L, Holland P. (2006). Hepatitis B virus blood screening: unfinished agendas. *Vox Sang* **91**, 1–12.

Hollinger FB. (2008). Hepatitis B virus infection and transfusion medicine: science and the occult. *Transfusion* **48**, 1001–1026.

Hourfar MK, Jork C, Schottstedt V. (2008). Experience of German Red Cross blood donor services with nucleic acid testing: results of screening more than 30 million blood donations for human immunodeficiency virus-1, hepatitis C virus, and hepatitis B virus. *Transfusion* **48**, 1558–1566.

Linauts S, Saldanha J, Strong DM. (2008). PRISM hepatitis B surface antigen detection of hepatits B virus minipool nucleic acid testing yield samples. *Transfusion* **48**, 1376–1382.

Velati C, Romanó L, Fomiatti L *et al.* (2008). Impact of nucleic acid testing for hepatitis B virus, hepatitis C virus, and human immunodeficiency virus on the safety of blood supply in Italy: a 6-year survey. *Transfusion* **48**, 2205–2213.

# CHAPTER 14

# Hepatitis C Screening of Donor Products

Krista L. Hillyer, MD

The incidence of transfusion-transmitted hepatitis C virus (TT-HCV) was ~10% of transfusions in 1974–1979. With the elimination of HIV high-risk donors and the addition of ALT and anti-HBc (the antibody against the core antigen of hepatitis B) testing in 1987, the risk of TT-HCV decreased by 70%. In 1990 anti-HCV testing was implemented, and in 1999 nucleic acid testing (NAT) for HCV became available. The current risk of TT-HCV is estimated to be 1:1,800,000 units transfused. Therefore, the mitigation of TT-HCV has been secondary to improved sensitivity of testing for the presence of HCV in donors and eliminating high-risk donors through the donor history questionnaire.

**Determination of Need and Requirement for Testing:** FDA requires testing of allogeneic donors for evidence of infection with HCV (21 CFR 610.40), and the AABB *Standards* specifically require the use of anti-HCV and HCV RNA to test donors for HCV infection (Standard 5.8.4). The HCV-positive units must be destroyed. In addition, all donors with positive infectious disease markers must be notified and educated, and the donors placed in a donor deferral registry. Lastly, the blood center must ensure that previous units donated by HCV-positive donors have not transmitted HCV. "Lookback" investigations require locating, notifying, and testing recipients of products donated previously by an HCV-positive donor, at a time when that donor did not test positive for HCV because of either having not yet acquired the infection or having markers of infection that were below the limits of test detection (i.e. the previous donation(s) fell within the window period).

## Approach to Testing:

**Anti-HCV:** The primary screening test for HCV is a third-generation antiglobulin enzyme immunoassay (EIA) that detects antibodies to HCV. The currently available EIA, made up of recombinant HCV polypeptides representing four viral sequences, has an estimated 98% sensitivity for detecting antibody to HCV approximately 10 weeks after infection with HCV. If the initial EIA for anti-HCV is positive, the test is repeated. If the second EIA is "repeat reactive" for anti-HCV (which occurs in 0.2% of US blood donors), a supplementary confirmatory test is performed.

Confirmatory/supplemental testing for a repeatedly reactive anti-HCV EIA is currently a recombinant immunoblot assay (RIBA) against HCV antigens, which is FDA-approved for supplemental testing. Of donors who are reactive for HCV by RIBA, 70–90% are also positive with HCV NAT testing. A donor with a reactive RIBA is permanently deferred from donating and is considered to be infected with HCV. Of EIA repeatedly reactive donors, 37% have a non-reactive or indeterminate RIBA 3.0 (third-generation) test result and are rarely infectious for HCV. Whether or not RIBA 3.0 is reactive, if a blood donor is repeatedly reactive for anti-HCV by EIA, the donation cannot be used for transfusion. However, a donor who has a non-reactive RIBA and reactive EIA (without HCV NAT positivity) can be considered for re-entry by application to the FDA.

An alternate confirmatory/supplemental testing algorithm is to first perform HCV NAT testing and then to perform RIBA testing only on non-reactive NAT samples. Data compiled from 1999 to 2003 on over 33 million donations screened are presented in Figure 14.1. This algorithm requires FDA variance because NAT testing has not been approved as a confirmatory/supplemental test.

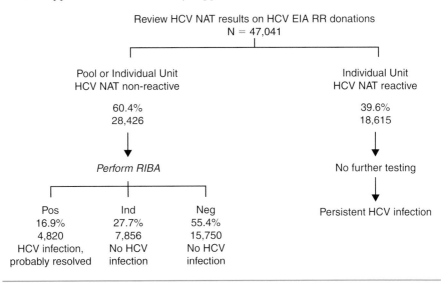

**FIGURE 14.1** HCV supplemental testing algorithm for HCV EIA-repeat-reactive (RR) donations incorporating NAT and RIBA results. RIBA results are positive (Pos), indeterminate (Ind) or negative (Neg). From Kleinman SH, Stramer SL, Brodsky NP *et al.* (2006). Integration of nucleic acid amplification test results into hepatitis C virus supplemental serologic testing algorithms: implications for donor counseling and revision of existing algorithms. *Transfusion* **46**, 695–702.

**HCV NAT:** Currently there are two commercially available, FDA-approved HCV NAT procedures: a PCR test performed on RNA tested by chemical methods on mini-pools (MP) of 16–24 donor blood samples; and a transcription-mediated amplification (TMA) test on nucleic acid preparations in a solid phase (probe-capture) method, also performed on MP of donor samples. Both test methods are sensitive and specific and perform equally well in identifying HCV RNA, particularly when compared with the third-generation EIAs that are currently commercially available (above). If a blood donor is HCV NAT positive, the likelihood that he or she has acute or chronic HCV infection approaches 100%. As reported by Stramer and colleagues (2004), since the inception of HCV NAT in 1999, the use of HCV MP-NAT is estimated to have prevented the transmission of 56 TT-HCV infections annually and has reduced the residual risk of TT-HCV to approximately 1 in 2 million blood transfusions.

**International Standards:** The use of HCV MP-NAT in Germany has resulted in a residual risk of 1 in 10.9 million transfusions. In Italy, HCV MP-NAT has reduced the risk to 2.5 HCV infected units per 1 million products.

## Recommended Reading

Kleinman SH, Stramer SL, Brodsky JP *et al.* (2006). Integration of nucleic acid ampli-fication test results into hepatitis C virus supplemental serologic testing algorithms: implications for donor counseling and revision of existing algorithms. *Transfusion* **46**, 695–702.

Stramer SL, Glynn SA, Kleinman SH *et al.* (2004). Detection of HIV-1 and HCV infections among antibody-negative blood donors by nucleic acid-amplification testing. *N Engl J Med* **351**, 760–768.

# CHAPTER 15

# West Nile Virus Screening of Donor Products

Krista L. Hillyer, MD

West Nile virus (WNV), a single-stranded RNA flavivirus primarily spread by mosquito bites, has a transmission cycle that involves birds and mosquitoes, and thus peaks in July through October. Humans are incidental hosts following mosquito bites. WNV was first reported in the US in 1999 in New York, and subsequently spread westward throughout the continental US, where it caused significant epidemics. While approximately 80% of human WNV infections are asymptomatic, symptomatic infections are not uncommon and result in fever, muscle pain and headache, and nausea and vomiting. About 1 in 150 WNV-infected individuals will have severe disease elements, such as meningitis and/or encephalitis (convulsion, coma, and paralysis) and, less frequently, death.

Transfusion-transmission of WNV was confirmed in 23 patients following red blood cell (RBC), platelet, and plasma transfusions in 2002. Of 23 recipients, 15 had WNV-associated illness (13 with meningoencephalitis and 2 with fever) and 7 patients died of WNV infection (6 had an unclear cause of death). Of the 16 donors of these 23 units, 9 had viral symptoms before or after donation and 5 were asymptomatic (2 were lost to follow-up).

In October 2002, the FDA issued guidance on WNV for deferral of donors with suspected or acute WNV infection, and for retrieval and quarantine of any blood or blood components from donors with post-donation illness that could be from WNV infection (deferral of donors based on donor questionnaire was retracted in 2005). By the summer of 2003 it was recognized that a screening test for WNV was needed, and a minipool (MP) nucleic acid test (NAT) was implemented. From 2003 to 2005, > 1000 viremic donors were documented and 7 cases of probable or confirmed transfusion-transmission occurred. It then became increasingly recognized that many donors had low viral copy numbers of circulating WNV, which results in false-negative testing by MP-NAT yet can be identified by individual (ID) NAT testing. Thus, the testing algorithm was changed such that ID-NAT was used selectively – that is, MP-NAT was employed until viremic donations were identified, then ID-NAT was used to increase sensitivity. In 2006, at least one case of transfusion-transmitted WNV was reported, resulting in further changes in the testing algorithm – most notably, changing the trigger used to instigate the change from MP- to ID-NAT. The trigger criteria continue to be modified as needed (*AABB Association Bulletin* 08-03). These steps have mitigated the risk of transfusion-transmission of WNV.

**Determination of Need and Requirement for Testing:** The AABB *Standards* requires screening of allogeneic blood donors for WNV RNA (seasonal criteria apply); further guidance for testing is supplied through AABB Association Bulletins (most recently #08-03) and FDA Guidance documents.

**Approach to Testing:** Blood transfusion-associated transmission of WNV infection during the 2002 US epidemic prompted rapid development of two investigational NAT assays to screen donated blood for WNV viremia, implemented under an IND in 2003. These tests are now approved by the FDA. As above, in general year-round MP-NAT testing is performed, and when an appropriate epidemiologic trigger is reached then ID-NAT is substituted.

Such a strategy was supported by the following studies: Stramer and colleagues (2005) in which NAT interdicted approximately 1000 infectious blood donations from the American Red Cross inventories during 2003 and 2004, when the frequency of WNV NAT-positive donations was found to be ~1.5/10,000 and ~0.4/10,000, respectively; Busch and colleagues, who implemented ID-NAT during times of high prevalence and Custer and colleagues, who suggested that using a moving trigger based on seasonal and geographic criteria would be more cost-effective. This strategy has come to be known as "selective" ID-NAT. WNV testing has been estimated to cost ~1.5 million US dollars per quality-adjusted life year gained. Both the AABB WNV Biovigilance Network and the CDC track regional WNV activity, which allows efficient data communication between donor centers. This communication provides the information needed for the implementation of ID-NAT.

The selective criteria were newly modified for the 2008 season secondary to continued false negatives in MP-NAT. The current recommended trigger for implementing ID-NAT is when two presumed viremic donors (PVD; defined as an initial reactive donor that repeats on the original sample) occur within a 7-day period; MP-NAT is resumed after 7 days without PVD donations (or longer if collecting samples from overlapping areas) or ongoing regional activity, or at the medical director's discretion. Conversion from MP- to ID-NAT should occur within 48 hours of collection time. The donor's zip code should be used for tracking.

**Confirmatory Test:** WNV infectivity of the donor is confirmed by either repeat NAT reactivity on a follow-up sample, or IgM and IgG antibody reactivity with neutralization testing on either the original (index) sample or a follow-up sample.

**Donor Deferral:** WNV NAT-reactive donations cannot be used for transfusion, and all in-date components from a positive donor should be quarantined. The donor is then deferred for at least 120 days.

**International Considerations:** In countries without the needed mosquito vector, or in which there is a less optimal climate or season, WNV strategies can vary. In Canada, testing is employed from June 1 to November 30. During periods of non-testing, Canadian donors are tested only if they have traveled to the US.

## Recommended Reading

Busch MP, Caglioti S, Robertson EF *et al.* (2005). Screening the blood supply for West Nile virus RNA by nucleic acid amplification testing. *N Engl J Med* **353**, 460–467.

Custer B, Busch MP, Marfin AA, Petersen LR. (2005). The cost-effectiveness of screening the US blood supply for West Nile virus. *Ann Intern Med* **143**, 486–492.

Pealer LN, Marfin AA, Petersen LR. (2003). Transmission of West Nile virus through blood transfusion in the United States in 2002. *N Engl J Med* **349**, 1236–1245.

Stramer SL, Fang CT, Foster GA *et al.* (2005). West Nile virus among blood donors in the United States, 2003 and 2004. *N Engl J Med* **353**, 451–459.

# CHAPTER 16

# Syphilis, HTLV and Chagas Testing of Donor Products

Chelsea A. Sheppard, MD and Krista L. Hillyer, MD

Testing of blood donors for syphilis, Human T-cell lymphotropic virus (HTLV) infection and Chagas' disease is either mandated or has become a *de facto* community standard in the US, though the cost–benefit of this practice remains under review. Syphilis was the first infection for which the blood supply was tested in the US, and in the past was not uncommonly transmitted by transfusion. HTLV is a retrovirus which can be transfusion transmitted, and donor testing has been mandated. Chagas' disease is an emerging infection within the US blood supply, secondary to increased immigration from endemic countries (primarily Latin America). Infection with *T. cruzi*, the agent responsible for Chagas' disease, is lifelong, and may result in severe chronic disease with cardiac and gastrointestional complications.

Syphilis: Syphilis was the most commonly recognized disease transmitted by transfusion pre-World War II. For over 50 years, the blood supply has been screened for evidence of infection by *Treponema pallidum*, the organism responsible for causing syphilis. The last reported case of transfusion-associated syphilis was in 1966. The disappearance of transfusion-associated syphilis is thought to be a result of multiple factors, including the decline in and control of syphilis infection in the US, no longer performing direct donor-to-patient transfusions, the rarity in blood donors of spirochetemia, cold storage of some blood components (*T. pallidum*, which is anaerobic, is not infectious after ~72 hours of refrigeration), better donor screening, and appropriate antibiotic use for syphilis infections.

Background: The first sign of syphilis is most often a chancre, which appears on average 21 days after primary infection. The organism then disseminates to the blood, and 6–8 weeks after the primary infection the infected individual will develop a rash and spirochetemia. At this stage, nearly all infected persons will seroconvert. If untreated, 20% of these individuals may go on to develop a recurrent fulminant infection within 2 years of the primary infection. Eventually, patients become immune to reinfection and treponemal antibody titers wane. Tertiary syphilis, or neurosyphilis, develops after a long latency period (usually years). Both treponemal and nontreponemal antibodies are present during this stage.

Determination of Need and Requirement for Testing: FDA and AABB *Standards* require serologic testing for syphilis (21 CFR 640.5 and Standard 5.8.4).

Approach to Testing:

*Screening Tests:* Syphilis screening is through antibody detection via either nontreponemal or treponemal tests. Most blood centers use treponemal tests, because they are automated and have better performance characteristics. Previously, non-treponemal tests, such as the rapid plasma reagin (RPR) test, were used, but they are non-specific. Non-treponemal tests identify active or recent infections and become negative after disease treatment, while treponemal tests identify current and distant infections, even after successful treatment. The frequency of reactive screening tests in the US is about 0.18%.

*Non-treponemal Antibody Tests (RPR or VDRL):* Non-treponemal tests detect antibodies which react against cardiolipin phospholipids in response to infected host tissue. Non-treponemal tests have a false positive rate (positive non-treponemal result with a negative FTA-ABS, see below) of 1–2% because they may be positive in multiple conditions, including pregnancy, other infections (e.g. HIV, mononucleosis, tuberculosis, rickettsial infection, other spirochetal infections and bacterial endocarditis), and disorders of immunoglobulin production (rheumatoid arthritis, ulcerative colitis and cirrhosis). The RPR test uses cardiolipin-coated carbon particles, which agglutinate upon addition of antibody-containing serum.

*Treponemal Antibody Tests (Fluorescent Treponemal Antibody Absorption Test (FTA-ABS), Particle Aggregation (TP-PA), Recombinant Antigen Tests):* Treponemal tests detect antibodies specific to *T. pallidum*. One method on an automated platform uses microhemagglutination for the detection of *T. pallidum* IgG and IgM antibodies. False positivity occurs in individuals with autoimmune disease, Lyme disease, Legionella, infectious mononucleosus and other conditions. Enzyme-linked immunosorbent assays (EIA) are also available.

*Confirmatory/Supplemental Tests:* FTA-ABS, treponemal EIAs and non-treponemal tests are used to confirm or supplement repeatedly reactive samples. Approximately 46% of screening tests are confirmed by FTA-ABS.

In 2002, Orton and colleagues performed NAT testing for *T. pallidum* on 169 positive and control samples and found no evidence of *T. pallidum* DNA and/or RNA; they concluded that serologically positive donors are unlikely to have circulating *T. pallidum*. This study has been used as the basis for the argument that syphilis testing should be discounted in the US. However, its use continues, at least in part, as a surrogate marker for HIV and other cross-reactive infectious agents as above.

**Donor Deferral:** Donor units that test positive for screening serologic tests for syphilis should not be used for allogeneic transfusion. Donors are deferred for 1 year after diagnosis and treatment, a reactive screening test, or a confirmed positive test for syphilis.

**International Considerations:** In some developing nations, both the prevalence of syphilis and the incidence of transfusion-transmitted syphilis are significant. Also, in certain countries it is not uncommon for blood to be transfused before it is refrigerated for a period sufficient to render *T. pallidum* non-infectious. Thus, testing strategies, despite their limitations, are commonly employed and considered to be of value.

**Human T-cell Lymphotropic Virus:** HTLV, a retrovirus, has a high transfusion-transmission efficiency (50% reported in Japan and 10–20% in the US). In 1988, universal screening for HTLV antibodies was implemented because of an estimated risk of HTLV transmission of 12 per 100,000 units and the known severity of HTLV-induced diseases (adult T-cell leukemia/lymphoma [ATLL] and tropical spastic paraparesis [TSP, also known as HTLV-1 associated myelopathy]). Currently the residual risk for HTLV infection is approximately 1 per 3 million transfused units, though the need for continued testing is debated.

Background: Most HTLV infections are asymptomatic, but there is a 2-4% risk of developing disease up to 40 years after infection with HTLV-1 (and a lesser risk with HTLV-2). This infection is endemic in the Caribbean, where the risk of ATLL in individuals *infected at birth* is 4% in their lifetimes. The risk of developing ATLL is lower in those infected during adulthood (i.e. those who acquire HTLV from transfusion). ATLL has a high mortality rate within 1 year of disease onset.

Additionally, HTLV-1 and HTLV-2 are associated with TSP (also known as HTLV-1 associated myelopathy [HAM]), which is a slowly progressive myelopathy characterized by spastic paraparesis of the lower extremities, hyperreflexia, and bowel and bladder symptomatology). The risk of TSP is about 2% in HTLV-1 positive individuals, with a similar or lower risk in HTLV-2 positive individuals. Other diseases associated with HTLV-1 or HTLV-2 infection include lymphocytic pneumonitis, uveitis, polymyositis, arthritis, bronchitis, and dermatitis.

HTLV-1 predominately infects CD4+ lymphocytes while HTLV-2 preferentially infects CD8+ lymphocytes (and, to a lesser extent, infects CD4+ lymphocytes, B lymphocytes and macrophages). The HTLV-1/2 seroprevalence in blood donors in the US is approximately 0.025%.

Determination of Need and Requirement for Testing: The AABB *Standards* and the FDA require that transfusion transmission of HTLV be prevented by screening all allogeneic blood donations for HTLV antibodies.

Approach to Testing:

*Screening Tests:* The FDA has approved the use of EIAs which detect both HTLV-1 and -2 antibodies. These tests typically use viral lysates as the capture reagent, and adherent donor antibodies are identified with an antiglobulin conjugate.

*Confirmatory/Supplemental Tests:* There are no licensed confirmatory or supplemental tests for the HTLV-1/2 EIAs. One approach to supplemental testing to confirm a reactive anti-HTLV test is to repeat the test using an alternate manufacturer's EIA. Western blot and radioimmunoprecipitation (RIPA) tests are available. In the American Red Cross, 9% of EIA repeatedly reactive samples are positive by Western blot, 19% are negative and 72% are indeterminate. In addition, NAT testing is being explored. One study demonstrated that 99.3% of repeatedly reactive anti-HTLV donors, who were confirmed by alternative serologic testing (p21 EIA, Western blot, or RIPA), were PCR positive; 1.4% of the confirmatory indeterminate donors were PCR positive; and none of the confirmed negative donors were PCR positive.

*Donor Deferral and Counseling:* Donors who are repeatedly reactive for anti-HTLV-1/2 on more than one occasion or who are repeatedly reactive with a second licensed screening test are indefinitely deferred. Because of the lack of FDA-approved/ licensed supplemental and/or confirmatory tests, and the unclear nature of disease in HTLV antibody positive blood donors, counseling donors with positive HTLV results is difficult. In one study, Orland and colleagues prospectively followed 138 HTLV-1 infected, 358 HTLV-2 infected, and 759 uninfected control US blood donors and, after adjusting for known and potential confounders, found that HTLV-2 infection was associated with increased mortality (hazard ratio 2.8), whereas no statistically significant increase was seen in HTLV-1 infected individuals.

**International Considerations:** In November 1988, the FDA recommended testing for HTLV-1 in US blood donors. At that time, HTLV screening was also introduced in Canada, France, the Netherlands and Sweden. Other countries adopted similar strategies, including Australia by 1992. As in the US, authorities in Australia and other locations such as Norway are re-considering the need for such large-scale HTLV testing, and analyzing its cost versus benefit, when prevalence is below 8 per 100,000. In locations such as Japan and the Caribbean, where HTLV prevalence is significantly higher, HLTV testing of blood donors remains both used and recommended.

**Chagas' Disease:** Interest in preventing TT-Chagas' disease (i.e. *T. cruzi* infection) is a relatively recent development within the US. With increased Hispanic immigration the seroprevalence in the US has steadily risen in the past three decades, where it has been estimated that 25,000 to 100,000 people and 1 in 25,000 blood donors may be infected; almost all are immigrants. The seroprevalence has been reported to be higher in Miami (1:9000) and Los Angeles (1:7500). Since 1989, seven cases of TT-Chagas' disease have been reported in the US and Canada.

**Description:** *Trypanosoma cruzi* is endemic to the Americas, although infected persons can be found worldwide. *T. cruzi* is transmitted by the bite of infected reduviid insects. In the acute phase of the disease, patients may demonstrate a chagoma with local lymphadenopathy, Romaña's sign, or painless unilateral edema of the palpebrae and periocular tissues, malaise, fever, anorexia, and edema of the face and lower extremities, in addition to generalized lymphadenopathy and hepatosplenomegaly. The acute phase may resolve spontaneously without the patient having realized he or she has been infected. The chronic phase of the disease may present years to decades following the initial infection in up to 30% of patients. Patients may develop arrythmias, cardiomyopathy leading to heart failure, or thromboembolism. Mega-esophagus and megacolon may be complicated by airway obstruction/aspiration pneumonitis, colonic obstruction, volvulus, septicemia and/or death.

**Determination of Need and Requirement for Testing:** Currently, neither the AABB *Standards* nor the Code of Federal Regulations (CFR) require allogeneic testing of donor blood for Chagas' disease in the US. A requirement for donor testing may be implemented in the future.

**Approach to Testing:**

*Screening Test:* A recently FDA-approved EIA (enzyme immunoassay) is now available to screen blood donors for antibodies to *T. cruzi*. In January 2007, the American

Red Cross and Blood Systems Inc. began screening all donations for *T. cruzi* infection. The AABB recommends that samples testing positive (repeatedly reactive by EIA) should be quarantined and removed from the blood supply. In addition, any in-date products from the same donor should be quarantined. The donor should be notified and indefinitely deferred. Recipient tracing should also be performed to identify all recipients exposed to potentially infected products (AABB *Association Bulletin* #06-08). The Chagas' Disease Biovigilance Network records screening and confirmatory results from the testing of donors for antibodies to *T. cruzi*.

At least one large US blood center has opted not to commence testing donors for infection with the agent of Chagas' disease. This decision may become more widely adopted as the cost–benefit for Chagas testing becomes further elucidated. Some authors have suggested that these determinations need to be site-specific (see Wilson *et al.*, 2008).

Confirmatory Tests: There is currently no FDA-approved confirmatory test for *T. cruzi* infection. Confirmatory testing can be performed using a second EIA test, a radioimmunoprecipitation assay (RIPA), or an immunofluorescence assay (IFA). From the AABB Chagas' Disease Biovigilance Network, there have been 2505 repeatedly reactive US donors, of which 654 are RIPA positive, 1791 are RIPA negative and 7 RIPA are indeterminate. False-positive EIA reactivity may occur secondary to *Leishmania* infection.

*Donor Deferral:* Donors who have a history of Chagas' disease or who are confirmed positive for *T. cruzi* infection are indefinitely deferred.

## Recommended Reading

Alter HJ, Stramer SL, Dodd RY. (2007). Emerging infectious diseases that threaten the blood supply. *Semin Hematol* **44**, 32–41.

Busch MP, Laycock M, Kleinman SH *et al.* (1994). Accuracy of supplementary serologic testing for human T-lymphotropic virus types I and II in US blood donors. Retrovirus Epidemiology Donor Study. *Blood* **83**, 1143–1148.

Dodd RY. (2007). Current risk for transfusion transmitted infections. *Curr Opin Hematol* **14**, 671–676.

Orland JR, Wang B, Wright DJ *et al.* (2004). Increased mortality associated with HTLV-II infection in blood donors: a prospective cohort study. *Retrovirology* **24**, 1–4.

Orton SL, Liu H, Dodd RY *et al.* (2002). Prevalence of circulating *Treponema pallidum* DNA and RNA in blood donors with confirmed-positive syphilis tests. *Transfusion* **42**, 94–99.

Stramer SL, Foster GA, Dodd RY. (2004). Effectiveness of human T-lymphotropic virus (HTLV) recipient tracing (lookback) and the current HTLV-I and -II confirmatory algorithm, 1999 to 2004. *Transfusion* **46**, 703–707.

Wilson LS, Ramsey JM, Koplowicz YB *et al.* (2008). Cost-effectiveness of implementation methods for ELISA serology testing of *Trypanosoma cruzi* in California blood banks. *Am J Trop Med Hyg* **79**, 53–68.

# CHAPTER 17

# Bacterial Detection Methods

Beth H. Shaz, MD

During the 1990s, substantial data accumulated supporting the belief that the room temperature, plasma-rich, oxygenated environment of platelet storage was leading to an unacceptability large number of septic transfusion reactions in recipients due to bacterial contamination of the product. Indeed, the septic reactions were associated with significant morbidity and mortality. In order to mitigate the risk associated with bacterial contamination the AABB created Standard 5.1.5.1, which requires methods to limit and detect bacteria in platelet products. The introduction of this requirement by culture methods for bacterial detection of platelet apheresis products has decreased the rate of septic reactions from a $\sim 1:40,000$ to $\sim 1:75,000$ and for fatalities $\sim 1:240,000$ to $\sim 1:500,000$. Therefore, the risk of a febrile, septic or fatal transfusion reaction due to undetected bacteria in platelet products continues to be a concern. This residual risk may be as frequent for mild reactions as $1:15,000$ apheresis platelet products. Thus, it appears that the implementation of bacterial screening has decreased the overall risk by approximately 50%, thus methods to reduce this risk further are needed. This chapter will address the methods used to limit and detect bacterial contamination in platelet products (Table 17.1). Bacterial contamination reactions are described in Chapter 59.

| TABLE 17.1 Methods to Reduce Bacterial Contamination | |
| --- | --- |
| Bacterial contamination *avoidance* | Donor eligibility<br>Phlebotomy techniques |
| Bacterial contamination *reduction* | Diversion pouch<br>Leukoreduction |
| Bacterial contamination *inhibition* | Cold storage**<br>Storage solutions* |
| Bacterial contamination *inactivation* | Pathogen inactivation* |
| Bacterial contamination *detection* | Culture<br>Nucleic acid testing**<br>Metabolic changes |

*Currently not available in the US. **Currently not available.

## Methods to Avoid Bacterial Contamination:

**Donor Screening:** Donor eligibility questions are used to exclude donors with fever or symptomatic infections, or those receiving antibiotics for infection. Donor screening does not eliminate donors with asymptomatic infections, which are commonly a result of gram-negative organisms and can result in severe septic transfusion reactions.

**Skin Preparation:** The skin is the source of contamination for the majority of bacterially contaminated units (~80%). Skin decontamination with iodine decreases bacterial contamination. If the donor is allergic to iodine, chlorhexidine should be used.

## Methods to Reduce Bacterial Contamination:

**Diversion Pouch:** Bacteria within hair follicles and scar tissue (which occurs in repeat donors) are not removed by skin preparation. A diversion pouch collects the skin plug and first few milliliters of blood, which adds to the reduction of bacterial contamination from the skin. Blood collected in the diversion pouch may be used to obtain samples for infectious-disease testing. Published studies show a decrease in *Staphylococcus* species bacterial contamination from 0.14% to 0.03% of units following implementation of this approach. Even though the skin is the major source of bacterially contaminated products, gram-negative bacteria as a result of asymptomatic donor bacteria result in the majority (83%) of fatalities from septic reactions. Thus, additional methods are needed to decrease bacterial contamination.

**Leukoreduction:** Some studies have demonstrated that leukoreduction filters decrease the rate of bacterial contamination, but this effect is inconsistent and thus leukoreduction should not be considered a reliable method.

## Methods to Inhibit or Inactivate Bacteria:

**Cold Storage:** Cold storage of platelets would decrease the growth of bacteria and, therefore, the risk of septic transfusion reactions. Recent data using a murine model have suggested that platelets could be stored at refrigerated temperatures if a sugar solution was added, and subsequently successfully transfused, but this was not reproducible with human platelets. Currently, no cold storage techniques for platelets are available, though this is a focus of continued investigation.

**Storage Solutions:** Platelet storage (additive) solutions decrease the amount of plasma in platelet products, and may therefore decrease the rate of bacterial contamination. Currently, these are available only outside the US.

**Pathogen Inactivation:** Pathogen inactivation technologies are available outside the US which eliminate bacteria within the product. These methods use photochemical energy to disrupt bacterial DNA.

## Methods to Detect Contaminated Products:

**Culture-based Methods:** There are currently two culture-based systems which are FDA approved for *quality control* testing in leukoreduced apheresis platelet products (the second method can also be used for leukoreduced whole blood derived platelet products). Both methods require an incubation period of at least 24 hours. These methods miss slow-growing bacteria, such as *Propionibacterium acnes*, which are usually of low pathogenicity. The false positive rate is ~1 : 5000 if culture is performed in a laminar flow hood, and ~1 : 500 if performed on a bench.

   The first method uses a minimum of 4 ml (up to 10 ml) to inoculate an aerobic bottle at least 24 hours after collection. Additionally, an anaerobic bottle can be used, but anaerobes are rarely implicated in bacterial contamination, and a second bottle increases sample collection volume and cost, and false-positivity rate. Typically, the product is not released for 24 hours after inoculation. The culture bottle is retained until the expiration date of the unit, until 24 hours prior to release of the unit, or until positive. A positive is determined by an increase in $CO_2$. This method detects $10^{1-2}$ CFU/ml of bacteria. Products tested with this method are released if the culture is negative to date, but the culture remains incubated. Therefore, these cultures can become positive after release from the blood collection center. Subsequently, the blood center should notify the transfusion service of a potentially contaminated product so the transfusion service can either notify the patient care team, if the product was transfused, or discard the product.

The second method uses a pouch to collect a filtered (to remove the platelets) ~5 ml sample. The sample is incubated for 24 hours and then read for a decrease in oxygen concentration (either PASS or FAIL). This method detects $10^{2-3}$ CFU/ml of bacteria. Products are released from the blood collection center after the PASS readout. A recently FDA-approved system allows for pre-storage leukoreduction, bacterial detection, and pooling of whole blood derived platelet units.

True-positive bacterial cultures then require identification of the organism, and, for gram-negative organisms, measurement of the endotoxin level in the product, if the product was transfused. The blood center may wish to notify a donor implicated in a contaminated unit with a non-skin flora organism, because of diseases associated with asymptomatic bacteremia. For example *Streptococcus bovis* and *Streptococcus* G have been reported in donors with occult malignancy, especially colon carcinoma.

## Methods Performed Just Prior to Product Release:

*Immunoassay:* This is a recently approved FDA adjunct device for quality control bacterial detection for leukoreduced apheresis platelet products, which must be used with another FDA-approved quality control bacterial detection method (such as a culture method described above). This is a point-of-care device which uses 500 µl of product, takes 30 minutes, and has a positive/negative color-change read-out. Its detection limit is $10^{3-5}$ CFU/ml of bacteria, which is a lower sensitivity rate than with culture, but is performed at a later time period in storage, where the number of bacteria has expanded.

*Staining:* Gram stain, accrine orange and Wright stain can be used to detect the presence of bacteria in samples from platelet products. These methods have a high false-positive rate and a detection limit of $10^6$ CFU/ml of bacteria.

*Metabolic Detection Methods:* There is a variety of methods, which are insensitive and non-specific, with a detection limit of $10^{6-7}$ CFU/ml of bacteria, based on metabolic changes which occur in the presence of bacterial contamination.

One method is to check the platelet product for the presence of *swirling* (this method per AABB does not qualify to be the sole method to detect bacterial contamination). Platelets are normally discoid, which results in swirling; when there is increased acid, platelets become round and lose their capacity to swirl.

A second method is to use either a urine reagent strip or automated technology to determine the glucose, pH and/or oxygen content of the platelet product. Different cut-off values are used depending on the product and method. While inexpensive, these methods are generally considered to be suboptimal due to their insensitivity.

<u>Future Considerations:</u> Other methods to detect bacteria are currently under development, including fluorescent cytometric techniques and nucleic acid (PCR) methods. In addition, strategies to optimize the sensitivity and specificity of culture techniques are being investigated. Lastly, pathogen inactivation methods are available outside the US.

## Recommended Reading

AABB (2003). *AABB Association Bulletin* **03–12**, October 1.

Blajchman MA, Beckers EA, Dickmeiss E *et al.* (2005). Bacterial detection of platelets: current problems and possible resolutions. *Transfus Med Rev* **19**, 259–272.

Fang CT, Chambers LA, Kennedy J *et al.* (2005). Detection of bacterial contamination in apheresis platelet products: American Red Cross experience, 2004. *Transfusion* **45**, 1845–1852.

Hillyer CD, Josephson CD, Blajchman MA *et al.* (2003). Bacterial contamination of blood components: risks, strategies, and regulation: joint ASH and AABB educational session in transfusion medicine. *Hematology (Am Soc Hematol Educ Program)*, 575–589.

Vostal JG. (2005). *Update on FDA Review of Bacterial Detection Devices for a Platelet Release Test Indication and Extension of Platelet Dating*. Washington, DC: US DHHS.

# CHAPTER 18

# The Role of the Transfusion Service Physician

Beth H. Shaz, MD and Christopher D. Hillyer, MD

The hospital transfusion service (TS) physician plays a central and leading role in a critical operation within the hospital by ensuring a reliable and adequate supply of safe and effective blood products, and ensuring optimal component choice and matching with recipients. As blood components are required 24 hours a day, 7 days a week for the practice of obstetrics, surgery, hematology/oncology and transplantation, for trauma/damage control resuscitation and other medical disciplines, hospital transfusion services require constant medical input and oversight, making the physician's roles and responsibilities both demanding and rewarding. In general, the TS physician provides leadership, oversight and clinical care, and strives for continued improvement in the practice of transfusion medicine.

**Organization Chart:** According to the AABB *Standards for Blood Banks and Transfusion Services* (Standard 1.0), the TS must have a clearly defined structure and documentation of the individuals responsible for key functions, including executive management. The TS must have a medical director who is responsible for all medical and technical policies, processes and procedures. The hospital TS, most commonly referred to as the hospital blood bank, is typically within the administrative hegemony of the departments of pathology and laboratory medicine (or clinical pathology) or the department of medicine, and includes medical, technical and administrative personnel. The TS and clinical laboratories function as part of the health system, and provide services that are vital to the overall hospital mission and function. An organization chart clarifies the lines of communication and the levels of authority within the organization (Figure 18.1).

**Specific TS Physician Roles:** The TS physician medical director typically has multiple roles, responsibilities and functions, including overseeing specific laboratories, laboratory and administrative policies and procedures, direct and indirect patient care, education and research (Figure 18.2).

**Transfusion Service:** The TS physician not only directs the traditional hospital blood bank, where pretransfusion testing and blood product dispensing occur, but may also direct a number of other laboratory and/or clinical services. Other laboratories under the TS physician may include HLA, HPC processing, and coagulation laboratories. Other covered services may include therapeutic apheresis and phlebotomy, infusion and in- and outpatient transfusion, whole blood and apheresis collection, perioperative services, and tissue banking. Each of these laboratories or services requires policies and procedures, and medical, administrative and technical oversight.

**Transfusion Service Management:** Management of the TS requires the establishment and pursuit of organizational goals in an effective and efficient manner. The initial

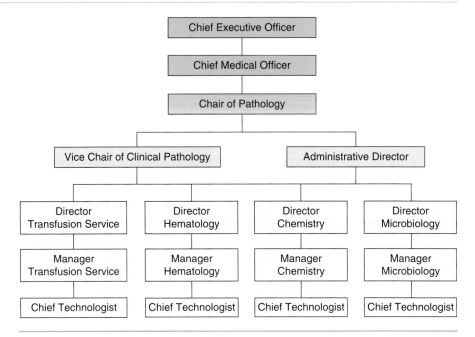

FIGURE 18.1  Organizational chart of a clinical laboratory.

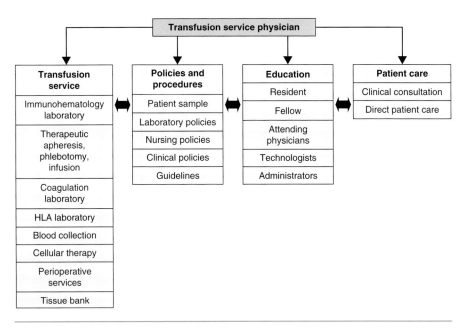

FIGURE 18.2  Transfusion service physician oversight.

step is to create goals, and, through the process of planning, organizing and implementing, these goals should be achieved. The major components of planning are the writing and acceptance of mission and vision statements, core values and strategic planning, including vision, goals and operational and time-line linkages (Table 18.1).The major components of organizing are prioritization, coordination, communication and collaboration. The major components of implementing are monitoring, outcomes evaluation, staff development and management, and continual adjustment and improvement.

| TABLE 18.1 Mission Statement |
| --- |
| The Transfusion Service is dedicated to the optimal transfusion of patients and the advancement of the field through outstanding clinical service, basic and clinical research, and education of our members and the next generation of leaders in transfusion medicine. |

The scope of TS physician includes executive, medical director and administrative functions, including strategic planning, leadership, project management, identification of new technologies, determination of the scope and complexity of services offered, information technology, clinical consultation, TS and laboratory representation to the hospital staff and administration, utilization review, formation and promulgation of clinical guidelines, regulatory compliance, quality systems and assurance, quality control, human resources and customer service.

**Quality Management:** The TS physician is usually, or will become, an expert in quality systems, as these systems have been a requirement in the AABB *Standards* for over 10 years. Quality systems provide a process and procedure structure for optimal pretransfusion testing, post-manufacture processing, crossmatching, and issuing and administration of products, and for validation, preventative maintenance, and quality control of technologies, to name a few of the elements incorporated into the quality plan.

**Patient Care:** The TS physician can provide both direct and consultative patient care. Direct patient care, where services can be reimbursed through either Evaluation and Management (E&M) or Current Procedural Terminology (CPT) codes, is performed through therapeutic apheresis, therapeutic phlebotomy and HPC collection, as well as through the interpretation of complex serologic and patient matching issues and evaluation of transfusion reactions and complications. Consultative services are often requested regarding provision of appropriate or specialized blood products, transfusion reactions, blood management, hemostasis and thrombosis-related test result interpretation, and consideration of optimal products for injection, administration, infusion or transfusion. Much of the TS (and other clinical laboratory) test consultation activities and services can be reimbursed through CPT codes and billing. Outside these roles, the TS physician improves patient care through policies and procedures that enhance quality, and through participation in hospital committees (such as the transfusion committee or patient safety committee). Transfusion committees oversee blood utilization and the creation of guidelines, as well as reviewing adverse transfusion outcomes.

**Transfusion Committee:** The Joint Commission (TJC) standards (PI.1.10) require hospitals to collect and monitor their performance regarding the use of blood products, which can be overseen by the hospital transfusion committee or its equivalent. The data must be analyzed and tracked over time to identify levels of performance, patterns, trends and variation, compared with external sources, and the results must identify opportunities to improve. Areas of improvement must be prioritized, actions taken to improve, changes evaluated to ensure that they result in improvement, and actions taken when improvement is not achieved or sustained. These requirements can be met by reviewing the ordering, distribution, handling, dispensing, administration and effects of blood products.

*Membership:* The members of the transfusion committee include representatives from the departments that utilize blood products, such as hematology, surgery, anesthesia and obstetrics; and blood bank, clinical laboratory, nursing and hospital administration.

*Goals:* The primary goal of the transfusion committee is to improve performance, defined as what is done and how well it is done. The committee must approve the blood supplier, and monitor blood utilization, adverse events, errors, accidents, and the blood bank quality plan. The committee should ensure that the blood bank has the resources it needs to carry out the responsibilities of the TS. The committee also oversees development of transfusion practice guidelines, criteria for blood product use, and blood administration procedures.

**Audit:** Audit of particular areas, such as turn-around time, specimen mislabeling, wastage and patient identification, is used to identify areas in need of improvement. The root cause of the deviation can be determined, and a corrective action created to ameliorate the problem. The audit is then repeated to ensure improvement and sustainability. An audit of blood administration may include specimen collection (use of two patient identifiers), blood bank testing (appropriate reagent use and product selection), product dispensing (patient identification, verification of patient history), and product administration (patient identification, appropriate informed consent, compliance with physician order, transfusion with no other fluid except normal saline and completed within 4 hours).

Audits can be *prospective*, occurring at the time of the event; *concurrent*, occurring within 24 hours of the event; or *retrospective*, occurring later than 24 hours after the event.

**Transfusion Practice Guidelines:** Clinical transfusion guidelines encompass both the transfusion of products (i.e. RBCs, plasma, platelets and cryoprecipitate) as well as transfusion in special clinical scenarios (i.e. sickle cell disease, massive transfusion and liver transplantation). The creation of clinical transfusion guidelines requires communication between the ordering physician and the TS physician, as well as an understanding of the steps involved in providing blood products to the patient. A model process is as follows:

1. An audit is performed in order to best understand current and established transfusion practices.
2. Draft guidelines are created by a multidisciplinary team based on evidence in the literature.

3. Education of the ordering physicians and other members of the healthcare team ensues, and is required to determined the understandability and likelihood of implementation and compliance.

4. The guidelines are revised, approved, adopted and implemented.

5. Periodically, repeat auditing is required to guarantee that the guidelines are being followed and continue to be appropriate.

**Maximum Surgical Blood Order:** The establishment of the maximum blood surgical order schedule (MSBOS) for common surgical procedures (e.g. 2 units of RBCs for total hip replacement, 2 units of RBCs for coronary artery bypass grafting) makes it possible to determine the amount of blood that will be made available for a particular type of operation, and enables the hospital TS to avoid the excess inventory and wastage that can result from excessive amounts of blood being ordered for surgical procedures.

**Informed Consent:** The TS physician is required by the College of American Pathologists (CAP) and the AABB *Standards* to participate in the development of policies, processes and procedures regarding recipient informed consent for transfusion. At a minimum, the procedure must communicate the risks and benefits for transfusion, the alternatives to transfusion, the right to refuse transfusion, and the ability to ask questions. In some hospitals, it may be helpful to have a separate and approved process for refusal of blood product administration.

**Inventory Management:** Although the overall US blood supply is usually adequate, the margin of supply and demand can at times be low, and regional and local differences can be extreme. The TS physician is significantly involved in inventory management, especially when the supply is low (such as during the winter holidays and the summer), and is responsible for the establishment of a communication plan to allow all stake holders to be apprised of significant blood and blood component shortages. The TS physician is also responsible for having a system for patient evaluation and distribution/triage of units during times of significant shortage or recalls. TS physicians also oversee product inventory management for specific patients or patient groups, including patients with multiple alloantibodies that may require large numbers of RBC products for transfusion.

**Education:** The TS physician educates many groups of individuals. Formal education of residents and transfusion medicine fellows is overseen through the ACGME (Accreditation Council for Graduate Education), which requires learning opportunities, assessment and outcomes in six core competencies: patient care, medical knowledge, system-based practice, practice-based learning, professionalism, and interpersonal and communication skills. In addition, the TS physician educates technologists, ordering physicians and hospital administrators. Education is via multiple methods, including presentations, patient cases, and one-on-one conversations. In order to improve the quality of patient care, continued education of residents, fellows, ordering physicians and hospital administrators is mandatory.

**Research:** It is possible to participate in and/or direct a broad range of opportunities and activities in the blood bank and transfusion service, ranging from NIH-funded to quality improvement projects. Large clinical trials currently funded by the National Heart, Lung, and Blood Institute through the Transfusion Medicine Hemostasis/ Clinical Trials Network are being conducted to evaluate new therapies and approaches in the transfusion therapy and treatment of patients with hemostatic disorders. In addition, ongoing basic science research is needed to better understand the beneficial and adverse effects of transfusion and to improve transfusion management. The collaboration between basic scientists and clinicians allows for the "bench to bedside and back" development of optimal patient care. Lastly, quality improvement projects affect patient care by highlighting areas of improvement, such as turn-around time, patient identification and blood utilization review.

**TS Medical Direction within Small- to Medium-Sized Community Hospitals:** The TS in small- to medium-sized community hospitals is frequently overseen by a general pathologist, who is responsible for other areas in the pathology practice as well, or by a board-certified transfusion medicine physician who has additional revenue-generating responsibilities in a private practice environment.

**Transfusion Medicine as a Career:** Blood banking and transfusion medicine (BB/TM) physicians work in a variety of settings, including blood centers, hospital transfusion services, industry, government, and large commercial laboratories. BB/TM physicians continue to participate in expanding areas outside the hospital transfusion service and donor program, such as HLA laboratory, HPC collection and processing laboratory, tissue banking, and coagulation laboratory direction. BB/TM physicians participate in research, including basic and translational investigations, cellular therapy, blood donation and transfusion issues, and clinical trials. A recent survey demonstrated that approximately 70% of BB/TM fellowship graduates from 1995–2004 had published articles within the previous 3 years, and 12% had published over 10 publications within that timeframe.

There are multiple career options in the field of BB/TM, with wide-ranging opportunities including clinical, teaching, research and administration. BB/TM physicians should be advocates for the best transfusion care of patients and the best care of donors, be teachers and mentors, and be devoted to the continual improvement of the field of BB/TM.

## Recommended Reading

Conry-Cantilena C, Klein HG. (2004). Training physicians in the discipline of transfusion medicine. *Transfusion* **44**, 1252–1256.

Jackson GN, Snowden CA, Indrikovas AJ. (2008). A prospective audit program to determine blood component transfusion appropriateness at a large university hospital: a 5-year experience. *Transfus Med Rev* **22**, 154–161.

McCullough J. (2006). The role of physicians in blood centers. *Transfusion* **46**, 854–861.

McCullough J, Benson K, Roseff S *et al.* (2008). Career activities of physicians taking the subspecialty board examination in blood banking/transfusion medicine. *Transfusion* **48**, 762–767.

Szczepiorkowski ZM, AuBuchon JP. (2006). The role of physicians in hospital transfusion services. *Transfusion* **46**, 862–867.

# CHAPTER 19

# Pretransfusion Testing

Beth H. Shaz, MD

Pretransfusion testing (or pretransfusion *compatibility* testing) includes the immunohematologic testing needed for proper patient blood typing, component selection and compatibility testing, which ensure optimal patient safety and transfusion efficacy.

AABB *Standards for Blood Banks and Transfusion Services* require that the following are performed:

1. Positive identification of patient and their corresponding blood sample;
2. ABO group and D typing of patient's sample;
3. Testing of the patient's serum/plasma (hereafter referred to as "plasma" for simplicity) for unexpected, clinically significant red blood cells (RBC) antibodies, which are defined as antibodies known to cause hemolytic transfusion reactions or hemolytic disease of the fetus and newborn;
4. Comparison of current findings with previous results;
5. Confirmation of ABO group of RBC components;
6. Confirmation of D type of D-negative RBC components;
7. Selection of ABO- and D-appropriate components for the patient;
8. Serologic or computer crossmatching; and
9. Labeling of the component for issue with the patient's identifying information.

The number of these tests, their frequency and their methodology differ based on the type of component to be issued.

RBC components (and other components containing >2 ml of RBCs) require a crossmatch, for which the patient's sample should be no more than 3 days old unless the patient has not been pregnant or transfused within the preceding 3 months. If the patient's history is uncertain or unavailable, compatibility testing must be performed on blood samples collected within 3 days of the transfusion event. Each institution should have a policy in place that defines the length of time samples may be used from patients who have not been pregnant or transfused within the last 3 months. Figure 19.1 shows the steps involved before a unit of RBCs can be issued for transfusion.

Plasma, platelet, and cryoprecipitate components do not require crossmatching; nor does ABO/D typing need to be confirmed on these components.

Transfusion requests must contain sufficient information for positive patient identification and determination of the patient's component needs, including the quantity of the specific blood component requested and any other special processing needed (for example, irradiation or leukoreduction), and the name of the responsible physician.

<u>Patient Identification:</u> Appropriate patient identification is critical to the safety of blood transfusion. *Mistransfusion*, a term that typically refers to erroneous

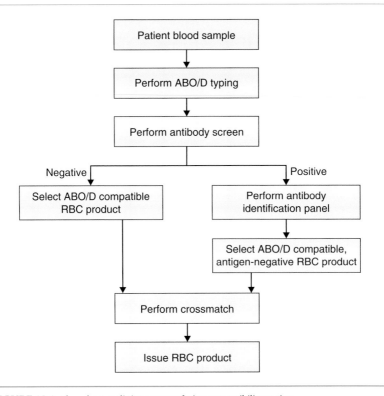

FIGURE 19.1   Flow chart outlining pretransfusion compatibility testing.

administration of an ABO-incompatible RBC unit, remains a significant source of morbidity and mortality. Mistransfusion is most often due to errors in sample or patient identification. Each facility must have patient identification and specimen collection policies that require positively identifying each patient prior to drawing the specimen to be used for pretransfusion testing. The specimen must be labeled with two independent patient identifiers and the date of collection while at the patient's side. In addition, there must be a mechanism to identify the phlebotomist. Once the specimen is received in the laboratory, the specimen label must match the information on the transfusion request. If there is any doubt about the identity of the patient, a new sample must be obtained. It is unacceptable to correct identifying information on an incorrectly labeled specimen. Each laboratory must have procedures and policies in place that define the required identifying information and describe how to document and handle mislabeled specimens.

**Specimen Requirements:** Pretransfusion testing can be performed on either serum or plasma.

**Method:** The classic foundation of serologic methods is the agglutination of RBCs when antigens on RBCs interact with antibodies in the plasma. Usually, agglutination is accelerated by brief centrifugation ("spinning"). Reactions are interpreted based on the degree of agglutination, and are graded as 0 (negative, no agglutination) to 4+ (one solid agglutinate). The reactions are also read for the presence of hemolysis; there are some antibodies which result in *in vitro* hemolysis, such as anti-Jk$^a$ and anti-Jk$^b$. The two major classes of antibodies to RBC antigens are IgM and IgG. Tyically, IgM antibodies result in agglutination in the immediate spin (IS) phase while IgG antibodies result in agglutination in the anti-human globulin (AHG) phase.

**Immediate Spin Phase:** IgM antibodies are pentamers which bind to corresponding antigens and directly agglutinate RBCs. Agglutination occurs after centrifugation, without additional reagents or extended incubation, and is known as "immediate spin" reactivity.

**Anti-human Globulin Phase:** IgG antibodies are monomers which do not directly result in agglutination (termed *non-agglutinating antibodies*). In order to detect IgG antibodies, AHG techniques must be used. The AHG phase (also termed the Coombs' phase) is based on the principle that AHGs obtained from immunized non-human species bind to human globulins such as IgG or complement attached to RBC antigens. The binding of the AHGs to the sensitized RBCs (RBCs covered with IgG and/or complement) results in visible agglutination following centrifugation. AHG techniques require additional reagents (potentiators) and extended incubation for optimal sensitivity. Alternative methods that do not require agglutination have also been developed, including solid phase and flow cytometry. In addition, some of these methods have been automated.

**AHG Reagents:** The AHG reagent can be polyclonal, monoclonal, or a blend of both, and either anti-IgG or polyspecific – which contains anti-IgG and anti-C3d, and may contain anti-C3b and other immunoglobulin and complement antibodies (Figure 19.2). A negative AHG test must be followed by a control system of IgG-sensitized cells, termed *check cells*, to confirm that the result is not a false negative. If the check cells do not agglutinate, then the test must be repeated.

**Detection Techniques:** A variety of techniques can be used for detection of RBC alloantibodies; in the US, tube and gel methods are most commonly used, followed by solid phase methods. Regardless of the method, test sensitivity is usually greater than 97–99% when performed under FDA-approved conditions. Each laboratory should determine how testing will be performed, including which methods are to be used as primary and alternative methods.

**Tube Test:** The tube test method (also known as liquid-phase tube or "wet" method) can be performed with no enhancement reagents (saline). More often an enhancement media is used to increase the sensitivity; these media include albumin and low ionic strength solution (LISS), which decrease the zeta potential (the repulsive electric potential between RBCs that prevents their aggregation) and bring the RBCs closer together. Polyethylene glycol (PEG), which removes water and concentrates the antibodies in the sample, is also used. Each enhancement media has its limitations. For example, LISS and albumin enhance cold autoantibodies and PEG enhances warm autoantibodies. In addition, different methods have varying sensitivity and specificity. The tube test is flexible and relatively inexpensive, but the reactions are unstable, the

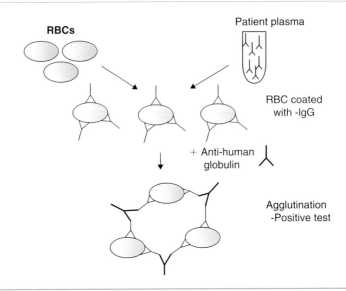

FIGURE 19.2   Antihuman globulin test. Modified from Hillyer CD, Silberstein LE, Ness PM *et al.* (eds). (2007). *Blood Banking and Transfusion Medicine: Basic Principles & Practice*, 2nd edition. Philadelphia, PA: Churchill Livingstone, Elsevier.

grading of reactions is subjective, the test requires increased technologist time, and there are test failures due to inadequate washing between steps.

**Gel Test:**  The gel method (also known as column agglutination) has a similar sensitivity to the PEG tube test, requires less technologist time, produces reactions that are stable for up to 24 hours, and allows for a more standardized grading of reactions. However, it is more expensive than the tube test, as special equipment must be purchased to accommodate the gel cards.

**Solid Phase Test:**  The solid phase method has benefits similar to those of the gel method, but the solid phase method requires that technologists are carefully trained to interpret reactions because, unlike the tube and gel methods, solid phase reactions are not based on agglutination.

**Automation:**  Both the gel and solid phase methods have been automated and require smaller sample volumes than the tube test method.

**ABO Blood Type:**  The ABO antigens and antibodies are the most clinically significant of all RBC antigens and antibodies in transfusion practice. The reciprocal antibodies (also called antithetical isohemagglutinins) are consistently present in the sera of individuals without previous RBC exposure because they are produced in response to environmental stimulants, such as bacteria. These antibodies may result in severe intravascular hemolysis after transfusion of ABO-incompatible RBCs. The prevention of ABO-incompatible transfusion is a primary objective of pretransfusion testing. In order to determine the ABO group of the patient using serology, the patient's RBCs must be tested with anti-A and anti-B (known as the front/forward type) and the patient's plasma with $A_1$ and B RBCs (known as the back/reverse type). These assays use FDA-licensed antisera and reagent RBCs (Table 19.1). ABO RBC typing reagents are currently manufactured as monoclonal antibodies (monoclonal reagents), while in the past they were derived from human sera (polyclonal reagents). Typing can be performed by tube, gel and solid phase methods.

### TABLE 19.1 ABO Typing

| Reaction With RBCs Tested With: | | Reaction With Plasma Tested With: | | Interpretation |
| --- | --- | --- | --- | --- |
| Anti-A | Anti-B | $A_1$ Cells | B Cells | ABO Group |
| 0 | 0 | + | + | O |
| + | 0 | 0 | + | A |
| 0 | + | + | 0 | B |
| + | + | 0 | 0 | AB |

**D Type:** After the ABO group system, D is the most clinically important RBC antigen in transfusion practice. In contrast to A and B antigens, an individual who lacks the D antigen does not form the corresponding antibody unless exposed to D-positive RBCs through pregnancy or transfusion. The D antigen has greater immunogenicity than all other RBC antigens; 20% of hospitalized D-negative individuals who receive a D-positive RBC component will form anti-D. Therefore, most countries where there is a significant population of D-negative individuals will test all patients and donors for the D antigen to ensure D-negative individuals receive D-negative RBC-containing components. The patient must be tested serologically using FDA-licensed anti-D reagents, which can be monoclonal, polyclonal, or monoclonal–polyclonal blends. If the D typing cannot be interpreted, then the patient should receive D-negative RBCs. Testing a patient's RBCs for weak D (where the D test is taken to the AHG phase) is not necessary; without this additional step weak D individuals are treated as D negative, and there is no harm in giving these individuals D-negative RBCs.

**D Type of a Blood Product:** In contrast to the transfusion service, the donor center must ensure that D-negative products are appropriately labeled, such that a recipient of a D-negative product does not form anti-D in response to transfused RBCs. Currently, donor centers use typing reagents to detect the presence of the D antigen, and to increase the sensitivity of these reagents the test is brought to the AHG phase (weak D testing) (see Chapter 10).

**Antibody Screen:** The objective of the antibody screen is to identify any clinically significant antibodies against foreign (non-self) RBC antigens (RBC alloantibodies). Each institution must decide which antibodies it considers clinically significant (examples are given in Chapter 20, Tables 20.1 and 20.2). The procedure should also minimize detection of clinically insignificant antibodies, and be performed in a timely manner. Because these alloantibodies are primarily IgG, AHG must be used for maximal sensitivity.

## Blood Component Selection:

**ABO Compatibility:** Blood components should be ABO-identical with the transfusion recipient, although situations frequently arise in which this is not possible (Table 19.2). If the component contains 2 ml or more of RBCs (i.e. RBC and granulocyte components), then the donor's RBCs must be compatible with the recipient's plasma (Table 19.3). Since plasma-containing components (i.e. plasma, platelets and whole blood) have ABO antibodies, the donor plasma should be compatible with the recipient's RBCs (Table 19.4). Whole blood must be ABO-identical to the recipient's ABO type.

### TABLE 19.2  ABO Selection of Blood Components

| Component | ABO Selection of the Component |
| --- | --- |
| Whole blood | Identical to the patient |
| RBCs | Compatible with the patient's plasma |
| Granulocytes pheresis | Compatible with the patient's plasma |
| Plasma | Compatible with the patient's RBCs |
| Platelet pheresis | All ABO groups acceptable; preferable to give components compatible with the patient's RBCs |
| Cryoprecipitate | All ABO groups acceptable |

### TABLE 19.3  ABO Group Selection for RBC Transfusion

| Patient ABO Group | RBC Containing Component ABO Group | | | |
| --- | --- | --- | --- | --- |
| | 1st Choice | 2nd Choice | 3rd Choice | 4th Choice |
| AB | AB | A | B | O |
| A | A | O | | |
| B | B | O | | |
| O | O | | | |

### TABLE 19.4  ABO Group Selection of Plasma Component Transfusion

| Patient ABO Group | Plasma Containing Component ABO Group | | | |
| --- | --- | --- | --- | --- |
| | 1st Choice | 2nd Choice | 3rd Choice | 4th Choice |
| O | O | A | B | AB |
| A | A | AB | | |
| B | B | AB | | |
| AB | AB | | | |

**D Compatibility:** Only RBC-containing components (RBCs, granulocytes, whole blood and platelets) need to be matched for the D-antigen. D-negative components should be reserved for D-negative recipients in order to prevent anti-D formation. Occasionally, D-negative components are not available for D-negative patients. In this situation, it is important to weigh the risk of alloimmunization against the urgency of transfusion. Formation of an anti-D is most detrimental to a female with child-bearing potential, but prevention of antibody production may also be important in patients who may require ongoing RBC support (such as patients undergoing transplantation or with chronic anemia). The risk of sensitization to the D antigen is highest with RBC components (80% in healthy volunteers and approximately 30% in hospitalized patients), and is very low in patients with malignancy undergoing chemotherapy and receiving apheresis platelet components. For platelet components and other low-volume RBC-containing components, Rh immune globulin (RhIg) can be given to prevent anti-D formation after transfusion of D-positive RBCs. It has been reported that RhIg can prevent anti-D formation after D-positive RBC transfusion in a D-negative patient, but there is a risk of hemolysis with large doses of RhIg in the presence of large volumes of D-positive RBCs. In emergency situations, there may not be adequate time to acquire a D-negative component for a D-negative patient.

**Crossmatch:** RBC components (including whole blood) must be crossmatched with the patient's sample prior to issue, except in emergency situations. The objective of the crossmatch is to demonstrate compatibility between antigens on the donor RBC and the recipient's plasma antibodies.

**Immediate Spin Crossmatch:** The immediate spin crossmatch is designed to detect incompatibility due to ABO antigens on donor RBCs, while an AHG crossmatch is used to detect incompatibility between alloantibodies in the patient's plasma and non-ABO antigens on donor RBCs. If the patient has no history of clinically significant alloantibodies and the current screen does not detect any clinically significant alloantibodies, then an AHG test is not required and only an immediate spin or computer crossmatch is necessary. The advantages of an immediate spin crossmatch are decreased workload, turnaround time and reagent costs.

**Electronic Crossmatch:** Electronic crossmatching can be used with a validated computer system in patients who have no previous or current clinically significant alloantibodies and whose ABO group has been determined at least twice (once on the current sample). The advantages of the electronic crossmatch are decreased workload, decreased volume requirement of patient samples, improved blood inventory, and decreased turnaround time for component issue. In the US, for non-alloimmunized patients, immediate spin tube testing is currently more commonly used than electronic crossmatch, but this may be changing.

**AHG Crossmatch:** For alloimmunized patients, AHG crossmatch must be performed. In the US, tube testing with LISS and gel are most commonly used, followed by tube testing with PEG, albumin or saline, and solid phase testing.

**Neonates:** An initial pretransfusion specimen is required to determine ABO/D type. Only the neonatal RBCs need to be tested with anti-A and anti-B (only the forward type needs to be performed), as neonates do not usually form anti-A and/or anti-B in their plasma for the first 4–6 months and the antibodies that are detected soon after birth are from placental transfer from their mother. If a non-group-O infant is to receive non-group-O RBCs that are not compatible with the maternal ABO group, then the infant's plasma must be tested for the presence of ABO antibodies using a method that includes the AHG phase. Plasma from the infant's mother or the infant can be used for unexpected RBC antibody detection and crossmatch. If no clinically significant antibodies are present, then it is not necessary to perform a crossmatch for the initial or subsequent transfusions. Repeat testing of an infant less than 4 months of age during any single hospital admission is not required.

**Labeling and Release of Blood Components:** The AABB *Standards* require that the following occur at the time of issue:
1. A tag or label with the patient's two independent identifiers, donor unit number, and compatibility testing results (if performed) must be securely attached to the container.
2. A final check of records (patient name and identification number, patient's ABO/D type, donor component number, donor's ABO/D type, interpretation of crossmatch [if performed], date and time of issue, and identification of special transfusion requirements [such as irradiation, antigen-negative, and CMV-reduced risk] must be performed).
3. There must be confirmation that identifying information on the request, records and components are in agreement. Prior to issue, the component must be acceptable for use and should be checked for abnormal color, leakage, and outdate.

Final identification of the patient and component must be performed by the transfusionist, who must identify the patient and donor unit and certify that all identifying information on forms, tags and labels are in agreement. A record of the transfusion should be made as part of the patient's medical record.

**Causes of Unexpected Test Results:**
- *Negative antibody screen and incompatible immediate spin crossmatch*: anti-A$_1$ is present in an A$_2$ or A$_2$B individual being crossmatched with A$_1$ component, patient has room temperature antibodies (anti-M, anti-I), there is rouleaux, or donor RBCs are polyagglutinable.
- *Negative antibody screen and incompatible antiglobulin crossmatch*: donor RBCs have positive direct antiglobulin test, donor RBCs have low-incidence antigen for which the patient has the corresponding antibody, donor RBCs have antigen for which the patient has the corresponding antibody but was missed on the antibody screen secondary to strength of the antigen expression, or passive antibody is present (ABO antibodies from the transfusion of ABO non-compatible plasma-containing components).
- *Positive antibody screen and compatible crossmatch*: antibodies dependent on reagent RBC diluent, component does not contain RBC antigen corresponding to the patient's antibody or it is at a lower dose where it does not result in agglutination, or the patient has autoanti-H or autoanti-IH (antibody reacts with group O RBCs but not type-specific RBCs).

- *Positive antibody screen, incompatible crossmatch, and negative auto control*: patient has RBC alloantibodies or there is unexpected interaction with the reagents.
- *Positive antibody screen, incompatible crossmatch, and positive auto control*: patient has RBC alloantibodies and is experiencing a delayed serologic or hemolytic transfusion reaction, passively transferred alloantibody (intravenous immune globulin or RhIg) reactive with the patient's RBCs, or cold or warm reactive autoantibodies, or there is rouleaux formation or unexpected interaction with the reagents.

Quality Assurance: The Clinical Laboratory Improvement Amendments of 1988 (CLIA '88) regulates ABO group and D typing, antibody detection and crossmatch. Proficiency testing must be performed at least twice a year. A program of quality control must be established to ensure that reagents, equipment and methods function as expected. Results need to be reviewed and, when appropriate, corrective action taken. Reagents and other materials must be stored and used in accordance with the manufacturer's written instructions.

## Recommended Reading

BCSH Blood Transfusion Force (Chairman: JAF Napier; Membership: JF Chapman, K Forman, P Delsey, SM Knowles, MF Murphy and JK Wood; Drafting group: JF Chapman (Convenor), T Balin, SM Knowles, CE Milkins and G Poole). (1996). Guidelines for pretransfusion compatibility procedures in blood transfusion laboratories. *Transfusion Med* **6**, 273–283.

Frohn C, Dümbgen L, Brand J *et al.* (2003). Probability of anti-D development in D− patients receiving D+ RBCs. *Transfusion* **43**, 893–898.

Shulman IA, Downes KA, Sazama K, Maffei LM. (2001). Pretransfusion compatibility testing for red blood cell administration. *Curr Opin Hematol* **8**, 397–404.

Shulman IA, Maffe LM, Downes KA. (2005). North American pretransfusion testing practices, 2001–2004: results from the College of American Pathologists Interlaboratory Comparison Program Survey Data, 2001–2004. *Arch Pathol Lab Med* **129**, 984–989.

Yazer MH. (2006). The blood bank "black box" debunked: pretransfusion testing explained. *Can Med Assoc J* **174**, 29–32.

# CHAPTER 20

# Antibody Identification

Beth H. Shaz, MD

Unexpected RBC antibodies (defined as all alloantibodies other than anti-A and anti-B) are found in 1.2% to 35% of the population, depending on the age (pediatric patients are less likely to have antibodies than adults), number of previous transfusions, and patient population (the highest rates of alloimmunization occur in patients with sickle cell disease). Immunization to RBC antigens that require RBC exposure usually results from pregnancy or transfusion. In contrast, there are some non-anti-A or -anti-B RBC alloantibodies that are naturally occurring (i.e. do not occur in response to known sensitization through RBC exposure, but rather may be examples of antigenic mimicry), such as anti-M or anti-Le$^a$.

Initial detection of unexpected RBC alloantibodies can occur when testing the patient's plasma for ABO type (ABO typing discrepancy), antibody screen, crossmatch, or through an eluate prepared from the patient's RBCs coated with the alloantibody (ABO typing and antibody screen are discussed in Chapter 19; the eluate is discussed in Chapter 21). Once an antibody is detected, its specificity should be determined and its clinical significance assessed. This further testing is termed *RBC antibody identification* (also known as antibody panel, identification of alloantibodies to RBC antigens, or serologic investigation of unexpected antibodies).

<u>Antibody Identification Panel:</u> If the antibody is present in the patient's plasma, then the specificity of the unexpected RBC antibody can be determined by testing the plasma against a panel of reagent RBCs. The antibody specificity can be determined by correlating the RBCs that do or do not react to the plasma with the antigen profile of those RBCs (e.g. if there is agglutination of K-positive RBCs but not of K-negative panel RBCs, then the patient's plasma contains an anti-K). Once the specificity of the antibody is determined, then the clinical significance can be assessed.

Determination of Clinical Significance: Clinically significant alloantibodies result in the decreased survival of transfused RBCs (i.e. hemolysis) and/or are associated with hemolytic disease of the fetus and newborn (HDFN) (Tables 20.1 and 20.2). If an antibody is determined to be clinically significant, then antigen-negative RBCs should be used for all present and future transfusions. If the antibody is associated with HDFN, then the appropriate prenatal testing needs to be performed, such as antibody titers. Therefore, knowing the patient's medical history is important to aid in antibody identification (history of recent or past RBC transfusion or pregnancy), and in assessing the clinical significance of the antibody and the increased likelihood of autoantibodies (medication or disease history or clinical presentation).

History of Antibody: Patients with previously identified antibodies are required to be tested for additional clinically significant antibodies per AABB *Standards*.

**TABLE 20.1 Summary of Major Blood Group Antigens**

| Antigens | Systems | IgM | IgG | Transfusion Reactions | HDFN | Prevalence (%) White | Prevalence (%) African American |
|---|---|---|---|---|---|---|---|
| A | ABO | X | X | Mild–severe | None–moderate | 40 | 27 |
| B | ABO | X | X | Mild–severe | None–moderate | 11 | 20 |
| D | Rh | X | X | Mild–severe | Mild–severe | 85 | 92 |
| C | Rh | | X | Mild–severe | Mild | 68 | 27 |
| E | Rh | X | X | Mild–moderate | Mild | 29 | 22 |
| c | Rh | | X | Mild–severe | Mild–severe | 80 | 96 |
| e | Rh | | X | Mild–moderate | Rare | 98 | 98 |
| K | Kell | X | X | Mild–severe | Mild–severe | 9 | 2 |
| k | Kell | | X | Mild–moderate | Mild–severe | 99.8 | >99 |
| Kp$^a$ | Kell | | X | Mild–moderate | Mild–moderate | 2 | <1 |
| Kp$^b$ | Kell | | X | None–moderate | Mild–moderate | >99 | >99 |
| Ja$^a$ | Kell | | X | None–moderate | Mild–moderate | <1 | 20 |
| Jp$^b$ | Kell | | X | Mild–moderate | Mild–moderate | <99 | 99 |
| Fy$^a$ | Duffy | | X | Mild–severe | Mild–severe | 66 | 10 |
| Fy$^b$ | Duffy | | X | Mild–severe | Mild | 83 | 23 |
| Jk$^a$ | Kidd | | X | None–severe | Mild–moderate | 77 | 92 |
| Jk$^b$ | Kidd | | X | None–severe | None–mild | 74 | 49 |
| M | MNS | X | | None | None–mild | 78 | 74 |
| N | MNS | X | | None | None | 70 | 75 |
| S | MNS | | X | None–moderate | None–severe | 52 | 31 |
| s | MNS | | X | None–mild | None–severe | 89 | 94 |
| U | MNS | | X | Mild–severe | Mild–severe | 100 | >99 |
| Le$^a$ | Lewis | X | | Few | None | 22 | 23 |
| Le$^b$ | Lewis | X | | None | None | 72 | 55 |
| Lu$^a$ | Lutheran | X | X | None | None–mild | 8 | 5 |
| Lu$^b$ | Lutheran | X | X | Mild–moderate | Mild | >99 | >99 |
| Do$^a$ | Dombrock | | X | Rare | +DAT/No HDFN | 67 | 55 |
| Do$^b$ | Dombrock | | X | Rare | None | 82 | 89 |
| Co$^a$ | Colton | | X | None–moderate | Mild–severe | >99.9 | >99 |
| Co$^b$ | Colton | | X | None–moderate | Mild | 10 | 10 |
| P$_1$ | P | X | | Rare | None | 79 | 94 |

| TABLE 20.2 Clinical Significance of Antibodies to the Major Blood Group Antigens | | | |
|---|---|---|---|
| **Usually Clinically Significant** | **Sometimes Clinically Significant** | **Clinically Insignificant if not Reactive at 37°C** | **Generally Clinically Insignificant** |
| A and B | At$^a$ | A$_1$ | Bg |
| Diego | Colton | H | Chido/Rogers |
| Duffy | Cromer | Le$^a$ | Cost |
| H in O$_h$ | Dombrock | Lutheran | JMH |
| Kell | Gerbich | M, N | Knops |
| Kidd | Indian | P$_1$ | Le$^b$ |
| P, PP$_1$P$^k$ | Jr$^a$ | Sd$^a$ | Xg$^a$ |
| Rh | Lan | | |
| S, s, U | LW | | |
| Vel | Scianna | | |
| | Yt | | |

Reproduced from Hillyer CD, Strauss RG, Luban NLC (eds). (2004). *Handbook of Pediatric Transfusion Medicine*. San Diego, CA: Elsevier Academic Press.

**Indication:** Antibody identification panels are required when a patient has an unexpected positive RBC antibody screen, ABO typing discrepancy, or unexpected positive crossmatch possibly due to RBC antibody.

**Specimen Requirements:** Antibody identification can be performed on either serum or plasma. Blood samples anticoagulated with EDTA avoid the problems associated with the *in vitro* uptake of complement components by RBCs, which may occur in clotted samples.

**Method:** The most common methods for antibody identification are based on the agglutination of RBCs. A variety of techniques can be used for detection of RBC agglutination; in the US, tube and gel techniques are most commonly used, followed by the solid phase method. The tube method can be performed with no enhancement reagents (saline), but more often one of a variety of enhancement media, such as albumin, low ionic strength solution (LISS) or polyethlene glycol (PEG), is used. Each enhancement media has its limitations – for example, LISS and albumin enhance cold autoantibodies and PEG enhances warm autoantibodies. Different techniques have varying sensitivity and specificity: one study found the automated solid phase method to be less sensitive (65%) but more specific (99%) than either manual solid phase or tube methods using PEG. Another study comparing the performance of a variety of methods determined that the sensitivities of the gel and solid phase systems were similar and higher than the tube method with LISS, yet all of the higher sensitivity methods had higher false-positive rates and identified more antibodies of minor clinical significance. Each laboratory should determine how testing will be performed, including which methods are to be used as primary and alternative methods and what phase of testing to use (immediate spin, 37°C, and/or antiglobulin phase).

**Panel of RBCs of Known Antigenic Composition:** Identification of an antibody requires testing the plasma against a panel of selected RBCs with known antigen composition for the major blood groups (Rh, Kell, Kidd, Duffy and MNS) (Figure 20.1).

- These panel RBCs are usually obtained from a commercial supplier, are usually group O, and are selected so a distinctive pattern results from positive and negative reactions for each of many antigens.
- A panel should make it possible to identify most of the common clinically significant alloantibodies, such as anti-D, -E, -K, and -Fy$^a$.
- Negative reactions are used to eliminate the presence of alloantibodies.
- Dosage occurs when an antibody reacts more strongly with a RBC with a double dose of antigen (homozygous for the allele), than a RBC with a single dose of antigen (heterozygous for the allele). An example is an anti-M that reacts more strongly with a RBC that is M positive and N negative than a RBC that is M positive and N positive.
- Ideally, antibodies should be ruled out on RBCs that carry a double dose of antigen, but this may not be possible for low-frequency antigens that are rarely carried homozygously (such as K).
- Additional selected RBCs are often required to complete the evaluation.
- Criteria for ruling in or ruling out antibodies vary by institution. One example is to test the patient's plasma with at least three antigen-positive RBCs and three antigen-negative RBCs which will result in a probability value of 0.05, indicating that the probability the results are due to chance alone is < 5%.

**RBC Phenotype:** Performing a phenotype of the patient's RBCs is useful in aiding antibody identification. If the patient lacks a RBC antigen, then he or she is capable of forming an antibody to that antigen. If the patient has a RBC antigen, then he or she usually does not make an antibody to that antigen. There are rare circumstances when this rule is broken; for example, individuals who are D positive and have the partial D phenotype may form an anti-D. RBC phenotyping can also be useful to help focus antibody identification by aiding in the determination of what antibodies a patient could make. RBCs from a patient sample prior to RBC transfusion are required (pretransfusion sample); transfused RBCs can circulate for up to 120 days after transfusion, and therefore their presence could produce a false-positive phenotype result.

*Phenotyping Direct Antiglobulin Test (DAT)-positive RBCs:* In patients with a positive DAT, the use of phenotyping reagents, which require anti-human globulin (AHG), may result in false-positive antigen typing results unless the IgG can be removed from the patient's RBCs prior to testing. Fortunately, many RBC typing reagents now consist of monoclonal antibodies, allowing for antigen detection in the immediate spin phase. If necessary, IgG can be removed from the DAT-positive RBCs by a variety of techniques, such as chloroquine or a combination of acid glycine and EDTA. Sometimes the antigen is destroyed by these removal methods (such as Kell system antigens when using the acid glycine/EDTA method) or the removal is not effective. In such cases, genotyping provides an acceptable alternative by allowing the prediction of a patient's RBC antigen status without the interference of the antibody.

| CELL | Rh | | | | | | | | MNS | | | | Lutheran | | P1 | Lewis | | Kell | | Duffy | | Kidd | | IS | LISS 37C | AHG | Check cells |
|---|---|---|---|---|---|---|---|---|---|---|---|---|---|---|---|---|---|---|---|---|---|---|---|---|---|---|---|
| | D | C | E | c | e | f | V | Cw | M | N | S | s | Lua | Lub | P1 | Lea | Leb | K | k | Fya | Fyb | Jka | Jkb | | | | |
| 1. r'r-2 | 0 | + | 0 | + | + | + | 0 | 0 | + | + | + | + | 0 | + | 0 | 0 | + | 0 | + | + | 0 | + | + | ∅ | ∅ | 2+ | |
| 2. R1wR1-1 | + | + | 0 | 0 | + | 0 | 0 | + | + | + | 0 | 0 | + | + | + | 0 | + | 0 | + | 0 | + | + | 0 | ∅ | ∅ | 2+ | |
| 3. R1R1-6 | + | + | 0 | 0 | + | 0 | 0 | 0 | 0 | + | + | + | 0 | + | + | + | 0 | + | + | 0 | + | + | 0 | ∅ | ∅ | 2+ | |
| 4. R2R2-8 | + | 0 | + | + | 0 | 0 | + | 0 | + | + | 0 | + | 0 | + | + | 0 | + | 0 | 0 | 0 | + | 0 | + | ∅ | ∅ | ∅ | ✓ |
| 5. r'r-3 | 0 | + | 0 | + | + | + | 0 | 0 | + | + | 0 | 0 | 0 | + | + | + | 0 | 0 | + | + | + | 0 | + | ∅ | ∅ | ∅ | ✓ |
| 6. rr-32 | 0 | 0 | 0 | + | + | + | 0 | 0 | + | + | 0 | 0 | 0 | + | 0 | 0 | 0 | 0 | + | 0 | + | + | 0 | ∅ | ∅ | ∅ | ✓ |
| 7. rr-10 | 0 | 0 | 0 | + | + | + | + | 0 | + | + | + | + | 0 | + | + | 0 | + | + | + | + | + | + | + | ∅ | ∅ | ∅ | ✓ |
| 8. rr-12 | 0 | 0 | 0 | + | + | + | 0 | 0 | 0 | 0 | 0 | + | 0 | + | + | 0 | + | 0 | + | 0 | 0 | 0 | + | ∅ | ∅ | ∅ | ✓ |
| 9. Ro-4 | + | 0 | 0 | + | + | + | 0 | 0 | + | 0 | + | + | 0 | + | 0 | + | 0 | 0 | + | 0 | 0 | 0 | + | ∅ | ∅ | ∅ | ✓ |
| Cord cell | / | / | / | / | / | / | / | / | / | / | / | / | / | / | / | 0 | 0 | / | / | / | / | / | / | | | | |
| Patient | | | | | | | | | | | | | | | | | | | | | | | | ∅ | ∅ | ∅ | ✓ |

FIGURE 20.1 An example of an antibody identification panel in a patient with anti-C.

*Phenotyping Patients who have been Recently Transfused:* Due to the presence of circulating donor RBCs, which may persist for weeks, determination of an individual's phenotype by traditional hemagglutination methods is complicated by the presence of a mixed field population. The RBC phenotype may be determined with the use of time-consuming and labor-intensive techniques, such as isolating reticulocytes or sickle cells (in individuals with SCD), but it may be inaccurate. Molecular methods performed on DNA obtained from peripheral blood leukocytes can overcome some limitations of serologic techniques.

*Genotyping:* RBC antigen identification can be performed through DNA testing. High-throughput molecular technologies in the field of transfusion medicine make it possible to determine multiple RBC antigen single nucleotide polymorphisms (SNPs) simultaneously. There are multiple mass-scale genotyping platforms in use and in development in North America and Europe, with a single platform currently FDA-approved. This platform includes genotypes for Rh (CcEe), Kell (K/k), Kidd ($Jk^a/Jk^b$), Duffy ($Fy^a/Fy^b$), Duffy-GATA, MNS (M/N, S/s), Lutheran ($Lu^a/Lu^b$), Diego ($Di^a/Di^b$), Colton ($Co^a/Co^b$), Dombrock ($Do^a/Do^b$, Ja(a+)/Jo(a−), Hy+ /Hy−), Landsteiner-Wiener ($LW^a/LW^b$), and Scianna (Sc1/Sc2) blood groups and Hemoglobin S.

**Prewarm Technique:** Cold-reactive antibodies may react in the AHG phase of testing. Reactivity can often be eliminated by removing the enhancement media, using saline-suspended RBCs, and performing a 30- to 60-minute incubation followed by the AHG test. If reactivity remains, the prewarm technique may be used. In prewarm testing, the patient serum and reagent RBCs are warmed to 37°C separately and then mixed. Although clinically significant antibodies that are weakly reactive can become negative in prewarm testing, the risk is considered minimal. A small percentage of patients with cold agglutinins will still have reactivity. In these patients, cold adsorption or adsorption with rabbit erythrocyte stroma (REST, which removes anti-I and anti-IH and some clinically significant alloantibodies, such as anti-B, -D, and -E) may be used.

**Cold Antibody Screen:** The cold antibody screen is used to confirm and identify the presence of cold autoantibodies. The patient's plasma is combined with appropriate reagent RBCs (such as $A_1$, B, O adult, O cord, and autologous RBCs) and then read for agglutination at varying temperatures (room temperature, 18°C, and 4°C). This technique can be useful in resolving antibody identification problems.

**Enzyme-treated RBCs:** Enzyme-treated RBCs can be useful in antibody detection either to confirm the presence of an antibody or to detect additional antibodies because enzymes destroy or alter RBC antigens. Ficin, papain, trypsin and bromelin are most commonly used, and destroy or alter the Duffy and MNS blood group antigens and $Xg^a$, JMH, Ch, Rg, S, $Yt^a$, Mg, $Mi^a$/Vw, $Cl^a$, $Je^a$, $Ny^a$, JMH, some Ge, and $In^b$ antigens, and enhance the Rh, Kidd, Lewis, and ABO blood group antigens, and the $P_1$ and I antigens. DTT destroys or alters Kell, Lutheran, Dombrock, and Cromer blood group antigens, and the $Yt^a$, JMH, $Kn^a$, $McC^a$, $Yk^a$, $LW^a$ and $LW^b$ antigens. ZZAP is a combination of DTT and proteolytic enzymes. When enzyme-treated RBCs are used for antibody identification, alloantibodies to the blood group antigens which are destroyed or altered with enzymes cannot be ruled out.

Neutralization: Neutralization of antibodies can be useful in confirming the antibody's identity or in identifying other antibodies. Anti-Sd$^a$ is neutralized by urine (Sd$^a$ substance); anti-Ch, anti-Rg (Chido and Rodgers substance), anti-Le$^a$ and anti-Le$^b$ (Lewis substances) are neutralized by plasma; and P$_1$ is neutralized by P$_1$ substance found in hydatid cyst fluid and pigeon egg whites. When using neutralized plasma for antibody identification, alloantibodies to the blood group antigens that were neutralized cannot be ruled out.

Adsorption: In patients with panagglutinating warm autoantibodies, special techniques are used to adsorb out the autoantibodies from the plasma in order to determine whether alloantibodies are present. These tests are not routinely performed in all laboratories, and may need to be performed in a reference laboratory.

Autologous Adsorption: If the patient has not been transfused within the last 3 months and has an adequate hematocrit, then an autologous adsorption can be performed. In this procedure, a whole blood sample is first separated into RBC and plasma fractions, and then heat or other elution techniques are used to remove autoantibody from the RBCs. The treated RBCs, now stripped of autoantibodies, are mixed with the plasma fraction in order to adsorb out any additional autoantibody present in the plasma. When performed correctly, the adsorbed plasma should be free of autoantibody, leaving only alloantibodies remaining.

Allogeneic Adsorption: If the patient has been recently transfused (within the last 3 months), then an allogeneic adsorption must be performed. In allogeneic adsorption, allogeneic RBCs (which may or may not be enzyme-treated) are used to adsorb out the autoantibody. However, because the adsorbing cells may differ from the patient's RBCs in alloantigen expression, they must be carefully selected so that they do not inadvertently adsorb alloantibodies as well as autoantibodies from the patient's plasma. Therefore, at least three phenotype-appropriate RBCs must be used for allogeneic adsorption in order to rule out the majority of alloantibodies. For example, a D-positive RBC used for adsorption cannot be used to rule out the presence of anti-D. The adsorbed plasma can then also be used for crossmatch.

Blood Component Selection: RBC components need to be negative for the corresponding antigen against which the patient has a clinically significant alloantibody. Each laboratory determines which antibodies require the use of antigen-negative RBCs and/or crossmatch using the AHG phase. Usually clinically significant alloantibodies require the use of antigen-negative RBCS, while antibodies that are only significant at 37°C may require crossmatching using the AHG phase, and clinically insignificant alloantibodies require immediate spin crossmatch only. Some RBC typing reagents are not routinely available, making it difficult to obtain RBCs negative for certain antigens. If RBC typing reagents are not available, then a crossmatch with the patient's plasma to detect the presence of an incompatibility can be used. If the patient's antibody titer is too low to be detected, and there is no method to ensure the components are antigen negative, a reference laboratory may be able to provide antigen-negative components either by serological (using non-commercial typing reagents) or by molecular techniques.

**Warm Autoimmune Hemolytic Anemia:** It is difficult to select RBCs which will have optimum survival for patients with autoantibodies, because the antibodies typically have a broad spectrum of reactivity. Even donor RBCs that lack antigens for which the patient has the corresponding alloantibody will likely be crossmatch incompatible. However, usually the destruction of the transfused RBCs is not in excess of that already occurring to the patient's own RBCs secondary to the autoimmune disease. If the autoantibody has specificity, RBCs lacking the corresponding antigen may provide improved survival, but the data to support this practice are lacking.

**Paroxysmal Nocturnal Hemoglobinuria (PNH):** Patients with PNH, which is a stem cell disorder manifested by complement-mediated hemolytic anemia, thrombophilia and marrow failure, should receive ABO-identical blood components in order to avoid complement activation.

**Crossmatch:** The objective of the immediate-spin crossmatch is to demonstrate ABO compatibility between donor and recipient before transfusion. An AHG crossmatch is required in patients who have clinically significant RBC alloantibodies, and is used to detect incompatibility due to the presence of an antigen on the donor RBCs against which the patient has formed the corresponding alloantibody.

**Warm Autoantibodies:** Adsorbed plasma can be used for crossmatch, if available. In patients with warm autoantibodies, unabsorbed plasma will often result in an incompatible crossmatch.

**Neonates:** Infants younger than 6 months of age will have passive antibodies from their mothers. In addition, passive antibodies can result from transfused plasma and intravenous immunoglobulin. RBCs that do not contain the corresponding antigen are required until the passive antibody is no longer detected. A maternal sample can be used for antibody detection until the neonate is 4 months of age.

**Quality Assurance:** The Clinical Laboratory Improvement Amendments of 1988 (CLIA '88) regulates antibody detection and crossmatch. Proficiency testing must be performed at least twice a year. A program of quality control must be established to ensure that reagents, equipment and methods function as expected. Results need to be reviewed and, when appropriate, corrective action taken. Reagents and other materials must be stored and used in accordance with the manufacturer's written instructions.

## Recommended Reading

Hillyer CD, Shaz BH, Winkler AM, Reid M. (2008). Integrating molecular technologies for RBC typing and compatibility testing into blood centers and transfusion services. *Transfus Med Rev* **22**, 117–132.

Shulman IA, Downes KA, Sazama K, Maffei LM. (2001). Pretransfusion compatibility testing for red blood cell administration. *Curr Opin Hematol* **8**, 397–404.

Weisbach V, Kohnhauser T, Zimmermann R *et al.* (2006). Comparison of the performance of microtube column systems and solid-phase systems and the tube low-ionic-strength solution additive indirect antiglobulin test in the detection of red cell alloantibodies. *Transfus Med* **16**, 276–284.

# CHAPTER 21

# Direct Antiglobulin Test

Beth H. Shaz, MD

The direct antiglobulin test (DAT; also known as the Direct Coombs test [DCT]; Figure 21.1) detects the presence of IgG and/or C3 (when IgM binds to RBCs it fixes complement, which is more readily detected than IgM) coating red blood cells (RBCs) *in vivo*. Reasons for a positive DAT, which may or may not be associated with a shortened RBC lifespan (i.e. less than 100–120 days; hemolytic anemia) are listed in Table 21.1. Small amounts of IgG (5–90 IgG molecules/RBC) and complement (5–97 C3d molecules/RBC) are found on RBCs of normal individuals. The DAT can detect a minimal level of 100–500 IgG molecules/RBC and 400–1100 C3d molecules/RBC, depending on the method and testing techniques.

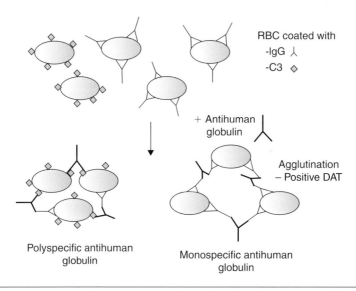

**FIGURE 21.1** Direct antiglobulin test (DAT). Modified from Hillyer CD, Silberstein LE, Ness PM *et al.* (eds). (2007). *Blood Banking and Transfusion Medicine: Basic Principles and Practice*, 2nd edition. San Diego, CA: Elsevier Academic Press.

Positive DATs are reported in 1 in 1000 to 1 in 14,000 healthy blood donors, and 1 in 100 to 1 in 17 in hospitalized patients without clinical manifestation of hemolysis or hemolytic anemia. Multiple diseases are associated with a positive DAT, not necessarily correlating with the presence of hemolytic anemia, such as sickle cell disease, thalassemia, renal disease, multiple myeloma, autoimmune disorders and HIV/AIDS.

| TABLE 21.1 Causes of a Positive DAT |
|---|
| Autoantibodies directed at RBC antigens (warm autoimmune hemolytic anemia [WAIHA], cold agglutinin syndrome [CAS]) |
| Alloantibodies in a patient who was recently transfused antigen-positive RBCs (acute or delayed hemolytic transfusion reaction [AHTR, DHTR]) |
| Passively transfused alloantibodies against the patient's RBCs resulting from plasma-containing components (platelet component) or a plasma derivative (intravenous immunoglobulin [IVIG] or Rh immune globulin [RhIg]) |
| Alloantibodies in the maternal circulation which cross the placenta and coat the fetal RBCs (hemolytic disease of the fetus and newborn [HDFN]) |
| Antibodies against drugs which bind to the RBC membrane (penicillin) |
| Absorbed proteins (IgG) which attach to altered RBC membrane or RBCs modified by drugs (cephalosporins) |
| Immune complex or complement binding to RBCs after drug administration secondary to a drug/anti-drug interaction (quinidine, phenacetin) |
| Nonspecific uptake of protein, usually IgG (patients with hypergammaglobulinemia or recipients of high dose IVIG) |
| Antibodies derived from passenger lymphocytes as a result of either solid organ or HPC transplantation |

Indication: The DAT is useful in the diagnosis of the following situations:

- *Antibody identification*: A DAT can be performed as part of the evaluation for unexpected antibodies. A DAT and, if positive, an eluate may aid in antibody identification (see Chapter 20).
- *Autoimmune hemolytic anemia*: The DAT is performed in order to determine the presence of IgG and/or complement coating the RBCs. A positive DAT with IgG (with or without complement) and a panagglutinin (see below) eluate is consistent with the diagnosis of warm autoimmune hemolytic anemia (WAIHA). A DAT positive with complement only is seen in patients with cold agglutinin disease (CAD) (see Chapter 44).
- *Drug-induced hemolytic anemia*: The DAT can be used to evaluate the presence of drug induced hemolytic anemia (see Chapter 44).
- *Hemolytic disease of the fetus and newborn (HDFN)*: A DAT is performed to evaluate the presence of HDFN. A neonatal sample should have a DAT performed if the mother has a positive antibody screen. It may also be indicated when the mother is ABO-incompatible with the neonate (e.g. the mother is group O and the neonate is group A) (see Chapter 43).
- *Transfusion reactions*: DAT is performed on the post-transfusion sample in order to evaluate a possible acute or delayed hemolytic transfusion reaction (see Chapters 55 and 56). If the post-transfusion DAT is positive, a DAT should also be performed on the pre-transfusion sample in order to assess whether the strength of the reaction has increased. If the DAT is positive for IgG, then an eluate should be performed to determine the specificity of antibody coating the RBCs. If the patient is non-group O, then the eluate should be tested with group O screening cells as well as group A and group B cells.

**Specimen Requirements:** Specimens must be blood samples anticoagulated with EDTA.

**Method:** DATs can initially be performed with a polyspecific anti-human globulin (AHG) reagent, which contains anti-IgG and anti-C3d and may also contain antibodies to other C3 determinants (C3dg, C3b, C3c) and other immunoglobulins (IgA, IgM). If positive with the polyspecific reagent, the sample can be retested with monospecific anti-IgG and anti-C3d/C3dg reagents to further characterize the reactivity. These reagents are licensed in the US by the FDA. DAT can be performed with a variety of serologic techniques, such as tube ("wet"), microtube columns ("gel") and solid phase methods. The sensitivity and specificity of the DAT performed by the gel test in comparison with the tube test has been reported to be 74–98% and 89–95%, respectively.

**Evaluation of a Positive DAT:** The extent to which a positive DAT is evaluated depends on the clinical context of the patient, such as the patient's underlying diagnosis, medication history, pregnancy, transfusion history, and presence of unexplained hemolytic anemia. The clinical significance of a positive DAT needs to be assessed in conjunction with the clinical and laboratory information (i.e. the presence of hemolysis). Further evaluation may be indicated in patients who have evidence of hemolytic anemia, patients with transfusion within the last 3 months, patients receiving medications associated with a positive DAT and RBC destruction, patients who have received a solid organ or HPC transplantation, and patients receiving IVIG or RhIg.

Additional tests for the evaluation of a positive DAT include testing the patient's plasma for the presence of clinically significant auto- or alloantibodies, and testing the eluate if there is IgG coating the RBCs.

**Eluate:** An eluate removes antibodies from the RBCs using a variety of techniques, such as heat, freeze–thaw, cold acid, digitonin acid, and dicholormethane glycine-HCl/EDTA. Commercial elution kits are available. The eluate is then used for antibody identification employing the same techniques used to test plasma (see Chapter 20). Eluate preparations concentrate the antibody, which may aid in antibody identification by increasing sensitivity.

**Panagglutinin:** When the eluate reacts with all RBCs tested (panagglutinin), then an autoantibody is most likely the reason for the positive DAT.

**Non-reactive Eluate:** A non-reactive eluate (i.e. no RBCs reacting) may occur if the sample is tested against RBCs that do not express the appropriate antigen(s). This may be the case with antibodies against low-frequency antigens. In addition, a non-reactive eluate may occur if the positive DAT was due to anti-A and/or anti-B, but the eluate was only tested against group O cells. This is an important consideration in patients who have received ABO-out of group components (e.g. a group A patient who received a group O platelet component). Approximately 80% of hospitalized patients with a positive DAT will have a non-reactive eluate, likely attributable to non-specific protein uptake by the RBCs.

**Enhancement of Reactivity:** Reactivity of eluates can be enhanced by using polyethylene glycol (PEG), concentrating the eluate, or other methods. Some eluate methods result in non-reactive results with certain antibodies; when this occurs, an alternate method should be used.

**False-negative DAT:** If the antibody causing the immune hemolysis is not IgG or IgM (e.g. IgA), it may not be identified if the polyspecific reagent used in testing does not have the appropriate specificity. The IgG bound to RBCs may also be at a concentration too low for the reagent to detect its presence. In this situation, more sensitive techniques can be used. In addition, incorrect washing or resuspending, or delayed testing, may result in a false-negative DAT.

**False-positive DAT:** RBCs from clotted specimens, especially when the specimens have been refrigerated, often have a positive DAT, usually due to false complement reactivity. Previously available polyclonal AHG reagents resulted in false-positive DATs prior to the use of monoclonal reagents. For example, anti-T in some reagents caused a positive DAT in patients with T activation, while anti-transferrin caused a false-positive DAT in patients with high reticulocyte counts.

**Clinical Significance of a Positive DAT:** In any single patient, the clinical significance of a positive DAT is difficult to assess. Among hospitalized patients, 0.7% have positive DATs, while 18% of AIDS patients have positive DATs. The clinical significance of a positive DAT is assessed based on clinical and laboratory findings suggestive of decreased RBC survival, such as anemia, jaundice, hematuria; elevated lactate dehydrogenase, reticulocyte count, and bilirubin (especially the indirect fraction); and decreased haptoglobin (see Chapter 44).

**Recommended Reading**

Das SS, Chaudhary R, Khetan D. (2007). A comparison of conventional tube test and gel technique in evaluation of direct antiglobulin test. *Hematology* **12**, 175–178.
Petz LD, Garraty G (eds). (2004). *Immune Hemolytic Anemias*, 2nd edition. Philadelphia, PA: Churchill Livingstone.

# CHAPTER 22

# ABO and H Blood Group Systems

Beth H. Shaz, MD

The ABO blood group system consists of the A; B; A,B and A antigens. Anti-A and anti-B antibodies are the most clinically significant in transfusion practice. These reciprocal antibodies are consistently present in the sera of the majority of individuals without previous red blood cell (RBC) exposure (e.g. anti-B antibodies in blood group A patients), and these antibodies may result in severe intravascular hemolysis after transfusion of ABO-incompatible blood components. The prevention of ABO-incompatible transfusion is the primary objective of pre-transfusion testing.

**Antigens:** The ABO antigens are carbohydrate structures that are synthesized in a stepwise fashion by glycosyltransferases that sequentially add specific monosaccharide sugars to glycoproteins and glycolipids. *H(FUT1)* gene, which is preferentially expressed in erythroid cells and is located on chromosome 19, encodes a fucosyltransferase which places a fucose in α(1,2) linkage on type 2 glycoproteins (where the terminal galactose has β(1,4) linkage to the N-acetylglucosamine) to form the H antigen on RBCs. The *Se(FUT2)* gene, which is preferentially expressed in epithelial cells and is located on chromosome 19, encodes a fucosyltransferase which adds fucose in α(1,2) linkage to a type I glycoprotein chains (where the terminal galactose has β(1,3) linkage to the N-acetylglucosamine). Individuals who carry the *Se* gene are termed ABO secretors, because they can form the H antigen on type 1 glycoproteins which are produced in epithelial cells and whose products reside on mucins in secretions, and subsequently can produce A and B antigens (given the presence of the appropriate transferases) in the secretions.

The H antigen formed in this way defines the O blood group, and is the precursor for A and B antigens. The A and B glycosyltransferases, encoded by the *ABO* genes on chromosome 9, add either N-acetylgalactosamine or galactose, respectively, to the H antigen, resulting in the A and B antigens, respectively (Figures 22.1 and 22.2). Group O individuals

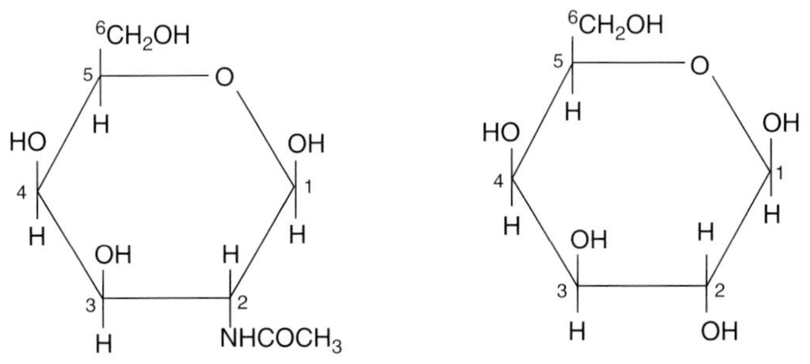


N-acetylgalactosamine
(A antigen)

Galactose
(B antigen)

FIGURE 22.1 Terminal carbohydrates that define the A and B antigens. The terminal galactose residues differ only in that the A antigen has substituted the amino-acetyl group on carbon number 2.

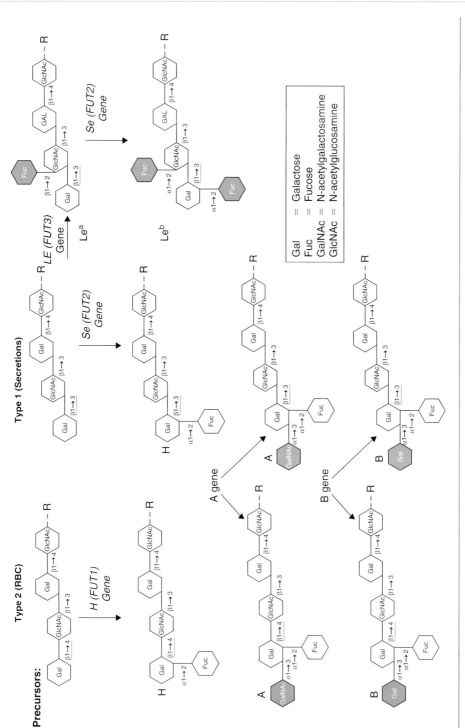

FIGURE 22.2 Synthesis of A, B, H, and Lewis antigens. Oligosaccharide precursor core type 1 and type 2 structures differ only in the linkage between the terminal galactose (Gal) and the N-acetylglucosamine (GlcNAc), shown underscored. Terminal carbohydrates that define the antigens are shown in blue. Modified from Hillyer CD, Silberstein LE, Ness PM et al. (eds). (2007). *Blood Banking and Transfusion Medicine: Basic Principles and Practice*, 2nd edition. San Diego, CA: Elsevier Academic Press.

lack the A or B glycosyltransferases, and therefore have large amounts of H antigen but no A or B antigens. Some H antigen remains on A and B RBCs, with $A_1B$ RBCs having the least and O the most amount of H antigen ($O >> A_2 > B > A_2B > A_1 > A_1B$).

The prevalence of ABO blood groups differs in various populations (Table 22.1). RBCs carry over 2 million ABO antigens. In addition, ABO antigens are found on other tissues, including endothelial and epithelial cells of the lung, gut, and urinary and reproductive tracts (and are therefore termed histo-blood group antigens). Hence they are important in solid organ transplantation, where ABO incompatibility may result in acute rejection. They are also found in secretions and fluids in individuals who carry the *Se* gene (secretor phenotype).

| TABLE 22.1  ABO Blood Group Prevalence | | | |
|---|---|---|---|
| | | Prevalence (%) | |
| ABO Group | White | African American | Asian |
| O | 45 | 49 | 43 |
| A | 40 | 27 | 27 |
| B | 11 | 20 | 25 |
| AB | 4 | 4 | 5 |

## ABO Variants:

*Transferases and ABO Genetic Alleles:* The transferases for A and B antigens differ by only 4 of 354 amino acids. Mutations in these transferase genes may result in weakened expression of the antigens and therefore in subgroup phenotypes. Mutations may also result in non-functional transferases and therefore group O RBCs. There are at least 41 different A subgroup alleles and 18 B subgroup alleles, as well as 61 different O alleles. This heterogeneity makes ABO typing with DNA technologies complicated. In addition, the same genotype gives rise to different phenotypes even within families, which further adds to the complexity.

*Inherited A and B Subgroups:* RBCs from some group A or B individuals that react weakly or not at all with standard anti-A or anti-B sera are termed *subgroups*; B subgroups are very rare and less frequent than A subgroups. These subgroups are secondary to variability in the ABO genes, which results in variations in the A or B antigen structure or the number of antigen sites per RBC. Most group A individuals are $A_1$, while $A_2$ is the major A subgroup; $A_2$ and $A_1$ together account for 99% of group A individuals with approximately 80% of group A individuals being $A_1$. The difference between $A_1$ and $A_2$ is both quantitative (there are fewer A antigens on $A_2$ than $A_1$ RBCs) and qualitative (structural differences between $A_1$ and $A_2$ antigens). The $A_2$ allele differs from the $A_1$ allele by only a single base pair. Because of the structural difference, $A_2$ individuals can form anti-$A_1$ (1–8% of $A_2$ individuals and 22–35% of $A_2B$ individuals have anti-$A_1$). Anti-$A_1$ does not usually result in hemolysis of group $A_1$ RBCs, although hemolytic anti-$A_1$ has been reported. Therefore, it is prudent to transfuse these individuals with group O RBCs (or other compatible RBCs that lack $A_1$). *Dolichos biflorus* lectin is used to distinguish $A_1$ from $A_2$, as it will agglutinate $A_1$ but not $A_2$ RBCs.

*Bombay Phenotype:* The H antigen on RBCs is encoded by the *H* gene, while the H antigen found in secretions is encoded by the *Se* gene, both located on chromosome 19. Both genes encode for a fucosyltransferase enzyme that adds fucose to a precursor oligosaccharide. Homozygosity for defective *Se* (*sese*) results in the non-secretor phenotype with an incidence of approximately 20%. The rare Bombay phenotype (group $O_h$) results from homozygosity of null alleles at *H* and *Se*. Bombay individuals lack the H antigen on RBCs and secreted proteins, and make a potent anti-H. These individuals must only be transfused with RBCs from other Bombay individuals. Para-Bombay denotes two different situations: individuals homozygous for the null allele at *H* but having at least one functional *Se* allele lack H antigen on their RBCs but have the H antigen in their secretions; alternatively, individuals who have minimal residual H production on their RBCs and are non-secretors.

**Expression:** ABO antigens are not fully developed at birth due to the linear structure of their oligosaccharides. Not until 2–4 years of age do the complex branching oligosaccharide structures appear on the RBCs.

**Antibodies:** Anti-A and anti-B are found in the plasma of individuals who lack the corresponding antigen (group O individuals form anti-A and anti-B; group A individuals form anti-B; group B individuals form anti-A; group AB individuals form neither antibody). They are produced in response to environmental stimulants, such as bacteria, and are therefore termed *naturally occurring antibodies*. Antibody production begins after birth and is usually detected by 4–6 months of age, reaches a peak at age 5–10 years, and then declines with increasing age. Immunodeficient patients may not produce detectable levels of anti-A and/or anti-B. The antibodies are a combination of IgM and IgG; the IgM content results in the agglutination at room temperature and makes these antibodies effective at activating complement, which in combination with the high amount of ABO antigens on RBCs are responsible for the severe transfusion reactions that may result from ABO-incompatible transfusions. Hemolytic disease of the fetus and newborn (HDFN) caused by ABO antibodies is usually mild because only IgG crosses the placenta, fetal ABO antigens are not fully developed, and ABO tissue antigens provide additional targets for the antibody. HDFN is most often seen in non-group O infants of group O mothers, because group O individuals often have significant amounts of IgG anti-A, anti-B, and another antibody that reacts with both A and B antigens (anti-A,B).

**Antibody Titers:** Anti-A and anti-B titer results are used in a variety of clinical situations, including the evaluation of ABO mismatched platelet components (especially when a group O platelet component is transfused to a group A or AB patient), ABO-incompatible solid organ transplantation, and ABO mismatched hematopoietic progenitor cell (HPC) transplantation. There currently is no uniform method for titering these antibodies; nor is there a uniform critical value. These antibodies may be tested in both the room temperature phase to detect IgM levels, and the anti-human globulin (AHG) phase at 37°C to detect the IgG levels.

**Plasma-rich Blood Components:** Fatal acute hemolysis has been reported in patients transfused with ABO-incompatible blood components containing high titers of anti-A or anti-B. Typically these are group O components, with significant amounts of plasma, such as platelet components. Because of inadequate platelet inventories, it is not always possible to transfuse ABO matched platelet components. Anti-A and/ or anti-B titers are performed on group O platelet components by some US institutions; a critical titer is used to identify components that are at higher risk for producing acute hemolysis. For example, the UK tests all group O blood donations at a 1:100 dilution for anti-A, and labels those with high titers (i.e. positive at this dilution) accordingly to be transfused to group O recipients only.

**Solid Organ Transplantation:** As a limited number of donor organs are available, ABO-incompatible solid organ transplantation are sometimes performed. Currently in the US, each institution develops a titer protocol, usually including both the IgG and the IgM phases of testing, and determines the critical titer for which ABO-incompatible transplantation can proceed. Some institutions will treat the plasma sample with dithiothreitol (DTT) to inactivate IgM, and therefore only have IgG present in the IgG phase of testing.

**Incompatible HPC Transplantation:** Major ABO incompatibly occurs when patients have antibodies against donor RBCs (for example, donor group A and patient group O); minor incompatibility occurs when donors have antibodies against patient RBCs (for example, donor group O and patient group A); and bidirectional incompatibility occurs when both major and minor incompatibilities are present (for example, donor group A and patient group B). Major incompatible transplantation is associated with immediate hemolysis of donor RBCs in the HPC component after infusion. This complication can be minimized by depleting RBCs from the HPC component or reducing the recipient's antibody titer, if necessary. In addition, there may be delayed production of donor-type RBCs secondary to the persistent presence of host antibody, which may require reducing the antibody titers in the patient by therapeutic plasma exchange. The risk of minor incompatibility transplantation is immediate hemolysis of recipient RBCs by the plasma contained within the HPC component, and therefore the HPC components should be plasma reduced. In addition, delayed hemolysis can occur when donor lymphocytes produce antibodies against the patient RBCs usually within 7–14 days after transplantation. This hemolysis can be severe and fatal, especially in group A patients with a group O donor, and is minimized in patients receiving methotrexate or similar medication. Antibody titers can be used in determining the need for HPC component modification, or the need to lower antibody levels in patients.

**Typing Discrepancies:** Any discrepancy in the results of ABO typing, which occurs when the RBC typing reactions (i.e. front type) do not match the plasma reactions (i.e. back type), should be resolved prior to issuing blood components to patients (or, if from a donor unit, then prior to release of the product). If transfusion is necessary prior to resolution, or results cannot be resolved, then group O RBCs and group AB plasma should be issued. The discrepancy may result from problems with the RBCs or plasma, test-related problems or technical errors. Negative results may be obtained when positive results are expected, or positive results may be obtained when negative results are expected. To resolve the discrepancy, the test must first be repeated after washing patient and reagent cells and also obtaining a history, such as previous transfusions or HPC transplantation.

**Resolving Discrepancies Due to Absence of Expected Antigens:** Acquired weakened A and B antigen expression can be seen in patients with hematologic diseases and in other conditions. In addition, somatic chromosomal deletion of the *ABO* locus can result in loss of antigen expression. There are a variety of methods to enhance the detection of weakly expressed antigens, such as 30-minute incubation at room temperature with washed RBCs, incubation at 4°C with appropriate controls (group O and autologous cells), treating RBCs with proteolytic enzymes (ficin, papain or bromelin), incubating RBCs with anti-A or anti-B to absorb antibody (with appropriate group O, A, and B RBC controls) and performing an elution, and testing the saliva (if the patient is a secretor) for the presence of ABO substances.

**Resolving Discrepancies Due to Unexpected Reactions with Anti-A/Anti-B:**

*B(A) Phenotype:* Some group B individuals have excessively high levels of the *B* gene associated galactosyltransferase, which results in the attachment of detectable levels of the A-determining sugar (GalNAc) to H antigen on RBCs. These RBCs are agglutinated by an anti-A reagent that contains murine monoclonal antibody MHO4. Using an alternate anti-A reagent (without MHO4) usually resolves the discrepancy.

*Acquired B Phenotype:* This phenotype arises, usually in $A_1$ patients, when microbial deacetylating enzymes modify the A-determining sugar (GalNAc) so that it resembles the B-determining sugar (galactose). Thus, the acquired B antigen develops at the expense of the A antigen. Usually the reactions with anti-B are weak, but they may be strong. This is most often observed with the monoclonal anti-B reagents containing the ES-4 clone. To confirm the presence of acquired B, the patient's history must be checked for a diagnosis that may be associated with bacteremia, such as intestinal obstruction or gastric or intestinal malignancy. In addition, the patient's RBCs or known acquired B RBCs must be tested with patient's plasma, which should not lead to agglutination. Additional approaches include use of an alternate anti-B reagent, use of human anti-B that is acidified, and, if the patient is a secretor, testing of their saliva for the presence of A and B antigens.

*Polyagglutination:* Polyagglutination refers to RBCs agglutinating with all human sera, which can result from genetic inheritance or infection. *T activation* occurs when a bacterium or virus produces an enzyme (neuraminidase) that cleaves the N-acetyl-neuraminic acid and exposes an otherwise hidden antigen (cryptantigen) on RBCs. All normal human sera contain anti-T, and therefore exposure of the T antigen results in polyagglutination. This is a transient condition which resolves upon elimination of the causal organism. $T_n$ *polyagglutination* is the defective synthesis of oligosaccharides exposing cryptantigens on the RBC surface. HPC mutations result in a permanent population of $T_n$-activated RBCs. If the RBCs are treated with proteolytic enzymes to degrade the cryptantigens, then the reactivity should cease. Polyagglutinable RBCs can be characterized by lectin typing. There are also other infectious and non-infectious causes for polyagglutination beyond the scope of this book.

*Mixed-field Agglutination:*  This occurs when there are two separate populations of RBCs in the patient's circulation. This can happen after transfusion of group O RBCs into a non-group O patient, or in an HPC transplant recipient who received a graft from a donor with a different ABO type, or in chimerism (intrauterine exchange of erythropoietic tissue by fraternal twin or mosaicism from dispermy). In addition, mixed-field agglutination can occur with an A subgroup, $A_3$ (uniquely if the agglutinated RBCs are removed from the non-agglutinated RBCs, and then the remaining RBCs are tested with anti-A the mix-field agglutination occurs again in this previously non-agglutinated population).

*Antibody-coated RBCs:*  RBCs heavily coated with IgG may agglutinate spontaneously in the presence of high protein concentration reagents. In this case, the IgG can be removed and the RBCs retested. In addition, RBCs from samples containing cold-reactive IgM autoagglutinins may agglutinate spontaneously. Incubating the RBCs at 37°C and then washing them with warm saline can usually remove the antibodies. Another option is to remove the IgM antibody with dithiothreitol (DTT).

## Resolving Discrepancies Due to Unexpected Serum Reactions:

1. Immunodeficient patients may not produce detectable levels of anti-A and anti-B; infants do not make these antibodies and they may be weak in the elderly.

2. High titers of anti-A or anti-B may result in the prozone effect (when there is excess of either antigen or antibody, resulting in the inability to form large antigen-antibody complexes required for visual agglutination) which can be resolved by diluting the plasma or treating it with EDTA.

3. Anti-$A_1$ in the plasma of $A_2$ or $A_2B$ individuals agglutinates $A_1$ RBCs. This can be resolved by using $A_2$ RBCs (with control group O and $A_1$ RBCs) and testing patient RBCs with *Dolichos biflorus* lectin to demonstrate that the patient is a non-$A_1$ subgroup.

4. Strongly reactive cold-reacting antibodies (autoantibodies or alloantibodies) can agglutinate reagent RBCs. For autoantibodies (for example, anti-I or anti-IH), warm the plasma and reagent RBCs to 37°C and continue testing at this temperature, remove the autoantibody by cold autoadsorption, or treat the serum with DTT to destroy the IgM autoantibody. For alloantibodies (for example, anti-$P_1$ or anti-M), identify the alloantibody, test the reagent RBC to determine if it carries the corresponding antigen, obtain RBCs that lack the corresponding antigen and use those for testing, raise the temperature of the reaction, or repeat testing on several $A_1$ and B RBCs.

5. Patients with abnormal concentrations of serum proteins, altered serum protein ratios, or who have received high molecular weight plasma expanders can aggregate reagent RBCs, mimicking agglutination. These reactions appear as rouleaux or irregularly shaped clumps. They can be corrected by diluting the plasma in saline or using a saline replacement method.

*Genotyping:*  RBC genotyping can be used to resolve typing discrepancies. The ability accurately to determine an individual's antigen status eliminates the use of group O RBCs and AB plasma for transfusion in the situation of ABO typing discrepancy or the loss of a component in the donor setting because of the inability to appropriately label it. Genotyping can aid in the differentiation between subgroup alleles and acquired weakened agglutination, and allows for proper ABO identification of both donors and patients.

## Recommended Reading

Harris SB, Josephson CD, Kost CB, Hillyer CD. (2007). Nonfatal intravascular hemolysis in a pediatric patient after transfusion of a platelet unit with high-titer anti-A. *Transfusion* **47**, 1412–1417.

Josephson CD, Mullis NC, Van Demark C, Hillyer CD. (2004). Significant numbers of apheresis-derived group O platelet units have "high-titer" anti-A/A,B: implications for transfusion policy. *Transfusion* **44**, 805–808.

MacLennan S. (2002). High titre anti-A/B testing of donors within the National Blood Service (NBS) (monograph on the Internet). London: NBS. Available from: http://hospital.blood.co.uk/library/pdf/hightit.pdf (accessed on June 30, 2008).

Olsson ML, Irshaid NM, Hosseini-Maaf B, Hosseini-Maaf B, Hellberg A, Moulds MK, Sareneva H, Chester MA. (2001). Genomic analysis of clinical samples with serologic ABO blood grouping discrepancies: Identification of 15 novel A and B subgroup alleles. *Blood* **98**, 1585–1593.

Rowley SD. (2001). Hematopoietic stem cell transplantation between red cell incompatible donor–recipient pairs. *Bone Marrow Transplant* **28**, 315–321.

Yazer MH. (2005). What a difference 2 nucleotides make: a short review of ABO genetics. *Transfus Med Rev* **19**, 200–209.

# CHAPTER 23
# Rh Blood Group System

Beth H. Shaz, MD

The Rh blood group system (C, c, E, e, D, and other antigens) is second to the ABO system in clinical importance because the Rh antigens, especially D, are highly immunogenic and the corresponding antibodies can result in hemolytic transfusion reactions (HTR) and hemolytic disease of the fetus and newborn (HDFN). The Rh family of proteins is involved in ammonia/ammonium transport, but the specific function of RhCE and RhD proteins has yet to be determined. Extensive reviews of the Rh blood group system have been published (see Recommended reading).

**Antigens:** The Rh system has at least 45 antigens, most notabily C, c, E, and e antigens, which are carried on the RhCE protein encoded by the *RHCE* gene, and the D antigen on the RhD protein encoded by the *RHD* gene (Table 23.1). The *RHD* and *RHCE* genes are 96% homologous and are adjacent to one another on chromosome 1; a third gene *RHAG* is 47% homologous to the other *RH* genes (Figure 23.1). The Rh antigens are carried on hydrophobic 12-pass transmembrane proteins. The RhAG protein does not carry Rh antigens, but is important for bringing the RhD and RhCE proteins to the membrane. Lack of RhAG results in the absence of Rh antigen expression (Rh null phenotype) or marked reduction of Rh antigen expression (Rh mod phenotype). The Rh and RhAG proteins are associated in the membrane and form the Rh core complex, which interacts with other proteins such as CD47, glycophorin B, LW, and possibly AE1/Band 3. The complex is linked to the membrane skeleton via Rh/RhAg-ankrin interaction and CD47-protein 4.2 association.

TABLE 23.1 Nomenclature and Prevalence of Rh Haplotypes

| Haplotype Based on Antigens Present | Shorthand for Haplotype | Prevalence (%) | | |
|---|---|---|---|---|
| | | White | African American | Asian |
| DCe | $R_1$ | 42 | 17 | 70 |
| DcE | $R_2$ | 14 | 11 | 21 |
| Dce | $R_0$ | 4 | 44 | 3 |
| DCE | $R_z$ | <0.01 | <0.01 | 1 |
| ce | r | 37 | 26 | 3 |
| Ce | r' | 2 | 2 | 2 |
| cE | r" | 1 | <0.01 | <0.01 |
| CE | $r^y$ | <0.01 | <0.01 | <0.01 |

Modified from Hillyer CD, Strauss RG, Luban NLC (eds). (2004). *Handbook of Pediatric Transfusion Medicine*. San Diego, CA: Elsevier Academic Press.

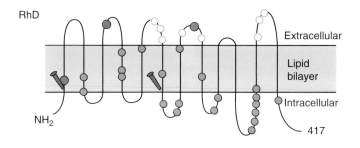

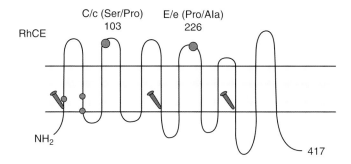

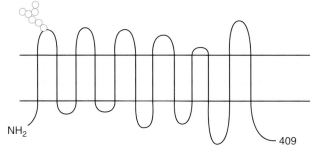

FIGURE 23.1 Predicted 12-transmembrane domain model of the RhD, RhCE and RhAG proteins in the red blood cell membrane. The amino acid differences between RhD and RhCE are shown as symbols. The eight extracellular differences between RhD and RhCE are indicated as *open circles*. The *nails* represent the location of possible palmitoylation sites. Positions 103 and 226 in RhCE that are critical for C/c and E/e expression, respectively, are indicates as *gray circles*. The N-glycan on the first extracellular loop of the Rh-associated glycoprotein is indicated by the *branched structure*. From Hillyer CD, Silberstein LE, Ness PM *et al.* (eds). (2007). *Blood Banking and Transfusion Medicine: Basic Principles and Practice*, 2nd edition. San Diego, CA: Elsevier Academic Press.

**D Antigen:**  It is the presence or absence of the D antigen that confers the Rh-positive or Rh-negative status used in lay and scientific parlance. In the US, 85% of the general population is D positive (Rh positive) and 15% is D negative (Rh negative). Most individuals, especially of European descent, who are D negative (known as Rh negative) do not carry the *RHD* gene. The D-negative phenotype can also occur as a result of various mutations in the *RHD* gene, including premature stop codons, insertions, *RHD/RHCE* hybrids, or other mutations. Most Asian D-negative individuals have a mutant *RHD* gene, while African American individuals usually either have a premature stop codon or hybrid genes.

**Weak D:** This term is applied to RBCs that carry weak (low antigen concentration) forms of D (formerly $D^u$). Weak D cannot be detected using direct agglutination with anti-D reagents at room temperature, but can be identified after prolonged incubation with the anti-D reagent or with the use of antiglobulin (AHG; with appropriate controls). If a patient has a positive direct anti-human globulin test (DAT), a false positive weak D test can result. The use of monoclonal anti-D reagents and improved polyclonal reagents, or monoclonal/polyclonal blends, has resulted in routine detection of D-positive RBCs that would have previously been considered weak D with less sensitive polyclonal reagents. Monoclonal reagents may result in direct agglutination of epitopes that previously required more sensitive test methods or failed to react.

The molecular basis of weak-D expression varies (over 50 mutations have been reported) and includes a variety of point mutations which are in the intracellular or transmembrane regions and not the outer surface of the RBC. Weak D can also result from the weakening of the D antigen expression by *RHC* gene in *trans* to the *RHD* gene (phenotype Dce/Ce). A very weak and rare form of D, $D_{el}$, detected by adsorption and elution of anti-D, is more prevalent in Hong Kong Chinese and Japanese than white individuals.

Weak D is less immunogenic than normal D-positive RBCs, but cases of anti-D stimulation are reported. Therefore, weak D components are labeled as D positive. Patients with weak D usually can receive D-positive RBCs without risk of immunization, but if the weak D expression is secondary to the partial-D phenotype (see below), then the patient may become immunized against D. Current licensed anti-D reagents detect most patients with weak D at direct agglutination. The few patients who require antiglobulin testing in order to demonstrate its presence can receive D-negative components without a problem. In contrast to donors, patient samples do not need to be tested for weak D.

**Partial D:** Individuals with partial D (previously termed D mosaic or D variant) lack a portion of the D protein, and when exposed to the D antigen on transfused RBCs can produce alloantibodies to the missing portion of D. Current D typing reagents in use in both donor centers and transfusion services may have difficulty in determining the partial D phenotypes. Typing of partial D individuals can result in strong, variable or weak expression of D. Patient typing is usually performed with an IgM monoclonal anti-D reagent that does not detect, in the direct phase, DVI, which is the most common form of partial D in whites. There are many partial D phenotypes (also known as D categories) which have arisen as a result of nucleotide exchange between *RHCE* gene and the *RHD* gene or from single-point mutations in *RHD*. The mutations are in the extracellular portion of the protein. Individuals who have the partial- D phenotype benefit from receiving D-negative RBCs for transfusion and potentially benefit from Rh immune globulin (RhIg) prophylaxis. In practice, these individuals are frequently typed as D positive and are recognized only after they form an anti-D alloantibody.

**C/c and E/e Antigens:** C and c differ by six nucleotide substitutions causing four amino acid changes. Only the Ser103Pro polymorphism strictly correlates with

C/c antigenicity, while Pro102 is critical to the c antigen. E and e differ by one nucleotide substitution, resulting in one amino acid difference.

**G Antigen:** The G antigen is expressed on both RhD and RhC proteins, and results from a serine at position 103.

**V and VS Antigens:** These low-frequency antigens are expressed on the red blood cells of ~30% of African Americans, and result from an amino acid substitution in the Rhce protein.

**Variation in e Antigens:** Individuals of African descent often have *RHce* genes that encode variant e antigens. These individuals' RBCs phenotype as e positive, but they form alloantibodies with e-like specificities, such as anti-hr$^S$, -hr$^B$, -RH18 and -RH34. Multiple molecular backgrounds are responsible for these phenotypes, resulting in variation in the antibodies formed, which makes it difficult to find compatible RBC components for these patients. Sometimes only D$--$ RBCs are compatible, but these Rhce variants can be inherited with partial D, so these individuals can also form anti-D.

**Rh-null:** Rh-null RBCs lack expression of Rh antigens, are stomatocytic and spherocytic, and affected individuals have variable degrees of anemia. The *regulator type* is caused by mutations in the *RHAG* gene so they have no Rh or RhAG proteins, and the *amorph type* is caused by mutations in the *RHCE* gene on a deleted *RHD* background so they have no Rh proteins and reduced RhAG proteins.

**D Typing Discrepancies:** In general, D reagents are configured differently for blood centers and for transfusion services, in order to determine weak D phenotypes as being D positive in the case of donors, and to show partial D phenotypes as D negative in recipients. Different typing reagents, which usually are blends of monoclonal anti-D antibodies, will type variant D phenotypes differently. Therefore, individuals may have discrepancy in their D type depending on the reagent used (e.g. an individual may type as D negative in one center and D positive in another center which uses an alternate reagent).

**RH Genotyping:** *RH* genotyping can be helpful in predicting the Rh phenotype of donors and patients. Currently, genotyping is not routinely available in the US and has not been approved by the FDA as a release test.

Situations where genotyping is useful include:
- Patients who have been recently transfused.
- IgG-coated RBCs, which may make RBC phenotyping reagents suboptimal.
- In the prenatal setting, to determine the Rh status of a fetus at risk for HDFN due to an antibody against a Rh antigen. Fetal DNA can be provided through chorio- and amniocentesis, or extracted from the maternal plasma.
- Where there are D typing discrepancies or determination of the presence of weak D type, partial D category, or D$_{el}$.
- For determination of antibody specificity and identification of appropriate RBC components for transfusion.

**Expression:** Rh proteins are restricted to erythroid and myeloid cells. The expression of Rh proteins occurs most actively during late erythropoiesis.

**Antibodies:** Most Rh antibodies are IgG (some have an IgM component) and are clinically significant, causing both HTRs and HDFN. Approximately 20% of D negative individuals who are exposed to D-positive RBC components in the hospital setting (primarily emergency room, operating room, intensive care unit and medicine ward) will form anti-D, which is in contrast to healthy male volunteers, 80% of whom can form anti-D after receiving as little as 0.5 ml of transfused D-positive RBCs. In the maternal setting, ABO-incompatibility between the mother and fetus has a partial protective effect on the mother forming anti-D, as the incompatible and D positive cells are rapidly removed by the antithetical isohemagglutinin. Anti-D and anti-c can cause severe HDFN, while anti-C, anti-E and anti-e usually cause no or mild HDFN.

**Autoantibodies:** Many autoantibodies in patients with warm autoimmune hemolytic anemia have specificity that appears to be directed towards a Rh antigen, most notably e. However, transfused antigen negative RBCs rarely survive better than antigen positive RBCs.

## Recommended Reading

Avent ND, Reid ME. (2000). The Rh blood group system: a review. *Blood* **95**, 375–387.

Flegel WA. (2006). How I manage donors and patients with a weak D phenotype. *Curr Opin Hematol* **13**, 476–483.

Lowmas-Francis C, Reid ME. (2000). The Rh blood group system: the first 60 years of discovery. *Immunohematology* **16**, 7–17.

Westhoff CM. (2007). Rh complexities: serology and DNA genotyping. *Transfusion* **47**, 17S–22S.

Yazer M, Triulzi DJ. (2007). Detection of anti-D in D− recipients transfused with D+ red blood cells. *Transfusion* **47**, 2197–2201.

# CHAPTER 24

# Kell and Kidd Blood Group Systems

Beth H. Shaz, MD

The Kell (K [Kell], k [Cellano], Kp$^a$, Kp$^b$, K$_0$, Js$^a$, Js$^b$ and other low- and high-frequency antigens) and Kidd (Jk$^a$ and Jk$^b$) blood group system antigens are carried on red blood cell (RBC) membrane glycoproteins. Antibodies to the Kell and Kidd blood group system antigens can cause hemolytic transfusion reactions (HTR) and hemolytic disease of the fetus and newborn (HDFN). Therefore, these are considered to be highly clinically significant.

**Kell Blood Group System:** The Kell blood group system contains over 25 antigens, but the major antigens are K (K1 [Kell]) and k (K2 [Cellano]). The Kell protein is a glycoprotein which is a zinc endopeptidase. It contains multiple intrachain disulfide bonds, and therefore is destroyed by reducing agents such as dithiothreitol (DTT). The *KEL* gene coding for the Kell protein is located on chromosome 7. Another protein, Kx, is essential for the expression of Kell system antigens. Kx, which has its locus on the X chromosome (*XK*), is linked by a disulfide bond to the Kell protein. This interaction shields the Kx protein such that anti-Kx does not react with RBCs that carry normal expressing Kell antigens. Individuals who lack Kx have the so-called McLeod phenotype, which is characterized by a marked reduction of the Kell antigens, anemia, and neuromuscular abnormalities (see below).

**Antigens:** A variety of nucleotide mutations that cause single amino acid substitutions in the K protein give rise to the majority of Kell antigens. There are five sets of high- and low-incidence antithetical antigens (**K** and k; Kp$^a$, **Kp$^b$**, and Kp$^c$; Js$^a$ and **Js$^b$**, **K11** and K17, and **K14** and K24 [high-incidence antigens denoted by bold]), in addition to other low- and high-incidence antigens (Table 24.1, Figure 24.1). K$_0$, a phenotype that lacks Kell antigens, results from an assortment of mutations, including nucleotide deletion, defective splicing and premature stop codons. Weak expression of Kell antigens results in "Kell-mod" phenotypes.

**McLeod Phenotype:** The McLeod phenotype arises through deletions and mutations of the *XK* locus, resulting in depressed expression of the Kell system antigens in addition to decreased RBC survival, deformability, and permeability to water, as well as acanthocytic morphology. Individuals with the McLeod syndrome, which has been reported in approximately 60 males, have the above RBC defects along with muscular and neurological defects, including skeletal muscle wasting, seizures, and cardiomyopathy with symptoms developing after the fourth decade of life. This phenotype has been found in individuals with chronic granulomatous disease, which results from a deletion of the X chromosome that includes both the *XK* and the *X-CGD* loci.

**Expression:** The Kell system antigens appear to be erythroid specific, and are most actively expressed very early during erythropoiesis. Kx, in contrast, is found in muscle, heart, brain and hematopoietic tissue.

### TABLE 24.1 Kell Blood Group System Phenotypes and Prevalence

|  | Prevalence (%) | |
| --- | --- | --- |
| Phenotype | White | African American |
| K−k+ | 91 | 98 |
| K+k+ | 8.8 | 2 |
| K+k− | 0.2 | Rare |
| Kp(a+b−) | Rare | 0 |
| Kp(a−b+) | 97.7 | 100 |
| Kp(a+b+) | 2.3 | Rare |
| Kp(a−b−c+) | 0.32 Japanese | 0 |
| Js(a+b−) | 0 | 1 |
| Js(a−b+) | 100 | 80 |
| Js(a+b+) | Rare | 19 |

Modified from Hillyer CD, Strauss RG, Luban NLC (eds). (2004). *Handbook of Pediatric Transfusion Medicine*. San Diego, CA: Elsevier Academic Press.

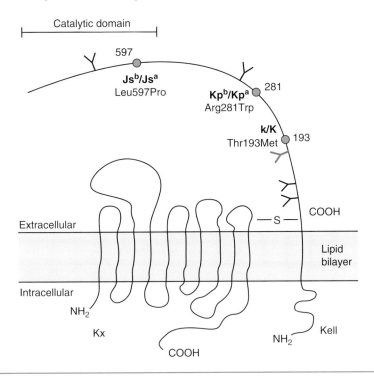

FIGURE 24.1 Kell and Kx proteins. Kell is a single-pass protein, but Kx is predicted to span the red blood cell membrane ten times. Kell and Kx are linked by a disulfide bond, shown as —S—. The amino acids that are responsible for the more common Kell antigens are shown. The N-glycosylation sites are shown as Y. The *hollow Y* represents the N-glycosylation site that is not present on the K (K1) protein. From Hillyer CD, Silberstein LE, Ness PM *et al*. (eds). (2007). *Blood Banking and Transfusion Medicine: Basic Principles and Practice*, 2nd edition. San Diego, CA: Elsevier Academic Press.

**Antibodies:** Because the K antigen is strongly immunogenic, anti-K is frequently formed in K-negative individuals who are exposed to K-positive RBCs. Rare cases of naturally occurring anti-K have been reported, but most anti-K alloantibodies are identified in individuals who have received RBC transfusion or have been pregnant. All other Kell system antibodies also result from RBC exposure. Anti-K and other antibodies to the Kell system antigens (including anti-Kp$^a$, anti-Kp$^b$, anti-Js$^a$, and anti-Js$^b$) can produce HTRs, both immediate and delayed, and HDFN. Neither maternal antibody titer nor amniotic fluid bilirubin levels are good predictors of the severity of HDFN. Because Kell antigens are expressed very early during erythropoiesis, anti-Kell antibodies can cause destruction of RBC precursor cells and thus clinically can suppress erythropoiesis. Therefore, these antibodies may result in anemia without hemolysis and elevated bilirubin.

$K_0$ individuals form anti-Ku, individuals with the McLeod phenotype with CGD make anti-Kx and anti-Km, while individuals with the McLeod phenotype without CGD make anti-Km.

**Kidd Blood Group System:** The Kidd blood group system has two antigens, Jk$^a$ and Jk$^b$, that are encoded by *HUT11* on chromosome 18. The Kidd protein is a urea transporter. Jk-null individuals have a reduced capacity to concentrate urine and their RBCs are more resistant to lysis by 2 M urea, but they display no other abnormalities.

**Antigens:** Jk$^a$ and Jk$^b$ result from a single amino acid substitution leading to three common phenotypes (Jk(a+b−), Jk(a−b+), and Jk(a+b+); Table 24.2, Figure 24.2). The final phenotype, Jk(a−b−), or null phenotype, is rare in whites and African Americans, but is found with increased prevalence in Asian and Polynesian individuals. The null phenotype arises from either a silent allele or inheritance of a dominant inhibitor gene, *In(Jk)*.

TABLE 24.2 Kidd Blood Group System Phenotypes and Prevalence

| Phenotype | Prevalence (%) | |
| --- | --- | --- |
| | White | African American |
| Jk(a+b−) | 26 | 52 |
| Jk(a−b+) | 23 | 8 |
| Jk(a+b+) | 50 | 40 |
| Jk(a−b−) | Rare | Rare |

Modified from Hillyer CD, Strauss RG, Luban NLC (eds). (2004). *Handbook of Pediatric Transfusion Medicine*. San Diego, CA: Elsevier Academic Press.

**Expression:** Kidd antigens are detected at 11 weeks gestation, and are well developed at birth. These antigens are also expressed on endothelial cells of vasa recta in the medulla of the human kidney.

**Antibodies:** Antibodies to the Kidd blood group system are not common, and are usually found in sera containing other antibodies. Because their titer often decreases below the limit of detection, they often react weakly, they may become undetectable

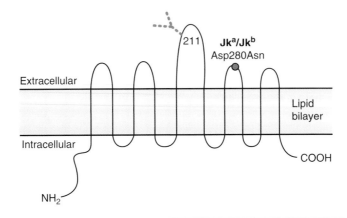

FIGURE 24.2  Predicted ten-transmembrane domain structure of the Kidd/urea transporter. The polymorphism responsible for the Kidd antigens and the site for the N-glycan are indicated. From Hillyer CD, Silberstein LE, Ness PM *et al.* (eds). (2007). *Blood Banking and Transfusion Medicine: Basic Principles and Practice*, 2nd edition. San Diego, CA: Elsevier Academic Press.

during storage, and they may react with only homozygous cells (show dosage), anti-Kidd antibodies can be difficult to identify. For these reasons, Kidd antibodies are responsible for about one-third of all delayed hemolytic transfusion reactions, which may be severe. The antibodies are mainly IgG, but can be partially IgM. Kidd antibodies rarely cause HDFN, and when they do it is not severe. Anti-Jk3 (also known as anti-Jk[ab]) can be produced by Jk(a−b−) individuals.

## Recommended Reading

Storry JR, Olsson ML. (2004). Genetic basis of blood group diversity. *Br J Hematol* **126**, 759–771.

# CHAPTER 25

# MNS and Duffy Blood Group Systems

Beth H. Shaz, MD and John D. Roback, MD, PhD

The MNS blood group system consists of the antigens M, N, S, s and U, which are determinants on glycophorins A and B. Antibodies to these antigens can be clinically significant, especially anti-S, anti-s, and anti-U, while anti-M is rarely clinically significant and anti-N is not clinically significant. The Duffy blood group system consists of the Fy$^a$, Fy$^b$, Fy3 and Fy6 antigens carried on the erythrocyte chemokine receptor known as the Duffy Antigen Receptor for Chemokines (DARC, also known as CD234). Antibodies to the Duffy antigens are clinically significant.

**MNS Blood Group System:** The MNS blood group system contains 43 antigens; the major ones being M, N, S, s and U, while the others are low-frequency antigens resulting from either amino acid substitutions or rearrangements between *GYPA* and *GYPB* (Table 25.1, Figure 25.1). The M and N antigens are located on glycophorin A (*GYPA*), while the S and s antigens are located on glycophorin B (*GYPB*). The genes for *GYPA* and *GYPB* are adjacent to each other on chromosome 4, and both encode single pass membrane sialoglycoproteins.

### TABLE 25.1 MNS Blood Group System Phenotypes and Prevalence

| Phenotype | Prevalence (%) | |
|---|---|---|
| | White | African American |
| M + N − S + s − | 6 | 2 |
| M + N − S + s + | 14 | 7 |
| M + N − S − s + | 10 | 16 |
| M + N + S + s − | 4 | 2 |
| M + N + S + s + | 22 | 13 |
| M + N + S − s + | 23 | 33 |
| M − N + S + s − | 1 | 2 |
| M − N + S + s + | 6 | 5 |
| M − N + S − s + | 15 | 19 |
| M + N − S − s − | 0 | 0.4 |
| M + N + S − s − | 0 | 0.4 |
| M − N + S − s − | 0 | 0.7 |

Modified from Hillyer CD, Strauss RG, Luban NLC (eds). (2004). *Handbook of Pediatric Transfusion Medicine.* San Diego, CA: Elsevier Academic Press.

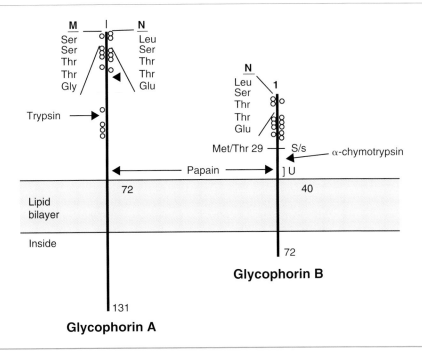

**FIGURE 25.1**  Diagram of glycophorin A and glycophorin B. From Brecher ME (ed.) (2005). *Technical Manual*, 15th edition. Bethesda, MD: AABB Press.

**Antigens:** The M and N antigens are the products of allelic genes, and as such are antithetical; N differs by two amino acids from the M antigen. The M and N antigens are sensitive to ficin, papain, trypsin and pronase, with variable sensitivity to sialidase (Figure 25.1 shows the proposed cleavage sites), which can aid in antibody identification (see Chapter 20). The S antigen is antithetical to the s antigen; the s antigen differs by only one amino acid from the S antigen. The S and s antigens are sensitive to α-chymotrypsin and pronase, with variable sensitivity to ficin, papain and pronase. The U antigen represents a conserved sequence of glycophorin B. The U− phenotype is seen in individuals with deletions of glycophorin B (S−s−U−). Additional null phenotypes in the MNS system include En(a−), which lacks the MN antigens, and M$^k$M$^k$, which lacks both MN and Ss antigens. Glycophorin A is a receptor (for bacteria, viruses and *Plasmodium falciparum*), a chaperone for Band 3 transport to the red blood cells (RBC) membrane, the major component contributing to the negatively-charged RBC glycocalyx, and may function as a complement regulator. However, the rare null phenotypes are not associated with any apparent health defects.

**Expression:** Cord RBCs express MNS blood group system antigens. These antigens are also expressed on renal endothelium and epithelium.

**Antibodies:** Anti-M is primarily IgM, but may have an IgG component. Anti-M reactivity at 37°C has rarely been associated with hemolytic transfusion reactions and hemolytic disease of the fetus and newborn (HDFN), and therefore patients with anti-M reactive at 37°C should be transfused with RBCs that are crossmatch compatible in the antiglobulin phase and should be assessed for the risk of HDFN. Anti-N is primarily IgM, and has not been associated with hemolytic transfusion reactions or HDFN. Anti-M and anti-N are typically naturally occurring alloantibodies, which do not require previous RBC antigen exposure for formation, and show a dosage effect (reacting to RBCs with a double dose of antigen more strongly than to those with a single dose of antigen). In contrast, anti-S, anti-s and anti-U are IgG antibodies that are formed in response to RBC stimulation and are associated with hemolytic transfusion reactions and HDFN. In addition, anti-En$^a$ is usually IgG, and may cause transfusion reactions or HDFN.

Autoantibodies with apparent M, N, S, s or U specificity are observed rarely. An autoanti-N has been found in patients on dialysis when equipment was sterilized with formaldehyde (anti-N$^f$).

**Duffy Blood Group System:** The protein containing the Duffy antigens is a multipass transmembrane glycoprotein with a protruding glycosylated amino terminal region (Figure 25.2). The antigens show a dosage effect, meaning there are twice as many Fy$^a$ antigens on RBCs from an individual who is homozygous for the Fy$^a$ allele than on RBCs from an individual who is heterozygous. The Duffy antigen frequency varies significantly between racial groups.

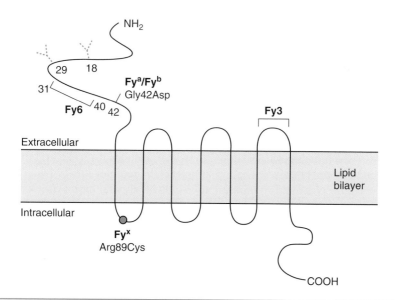

FIGURE 25.2 The predicted seven-transmembrane domain structure of the Duffy protein. The amino acid change responsible for Fy$^a$/Fy$^b$ polymorphism, the mutation responsible for Fy$^x$ glycosylation sites and the regions where Fy3 and Fy6 map are indicated. From Hillyer CD, Silberstein LE, Ness PM *et al.* (eds) (2007). *Blood Banking and Transfusion Medicine: Basic Principles and Practice*, 2nd edition. San Diego, CA: Elsevier Academic Press.

**Antigens:** $Fy^a$ and $Fy^b$ antigens differ by a single amino acid, are encoded by two alleles, *FYA* and *FYB*, and are responsible for the phenotypes Fy(a+b−), Fy(a−b+), and Fy(a+b+) (Table 25.2). The $Fy^x$ antigen results from weak expression of $Fy^b$, is found in whites, and is due to a single mutation in the *FYB* gene. The Fy(a−b−) phenotype found in African Americans is caused by a mutation in the promoter region of *FYB*, which disrupts a binding site for the erythroid transcription factor GATA-1 and results in the loss of Duffy expression on RBCs, but not its expression on endothelium. Fy(a−b−) individuals, who express no Duffy glycoprotein, form anti-Fy3, which reacts with all RBCs except Fy(a–b–) RBCs. $Fy^a$ and $Fy^b$ (but not Fy3) antigens are sensitive to proteolytic enzyme treatment.

### TABLE 25.2 Duffy Blood Group System Phenotypes and Prevalence

| | Prevalence (%) | |
| --- | --- | --- |
| Phenotype | White | African American |
| Fy (a+b−) | 17 | 9 |
| Fy (a−b+) | 34 | 22 |
| Fy (a+b+) | 49 | 1 |
| Fy (a−b−) | Rare | 68 |
| $Fy^x$ | 1.4 | 0 |

Modified from Hillyer CD, Strauss RG, Luban NLC (eds). (2004). *Handbook of Pediatric Transfusion Medicine*. San Diego, CA: Elsevier Academic Press.

The *FY* gene encodes an erythrocyte chemokine receptor (DARC; CD234), which is postulated to scavenge excess chemokines released in the circulation. Chemokines are chemotactic cytokines that attract white blood cells to sites of inflammation. DARC is known as a promiscuous chemokine receptor, because it binds chemokines from both the C-X-C (which includes IL-8) and C-C (which includes RANTES and monocyte chemotactic protein-1) classes. The Fy3 antigen is the receptor for the malarial parasites *Plasmodium vivax* and *P. knowlesi*; thus individuals who have Fy(a−b−) RBCs are resistant to infection by these malarial organisms.

**Expression:** The Duffy antigens are detected as early as 6–7 weeks' gestation, and are well developed at birth. Duffy antigens are expressed on endothelial cells, postcapillary venules of the kidney, spleen, heart, lung, muscle, duodenum, pancreas and placenta, and Purkinje cell neurons in the brain.

**Antibodies:** Duffy antibodies are almost always IgG, rarely IgM (which binds complement), and result from RBC stimulation. Anti-$Fy^a$ may cause mild to severe hemolytic transfusion reactions and HDFN. $Fy^b$ is a poor immunogen, and therefore anti-$Fy^b$ is rarely formed; when anti-$Fy^b$ is present, it rarely causes transfusion reactions or HDFN.

## Recommended Reading

Afenyi-Annan A, Kail M, Combs MR *et al.* (2008). Lack of Duffy antigen expression is associated with organ damage in patients with sickle cell disease. *Transfusion* **48**, 917–924.

Palacajornsuk P. (2006). Review: molecular basis of MNS blood group variants. *Immunohematology* **22**, 171–182.

Storry JR, Olsson ML. (2004). Genetic basis of blood group diversity. *Br J Hematol* **126**, 759–771.

# CHAPTER 26
# Lewis, I and P Blood Group Systems

Beth H. Shaz, MD

The antigens in the Lewis (Le$^a$, Le$^b$), I (I, i), P (P$_1$, P$_2$), and GLOB (P, P$^k$, LKE) blood group systems are composed of terminal carbohydrate moieties added to protein (glycoproteins) and/or lipid (glycolipids) backbones in a configuration similar to that of the ABO antigens. Some of these antigens are expressed on tissues, and in some individuals soluble forms can be found in secretions and excretions.

**Lewis Blood Group System:** Le$^a$ and Le$^b$ antigens are synthesized by two independent fucosyltransferases (Figure 26.1). The corresponding antibodies are usually not of clinical significance.

**Antigens:** Lewis antigens (Le$^a$ and Le$^b$) are not intrinsic to the red blood cell (RBC) membrane, but are synthesized by intestinal epithelial cells, circulate in the plasma either free or bound to liporoteins, and are then passively adsorbed onto the RBC membrane. Le$^a$ and Le$^b$ are synthesized in a stepwise fashion by two separate fucosyltransferases (enzymes which add fucose) acting on type I glycoprotein chains; these enzymes are encoded by the *Se(FUT2)* and *LE(FUT3)* genes, which are preferentially expressed in epithelial cells, located on chromosome 19; *H(FUT1)* gene, which is preferentially expressed in erythroid cells, places a fucose on type 2 glycoproteins to form the H antigen on RBCs, which is also on chromosome 19.

The *LE* product adds fucose to type 1 precursors creating Le$^a$, resulting in the Le(a+b−) phenotype in individuals who carry *LE* but not *Se* gene. Individuals who carry the *Se* gene and the *LE* gene add a second fucose which converts the Le$^a$ to Le$^b$, resulting in the Le(a−b+) phenotype. Individuals who do not carry *LE* are Le(a−b−) regardless of the presence of the *Se* gene.

Individuals who carry the *Se* gene are termed ABO secretors, because they can form the H antigen on type 1 glycoproteins which are produced in epithelial cells and whose products reside on mucins in secretions, and subsequently can produce A and B antigens (given the presence of the appropriate transferases) in the secretions.

Individuals who have reduced activity of the second fucosyltransferase (*Se$^w$*) have weak expression of both Le$^a$ and Le$^b$, and are termed partial secretors; this phenotype is principally found in Taiwanese individuals (Table 26.1).

**Expression:** Lewis antigens are not present on cord RBCs. Le$^a$ appears prior to Le$^b$, and usually within the first few months of life. The Lewis antigens do not reach adult levels until ~six years of age. Prior to that age, the Le(a+ b+) phenotype is not uncommon in individuals who will later become Le(a− b+). Lewis antigen levels on RBCs are often diminished during pregnancy and various diseases; the diminished antigens on RBCs are likely secondary to changes in the endothelial secretion or changes in the lipoprotein content in the plasma. Lewis antigens are also adsorbed onto platelets, lymphocytes and other tissues, as well as RBCs.

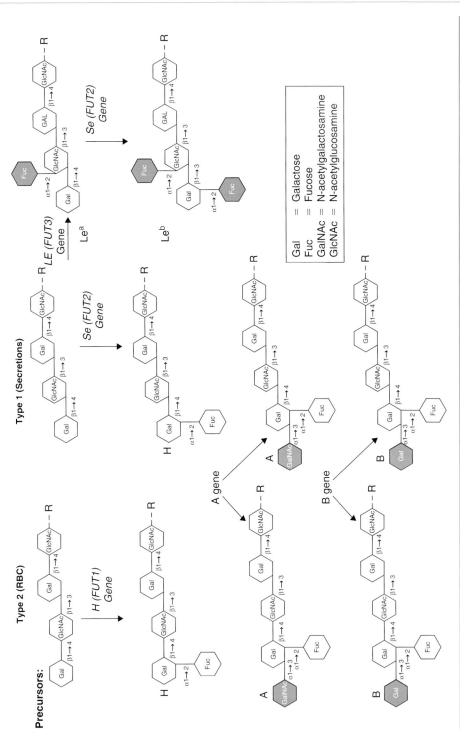

FIGURE 26.1 Synthesis of A, B, H, and Lewis antigens. Oligosaccharide precursor core type 1 and type 2 structures differ only in the linkage between the terminal galactose (Gal) and the N-acetylglucosamine (GlcNAc), shown underscored. Terminal carbohydrates that define the antigens are shown in blue. Modified from Hillyer CD, Silberstein LE, Ness PM et al. (eds) (2007). *Blood Banking and Transfusion Medicine: Basic Principles and Practice*, 2nd edition. San Diego, CA: Elsevier Academic Press.

| | Prevalence (%) | | |
|---|---|---|---|
| Phenotype | White | African American | Asian |
| Le(a− b+) | 72 | 55 | 72 |
| Le(a+ b−) | 22 | 23 | 22 |
| Le(a− b−) | 6 | 22 | 6 |
| Le(a+ b+) | Rare | Rare | 3 |

TABLE 26.1 Lewis Blood Group Phenotypes and Prevalence

Modified from Hillyer CD, Strauss RG, Luban NLC (eds). (2004). *Handbook of Pediatric Transfusion Medicine*. San Diego, CA: Elsevier Academic Press.

**Antibodies:** Antibodies to the Lewis antigens are primarily IgM, naturally occurring (present without previous exposure to antigen-positive RBCs), and not usually clinically significant. Rare cases of hemolytic transfusion reactions secondary to Lewis antibodies have been reported, more commonly due to antibodies against Le[a] than Le[b]. Lewis antibodies are usually not hemolytic because the antibody does not react at 37°C. Furthermore, Lewis antigens in the donor's plasma neutralize the antibody, and the Lewis antigens elute from RBCs into the plasma. Patients with Lewis antibodies may be transfused with RBCs that are crossmatch-compatible at 37°C. Lewis antibodies are of the IgM class, and thus do not cause hemolytic disease of the fetus and newborn because they are not able to cross the placenta. In addition, Lewis antigens are not present on fetal RBCs, as above.

**Ii Antigens:** I and i antigens, which are reciprocal to each other (meaning the I antigen is synthesized from the i antigen by the action of a transferase), are on the same carbohydrate chains that carry RBC ABO antigens (Figure 26.2). The corresponding

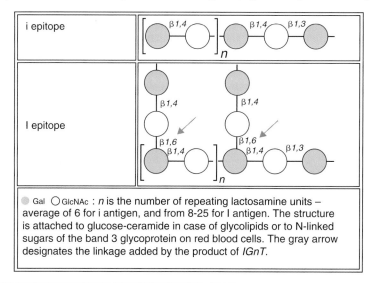

FIGURE 26.2 The I and i antigens. From the National Library of Medicine (NLM) web pages.

antibodies are most commonly benign autoantibodies; rare individuals with the i phenotype can form an alloanti-I antibodies which are only rarely hemolytic.

**Antigens:** The I and i antigens differ in their branching structure. The i antigen is found predominantly on fetal and infant RBCs, where a disaccharide unit (lactosasamine) is linked in a straight chain. During the first six years of life, increased expression of an acetyl glucosamine transferase, encoded by the *IGnT(GCNT2)* gene, results in increased branching of the carbohydrate structure, leading to the formation of the I antigen. Therefore, the formation of I antigen is at the expense of the i antigen (hence the term *reciprocal* for the I and i antigens).

In some adults the i antigen is not converted into the branched chain I antigen secondary to mutations in the *IGnT* gene, leading to lack of transferase activity and the rare i ("adult little i") phenotype.

**Expression:** Newborns have predominately i expression, while adults have predominantly I antigen. Adult RBCs vary in the amount of I antigen expressed. The Ii antigens are found on the surface of most cells and on soluble glycoproteins in saliva, plasma and other fluids, and therefore they are known as histo-blood group antigens.

**Antibodies:**

*Autoantibodies:* Cold agglutinins are IgM autoantibodies that react optimally in the cold (4°C) with RBCs and are present in sera of all adults at low titers. When these antibodies react at room temperature they can interfere with pre-transfusion testing, which can usually be circumvented by prewarming the reactions. Usually these antibodies are benign; rarely these antibodies, when they are high titer ($>1:1000$) and react at body temperature (37°C), can result in hemolytic anemia (termed *cold agglutinin disease*).

Cold agglutinin disease (CAD) can be transient (also referred to as secondary) as a result of an infection, or chronic (also referred to as primary or idiopathic) as a result of clonal B-cell expansion. The antigen specificities of the IgM autoantibody include anti-I (most commonly), anti-i, and anti-Pr (rarely; anti-Pr can be IgG or IgA). Cold agglutinins can be secondary to viral and bacterial infections. *Mycoplasma pneumoniae* infection is associated with CAD secondary to anti-I, and infectious mononucleosis is associated with CAD secondary to anti-i. The pathogenesis of the autoantibody formation is unknown; theories include (1) immune dysfunction, (2) antigens sharing between the infectious agent and RBC (known as antigen mimicry), and (3) infection-induced antigenic changes resulting in increased antigenicity.

*Alloantibodies:* Adults with the i phenotype typically make anti-I, which is usually IgM and only active at low temperatures ($<32$°C). Rare examples of hemolytic alloanti-I have been reported which have demonstrated activity and thus a thermal range to 37°C.

*Anti-IH:* Anti-IH, which is an autoantibody most commonly seen in group $A_1$ (less commonly group AB) individuals, agglutinates RBCs carrying both H and I determinants (i.e. adult group O RBCs which carry the I antigen and the most amount of the H antigen). The I and H antigens are carried on the same glycoproteins and glyocolipids. Autoanti-IH is usually not clinically significant, but hemolytic transfusion reactions resulting from it have been reported, usually when it is reactive at 37°C.

## P and GLOB Blood Group Systems:
The P ($P_1$, $P_2$) and GLOB (P, $P^k$, LKE) antigens are defined by sugars added to precursor glycosphingolipids (Figures 26.3 and 26.4). Their synthesis occurs by sequential addition of monosaccharides to the precursor molecule by glycosyltransferases. Two separate synthetic pathways act on a single precursor (lactosylceramide).

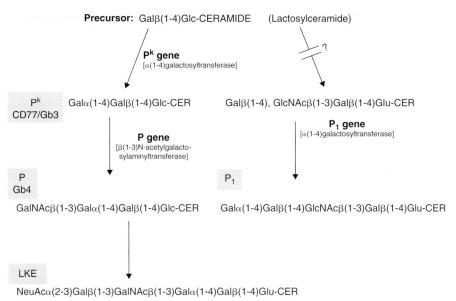

**FIGURE 26.3** Biosynthyesis of the $P_1$, $P^k$, P and LKE antigens. Glu, glucose; Cer, ceramide. From Hillyer CD, Silberstein LE, Ness PM *et al.* (eds) (2007). *Blood Banking and Transfusion Medicine: Basic Principles and Practice*, 2nd edition. San Diego, CA: Elsevier Academic Press.

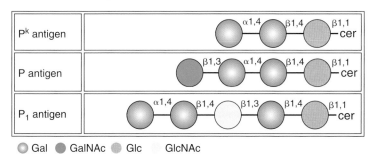

**FIGURE 26.4** The P, $P^k$ and $P_1$ antigens. From the National Library of Medicine (NLM) web pages.

**Antigens:** In the first pathway, lactosylceramide is converted to $P^k$ (globotriosylceramide; CD77), which is then converted to P (globoside) by a second enzyme, which adds acetylgalactosamine and is encoded by the *P* gene. The P antigen is found on the majority of RBCs, except for the rare p phenotype, which lacks CD77 and therefore cannot synthesize $P^k$, and the rare $P^k$ phenotype, which lacks the *P* gene and therefore cannot synthesize P. LKE is formed when first galactose is added, forming galactosylgloboside, and then sialic acid is added to the P antigen.

In the second pathway, lactosylceramide is converted to $P_1$ by the addition of a galactose. The absence of $P_1$ results in the $P_2$ phenotype (i.e. $P_2$ is the null-phenotype of $P_1$). The prevalence of the $P_1$ phenotype is 80% in whites and 96% in African Americans.

The P antigen is the receptor for the B19 parvovirus, which causes erythema infectiosum (fifth disease) and occasionally more severe disease, such as RBC aplasia. Individuals with the p phenotype lack P and $P^k$, and are therefore resistant to parvovirus B19 infection.

**Expression:** $P_1$ expression is weaker in children, and does not reach adult levels until approximately seven years of age. Individuals show varied $P_1$ expression. $P_1$ is also expressed on lymphocytes, granulocytes and monocytes and in a wide variety of organisms.

## Antibodies:

*Anti-$P_1$:* $P_1$-negative individuals frequently form anti-$P_1$, which is a cold-reactive IgM antibody, and usually naturally occurring. It does not cross the placenta, and therefore does not result in hemolytic disease in the fetus and newborn. It has, rarely, been reported to cause hemolytic transfusion reactions, and in those cases the antibody shows reactivity up to 37°C. Anti-$P_1$, as well as anti-$P^k$, can be neutralized with $P_1$ substance, hydatid cyst fluid (*Echinococcus* cyst fluid) or pigeon egg white.

**Autoanti-P:** Paroxysmal cold hemoglobinuria (PCH) is a rare autoimmune hemolytic anemia that typically occurs in young children following a viral infection. These children have a biphasic IgG antibody, termed the Donath-Landsteiner antibody (usually of anti-P specificity). The term *biphasic antibody* is used because the antibody binds to the RBCs at low temperature, but RBC lysis does not occur until the RBC with the antigen–antibody complex warms up. The Donath-Landsteiner test is used to diagnose PCH; a positive test result is when hemolysis does not occur if the specimen is keep at 4°C or 37°C, but occurs only when the specimen is first placed at 4°C and then warmed to 37°C.

## Recommended Reading

Daniels GL, Fletcher A, Garratty G *et al.* (2004). Blood group terminology 2004: from the International Society of Blood Transfusion committee on terminology for red cell surface antigens. *Vox Sang* **87**, 304–316.

Garratty G, Dzik W, Issitt PD *et al.* (2000). Terminology for blood group antigens and genes–historical orgins and guidelines in the new millennium. *Transfusion* **40**, 477–489.

International Society for Blood Transfusion: Committee on Terminology for Red Cell Surface Antigens. Available at: http://ibgrl.blood.co.uk/default.htm (accessed July 10, 2008).

Logdberg L, Reid ME, Lamont RE, Zelinski T. (2005). Human blood group genes 2004: Chromosomal locations and cloning strategies. *Transfus Med Rev* **19**, 45–57.

NCBI: dbRBC: Blood Group Antigen Gene Mutation Database. Available at: http://www.ncbi.nlm.nih.gov/projects/gv/mhc/xslcgi.cgi?cmd=bgmut/systems (accessed July 10, 2008).

# CHAPTER 27

# Other Blood Group Systems, Collections and Antigens

Beth H. Shaz, MD

This chapter will discuss blood group systems, collections, and low- and high-frequency antigens which are not discussed in other chapters (Figure 27.1). An exhaustive list of all red blood cell (RBC) antigens is beyond the scope of this book.

The International Society of Blood Transfusion (ISBT) has classified RBC antigens within systems and collections, and as low-frequency and high-frequency antigens. Per ISBT, a *blood group system* "consists of one or more antigens controlled at a single gene locus, or by two or more very closely linked homologous genes with little or no observable recombination between them." *Collections* "consist of serologically, biochemically or genetically related antigens, which do not fit the criteria required for system status." *Low-frequency antigens* have "an incidence of less than 1% and which cannot be included in a system or collection," while *high-frequency antigens* have "an incidence of greater than 90% and cannot be included in a system or collection."

## Blood Group Systems:

**Chido/Rogers Blood Group System:** The Chido (Ch) and Roger (Rg) antigens are high-incidence antigens present on the isotypes (C4A carries Ch and C4B carries Rg) of the 4th component of complement (C4), which serves as a platform for interaction of the antigen–antibody complex and the complement proteins. These antigens are absorbed onto the RBC from the plasma, and can be destroyed by papain and ficin treatment. Antibodies to Ch and Rg are usually clinically insignificant, and can be neutralized by plasma from antigen-positive individuals.

**Colton Blood Group System:** The Colton blood group antigens reside on the RBC water channel protein Aquaporin-1 (AQP-1), which may also serve as an ion channel. The Colton system consists of $Co^a$ (a high-incidence antigen), $Co^b$ (a low-incidence antigen) and Co3 (which is a product of either $Co^a$ or $Co^b$ antigen). Antibodies to the Colton system are rare, but have been implicated in hemolytic transfusion reactions (HTR) and hemolytic disease of the fetus and newborn (HDFN).

**Cromer Blood Group System:** There are ten antigens in the Cromer system; seven are high incidence ($>99\%$ incidence) and three are low incidence ($<1\%$ incidence) antigens (**Tc**$^a$ antithetical to Tc$^b$ [African Americans] or Tc$^c$ [whites], WES$^a$ antithetical to **WES**$^b$, **Cr**$^a$, **Dr**$^a$, **ES**$^a$, **IFC**, **UMC**; high-incidence antigens in bold). These antigens are located on the complement regulator protein decay-accelerating factor (DAF), which is encoded by the gene *DAF*. DAF (also known as CD55), which is anchored to the cell membrane by glycosylphosphatidylinositol (GPI), serves to regulate complement activation and is a receptor for a number of microbes (e.g. *Escherichia coli* and Hantavirus). Cromer/DAF antigens are present on leukocytes, platelets and placental

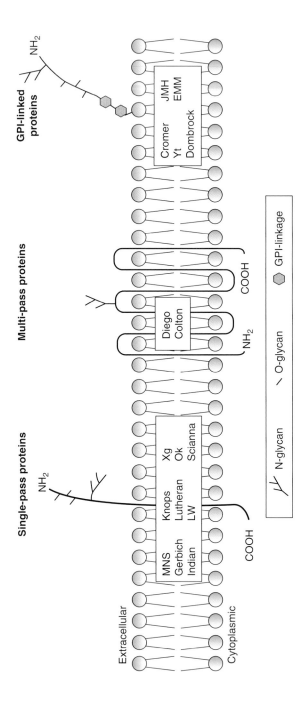

FIGURE 27.1 Diagram of the red blood cell membrane illustrates the type of membrane components that carry the blood group antigens. The figure does not show components carrying Chido/Rodgers antigens, because they are not integral membrane components, or most of the components carrying the blood group collections, and high- and low-incidence antigens, because their structure is unknown. From Hillyer CD, Silberstein LE, Ness PM et al. (eds) (2007). *Blood Banking and Transfusion Medicine: Basic Principles and Practice*, 2nd edition. San Diego, CA: Elsevier Academic Press.

trophoblasts and, in a soluble form, in plasma and urine. DAF is missing from the RBCs of patients with paroxysmal nocturnal hemoglobinuria because they are deficient in all GPI-linked proteins. Anti-Cromer antibodies are formed as a result of previous RBC exposure and have variable clinical significance, some of which may cause decreased RBC survival; however, they are not known to cause HDFN.

**Diego Blood Group System:** The Diego system antigens are on band 3 (AE-1) of the RBC, which functions at least as a structural membrane protein and as a chloride–bicarbonate counter transporter. Diego antigens consist of two pairs of antithetical antigens ($Di^a/Di^b$ and $Wr^a/Wr^b$) and other low-incidence antigens. The expression of $Wr^b$ is dependent on the presence of glycophorin A. Anti-$Di^a$ and anti-$Di^b$ may cause HTRs or HDFN. Anti-$Wr^a$ is reported in 1–8% of normal donors and 30% of patients with autoimmune hemolytic anemia, is IgG and/or IgM, and can occur with or without RBC exposure and rarely causes HTRs or HDFN. Anti-$Wr^b$ is rare, and is potentially clinically significant.

**Dombrock Blood Group System:** The Dombrock system includes $Do^a$ and $Do^b$, and the high-incidence antigens $Gy^a$, Hy and $Jo^a$. The null phenotype is Gy(a−). Antibodies to $Do^a$ and $Do^b$ are uncommon, and are usually found in sera containing other RBC alloantibodies. Anti-$Do^a$ has resulted in HTRs and HDFN, and anti-$Do^b$ has resulted in HTRs but not HDFN. Antibodies to $Gy^a$, Hy and $Jo^a$ may result in a mild HTR or mild HDFN. The antigen–antibody reactions are enhanced by papain- or ficin-treated RBCs, but are destroyed by sulfhydryl reagents.

**Gerbich Blood Group System:** The Gerbrich system includes seven antigens: three high incidence (Ge2, Ge3 and Ge4) and four low incidence (Wb, $Ls^a$, $An^a$ and $Dh^a$). These antigens are carried on glycophorin C and glycophorin D, which are products of a single gene (*GYPC*) and interact with band 4.1. Deficiencies of either band 4.1 or glycophorin C or D results in elliptocytosis. Gerbrich phenotypes included the Yus type (Ge:-2, 3, 4; individuals with this phenotype can form anti-Ge2), the Gerbrich type (Ge:-2, -3, 4; individuals with this phenotype can form anti-Ge2 or -Ge3) and the Leach type (Ge:-2, -3, -4; individuals with this phenotype can form anti-Ge2, -Ge3 or -Ge4). These antibodies can be RBC stimulated or naturally occurring, and have variable clinical significance. Antibodies to the low-incidence antigens rarely result in HDFN.

**Indian Blood Group System:** The Indian system antigens, $In^a$ (low incidence) and $In^b$ (high incidence), are located on CD44, a widely distributed cell adhesion molecule. The antigens are destroyed by papain, ficin and sulfhydryl reagents.

**John Milton Hagen Blood Group System:** Antibodies to the John Milton Hagen (JMH) antigens are not considered clinically significant.

**Knops Blood Group System:** The Knops antigens are located on the C3b/C4b receptor, CR1, which is the primary complement receptor on RBCs. All of the antigens are high incidence ($Kn^a$, $McC^a$) except for $McC^b$. These antigens are weakened or destroyed by sulfhydryl reagents. Antibodies to the Knops antigens show variable weak reactivity, but may continue to react even at high dilutions. These antibodies are not clinically significant.

**Lutheran Blood Group System:** The Lutheran system contains $Lu^a$ and $Lu^b$ in addition to a number of high-incidence antigens (Lu4, Lu5, Lu6, Lu7, Lu8, Lu11, Lu12, Lu13, Lu16, Lu17, Lu18) and low-incidence antigens (Lu9 and Lu14), as well as $Au^a$ (Lu18) and $Au^b$ (Lu19). The Lu(a−b−) phenotype can arise from an amorphic Lutheran gene (*Lu*), from an inhibitor gene (*In(Lu)*) which prevents normal expression of Lutheran and other blood group antigens ($P_1$, I, AnWj, $In^a$ and $In^b$), or from an X-borne suppressor (*XS*). The antigens are destroyed by trypsin, $\alpha$-chymotrypsin and sulfhydryl reagents. Anti-$Lu^a$ and anti-$Lu^b$ are usually formed in response to previous RBC exposure, but have been reported to be naturally occurring. Lutheran antigens are poorly developed at birth. Anti-$Lu^a$ has not been associated with HTRs, and has rarely been associated with mild HDFN. Anti-$Lu^b$ has been reported to cause mild HTRs and none to mild HDFN. Anti-$Au^a$ and anti-$Au^b$ may cause mild HTRs, but have not been associated with HDFN. Anti-Lu3 is found in immunized individuals with the Lu(a−b−) phenotype, and may result in delayed HTRs or HDFN.

**LW Blood Group System:** The LW system consists of $LW^a$ (high incidence) and $LW^b$ (low incidence). The antigens are destroyed by sulfhydryl reagents and pronase. $LW^a$ is more strongly expressed on D-positive than D-negative RBCs. Individuals who have the $Rh_{null}$ phenotype are LW(a−b−). Antibodies to LW antigens are not usually associated with HTRs or HDFN.

**Ok Blood Group System:** The Ok system consists of a single high-incidence antigen, $Ok^a$. Rare Ok(a−) Japanese individuals may form anti-$Ok^a$, which has resulted in HTRs but not HDFN.

**Raph Blood Group System:** The Raph system consists of a single antigen, MER2. Anti-MER2 has been reported in three Israeli Jews on renal dialysis, and has not caused HTRs or HDFN.

**Scianna Blood Group System:** The Scianna system consists of four antigens; Sc1, Sc2, Sc3 and Rd. Sc1 (high incidence) and Sc2 (low incidence) are antithetical antigens, and Sc3 is present on any RBC that contains Sc1 or Sc2. The Rd antigen is low incidence. Antibodies to these antigens are rare. Anti-Sc1 has not been reported to cause HTRs or HDFN, and anti-Sc2 has resulted in mild HDFN.

**Yt Blood Group System:** The Yt (Cartwright) system consists of $Yt^a$ and $Yt^b$ antigens, which are located on RBC acetylcholinesterase. Anti-$Yt^a$ has been reported to result in accelerated destruction of $Yt^a$-positive RBCs, and is not associated with HDFN. Anti-$Yt^b$ has not been associated with HTRs or HDFN.

**Xg Blood Group System:** The $Xg^a$ antigen gene is located on the X chromosome. $Xg^a$ is destroyed by enzymes. Anti-$Xg^a$ is uncommon, may either be RBC stimulated or naturally occurring, and has not been implicated in HTRs or HDFN.

## Blood Group Collections:

**Cost Blood Group Collection:** The Cost collection consists of $Cs^a$ and $Cs^b$. Antibodies to these antigens are not clinically significant.

**Er Blood Group Collection:** The Er collection consists of two antigens: $Er^a$ (high-incidence) and $Er^b$ (low-incidence), which give rise to four phenotypes: (Er(a+b−), Er(a+b+), Er(a−b+), and Er(a−b−)).

## High-incidence RBC Antigens:

**Vel Antigen:** Anti-Vel is formed as a result of RBC stimulation, is usually IgM, binds complement, and is associated with HTRs, but not HDFN.

**Lan Antigen:** Anti-Lan is IgG, may bind complement, may cause HTRs and has been associated with mild HDFN.

**$Jr^a$ Antigen:** Anti-$Jr^a$ may result in decreased RBC survival of transfused antigen-positive RBCs, but has not been associated with HDFN.

**AnWj Antigen:** AnWj is carried on CD44 and is the receptor for *Haemophilus influenzae*. Anti-AnWj has been implicated in severe HTRs but not in HDFN, because it is not present on umbilical cord RBCs.

**$Sd^a$ Antigen:** $Sd^a$ expression on RBCs is variable and may diminish during pregnancy. Anti-$Sd^a$ is usually IgM and is not considered clinically significant, but examples of HTRs with RBCs with strong $Sd^a$ expression have been reported.

**Low-incidence RBC Antigens:** These antigens occur in less than 1 in 500 individuals, and include Batty (By), Biles (Bi), Box ($Bx^a$), Christiansen ($Chr^a$), HJK, HOFM, JFV, JONES, Jensen ($Je^a$), Katagiri (Kg), Livesay ($Li^a$), LOCR, Milne, Oldeide ($Ol^a$), Peters ($Pt^a$), Rasmussen (RASM), Reid ($Re^a$), REIT, SARA, Torkildsen ($To^a$), Traversu ($Tr^a$), and others. Antibodies to these antigens may cause HTRs or HDFN.

*Bg (Bennett-Goodspeed) antigens* correspond to human leukocyte antigen (HLA) antigens: $Bg^a$ corresponds to HLA-B7, $Bg^b$ corresponds to HLA-B17, and $Bg^c$ corresponds to HLA-A28. These antigens are expressed variably on RBCs.

# CHAPTER 28

# Red Blood Cells and Related Products

Christopher D. Hillyer, MD

**Product Names:** Red blood cells (RBCs) are also known as packed RBCs (pRBCs) and red cells. Modified RBC products include frozen, deglycerolized, leukoreduced, rejuvenated, washed, irradiated, and volume-reduced RBCs. Many textbooks will list whole blood as a RBC product, though the use of whole blood is uncommon and inefficient secondary to the loss of platelet function, because platelets must be stored at room temperature and once refrigerated lose their function, and the deterioration of coagulation protein function over time. In this book, the term RBC or RBC component or product refers to pRBCs or apheresis RBC products (which are concentrated during the collection process).

**Description:** The majority of RBC products are made from whole blood donated as 450 ml or 500 ml into an anticoagulant-preservative solution; about 5% of the RBC products are collected through automated RBC apheresis. The initial step of component manufacturing of whole blood is centrifugation, which leads to sedimentation or packing of the iron-laden RBCs. Next the platelet rich plasma is expressed, resulting in the RBC product (i.e. pRBCs). This product can then undergo a number of modifications, including leukoreduction, freezing, rejuvenation, washing, irradiation and/or volume reduction.

Approximately 15 million units of whole blood are manufactured into the ~13 million RBC products that are transfused each year in the US. Worldwide, it is estimated that ~70 million units of whole blood are collected and >65 million RBC products are transfused. General characteristics of RBC products include: ~130–240 ml of RBCs, ~50–80 g of hemoglobin (Hbg) and ~150–250 mg of iron. The total volume, ~250–350 ml, and hematocrit, 55–80%, vary depending on the anticoagulant-preservative solution used; higher volume and lower hematocrit are found in additive-solution (AS) containing products. Small amounts of plasma, platelets and leukocytes remain in RBC products, unless the latter have been removed by leukoreduction, which is typical in about 85% of the RBC products used in the US.

**RBC Storage Lesion:** The *RBC storage lesion* is the term that collectively refers to a number of metabolic and physical changes occurring in the RBCs themselves, as well as the resultant changes in the entire RBC product (the RBCs and surrounding plasma) during storage (Table 28.1). The decreased 2,3-DPG results in a shift in the oxygen dissociation curve to the left, which leads to less oxygen release than from normal RBCs at the same partial pressure $O_2$. In addition to biochemical changes, RBCs change from a deformable biconcave disk, to reversibly deformed echinocytes, to irreversibly deformed spherechinocytes with increased membrane stiffness. These morphologic changes may also result in decrease oxygen transport because of the inability of the RBC to flow through the microcirculation, and the increase in RBC and vascular

TABLE 28.1 Biochemical Changes in Stored Non-Leukocyte-Reduced Red Blood Cells

| Variable | CPD | | CPDA-1 | | | | AS-1 | AS-3 | AS-5 |
|---|---|---|---|---|---|---|---|---|---|
| | Whole Blood | | Whole Blood | Red Blood Cells | Whole Blood | Red Blood Cells | Red Blood Cells | Red Blood Cells | Red Blood Cells |
| Days of Storage | 0 | 21 | 0 | 0 | 35 | 35 | 42 | 42 | 42 |
| %Viable cells (24 hours posttransfusion) | 100 | 80 | 100 | 100 | 79 | 71 | 76 (64–85) | 84 | 80 |
| pH (measure at 37°C) | 7.20 | 6.84 | 7.60 | 7.55 | 6.98 | 6.71 | 6.6 | 6.5 | 6.5 |
| ATP (% of initial value) | 100 | 86 | 100 | 100 | 56 ($\pm$16) | 45 ($\pm$12) | 60 | 59 | 68.5 |
| 2,3 DPG (% of initial value) | 100 | 44 | 100 | 100 | <10 | <10 | <5 | <10 | <5 |
| Plasma K+ (mmol/L) | 3.9 | 21 | 4.20 | 5.10 | 27.30 | 78.50* | 50 | 46 | 45.6 |
| Plasma hemoglobin | 17 | 191 | 82 | 78 | 461 | 658.0* | N/A | 386 | N/A |
| % Hemolysis | N/A | N/A | N/A | N/A | N/A | N/A | 0.5 | 0.9 | 0.6 |

*Values for plasma hemoglobin and potassium concentrations may appear somewhat high in 35-day stored RBC units; the total plasma in these units is only about 70 mL. From Brecher ME (ed.). (2005). *AABB Technical Manual*, 15th edition. Bethesda: AABB Press.

endothelial interactions. Recent data suggest that stored RBCs have reduced nitric oxide (NO) bioavailability, possibly via decreased levels of SNO-Hb (S-nitrosohemoglobin). NO plays a vital role in the vasodilation of blood vessels, and therefore in oxygen delivery to tissues. It has been shown in experimental models that stored RBCs, when transfused, do not allow appropriate NO-mediated hypoxic vasodilation.

**Indications:** Indications are listed in Table 28.2.

---

**TABLE 28.2  Indications and Contraindications for Transfusion of Red Blood Cells**

**Indications – red blood cells may be used:**
- To mitigate tissue hypoxia due to decreased oxygen-carrying capacity associated with an inadequate red blood cell mass.
- As a source of replacement red blood cells during red blood cell exchange (erythrocytopheresis).
- To ensure optimal tissue oxygenation in anemic patients undergoing radiation therapy.

**Contraindications – red blood cells should not be used:**
- To correct anemia due to iron deficiency.
- As a souce of nutritional supplementation.
- For volume expansion or to increase oncotic pressure.

---

**Signs and Symptoms of Anemia:** RBCs are transfused to mitigate the signs and symptoms of anemia, which reflect a significant deficiency of oxygen-carrying capacity and/or tissue hypoxia due to an inadequate circulating RBC mass, which typically occurs from disorders leading to decreased bone marrow erythropoiesis (malignancy and cancer chemotherapy), loss of erythrocytes (trauma, surgery and parturition) or increased clearance or destruction of erythrocytes (immune and non-immune hemolysis, including autoimmune hemolytic anemia, malaria and sickle cell disease).

**RBC Exchange:** RBCs are used during RBC exchange (erythrocytapheresis), either therapeutically or prophylactically. RBC exchange can be required in hemolytic disease of the fetus and newborn for the prevention of kernicterus, and in patients with sickle cell disease either therapeutically for the treatment of severe complications (stroke or acute chest syndrome) or prophylactically for stroke prevention or prior to surgery.

**As an Adjunct to Radiation Therapy:** Some practitioners will transfuse anemic cancer patients during periods of radiation therapy in order to ensure that there is optimal oxygen delivery to the treatment site and the formation of oxygen radicals, which help potentate the efficacy of the radiotherapy.

**Laboratory Values as Transfusion Triggers and Guidelines for RBC Administration:**

*The Transfusion Trigger:* Because there are no precise indicators of tissue hypoxia, nor objective, measurable indicators of symptomatic anemia, it has become common practice to administer RBC products based on laboratory parameters including hemoglobin and hematocrit. These numbers constitute "transfusion triggers" at which transfusion is generally considered appropriate and above which it is not. While the use of transfusion triggers is helpful in considering a patient's general condition, there is no universal transfusion trigger, and therefore the clinical assessment of each patient is imperative so that unnecessary transfusion can be avoided in patients who have well adapted to their current level of anemia, and so that transfusion is not withheld (i.e. undertransfusion) when appropriate and needed.

*Guidelines for RBC Administration:* A number of countries and blood systems around the world have produced guidelines regarding administration of RBCs. These are helpful both to the practitioner and during audits of RBC use. Many guidelines do not give specific transfusion triggers, but rather describe clinical conditions in which RBC transfusion is likely to be appropriate and needed. One example of guidelines for RBC transfusion can be found in Figure 28.1. As a general concept only, transfusion of RBC products in adults is rarely indicated when the Hbg concentration is $> 10\,g/dl$, and is almost always considered to be indicated when the Hbg concentration is $< 6\,g/dl$.

*Restrictive Versus Liberal Transfusion Strategies:* There is some evidence that restrictive (versus liberal) transfusion policies result in a decrease in patient complications following allogeneic RBC product transfusion. In particular, Hébert and colleagues reported the results of a multicenter, randomized controlled trial of 838 stable but critically ill patients with $Hgb < 9.0\,g/dl$ within 72 hours of intensive care unit (ICU) admission randomized to a "restrictive-strategy" (Hgb of $7\,g/dl$) or a "liberal strategy" (Hgb of $10.0\,g/dl$) for transfusion. Though 30-day mortality was similar in the two groups, the overall mortality rate during hospitalization was significantly lower in the restrictive strategy group (22.2% versus 28.1%; $P = 0.05$). Similarly, Lacroix and colleagues reported the results of a non-inferiority trial of 637 stable but critically ill children with $Hbg < 9.5\,g/dl$ within 7 days of ICU admission randomized to a "restrictive strategy" (Hbg of $7\,g/dl$) or a "liberal strategy" (Hbg of $9.5\,g/dl$) for transfusion. Children who were transfused based on the restrictive transfusion "trigger" had 44% fewer transfusions ($P < 0.001$) without an increase in adverse outcomes, including multiple organ dysfunction or death.

In the past decade, data have accumulated suggesting that RBC product transfusion itself may be associated with, or an independent predictor or risk factor for, increased of morbidity in the recipient (including particularly acute respiratory distress syndrome [ARDS] and multi-organ failure) and mortality. The reasons for this are unclear, but current hypotheses include elements of the storage lesion, including the inability of stored RBCs to maintain their NO bioavailability leading to poor oxygen delivery and absence of NO-driven vasodilation in response to tissue hypoxia. Decreased RBC deformability of stored RBCs may also contribute to the observed phenomenon. In multiple clinical studies, the use of RBC transfusion has been associated with increase morbidity and mortality. In addition, the use of older (>14 days of storage) versus fresh (<14 days of storage) product has been associated with increased morbidity and mortality in one study that needs to be confirmed in other trials. Most of the studies referred to above have been retrospective, and performed in cardiac surgery, trauma and other acutely ill patients.

**Special RBC Products and Circumstances:** Details of each of the special blood product processes and modifications are discussed elsewhere in this book. The listing

South Glasgow University Hospitals NHS Trust
Southern General Hospital

**NHS**
Greater
Glasgow

## GUIDELINES FOR BLOOD TRANSFUSION

These guidelines promote best practice regarding blood use within the Southern General Hospital. Recent audit revealed widespread differences in practice and confusion as to when and how to prescribe blood.

### INDICATIONS FOR TRANSFUSION
#### 1. Acute Blood Loss

**An acute blood loss of greater than 20% of blood volume (about 1000 ml blood)** will often need a tranfusion. Do not delay ordering blood in situations where blood loss is acute and rapid. If blood loss is very rapid, the hospital **Major Hemorrhage Protocol** should be activated by dialing 3333.

#### 2. For Surgical Patients

Consider transfusion if:
* Preoperative hemoglobin is less than 8 g/dl and the surgery is associated with the probability of major blood loss.
* Post-operative hemoglobin falls below 7 g/dl
* Preoperative anemia MUST be investigated, as medical management may be more appropriate than transfusion.

#### 3. Anemia In Active Myocardial Infarction (Hb below 10 g/dl)

These patients are among the few who may benefit from a Hb above 80. Transfusion to an Hb of 10 g/dl is acceptable but to overshoot to 11 may be excessive. Evaluate effect of each unit as it is given.

#### 4. Anemia In Other Patients (Hb below 10 g/dl but above 7 g/dl)

Consider transfusion in normovolemic patients ONLY if they have symptomatic anemia. Symptoms and signs of anemia include:

* Shortness of breath for no other reason
* Angina
* Syncope/postural hypotension
* ST depression on ECG
* Tachycardia for no other reason

Transfusion above 10 g/dl is very rarely indicated and WILL be questioned by hematology staff.

* Think before transfusion. Blood is expensive and potentially dangerous if used inappropriately.
* Reasses after each unit is given. Do you need to give more?
* Stop if symptoms/signs shown above resolve.
* Stop if you have reached an adequate Hb i.e. above 8 g/dl in symptomless patients (10 g/dl in acute MI).

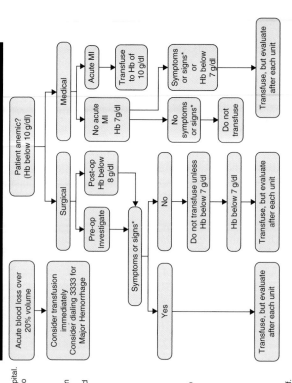

FIGURE 28.1 An example of RBC transfusion guidelines. From Garrioch M, Sandbach I, Pirie E *et al.* (2004). Reducing red cell transfusion by audit, education and a new guideline in a large teaching hospital. *Transfusion Med* **14**, 25–31.

below provides a framework to understand the array of products available, and the setting(s) in which they are used.

*Autologous Whole Blood:* Autologous products are RBC products donated by an individual for his or her own use in the future, most commonly for a scheduled surgical procedure. These autologous RBC products are often stored as whole blood because the plasma and platelets are usually not of clinical use because they cannot be used in the allogeneic setting. Therefore, in order to decrease and the cost of preparation, blood components are not manufactured from autologous whole blood units.

*Sickle Cell Disease:* Because patients with sickle cell disease often require serial and ongoing RBC transfusion, and because some of these patients can make a large number of alloantibodies to RBC antigens, specific guidelines for the transfusion of such patients include more extensive RBC antigen matching and the use of hemoglobin S negative products.

*Leukoreduced RBCs:* RBCs leukoreduced have $<5 \times 10^6$ residual leukocytes per product (US) and $<1 \times 10^6$ residual leukocytes per product (EU). Leukoreduction mitigates leukocyte-associated complications of blood transfusion, which can include febrile non-hemolytic transfusion reactions, HLA alloimmunization, CMV and EBV (and possibly HHV-8) transmission, and transfusion-related immunomodulation.

*Irradiated RBCs:* RBC products are irradiated to prevent the occurrence of transfusion associated graft-versus host disease (TA-GVHD) in appropriate recipients. Patients at highest risk for TA-GVHD thus need irradiated cellular blood products, including RBCs. Such patients include those with congenital immunodeficiencies or hematologic malignancies, and those receiving directed donor products (i.e. from a relative) and hematopoietic progenitor cell transplantation, and fludarabine and other purine analogues.

*Frozen RBCs:* Products with unusual RBC phenotypes can be frozen within 6 days of collection or after rejuvenation to create an inventory of rare products. After thawing and removal of the cryoprotectant glycerol, these products can be used for transfusion following the indications provided above in patients with related alloantibodies. Autologous products can also be frozen in advance of an anticipated surgery, or for patients with very rare RBC phenotypes, for their own use.

*Rejuvenated RBCs:* RBCs that are within 3 days of expiration can be rejuvenated and then frozen, if indicated; this process requires specialized rejuvenation solutions that contain pyruvate, inosine, phosphate and adenine, which restore the levels of ATP and 2,3-DPG to levels approaching those of a freshly drawn unit.

*Washed RBCs:* Patients with a history of severe allergic or anaphylactic reactions may require washed RBCs to remove the plasma proteins. Washing is also required for the use of rejuvenated or frozen RBC products, to remove the cryoprotectant and rehydrate the cells.

*Volume-reduced RBCs:* Volume reduction of a product may be required for the removal of the plasma and supernatant in patients who are extremely volume-sensitive or who are sensitive to the additive solution and/or increased potassium load (e.g. neonates) within the supernatant following prolonged storage or irradiation.

**Contraindications:** RBCs should not be used to treat an anemia that is asymptomatic or can be treated without transfusion-based therapy, such as iron deficiency. Indeed, in one study of hospitalized elderly patients, ~40% of transfusions were given for "anemia," the majority of which were due to iron deficiency. Furthermore, RBCs should not be used for volume expansion or to increase oncotic pressure. Finally, RBCs are not indicated in the treatment of any nutritional deficiencies; nor should this product be used to speed recovery, increase wound healing or improve a sense of wellbeing.

**Processing and Storage:** As above, RBC products are made from 450- or 500-ml collections of whole blood donated into 63 or 70 ml of anticoagulant-preservative solution. Initially the whole blood unit is centrifuged and the resulting supernatant, usually platelet-rich plasma, is decanted, leaving pRBCs. Anticoagulant and additive solutions used for RBC storage include ACD-A, CPDA-A, CPD, CP2D, AS-1, AS-3 and AS-5. These solutions vary in their constituents, which can include citrate or trisodium citrate, phosphate, adenine, monobasic sodium phosphate, and mannitol. The allowed storage duration depends on the anticoagulant-preservative and additive solution chosen used: there is a 21-day expiration period for ACD, CPD and CP2D, a 35-day period for CPDA-1, and a 42-day period for AS-1, AS-3 and AS-5. Storage of liquid RBC products is at 1–6°C.

Products that are to be frozen require the addition of a cryoprotectant, usually in the form of 40% glycerol which protects the erythrocytes from dehydration and ice-particle formation during the freezing process. Frozen products are stored at −65°C or below and have an approved shelf-life of 10 years, though products >20 years old have been used without complication. Thawing of frozen products is followed by a process of deglycerolization using hypertonic saline solutions. Once thawed and deglycerolized, the products must be used within 24 hours.

A number of additional manufacturing and/or processing steps may also be employed. These include pre-storage leukoreduction via filtration and gamma irradiation. Pre-storage irradiation modifies the product's shelf-life to the earlier of 28 days or the outdate based on the date of initial collection. This precaution was put in place in order to ensure that the plasma potassium concentration in the RBC product is not excessive, as the radiation enhances the release of potassium and decreases the post-transfusion survival of the RBCs. Post-storage processing can include washing or volume reduction to remove the excess potassium in at-risk individuals. These are accomplished via open systems, and thus the product has only 24 hours of approved shelf-life following such washing or volume reduction. For pediatric transfusion, RBC aliquots can be made from individual products, thus allowing the re-use of a given product for multiple transfusion episodes for a given recipient.

**Preparation and Administration:** RBCs are administered only with a physician's order. It is also recommended that a process for obtaining informed consent for RBC (and other blood product) administration is in place. Ideally, a process should be available that allows individuals who will not accept RBC transfusion to document their refusal; this is especially important for Jehovah's Witnesses, who may object to transfusion of RBC-containing products on religious grounds.

RBC products are transfused through tubing that contains a micro-aggregate filter capable of removing particulate matter >170 μm in size. In general it is not necessary to pre-warm the products, though some patients with cold agglutinin disease may require the use of blood warmers and patients who are transfused at rates >100 ml/min are at increased risk of cardiac events and hypothermia unless the blood is warmed.

ABO/D Compatibility: RBC products must be ABO compatible with the recipient's plasma to avoid potentially immediate and life-threatening immune hemolysis. Ideally, crossmatched ABO-type specific or ABO-type compatible products (if type-specific products are not available) are issued. In emergent circumstances, group O products can be issued without crossmatch or knowledge of the recipient's ABO type.

In order to minimize alloimmunization to the D antigen, and thus limit the risk of future hemolytic disease of the fetus and newborn or hemolytic transfusion reactions, it is expected that D-compatible products will be administered. In emergent situations, where the D typing is unknown, D-negative products should be given to all recipients; this is especially important in D-negative women of child-bearing potential. In situations where the D-negative inventory is inadequate to meet the transfusion needs, transfusion of D-positive products to D-negative individuals should be directed by transfusion service policy and medical director decision by weighing the risk of alloimmunization against the risk of withholding transfusion.

Quality Assurance: The requirements for RBC product preparation are dictated by AABB *Standards* and the *Code of Federal Regulations* (*CFR*). In the US, current regulations focus primarily on the volume of blood collected and ensuring that at least 75% of erythrocytes are viable 24 hours following transfusion. In some developed nations, it is recommended that the amount of free hemoglobin not exceed 1% of the RBC mass. Current US regulations or standards do not address post-manufacturing RBC product hemoglobin concentration, though some authorities have suggested that this would be appropriate.

Dose: In adult recipients, RBC products are typically given as 1–2 units per transfusion episode. In an average-sized adult without ongoing bleeding or hemolysis, each RBC product can be expected to increase the Hgb by 1 g/dl or the Hct by ~3%. In children, dosing is often based on recipient weight, and a dose of 10–15 ml/kg is not atypical.

Adverse Events: Acute hemolytic transfusion reactions can result from inadvertent ABO-incompatible RBC product transfusion, and may result in hypotension, DIC, renal failure and death. Indeed, this is most commonly termed *mistransfusion*, and is considered to be one of the leading causes of death following transfusion. Alloimmunization to minor RBC antigens can also occur, and may result in delayed hemolytic transfusion reactions upon subsequent transfusion with antigen-positive cells. As RBC products are derived from human whole blood donation, transmission of infectious diseases is possible. Further details about adverse events of blood transfusion can be found in Chapters 52–66 of this book.

Alternatives and Future Considerations: Alternatives to standard allogeneic include preoperative autologous donation, intraoperative and post-operative autologous RBC salvage, and acute normovolemic hemodilution. The administration of erythropoietin can reduce the need for allogeneic RBC transfusion in some clinical settings. Pharmacologic agents to minimize bleeding can also be helpful in minimizing RBC transfusion (see Chapter 51). Hemoglobin-based oxygen carriers (HBOCs), hemoglobin solutions in a number of formulations, are under investigation. Currently none of these is approved for use in the US. Pathogen-inactivated RBC products are also under investigation, though no process is currently in use for RBCs around the world. In addition, enzymatic removal of the terminal sugars of the A and B antigen molecules is under investigation in order to produce chemically modified, universal group O RBC products.

**International Considerations:** In countries where the risk of transfusion-transmitted infectious diseases is significantly higher than in the US and the EU RBC transfusion strategies have evolved that are quite divergent from those in more developed countries. In areas highly endemic for malaria, and thus severe malarial anemia, RBC transfusions are not typically initiated until the Hbg is 3 g/dl or below due to blood inventory limitations and the high prevalence of HIV in some donor populations. As there are significant shortages, the use of umbilical cord blood for transfusion is being considered for RBC transfusion in neonates and small children in under-resourced countries.

In some countries the volume of whole blood collection varies. For example, the standard whole blood unit donation in Japan is 200–400 ml. Many countries outside of the US use stored RBC products in an additive solution termed SAG-M, which contains saline (sodium chloride, for isotonicity), adenine (for ATP generation), glucose (for metabolism) and mannitol (for decreasing RBC lysis).

## Recommended Reading

American Society of Anesthesiologists Task Force on Perioperative Blood Transfusion and Adjuvant Therapies. (2006). Practice guidelines for perioperative blood transfusion and adjuvant therapies: an updated report by the American Society of Anesthesiologists Task Force on Perioperative Blood Transfusion and Adjuvant Therapies. *Anesthesiology* **105**, 198–208.

Garrioch M, Sandbach J, Pirie E *et al.* (2004). Reducing red cell transfusion by audit, education and a new guideline in a large teaching hospital. *Transfusion Med* **14**, 25–31.

Hardy JF. (2004). Current status of transfusion triggers for red blood cell concentrates. *Trans Apheresis Sci* **31**, 55–66.

Hasall O, Ngina L, Konga W *et al.* (2008). The acceptability to women in Mombasa, Kenya, of the donation and transfusion of umbilical cord blood for severe anaemia in young children. *Vox Sang* **94**, 125–131.

Hébert PC, Wells G, Blajchman MA *et al.* (1999). A multicenter, randomized, controlled clinical trial of transfusion requirements in critical care. *N Engl J Med* **340**, 409–417.

Hogman CF, Merryman HT. (2006). Red blood cells intended for transfusion: quality criteria revisited. *Transfusion* **46**, 137–142.

Ikeda H. (2002). Blood program in Japan: current status. *Trans Apheresis Sci* **26**, 129–133.

Josephson CD, Su LL, Hillyer KL, Hillyer CD. (2007). Transfusion in the patient with sickle cell disease: a critical review of the literature and transfusion guidelines. *Transfus Med Rev* **21**, 118–133.

Koch CG, Li L, Sessler DI *et al*. (2008). Duration of red-cell storage and complications after cardiac surgery. *N Engl J Med* **358**, 1229–1239.

Lacroix J, Hébert PC, Hutchison JS *et al*. (2007). Transfusion strategies for patients in pediatric intensive care units. *N Engl J Med* **356**, 1609–1619.

Tinmouth A, Fergusson D, Yee IC *et al*. (2006). Clinical consequences of red cell storage in the critically ill. *Transfusion* **46**, 2014–2027.

# CHAPTER 29

# Plasma Products

Cassandra D. Josephson, MD and Christopher D. Hillyer, MD

**Product Names:** Fresh frozen plasma (FFP), plasma frozen within 24 hours after phlebotomy (FP24 [sometimes called F24 or PF24]), cryoprecipitate-reduced plasma (CRP, cryo-poor plasma and cryosupernatant) and thawed plasma are plasma-based products primarily used in the clinical setting. Recovered plasma (usually converted from "plasma" and liquid plasma) is usually used for further manufacturing into plasma derivatives, including albumin, immunoglobulin and factor concentrates.

In this book the term *plasma products* refers to fresh frozen plasma (FFP), plasma frozen within 24 hours (FP24) and thawed plasma, which can be used, in the majority of patients, interchangeably.

**Description:** Plasma is the acellular, fluid compartment of blood, and consists of 90% water, 7% protein and colloids, and 2–3% nutrients, crystalloids, hormones and vitamins. The protein fraction contains the soluble clotting factors, fibrinogen, Factor XIII, von Willebrand Factor (VWF) and Factor VIII primarily bound to its carrier protein VWF, and the vitamin K-dependent coagulation Factors II, VII, IX and X. Plasma can be manufactured from whole blood after centrifugation and RBC removal, or collected by apheresis as a single product or via multicomponent apheresis collections (often referred to as concomitant plasma). Clotting factors are the constituents for which transfusion of plasma is most often required.

**Fresh Frozen Plasma:** FFP contains approximately 1 IU per ml of each clotting factor, is prepared from whole blood (or apheresis), is frozen at $-18°C$ or colder within 8 hours of collection (or within 6 hours with the use of some storage bags after apheresis collection), and has a frozen shelf-life of 1 year.

Prior to transfusion, FFP must be thawed at 30–37°C, which takes 20–30 minutes (termed FFP Thawed). Thawed FFP should be transfused immediately, or stored at 1–6°C for up to 24 hours. If storage of thawed FFP must be extended to prevent wastage, it becomes *thawed plasma* (see below) and must be relabeled as such. Thawed plasma can stored for up to 5 days (in total from the time of thawing) at 1–6°C.

**Plasma Frozen within 24 hours After Phlebotomy:** FP24 is plasma frozen within 24 hours of phlebotomy, is stored at $-18°C$ or colder, and has a frozen shelf-life of 1 year. Like FFP, FP24 is thawed at 30–37°C and either transfused immediately or stored for up to 24 hours at 1–6°C, and also can be stored for up to 5 days (again as *thawed plasma*).

FFP and FP24 are considered to be equivalent products for most patient populations; compared to FFP, FP24 has approximately 20% lower Factor VIII levels. However, because Factor VIII is an acute-phase reactant, its levels are quickly replenished in recipients without hemophilia A. Therefore, FFP and FP24 are generally considered to be interchangeable. The manufacture or FP24 has significant advantages over FFP for the blood collection facility, including improved logistics (and thus cost-efficient use) of whole blood donated at distant locations; the interchangeable use of FP24 has facilitated the implementation of male-only plasma in the US as TRALI mitigation strategy (see Chapter 58).

**Thawed Plasma:** Thawed plasma originates from either FFP or FP24 that has been thawed and stored at 1–6°C. Thawed plasma may be stored at 1–6°C up to 5 days after the original product was thawed. Thawed plasma has lower levels of Factor VIII (reduced by 35–41% by day 5, resulting in 41–63% activity), Factor V (reduced by 9–16%, but reported as high as 50%) and Factor VII (reduced by 4–20%), which are usually not of clinical significance.

**Recovered Plasma (Plasma for Manufacture):** Plasma products (liquid plasma and "plasma") that are derived from whole blood and intended for further manufacturing (i.e. fractionation into albumin, intravenous immunoglobulin and factor concentrates) are sent to a manufacturer from a collection facility through a "short supply agreement." *Liquid plasma* is defined as plasma that is separated from whole blood at any time during storage at 1–6°C, up to 5 days after the whole blood expiration date. Another product, termed "plasma," is defined as liquid plasma that is frozen at −18°C or colder, with a frozen shelf-life of 5 years.

**Source Plasma:** Source plasma, a FDA-licensed product, is collected by apheresis and is intended for further manufacturing. The Plasma Protein Therapeutics Assocation promotes safe collection of and manufacturing practices for plasma derivatives.

**Cryoprecipitate-reduced Plasma:** CRP is a plasma product consisting of the supernatant expressed during the manufacture of cryoprecipitate from FFP. CRP is therefore deficient in Factor VIII, VWF, fibrinogen, cryoglobulin and fibronectin. CRP is refrozen and stored at −18°C or colder for up to 1 year. CRP is used only for transfusion or plasma exchange therapy in patients with thrombotic thrombocytopenic purpura (TTP).

**Solvent-detergent Plasma (SD Plasma):** SD plasma is a product manufactured from ≤ 2500 pooled plasma products that has been treated with solvent (tri-η-butyl phosphate)/detergent (triton X-100) to inactivate lipid-enveloped viruses (HIV, hepatitis B, hepatitis C). The product is distributed in 200 ml containers, frozen at −18°C, with a shelf-life of 1 year. This product has an approximately 50% loss of protein S and approximately 10% loss of other factors. SD plasma is approved for use by the FDA, but is no longer sold in the US; it is available in other countries.

**Indications:** Plasma transfusions are usually used in either the prevention of bleeding or for the treatment of hemorrhage in patients with either a congenital or acquired coagulation defect. Coagulation defects are routinely determined by the measurement of the prothrombin time (PT) or International Normalized Ratio (INR), and the partial thromboplastin time (PTT). There are no evidence-based studies that define a laboratory value that can serve as a "trigger" for plasma administration, though many clinicians will consider plasma transfusion when there is an INR of 2.0 or greater, depending on the presence, or risk, of bleeding. Audits of transfusion practices have consistently suggested that plasma use is inappropriately high (reportedly only 60% of plasma usage in the UK complies with the current UK guidelines). Specifically, plasma should not be used as a simple volume expander, or as a source of nutrients. Indications and contraindications for plasma products are described in Table 29.1.

---

**TABLE 29.1  Administration of Plasma Products**

**Indications**

- Congenital coagulation defects where no factor concentrates are available*
- Disseminated intravascular coagulopathy*
- Liver disease*
- Massive transfusion*
- Plasma infusion or exchange for thrombotic thrombocytic purpura (TTP)
- Plasma exchange in other conditions*
- Rapid reversal of warfarin effect*
- Other causes of multiple acquired coagulation factor deficiency*

**Contraindications**

- Immunodeficiency
- Burns
- Wound healing
- Volume expansion
- Source of nutrients

*Transfusion only for a risk of bleeding or in the presence of bleeding.

---

**Liver Disease and Transplantation:** Patients with severe liver disease may have low levels of the vitamin K-dependent clotting factors (II, VII, IX and X), and such patients can develop a prolonged PT/ INR and PTT. Additionally, the thrombin time (TT) may be prolonged, fibrin split products may be elevated, and in later stages the fibrinogen levels decrease. Hemorrhage, most often secondary to an anatomic lesion, may be complicated by the coagulopathy resulting from these abnormalities. Plasma infusion is indicated for the treatment of major bleeding and the prevention of bleeding prior to major procedures. Plasma products are not recommended prophy-lactically before minor surgical procedures. Increased PT/INR and PTT are poor pre-dictors of surgical bleeding. Transfusion should be guided by clinical assessment, and the evidence and degree of bleeding in conjunction with coagulation test results.

**Massive Transfusion:** The administration of plasma products to patients receiving massive transfusion is recommended to treat and mitigate the effects of both the early trauma-induced coagulopathy and the dilutional coagulopathy secondary that occurs with colloid and RBC resuscitation. Plasma, platelet and cryoprecipitate products can be administered either at set doses and time-points through a massive transfusion protocol, or as determined by coagulation laboratory values. Recent data demonstrate that early and aggressive use of plasma and platelets in trauma patients undergoing massive transfusion decreases mortality (see Chapter 50).

**Rapid Reversal of Warfarin:** Warfarin inhibits the hepatic synthesis of vitamin K-dependent clotting Factors (II, VII, IX and X) by blocking the recovery of the active form of vitamin K. Vitamin K administration typically corrects the coagulopathy in 12–18 hours and should be used if time is available and bleeding is not problematic or in combination with plasma products. In patients anticoagulated with warfarin who have active bleeding or require emergency surgery, the warfarin effect can be immediately reversed by prothrombin complex concentrates (PCC), activated Factor VII or plasma administration. PCC or Factor VIIa should be first-line therapy, especially in volume-sensitive patients (require less volume), and for emergent reversal (administration may be faster). The ACCP evidence-based warfarin reversal guidelines are presented in Table 29.2.

**TABLE 29.2 Recommendations for Managing Elevated INRs or Bleeding in Patients Receiving Warfarin**

| Condition | Intervention |
|---|---|
| INR more than therapeutic range but <5.0; no significant bleeding | Lower dose or omit dose; monitor more frequently and resume at lower dose when INR therapeutic; if only minimally above therapeutic range, no dose reduction may be required. |
| INR ≥ 5.0, but <9.0; no significant bleeding | Omit next one or two doses, monitor more frequently, and resume at an appropriately adjusted dose when INR in therapeutic range. Alternatively, omit dose and give vitamin K (1–2.5 mg po), particularly if at increased risk of bleeding. If more rapid reversal is required because the patient requires urgent surgery, vitamin K (≤5 mg po) can be given with the expectation that a reduction of the INR will occur in 24 h. If the INR is still high, additional vitamin K (1–2 mg po) can be given. |
| INR ≥ 9.0; no significant bleeding | Hold warfarin therapy and give higher dose of vitamin K (2.5–5 mg po) with the expectation that the INR will be reduced substantially in 24-48 h. Monitor more frequently and use additional vitamin K if necessary. Resume therapy at an appropriately adjusted dose when INR is therapeutic. |
| Serious bleeding at any elevation of INR | Hold warfarin therapy and give vitamin K (10 mg by slow IV infusion), supplemented with FFP, PCC, or rVIIa, depending on the urgency of the situation; vitamin K can be repeated ql2h. |
| Life-threatening bleeding | Hold warfarin therapy and give FFP, PCC, or rVIIa supplemented with vitamin K (10 mg by slow IV infusion). Repeat, if necessary, depending on INR. |
| Administration of vitamin K | In patients with mild to moderately elevated INRs without major bleeding, give vitamin K orally rather than subcutaneously. |

If continuing warfarin therapy is indicated after high doses of vitamin $K_1$, then heparin or low-molecular-weight heparin can be given until the effects of vitamin $K_1$ have been reversed, and the patient becomes responsive to warfarin therapy. It should be noted that INR values >4.5 are less reliable than values in or near the therapeutic range. Thus, these guidelines represent an approximate guide for high INRs.
From Ansell J, Hirsh J, Hylek E *et al.* (2008) Pharmacology and management of the vitamin K antagonists: American College of Chest Physicians Evidence-Based Clinical Practice Guidelines (8th edition). *Chest* **133**, 160S–198S.

**Disseminated Intravascular Coagulopathy:** Disseminated intravascular coagulopathy (DIC) is typically secondary to a variety of diseases, including sepsis, liver disease, hypotension, surgery-associated hypoperfusion, trauma, obstetric complications, leukemia (usually promyelocytic) and underlying malignancy. Successful treatment of the underlying cause is critical. Patients with DIC, demonstrable multiple coagulation factor deficiencies, thrombocytopenia and bleeding (or high risk of bleeding) should be transfused with plasma products in amounts sufficient to correct or mitigate the hemorrhagic diathesis. Normalization of coagulation values will likely not be achieved with the plasma infusion.

**Plasma as a Replacement Fluid for Plasma Exchange:** Plasma can be used as replacement fluid for patients undergoing therapeutic plasma exchange (TPE), either alone or in combination with saline or albumin. Patients with thrombotic thrombocytopenic purpura (TTP) require plasma to be used as the replacement fluid, as it provides the missing enzyme ADAMTS-13 (see below). The goal of TPE in Refsum's disease is to decrease the level of phytanic acid, and not to replace the missing enzyme, phytanoyl-CoA hydroxylase; therefore, albumin is primarily used for replacement fluid. For patients with other diagnoses undergoing TPE, plasma may be added as a replacement fluid when the patient is actively bleeding secondary to the dilutional effects of ongoing plasma-exchange therapy or a concomitant coagulopathy exists.

**Thrombotic Thrombocytopenic Purpura:** The treatment of idiopathic TTP is TPE with plasma as the replacement fluid. TPE has decreased the mortality of TTP from $>90\%$ to $<10\%$. If TPE is not immediately available, plasma infusion is recommended. In addition, patients with familial TTP may require simple plasma infusion only.

**Congenital Coagulation Factor Deficiencies:** There are rare factor deficiencies for which purified plasma derivatives are not available in the US, such as C1-esterase inhibitor, and Factors V, X or XI. In these factor deficiencies, the use of plasma may be indicated in situations of bleeding, or as prophylaxis to prevent bleeding. Plasma is indicated for the treatment of C1-esterase inhibitor deficiency (also called hereditary angioedema) prior to surgical or dental procedures in order to prevent the onset of angioedema.

**Other Multiple Coagulation Defects:** There are other circumstances which result in acquired coagulation defects, such as cardiac surgery or extracorporeal membrane oxygenation, where plasma administration may be indicated.

**Dosage:** The appropriate dose and frequency of plasma administration may be estimated from the plasma volume, the desired increment of factor activity, and the expected half-life of the factor being replaced. The typical plasma dosage is 10–20 ml/kg. A single plasma unit volume is 200–300 ml and contains, on average, 0.7–1.0 IU of activity of each coagulation factor per ml of product and 1–2 mg/ml of fibrinogen. At 10–20 ml/kg dosing, clotting factor activities will increase by at least 30% in the absence of rapid and ongoing consumption.

Ideally plasma products should be ordered as the number of milliliters to be infused, but this practice is uncommon in adults. Common practice is to transfuse

2–4 plasma units in adults. The frequency of administration depends on the clinical response to the infusion, and moderation of corresponding laboratory parameters. Normalization of coagulation laboratory values will likely not be achieved with plasma infusion. In addition, minimal elevation of the PT/INR (e.g. INR < 1.5) is usually not corrected with plasma administration.

**Compatibility:** Plasma is screened for unexpected RBC alloantibodies at the time of collection, and should be ABO-type compatible for transfusion (see Chapter 19). Tests for serologic compatibility, major and/or minor crossmatching, are not performed before administration.

**Adverse Events:** Allergic reactions and volume overload (i.e. transfusion-associated circulatory overload [TACO]) are the most frequent adverse events of plasma transfusion. Rarely, other more serious events occur, including anaphylaxis and transfusion-related acute lung injury (TRALI).

**Alternatives to Plasma Products:** Recombinant Factors VIII and IX are available and may be used (and in some circumstances preferentially) in patients with hemophilia A and hemophilia B, instead of the plasma derived concentrates. Recombinant Factor VIIa is available for the treatment of patients with Factor VIII inhibitors and Factor VII deficiency. Plasma derived factor concentrates which are available in the US include Factor VIII, Factor IX, antithrombin III and von Willebrand Factor, which mitigate the need for plasma transfusion for the replacement of these specific factors. Thus, in these circumstances plasma infusion is contraindicated. Outside of the US, C1-esterase inhibitor, protein C, fibrinogen, and Factors V, VII, XI and XIII concentrates are available, and should be used as appropriate in place of plasma product infusion.

**International Considerations:** Outside the US, pathogen-inactivated plasma products treated with pathogen-reduction technologies, including riboflavin, S-59, solvent/detergent and methylene blue, are available.

## Recommended Reading

Abdel-Wahab OI, Healy B, Dzik WH. (2006). Effect of fresh-frozen plasma transfusion on prothrombin time and bleeding in patients with mild coagulation abnormalities. *Transfusion* **46**, 1279–1285.

Ansell J, Hirsh J, Hylek E et al. (2008). Pharmacology and management of the vitamin K antagonists: American College of Chest Physicians Evidence-Based Clinical Practice Guidelines (8th edition). *Chest* **133**, 160S–198S.

Chowdhury P, Saayman AG, Paulus U et al. (2004). Efficacy of standard dosing and 30 ml/kg fresh frozen plasma in correcting laboratory parameters of haemostasis in critically ill patients. *Br J Haematol* **125**, 69–73.

Eder AF, Sebok MA. (2007). Plasma components: FFP, F24, and thawed plasma. *Immunohematology* **23**, 150–157.

Ketchem L, Hess JR, Hiippala S. (2006). Indications for early fresh frozen plasma, cryoprecipitate, and platelet transfusion in trauma. *J Trauma* **60**, S50–S51.

O'Shaughnessy DF, Atterbury C, Bolton Maggs P et al. (2004). Guidelines for the use of fresh-frozen plasma, cryoprecipitate and cryosupernatant. *Br J Haematol.* **126**, 11–28.

# CHAPTER 30
# Platelet Products

Beth H. Shaz, MD

**Product Names:** Platelet products include those manufactured from whole blood and those manufactured by apheresis. An official FDA nomenclature exists for each one, but there are also many other names that are used in common parlance, occasionally giving rise to significant confusion in published papers and with ordering physicians. Platelets derived from whole blood are called *platelets* by the FDA; common names include *whole blood derived platelets, random donor platelets* and *platelet concentrates*. Platelets produced by apheresis are called *platelets,pheresis* by the FDA; common names include *single donor platelets, apheresis platelets* and *plateletpheresis*. The method of collection does not reflect or define a platelet *dose*, which can vary given the clinical circumstance but usually approximates $3–4 \times 10^{11}$ platelets for an adult. As whole blood derived platelet units typically contain $0.5 \times 10^{11}$ platelets, 5–6 units must be "pooled" to make a dose. Many apheresis derived platelet collections contain two or three times the required minimum of $3 \times 10^{11}$ platelets, and thus they are "split" to make multiple platelet doses. The 2007 FDA guidelines for platelets make QA determinations based on "dose" and "collection." Clinically, there is no single number that represents, or is accepted as, the optimal dose.

**Description:** Platelets are an essential component of hemostasis, and deficiencies in platelet number or function can result in bleeding and hemorrhage. Medications, liver or kidney disease, sepsis, disseminated intravascular coagulopathy (DIC), hematologic diseases and associated therapies, massive transfusion, and cardiac bypass may cause low platelet counts (thrombocytopenia) and/or platelet dysfunction. Symptoms of thrombocytopenia and platelet dysfunction include petechiae, easy bruising, or mucous membrane bleeding. The average *in vivo* lifespan of a platelet is 10 days, but that of a transfused platelet is 4–5 days. A platelet's lifespan is shortened by bleeding, DIC, splenomegaly, platelet antibodies, drugs, sepsis, endothelial cell or platelet activation, and thrombocytopenia.

**Indications:** Platelet transfusions are used for prophylaxis to prevent bleeding, or as treatment of bleeding in patients who have thrombocytopenia or qualitative defects in their platelet function (inherited or acquired by disease or anti-platelet medications).

**Prophylactic Platelet Transfusion:** The threshold to transfuse platelets is determined based on the patient's platelet count in addition to the underlying disease, presence of fever or sepsis, medications, and coagulation status. For most patients, the platelet transfusion threshold of 10,000/μl is safe in the absence of fever and sepsis, in which case a higher threshold may be used. Typically, platelets are increased to >50,000/μl prior to lumbar puncture, indwelling catheter insertion, thoracentesis, liver biopsy or transbronchial biopsy. For procedures in critical locations, such as the eye or brain, and for major surgery, maintenance of platelet count >100,000/μl should be considered.

**Therapeutic Platelet Transfusion:** Platelet transfusions should be considered in actively bleeding patients with platelet counts <50,000/μl or in those whose platelets are dysfunctional (e.g. secondary to medications). In patients with brain injury or surgery, platelet transfusion should be considered with platelet counts <100,000/μl. In the perioperative setting of patients undergoing cardiac surgery, platelet transfusions should be considered in patients experiencing excessive postoperative bleeding without a surgical cause.

**Transfusion for Platelet Dysfunction:** If possible, the underlying cause of the platelet dysfunction (e.g. removal of medication) should be corrected. Other treatments include increasing the hematocrit to >30%, which appears to aid in the treatment of bleeding by moving the platelets closer to the subendothelium, or administering DDAVP (1-deamino-8-D-arginine vasopressin desmopressin), which increases circulating Factor VIII and von Willebrand Factor levels and may be used in certain inherited diseases (e.g. storage pool disease). Treatment of platelet dysfunction secondary to uremia includes DDAVP, cryoprecipitate and conjugated estrogens (see Chapter 108). Platelet transfusions should be considered if these therapies are not appropriate, or are ineffective.

**Platelet Transfusion for Neonates:** Platelet transfusion triggers for neonates are not clearly defined, with practices varying by institution and provider, depending on the clinical situation and gestational age of the neonate. Generally, platelet counts are kept >50,000/μl in bleeding neonates or those undergoing an invasive procedure, with counts kept >100,000/μl in extremely ill, premature infants. Prophylactic transfusions are generally given for platelet counts <20,000/μl in neonates, and for platelet counts <50,000/μl in extremely preterm and/or critically ill neonates.

**Relative Contraindication for Platelet Transfusions:** Traditionally platelet transfusions have been contraindicated in patients with thrombotic thrombocytopenic purpura (TTP) or heparin-induced thrombocytopenia (HIT) unless there is a severe or life-threatening hemorrhage, because there are anecdotal reports of clinical worsening after platelet transfusion. The proposed theory is that the transfused platelets contribute to the ongoing platelet aggregation, resulting in increased microthrombi formation (see Chapters 90 and 93).

Platelet transfusion in patients with immune thrombocytopenic purpura (ITP) may yield lower platelet increments secondary to the platelet autoantibody, which may also react with the transfused platelets.

**Preparation and Administration:** Platelet products are either derived from whole blood collections or collected through apheresis as above. In the US, approximately 75% of platelet transfusions are collected by apheresis. The use of apheresis platelet products has increased in the US because whole blood derived platelet units are usually pooled prior to transfusion, most apheresis collection methods create a leukoreduced product and therefore do not require further manufacturing, and bacterial testing is simpler to perform on apheresis products.

**Whole Blood Derived:**  Whole blood is drawn into a bag containing an anticoagulant-preservative solution (citrate-phosphate-dextrose [CPD or CP2D] without or with adenine [CPDA-1]). The whole blood unit undergoes a low speed centrifugation step (soft spin), after which the platelet-rich plasma (PRP) is expressed and the red blood cells (RBCs) remain. Then a second centrifugation step separates the platelets from the plasma and the plasma is expressed, leaving the platelet concentrate. This method is used predominately in the US. The EU use a method that prepares platelet from the buffy coat and is described below. The CFR and AABB *Standards* require these platelets must contain $\geq 5.5 \times 10^{10}$ platelets in 90% of the units tested.

Prior to issuing whole blood derived platelets, the platelet products need to be pooled to make a suitable adult dose of platelets. This typically occurs in the hospital transfusion service immediately prior to release to the recipient. However, a system for pre-pooling, leukoreducing and bacterial testing of platelet concentrates has recently been approved by the FDA; these are referred to as "pool and store" platelets.

**Apheresis:**  Apheresis platelets are collected into an ACD-A (citric acid, trisodium citrate, dextrose) solution either as platelet-rich plasma or as platelet-pellet that requires re-suspension in concurrently collected plasma. Apheresis derived platelets are collected as leukoreduced (via process leukoreduction). CFR and AABB *Standards* require these platelets to contain $\geq 3.0 \times 10^{11}$ platelets in 90% of the products tested. Many products collected by apheresis contain platelet numbers well in excess of the required minimum and can be split into two or three dose-sized products; 95% of these products must contain $\geq 3.0 \times 10^{11}$ platelets per CFR.

**Storage:**  Platelet products are stored at 20–24°C with continuous gentle agitation for up to 5 days. In the US, platelets must be stored in plasma, as no platelet storage solutions (see below) are FDA approved. Plasma is not considered by many authorities to be optimal as it is associated with a number of adverse events, including allergic reactions, transfusion-related acute lung injury and ABO-incompatible hemolysis.

**Storage Containers:**  Platelets must be stored in oxygen-permeable containers, because in anoxic conditions platelet metabolism shifts to using the anaerobic glycolytic pathway, which produces lactic acid; this results in acidosis, leading to platelet death during storage. Therefore, adequate oxygenation is required to allow platelets to use aerobic mitochondrial oxidative phosphorylation.

**Bacterial Testing:**  During the 1990s, substantial data accumulated supporting the theory that the room temperature, plasma-rich, oxygenated environment of platelet storage was leading to bacteria growth and resulting in an unacceptably large number of reactions in recipients, including fever, sepsis, shock and death. Indeed, it was reported as many as 1 in 3000 plateletpheresis products had evidence of bacterial contamination. Furthermore, late in the 5-day storage period, the bacteria entered an exponential growth phase. It was indeed data of this type that had led the FDA to stop allowing 7-day storage in the preceding decades. In order to mitigate the risk associated with bacterial contamination, the AABB created Standard 5.1.5.1, which requires methods to limit and detect bacteria in platelet products. Currently there are several methods to limit and detect bacterial contamination in platelet products; bacterial testing is discussed in detail in Chapter 17.

The introduction of the requirement for bacterial testing of platelet products has decreased the rate of septic reactions from ~1:40,000 to ~1:75,000 and of fatal septic reactions from ~1:240,000 to ~1:500,000 apheresis platelet products transfused. These numbers are best considered as general estimates, as the true residual risk in tested platelets is not known. Overall, preliminary data suggests that culture methods have decreased the bacteria-related risks by one-half.

**TRALI Reduction Strategies:** In order to decrease the incidence of transfusion related acute lung injury (TRALI) from high plasma volume-containing products, donors who are at high risk of having HLA antibodies should be deferred from donating these products. Current strategies to limit these high-risk donors include male-only donors, exclusion of multiparous women donors, or testing and eliminating donors who have leukocyte antibodies (see Chapter 58).

**Leukoreduction:** Prestorage leukoreduction results in a decrease in febrile transfusion reactions by minimizing the levels of cytokines which are released from white blood cells during storage. In addition, leukoreduction reduces the risk of CMV transmission, HLA alloimmunization and the resulting risk of platelet refractoriness. Poststorage leukoreduction (i.e. the use of bedside leukoreduction filtration) does not decrease the incidence of febrile transfusion reactions and it is difficult to maintain adequate quality assurance with bedside leukoreduction. Therefore, the majority of institutions use prestorage leukoreduction. In addition, since the majority of patients who receive platelet products are hemotology/oncology patients, most institutions use only prestorage leukoreduction platelet products (see Chapter 37).

**Irradiation:** Irradiation of platelet products prevents transfusion-associated graft versus host disease. Patients at risk include those receiving products donated from a family member, products that are HLA matched or crossmatched (where the recipient's serum is tested against the donated platelets and those platelet products which do not react [i.e. are crossmatch compatible] are issued for transfusion), and those patients with congenital immunodeficiency syndromes, undergoing hematopoietic progenitor cell (HPC) transplantation, and with other diseases with congenital or acquired immunodeficiency states (see Chapter 36, Table 36.1). Some institutions irradiate all platelet products (universal irradiation) because the majority of platelet products are transfused to hematology/oncology patients and to neonates, some of whom will have as yet diagnosed immunodeficiency syndromes.

**Washing or Volume Reduction:** Washing or volume reduction may be used when transfusing ABO-incompatible platelet products, especially a group O product being transfused to a group A individual. In addition, volume reduction or washing will remove the antibody contained within the plasma for which the recipient carries the corresponding antigen – for example, maternal platelets being transfused to a neonate with neonatal alloimmune thrombocytopenia (NAIT), in which removal of the maternal platelet antibody is required (as well as irradiation) prior to transfusion. Lastly, volume reduction may be used in patients who are sensitive to extra volume. Washing or volume reduction results in loss of 5–20% of the platelets (see Chapters 40 and 41), and thus should be used only in the most necessary circumstances.

**Aliquots:** Platelet products are often dispensed in small aliquots via syringe for neonatal transfusions; these products are acceptable for 4–6 hours in the syringe. The lowest acceptable volume remaining in the platelet product for further aliquot removal is not known.

**Quality Assurance:** Per AABB *Standards* and CFR: 90% of whole blood derived platelets must contain $>5.5 \times 10^{10}$ platelets and have a pH $\geq 6.2$. Leukoreduced whole blood derived platelet units must have 95% $<0.83 \times 10^6$ residual white blood cells, and 75% of the units must have $>5.5 \times 10^{10}$ platelets; when the products are pooled they must have $<5.0 \times 10^6$ white blood cells. Regarding apheresis derived platelets, 90% must contain $>3.0 \times 10^{11}$ platelets and have a pH of $\geq 6.2$; to be considered leukoreduced, 90% of the products tested must have $<5.0 \times 10^6$ white blood cells. Per the CFR, apheresis collections can be split into up to three products: 95% of each product must have $\geq 3.0 \times 10^{11}$ platelets and 95% must have pH $\geq 6.2$. For leukoreduction, 95% of single collections must have $<5.0 \times 10^6$ white blood cells; 95% of double collections must have $<8.0 \times 10^6$ white blood cells (and 95% of the products must have $<5.0 \times 10^6$ white blood cells); and 95% of triple collections must have $<12 \times 10^6$ white blood cells (and 95% of the products must have $<5.0 \times 10^6$ white blood cells).

**Dose:** Daily platelet loss is through platelet senescence (half-life of approximately 10 days) and secondary to the requirements of platelets to maintain the vascular integrity. Together, these amount to $\sim 7000$–$10,000/\mu l/day$. In addition, splenomegaly, sepsis, drugs, DIC, auto- or alloantibodies, endothelial cell activation, and platelet activation through cardiac bypass or other circulatory devices may increase daily platelet consumption and, therefore, platelet need.

**Prophylactic Transfusion:** The optimal dose of platelets for prophylactic use is currently unknown. Currently, most ordering physicians will transfuse a single platelet dose in non-bleeding thrombocytopenic cancer patients at platelet counts below $10,000/\mu l$. Some physicians prefer a value of $20,000/\mu l$ in this situation.

Platelet administration using smaller platelet dose numbers meets the daily need to replace ongoing platelet loss, but requires more frequent platelet transfusions. This is in contrast to larger platelet doses per transfusion, which result in higher posttransfusion platelet counts and thus less frequent transfusions.

A recently completed trial (Optimal Platelet Dose Strategy for Management of Thrombocytopenia [PLADO]) by the NHLBI, where thrombocytopenic patients were randomized to receive low, medium or high doses based on the patient body surface area, demonstrated no clinically significant difference in bleeding between the three arms and will likely lead to a reduction in the dose of platelets which are prophylactically administered in non-bleeding thrombocytopenic hematology/oncology patients, and a change to per kg body-weight platelet dosing.

**Therapeutic Transfusion:** In adults, a typical platelet dose is a single apheresis product or 4–6 whole blood derived platelet units pooled, which will represent a platelet dose of $3$–$4 \times 10^{11}$ platelets and can be expected to increase the patients' platelet count by $30,000$–$60,000/\mu l$. In neonates and pediatrics, the dose of apheresis platelets is usually 10 ml/kg or 1 whole blood derived unit/10 kg with an expected increase in platelet count of 50,000–100,000/µl. As above, a significant number of clinical conditions (fever, sepsis, DIC, bleeding, etc.) can be expected to decrease the expected post-transfusion increment.

Response to platelet transfusions (i.e. platelet increment) can be assessed by improvement of bleeding and through measuring the posttransfusion platelet increment typically collected 15 to 60 minutes posttransfusion. Each transfused product should increase an adult's platelets by at least 10,000/µl, and if the platelet increment is below 10,000/µl on at least two occasions then the cause for platelet transfusion refractoriness should be sought (see Chapter 49). A more formal evaluation of the platelet increment can be performed, usually using the corrected count increment (CCI); the CCI takes the patient's size and the number of platelets in the product into account (Table 30.1).

---

**TABLE 30.1  Calculation of Corrected Count Increment (CCI)**

$$CCI = \frac{\text{Body surface area (m}^2) \times \text{platelet count increment} \times 10^{11}}{\text{Number of platelets transfused}}$$

*Example*: If $5 \times 10^{11}$ platelets are transfused to a patient whose body surface area is 2.0 m$^2$ and the increase in post-transfusion platelet count is 30,000/µl, then:

$$CCI = \frac{2.0\,\text{m}^2 \times 30,000/\mu l \times 10^{11}}{5 \times 10^{11}}$$

$$= 12,000 \text{ platelets} \times \text{m}^2/\mu l$$

---

## Product Selection:

**ABO Compatibility:** In general, ABO type-specific platelet transfusions should be administered. However, platelet supply is often suboptimal, and selection of out-of-group platelets for transfusion is not uncommon. As ABO antigens are present on the surface of platelets, lower recovery of ABO-incompatible platelets will be observed compared to compatible platelets (i.e. group A product being transfused to a group O recipient versus a group O product to a group O recipient), but this difference is not typically of clinical significance.

More important, however, is the co-administration of ABO-incompatible plasma present in the platelet product. As some O donors, and thus O plateletpheresis products, have high titer anti-A, often of both IgG and IgM classes, a positive direct antiglobulin test can result or, in some cases immediate hemolysis can result which can be fatal. It is thus recommended that the anti-A titer of group O platelets be determined, and only those with low titers be administered to group A patients.

In addition, ABO-incompatible plasma transfused into patients after HPC transplantation has been associated with deceased patient survival. This finding requires further validation.

Neonates and infants should receive ABO type-specific products. If these are not available, it is recommended that the product be washed or volume reduced.

**D Compatibility:** The D antigen is not on platelets but is present on any residual RBCs within the product. When present in sufficient dose, these D-positive RBCs can result in anti-D formation and future risk of HDFN. If a D-negative product is not available for a female with child-bearing potential, Rh immune globulin (RhIg) should be administered to prevent immunization. Hematology/oncology patients are at lower risk of forming anti-D in this circumstance, as they are significantly immuno-suppressed. Clinical consultation may be required.

**Adverse Events:** Platelet products, like other blood components, may result in adverse events, including infectious disease transmission (e.g. HIV, hepatitis C, and hepatitis B) and non-infectious hazards. The non-infectious hazards of transfusion include hemolytic transfusion reaction (usually from ABO-incompatible plasma within the product, especially a group O platelet product being transfused to a group A recipient), febrile non-hemolytic transfusion reaction, allergic reaction, and TRALI. Septic reactions related to bacterial contamination do occur as described above.

**Alternatives and Future Considerations:** Methods that replace plasma with appropriately constituted additives prior to storage would likely decrease hemolytic transfusion reactions, TRALI and potentially residual bacterial risk, and might also improve platelet survival and function following storage by improving metabolic parameters. These platelet additive solutions (PAS) are under investigation in the US, and have the potential to allow for prolonged platelet storage of up to 10 days. Storage of platelet products at 4°C would significantly decrease bacteria-related complications. However, at present no storage medium has been reproducibly found to accomplish this objective. The addition of some forms of sugar molecules may allow this practice, and is under investigation. Both lyophilized platelets and infusible platelet membranes appear to have some ability to function as "platelet substitutes," but neither is at the point of regulatory approval. Finally, pathogen-inactivation technologies by definition include removal of plasma as the storage solution, and thus may have the benefits of a PAS as above. Again, these are investigations only in the US at this time.

**International Standards:**

**Buffy Coat Prepared:** Many countries outside the US use buffy coat prepared platelets derived from whole blood. Whole blood first goes through a hard spin, after which the RBCs and plasma are removed, leaving the buffy coat. The buffy coat is then softly spun, after which the residual white blood cells are removed (i.e. leukoreduction), leaving the platelet concentrate. These buffy coat platelet concentrates can be pooled (prestorage) and stored in a single donor's plasma.

**Platelet Solutions:** In a variety of locations around the world including Europe, a variety of platelet solutions is approved for use. These PAS contain sodium chloride, potassium chloride, magnesium chloride, sodium acetate, and sodium gluconate or phosphate at present. The electrolytes in PAS inhibit platelet activation and aggregation, and acetate is used as a metabolic fuel and buffer.

**Pathogen-reduction/-inactivation Technologies:** Outside the US, pathogen-inactivation technologies are available, which minimize, or nearly eliminate, the risk of bacterial, viral and protozoal transmission. In addition, some methods inactivate white blood cells, which may eliminate the need for irradiation to prevent transfusion-associated graft versus host disease. As above, these technologies remove plasma from the product and thus mitigate hemolytic transfusion reactions and TRALI.

**Quality Assurance:** In UK, whole blood and apheresis derived platelets must contain $>2.4 \times 10^{11}$ platelet, and 90% of products must contain $<1.0 \times 10^6$ white blood cells (99% must contain $<5.0 \times 10^6$ white blood cells) in order to be labeled leukoreduced.

## Recommended Reading

British Committee for Standards in Haematology, Blood Transfusion Task Force (Chairman P. Kelsey). (2003). Guidelines for the use of platelet transfusions. *Br J Haematol* **122**, 10–23.

Hedges SJ, Dehoney SB, Hooper JS *et al.* (2007). Evidence-based treatment recommendations for uremic bleeding. *Natl Clin Pract Nephrol* **3**, 138–153.

Kaufman RM. (2006). Platelets: testing, dosing and the storage lesion-recent advances. *Hematology Am Soc Hematol Educ Program*, 492–496.

Prowse C. (2007). Zero tolerance. *Transfusion* **47**, 1106–1109.

# CHAPTER 31

# Cryoprecipitate

Beth H. Shaz, MD and Christopher D. Hillyer, MD

**Product Names:** Cryoprecipitated Antihemophilic Factor (thawed cryoprecipitated AHF; cryoprecipitated AHF, pooled; also called cryoprecipitate and cryo).

**Description:** Cryoprecipitate is made from human plasma. When fresh frozen plasma (FFP) is thawed in the cold (1–6°C), a precipitate forms (the cryoprecipitate) which, after the supernatant (cryosupernatant, cryoprecipitate-poor or cryoprecipitate-reduced plasma) is removed, is refrozen. Its main constituents are fibrinogen, fibronectin, Factor VIII, von Willebrand Factor (VWF) and Factor XIII.

**Indications:** Originally, cryoprecipitate was used as Factor VIII replacement for hemophilia A patients, but now there are more purified and virally inactivated or recombinant products available. Thus, use of this product for this reason is essentially contraindicated in the developed world. Similarly, this product should not be used to treat von Willebrand disease, as more purified and virally inactivated products containing VWF and Factor VIII are available. Currently, cryoprecipitate is used primarily for fibrinogen replacement, occasionally as Factor XIII replacement, and in the manufacturing of fibrin sealants and glue (see Table 31.1).

TABLE 31.1 Primary and Secondary Indications, Common Misuses, and Underutilization of Cryoprecipitate

**Primary indications**

Acquired/congenital hypofibrinogenemia
Massive transfusion with bleeding
As a component of fibrin glue/sealants
Factor XIII deficiency
Reversal of thrombolytic therapy with bleeding

**Secondary indications**

Hemophilia A
von Willebrand disease
Uremic coagulopathy

**Common misuses**

Replacement therapy in patients with normal fibrinogen measurements
Reversal of warfarin therapy
Treatment of bleeding without evidence of hypofibrinogenemia
Treatment of hepatic coagulopathy

**Common underutilization**

Massive transfusion with dilutional coagulopathy and bleeding

**Fibrinogen Replacement:** Hypofibrinogenemia occurs in patients with disseminated intravascular coagulopathy (DIC), with liver failure, during the anhepatic phase of liver transplantation surgery, and during massive transfusion via a dilutional coagulopathy. In an actively bleeding patient or prior to surgery, cryoprecipitate should be given when fibrinogen levels fall below 100 mg/dl. When the fibrinogen does fall below 100 mg/dl, there is a prolongation of the prothrombin test (PT) and activated partial thromboplastin time (PTT) that cannot be corrected by the infusion of FFP. Once the PT and PTT are critically abnormal, with significant bleeding in the patient, it can be more important to intervene with the transfusion of a pool of cryoprecipitate than to await laboratory results.

Cryoprecipitate is also used for fibrinogen replacement in patients with congenital or acquired abnormalities in fibrinogen, such as dysfibrinogenemia or hypofibrinogenemia. Low fibrinogen levels are detected by measuring fibrinogen, but dysfibrinogenemia, where fibrinogen is present but functionally defective, can be best detected by the thrombin time.

**Massive transfusion:** Massive transfusion, defined as the replacement of one blood volume with RBC products (i.e. 10 products in an adult), is often complicated by coagulopathy secondary to dilution and consumption of platelets and coagulation factors. Fibrinogen decreases to 100 mg/dl after approximately two blood volumes (i.e. 20 RBC products in an adult) are replaced. Therefore, cryoprecipitate use should be incorporated into the treatment of massively transfused patients, either as part of a massive transfusion protocol or as replacement once fibrinogen levels begin to decline (see Chapter 50).

**Fibrin Glue or Fibrin Sealant:** Fibrin sealant is the combination of thrombin and fibrinogen mixed with calcium to form fibrin, which is used as a topical hemostatic agent. Products may contain antifibrinolytics (i.e. aprotinin) to reduce fibrinolysis, or Factor XIII to increase the strength of the clot. A variety of commercially FDA-approved and individually produced fibrin sealants are available. Bovine thrombin products are commonly used, but have the risk of severe allergic reactions and antibody formation. Antibodies to bovine thrombin can cross-react with human Factor V, leading to acquired Factor V deficiency and a risk of hemorrhage. In contrast, human thrombin, though virally inactivated, has the small risk of transfusion-transmitted disease. Currently there are two fibrin sealant products which are commercially available in the US. Alternatively, automated devices also exist to produce fibrin sealant from autologous plasma. Another option is autologous fibrin glue prepared from the cryoprecipitated portion of autologous plasma; after thawing, this material is mixed with bovine thrombin immediately before application to the surgical field site. A disadvantage of fibrin sealant is that it takes time to prepare, especially in the autologous setting, and also time for the clot to form.

**Factor XIII Deficiency:** Factor XIII deficiency is a rare autosomal recessive congenital deficiency. Factor XIII plays an important role in the cross-linking of polymerized fibrin. Patients present with bleeding and delayed wound healing, usually first noted at the umbilical stump or following circumcision. They have normal PT and

PTT tests, but increased clot solubility. The half-life of Factor XIII is 9–15 days. Until Factor XIII concentrates are FDA approved, cryoprecipitate is the appropriate blood product for Factor XIII replacement.

**Bleeding Complications After Thrombolytic Therapy:** Approximately 1% and 4% of patients who receive thrombolytic therapy for either myocardial infarction and stroke, respectively, have intracranial hemorrhage. Cryoprecipitate has been used in the algorithm to treat these life-threatening bleeds; it is especially indicated if the fibrinogen is <100 mg/dl.

**Uremic Bleeding:** Cryoprecipitate has been reported to shorten the bleeding time in some uremic patients, and it has a variable hemostatic effect. Therefore, in addition to DDAVP, platelet transfusion and dialysis, it may have an adjunctive role.

**Processing and Storage:** A single unit of cryoprecipitate is manufactured from one unit of FFP which has been frozen to −18°C within 8 hours of collection if collected in CPD, CP2D or CPDA-1 anticoagulants, or within 6 hours if collected in ACD anticoagulant. Cryoprecipitate can be made from FFP within 12 months of the FFP collection. To make cryoprecipitate, FFP is slowly thawed to 1–6°C; the resulting cold-insoluble proteins are centrifuged to the bottom of the bag, and the supernatant is expressed. The remaining proteins are refrozen within 1 hour at –18°C in a small amount of remaining plasma/anticoagulant solution. The supernatant plasma, also known as plasma cryoprecipitate-reduced or cryoprecipitate-poor supernatant, is also refrozen for storage at −18°C, and later may be used in the treatment of thrombotic thrombocytopenic purpura (TTP). The volume of a single unit of cryoprecipitate is 10–15 ml, and for dosing purposes is usually pooled, post-thaw, into a single bag of 10 pooled units for transfusion. Some blood centers now supply cryoprecipitate in pre-pooled units from five donors. Cryoprecipitate must be used within 12 months of collection.

## Preparation and Administration:

**ABO/D Compatibility:** Since cryoprecipitate contains negligible amounts of RBCs and minimal isohemagglutinins, anti-A and/or anti-B, choosing units with ABO or D compatibility is not necessary for most adult and pediatric patients. Neonatal transfusion guidelines often recommend the use of ABO-compatible cryoprecipitate.

**Thawing and Pooling:** Cryoprecipitate takes 10–15 minutes to thaw in a 30–37°C water bath. Prior to administration, cryoprecipitate is usually pooled for easier infusion; this takes an additional 10–15 minutes. Therefore, it can take more than 30 minutes to prepare pooled cryoprecipitate. Some blood centers and transfusion services have started to manufacture and store pre-pooled cryoprecipitate, which eases the process of thawing and administration, especially in times of emergency such as trauma and liver transplantation. After thawing, the cryoprecipitate is maintained between 20–24°C.

**Expiration:** Once thawed, a single unit of cryoprecipitate expires in 6 hours; pooled units or open system units expire in 4 hours.

**Quality Assurance:** Per AABB *Standards*, each unit of cryoprecipitate must contain a minimum of 150 mg of fibrinogen (it usually contains 250–350 mg) and a minimum of 80 IU of Factor VIII (it usually contains 80–120 IU). In addition, cryoprecipitate contains 30–60 mg of fibronectin, 40–60 IU of Factor XIII, and ~80 IU of VWF.

## Dose:

**Fibrinogen Deficiency States:** One unit of cryoprecipitate will increase the fibrinogen concentration by approximately 50 mg/dl per 10 kg of body weight. The dose of cryoprecipitate for fibrinogen replacement can be calculated by first determining the mg of fibrinogen required (desired fibrinogen level (mg/dl) − initial fibrinogen level (mg/dl)) × plasma volume (ml)/100 ml/dl. The number of units of cryoprecipitate is the mg of fibrinogen required divided by 250 mg of fibrinogen/unit. While the above calculation is the most precise, most clinicians will order a single dose of a 10-unit pool of cryoprecipitate (an equivalent of one blood volume's worth), and re-measure the fibrinogen level.

**Factor XIII Deficiency States:** One unit of cryoprecipitate for every 10–20 kg given every 3–4 weeks prophylactically, and higher doses more frequently in the setting of active bleeding, is a common strategy, with laboratory measurements obtained as indicated.

**Adverse Events:** Cryoprecipitate is a human plasma derived product, and currently there are no available virally inactivated products that replace cryoprecipitate or its fibrinogen component. Adverse reactions include fever, chills and allergic reactions. Large volumes of ABO-incompatible cryoprecipitate may cause a positive direct antiglobulin test and have, rarely, been reported to cause mild hemolytic transfusion reactions.

**International Standards:** In the UK, cryoprecipitate is manufactured from a single unit of FFP by rapid freezing to less than −30°C then thawing slowly to 4°C. Current guidelines require that 75% of the units of cryoprecipitate include at least 140 mg of fibrinogen and 70 IU/ml of Factor VIII with a 24-month storage period.

## Recommended Reading

Hedges SJ, Dehoney SB, Hooper JS *et al.* (2007). Evidence-based treatment recommendations for uremic bleeding. *Natl Clin Pract Nephrol* **3**, 138–153.

Stinger HK, Spinella PC, Perkins JG *et al.* (2008). The ratio of fibrinogen to red cells transfused affects survival in casualties receiving massive transfusions at an army combat support hospital. *J Trauma* **64**, S79–85.

# CHAPTER 32

# Granulocyte Products

Christopher D. Hillyer, MD and Lawrence B. Fialkow, DO

**Description:** Granulocytes are the immune system's main cellular defense against bacterial and fungal infections. Transfusion of granulocytes is considered a therapeutic modality for severe bacterial and fungal infections in patients with prolonged neutropenia and with functional neutrophil disorders. Good theoretical and experimental evidence demonstrating granulocyte transfusion efficacy exists in preventing and treating severe infection. Clinical evidence has been more difficult to interpret, with efficacy equivocal in many studies, and further trials hindered by limitations in collecting "adequate doses" of leukocytes from healthy donors. However, the recent development and use of granulocyte-colony stimulating factor (G-CSF) to stimulate normal donors has generated renewed interest in granulocyte transfusions. Preliminary clinical evidence, when granulocyte dose per patient body weight is optimized, suggests efficacy; however, well-designed randomized clinical trials are necessary to definitively establish granulocyte transfusions as a viable therapeutic modality in the treatment of severe bacterial and fungal infections in patients with functional neutrophil disorders or neutropenia. Such a trial is underway as of this writing.

Granulocyte products typically:

- contain $> 1 \times 10^{10}$ granulocytes, 10–30 ml of RBCs and $1$–$6 \times 10^{10}$ platelets,
- must undergo crossmatch and an emergency release procedure as results of infectious disease testing may be unknown at the time of administration,
- should be irradiated and CMV safe, and
- should be transfused within 24 hours of collection.

**Indications:** Data from remote and recent trials do not allow for a definitive set of indications, or contraindications, for granulocyte transfusion. Data from the 1970–1980s supported efficacy, but patients in those reports did not have access to current antimicrobial (antibiotic and antifungal) medications. In the 1990s, investigators considered that even with more modern antimicrobial medications granulocyte transfusion might be clinically helpful because mobilization strategies using G-CSF allow higher doses of neutrophils to be collected and infused. However, no definitive, randomized trial has been accomplished. Currently (2008) a study (The RING Study [Resolving Infection in Neutropenia with Granulocytes]) is underway through the Transfusion Medicine/Hemostasis Clinical Trials Network funded by the NHLBI, which is investigating the use of G-CSF/dexamethasone mobilized high-dose granulocytes in infected neutropenic patients. Results are not likely to be available until 2010.

**Neonatal Sepsis:** Neonates are at particular risk for bacterial and fungal infection because of their diminished storage pool of neutrophils, which can be rapidly depleted; neonates also have a qualitative defect of neutrophil function. In 1992, Cairo and colleagues performed a clinical trial involving 35 neonates with neutropenia and sepsis randomized to receive either intravenous immunoglobulin (n = 14) or granu-locyte transfusion (n = 21); survival was significantly higher in the granulocyte trans-fusion group (100% survival) versus in the immunoglobulin group (64% survival). Vamvakas and colleagues performed a meta-analysis of seven clinical studies in adults and five clinical trials in neonates with sepsis, and reported on the efficacy of granu-locyte transfusions in the treatment of bacterial sepsis. These investigators found that the most significant factor in all of the studies showing favorable outcomes was the dose of granulocytes transfused.

**Neutrophil Function Defects:** Patients whose neutrophils have defects in adhe-sion and motility generally develop cutaneous abscesses with common pathogens such as *Staphalococus aureus*, or have mucous membrane lesions caused by *Candida albicans*; they can also develop sepsis from these common pathogens or from other gram-positive or gram-negative bacteria. Disorders of phagocytic microbicidal activ-ity such as chronic granulomatous disease (CGD) are also associated with cutaneous abscesses, lymphadenitis, pulmonary infections and sepsis. These patients also tend to have deep-seated abscesses in the liver, lungs and gastrointestinal tract. Granulocyte transfusions in CGD are supported by the observation that a small number of normal neutrophils may be able to compensate for the metabolic defect in CGD neutrophils.

Numerous case reports document the potential benefit of granulocyte transfusions in this patient population. For example, in 2000 Bielorai and colleagues published a case report demonstrating the successful treatment of invasive pulmonary aspergil-losis and osteomyelitis in a 4-year-old boy with X-linked CGD undergoing stem cell transplantation, and a 2001 case report by Watanabe and colleagues described the suc-cessful treatment of a 20-year-old CGD patient with antibiotic-resistant pneumonitis and Aspergillus osteomyelitis undergoing hematopoietic progenitor cell (HPC) trans-plantation. However, the value of granulocyte transfusions in patients with these dis-eases has not been studied in a prospective controlled clinical trial.

**Neutropenic Patients:** In 1995, Strauss reviewed 32 papers that studied the efficacy of granulocyte transfusions in neutropenic patients and found that about 60% of the more than 200 patients with bacterial sepsis showed improvement and recovery from their infection. In 2000, Price and colleagues reported that of 11 patients with bacterial and candidial infections, 8 cleared their infections; however, none of the patients with invasive aspergillus cleared their infections.

In 2005, Stanworth and colleagues reported a meta-analysis performed on eight parallel randomized controlled trials which included 310 patient episodes. Mortality, extracted from six trials using standard dosing, favored granulocyte transfusion (RR = 0.64), while four studies that transfused $>1 \times 1^{10}$ granulocytes indicated a significant summary (RR = 0.37). Still, the authors suggested that there was incon-clusive evidence to definitively support a positive role for granulocyte transfusion therapy.

Also in 2005, Mousset and colleagues studied the effects of granulocyte transfu-sions on acute and recurrent infections in patients undergoing HPC transplantation and reported that granulocyte transfusion achieved control of acute life-threatening infections in 82% of patients (36/44). In 2006, Oza and colleagues studied 151 HPC transplant patients; compared with the control group, patients who received granulocytes demonstrated a significant reduction in the percentage with fever (64.2% versus 82.7%, $p = 0.03$) and days on antibiotics (9 versus 11, $p = 0.03$). However, no difference was noted in length of stay or 100-day survival. Also in 2006, Grigull and colleagues published a retrospective review of 32 neutropenic children with sepsis who received granulocyte transfusions. Survival rates of 73% (bacterial infection), 57% (fungal), 25% (viral) and 60% (fever only) were observed. Granulocyte transfusions were felt to be safe, efficacious treatment for neutropenic pediatric patients with sepsis.

## Processing and Storage:

**Collection:** There are two main methods of collecting granulocytes from donors: filtration and continuous flow centrifugation leukapheresis. The filtration technique employs nylon fibers and procures a large yield of cells; however, the fibers of the filter damage and activate the leukocytes, promoting granule release and activation of the complement system. Therefore, most blood collection centers employ the continuous flow centrifugation separation apheresis method for collecting granulocytes.

Apheresis employs a commercial blood cell separator which selectively removes the leukocyte fraction of the blood and returns most of the RBC and platelet fractions to the donor. Final granulocyte yield is dependent on the total volume of blood processed, as well as the white blood cell count of the donor at the time of donation. The donor's erythrocyte sedimentation rate may be used to predict collection efficiency.

Collection efficiency may be enhanced by the addition of pentastarch, hetastarch and/or hydroxyethyl starch (HES), all of which increase density gradients in the apheresis device, thus improving collection by facilitating separation of the leukocytes from RBCs. Donor side-effects, such as allergic reactions, edema and weight gain, may occur with these agents.

Donor stimulation with G-CSF and/or corticosteriod increases the granulocyte product yield as stated above. G-CSF and/or dexamethasone are administered 8–16 hours prior to product collection. Side-effects include bone-pain, headache, insomnia, fluid accumulation and weight gain.

**Storage:** Granulocyte storage should be at 20–24°C without agitation. Granulocyte products have been stored in this manner for up to 8 hours without any reduction of their chemotactic or adhesion properties. However, at between 8 and 24 hours of storage the ability of the transfused granulocytes to migrate to areas of inflammation may be decreased. More recent data by Hubel and colleagues in healthy volunteer donors that were treated with G-CSF and dexamethasone 12 hours prior to collection demonstrated that L-selectin expression declined, but CD11b, CD18, CD14, CD16, CD32 and CD64 expression were maintained. Bactericidal and fungicidal activities were sustained during storage for 48 hours at 10°C. Additional studies have shown prolonged neutrophil function after G-CSF donor stimulation.

## Patient Selection, Dose Preparation, Administration and Toxicities:

**Patient Selection:** Typically, candidates for granulocyte transfusion have neutropenia with an ANC $< 500/\mu l$, and fungemia, bacteremia, and/or proven or probable invasive tissue fungal or bacterial infection or proven neutrophil dysfunction.

**Dose:** It is generally recommended that the dose of granulocytes should be $\geq 4 \times 10^{10}$, once daily, either until significant clinical improvement is noted or the endogenous ANC is $>500/\mu l$, or 42 days, whichever comes first. Most often, 5 days of granulocyte products are ordered from the blood center so as to have donors recruited, stimulated and ready to donate; at 5 days the clinical picture is reassessed, and additional days of granulocytes are ordered as needed.

**Preparation:** Most authorities recommend that the granulocyte product be irradiated in order to prevent TA-GVHD, and some authorities prefer granulocyte collection from an HLA-matched donor, although this is not typically feasible when the patient is receiving granulocytes on a daily basis. CMV-seronegative granulocyte donors are required for CMV-seronegative recipients (since granulocyte products cannot be leukoreduced). Because granulocytes often have a greater than 0.5% hematocrit, ABO compatibility is required and a crossmatch must be performed.

**Administration:** Infusion of a granulocyte product should occur within 24 hours of the beginning of the collection. Given this short shelf-life, the results of most infectious disease testing for granulocyte donors may not be known at the time of product release for transfusion. Thus, an emergency release procedure is usually required by the blood center or collecting facility. Granulocytes should not be transfused through leukoreduction or microaggregate filters. One study by Wright and colleagues demonstrated increased pulmonary toxicity when granulocytes were co-administered with amphotericin B; thus generally this co-administration is avoided when possible.

**Toxicities:** Fever and chills are relatively common symptoms that occur during granulocyte transfusions. More severe side-effects, such as respiratory distress and hypotension, are rare events, and have been associated with the concomitant administration of amphotericin B and granulocytes. As with other blood products, granulocyte transfusion has the potential risk of transmitting infectious agents, such as hepatitis viruses, HIV and CMV.

**Quality Assurance:** Granulocyte products are not licensed by the Food and Drug administration (FDA). AABB *Standards* requires that at least 75% of products contain at least $1 \times 10^{10}$ granulocytes.

**International Issues:** The UK National Blood Service has guidelines for granulocyte transfusion which include a method for preparing granulocytes by pooling buffy coats for adults or children. Each buffy coat is 50 ml with a 45% hematocrit, and contains $1–2 \times 10^9$ WBCs and $9 \times 10^{10}$ platelets. A pool of 10 is usually used for adults; children $<50$ kg are dosed at 10 ml/kg. The guidelines state that "granulocyte transfusions can be used as supportive therapy in patients with life-threatening neutropenia caused by bone marrow failure or neutrophil dysfunction" and the benefits should outweigh the risks.

In a 2003 publication by the French Health Products Agency, the minimum number for granulocytes in products was defined as $2 \times 10^{10}$ with a maximum regulatory storage period of 12 hours. The indications for transfusion were noted to be "rare," and granulocyte products were not validated for use in neonatal patients. It is likely that a positive result from the RING trial described above will affect practice guidelines for granulocyte transfusion worldwide.

## Recommended Reading

Price TH. (2007). Granulocyte transfusion: current status. *Semin Hematol* **44**, 15–23.

Schiffer CA. (2006). Granulocyte transfusion therapy 2006: the comeback kid? *Medical Mycol* **44**, S383–386.

Stanworth SJ, Massey E, Hyde C *et al.* (2005). Granulocyte transfusions for treating infections in patients with neutropenia or neutrophil dysfunction. *Cochrane Database Syst Rev* (3):CD005339.

# CHAPTER 33

# Albumin and Related Products

Anne M. Winkler, MD and Beth H. Shaz, MD

Albumin, the most abundant protein in human plasma (50–60% of the total plasma protein content), accounts for 80–85% of the osmotic pressure of plasma and therefore maintains and regulates plasma volume. Albumin also acts as a carrier for other physiologic molecules and administered drugs. In the US, albumin is available in 5% and 25% solutions, with albumin comprising 96% of the total protein content. Albumin infusions draw fluids from the extravascular space into the intravascular space, and thus are used primarily to maintain intravascular volume and increase osmotic pressure; 25% albumin solution will draw 3.5 times the amount of volume administered into the intravascular space.

**Indications:** Albumin was initially used as a treatment for shock during World War II, but its use has expanded to the treatment of various conditions such as hypoalbuminemia, malnutrition, hypotension and fluid replacement, though many of these uses are not supported by data from randomized clinical trials. As the use of albumin expanded, its cost increased and periodic shortages in supply were observed. This, in part, led the University Health System Consortium to establish guidelines for albumin administration in 2000; also, the Cochrane Collaboration has published systematic reviews of, and recommendations for, albumin use in specific clinical circumstances.

Importantly, the advantages and disadvantages of albumin infusion versus the use of non-protein colloid (such as dextran or hetastarch) or crystalloid solutions have not been well addressed in appropriately designed studies. Thus, albumin administration should be based on an individual patient's clinical status. Clinical situations where albumin is commonly administered are described below.

**Therapeutic Plasma Exchange:** Albumin is the primary replacement fluid used in therapeutic plasma exchange (TPE), except in clinical disorders which require replacement with specific factors present in plasma. For example, patients with thrombotic thrombocytopenic purpura (TTP) require plasma products (i.e. fresh frozen plasma [FFP] and cryoprecipitate reduced plasma) as the replacement fluid to restore levels of the deficient metalloprotease (ADAMTS13) (see Chapter 93).

The use of albumin alone or in combination with another replacement fluid such as normal saline, rather than plasma, reduces adverse events such as viral transmission, allergic reactions and transfusion-related acute lung injury (TRALI), and is considered by many to be the most appropriate and indicated use of albumin solutions; however, some cautions apply.

Specifically, in patients receiving angiotensin converting enzyme (ACE) inhibitor therapy and in whom TPE is accomplished, albumin use can cause hypotension, bradycardia and flushing. This reaction is due to the patient's inability to metabolize (secondary to ACE inhibition) the bradykinin that is contained in albumin preparations, and that produced during TPE. The use of FFP or colloid starches instead of albumin as the TPE replacement fluid may be considered in these patients. Another option is to use albumin, but to monitor the patient more closely for hypotension, substituting FFP or colloid starches if an untoward reaction does occur. A third possibility is to delay commencement of TPE if possible. Slow infusion rates are generally recommended if albumin solutions are chosen in patients on ACE therapy.

Ovarian Hyperstimulation Syndrome: Ovarian hyperstimulation syndrome (OHSS) is usually a result of iatrogenic administration of human chorionic gonadotrophin (hCG) to induce ovulation. OHSS is typified by enlarged ovaries which release vascular endothelial growth factor that can result in increased capillary permeability. This in turn leads to a fluid shift out of the intravascular compartment to the abdominal/pleural spaces, resulting in ascites and hypovolemia. In the most severe form, the patient can develop tense ascites, oliguria, dyspnea, hemodynamic instability and thromboembolism. Treatment includes fluid restriction, analgesics, and close monitoring; occasionally hospitalization may be necessary.

Mild OHSS occurs in approximately one-third and moderate to severe in approximately 5% of women receiving exogenous hCG. Increased risks of OHSS include young age, low body weight, polycystic ovarian syndrome, high-dose hCG, high or rapid rise in estradiol level, and previous history of OHSS. In addition, the risk is proportional to the number of developing follicles and number of oocytes retrieved. Moderate to severe OHSS can be mitigated by closely monitoring women during treatment and subsequently withholding or reducing hCG administration when there is a large number of intermediate-sized developing follicles present, or when estradiol levels are elevated.

The Cochrane Collaboration systematically reviewed five randomized clinical trials of albumin administration in OHSS, and concluded that there is a decreased incidence and severity of OHSS when albumin is administered during oocyte retrieval in high-risk women. In contrast, Bellver and colleagues published a large randomized trial which demonstrated no difference in moderate to severe OHSS when 40 g of albumin was administered after the retrieval of 20 or more oocytes. Therefore, current data are inconclusive regarding the use of albumin in the prevention of OHSS.

Cirrhosis with Spontaneous Bacterial Peritonitis: Spontaneous bacterial peritonitis is complicated by renal failure in ~30% of patients. The pathophysiology is thought to be secondary to an increase in intraperitoneal nitric oxide leading to systemic vasodilation and effective hypovolemia, which results in renal vasoconstriction and subsequent renal failure. In a single randomized clinical trial, albumin infusion with antibiotics decreased the risk of renal failure and improved mortality in cirrhotic patients with spontaneous bacterial peritonitis, when compared to administration of antibiotics alone. In this trial, 1.5 g/kg body weight of albumin was administered at the time of detection and then 1.0 g/kg was administered at day 3.

Large-volume Therapeutic Paracentesis: Albumin is often used in cirrhotic patients undergoing large-volume paracenteses (>5 liters) for diuretic-refractory ascites in order to mitigate renal and circulatory system dysfunction. A randomized control trial compared the use of diuretics alone versus diuretics plus albumin in this patient population, and demonstrated a decrease in hospital stay by 3 days. In addition, patients who received albumin in this trial had a lower probability of developing ascites at 12, 24 and 36 months, and a lower probability of readmission. While some studies have shown no survival difference upon comparison of crystalloid (or other non-protein colloids) administration and albumin administration (typically at a dose of 5–10 g of albumin per liter of ascites removed), albumin administration remains a not uncommon practice after large-volume paracentesis.

**Nephrotic Syndrome:** Nephrotic syndrome is secondary to increased permeability of the glomerular capillary basement membranes leading to increased urinary protein excretion, which results in hypoalbuminemia, edema, renal failure and hyperlipidemia. Treatment is focused on mitigating the underlying cause of the nephrotic syndrome, on diuretic therapy and on a sodium-restricted diet. Albumin administration (replacement) combined with diuretics has been used with the intent of increasing vascular pressure at the glomerulus, and thus to increase diuresis. One earlier study did reveal some increased natriuresis; however, several more recent studies have shown no benefit to this combination, or worse outcomes including hypertension, respiratory distress and electrolyte abnormalities. For this reason, albumin replacement in nephrotic patients is currently considered a second-line therapy for refractory cases when standard treatment with diuretics and diet restriction has failed.

**Hypoalbuminemia:** Hypoalbuminemia occurs secondary to decreased production (such as in liver disease) or increased loss (such as nephrotic syndrome) of albumin, which leads to intravascular volume depletion and peripheral or pulmonary edema. A recent small randomized trial demonstrated a decrease in organ dysfunction in hypoalbuminemic critically ill patients with the administration of albumin, compared to no albumin administration. Albumin infusion for this indication is considered to be controversial.

**Contraindications:** Albumin may be contraindicated in any disease state that would be exacerbated by volume expansion, including (but not limited to) severe anemia, congestive heart failure and pulmonary edema. In addition, albumin is contraindicated in patients who have experienced previous anaphylactic reactions after receiving it.

**Adverse Effects:** Adverse effects of albumin administration are rare; these include changes in vital signs (heart rate, blood pressure and respiration), nausea, fever/chills, and allergic reactions. Furthermore, since negatively-charged albumin binds calcium, administration can lead to hypocalcemia and related complications. Albumin solutions may also contain trace amounts of aluminum, which can cause toxicity in infants and in patients with chronic renal failure. Because albumin acts to increase osmotic pressure, rapid infusion can lead to significant shifts in intravascular volume, and resultant circulatory overload (including pulmonary edema) is possible. These shifts can also cause a dilutional anemia and electrolyte imbalances.

**Cost and Usage:** The choice of replacement fluid has considerable cost implications. Albumin is over 30 times more expensive than crystalloids, and is often used in situations where randomized control trials have not shown a difference in patient mortality. The cost incurred due to inappropriate albumin use is substantial. In a retrospective US hospital audit of 15 academic health centers, over $200,000 was spent on albumin and non-protein colloid solutions at all participating institutions in less than 1 month. Of that amount, only $49,702 (24%) was spent on "appropriate indications"; the remainder (76%) was spent on either "inappropriate" or "unevaluated" indications. All indications were deemed "appropriate" or "inappropriate" by retrospective evaluation of guideline-based algorithms for various clinical situations. Another study has shown albumin over-usage rates of 57.8% in adults and 52.2% in pediatrics. In this study, the most common inappropriate uses were for intradialytic blood pressure support and hypoalbuminemia. Similar trends have also been reported in Europe, which has initiated the development of guidelines for colloid administration.

**Manufacturing:** Although recombinant albumin is presently under study, current preparations of albumin solutions are purified from human sources (manufactured from either whole blood or plasmapheresis donations). The preparations must have 96% of the protein composition consisting of albumin. Additionally, the product contains non-albumin proteins ($< 4\%$), endotoxins, trace metals (aluminum), prekallikrein activator, bradykinin, sodium, potassium and stabilizers (sodium caprylate and/or sodium acetyltryptophanate).

The manufacturing process includes steps to prevent transfusion-transmitted diseases, including heat treatment (60°C for 10 hours) and cold ethanol fractionation. To date, no transmission of HIV or hepatitis C has been reported due to albumin. There has been a single lot of plasma protein fraction (PPF) product that transmitted HBV (see PPF section below); however, this was a single isolated event, and no HBV transmission has been reported for purified albumin. While processing does reduce the risk of contamination, the risk of microbial contamination does still remain. In fact, seven patients developed *Pseudomonas* bacteremia after receiving albumin from the same lot in the 1970s.

The US Food and Drug Administration (FDA) mandates testing on final albumin products, which includes determination of protein concentration, protein composition, pH, sodium concentration, potassium concentration, and visual comparison to controls after heat treatment. Moreover, this information must be provided by the manufacturer on the product label. Each manufacturer must comply with the standards of the US Pharmacopeia (USP), and perform additional testing such as nucleic acid testing (NAT) for HBV, HCV and HIV.

**Storage:** Albumin is stored at room temperature in either glass or specialized plastic containers for up to 2 years. Albumin solutions are inspected for turbidity by the manufacturer, and prior to use as a quality control measure to detect potential bacterial contamination.

**Preparation and Administration:** Preparation and products vary, but albumin is most commonly provided in 5% and 25% concentrations in the US; the former is slightly hyperoncotic to human plasma and therefore may result in a dilutional anemia, while the latter is significantly hyperoncotic and therefore may result in dilutional anemia, pulmonary edema and circulatory overload. The concentration to be used is determined based upon the patient's clinical volume status. Typically, 5% albumin expands the volume equal to the volume of albumin infused; whereas 25% albumin will expand the volume by 3.5 times by drawing fluid into the intravascular space.

Thus, if 25% albumin is used in dehydrated patients, additional fluids are indicated to avoid further exacerbating dehydration of tissues.

It is possible to dilute 25% albumin, but it must be diluted with normal saline only. Dilution with sterile water can lead to hemolysis of patient RBCs due to hypotonicity, and has resulted in death. Smaller doses may also be diluted with 5% dextrose in water (D5W), but large volumes of D5W-diluted albumin may lead to hyponatremia with resultant sequelae, including cerebral edema. In any case, clinical factors surrounding fluid status should guide product choice in the context of the above considerations.

## Albumin Dosing:

**Adult Dosing:** A typical initial adult does is 25 g, which may be repeated in 15–30 minutes, depending on the patient response. In a 48-hour period, a maximum of 250 g of albumin can be infused; however, no standard dose of albumin is applicable to all clinical situations, and clinical parameters must be used to determine appropriateness of response. Infusion rates vary depending on the albumin concentration, to prevent complications of rapid volume expansion; as a guideline, 5% albumin solutions are started at a rate of 1–2 ml/min and increased to a maximum rate of 4 ml/min, while 25% albumin solutions are not infused at rates $> 1$ ml/min. However, infusion rates may be increased in patients with hypoproteinemia.

**Pediatric Dosing:** Dosing in pediatrics is dependent on the clinical indication for administration. A typical dose for hypoproteinemia is 0.5–1.0 g/kg per dose of 25% albumin that can be repeated every 1 to 2 days. For infants and children with hypovolemia, 0.5–1.0 g/kg per dose can be administered up to a maximum of 6 g/kg per day. For neonates with hypovolemia the typical dose is 0.25–0.5 g/kg as a 5% solution, and the 25% solution should be avoided. Lastly, dosing for infants and children with nephrotic syndrome varies from 0.25 to 1.0 g/kg per dose of 25% albumin.

## Other Colloid Solutions:
Alternatives to albumin for plasma expansion include crystalloids (e.g. 0.9% sodium chloride [normal saline], Ringer's lactate), alternate protein colloids (e.g. plasma protein fraction [PPF]) and non-protein colloids (e.g. dextran, geletan and starches). Crystalloids and non-protein colloids have not demonstrated a benefit over albumin, but are less expensive. Non-protein colloids have been associated with side-effects, including coagulopathy, pruritis, and head and back pain.

**Plasma Protein Fraction:** Plasma protein fraction (PPF), an alternative to albumin for plasma expansion, is seldom used because it is associated with more frequent hypotensive and allergic reactions than albumin. Like albumin, it is derived from human plasma and comes as a 5% solution. PPF must contain at least 83% albumin and $< 1\%$ can be gamma globulin. PPF was associated with a single outbreak of hepatitis B secondary to failure of heating in the manufacturing process in 1973. No further documentation of transfusion-transmitted diseases in PPF or albumin has occurred.

**Dextrans:** Dextran is a synthetic colloid consisting of a mixture of glucose polymers derived from the action of *Leuconostoc mesenteroides* on sucrose, and is currently available in 10% dextran 40 (40 kDa) and 6% dextran 70 (70 kDa) formulations. Dextrans have a high water-binding capacity; for example, 1 g of dextran 40 retains

30 ml of water, while 1 g dextran 70 retains 20–25 ml of water. Dextran is mostly eliminated by the kidneys and, as a result, should be avoided in patients with impaired renal function. Moreover, renal dysfunction has been reported after dextran infusion. Other side-effects of dextrans include anaphylactic/anaphylactoid reactions, and significant effects on coagulation. Dextrans can induce an acquired von Willebrand's syndrome by decreasing the activity of von Willebrand factor (VWF) and Factor VIII and enhancing fibrinolysis, resulting in potential bleeding sequelae following dextran administration. Dextrans were historically used to maintain circulation in shock and the setting of reperfusion injury, due to its ability to reduce endothelial cell damage from activated leukocytes. However, the use of dextrans is declining owing to the abovementioned adverse reactions.

**Gelatins:** Gelatins are synthetic colloids composed of polypeptides produced from the degradation of bovine collagen. Gelatins are sterile, pyrogen free, do not contain preservatives and, when stored at room temperature, have a 3-year expiration date. The molecular size varies, and high molecular weight products have greater oncotic effect and increased blood viscosity. However, the increase in blood volume is generally less than the amount administered due to the passage of gelatins into the interstitial space and rapid renal clearance. As a result, repeated infusions are necessary to maintain intravascular volume. Similar to dextrans, gelatins also affect hemostasis by interfering with platelet function. Therefore, due to the inefficiency and potential bleeding complications, gelatins are a less commonly used colloid.

**Hydroxyethylene Starch:** Hydroxyethylene starch (HES) is a class of synthetic colloids derived from amylopectin, a starch obtained from maize or potatoes, which is a similar polysaccharide to glycogen. HES is commercially available in multiple concentrations (3%, 6% and 10%) and a variety of molecular weights; it is important to recognize this because of the effects on intravascular volume expansion and the dose-dependent adverse effects on hemostasis and renal function. In general, HES solutions can retain 20–30 ml of water per gram; however, 50% of HES is degraded and excreted renally within 24 hours of administration. HES is being increasingly used for volume expansion in the critically ill, especially in Europe where the use of albumin is declining.

## Recommended Reading

Aboulghar M, Evers JH, Al-Inany H. (2002). Intra-venous albumin for preventing severe ovarian hyperstimulation syndrome. *Cochrane Database Syst Rev* (2): CD001302.

Alderson P, Bunn F, Li Wan Po A *et al.* (2004). Human albumin solution for resuscitation and volume expansion in critically ill patients. *Cochrane Database Syst Rev* (4): CD001208.

Bellver J, Muñoz EA, Ballesteros A *et al.* (2003). Intravenous albumin does not prevent moderate-severe ovarian hyperstimulation syndrome in high-risk IVF patients: a randomized controlled study. *J Hum Reprod* **18**, 2283–2288.

Bunn F, Trivedi D, Ashraf S. (2008). Colloid solutions for fluid resuscitation. *Cochrane Database Syst Rev* (1): CD001319.

Chalidis B, Kanakaris N, Giannoudis PV. (2007). Safety and efficacy of albumin administration in trauma. *Expert Opin Drug Safety* **6**, 407–415.

Knoll G, Grabowski J, Dervin G, O'Rourke K. (2004). A randomized, controlled trial of albumin versus saline for the treatment of intradialytic hypotension. *J Am Soc Nephrol* **15**, 487–492.

Laffi G, Gentilini P, Romanelli RG, La Villa G. (2003). Is the use of albumin of value in the treatment of ascites in cirrhosis? The case in favour. *Dig Liver Dis* **35**, 660–663.

Matejtschuk P, Dash C, Gascoigne E. (2000). Production of human albumin solution: a continually developing colloid. *Br J Anaesth* **85**, 887–895.

Mathur R, Kailasam C, Jenkins J. (2007). Review of the evidence base of strategies to prevent ovarian hyperstimulation syndrome. *Hum Fertil (Camb)* **10**, 75–85.

Perel P, Roberts I. (2007). Colloids versus crystalloids for fluid resuscitation in critically ill patients. *Cochrane Database Syst Rev* (4): CD000567.

SAFE Study Investiagors. (2004). A comparison of albumin and saline for fluid resuscitation in the intensive care unit. *N Engl J Med.* **350**, 2247–2256.

Sort P, Navasa M, Arroyo V *et al.* (1999). Effect of intravenous albumin on renal impairment and mortality in patients with cirrhosis and spontaneous bacterial peritonitis. *N Engl J Med* **341**(6): 403–409.

Tanzi M, Gardner M, Meggelas M *et al.* (2003). Evaluation of the appropriate use of albumin in adult and pediatric patients. *Am J Health Syst Pharm* **60**, 1330–1335.

Wong F. (2007). Drug insight: the role of albumin in the management of chronic liver disease. *Natl Clin Pract Gastroenterol Hepatol* **2**, 43–51.

# CHAPTER 34

# Human Immunoglobulin Preparations

Beth H. Shaz, MD

Human immune globulin (Ig), albumin, and coagulation factor concentrates are commonly prepared by cold ethanol fractionation (Cohn fractionation) from large pools of whole blood or apheresis derived plasma, and are typically referred to as plasma derivatives (Figure 34.1). Ig preparations are concentrated, purified and sterilized, making the risk of infectious disease transmission from Ig virtually zero. Ig preparations can be made for intramuscular (IM) or intravenous (IV) administration, and the clinical indications for Ig are many and varied. Below, a description, indications, dosing, mechanism of action, preparation and storage, potential adverse events and

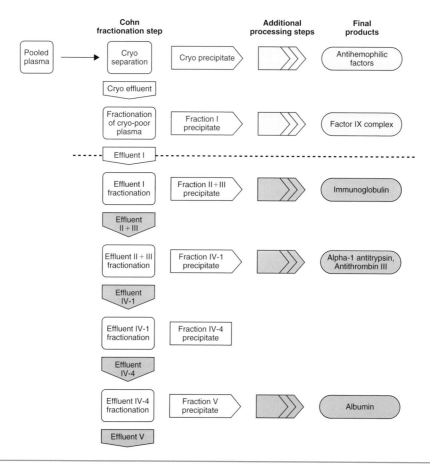

FIGURE 34.1 Cohn plasma fractionation process. Modified from Burdick MD, Pifat DY, Petteway SR, Cai K (2006). Clearance of prions during plasma protein manufacture. *Transfus Med Rev* **20**: 57–62.

international considerations are presented for intravenous immune globulin (IVIG) and hyperimmune globulin preparations available or in common use. RhIg (Rh immune globulin) and other plasma derivatives are discussed in Chapters 35 and 143–149 respectively.

## Intravenous Immune Globulin (IVIG):

**Description:** IVIG is derived from human plasma, and is a highly purified product that consists mostly of IgG and has a half-life of 21–28 days. IVIG is used to replenish IgG in patients with congenital hypogammaglobulinemia. It also has immunomodulatory properties, resulting in an increasing list of both FDA-approved and non-approved indications.

**Mechanism of Action:** For patients with acquired or congenital Ig deficiency (hypo- or agammaglobulinemia), IVIG supplements or replaces the missing antigen-specific humoral component of their immune system. Since IVIG contains a diverse collection of antibody specificities, this protects patients from increased susceptibility to infections by classical elimination of opsonized infectious organisms by antibody-dependent cell-mediated cytotoxicity or by complement activation. These processes are followed by lysis and/or neutralization of soluble infectious proteins by immuno-complex formation and elimination through the reticuloendothelial system.

In the treatment of autoimmune disorders or other diseases associated with antibodies, IVIG results in immunomodulation and may alleviate the symptoms of the disease. The mechanisms of Ig-induced immunomodulation are incompletely understood, but include:

- macrophage Fc receptor blockage by immune complexes formed between the IVIG and native antibodies,
- modulation of complement,
- suppression of antibody production,
- suppression of inflammatory cytokines and chemokines, and/or
- anti-idiotypic regulation of autoreactive B lymphocytes or antibodies.

**Indications and Dose:** There are currently six FDA-approved indications for IVIG, and an expanding list of off-label uses (Table 34.1). The FDA-approved indications and some of the non FDA-approved indications are listed below, an extensive list of all diseases for which IVIG could potentially be used is beyond the scope of this book. There are multiple different products of IVIG, and these products may be reconstituted in different concentrations and with different liquids (sterile water, 5% dextrose, and 0.9% saline), according to the manufacturers' instructions. Not all *products* have been FDA-approved for each of the listed indications, and each disease indication has an individual recommended dose. Many clinicians use the different IVIG products interchangeably.

### FDA-approved Indications:

**Primary Immune Deficiency:** Patients with primary immunodeficiency syndromes have decreased levels of IgG and increased susceptibility to infections. Prophylactic administration of IVIG reduces the number and duration of infections. The typical

TABLE 34.1 Some of the Indicators for IVIG Administration

| FDA-approved Indications | Off-label Indications |
|---|---|
| Primary immune deficiency | Aplastic anemia secondary to parvovirus |
| Secondary immune deficiency | Chronic inflammatory demyelinating polyradiculoneuropathy |
| Idiopathic thrombocytopenic purpura | Dermatomyositis |
| Kawasaki disease | Guillan-Barré syndrome |
| | Hemolytic disease of the fetus and newborn |
| | Hypogammaglobulinemia associated with multiple myeloma |
| | IgM paraproteinemic demyelinating neuropathy |
| | Lambert-Eaton myasthenic syndrome |
| | Multifocal motor neuropathy |
| | Multiple sclerosis |
| | Myasthenia gravis |
| | Neonatal alloimmune thrombocytopenia |
| | Organ transplantation |
| | Posttransfusion purpura |
| | Sepsis and septic shock in adults |
| | Stiff-person syndrome |

maintenance dose for adults is 400–600 mg/kg every month (recommended doses per manufacturer range from 200 to 600 mg/kg every 3–4 weeks).

**Secondary Immune Deficiency:** Patients with secondary immunodeficiency syndromes have acquired disorders of the immune system, which may be caused by hematologic malignancy (e.g. chronic lymphocytic leukemia [CLL]), infection (e.g. Human Immunodeficiency Virus [HIV]) or immunosuppressive therapy (e.g. related to human progenitor cell [HPC] transplantation). IVIG is effective at preventing infection in patients with hypogammaglobinemia associated with CLL, but the cost-effectiveness of this treatment has been debated. The efficacy of IVIG in preventing bacterial infection in patients with HIV infection in the current era of HARRT and prophylactic trimethoprim-sulfamethoxazole is unknown. In the past, IVIG was used in HPC transplant patients to decrease infection and acute graft-versus-host disease, but it has been shown to result in delayed recovery of recipient humoral responses; therefore, the clinical benefit of IVIG use in HPC recipients is unclear. The typical dose of IVIG is 400 mg/kg every 3–4 weeks, but a wide variety of doses has been used in clinical trials assessing its usefulness in secondary immunodeficiency states.

**Idiopathic Thrombocytopenic Purpura:** Idopathic thrombocytopenic purpura (ITP) is a bleeding disorder characterized by immune-mediated platelet destruction and resultant thrombocytopenia. The most effective pharmacologic therapies for acute ITP currently are corticosteroids, IVIG and RhIg. RhIg may only be used in D positive individuals who have not undergone splenectomy. Both RhIg and IVIG have a rapid but temporary response. The typical dose of IVIG is 2000 mg/kg given over 2–5 days (see Chapter 87 ITP).

**Kawasaki Disease:** Kawasaki disease is an acute, self-limited childhood disorder manifested by fever, bilateral conjunctivitis, rash, and cervical lymphadenopathy. Kawasaki disease is associated with systemic vasculatis, which may result in coronary artery aneurysms in 15–25% of untreated children. In order to prevent coronary artery aneurysms, a single infusion of IVIG is given with aspirin within 7 days of illness presentation. The typical dose is 2000 mg/kg given as a single infusion.

*Off-label Indications:*

**Aplastic Anemia Secondary to Parvovirus:** Parvovirus B19 infection can result in severe anemia and reticulocytopenia, especially in immunocompromised individuals or individuals with sickle cell disease or thalassemia, and the use of IVIG may be beneficial in the treatment of these patients (typical dose 1000 mg/kg over 1–2 days).

**Chronic Inflammatory Demyelinating Polyradiculoneuropathy:** Chronic inflammatory demyelinating polyradiculoneuropathy (CIDP) is a chronic disorder resulting in demyelination of peripheral nerves, which leads to weakness and sensory changes. Equivalent outcomes have been observed in the treatment of CIDP with IVIG (reported dose 2000 mg/kg over 2–5 days), plasma exchange or glucocorticosteriods. The decision as to which treatment to use is made on an individual basis, balancing the risks and benefits of each treatment modality.

**Dermatomyositis:** Dermatomyositis is a chronic inflammatory disorder which results in progressive weakness and rash. IVIG (typical dose 2000 gm/kg per month for three months) results in improved muscle strength and neuromuscular symptoms.

**Guillain-Barré Syndrome (Acute Inflammatory Demyelinating Polyneuropathy):** Guillain-Barré syndrome is an acute demyelinating peripheral neuropathy affecting both motor and sensory nerves. IVIG (typical dose 400 mg/kg per day for 5 days) is likely equivalent to TPE in improving disability and shortening the time to improvement.

**Hemolytic Disease of the Fetus and Newborn:** Hemolytic disease of the fetus and newborn (HDFN) results from maternal RBC alloantibodies binding to fetal/neonatal RBCS and may result in hemolysis. IVIG has been used at a dose of 1000 mg/kg to treat newborns with HDFN. In addition, maternal IVIG infusion (with or without therapeutic plasma exchange) has been used in severe cases of HDFN where treatment must occur prior to the ability to perform intrauterine transfusion.

**Hypogammaglobulinemia Associated with Multiple Myeloma:** Multiple myeloma is a monoclonal B-cell (plasma cell) disorder with clinical symptoms arising as a result of plasma cell infiltration of the bone marrow, monoclonal Ig in the blood and urine, and immunosuppression. IVIG has been shown to be beneficial in preventing serious infections in plateau-phase multiple myeloma, where the patients have hypogammaglobinemia, at doses of 300–400 mg/kg monthly.

**IgM Paraproteinemic Demyelinating Neuropathy:** Paraproteinemic demyelinating neuropathy is a chronic disorder resulting in decreased sensory and motor function, similar to CIDP, in association with monoclonal immunoglobulins. IVIG (total dose of 2000 mg/kg) may improve overall disability.

**Inclusion Body Myositis:** Inclusion body myositis is an inflammatory myopathy resulting in chronic muscular weakness. IVIG (dose 2000 mg/kg per month for 3 months) appears to result in better strength scores and duration of improved swallowing, but is equivalent to treatment with glucocorticosteroids in small clinical studies.

**Lambert-Eaton Myasthenic Syndrome:** Lambert-Eaton myasthenic syndrome results from antibodies to the neuromuscular junction, leading to autonomic dysfunction. IVIG (dose 1000 mg/kg for 2 days) improves strength and decreases serum calcium channel antibody titers.

**Multifocal Motor Neuropathy:** Multifocal motor neuropathy is a chronic progressive disorder resulting in primarily hand weakness. IVIG improved strength at a dose of 400 mg/kg per day for 5 days.

**Multiple Sclerosis:** Multiple sclerosis is a chronic progressive or relapsing and remitting disorder characterized by brain white-matter demyelination. There are four randomized, double-blind clinical trials in patients with relapsing-remitting multiple sclerosis, using a wide range of IVIG doses, that have demonstrated the success of IVIG in reducing the number of exacerbations and the disability in patients with this form of multiple sclerosis.

**Myasthenia Gravis:** Myasthenia gravis is a chronic neurologic autoimmune disorder, characterized by weakness and fatigue upon repetitive use which improves with rest. IVIG used for acute exacerbation of myasthenia gravis at a dose of 1000–2000 mg/kg appears comparable to plasma exchange.

**Neonatal Alloimmune Thrombocytopenia:** Neonatal alloimmune thrombocytopenia (NAIT) results from maternal platelet alloantibodies against the fetal/neonatal platelets resulting in neonatal/fetal thrombocytopenia. The treatment of NAIT during pregnancy is maternal administration of 1000 mg/kg IVIG weekly, beginning at 20–24 weeks of gestational age, with or without the use of glucocorticosteroids. Moreover, the neonate may need to receive IVIG and platelet transfusions after delivery to increase fetal platelet counts to prevent intracerebral hemorrhage, at a dose of 1000 mg/kg (see Chapter 81).

**Organ Transplantation:** HLA antibodies result in antibody-mediated rejection of solid organ transplants, especially kidney and cardiac transplants. In order for some patients who have HLA antibodies to undergo transplantation, these antibodies must be removed or decreased with immunosuppressive agents, IVIG and/or plasma exchange. In addition, IVIG with or without plasma exchange is used in the treatment of antibody-mediated rejection.

**Posttransfusion Purpura:** Posttransfusion purpura (PTP) is a rare complication of transfusion resulting in acute, profound thrombocytopenia, secondary to platelet antibodies which destroy both transfused and autologous platelets. PTP treatment with IVIG at a dose of 2000 gm/kg over 2–5 days results in a rapid increase in platelet count (see Chapter 61).

**Sepsis and Septic Shock in Adults:** The use of IVIG in adult patients with bacterial sepsis or septic shock is potentially beneficial, but further studies are needed.

**Stiff-person Syndrome:** Stiff-person syndrome is a neurologic disorder associated with truncal and limb rigidity and heightened sensitivity. IVIG (dose 2000 gm/kg per month for 3 months) results in a significant decrease in stiffness and heightened sensitivity.

## Preparation and Administration:

*Production:* IVIG production is regulated by the IUIS/WHO (International Union of Immunological Societies/World Health Organization), which require the following:

- the source material must be plasma obtained from a minimum pool of 10,000 donors;
- the product must be free of prekallikrein activator, kinins, plasmin, preservatives, or other potentially harmful contaminants;
- IgA content and IgG aggregate levels need to be as low as possible;
- the product must contain at least 90% intact IgG;
- the IgG should maintain opsonin activity, complement binding, and other biological activities;
- the IgG subclasses should be present in similar proportions to those in normal pooled plasma;
- antibody levels against at least two species of bacteria (or toxins) and two viruses should be determined;
- the product must demonstrate at least 0.1 IU of hepatitis B antibody per mL, and a hepatitis A radioimmunoassay titer of at least 1 : 1000; and
- the manufacturer should specify the contents of the final product, including the diluent and other additives, and any chemical modification of the IgG.

*Plasma Collection:* IVIG is derived from plasma collected either by whole blood donations as recovered plasma (20%), or by apheresis as source plasma (80%).

*Processing:* Manufacturers differ in the steps used to fractionate, purify and stabilize Ig, in the methods used to inactivated and/or remove viruses, and in the formulation of the final product. Cold ethanol is commonly used for fractionation, and the product is then purified by filtration (nanofiltration, ultrafiltration, depth filtration), chromatography and/or precipitation. Viral inactivation is achieved by heat (pasteurization) and chemical/enzymatic methods (incubation at low pH with or without enzymatic treatment, treatment with methylene blue, psoralens, riboflavin, caprylate, or solvent detergents). To limit IgG aggregates, ion exchange chromatography, treatment with pepsin at a pH of 4, polyethylene glycol and/or stabilizers (such as sucrose, glucose, glycine, maltose, sorbitol and/or albumin) are used.

*Selecting a Product:* Products are available in liquid or lyophilized forms. The lyophilized forms can be reconstituted to a variety of different concentrations and osmolarities, depending on the amount of liquid used and the choice of liquid (sterile water, 5% dextrose, or 0.9% saline) conditional on the manufacturers' instructions. No other medications or fluids should be mixed with IVIG. Different IVIG products have differences in concentration of additives, IgA content, osmolarity, osmolality and pH (Table 34.2). The antibody titers and biologic function of different products

**TABLE 34.2 Variables to Consider When Choosing an IVIG Product**

| Variable | Clinical significance |
| --- | --- |
| Sucrose | The FDA issued a warning letter stating that the administration of sucrose-containing products may increase the risk of development of acute renal failure. Patients at increased risk include those with any degree of pre-existing renal insufficiency, diabetes mellitus, age >65 years, volume depletion, sepsis, paraproteinemia, and concomitant nephrotoxic drugs. |
| Sorbitol | Patients with hereditary fructose intolerance who receive sorbitol- or fructose-containing products may develop irreversible multi-organ failure. |
| Glucose | Glucose-containing products should be used with caution in patients with diabetes or renal dysfunction and the elderly. |
| Glycine | Glycine-containing products are associated with increased frequency of vasomotor events. |
| Maltose | Some glucose monitors may interpret maltose as glucose and give falsely elevated results, which may result in iatrogenic insulin overdose. |
| Sodium | High sodium products should be cautiously given to patients with heart failure or renal dysfunction, to neonates, young children, the elderly, and those at risk for thromboembolism. |
| pH | Low-pH products should be administered cautiously to those with compromised acid-base compensatory mechanisms, such as neonates or those with renal dysfunction. |
| Osmolality and osmolarity | The osmolality and osmolarity should be considered in patients with heart disease or renal dysfunction, young children, elderly, and those at risk for thromboembolism. |
| Volume | The volume should be considered when administering IVIG to volume-sensitive patients, including patients with renal dysfunction, heart disease, elderly, neonates, and small children. |

Modified from Hillyer CD, Silberstein LE, Ness PM *et al.* (eds). (2007). *Blood Banking and Transfusion Medicine: Basic Principles & Practice*, 2nd edition. San Diego, CA: Elsevier.

are not typically tested, even though there may be a potentially clinically significant difference between products. Most hospital pharmacies only stock a limited number of IVIG products.

*Administration:* The FDA has approved IM and IV Ig products; subcutaneous administration is an off-label use that may be helpful when venous access is difficult, or for home self-administration. The rate of infusion for those not previously exposed to IVIG is low (0.01 mL/kg per minute), but can then be gradually increased to 0.03–0.06 mL/kg per minute, and to a maximum rate of 0.06–0.10 mL/kg per minute, depending on the product (see manufacturers' instructions). Infusion in patients at risk for renal dysfunction or thrombosis should also be slow. Vital signs should be monitored every 15 minutes for the first hour, and then every 30–60 minutes. The most common adverse reactions, such as headache, nausea, vomiting, chills, fever and malaise, seem to be related to the rate and dose of infusion.

*Adverse events:* Adverse effects occur in approximately 2–10% of infusions, and include erythema, phlebitis, eczema, fever, chills, myalgias, malaise, flushing, rash, diaphoresis, pruritus, bronchospasm, chest pain, back pain, dizziness, blood pressure changes, nausea, vomiting, and headache. Reactions are the result of either allergy or the rate/dose of the infusion. Those that are dose-related may be ameliorated by decreasing the rate or dose of infusion. In addition, adverse reactions differ among different preparations, such that patients may tolerate one product better than another.

*Anaphylactic Reactions:* Individuals who are IgA deficient and have anti-IgA may have anaphylactic reactions to IVIG products. There are products with low IgA content ($\leq 2.2\,\mu g/mL$) for use in IgA deficient individuals.

*Aseptic Meningitis:* This is characterized by severe headache, nuchal rigidity, drowsiness, fever, photophobia, painful eye movements, nausea and vomiting, beginning 6–48 hours after infusion. Patients with a history of migraine and who have received high-dose Ig treatment appear more susceptible. The cerebrospinal fluid demonstrates pleocytosis and elevated protein. The symptoms resolve in hours to days.

*Renal Failure:* The FDA issued a warning letter in 1998 regarding the association between acute renal failure and the administration of IVIG (see Table 34.2).

*Thromboembolic Events:* IVIG has been associated with deep venous thrombosis, myocardial infarction, cerebrovascular accidents, transverse sinus thrombosis and pulmonary embolism, possibly related to increase in blood viscosity after IVIG administration. Patients who receive large doses rapidly, as well as elderly, overweight or immobilized patients and patients with cardiovascular disease, are thought to be at highest risk for this complication.

*Passively Acquired Antibodies:* Patients who receive IVIG may passively acquire a variety of antibodies, including anti-HBc and anti-CMV, and therefore this may result in false-positive serologic testing. The testing can either be repeated at a later time interval, or non-serologic methods can be used to determine the presence of the infectious agent. Blood group antibodies may also be passively acquired, particularly anti-A and/or anti-B, resulting in a positive direct antiglobulin test or, rarely, significant hemolysis.

### TABLE 34.3 Hyper-Ig Preparations, Indications and Dose

| Hyper-Ig | Indication | Dose |
|---|---|---|
| Botulism Ig | Botulism in infants only | 1 mL/kg (50 mg/kg) IV |
| Cytomegalovirus (CMV) Ig | No clear recommendations | |
| Hepatitis A Ig | Unvaccinated individuals at high risk of infection | 0.02 mL/kg IM prior to or within 2 weeks of exposure |
| Hepatitis B Ig | Unvaccinated individuals at high risk of infection | 0.6 mL/kg IM within 24 hours |
| | Infants born to hepatitis B positive mothers | 0.5 mL IM |
| | Prevention of recurrent hepatitis B in liver transplantation recipients | 20,000 IU IV concurrent with graf-ting of the transplanted liver, then 20,000 IU/day IV on days 1–7 post-operatively, then 20,000 IU IV every 2 weeks starting on day 14 post-operatively, then 20,000 IU IV every month starting on month 4 postoperatively; target serum anti-HBs concentration >500 IU/l; if the serum anti-HBs concentration is <500 IU/l within the first week of transplantation, increase the dose to 10,000 IU IV every 6 hours until the target is reached |
| Rabies Ig | Rabies exposure | 20 IU/kg IM up to 7 days after exposure |
| Respiratory syncytial virus (RSV) Ig | Prophylaxis for high-risk infants (children under 24 months of age with bronchopulmonary dysplasia, chronic lung disease or history of premature birth) | 750 mg/kg monthly during the RSV season; alternate treatment is palivizumab, which is a monoclonal antibody to RSV. |
| Tetanus Ig | Prophylaxis of tetanus infection in patients with traumatic injuries, other than a minor, clean wound, who have not been immunized within the last 5 years | 250 U IM or for children under 7 years of age either 4 U/kg or 25 U IM; with concomitant vaccination |
| | Tetanus infection | 3000–6000 U IM within 24 hours of infection |
| Vaccinia Ig | Serious adverse reactions to smallpox vaccine | Obtained from the Center for Disease Control or the US Department of Defense |
| Varicella-zoster virus Ig | Unvaccinated immunocompromised individuals or neonates who are exposed to varicella | 125 U/10 kg IM (minimum dose of 125 U and maximum dose of 625 U within 96 hours of exposure) |

*Hemolytic Transfusion Reactions:* A recent review demonstrated a 1.6% incidence of decreased hemoglobin after IVIG administration, especially in non-group O women in an inflammatory state receiving large doses of IVIG. The proposed mechanism is from passively transfused anti-A and/or anti-B antibodies within the product.

*Transfusion Related Acute Lung Injury:* There has been a single case report of possible transfusion related acute lung injury (TRALI) occurring after IVIG administration.

*Infectious Disease Transmission:* The risk of infectious disease transmission is near zero, as above, due to donor interview and testing, fractionation, and additional safety steps (such as ultrafiltration) resulting in viral reduction.

**Hyperimmune Globulin Products:** Hyperimmune globulin (Hyper-Ig) products are manufactured from donors with high titers of the Ig specificity of interest. The high titers of the donors can be achieved by natural immunity, prophylactic immunizations, or through target immunizations. Hyper-Ig products should contain at least five-fold increased titers compared to standard preparations of IVIG, according to IUIS/WHO (Table 34.3).

**Pathophysiology:** Hyper-Ig products transfer a specific passive immunity for a variety of conditions. Selected opsonized cellular targets are lysed through complement activation or by antibody-dependent cell-mediated cytotoxicity, while soluble target antigens are bound in immunocomplexes and eliminated by the reticuloendothelial system.

**Preparation and Administration:** Hyper-Ig products are manufactured in a similar manner to IVIG.

**Adverse Events:** Local adverse reactions such as tenderness, pain, soreness, or stiffness of the muscle may occur at the site of intramuscular injection. In addition to the adverse events associated with IM administration, adverse events similar to those with IVIG are associated with hyper-Ig.

## Recommended Reading

Alejandria MM, Lansang MA, Dans LF, Mantaring JBV. (2002). Intravenous immunoglobulin for treating sepsis and septic shock. *Cochrane Database Syst Rev* (1): CD001090.

Burdick MD, Pifat DY, Petteway SR, Cai K. (2006). Clearance of prions during plasma protein manufacture. *Transfus Med Rev* **20**: 57–62.

Chevalier I, L'Italien C, David M, Lacroix J. (2003). Steroids versus immunoglobulin for pediatric acute idiopathic thrombocytopenic purpura. *Cochrane Database Syst Rev* (1): CD004140.

Dalakas MC. (2004). Intravenous immunoglobulin in autoimmune neuromuscular diseases. *J Am Med Assoc* **291**, 2367–2375.

Darabi K, Abdel-Wahab O, Dzik WH. (2006). Current usage of intravenous immune globulin and the rationale behind it: the Massachusetts General Hospital data and a review of the literature. *Transfusion* **46**, 741–753.

Gajdos P, Chevret S, Toyka K. (2008). Intravenous immunoglobulin for myasthenia gravis. *Cochrane Database Syst Rev* (1): CD002277.

Hodson EM, Jones CA, Strippoli GFM *et al.* (2007). Immunoglobulins, vaccines or interferon for preventing cytomegalovirus disease in solid organ transplant recipients. *Cochrane Database Syst Rev* (2): CD005129.

Ohlsson A, Lacy JB. (2004a). Intravenous immunoglobulin for preventing infection in preterm and/or low birth weight infants. *Cochrane Database Syst Rev* (1): CD000361.

Ohlsson A, Lacy JB. (2004b). Intravenous immunoglobulin for suspected or subsequently proven infection in neonates. *Cochrane Database Syst Rev* (1): CD001239.

van Schaik IN, van den Berg LH, de Haan R, Vermeulen M. (2005). Intravenous immunoglobulin for multifocal motor neuropathy. *Cochrane Database Syst Rev* (2): CD004429.

# CHAPTER 35

# Rh Immune Globulin

Beth H. Shaz, MD

Rh immune globulin (RhIg) is a human-plasma derived product consisting of IgG antibodies to the D antigen. It is most commonly used to prevent immunization to the D antigen in D-negative individuals, and for the treatment of idiopathic thrombocytopenic purpura (ITP). Perinatal administration has decreased the risk of forming anti-D in D-negative women carrying D-positive infants from approximately 13% to 0.1%, and substantially reduced the risk of Rh hemolytic disease of the fetus and newborn (HDFN).

Each dose of RhIg suppresses the immune response ("masks the D antigen") for up to a certain amount of whole blood or D-positive red blood cells (RBCs) (Table 35.1). The measurement of the amount of D-positive RBCs within the patient is imprecise, and therefore the dose of RhIg recommended is greater than the exact dose calculated (see below). The half-life of RhIg is 21 days.

| TABLE 35.1 RhIg Vial Size and the Corresponding Amount of Whole Blood or RBCs Neutralized | | | |
| --- | --- | --- | --- |
| Vial size | IU | Whole blood (ml) | RBCs (ml) |
| 50 μg | 250 | 5 | 2.5 |
| 120 μg | 600 | 12 | 6 |
| 300 μg | 1500 | 30 | 15 |
| 1000 μg | 5000 | 100 | 50 |

The mechanism of action for the prevention of D immunization is likely due to the anti-D neutralizing the D antigen on D-positive RBCs, but the exact mechanism is not known. RhIg is used most commonly in the antenatal/postnatal prevention of anti-D formation in D-negative women, and the prevention of anti-D formation in D-negative recipients of D-positive blood products.

## Indications in the Prevention of Anti-D Formation:

**Perinatal Administration:** If a pregnant woman is D-negative and has not been previously sensitized to the D antigen, then she should receive RhIg during pregnancy to prevent anti-D formation and the risk of HDFN in future pregnancies. Antepartum and postpartum dosing is usually with 300 μg of RhIg, though some institutions prefer to give a vial of 50 μg at ≤12 weeks gestational age, but there is a risk of inadvertent misadministration of the lower dose. Outside of the US, varying doses from 100–300 μg of RhIg are administered antepartum and postpartum.

RhIg is administered:

- at 28 weeks gestational age (dose 300 µg),
- at delivery, if the neonate is D-positive, weak-D positive, or D untested (minimum dose of 300 µg, further dosing determine by fetomaternal hemorrhage [FMH] testing), and
- following perinatal events associated with FMH, such as abortion, ectopic pregnancy, amniocentesis, chorionic villus sampling, external cephalic version, abdominal trauma, and antepartum hemorrhage (minimum dose of 300 µg, further dosing determine by FMH testing if >20 weeks gestational age).

After a perinatal event or delivery beyond 20 weeks gestational age, when the fetal blood volume exceeds 30 ml, quantification of FMH is recommended. Multiple methods for determining the FMH volume are described in Chapter 43. If a FMH is detected, then the dose of RhIg should be calculated as shown in Table 35.2.

### TABLE 35.2  Calculation of RhIg Dosing

% Fetal RBCs in maternal circulation is determined by Kleihauer-Betke test or flow cytometry.

Maternal blood volume (ml) = 70 ml/kg × maternal weight (kg), or 5000 ml if maternal weight unknown

Fetal bleed (ml) = % fetal RBCs × maternal blood volume

Dose of RhIg = fetal bleed (ml)/30 ml per dose (300-µg vial)

If the number to the right of the decimal point is <5, round down and add one dose of RhIg (e.g. 2.3 → 2 + 1 = 3 vials).

If the number to the right of the decimal point is ≥5, round up and add one dose of RhIg (e.g. 2.6 → 3 + 1 = 4 vials)

In all situations, RhIg should be administered within 72 hours of the event. If a dose is not administered during that timeframe, then it should be administered as soon as possible, even up to 28 days after the event.

D-positive Blood Product Transfusion into a D-negative Recipient: Individuals who receive D-positive blood products can be given RhIg to prevent D alloimmunization. Reasons to prevent D sensitization include women with childbearing potential, to prevent future risk of HDFN, or individuals who require ongoing transfusion support. If an entire RBC product (250 ml RBCs) is administered, then multiple doses of RhIg can be administered; however, there is a potential risk of hemolysis as a result of the RhIg itself. When multiple D-positive RBC products are transfused to a D-negative female with child-bearing potential, case reports have used a combination of RBC exchange with replacement using D-negative RBCs and RhIg to prevent D sensitization. Whole-blood derived platelets contain approximately 0.5 ml of RBCs, and an apheresis platelet product contains less than 2 ml (usually significantly less). Given the small volume of RBCs in platelet products, a single 300 µg dose of RhIg will prevent immunization for up to 30 whole-blood derived platelet products and 7 apheresis platelet products. Granulocyte transfusions contain 10–30 ml of RBCs, and therefore, to prevent anti-D formation, a dose of at least 600 µg should be administered.

**Indications in Idiopathic Thrombocytopenia Purpura:** ITP is a bleeding disorder characterized by immune-mediated platelet destruction and resultant thrombocytopenia. The most effective pharmacologic therapies for acute ITP currently are corticosteroids, intravenous immune globulin (IVIG) and RhIg. RhIg can be used to treat ITP in D-positive patients with intact spleens. It is thought to work by binding the D antigen on D-positive RBCs and thus blocking receptor-mediated phagocytosis in the spleen; additional mechanisms include modulation of Fcγ receptor expression and immunomodulation.

For ITP, dosing is typically based on the patient's hemoglobin (Hgb): 50 µg/kg if the Hgb is ≥10 g/dl and 25–40 µg/kg when Hgb is 8–10 µg/dl. The administration of RhIg results in a decrease in Hgb, and therefore should be used with caution in patients with Hgb <8 g/dl. A second dose may be given if there is no response to the first dose, and should be adjusted to the Hgb level (see Chapter 87).

**Preparation and Administration:** RhIg is prepared from pooled plasma after cold alcohol fractionation and subsequent purification and infectious disease reduction technologies. RhIg is formulated in intravascular and/or intramuscular preparations, depending on the product; IV injections are more comfortable for patients than IM injections, if the patient has venous access. The IV formulation should be used in ITP patients or other patients with thrombocytopenia in order to prevent intramuscular bleeding.

**Adverse Events:** Reactions to low-dose RhIg include fever, chills, pain at the injection site and, rarely, hypersensitivity reactions. Administration of large doses of RhIg for the treatment of ITP can result in a mild hemolytic reaction in approximately 20% of infusions, and in intravascular hemolysis in approximately 0.7%, with complications including hemoglobinuria, pallor, hypotension, sinus tachycardia, oliguria, anuria, edema, dyspnea, ecchymosis, prolonged bleeding time and, rarely, death. Transfusion related acute lung injury has been associated with the administration of RhIg in the treatment of ITP.

The risk of infectious disease transmission is limited by donor testing, fractionation and additional safety steps (such as ultrafiltration) resulting in viral reduction. Due to improvements in viral reduction, no reported case of infectious disease transmission has occurred since the mid-1990s, when there was an outbreak of hepatitis C transmission from RhIg.

**International Standards:** There is no universal policy regarding the dosing of perinatal prophylactic RhIg, and the dosing varies throughout the world from 100 µg in the UK to 100–120 µg in Canada, and 200–250 µg in many European countries.

## Recommended Reading

Andemariam B, Bussel J. (2007). New therapies for immune thrombocytopenic purpura. *Curr Opin Hematol* **14**, 427–431.

Crow AR, Lazarus AH. (2008). The mechanisms of action of intravenous immunoglobulin and polyclonal anti-D immunoglobulin in the amelioration of immune thrombocytopenic purpura: what do we really know? *Transfus Med Rev* **22**, 103–116.

Engelfriet CP, Reesink HW, Judd WJ *et al.* (2003). Current status of immunoprophylaxis with anti-D immunoglobulin. *Vox Sang* **85**, 328–337.

Judd WJ, for the Scientific Section Coordinating Committee of the AABB. (2001). Practice guidelines for prenatal and perinatal immunohematology, revised. *Transfusion* **41**, 1445–1452.

# CHAPTER 36

# Irradiation of Blood Products

Beth H. Shaz, MD and Christopher D. Hillyer, MD

Irradiation, usually gamma irradiation, of blood products is performed to abrogate the risk of transfusion associated graft versus host disease (TA-GVHD), which is a rare and almost universally fatal complication of blood transfusion with no successful treatment options. Radiation results in the generation of electrons which damage the DNA of lymphocytes, and therefore make the lymphocytes unable to proliferate. A few institutions and countries practice universal irradiation of cellular blood products (see below), but most choose to irradiate cellular blood products only for those patients at increased risk for development of TA-GVHD (see Chapter 62).

Risk of TA-GVHD: It is difficult to quantify the incidence of TA-GVHD for any patient population because the number of patients with the disease, the number of patients with the disease who are transfused and the number of transfusions or type of products each patient receives are not known. The risk is therefore derived from case reports or hemovigilance data, which may be biased by under-recognition, misdiagnosis and under- and passive reporting of TA-GVHD. Still, TA-GVHD is very rare; the most accurate incident data can be obtained from hemovigilance systems, such as the UK SHOT (Serious Hazards of Transfusion), which reported 13 cases from 25 million platelet and RBC products transfused from 1996 to 2005, or a risk of 1:2,000,000 products transfused. Even though TA-GVHD is rare, it is, as above, a lethal condition, and irradiation of products is highly effective, non-toxic in general to the irradiated cells in the product, and relatively inexpensive; irradiation is therefore highly recommended in indicated individuals.

Indications for Irradiated Products: Table 36.1 lists the clear indications, the indications deemed appropriate by most authorities, and the indications for which irradiation is considered unwarranted by most authorities for irradiated blood products. Even in centers of excellence, some divergence of indications for irradiation does occur, so each blood and transfusion service should develop its own criteria and indications in concert with ordering physicians.

Guidelines and Standards for Irradiation and Mitigation of TA-GVHD:
AABB Standard 5.17.3 states that

> "the transfusion service shall have a policy regarding the transfusion of irradiated components; at a minimum cellular components shall be irradiated when a patient is identified as being at risk for TA-GVHD, the donor of the component is a blood relative of the recipient, and the donor is selected for HLA compatibility, by typing or crossmatching."

The US currently does not have established or widely adopted guidelines or indications for irradiation of blood components, and thus the indications vary from

TABLE 36.1   Indications for Irradiated Cellular Blood Products to Prevent TA-GVHD

*Indications for which irradiation is considered to be required*

Congenital immunodeficiency syndromes (suspected or known)

Allogeneic and autologous hematopoietic progenitor cell transplantation

Transfusions from blood relatives

HLA-matched or partially HLA-matched products (platelet transfusions)

Granulocyte transfusions

Hodgkin's disease

Treatment with purine analogue drugs (fludarabine, cladribine and deoxycoformycin)

Treatment with Campath (anti-CD52) and other drugs/antibodies that affect T-lymphocyte number or function

Intrauterine transfusions

*Indications for which irradiation is deemed appropriate by most authorities*

Neonatal exchange transfusions

Pre-term infants/ low birth-weight infants

Infant/ child with congenital heart disease (secondary to possible Di George's syndrome)

Acute leukemia

Non-Hodgkin's lymphoma and other hematologic malignancies

Aplastic anemia

Solid tumors receiving intensive chemotherapy and/ or radiotherapy

Recipient and donor pair from a genetically homogeneous population

*Indications for which irradiation is considered unwarranted by most authorities*

Solid organ transplantation

Healthy newborns/ term infants

Patients with HIV/AIDS

From Shaz BH and Hillyer CD. "Transfusion-associated graft-versus-host disease." In M Murphy, D Pamphilon (eds) (2008). *Practical Transfusion Medicine*, 3rd edition. Oxford: Blackwell Synergy.

institution to institution, as suggested above. Results of a survey performed by the AABB Transfusion Practice Committee of irradiation practices in 1990 demonstrated that: 88% of institutions provided irradiated components for patients with allogeneic hematopoietic progenitor cell (HPC) transplantation, 81% for autologous HPC transplantation, 62% for congenital immunodeficiency syndrome, 54% for premature newborn, 51% for leukemia, 40% for organ transplantation, 34% for Hodgkin's disease (HD), 32% for Non-Hodgkin's lymphoma (NHL), 31% for HLA matched product, 24% for AIDS, 24% for term newborn, and 20% for solid tumor.

Table 36.2 allows comparison of a number of elements related to irradiation, including dose, products, and expiry date changes for the US, the United Kingdom and Japan.

Universal Irradiation:  As: 1) TA-GVHD can occur in immunocompetent patients secondary to the donor having a homozygous HLA haplotype for which the recipient

| | United States | United Kingdom | Japan |
|---|---|---|---|
| TABLE 36.2 Comparison of Gamma-irradiation Product Requirements | | | |
| Techniques | **Gamma-irradiation** | **Gamma-irradiation** | **Gamma-irradiation** |
| Dose | 2500 cGy at center of product; minimum 1500 cGy at any point; maximum 5000 cGy | Minimum 2500 cGy; no part >5000 cGy | Between 1500 cGy and 5000 cGy |
| Type of product | All cellular products: whole blood, RBCs, platelets, granulocytes | All cellular products: whole blood, RBCs, platelets, granulocytes | All cellular products: whole blood, RBCs, platelets, granulocytes, fresh plasma |
| Age of product | RBCs, platelets, granulocytes any time | RBCs < 14 d after collection; platelets, any time during 5 d storage; for exchange or intrauterine transfusion <24 h | RBCs ≤3 d regardless of recipient; ≤14 d, if clinically indicated; at any time if patient is immunocompromised |
| Expiration | RBCs stored up to 28 d after irradiation or original outdate, whichever is sooner | RBCs stored 14 d after irradiation | RBCs stored up to 3 weeks after collection |
| General | All blood from relatives; all HLA matched products | All blood from relatives; all HLA matched products; all granulocytes | All blood from relatives; all HLA matched products |

Modified from Schroeder ML (2002). Transfusion-associated graft-versus-host disease. *Br J Haematol* **117**, 275–287.

is heterozygous (possible random donor and recipient partial HLA matching has been estimated to be possible in as many as 1 in 7174 in the US) or from receiving a product from a relative, and in individuals whose degree of immunocompromise was not known or properly identified prior to transfusion; 2) TA-GVHD is almost universally fatal; 3) the adverse effects of radiation on the blood product and its constituents are minimal, and 4) the cost of irradiating products is modest, universal irradiation has been recommended by some authors and indeed is practiced in some institutions within the US and the world (e.g. Japan, where the risk of inadvertent, non-relative partial matching of HLA haplotypes approaches 1 : 1000).

**Blood Products Requiring Irradiation:** All cellular blood products, defined as red blood cells, platelets, granulocytes, whole blood and fresh plasma (not fresh frozen plasma), contain viable T lymphocytes that are capable of causing TA-GVHD. These products should at a minimum be irradiated for patients at increased risk for TA-GVHD (Table 36.1). Two special situations apply: granulocyte transfusions and transfusion to HLA-matched individuals, especially to recipients who are first-degree relatives of the donor.

Granulocyte transfusions are a very high-risk product for TA-GVHD because they are given soon after collection, have a high lymphocyte count, and are administered to neutropenic and immunosuppressed patients. Therefore, it is recommended that all granulocyte transfusions undergo irradiation prior to transfusion.

All cellular blood products transfused to a relative of the donor, and all HLA-matched products (including both HLA-matched/selected and crossmatched platelet products) should be irradiated because the viable donor lymphocytes within the product may be homozygous for a HLA haplotype for which the recipient is heterozygous. This results in the inability of the recipient to recognize the donor lymphocytes as foreign, and therefore the donor lymphocytes can proliferate and cause TA-GVHD.

## Processing and Storage:

**Sources of Irradiation:** Both gamma rays and X-rays can be used to irradiate blood products and cause adequate T-lymphocyte inactivation at the doses described. Usually gamma rays originate from cesium 137 or cobalt 60, while X-rays are generated by linear accelerators.

After the September 11, 2001 attacks, there was increased regulation of blood irradiators (1500 and 2500 cGy) by the US Nuclear Regulatory Commission (NRC). One initiative through the Energy Policy Act is to find alternative technologies that either do not use radionucleotides, such as electricity, or use lower-risk sources.

**Dose:** The dose of irradiation must be sufficient to inhibit lymphocyte proliferation, but not so high as to significantly damage RBCs, platelets and granulocytes or their function. Assays to assess the effect of irradiation on T-lymphocyte proliferation include the mixed lymphocyte culture (MLC) assay and the limiting dilution analysis (LDA). The recommended dose varies between 15 and 50 Gy (1500 and 5000 cGy) (Table 36.2); the US requires a dose of 25 Gy (2500 cGy) at the center of the product and a minimum of 15 Gy (1500 cGy) and maximum of 50 Gy (5000 cGy) at any point within it. Of note, there have been three patients transfused with irradiated blood products, two at doses of 20 Gy (2000 cGy) and one at 15 Gy (1500 cGy), who developed TA-GVHD. It is unknown whether these events were due to a process or dose failure.

**Storage and Expiration:** In the US, RBC product outdates are shortened to 28 days after irradiation; RBC product outdate is variably shortened to 14–28 days after irradiation in other countries (Table 36.2). These changes in outdating are due to the small but not insignificant effects of radiation on erythrocyte membranes leading to increased potassium accumulation and accelerated cell death over time during the storage period. There is no change in outdate of other blood products.

**Quality Assurance:** Quality-related measures for blood product irradiation include those focused on the irradiator itself and those focused on the product. AABB *Standards* states that

> "*irradiated blood and components shall be prepared by a method known to ensure that irradiation has occurred. A method shall be used to indicate that irradiation has occurred with each batch. Alternate methods shall be demonstrated to be equivalent.*"

Quality assurance measures should be performed on the irradiator, including dose mapping, adjustment of irradiation time to correct for isotopic decay, ongoing detection for radiation leakage, timer accuracy, turntable operation, and preventive maintenance. Each batch of irradiated products should have attached a qualitative radiation dosimeter; usually a label is placed on individual products, and the label physically changes for the proper dose of radiation to demonstrate that the blood product has in fact been properly irradiated.

**Adverse Events:** At recommended doses, radiation causes a very low level of oxidation and damage to lipid components of membranes which occurs over time. Products, and the constituent cells within, irradiated immediately prior to transfusion appear to be unaffected, and have virtually normal function. The effects of radiation are most significant on RBC products, and include increase in extracellular potassium and decrease in post-transfusion RBC survival. The *in vivo* viability of irradiated RBCs evaluated at 24-hour recovery is reduced by 3–10% compared to non-irradiated RBCs.

The increase in extracellular potassium is usually not of clinical significance because of posttransfusion dilution of the potassium. However, there may be certain patients where attention should be paid to the increased potassium, such as premature infants; infants receiving large RBC volumes; and infants receiving intrauterine transfusions (IUT), neonatal exchange transfusions, or intracardiac transfusions via central line catheters. The potassium increase can be prevented by either irradiating the RBC product shortly before transfusion (usually within 24 hours) or by washing or volume-reducing the RBC product prior to transfusion. Newer methods for potassium removal include filters that specifically remove this element from blood products. At present, these filters are investigational in the US.

**International Considerations:** There are differences in the dose of gamma-irradiation, product requirements, product expiration, and indications between countries (Table 36.2). The UK has developed criteria for receiving irradiated cellular blood components, and Japan irradiates all cellular blood components because of the similarity of HLA haplotypes within the Japanese population (as stated above). In addition, some countries use pathogen-reduction technologies for platelets, which are currently not FDA approved in the US, that interfere with the replication of nucleic acids and thus may replace the need for irradiation to prevent TA-GVHD.

## Recommended Reading

BCSH Blood Transfusion Task Force. (1996). Guidelines on gamma irradiation of blood components for the prevention of transfusion-associated graft-versus-host disease. *Transfus Med* **6**, 261–271.

Hume HA, Preiksaitis JB. (1999). Transfusion associated graft-versus-host disease, cytomegalovirus infection and HLA alloimmunization in neonatal and pediatric patients. *Transfus Sci* **21**, 73–95.

Moroff G, Luban NLC. (1997). The irradiation of blood and blood components to prevent graft-versus-host disease: Technical issues and guidelines. *Transfus Med Rev* **11**, 15–26.

Moroff G, Leitman SF, Luban NL. (1997). Principles of blood irradiation, dose validation, and quality control. *Transfusion* **37**, 1084–1092.

Przepiorka D, LeParc GF, Stovall MA *et al.* (1996). Use of irradiated blood components: practice parameter. *Am J Clin Pathol* **106**, 6–11.

Schroeder ML. (2002). Transfusion-associated graft-versus-host disease. *Br J Haematol* **117**, 275–287.

Williamson LM, Stainsby D, Jones H *et al.* (2007). The impact of universal leukodepletion of the blood supply on hemovigilance reports of posttransfusion purpura and transfusion-associated graft-versus-host disease. *Transfusion* **47**, 1455–1467.

# CHAPTER 37

# Leukoreduction of Blood Products

James C. Zimring, MD, PhD

Although routine blood fractionation procedures allow for red blood cell (RBC), platelet (PLT) and plasma components to be manufactured, white blood cells (WBC) remain in these components and are referred to as residual leukocytes. Residual leukocytes have the potential to cause deleterious effects in the transfusion recipient, including febrile non-hemolytic transfusion reactions (FNHTR), HLA-alloimmunization and transmission of cytomegalovirus (CMV). Additional possible effects include transmission of other leukocyte-associated herpesviruses, and transfusion-related immunomodulation (TRIM). The effects that the residual WBCs have is dependent on the number in each component, the storage temperature of the component, and the clinical status of the recipient, including previous and latent infection with the viruses above and relative degree of immunosuppression. As residual leukocytes play a central role in these post-transfusion complications, it was predicted that WBC removal would decrease their incidence.

Beginning prior to the 1980s, it was recognized that residual leukocytes could be removed by washing components or by some membrane and fiber methods. These later methods were further developed into what is termed "filter leukoreduction." Also, as technologies for apheresis collections of platelets advanced, only trace numbers of WBCs were concomitantly collected. Thus, apheresis-derived platelets are leukoreduced as a matter of course on most modern apheresis devices, and this method of WBC "removal" is known as "process leukoreduction."

The use of leukoreduction methodologies has become commonplace in most developed nations. Ongoing clinical studies have established a clear benefit for leukoreduction in some settings, and a likely benefit in others. However, there are additional proposed indications for which clear evidence of efficacy is lacking.

## Indications for Leukoreduction:

**Decreasing Incidence of Febrile Non-Hemolytic Transfusion Reactions (FNHTR):** The frequency of FNHTRs is significantly decreased when leukoreduced products are transfused. There are two proposed mechanisms by which leukocytes in transfused units contribute to FHNTRs. The first mechanism involves leukocyte-derived cytokines, and is most commonly associated with PLT products. Leukocyte-derived cytokines (e.g. interleukins [IL-1, IL-6, IL-8] and tumor necrosis factor alpha [TNF-$\alpha$]) accumulate in the supernatant during room-temperature storage of platelet products. Because the cytokines have already been released during storage, bedside (or post-storage) leukoreduction does not result in decreased incidence of FNHTRs from platelet products. The second mechanism involves anti-WBC antibodies, and is most commonly associated with RBC products. Anti-HLA and anti-HNA antibodies in the transfusion recipient's plasma directed against the transfused WBCs leads to a release of inflammatory molecules that induces fever and other symptoms.

**Decreasing Incidence of HLA Alloimmunization:** Leukoreduction of RBCs and PLTs has been shown to decrease the incidence of HLA alloantibody formation in transfusion recipients. This is of primary importance in patients who require ongoing PLT transfusion support, as anti-HLA antibodies can lead to platelet refractoriness by binding to the corresponding antigens (major histocompatibility complex) on transfused PLTs. In addition, HLA immunization can contribute to rejection of subsequent organ transplantation, and therefore potential transplantation recipients should receive leukoreduced products (especially potential kidney and heart recipients). The TRAP study in 1997 assessed reduction in HLA alloimmunization by randomized controlled trial; in addition, a number of smaller studies had previously addressed the same issue. In aggregate, meta-analysis of these studies showed a relative risk reduction of HLA immunization of approximately one-third for patients not previously sensitized to HLA antigens.

**Decreasing Cytomegalovirus (CMV) Transmission:** Except in cases of active viremia, CMV remains in a latent state in circulating leukocytes. Such latent CMV can reactivate upon transfusion and infect the naïve recipient. Thus, leukoreduction can substantially decrease the inoculum of latent CMV genomes a recipient receives. Accordingly, transmission of CMV by blood transfusion is substantially decreased by leukoreduction (from as high as 30% in susceptible patients to approximately 2.5%). It has been argued that leukoreduced blood has rates of CMV transmission practically as low as seronegative products; however, some patient populations may still benefit from CMV seronegative products (see Chapter 38).

## Potential Indications:

**Decreasing other Human Herpesvirus Transfusion Transmitted Infections:** Like CMV, other human herpes viruses can likewise be associated with leukocytes. Thus, transmission of such viruses (i.e. EBV, HHV-6, HHV-7 and HHV-8) may in theory be reduced through the use of leukoreduced cellular blood components.

**Prion Disease:** Clinical cases consistent with the transmission of variant Creutzfeldt-Jakob disease (vCJD) by transfusion have now been reported. As prions can be associated with leukocytes, leukoreduction technology has been assessed for the ability to prevent transfusion transmission of vCJD. Prions can be decreased by passing blood products through existing filters, and modified filters have also been described. Whether such filters will decrease transmission of disease in humans is undetermined; however, a near-50% decrease in transmission was observed in an animal model.

## Controversial Indications:

**Transfusion Related Immunomodulation (TRIM):** TRIM is defined as effects of transfusion on the recipient immune system, including potential downregulation of cellular immunity, induction of humoral immunity, and altered inflammatory responses (see Chapter 64). The TRIM effect may be not only secondary to the transfused WBCs, but also the result of the other blood constituents, such as proteins, lipids and inflammatory mediators. However, as TRIM can be mediated in large part by transfused leukocytes, leukoreduction is predicted to decrease TRIM; nevertheless, the clinical benefit of leukoreduction for this purpose is unestablished.

**Contraindications:** Bedside leukoreduction should be avoided in patients on ACE inhibitors, as hypotensive episodes may be induced.

**Methods of Leukoreduction:** Traditionally, some leukoreduction was performed by centrifuging the product and removing the buffy coat. However, this process is inefficient and highly variable.

**Filtration Leukoreduction:** Currently, there is a variety of leukoreduction filters (termed *filter leukoreduction*) which remove three logs of WBCs, available from several different manufacturers. These filters use a combination of barrier filtration (i.e. pore size) and cell adhesion to decrease leukocyte content. Although each must conform to the established requirements for the numbers of residual leukocytes in the product, different filters may not generate equivalent products. While the numbers of residual leukocytes may be equivalent, the subset composition of the leukocytes may differ. In addition, filter leukoreduction of RBCs from hemoglobin AS (sickle trait) patients is often less effective secondary to reduced RBC deformability leading to clogging or decreasing the area of the filter.

**Process Leukoreduction:** Apheresis technologies are able to collect PLT products that are leukoreduced during collection, termed *process leukoreduction*.

**Quality Assurance:** In the US, the requirements for leukoreduction of RBCs and apheresis PLTS are $<5 \times 10^6$ leukocytes per product, and for whole blood derived platelets $< 0.83 \times 10^6$ leukocytes per product. At least 95% of the products sampled must meet this specification.

**Pre-storage versus Bedside Leukoreduction:** Leukoreduction can be performed at the time of collection or at the bedside; however, bedside leukoreduction has several substantial problems. First, patients on ACE inhibitors may have severe hypotensive episodes due to bradykinin activation, which is not observed for pre-storage leukoreduction. Second, as leukocytes elaborate inflammatory molecules during storage, which are not removed by the filter, leukoreducing at the bedside may have substantially decreased benefits regarding febrile reactions. Third, the quality control of bedside leukoreduction is not equivalent to that of in-laboratory pre-storage leukoreduction.

**Universal versus Diagnosis-specific Leukoreduction:** Whether it is safe, appropriate and cost-effective to leukoreduce all blood products prior to storage, or to reserve leukoreduced products for particular patient subsets, is currently a matter of debate. Practices differ by blood supplier, transfusion service and region.

**International Differences:** In Europe, the requirements for leukoreduction of RBCs and apheresis derived platelets are fewer than $1 \times 10^6$ leukocytes per product.

## Recommended Reading

Blajchman MA. (2006). The clinical benefits of the leukoreduction of blood products. *J Trauma* **60**, S83–90.

Cervia JS, Sowemimo-Coker SO, Ortolano GA *et al.* (2006). An overview of prion biology and the role of blood filtration in reducing the risk of transfusion-transmitted variant Creutzfeldt-Jakob disease. *Transfus Med Rev* **20**, 190–206.

The Trial to Reduce Alloimmunization to Platelets Study Group (1997). Leukocyte reduction and ultraviolet B irradiation of platelets to prevent alloimmunization and refractoriness to platelet transfusions. The Trial to Reduce Alloimmunization to Platelets Study Group. *N Engl J Med* **337**, 1861–1869.

Vamvakas EC. (1998). Meta-analysis of randomized controlled trials of the efficacy of white cell reduction in preventing HLA-alloimmunization and refractoriness to random-donor platelet transfusions. *Transfus Med Rev* **12**, 258–270.

# CHAPTER 38

# CMV-safe Blood Products

John D. Roback, MD, PhD

Cellular blood components (RBCs, platelets, and granulocytes) can transmit cytomegalovirus (CMV) infection to susceptible transfusion recipients. Two approaches are routinely available to reduce the risk of CMV transmission by these components: (1) selection of CMV-seronegative components; and (2) leukoreduction of red blood cell (RBC) and platelet components. In addition, broadly-active pathogen reduction methods in use in some countries outside the US eliminate CMV infectivity.

**Description:** CMV infection is typically of little consequence to individuals with normally functioning immune systems. In fact, 40–80% of the population (depending on the community) has been infected and harbors CMV, which persists in a relatively innocuous latent state. In contrast, CMV infection of immunocompromised patients has devastating consequences, including death. One route of transmission to at-risk patients is by transfusion, and prevention of transfusion-transmitted CMV infection (TT-CMV) is an important concern in transfusion medicine.

**Indications:** Most studies suggest that 13–37% of immunocompromised patients will contract TT-CMV from transfusion of cellular blood components (RBCs, platelets, and granulocytes) that are not screened for CMV antibody or filtered. The most well-established immunocompromised groups at-risk for TT-CMV include premature low birthweight infants (<1250–1500 g) born to seronegative mothers, seronegative recipients of seronegative allogeneic or autologous hematopoietic progenitor cells (HPC) transplantation, seronegative recipients of seronegative solid organ transplants, and seronegative patients with HIV-infection (Table 38.1). In these patients, primary CMV infection can progress to disseminated tissue-invasive CMV disease including CMV hepatitis, retinitis, interstitial pneumonitis, encephalitis and gastroenteritis, and can also predispose to additional complications, including graft-versus-host disease in transplant recipients, accelerated solid organ graft rejection, and other opportunistic infections.

---

**TABLE 38.1 CMV-Seronegative Patient Populations at High Risk of Serious Illness/Death After CMV Infection**

- Patients receiving or who may receive an allogeneic or autologous HPC transplant
- Patients receiving chemotherapy which produces severe neutropenia
- Solid-organ transplant recipients
- Pregnant women
- Fetuses (intrauterine transfusion) + low birthweight (<1250−1500 g) neonates
- HIV-infected patients

Modified from Blajchman MA, Goldman M, Freedman JJ, Sher GC (2001). Proceedings of a Consensus Conference: Prevention of Post-Transfusion CMV in the Era of Universal Leukoreduction. *Transfusion Med Rev* **15**: 1–20.

There are also additional groups that may benefit from transfusion of CMV-safe blood. These include pregnant women who are seronegative (to prevent primary infection with transmission to the fetus), and seronegative patients who are *candidates* for HPC transplantation.

## Processing and Storage:

**CMV Seronegative Products:** The standard approach for identifying blood donors who have not been infected with CMV is through screening for anti-CMV antibodies, using one of a variety of serological methodologies (solid-phase fluorescence immunoassay, enzyme immunoassay, latex or particle agglutination, and solid phase red cell adherence). Different configurations of CMV antibody test kits exist, including one that detects combination IgM/IgG, and an IgG-only kit.

The determination that a blood donor is truly seronegative can be problematic in some individuals. Specifically, anti-CMV antibodies may not be detected by serology until 6–8 weeks following primary infection (the window period), leaving a small risk that some donors who test "seronegative" are in fact newly infected, and thus can transmit infectious CMV. Also, in some remotely-infected donors anti-CMV antibodies can decrease to undetectable levels, although the continued presence of CMV T-cells shows that at one time they had a CMV infection.

Exclusive use of CMV-seronegative components for transfusion markedly decreases the incidence of TT-CMV. For example, in one study none of 90 seronegative infants transfused with seronegative components contracted CMV infections, as compared to 13.5% of those receiving seropositive transfusions. The exclusive use of seronegative components in immunocompromised adult seronegative recipients of seronegative HPCs for transplantation also decreases the incidence of TT-CMV and severity of resulting CMV disease as compared to the use of unscreened blood products.

*CMV Safe by Leukoreduction:* Because white blood cells (WBC) latently infected with CMV are the primary vector for TT-CMV (transfusion of products from a recently infected donor with infectious CMV particles in the plasma is less common), removal of WBCs from components also mitigates TT-CMV. Current filtration and apheresis technologies can produce blood components that have well below the maximal limit of $5 \times 10^6$ WBCs mandated for leukoreduced products in the US. This level of leukoreduction significantly reduces the incidence of TT-CMV. For example, in the largest prospective randomized controlled clinical trial to address this issue to date, 502 seronegative recipients of seronegative HPC transplants received either leukoreduced or CMV-seronegative components. The probabilities of developing CMV infection were similar in patients receiving filtered (2.4%) versus seronegative components (1.3%; $P = 1.00$). The probabilities of developing CMV disease were also similar (2.4% versus 0%, respectively; $P = 1.00$). However, because of a statistically greater progression to CMV disease (including lethal CMV pneumonia) in the group receiving filtered as compared to seronegative components (2.4% versus 0%, respectively; $P = 0.03$), there is a possibility that seronegative components may be slightly safer than filtered components with respect to the risk of TT-CMV. In the results of a Canadian Consensus Conference published in *Transfusion* in 2001, both CMV-seronegative components as well as those that were rendered CMV-safe by leukoreduction were considered to be effective, although both methods had some failures. Nonetheless, the panel concluded that at-risk patients may still benefit from seronegative products despite concomitant leukoreduction of all components in Canada. CMV NAT does not appear to be effective for further reducing the risk of TT-CMV, since in a recent large-scale study CMV NAT did not identify CMV-positive donors that were not already identified by serology. However, its utility is under investigation in very low birth-weight infants born to CMV-seronegative mothers, and this may guide future decision-making.

**CMV Seronegative versus CMV-safe by Leukoreduction:** Based in part on the data presented above, some authorities argue for the exclusive use of CMV-seronegative blood components for the transfusion of seronegative patients at risk for TT-CMV. In contrast, other authorities have determined that leukoreduced products are adequate because of prospective monitoring in high-risk patients with the pp65 CMV antigenemia test or CMV PCR, and pre-emptive treatment with CMV antiviral agents (e.g. ganciclovir). Additionally, most institutions employ universal leukoreduction and thus can use the potentially additive (but not proven) approach of combining CMV-serology and leukoreduction to reduce the risks of TT-CMV. Each institution should weigh the risks and benefit of the use of CMV-seronegative versus leukoreduced cellular blood components, and the magnitude of clinical CMV disease.

## Quality Assurance:

**Antibody Tests:** CMV donor screening assays that detect total antibody (IgG + IgM) to CMV are generally in use and are "cleared" by the FDA. Donor screening is not required by the FDA for blood component preparation; however, donors of viable, leukocyte-rich HPCs must be tested for evidence of infection due to CMV using a method to "adequately and appropriately reduce the risk of transmission" (§1271.85(b)(2)). Such services must establish and maintain an SOP governing the release of HPCs from CMV-positive donors (§1271.85(b)(2)). It is notable that a positive test for CMV does not necessarily make that donor ineligible.

   Leukoreduction quality control can be achieved by performing leukocyte counts on either the product or the process used to perform the leukoreduction. Nageotte chamber manual counting is commonly employed, although it is labor-intensive and not easy to control. Nor does it have commercially available standards for comparison of technologist reading. Some manual flow-cytometric based assays are available, but are again expensive. Fully automated rapid fluorescent cytometry is under development for large-scale component-by-component testing, but is not FDA cleared at this point.

**Adverse Events:** Except the residual risk of TT-CMV, as noted above, no additional adverse effects are associated with transfusion of CMV-seronegative or leukoreduced blood components.

**International Differences:** As with Canada, and referred to above, the UK Blood and Transfusion Service in a joint position paper put forward three statements:

1. The balance of evidence suggests that TT-CMV can be significantly mitigated by pre-storage leukoreduction

2. Leukoreduced components are likely comparable to seronegative components in providing CMV-safe transfusion

3. At present, due to the lack of adequate randomized controls clinical trials, they could not firmly recommend the discontinuation of CMV serologic testing.

Finally, as opposed to serological and leukoreduction approaches, which are associated with breakthrough TT-CMV cases, pathogen-inactivation (PI) technology carries the potential for completely preventing TT-CMV, as well as eliminating the transmission of other infectious agents. Outside the US, including in some European countries, these approaches have now entered clinical use for platelets. At present, there is no approved PI technology for RBCs.

## Recommended Reading

Blajchman MA, Goldman M, Freedman JJ, Sher GD. (2001). Proceedings of a Consensus Conference: prevention of post-transfusion CMV in the era of universal leukoreduction. *Transfusion Med Ref* **15**, 1–20.

Drew WL, Roback JD. (2007). Editorial Prevention of TT-CMV: reactivation of the debate? *Transfusion* **47**, 1955–1958.

Laupacis, Brown J, Costello B *et al.* (2001). Prevention of posttransfusion CMV in the era of universal WBC reduction: a consensus statement. *Transfusion* **41**, 560–569.

Roback JD. (2002). CMV and blood transfusions. *Rev Med Virology* **12**, 211–219.

Roback JD, Drew WL, Laycock ME *et al.* (2003). CMV DNA is rarely detected in healthy blood donors using validated PCR assays. *Transfusion* **43**, 314–321.

Vamvakas EC. (2005). Is white blood cell reduction equivalent to antibody screening in preventing transmission of cytomegalovirus by transfusion? A review of the literature and meta-analysis. *Transfusion Med Rev* **19**, 181–199.

# CHAPTER 39

# Frozen Blood Products

Cassandra D. Josephson, MD and Beth H. Shaz, MD

Blood products are frozen in order to lengthen their storage time. RBC products are cryopreserved in glycerol, which must be removed prior to transfusion. The primary indications for freezing RBC products are to increase the storage length of rare and/or autologous products; possible uses include inventory management in times of blood shortages and disasters. Currently, the cryopreservation of platelet products is under investigation only. Autologous hematopoietic progenitor cell (HPC) products and umbilical cord blood HPCs are routinely cryopreserved.

**Cryopreservation of RBC Products:** Freezing of RBC products not only increases the duration of storage to 10 years, but also provides a product with restored levels of ATP and 2,3 DPG. Frozen RBC products are indicated for rare RBC products (e.g. RBCs which are negative for high-frequency antigens), autologous products and, potentially, for disaster management. In addition, because of the removal of the supernatant plasma and storage solution, and the restored levels of ATP and 2,3-DPG, frozen and deglycerolized RBC products may be preferred in some situations, such as neonatal RBC exchange or intrauterine transfusion.

Cryopreservation with Glycerol: RBCs must be protected during freezing to prevent cellular dehydration and mechanical trauma as a result of intracellular ice formation. Glycerol is a penetrating cryoprotective agent which crosses the cell membrane into the cytoplasm, providing an osmotic force that prevents water from migrating outward as extracellular ice is formed and thus ultimately preventing cellular dehydration. A high-concentration glycerol method (40% glycerol concentration) is used in most blood banks, which can be automated. Glycerol must be introduced slowly, as rapid introduction itself can result in hypertonic damage to the RBCs and hemolysis. A slow freezing rate with an initial freezing temperature of $-80°C$ and storage temperature of $\leq -65°C$ is employed. Polyolefin plastic is preferred over polyvinyl chloride (PVC) storage bags because PVC storage bags result in a higher degree of hemolysis and breakage.

RBCs used for freezing are collected in CPD or CPDA-1, and stored as liquid whole blood or RBC products. These RBCs must be frozen within 6 days of collection; if RBC products need to be frozen and are older than 6 days, they must first be rejuvenated with a FDA-approved solution (see below) and then frozen. Per the FDA, frozen RBC products can be stored for 10 years, but RBCs stored over 20 years have been used successfully.

Rejuvenation: FDA-approved rejuvenation solutions containing pyruvate, inosine, phosphate and adenine, which restores levels of ATP and 2,3-DPG to that of a freshly-drawn unit. RBC products can be rejuvenated up to 3 days after expiration.

Rejuvenated RBCs must be frozen or washed to remove the inosine if transfused within 24 hours.

**Thawing of Frozen RBC Product:** Prior to transfusion, the glycerol must be removed completely (deglycerolized). Glycerol must be gradually reduced to avoid hemolysis. One method adds 150 mL of 12% saline then washes with 1 L of 1.6% saline, followed by 1 L of 0.9% saline with 0.2% dextrose. Other automated methods are available depending on the instrument used.

Frozen RBCs from donors with sickle cell trait (hemoglobin SA) will form a jelly-like mass and hemolyze upon deglycerolization. Modified wash procedures are available to prevent the hemolysis, but many institutions screen donors for hemoglobin S prior to freezing because the specialized wash procedures are not routinely available.

RBC survival 24 hours post-transfusion of deglycerolized RBCs is >75%. Most institutions use an open system to deglycerolize the RBC product, which is not only time-consuming (approximately 30 minutes); the product also then has a 24 hour outdate. Closed system, automated deglycerolization is also possible, allowing post-thaw storage in AS-3 for 14 days or SAGM for 7 days.

**Refreezing of Thawed RBC Products:** Refreezing may be indicated for unique RBC products that were thawed unintentionally or unexpectedly not used. Products that are deglycerolized, stored ≤20 hours at 1–6°C and then reglycerolized, refrozen and rethawed when needed demonstrate no loss in survival. This should not be considered routine practice.

**Indications:** Indications to freeze RBC products include the following.

*Rare RBC Inventory Management:* RBC products with a rare phenotype may be frozen for transfusion to patients with antibodies to high-frequency antigens or with multiple antibodies resulting in difficulty in finding compatible blood. The AABB and American Red Cross maintain the American Rare Donor Program (ARDP) and ISBT maintains the International Blood Group Reference Laboratory (IBGRL), which aid in the identification of rare RBC products. These products, which are frequently frozen, can be transported between blood centers for the treatment of these patients throughout the world. In addition, those patient with alloantibodies may store frozen autologous RBCs for future needs.

*Autologous Donations:* Cryopreserved autologous RBC products may be used either when a large number of autologous products is needed for surgery and the duration of time to collect all of the products needed exceeds the outdate of the product, or when the date of the surgery changes and the products would outdate prior to the rescheduled surgery date.

*Inventory Management in Times of Blood Shortages and Disasters:* Some institutions have implemented large-scale storage of group O units to provide a source of RBCs during periods of severe blood shortages. A frozen inventory is used by the US Department of Defense, which has the ability to thaw large amounts of products relatively quickly with an extended shelf-life. In the civilian setting, clear benefit of a frozen inventory is lacking because it is expensive and logistically challenging. Further, the inventory would periodically require rotating of the products, and standard thawing methodologies require a 24-hour outdate. In addition, when new donor requirements are implemented, the frozen products cannot undergo retrospective donor screening.

**Cryopreservation of HPC Products:** Because the storage time of non-frozen HPC products is usually less than 72 hours, all autologous HPC products and umbilical-cord blood products must be cryopreserved. In addition, some allogeneic bone marrow and peripheral blood HPC products are frozen.

**Cryopreservation with DMSO:** In order to maximize viability of frozen HPC, the cryoprotectant solution, freezing kinetics, and frozen storage conditions must be optimized. Addition of dimethyl sulfoxide (DMSO) to HPCs prior to freezing effectively prevents cellular dehydration as well as the formation of ice crystals. When DMSO (at a final concentration of between 5 and 10%) is used in combination with controlled-rate freezing, the HPCs retain high viability after thawing. Computer-controlled freezers are available for precise regulation of the HPC freezing rate. Once frozen, HPCs can be maintained at $-70°C$ or less in a mechanical freezer, or in a liquid nitrogen freezer. If the second option is chosen, the product can be stored beneath the surface of the liquid nitrogen ($-196°C$) or in the vapor phase above the liquid nitrogen surface ($-150°C$ or less). Liquid phase storage has the advantage of reducing temperature fluctuations. However, the vapor phase is usually chosen for storage of potentially infectious HPC components to limit the risks of contamination of other components. There is no defined expiration date for frozen HPC components, and components frozen for over 10 years have been used successfully in transplantation.

**Thawing of HPC Product:** Frozen HPC products are typically thawed at 37°C at or near the patient's bedside, and then immediately infused at a rate of approximately 5–20 mL/min. In most cases cryopreserved HPC products are not washed to remove DMSO, although infusion of large volumes of DMSO has been associated with some adverse effects. These include hypertension, hypotension, nausea, vomiting, abdominal pain, diarrhea and, rarely, cardiac problems. Children may be more susceptible to these symptoms. As a general role, DMSO infusions should be limited to less than 1 g/ kg per day. Side-effects can often be managed by slowing or briefly halting the HPC product infusion. While HPC products can be washed in the laboratory to remove DMSO prior to infusion, the general belief is that washing can lead to an unacceptable level of loss of HPCs and reduced viability of remaining HPCs, which could impair engraftment.

### Recommended Reading

Popovsky MA. (2001). Frozen and washed red blood cells: new approaches and applications. *Transfus Apher Sci* **25**, 193–194.

Valeri CR, Ragno G, Van Houten P *et al.* (2005). Automation of the glycerolization of red blood cells with the high-separation bowl in the Haemonetics ACP 215 instrument. *Transfusion* **45**, 1621–1627.

# CHAPTER 40

# Washed Blood Products

Cassandra D. Josephson, MD and Beth H. Shaz, MD

Washing refers to a process that removes the non-cellular fluid in RBC, platelet and other products and replaces it, typically with saline. This removes more than 99% of the plasma proteins (including antibodies) and the original supernatant, which may contain unwanted substances (i.e. cytokines) or levels of substances (i.e. potassium). Washing may be indicated for patients with recurrent severe allergic, anaphylactoid or anaphylactic reactions, and in situations similar to volume reduction, such as removal of the plasma-containing anti-A and/or anti-B in ABO out-of-group transfusions (i.e. group O platelet product to a group A recipient) and removal of supernatant in neonatal transfusions (see Chapter 42).

**Washing of RBC Products:** RBC products can be washed with 1–2 liters of normal saline using a manual or automated method. Approximately 90% of the leukocytes are removed, but insufficient amounts to label the product as leukoreduced. Washing may result in loss of up to 20% of the RBC mass of the product. Washed products must be used within 24 hours, as they are washed in an open system.

**Washing of Platelet Products:** Platelets can be washed with normal saline, saline buffered with ACD-A or citrate, or platelet storage solutions (currently none are approved in the US), using a manual or automated method. Washing results in approximately 33% loss of the platelets, but does not result in leukocyte reduction. Washed platelets must be used within 4 hours.

**Indications:** Indications for washed products are presented in Table 40.1.

**Prevention of Recurrent Severe Allergic/Anaphylactic Transfusion Reactions:** Anaphylactic reactions can be secondary to anti-IgA; to prevent these reactions in the future, IgA deficient products can be used for transfusion. However, these products may not always be readily available. Therefore, deglycerolized and/or washed RBC products can be used to remove the IgA in the RBC product. Deglycerolized and/or washed RBCs can also be used to prevent recurrent severe allergic/anaphylactoid/anaphylactic reactions not secondary to anti-IgA, by removing residual plasma proteins.

**Neonatal Alloimmune Thrombocytopenia:** Neonatal alloimmune thrombocytopenia (NAIT) is secondary to maternal alloantibodies (usually anti-HPA-1a) against the platelet antigen in the fetus or neonate, which may result in fetal/neonatal thrombocytopenia, hemorrhage and death. Maternal platelets are serologically compatible and are therefore a potential source of platelets for transfusion to the fetus/neonate. Prior to transfusion the maternal platelets must be washed to remove the antibody contained within the plasma, and irradiated to prevent transfusion-associated graft-versus-host disease.

**TABLE 40.1 Clinical Indications and Rationale for Washed Blood Products**

| Clinical Indication | Rationale |
| --- | --- |
| Allergic or anaphylactic reaction | Removes allergenic plasma proteins |
| NAIT | Removes maternal antibodies from maternal platelet product |
| Large-volume or rapid transfusion | Newborns and small children to decrease the risk of hyperkalemia and cardiac arrhythmia |
| Following irradiation and storage | Decreases potassium |
| T-activation | During ongoing hemolysis, reduces further hemolysis by eliminating IgM anti-T |
| PNH | No longer recommended |

NAIT, neonatal alloimmune thrombocytopenia; PNH, paroxymal nocturnal hemoglobinuria.
Modified from Hillyer CD, Silberstein LE, Ness PM et al. (eds.) (2007). *Blood Banking and Transfusion Medicine: Basic Principles and Practice*, (2nd edition). San Diego, CA: Elsevier.

**Large-volume or Rapid Transfusion into Neonates and Small Children:** Washing may be used to remove potassium, anticoagulant-preservative solution (particularly additive solutions) and other substances in the supernatant in RBC products. This is especially indicated in large-volume (greater than 25 ml/kg) or rapid transfusion in neonates and small children to decrease the risk of hyperkalemia and cardiac arrhythmias. Volume reduction can achieve the same goal.

**Irradiated RBC Products:** Potassium leakage out of the RBC increases after irradiation. Therefore, patients who are sensitive to increase(s) in potassium concentration may benefit from washed or volume-reduced RBC products.

**Patients with T-Activation:** T-activation occurs when bacterial neuraminidase removes N-acetyl neuraminic acid and exposes RBC T crypt antigens. The exposed T crypt antigen binds with IgM anti-T, which is present in adult plasma, resulting in hemolysis. T-activation has been associated in children with necrotizing enterocolitis, septiciema, hemolytic uremic syndrome and *Streptococcus pneumoniae* infection. T-activation is detected by ABO typing discrepancies caused by polyagglutination or *in vitro* hemolysis. Confirmation testing is performed by agglutination using *Arachis hypogaea* and *Glycine soja* lectins. Some physicians recommend washing cellular blood components to prevent transfusion of the anti-T within the plasma and to avoid transfusion of plasma-containing products, while others question the clinical importance of this step.

**Paroxysmal Nocturnal Hemoglobinuria:** Paroxysmal nocturnal hemoglobinuria (PNH) is manifested by complement-mediated hemolytic anemia, thrombophilia, and marrow failure. Washing of RBC products for PNH patients was advocated, but recent data do not support their use.

## Recommended Reading

Popovsky MA. (2001). Frozen and washed red blood cells: new approaches and applications. *Transfus Apher Sci* **25**, 193–194.

Valeri CR, Ragno G, Van Houten P *et al.* (2005). Automation of the glycerolization of red blood cells with the high-separation bowl in the Haemonetics ACP 215 instrument. *Transfusion* **45**, 1621–1627.

# CHAPTER 41

# Volume-reduced Products

Cassandra D. Josephson, MD

Volume reduction (also known as hyperpacking or hyperconcentrating) – a process performed after a blood component has been manufactured – is the removal of the fluid portion of a cellular blood product RBC or platelet product. The fluid phase contains residual plasma and anticoagulant-preservative solution. Volume reduction results in a more concentrated cellular product. This procedure is performed more often in the pediatric than the adult setting. In general, volume reduction requires a centrifugation step followed by expression of the supernant fluid, and is performed on request immediately prior to issue from the hospital blood bank to the patient for administration. Most transfusion authorities suggest that this process modification only be used in explicitly indicated circumstances.

## Red Blood Cell Products:

**Methods:** Volume-reduced pediatric products are manufactured using a modification of the RBC component preparation method described in Chapter 28. The volume-reduced method was described by Strauss and colleagues, and enables the manufacture of small aliquots of RBCs for neonates with hematocrit >90% using a single RBC product until expiration on day 42.

Briefly, donated whole blood collected in CP2D is centrifuged at 5000 g for 5 minutes, the supernatant platelet-rich plasma is removed, and 100 ml of extended storage media, preferably AS-3 (because it does not have mannitol), is added to and mixed with the RBCs. The product is drained by gravity through a leukoreduction filter into a primary storage bag. Attached to this primary storage bag, by way of a sterile connecting device, are a cluster of small-volume bags. When an aliquot for transfusion is ordered, the storage bag is centrifuged in an inverted position to pack the RBCs to a hematocrit of approximately 90%. The volume of RBCs requested flows out into one of the attached small-volume bags, which is subsequently disconnected. The remainder of the AS-3 product is mixed and returned to storage. In addition, the AS-3 storage bag is mixed thoroughly each week.

A less complicated method of concentrating RBC products is by *inverted gravity sedimentation*. This method stores RBC products in the refrigerator "upside down," which will concentrate an additive solution product to a hematocrit of around 70–90% within 72 hours. In addition, this method does not require a refrigerated centrifuge.

**Indications:** Two primary indications exist for the volume reduction of RBC products, including prevention of transfusion of unnecessary fluid to a volume-sensitive patient, and for the removal of the potassium-containing supernatant to prevent hyperkalemia in an at-risk patient. Volume reduction of RBC products to prevent passive transfusion of potentially hazardous concentrations of potassium applies to large volume transfusions (>20 ml/kg). Safety with small-volume transfusions has been established by Strauss. The concentration of extracellular potassium in the supernatant depends upon the age of the RBC product and when/if the product was irradiated. The extracellular potassium increases from ~4 mmol/l at day 0 to ~60 mmol/l at day 42 in an additive solution RBC product, or to 80 mmol/l at day 35 in a CPDA-1 RBC product. Therefore, a 1 kg infant receiving a 15 ml transfusion would only receive 0.3–0.4 mEq of potassium, which is a small amount compared to the daily requirement of potassium of 2–3 mEq per kg. There have been several studies confirming the lack of hyperkalemia or other untoward events for small-volume transfusions of 42-day storage additive solution RBC products into neonates. There are individual patients, such as patients with or at risk for hyperkalemia (e.g. small infants with only one vascular access point and the tip of the catheter near the right atrium) and patients receiving large-volume transfusion, who would be at an increased risk of transfusion-related hyperkalemia and may benefit from volume-reduced RBC products.

**Adverse Effects:** Volume reduction of RBC products does not appear to harm the RBCs. The removal of supernatant can be performed in a closed system, so the expiration date of the product does not change.

## Platelet Products:

**Methods:** Pooled individual platelet concentrates, and apheresis platelets, may be volume reduced if necessary. The stored platelets may be centrifuged at 20–24°C at 580 g for 20 minutes, or at 2000 g for 10 minutes, or at 5000 g for 6 minutes; optimal centrifugation rates and times have not been clearly determined. Following centrifugation, the platelets must rest without agitation for 20–60 minutes prior to resuspension and eventual transfusion. Because volume reduction is performed in an open system, the expiration date of the platelet products must be changed to 4 hours, starting from the time the product was entered for processing.

## Indications:

*ABO Out-of-group Platelet Transfusion:* When ABO-type specific platelets are unavailable, volume reduction to remove anti-A and/or anti-B in the platelet product plasma may be indicated if these antibodies are incompatible with the recipient's RBCs. The ABO-incompatible plasma most commonly causes a weakly positive direct antiglobulin test, but may result in severe acute intravascular hemolysis, which in rare circumstances is fatal. This possibility is especially true in small children, who have a higher transfusion volume to total blood volume ratio. The situation with the highest risk of hemolysis is a group O apheresis product containing high-titer anti-A (usually considered > 1 : 64 IgM or > 1 : 256 IgG) that is transfused to a group A or AB recipient. Alternative strategies to mitigate the risk of a hemolytic transfusion reaction include: (1) washing the product, or (2) titering group O platelet products for anti-A, and using only those with non-high titer anti-A for transfusion to group A or group AB individuals.

*Prevention of Transfusion-associated Cardiovascular Overload:* Transfusion-associated cardiovascular overload (TACO) may be minimized by employing volume reduction in platelet products. This is especially helpful in those individuals who are sensitive to volume, such as low birth-weight newborns.

*Decrease in Febrile Non-hemolytic Transfusion Reactions:* In adults, plasma-reduced platelet components decrease the incidence of FNHTRs. This effect is secondary to a decrease in leukocyte derived cytokines which accumulate in the plasma during storage. Decrease in FNHTRs is currently more often attained by pre-storage leukoreduction.

**Adverse Effects:** The centrifugation process has untoward effects on the platelets themselves, which may result in platelet loss, clumping and dysfunction. Thus, the AABB Committee on Pediatric Hemotherapy has recommended that "volume reduction of platelets should be reserved for special infants for whom marked reduction of all intravenous fluids is truly needed."

**Hematopoietic Progenitor Cell Products:** The indication for volume reduction of hematopoietic progenitor cell (HPC) products, particularly bone marrow and peripheral blood HPC derived products, is ABO incompatibility between the donor and recipient, where the donor has antibodies against the recipient RBCs. The HPC product can be centrifuged and the plasma decanted prior to transfusion.

## Recommended Reading

Josephson CD, Mullis NC, Van Demark C, Hillyer CD. (2004). Significant numbers of apheresis-derived group O platelet units have "high-titer" anti-A/A,B: implications for transfusion policy. *Transfusion* **44**, 802–804.

Larrsson LG, Welsh VJ, Ladd DJ. (2000). Acute intravascular hemolysis secondary to out-of-group platelet transfusion. *Transfusion* **40**, 902–906.

Sherwood WC, Donato T, Clapper C, Wilson S. (2000). The concentration of AS-1 RBCs after inverted gravity sedimentation for neonatal transfusions. *Transfusion* **40**, 618–619.

Strauss RG. (2000). Data-driven blood banking practices for neonatal RBC transfusions. *Transfusion* **40**, 1528–1540.

Strauss RG. (2008). How I transfuse red blood cells and platelet to infants with the anemia and thrombocytopenia of prematurity. *Transfusion* **48**, 209–217.

Strauss RG, Villhauser PJ, Cordle DG. (1995). A method to collect, store and issue multiple aliquots of packed red blood cells for neonatal transfusions. *Vox Sang* **68**, 77–81.

# CHAPTER 42

# Neonatal and Pediatric Transfusion Medicine

Jeanne E. Hendrickson, MD and Cassandra D. Josephson, MD

Transfusion of pediatric patients and neonates is complicated by the physiologic changes that occur from the fetal period to infancy and beyond. Specifically, the patients' (1) small size, (2) immature coagulation and immune system, (3) unique conditions requiring transfusion therapy, and (4) potential metabolism concerns due to transfusion therapy all contribute to the complexities of transfusion therapy in these populations.

## Red Blood Cell Transfusions:

**RBC Transfusion Considerations in Neonates:** Neonatal RBC transfusion thresholds vary from institution to institution, with the gestational age, the postnatal age, and the clinical condition of the patient being important considerations. Given the conflicting results of two prospective randomized studies comparing liberal versus restrictive transfusion strategies in neonates, no definite recommendations for transfusion threshold guidelines in neonates exist at this time. Generally, RBCs are transfused at a hemoglobin (Hgb) approaching 13 g/dl in infants with severe cardiopulmonary disease, at levels of 10 g/dl in infants with moderate cardiopulmonary disease or those undergoing major surgery, and at levels of 8 g/dl in infants with uncomplicated symptomatic anemia.

**RBC Product Selection for Neonates:** The type of anticoagulant-preservative solution most likely does not pose a risk to premature infants and neonates in the context of a *small*-volume transfusion (i.e. 10–15 ml/kg of RBCs). Some hospitals utilize CPDA-1 RBC products for neonates due to the higher hematocrit (Hct), lower anticoagulant-preservative:RBC ratio, and lack of mannitol (which is in AS-1 and AS-5 solutions). Others utilize additive solution (AS) RBC products; still others volume-reduce AS RBC aliquots prior to transfusion to increase the Hct of the unit. The relative age of the products transfused to neonates also varies by institution, with some transfusing RBC products <7 days old to neonates, and others using a dedicated unit for a particular neonate until the outdate to reduce donor exposure.

Although the rise in plasma potassium following a small-volume transfusion is relatively slight, there is a risk of hyperkalemia in neonates receiving *large*-volume transfusions. Thus, consideration should be given to volume-reducing products for infants requiring >20–25 ml/kg of RBCs. An additional consideration is the citrate load in CPDA-1 units, which can lead to hypocalcemia in infants following large-volume transfusions. Finally, the decrease in intracellular 2,3-diphosphoglycerate (DPG) with aging of RBCs may be a consideration, especially in the most critically ill neonates. Thus, some institutions reserve the freshest RBC products for these patients.

*Leukoreduction:* Pre-storage leukoreduced RBC products are typically provided for neonates in the US. Leukoreduction decreases the risk of febrile transfusion reactions,

HLA alloimmunization, and CMV transmission (leukoreduced RBCs are considered "CMV safe").

*CMV Seronegative:* A number of institutions provide blood from CMV-seronegative donors ("CMV negative") for low birth-weight infants (LBWI) (<1200 g) who are at highest risk for contracting CMV. The risk of CMV transmission may be slightly lower following transfusion of "CMV-negative" as compared to "CMV-safe" (i.e. leukoreduced) products, but this has not been proven. The risk of acquiring CMV from CMV-seronegative donors, however, is not zero, due to the pre-seroconversion "window period."

*Irradiation:* The use of irradiated RBC products also varies by country and institution. Given the potential for an undiagnosed cellular immune deficiency which would increase the neonate's risk for transfusion-associated graft-versus-host disease (TA-GVHD), some institutions opt to irradiate all blood products given to infants under a certain age (i.e. ≤6 months of age). Irradiated RBCs must be provided to neonates with congenital immunodeficiencies or malignancies, those undergoing HPC transplantation, those requiring treatment with fludarabine or other purine analogues therapy, and those receiving products from relatives.

RBC products should be irradiated close temporally to the transfusion for patients who are receiving large-volume transfusions or are more sensitive to increased levels of potassium. Alternatively, the product can be volume reduced or washed to decrease the potassium load.

**Neonatal RBC Compatibility Testing:** Given that alloimmunization to transfused RBCs in infants under 4 months of age is extremely unusual, AABB *Standards* (Standard 5.16) require limited pretransfusion testing for neonates. ABO and D forward typing must be done once, and neonatal plasma/serum must initially be tested for passively acquired maternal anti-A or anti-B using the antiglobulin (AHG) phase if non-group O RBCs are to be transfused. Additionally, one antibody screen must be performed on a sample from either the neonate or the mother to ensure maternal alloantibodies are not present. During a single hospitalization, repeat antibody screens are not necessary if the initial screen is negative, until 4 months of age. Furthermore, compatibility testing and repeat ABO/D typing are necessary only in certain circumstances, such as an initial positive antibody screen or passively acquired maternal anti-A/anti-B.

**Neonatal RBC Exchange Transfusion Considerations:** Exchange transfusion is utilized primarily to decrease bilirubin levels, in order to avoid kernicterus. In cases of hemolytic disease of the fetus and newborn (HDFN), maternal alloantibodies and neonatal antigen positive RBCs are removed, and replaced with RBCs negative for the relevant antigen. Cord bilirubin levels >5 mg/dl, bilirubin levels that rise >1 mg/dl/hour, or indirect bilirubin >20 mg/dl are all potential indications for exchange transfusion. A double volume exchange transfusion removes approximately 75–90% of circulating RBCs and 25% of the total bilirubin (which quickly equilibrates between the extravascular tissues and the intravascular space following the procedure). RBCs are typically volume reduced, washed, or deglycerolized, and reconstituted with fresh frozen plasma (FFP) to a hematocrit of about 50% for an exchange transfusion. Post-procedure labs must be checked to ensure Hgb, platelet count, PT, PTT and fibrinogen levels are appropriate (see Chapter 43).

### RBC Transfusions in Children and Adolescents:

RBC transfusions are dosed by weight for neonates and children, with 10–15 ml/kg of RBCs increasing Hgb approximately 2–3 g/dl. Children are generally better able to compensate for anemia than adults, and may be transfused at lower Hgb levels than adults. A recent study of RBC transfusion thresholds (Lacroix *et al.*, 2007) in a pediatric intensive care unit setting found that a Hgb threshold of 7 g/dl for RBC transfusion can decrease transfusion requirements without increasing adverse outcomes in stable, critically ill children. However, higher Hgb levels are necessary for adequate growth and development in children with chronic anemias, such as thalassemia major.

### RBC Product Selection in Children and Adolescents:

The indications for irradiation, leukoreduction and CMV-seronegative products are generally similar to those for adults, and are discussed in Chapters 36, 37 and 38, respectively. An exception is that some institutions irradiate cellular products for young children to avoid the risk of TA-GVHD in patients with undiagnosed cellular immune deficiencies.

### Platelet Transfusions:

### Platelet Transfusions in Neonates:

Thrombocytopenia can occur in neonates due to decreased production or accelerated destruction of platelets. Most cases of neonatal thrombocytopenia are multifactorial, with prematurity and infection accounting for a large percentage of them. The diagnosis of neonatal alloimmune thrombocytopenia (NAIT) or maternal idiopathic thrombocytopenic purpura (ITP) should be considered in otherwise healthy infants with thrombocytopenia, as should congenital platelet disorders. Both NAIT and maternal ITP result in passively transferred maternal antibodies (alloantibodies in the case of NAIT and autoantibodies in the case of maternal ITP) against fetal/neonatal platelet antigens.

Platelet transfusion triggers for neonates are not clearly defined, with practices varying by institution and provider, depending on the clinical situation and gestational age of the neonate. Generally, platelet counts are kept above 50,000/μl in bleeding neonates or those undergoing an invasive procedure, with counts kept above 100,000/μl in extremely ill, premature infants. Prophylactic transfusions are generally given for platelet counts below 20,000/μl in neonates. Transfusion of 10 ml/kg of platelets generally leads to an expected increase in platelet counts of 50,000–100,000/μl.

### Platelet Transfusions in Children:

Indications for platelet transfusions in children are similar to those in neonates. As in neonates, platelets are dosed by volume in children (approximately 10 ml/kg per dose). Ongoing studies are investigating dosing by platelet number per kilogram, rather than by volume.

### Platelet Product Selection:

*Irradiation:* Irradiation of platelets is indicated for situations similar to those outlined in the RBC section above. One notable indication for irradiation is for maternal platelets being transfused to neonates with NAIT. This is necessary to prevent TA-GVHD, given the HLA similarity between mother and neonate. In addition, these maternal platelets must be washed prior to transfusion to remove the antibody-containing plasma.

*ABO/Rh Compatibility:* Platelet components should be ABO/Rh compatible, if possible. The transfusion of group O platelets to group A or B recipients should be avoided, due to the possibility of the passive transfer of anti-A or anti-B. These antibodies have been reported to cause hemolysis in children. If needed, ABO-incompatible products can be volume reduced or washed prior to transfusion. Should Rh(D)-positive platelets have to be given in instances of blood shortages to Rh(D)-negative individuals, RhIg should be considered to prevent alloimmunization.

*Leukoreduction:* Pre-storage leukoreduced platelet products are typically provided for infants and children in the US, as outlined in the RBC section above.

*CMV Seronegative:* As with RBCs, leukoreduced "CMV-safe" platelets are generally provided for infants and children in the US. When possible, platelets from CMV seronegative donors may be provided for low-birth-weight infants.

*Volume Reduction:* Volume reduction is not recommended, except in exceptional cases, due to the platelet loss (both in number and function) that occurs as a result of the procedure.

**Plasma Transfusions:** Reference values of coagulation factors for children under 6 months of age are different for children and adults. However, indications for plasma transfusions are similar for neonates, children and adults. Dosing of plasma for neonates and children is weight-based. Most coagulation factor levels will be raised by approximately 20%, using 10–20 ml/kg of fresh frozen plasma (FFP).

**Plasma Product Selection:** FP24 (plasma frozen within 24 hours of collection) has lower levels of Factor VIII than FFP, which is frozen within 8 hours of collection. FP24 should not be chosen to treat Factor VIII deficiencies; many institutions use these products interchangeably for other indications.

**Cryoprecipitate Transfusions:** Indications for cryoprecipitate transfusions are similar for neonates, children and adults, where cryoprecipitate is primarily transfused to increase fibrinogen. Cryoprecipitate is dosed by weight in neonates and children, with 1–2 units transfused per 10 kg.

**Cryoprecipitate Product Selection:** It is generally recommended that ABO-compatible cryoprecipitate is transfused to neonates and children whenever possible. However, cryoprecipitate contains smaller amounts of anti-A and anti-B than other plasma products.

**Granulocyte Transfusions:** Transfused granulocytes may be more efficacious in neonates and children than in adults, due to the size of the recipient allowing for a larger dose per kg. Bacterial sepsis unresponsive to antibiotics in patients with severe neutropenia (ANC <500/µl for children, with higher ANC counts in infants) is a potential indication for granulocyte therapy. Likewise, severe infection unresponsive to antibiotics in patients with qualitative neutrophil defects (i.e. chronic granulomatous disease) is a potential indication for granulocyte therapy. Refractory fungal sepsis in neutropenic oncology patients is a third potential indication for granulocyte therapy (see Chapter 32).

**Granulocyte Product Selection:** Granulocytes should not be transfused through a leukoreduction filter. CMV-seronegative recipients should receive granulocytes from CMV-seronegative donors, when possible. All granulocyte products transfused to immunocompromised recipients must be irradiated. Additionally, due to significant RBC contamination, the product must be ABO/D compatible and crossmatch compatible with the recipient.

## Recommended Reading

Bell EF, Strauss RG, Widness JA et al. (2005). Randomized trial of liberal versus restrictive guidelines for red blood cell transfusion in preterm infants. *Pediatrics* **115**, 1685–1691.

Kirpalani H, Whyte RK, Andersen C et al. (2006). The Premature Infants in Need of Transfusion (PINT) study: a randomized, controlled trial of a restrictive (low) versus liberal (high) transfusion threshold for extremely low birth-weight infants. *J Pediatr* **149**, 301–307.

Lacroix J, Hebert PC, Hutchison JS et al. (2007). Transfusion strategies for patients in pediatric intensive care units. *N Engl J Med* **356**, 1609–1619.

Luban NL. (2002). Neonatal red blood cell transfusions. *Curr Opin Hematol* **9**, 533–536.

Roseff SD, Luban NL, Manno CS. (2002). Guidelines for assessing appropriateness of pediatric transfusion. *Transfusion* **42**, 1398–1413.

Strauss RG. (2008). How I transfuse red blood cells and platelets to infants with the anemia and thrombocytopenia of prematurity. *Transfusion* **48**, 209–217.

# CHAPTER 43
# Perinatal Transfusion Medicine

Beth H. Shaz, MD

Transfusion management of the pregnant woman and fetus requires special consideration. This chapter will address the following related issues: 1) routine prenatal and neonatal transfusion testing in relationship to maternal alloimmunization, 2) hemolytic disease of the fetus and newborn (HDFN), 3) neonatal alloimmune thrombocytopenia (NAIT), and 4) treatment of HDFN and NAIT, including proper selection of blood components.

The management of HDFN and NAIT includes maternal, fetal and neonatal testing and treatment. HDFN occurs when maternal plasma contains a RBC alloantibody against an antigen carried on the fetal RBCs, resulting in hemolytic anemia. The administration of Rh immune globulin (RhIg) perinatally has dramatically decreased the incidence of HDFN due to anti-D. The primary goals of prenatal testing are to determine which women would benefit from RhIg prophylaxis and which women/fetuses require further monitoring/treatment for HDFN. NAIT occurs as a result of a maternal platelet alloantibodies directed against an antigen on fetal platelets resulting in thrombocytopenia, which is more extensively discussed in Chapter 81.

Other obstetrical issues associated with transfusion medicine that may occur are discussed in different chapters of this book. Severe hemorrhage, which requires massive transfusion, results from placenta previa, or uterine atony or rupture, and DIC can lead to hysterectomy and loss of future reproductive capacity, and/or loss of the mother, child, or both (Chapter 50). In addition, thrombocytopenia can occur in pregnancy, and may be secondary to immune thrombocytopenic purpura (ITP); thrombotic thrombocytopenic purpura (TTP); hemolysis, elevated liver enzymes, and low platelets (comprising the HELLP syndrome); and acute fatty liver of pregnancy (Chapter 103). Lastly, transfusion management of pregnant patients with hemoglobinopathies, such as sickle cell disease and thalassemia, is discussed in Chapter 45.

**Hemolytic Disease of the Fetus and Newborn:** HDFN occurs when maternal plasma contains an alloantibody against an antigen carried on the fetal RBCs. The maternal IgG crosses the placenta and coats the fetal RBCs. The sensitized RBCs are removed from circulation through splenic macrophages, which leads to fetal anemia. In an effort to compensate for the RBC loss, the bone marrow releases immature RBCs (erythroblastosis fetalis). When the bone marrow fails to produce enough RBCs, then erythropoiesis occurs outside of the bone marrow in the spleen and liver. As the spleen and liver enlarge (hepatosplenomegaly), this results in hepatocellular damage where the hepatocytes can no longer produce enough plasma proteins, leading to high-output cardiac failure with generalized edema, effusions and ascites (hydrops fetalis). Hydrops fetalis, which can develop as early as 17 weeks' gestational age, was previously uniformly fatal, but with the use of intrauterine transfusions (IUT) and other therapeutic modalities there is a 74% survival rate. Severe non-hydropic HDFN, requiring IUT, has a 90% survival rate.

If severe anemia and/or hydrops develops prior to the ability to perform IUT (prior to 18 weeks gestational age), a combination of maternal treatment of plasma exchange and intravenous immunoglobulin has been used until IUT is possible.

*In utero*, the bilirubin released from hemolyzed RBCs is cleared by the placenta. After birth, the neonatal liver has limited capacity to conjugate the bilirubin. When increased levels of unconjugated bilirubin exceed the binding capacity for albumin, the unbound, unconjugated bilirubin crosses the blood–brain barrier and results in neuronal cell death in the basal ganglia and brain stem (known as kernicterus). Kernicterus is prevented by treating the neonatal hyperbilirubinemia by phototherapy and, if needed, RBC exchange. Guidelines for phototherapy and RBC exchange transfusion are published by the American Academy of Pediatrics Subcommittee on Hyperbilirubinemia.

The severity of HDFN is influenced by the antibody titer and specificity (anti-D, anti-c, and anti-K have the highest likelihood of severe HDFN), the immunoglobulin class, and the number of antigenic sites on the RBC. In general, the severity of HDFN increases with subsequent pregnancies. Anti-A and/or anti-B antibodies are the most common antibodies associated with HDFN, but the disease is usually mild.

Immune sensitization to RBC antigens occurs after fetomaternal hemorrhage (FMH) during pregnancy or delivery, or through previous RBC transfusion. As little as 0.1 ml of D-positive RBCs may result in D sensitization. The incidence of maternal sensitization decreases with ABO incompatibility between the fetus and mother. Due to the use of RhIg prophylaxis, the incidence of anti-D formation has decreased from 14% to 0.1% of D-negative mothers.

**Prenatal Testing:** At the first prenatal visit, usually at 12 weeks' gestational age, the maternal ABO/D type and antibody screen are performed. Testing for weak D is not required. If the pregnant woman is D positive and has no alloantibodies, then no further testing is recommended. If the pregnant woman is D negative and is not sensitized to the D antigen, then she should receive RhIg perinatally. If the pregnant woman is sensitized to a RBC antigen, then further blood bank testing is required and, potentially, monitoring or treating for HDFN.

**D Positive, No Alloantibodies:** No further testing is recommended if transfusion is unlikely.

**D Negative, Unsensitized to the D Antigen:** If the mother is D negative and has not been previously sensitized to the D antigen, then she should receive RhIg during pregnancy. If a woman is D positive or weak D positive, then she does not need to receive RhIg to prevent sensitization to the D antigen. Because 10% of individuals typed as weak D are partial D (who potentially can form anti-D), some authorities will either recommend further testing to determine if the woman is a partial D or recommend the use of RhIg. Yet it is unknown if women with partial D benefit from the administration of RhIg. In the US, most transfusion services do not perform additional D testing to determine if a woman is partial D (see Chapter 23).

*Rh Immune Globulin:* Antepartum and postpartum dosing is usually with 300 μg of RhIg (which neutralizes 30 ml of whole blood or 15 ml of D-positive RBCs), though some institutions prefer to give a vial of 50 μg (which neutralizes 5 ml of whole blood or 2.5 ml of D-positive RBCs) at ≤ 12 weeks gestational age, but there is a risk of inadvertent misadministration of the lower dose (see Chapter 35).

RhIg is administered:

- at 28 weeks' gestational age (dose 300 μg),
- at delivery, if the neonate is D positive, weak D positive or D untested (minimum dose of 300 μg, further dosing determined by FMH testing), and
- following perinatal events associated with FMH, such as abortion, ectopic pregnancy, amniocentesis, chorionic villus sampling, external cephalic version, abdominal trauma and antepartum hemorrhage (minimum dose of 300 μg, further dosing determine by FMH testing, if >20 weeks' gestational age).

After a perinatal event or delivery beyond 20 weeks gestational age, when the fetal blood volume exceeds 30 ml, quantification of FMH is recommended. In all situations, RhIg should be administered within 72 hours of the event. If a dose is not administered during that timeframe, then it should be administered as soon as possible, even up to 28 days after the event.

*Fetomaternal Hemorrhage Testing:* The goal of FMH testing is to determine an adequate dose of RhIg to neutralize a FMH. A sample for FMH testing should be obtained approximately one hour after delivery on all D negative women who deliver a D positive infant (or one hour after an event as described above). FMH testing can be performed by a screening test or rosette test and, if positive, quantification is typically performed by an acid elution test (Kleihauer-Betke test) or flow cytometry. Other FMH detection methodologies are in use, such as gel agglutination and enzyme linked antiglobulin test.

**Rosette Test:** The rosette test demonstrates the number of D positive cells in a D-negative suspension using an anti-D reagent. The anti-D binds to D positive RBCs, and when indicator D-positive RBCs are added rosettes are formed. This method has a FMH detection limit of about 10 ml. If the infant is weak D phenotype (or D unknown), the rosette test may result in a false negative and therefore a test that detects fetal hemoglobin should be used, such as the Kleihauer-Betke or flow cytometry test.

**Kleihauer-Betke Test:** The Kleihauer-Betke test has numerous limitations, including low sensitivity, poor reproducibility, and tendency to overestimate the FMH volume, yet it is the most commonly used test to quantify FMH in the US. The Kleihauer-Betke test is performed on a maternal blood smear treated with acid and then stained so the fetal cells remain red and the maternal cells appear as ghosts; 2000 cells are counted, and the percentage of fetal cells is determined.

**Flow Cytometry:** Flow cytometry techniques quantitate the amount of hemoglobin F or D-positive RBCs, and are simpler, more precise and more reliable than the Kleihauer-Betke test. Flow cytometry techniques are not routinely available throughout the US.

*RhIg Dosing in the Presence of a FMH:* The calculation of RhIg dosing used by the AABB method (other methods exist) is presented in Table 43.1.

---

**TABLE 43.1  Calculation of RhIg Dosing**

% Fetal RBCs in maternal circulation is determined by Kleihauer-Betke test or flow cytometry.
Maternal blood volume (ml) $= 70\,\text{ml/kg} \times$ maternal weight (kg), or 5000 ml if maternal weight unknown
Fetal bleed (ml) $=$ % fetal RBCs $\times$ maternal blood volume
Dose of RhIg $=$ fetal bleed (ml)/30 ml per dose (300-mg vial)
If the number to the right of the decimal point is $< 5$, round down and add one dose of RhIg (e.g. $2.3 \rightarrow 2 + 1 = 3$ vials)
If the number to the right of the decimal point is $\geq 5$, round up and add one dose of RhIg (e.g. $2.6 \rightarrow 3 + 1 = 4$ vials).

---

**Alloantibodies Currently or Previously Detected:**  In one series, 1% of pregnancies were complicated by alloimmunization to RBC antigens which were associated with HDFN (40% anti-D, 30% antibodies to other Rh antigens, and 30% to non-Rh antigens). The goals of blood bank prenatal testing are to identify the alloantibody (Chapter 20), to determine the clinical significance of the alloantibody (Table 43.2), and to perform antibody titrations to guide the clinicians as to when fetal monitoring is required.

**TABLE 43.2  Probability of Causing Severe HDFN Associated with RBC Antibodies**

| Blood Group | Highest Likelihood of Severe HDFN | Rare Cases of Severe HDFN | Usually Associated with Mild Disease | Not a Cause of HDFN |
|---|---|---|---|---|
| MNS | | M, S, s, U, Mi[a], Vw, Mur, Mt[a], Hut, Hil, M[v], Far, s[D], En[a], MUT | M, S, s, U, Mt[a], Mit | N |
| Rh | D,c | C, E, f, Ce, C[w], C[x], E[w], G, Hr$_0$, Hr, Rh29, Go[a], Rh32, Be[a], Evans,Tar, Rh42, Sec, JAL, STEM | E, e, f, C[x], D[w], Rh29, Riv, LOCR | |
| Lutheran | | | Lu[a] (rar), Lu[b] | |
| Kell | K | k, Kp[a], Kp[b], Ku, Js[a], Js[b], Ul[a], K11, K22 | Ku, Js[a], K11 | K23, K24 |
| Lewis | | | | Le[a],Le[b] |
| Duffy | | Fy[a] | Fy[b] (rare), Fy3 (rare) | |
| Kidd | | Jk[a] | Jk[b](rare), Jk3 | |
| Other | | Di[a], Wr[a], Rd, Co[a], Co3, PP$_1$P[k], Vel, MAM, Bi, Kg, JONES, HJK, REIT | Di[b], Sc3, Co[b], Ge2 (rare), Ge3, Ls[a], Lan, At[a], Jr[a] JFV, HOFM | P$_1$, Wr[b],Yt[a], Yt[b], Sc1, Sc2, CH/RG, CROM, KN, JMH, I, Jr[a], HLA: Bg[a], Bg[b], Bg[c] |

[*]For some of the antibodies listed, the information is based on a very small number of example, sometimes only one, resulting in overlap between categories. Eder AF (2006). Update on HDFN: new information on long-standing controversies. *Immunhematology* **22**, 188–195.

*Phenotype/genotype the Father or Fetus:* If paternity is assured, the father may be phenotyped to help in assessment of the risk of HDFN of the fetus. If the father's RBCs do not carry the antigen, then no further work-up needs to be performed. If the father carries a double dose of the antigen (i.e. a mother has anti-Jk$^a$ and the father types as Jk(a + b−)), then the fetus is at risk for HDFN. If the father carries a single dose of the antigen (i.e. Jk(a + b+)), then the fetus has a 50% chance of being at risk. Fetal genotyping or phenotyping may be performed to determine the antigen status of the fetus. Fetal DNA can either be obtained through amniocentesis or extracted from the maternal plasma, but currently the latter is not routinely available in the US.

*Antibody Titration:* Antibody titers should be performed using the saline tube or gel technique in the antihuman globulin phase with a previous sample (frozen at −20°C or colder), if available, run in parallel to the current sample to compare the change in titration from one sample to the next. Some institutions use RBCs with the strongest dose of the corresponding antigen (R$_2$R$_2$ cells in the case of anti-D) and some institutions use RBCs with a single dose of antigen; in either case the laboratory should consistently use RBCs of that phenotype. Antibody titers are performed by serially diluting the patient's plasma/serum against the appropriate RBCs and determining the titer where 1+ reaction still occurs (AABB method). The titration endpoint is reported as a reciprocal of the titer (i.e. 1:16 is reported as 16). The use of gel technology or enhancement strategies likely changes the critical titer.

The critical titer varies between laboratories, but most institutions use 8 to 32 for anti-D. Critical titers for other antibody specificities remain unclear. This is especially true for anti-K, where low titers of 4 have been associated with HDFN. With the advent of non-invasive measurements of fetal anemia through Doppler assessment of the middle cerebral artery peak systolic velocity, the role of antibody titration, as well as more invasive techniques (such as amniocentesis and cordocentesis), is limited. Titrations should be performed every 2–4 weeks after 18 weeks gestational age. Once the critical titer is reached no further titration studies are necessary, and the pregnancy should be monitored by middle cerebral artery Doppler. In addition, if the mother has had a previously affected fetus or infant, maternal antibody titers are not helpful.

*Middle Cerebral Artery Peak Systolic Velocity:* A recent study demonstrated equivalency between Doppler measurements of the peak velocity of systolic blood flow in the middle cerebral artery to serial $\Delta OD_{450}$ measurements using amniocentesis. Fetal anemia requiring transfusion was predicted by ≥1.5 multiples of the median. MCA blood flow has 88% sensitivity and 82% specificity in predicting fetal anemia.

*Amniocentesis:* Prior to the use of middle cerebral artery peak systolic velocity, serial amniocentesis was used to determine the severity of hemolysis. For the majority of antibodies, the most notable exception is anti-K (because anti-K affects early RBC precursors, which results in both suppression of erythropoiesis and hemolytic anemia); amniotic fluid bilirubin levels correlate with the severity of hemolysis. The bilirubin level is quantified by spectrophotometry and measured as the changed in the optical density at the wavelength 450 nm ($\Delta OD_{450}$). The $\Delta OD_{450}$ is then plotted on the Liley chart, which is most accurately used after 27 weeks gestational age, to determine the severity of hemolysis (zone I predicts mild or no disease, zone II predicts moderate disease and zone III predicts severe disease). Serial measurements are used to follow the severity of disease; when values are unchanging or increasing, then the disease is worsening. Amniocenesis can result in fetal injury or death.

*Periumbilical Cord Blood Sampling:* Fetal blood can be sampled through percutaneous umbilical blood samplings (PUBS, also known as cordiocentesis), which has a 1–2% risk of fetal loss. This sample can be tested for hemoglobin level, hematocrit and blood type, and the direct antiglobulin test (DAT) can be performed.

## Intrauterine Transfusions: 
IUT is performed after 18–20 weeks gestational age. With advances of ultrasonography, allogeneic blood is directly infused into the umbilical cord vein. An alternative method is injection of RBCs into the fetal peritoneal cavity, from where the RBCs are absorbed into circulation through the lymphatic system. A combination of intravascular and intraperitoneal transfusions can also be used to prolong the interval between transfusions. There is a 1–2% risk of fetal mortality with this procedure.

RBC Transfusions: The indications for intrauterine RBC transfusion are fetal anemia secondary to parvovirus infection, HDFN, twin-to-twin transfusion, large-volume fetal hemorrhage, and homozygous alpha thalassemia. Fetal anemia can be detected by amniotic fluid $\Delta OD_{450}$ or Doppler middle cerebral artery measurements, cordocentesis blood sample (hemoglobin $<10\,g/dl$ or hemoglobin two standard deviations below the mean for gestational age), or fetal hydrops noted on ultrasound examination.

For HDFN, the most common antibodies requiring the use of IUT include anti-D alone or with another antibody(ies), anti-K, and (rarely) anti-C. The transfusion volume is suggested to be between 20 and 50% of the fetoplacental blood volume. The normal fetus can tolerate rapid infusion of large blood volumes (5–7 ml/min), but this is not the case with a hydropic fetus, where the transfusion volume should be limited and the posttransfusion hematocrit should be lower for the first procedure. A hydropic fetus requires more frequent transfusions of smaller volumes. Typically, IUT is repeated every 1–3 weeks to maintain a fetal hematocrit at 27–30% (fetal hematocrit decreases 1–2%/day) until 35 weeks gestational age, with delivery at 37–38 weeks after confirmation of fetal lung maturity by amniocentesis. Intravascular exchange transfusions can be performed instead of simple intravascular transfusions, but the simple transfusion is preferred owing to the shorter procedure time and it being technically easier.

Neonates who have received IUT may have suppressed erythropoiesis, which can result in hypoproliferative anemia (low levels of erythropoietin and low reticulocyte counts). This may require RBC transfusions several weeks after birth, and/or the use of exogenous erythropoietin.

*RBC Product Selection:* The RBCs should be group O, D negative, CMV safe (some centers may decide to provide CMV seronegative, especially if the mother is CMV seronegative), leukoreduced, irradiated and hemoglobin S negative. RBC products should be antigen negative for the corresponding antibody(ies) and crossmatch compatible with the maternal plasma. Some recommend products that are less than 7 days old because they have improved levels of 2,3-DPG and lower levels of potassium. An alternative is to use frozen, deglycerolized RBCs, which also have near-normal levels of 2,3-DPG. If a RBC product which has been stored for >7 days needs to be used, it can be washed to remove the potassium and other electrolytes. Washing is also indicated for RBC products stored in additive solution (AS), to remove the additive. In some circumstances, especially when antigen-compatible allogeneic blood is unavailable, maternal RBCs can be used, even if they are ABO mismatched. The maternal RBC product should be washed to remove the antibody-containing plasma. The RBC product should have a high hematocrit, 75–85%, to minimize volume overload. The desired fetal hematocrit after the transfusion is usually 40–45%. IUT volume can be calculated by a variety of methods; Table 43.3 provides one of the calculation methods used.

---

**TABLE 43.3  Intrauterine Transfusion Volume**

**Intraperitoneal transfusion volume:**
  Volume to transfuse (ml) = (gestation in weeks – 20) × 10 ml
**Intravenous transfusion volume:**
  Fetoplacental volume (FPV) (ml) = fetal weight (g) × 0.14

$$\text{Volume to transfuse (ml)} = \frac{\text{FPV} \times (\text{Hct post} - \text{Hct pre})}{\text{Hct of donor product}}$$

Hct post = the desired fetal hematocrit after the transfusion, Hct pre = the fetal hematocrit prior to transfusion, and Hct of donor product = the hematocrit of the RBC donor product

---

**Platelet Transfusions:** The primary indication for intrauterine platelet transfusions is fetal thrombocytopenia secondary to NAIT when the platelet count is < 50,000/μl. The prenatal management of NAIT with intravenous immunoglobulin with or without steroids has decreased the use of intrauterine platelet transfusions in the management of this disease. The use of intrauterine platelet transfusion in thrombocytopenic fetus carries a risk of serious bleeding (see Chapter 81).

*Platelet Product Selection:* The platelet product should be irradiated and leukoreduced (CMV safe), and can be of maternal origin or negative for the corresponding antigen for which the maternal antibody is directed. If maternal platelets are used they must be washed to remove the plasma containing the antibody, and the platelets resuspended in AB plasma (or plasma compatible with both the mother and fetal ABO type) or saline.

**Neonatal Testing:** The cord blood sample should be tested for ABO/D, including weak D, and the corresponding antigen if the mother has a RBC antibody. A false-negative D type can occur if the fetal RBCs are coated with IgG anti-D, known as "blocked D." In this situation, an eluate of these cells will demonstrate anti-D. If there is ABO incompatibility between the mother and infant, or the mother has a RBC antibody, a DAT should be performed using antihuman globulin. The DAT is important in the diagnosis of HDFN, but the strength does not correlate with the severity of disease.

**Treatment of the Neonate with HDFN:** Phototherapy with ultraviolet light can be used in infants with mild to moderate HDFN. In severe cases, RBC exchange transfusion may be necessary to prevent kernicterus. Indications for exchange include cord hemoglobin $\leq 10\,g/dl$, cord bilirubin $\geq 5.5\,mg/dl$, rising bilirubin despite photo-therapy $\geq 0.5\,mg/dl$ per hour, or bilirubin $>20\,mg/dl$ in a term infant (lower bilirubin levels in premature infants or infants with hypoxemia, acidosis and hypothermia). The benefits of RBC exchange transfusion include removal of the bilirubin, removal of sensitized RBCs, removal of maternal antibody, provision of antigen-negative RBCs, and suppression of erythropoiesis. Intravenous immunoglobulin (dose $0.5\,g/kg$) has been shown to decrease the need for RBC exchange in neonates with HDFN.

**RBC Exchange Transfusion:** Group O, D negative, CMV seronegative, irradiated, and hemoglobin S negative products, which are less than 7 days old are used (alternately, frozen and deglycerolized RBC products can be used). In addition, RBC products should be antigen negative for the corresponding antibody(ies) and crossmatch compatible with the maternal plasma. If a maternal sample is not available, then the infant's plasma and/or an eluate from the infant's RBCs can be used. The Group O plasma should be removed and replaced with group AB fresh frozen plasma (FFP) to dilute the RBCs to an appropriate hematocrit (40–50%). In some circumstances, especially when antigen-compatible allogeneic blood is unavailable, maternal RBCs can be used (with the plasma removed) even if they are ABO mismatched. Typically, the volume exchanged is twice the blood volume of the infant (Table 43.4).

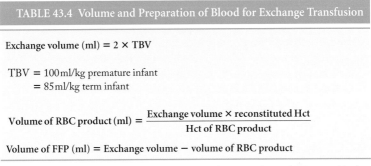

**TABLE 43.4  Volume and Preparation of Blood for Exchange Transfusion**

Exchange volume (ml) = 2 × TBV

TBV = 100 ml/kg premature infant
    = 85 ml/kg term infant

$$\text{Volume of RBC product (ml)} = \frac{\text{Exchange volume} \times \text{reconstituted Hct}}{\text{Hct of RBC product}}$$

Volume of FFP (ml) = Exchange volume − volume of RBC product

Reconstituted Hct is the desired Hct of the product once the RBC product and FFP are combined

## Recommended Reading

American Academy of Pediatrics, Subcommittee on Hyperbilirubinemia. (2004). Management of hyperbilirubinemia in the newborn infant 35 or more weeks gestation. *Pediatrics* **114**, 297–316.

Eder AF. (2006). Update on HDFN: new information on long-standing controversies. *Immunhematology* **22**, 188–195.

Engelfriet CP, Reesink HW, Judd WJ *et al.* (2003). Current status of immunoprophylaxis with anti-D immunoglobulin. *Vox Sang* **85**, 328–337.

Gottstein R, Cooke RW. (2003). Systematic review of intravenous immunoglobulin in haemolytic disease of the newborn. *Arch Dis Child Fetal Neonatal Ed* **88**, F6–F10.

Judd WJ for the Scientific Section Coordinating Committee of the AABB. (2001). Practice guidelines for prenatal and perinatal immunohematology, revised. *Transfusion* **41**, 1445–1452.

Koelewijn JM, Vrijkotte TG, van der Schoot CE *et al.* (2008). Effect of screening for red cell antibodies, other than anti-D, to detect hemolytic disease of the fetus and newborn: a population study in the Netherlands. *Transfusion* **48**, 941–952.

Oepkes D, Seaward G, Vandenbussche FPHA *et al.* (2006). Doppler ultrasonography versus amniocentesis to predict fetal anemia. *N Engl J Med* **355**, 156–164.

# CHAPTER 44
# Autoimmune Hemolytic Anemias

Beth H. Shaz, MD and Christopher D. Hillyer, MD

Autoimmune hemolytic anemias (AIHA) refers to a group of disorders where autoantibodies are directed against red blood cell (RBC) membrane antigens and result in shortened RBC survival (normally 100–120 days) through activation of the complement system and/or removal within the reticuloendothelial system (RES). Classification of AIHAs includes warm autoimmune hemolytic anemia (WAIHA), cold agglutinin disease (CAD), mixed type AIHA, paroxysmal cold hemoglobinuria (PCH), and drug-induced hemolytic anemia. The diagnosis of AIHA is dependent on the clinical features and the serologic work-up, primarily the direct antiglobulin test (DAT; see Chapter 21).

In the AIHAs, the degree of hemolysis depends on the antibody class and characteristics such as concentration and "antigen" affinity, ability to fix complement, and thermal amplitude. These antibodies typically react as *panagglutinins*, meaning that they show *in vitro* reactivity with all tested RBCs. Thus, these antibodies must be binding to a ubiquitous antigen or have non-specific binding. In WAIHA, IgG autoantibody most commonly reacts to Rh group proteins as panagglutinin antibodies. These IgG autoantibodies "coat" RBCs, and these coated RBCs are removed by phagocytosis by Fcγ receptors primarily within the spleen, thus achieving extravascular hemolysis.

In contrast to IgG alloantibodies, IgM molecules efficiently activate the complement cascade (starting with C1q) and can result in intravascular hemolysis. The degree of intravascular hemolysis is affected by antibody concentration, thermal amplitude, and amount and inactivation by complement regulatory proteins, such as decay accelerating factor and others. CAD is caused by IgM autoantibodies, usually of i or I specificity. In severe cases CAD can result in intravascular hemolysis, but also can result in extravascular hemolysis because RBCs are in the cool peripheral circulation for a short time-period and then the IgM dissociates when the RBC is warmed in the central circulation. The short time-period may only be sufficient to activate the complement cascade to the C3b stage and not to the membrane attack complex stage. The C3b-coated RBCs are cleared by hepatic macrophages with receptors specific for C3b if there is a sufficient quantity of C3b molecules on the RBCs (>500–800/RBC). C3b-coated RBCs, which are not removed from the hepatic circulation, are unharmed and the C3b is degraded to C3dg; C3dg-coated RBCs have near-normal survival.

**Clinical Manifestations:** The presence of anti-erythrocyte autoantibodies does not always cause hemolysis and thus anemia. In fact, most patients with low concentrations of IgG autoantibodies will have laboratory findings such as a positive DAT but will not have anemia or related symptoms. When hemolysis is of a significant degree, anemia, jaundice and splenomegaly can occur. Laboratory testing should include a complete blood count, reticulocyte count, serum bilirubin level, serum lactic dehydrogenase (LDH) level, serum haptoglobin value and peripheral blood smear. Hemolytic anemia results in anemia, elevated reticulocyte count, increased bilirubin levels (especially the indirect fraction), elevated LDH level and decreased haptoglobin level. The peripheral smear may demonstrate spherocytes (WAIHA) or RBC agglutination (CAD), which may aid in the diagnosis of hemolytic anemia. Urinalysis will show hemoglobinuria when intravascular hemolysis is present.

**Warm Autoimmune Hemolytic Anemia:** WAIHA is secondary to IgG (rarely IgA or IgM) warm autoantibodies with broad specificity, which typically react at 37°C and result primarily in extravascular hemolysis. Extravascular hemolysis is where the hemolysis occurs in the RES, resulting in increase in serum bilirubin, but usually not hemoglobinemia or hemoglobinuria. IgG-sensitized RBCs are destroyed or damaged within the RES, predominantly by splenic macrophages.

**Pathogenesis:** WAIHA may be idiopathic or secondary to a variety of diseases, such as chronic lymphocytic leukemia, lymphoma, systemic lupus erythematosus and AIDS (Table 44.1). IgG-coated RBC is detected serologically by a positive DAT with IgG; in addition, the RBC may sometimes be coated with complement and therefore the DAT may also be positive for complement (see Chapter 21). IgG, when eluted from RBCs, typically reacts as a panagglutinin, and thus does not demonstrate specificity. Of individuals with WAIHA, 80% have the autoantibody present in the plasma as well as on the RBC membrane when sufficient antibody titers are present (see Chapter 20). The plasma-phase autoantibody usually reacts as a panagglutinin in the indirect antiglobulin test (i.e. antibody screen/antibody identification panel). Occasionally antibody identification demonstrates specificity to antigens in the Rh system (usually c or e) or other RBC antigens, and thus is said to have "apparent specificity."

| TABLE 44.1 Immune Hemolytic Anemias | | | | | |
|---|---|---|---|---|---|
|  | **WAIHA** | **CAD** | **Mixed AIHA** | **PCH** | **Drug-induced** |
| Percent of cases | 48–70% | 16–32% | 7–8% | Rare in adults; 32% in children | 12–18% |
| DAT | IgG 20–66%; IgG + C3 24–63%; C3 7–14% | C3 91–98% | IgG + C3 71–100% | C3 94–100% | IgG 94%; IgG + C3 6% |
| Ig type | IgG (rare IgA or IgM) | IgM | IgG + IgM | IgG | IgG |
| Eluate | Panagglutinin | Non-reactive | Panagglutinin | Non-reactive | Panagglutinin |
| Plasma | Panagglutinin (IgG) | Panagglutinin (IgM) | Panagglutinin (IgG + IgM) | IgG biphasic hemolysin | Panagglutinin (IgG) |
| Antibody specificity | Rh | I (rare i or P^r) |  | P | Rh |

Modified from Brecher ME (ed.). (2005). *AABB Technical Manual*, 15th edition. Bethesda, MD: AABB Press.

Treatment: Treatment for WAIHA depends upon the clinical severity of the disease. First-line treatment is usually corticosteroids; other treatments include splenectomy, rituximab, intravenous immunoglobulin and alternate immunosuppressive medications. In secondary WAIHA, treatment should also target mitigating the underlying illness. Severe anemia with cardiac or cerebral dysfunction requires urgent management, which may include RBC transfusion. Patients with severe disease who are not responsive to RBC transfusion and other immunomodulatory treatments, secondary to rapid RBC destruction, may occasionally benefit from plasma exchange for removal of the pathogenic antibodies (see below).

Blood Bank and Transfusion Management: During an acute presentation in a patient with newly diagnosed AIHA, finding the appropriate RBC product for transfusion can be a challenge, and close communication between the transfusion service and the treating physician is necessary. Owing to the presence of a strong autoantibody, both the direct and indirect antiglobulin tests (and thus the antibody screening tests and panels) will be positive as the autoantibody, reacting as a panagglutinin, will react with all tested cells. The presence of positive reactions (i.e. *in vitro* agglutinin) in all tested cells makes the serologic identification of coexisting alloantibodies difficult, as their identification typically requires patterns of differential reactivity. Absorption techniques using either donor or patient RBCs are available at some hospital laboratories but usually reference laboratories perform these specialized tests, which are time-consuming. As 12–40% of patients with warm autoantibodies also have clinically significant alloantibodies, it is imperative that appropriate methods for determining the presence of coexisting alloantibodies be used. It is also helpful to recognize that the presence of a panagglutinin will create positive crossmatch results, and thus the crossmatch will not be helpful in determining compatibility with underlying alloantibodies.

Performing RBC phenotyping in patients with autoantibodies can be helpful because it focuses the antibody work-up on the possible alloantibodies the patient is capable of forming. In addition, if a complete phenotype can be determined, then the transfusion service can provide phenotype-matched RBCs, which may prevent future alloimmunization and delayed hemolytic transfusion reactions (DHTR) as well as circumventing the need for exhaustive absorption studies (see Chapter 20).

The Johns Hopkins Hospital published their approach to patients with warm autoantibodies, which included a phenotype for C, E, c, e, K, Jk$^a$, Jk$^b$, Fy$^a$, Fy$^b$, S and s, if they were able to perform the phenotype, and providing phenotype-matched as well as antigen-negative for any identified alloantibodies. During analysis of subsequent samples, if the serologic results were consistent with previous findings, then phenotype-matched products were provided. Twelve of the 20 patients studied could be fully phenotyped, and eight could be partially phenotyped or phenotyping was indeterminate. The patients received between 2 and 39 products, and none developed new alloantibodies during the study period of 13 months.

Due to the presence of the warm, IgG autoantibody in these patients, it is typically not possible to identify crossmatch-compatible units. This necessitates the administration of incompatible blood. Most blood banks and clinicians will select so-called "least incompatible" blood, meaning that they will choose those units for which the crossmatch demonstrates the least reactivity. The predicative value of this approach for *in vivo* hemolysis is not known. When the autoantibody does demonstrate apparent specificity (e.g. anti-e), selection of antigen-negative units (e.g. e negative) seems an appropriate decision, although these antibodies will still react with the patient's RBCs. Arguments against this approach include that there are scant data to support this approach, and that this practice increases the likelihood that an alloantibody will be induced to the antithetical antigen (e.g. E) if it is not carried on the patient's RBCs. In addition, the apparent antigen specificity is usually to a high-frequency antigen for which it is difficult to obtain antigen-negative RBCs (e.g. 98% of the population is e positive), such that transfusion may be delayed while acquiring antigen-negative RBCs.

Risks of transfusion in patients with WAIHA are increased hemolysis, resulting from the increase in RBC mass or inability to recognize and respect underlying alloantibodies, and congestive heart failure secondary to circulatory overload. It is recommended in severe cases that transfusion commence using small volumes of RBCs with close clinical observation.

Therapeutic plasma exchange (TPE) has been used in the management of WAIHA in severe cases unresponsive to RBC transfusion and other immunomodulatory therapies. WAIHA is an ASFA category III indication for TPE secondary to the rare and conflicting reports of its successfulness (see Chapter 68). Theoretically, TPE would be less effective in WAIHA because the IgG intravascular distribution is 45% (IgM is 80%) and the pathogenic IgG antibody is coating the RBCs and is less available in the plasma; therefore, TPE would result in inefficient antibody removal.

**AIHA Associated with a Negative DAT:** A patient may have AIHA with a negative DAT secondary to the RBC-bound IgG being below the threshold of detection (<150 IgG molecules/RBC), RBC-bound IgA and IgM being responsible (these are not detected by most routine reagents), or low-affinity IgG which is washed off the RBCs.

*Alternative Testing Techniques:* Reference laboratories may have alternative techniques for evaluating patients with WAIHA associated with a negative DAT. Flow cytometry can be used to detect small amounts of RBC-bound IgG, IgA and IgM. In addition, non-licensed anti-IgA and anti-IgM reagents to detect RBC-bound IgA or IgM, respectively, are available. Cold low-ionic strength saline wash DAT may be helpful in detecting low-affinity autoantibodies.

**Cold Agglutinin Disease:** CAD may be idiopathic or secondary to a variety of disease, such as *Mycoplasma pneunomiae* or mononucleosis infection, or lymphoproliferative disorders. When associated with an acute infection, CAD is usually transient; when associated with lymphoproliferative disorders, it is usually chronic. The clinical severity depends on the thermal amplitude, the antibody's ability to fix complement, and antibody titer. The typical clinical manifestations are a moderate chronic hemolytic anemia, which is exacerbated in the cold, in a middle-aged or elderly person with a good prognosis. In severe cases, the IgM autoantibody can result in life-threatening intravascular hemolysis.

CAD is characterized by a high titer (>10,000 at 4°C) IgM autoantibody that is reactive at >30°C (high thermal amplitude) with I or i specificity (rarely P$^r$) (Table 41.1).

The DAT is positive for C3 only (see Chapter 21). CAD secondary to *Mycoplasma pneumoniae* is associated with an IgM with anti-I specificity. The specificity of the antibody associated with infectious mononucleosis is anti-i. The pathogenesis of the autoantibody formation is unknown; theories include: 1) immune dysfunction, 2) antigens sharing between the infectious agent and RBC (known as antigen mimicry), and 3) infection induced antigenic changes resulting in increased antigenicity.

**Pathogenesis:** CAD results from IgM cold autoantibodies (also known as cold agglutinins) which lead to RBC agglutination at cold temperatures (4–18°C) *in vitro* and hemolysis *in vivo* when the antibody is present in high titer (>1:10,000) and reactive at warm temperatures (~37°C). The IgM autoantibodies, when present in sufficient quantity and able to react at near 37°C, activate the classic complement pathway on the RBC membrane, beginning with calcium-dependent C1q binding, under appropriate conditions, and ending with full assembly of the membrane (or terminal) attack complex. The membrane attack complex breaches the RBC membrane and leads to physical, intravascular hemolysis resulting in hemoglobinemia, hemoglobinuria and hemosiderinuria. Plasma-free hemoglobin binds haptoglobin, and thus free haptoglobin measurements are low. In some clinical situations complement activation is not sufficient to activate full assembly of the membrane attack complex, due to antibody concentration and reactivity at required temperatures and complement regulatory proteins, and the complement activation is only taken through the C3 stage, resulting in C3b-coated RBCs. C3b-positive RBCs can be phagocytized by macrophages with C3b receptors in the liver Kupfer cells, resulting in extravascular hemolysis. Because the RBCs must be coated with at least 500–800 C3b molecules for RBC clearance, many C3b-coated RBCs survive normally. In addition, C3b is degraded to iC3b or Cdg by Factor I; the former is formed if factor H, membrane cofactor protein or complement receptor CR1 is present.

**Treatment:** Treatment of CAD depends on the severity and rapidity of onset of the symptoms, and the underlying etiology for the formation of the cold agglutinin. The typical treatments for WAIHA (i.e. corticosteroids and splenectomy) are less effective in CAD. In the majority of cases the anemia is mild; thus cold avoidance is the sole treatment used to prevent exacerbations. Additional treatments for chronic CAD include rituximab, interferon-α, chlorambucil and cyclophasphamide. As with secondary WAIHA, treatment for secondary CAD should mitigate any underlying disease (e.g. lymphoma). Patients with transient CAD secondary to infection (e.g. mycoplasma pneumonia) usually require supportive measures only (transfusion and cold avoidance).

**Blood Bank and Transfusion Management:** Prewarming techniques may be used to mitigate the reactivity of the autoantibody, and thus can aid in the identification of underlying alloantibodies in patients with CAD (see Chapter 20). In less than 10% of patients with CAD, an alternate technique, such as cold adsorption, will be required.

Severe anemia can result from CAD and cardiovascular compromise, and death can follow; such cases may require RBC transfusions and TPE. These patients can usually be transfused with RBCs crossmatch compatible at 37°C; some suggest using a blood warmer, especially in severe cases, in addition to keeping the patient warm and transfusing the product slowly. TPE may be beneficial in severe cases because TPE efficiently removes the primarily intravascular IgM pathogenic antibody, but its effect is usually temporary. CAD is an ASFA category III indication for TPE secondary to the paucity of data (see Chapter 68).

**Combined Cold and Warm AIHA:** Combined cold and warm AIHA (mixed AIHA) occurs when a patient has serologic findings characteristic of WAIHA and has a cold agglutinin of high titer and thermal amplitude, and therefore has both WAIHA and CAD (Table 44.1). The IgG warm antibody is usually the more pathogenic antibody, and therefore treatments are similar to those for WAIHA. Depending on the titer and thermal amplitude of the cold autoantibody, a combination of the serologic techniques used for samples from patients with WAIHA and CAD can be employed.

**Paroxysmal Cold Hemoglobinuria:** PCH is a disorder that occurs due to the formation and activity of the Donath-Landsteiner antibody, a biphasic IgG antibody with P specificity. This biphasic antibody binds to autologous P-positive RBCs at low temperature and initiates complement activation, but intravascular hemolysis does not occur until the RBC with the antigen–antibody complex warm to 37°C. PCH may be idiopathic, or secondary to viral infections or syphilis. The clinical manifestation of PCH is an acute intravascular hemolytic anemia. It is typically self-limited, with rare cases of chronic PCH. The prognosis is excellent, and treatment may include steroids. The DAT is usually negative, but DAT positive with complement can be present (Table 44.1). Treatment for PCH is primarily supportive, as corticosteroids have not been effective. Supportive care includes RBC transfusion and cold avoidance. The treatment of syphilis usually eliminates PCH associated with syphilis.

There is some suggestion that transfused p or $P^k$ RBCs will have better *in vivo* survival, but the patient is likely to require transfusion prior to these RBC products becoming available. Therefore the use of RBCs of common P types should be provided, which usually provides adequate RBC support. In addition, RBCs should be transfused through a blood warmer and the patient maintained at a warm temperature.

**Drug-induced Hemolytic Anemia:** Administration of a number of commonly used medications can lead to hemolysis, typically occurring secondary to the mechanisms described below. Drugs sometimes induce the formation of antibodies, either against the drug itself or against RBC antigens that may result in a positive DAT with or without hemolysis. Some of the antibodies produced require the presence of the drug (i.e. are drug dependent) for their detection and/or destruction, while others do not (i.e. are drug independent). The clinical manifestations are variable. The disease has an excellent prognosis, and treatment is usually to discontinue the medication. Drug-induced hemolytic anemia is most often associated with an IgG antibody with Rh specificity. The drug is adsorbed onto the RBC membrane and antibodies are made to the drug, membrane components, or part-drug, part-membrane (Table 44.1, Figure 44.1).

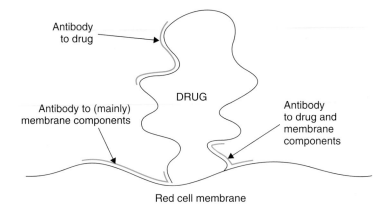

FIGURE 44.1 Proposed unifying theory of drug-induced antibody reactions (based on a cartoon by Habibi as cited by Garratty). The blue lines represent antigen-binding sites on the F(ab) region of the drug-induced antibody. Drugs (haptens) bind loosely, or firmly, to cell membranes and antibodies may be made to: (a) the drug (producing *in-vitro* reactions typical of a drug adsorption [penicillin-type] reaction); (b) membrane components, or mainly membrane components (producing *in-vitro* reactions typical of autoantibody); or (c) part-drug, part-membrane components (producing an *in-vitro* reaction typical of the so-called immune complex mechanism). From Brecher ME (ed) (2005). *AABB Technical Manual*, 15th edition. Bethesda, MD: AABB Press.

From a blood bank and transfusion service perspective, it is imperative to consider drug-induced hemolytic anemias under the appropriate clinical circumstances. Serologically, drug-induced hemolytic anemia may be difficult to demonstrate until the requisite medication is introduced into the test system. Specialized referral laboratories are able to perform these assays.

**Drug Adsorption (Hapten) Hypothesis:** In the drug adsorption hypothesis, the drug binds to the RBC membrane and the antibody is largely directed against the drug itself. The DAT is positive with IgG and possibly complement, and occasionally results in extravascular hemolysis. The antibody in serum/plasma and elute only reacts with drug-treated RBCs (not untreated RBCs). Approximately 3% of patients receiving large doses of penicillin and 4% of patients receiving cephalosporins develop a positive DAT. The dose of the drug required to cause hemolysis varies between drugs – it is millions of units daily for weeks in the case of penicillin. Treatment of the hemolytic anemia is discontinuation of the medication.

**Immune Complex Hypothesis:** Hemolysis occurs as a result of the drug/antidrug immune complex binding to the RBC. Cephalosporins may react by this mechanism. Usually the DAT is positive for complement only, but IgG can be present. The drug must be present to demonstrate the antibody in the patient's plasma. Only a small amount of drug is required to result in hemolysis. This mechanism results in intravascular hemolysis with hemoglobinemia, hemoglobinuria, renal failure and, occasionally, death. Treatment of the hemolytic anemia is discontinuation of the medication and transfusion support.

**Autoantibody Induction by Drugs:** Drugs may also cause a positive DAT, by inducing a drug-independent immune response, although the drug initiated the immune response. The proposed mechanism is through drug effects on immunoregulatory T cells. Serologically, this autoantibody is indistinguishable from a non-drug induced warm autoantibody. Methyldopa is the classic example of this mechanism, but fludarabine, cephalosporins, procainamide and non-steroidal anti-inflammatory drugs have also been associated with it. Treatment of the hemolytic anemia is discontinuation of the drug. If the hemolytic anemia is severe, it may be treated like a WAIHA.

**Non-immunologic Protein Adsorption (Membrane Modification):** Some drugs result in non-immunologic adsorption of proteins and in a positive DAT without hemolysis. Cephalosporins, diglycoaldehyde, suramin, cisplatin, clavulanate, sulbactam and taxobactam are associated with this mechanism.

## Recommended Reading

Garratty G, Arndt PA. (2007). An update on drug-induced immune hemolytic anemia. *Immunohematology* **23**(3), 105–119.

Garratty G, Petz LD. (2002). Approaches to selecting blood for transfusion to patients with autoimmune hemolytic anemia. *Transfusion* **42**, 1390–1392.

Johnson ST, Fueger JT, Gottschall JL. (2007). One center's experience: the serology and drugs associated with drug-induced immune hemolytic anemia – a new paradigm. *Transfusion* **47**(4), 697–702.

McLeod BC. (2007). Evidence based therapeutic apheresis in autoimmune and other hemolytic anemias. *Curr Opin Hematol* **14**(6), 647–654.

Shirey RS, Boyd JS, Parwani AV *et al.* (2002). Prophylactic antigen-matched donor blood for patients with warm autoantibodies: an algorithm for transfusion management. *Transfusion* **42**, 1435–1441.

# CHAPTER 45

# Transfusion Management in Patients with Hemoglobinopathies

Beth H. Shaz, MD

Patients with hemoglobinopathies require special consideration by blood center and hospital transfusion medicine physicians, for the following reasons: 1. They often require acute and chronic, simple and exchange transfusions, 2. They often require phenotypically matched transfusions, 3. They may develop allo- and autoantibodies that can be challenging to identify and occur with an increased incidence as compared to the general hospitalized population, 4. They can require specialized procedures such a prophylactic or urgent RBC exchange, 5. They may undergo HPC transplantation with curative intent.

This chapter addresses the above considerations for the two most common hereditary chronic anemia/hemoglobinopathy patient groups that require lifelong RBC transfusion support, namely those with sickle cell disease (SCD) or thalassemia.

**Sickle Cell Disease:** SCD is the most prevalent genetic disorder in the African American (AA) population. SCD patients are usually homozygous for hemoglobin (Hgb) S, but can also carry Hgb S along with β-thalassemia, Hgb C, or other hemoglobinopathies. SCD affects approximately 80,000 people in the US (1 in 400 AAs are Hgb SS, and 1 in 12 are Hgb SA). Present data suggest that SCD patients now live to an average of 45 years of age. A key component of the treatment of SCD patients is RBC transfusion.

**Pathophysiology:** When sickle Hgb deoxygenates, it forms polymers within the erythrocyte that distort its shape and decrease its deformability, leading to vaso-occlusion. These vaso-occlusive events can cause recurrent episodes of severe pain. Additionally, the chronic vaso-occlusive events can cause end-organ damage, especially in the CNS, lungs, spleen and kidney. Transfusion improves or prevents many of these complications.

The goals of RBC transfusion in SCD patients are to improve oxygen-carrying capacity by increasing Hgb AA concentration, decreasing blood viscosity, improving blood flow by decreasing or diluting the Hgb S containing RBCs, and suppressing endogenous erythropoiesis.

**Methods:**

*Simple versus Exchange Transfusion:* SCD patients can either receive simple (i.e. when RBCs are transfused without removal of the patient's RBCs) or exchange transfusions (i.e. when RBCs are transfused with removal of the patient's RBCs either by automated [known as erythrocytapheresis] or manual methods, which are used when there is not access to an automated instrument or if the patient is too small for erythrocytapheresis). Limitations of simple transfusion include volume overload and reduced blood flow due to increased blood viscosity. The advantages of exchange transfusion over simple transfusion include a rapid increase in hematocrit and decrease in Hgb S level with less risk of iron and fluid overload. The disadvantages of chronic erythrocytapheresis over chronic simple transfusion include increased RBC product exposure, requirements for venous access (erythrocytapheresis either requires two large-bore peripheral needles or a central catheter) and increased cost (which may not hold true once the cost of chelation therapy is included). For all transfusions in SCD patients, the goal should be to maintain a hematocrit of 30% or less in order to prevent increased blood viscosity.

*Acute versus Chronic Transfusion:* Indications for transfusion can either be in the acute (or intermittent) setting or the chronic setting, and the latter is typically used for prophylaxis. Acute simple transfusions are used when there is a need for an immediate increase in oxygen-carrying capacity. Chronic simple or exchange transfusions not only increase oxygen-carrying capacity, but can also significantly decrease the level of Hgb S and suppress production of Hgb S RBCs.

*Calculations:* Formulas for calculating the volume of RBCs to be transfused in patients using the acute simple transfusion and chronic transfusion methods are given in Table 45.1.

---

**TABLE 45.1  Calculation of the Volume of RBCs to be Transfused**

Total blood volume (ml) = 85 ml/kg child, 70 ml/kg adult

Estimated RBC volume (ml) = Hct × TBV

Acute simple transfusion:

$$\text{Volume of replacement RBCs (ml)} = \frac{[(\text{Hct}_d - \text{Hct}_i) \times \text{TBV}]}{\text{Hct}_r}$$

Chronic simple transfusion:

$$\%\text{HbS}_d = \left[1 - \frac{(\text{RBC}_t\ (\text{ml}) \times \text{Hct}_r)}{(\text{TBV} - \text{Hct}_i) + (\text{RBC}_t\ (\text{ml}) \times \text{Hct}_r)}\right] \times \text{HbS}_i$$

---

$\text{HbS}_d$ = desired hemoglobin S; $\text{HbS}_i$ = initial hemoglobin S; $\text{Hct}_d$ = desired hematocrit; $\text{Hct}_i$ = initial hematocrit; $\text{Hct}_r$ = hematocrit of replacement product; $\text{RBC}_t$ = transfused RBCs; TVB = total blood volume.

Multiple methods for manual RBC exchange in adults are available. The method used in the NIH *Management of Sickle Cell Disease* is described here:

1. 500 ml of whole blood is removed by phlebotomy,
2. 500 ml of normal saline is infused,
3. 500 ml of whole blood is again removed followed by the transfusion of 2 units packed RBCs.
4. These steps can be repeated until the volume to be exchanged has been reached, which is calculated as:

   Total volume exchanged = 1.5 × RBC volume

Manual exchange transfusion can utilize a stopcock for removal and infusion of saline or aliquots of pRBCs, and can be of high utility in very small patients or at times of considerable urgency.

For automated erythrocytapheresis, the apheresis device will calculate the volume of RBCs required for the replacement fluid. The required input information includes the Fraction of Cells Remaining (FCR), which is the target percentage of the original RBC volume that will remain after the procedure, as well as the patient's gender, height, weight and pretransfusion hematocrit. Lastly, the final desired hematocrit, average hematocrit of the replacement product (additive solution RBC products ~60% and CPDA1 RBC products ~75%) and the final fluid balance are required. By dividing the ml of replacement fluid by 350 ml for additive solution and 250 ml for CPDA1 products, the number of RBC products can be determined (see Chapter 69).

Indications for Acute Transfusions: The indications for acute simple and/or exchange transfusion are listed in Table 45.2, and each is described in more detail below.

TABLE 45.2  Clinical Indications for RBC Transfusion in Adults and Children With SCD

| Type of transfusion | Method | Indication |
| --- | --- | --- |
| Therapeutic/intermittent | Simple and/or exchange | Acute symptomatic anemia<br>Aplastic crisis<br>Acute splenic or hepatic sequestration<br>Acute stroke<br>Acute chest syndrome<br>Acute multiorgan system failure<br>Severe infection with symptomatic anemia<br>Before surgery requiring general anesthesia<br>Before eye surgery |
| Prophylactic/chronic | Simple and/or exchange | Prevention of recurrent stroke in children<br>Prevention of first stroke in children<br>Complicated pregnancy<br>Chronic renal failure |
| Controversial indications for intermittent or chronic transfusion | Simple and/or exchange | Frequent pain episodes<br>Acute pain crisis<br>Recurrent acute chest syndrome<br>Prevention of pulmonary hypertension/ cor pulmonale<br>Priapism<br>Normal pregnancy<br>Leg ulcers |

From Josephson CD, Su LL, Hillyer KL, Hillyer CD. (2007). Transfusion in the patient with sickle cell disease: a critical review of the literature and transfusion guidelines. *Transfus Med Rev* **21**, 118–133.

*Acute Chest Syndrome:* Acute chest syndrome is defined as a new pulmonary infiltrate on chest X-ray in a SCD patient with respiratory symptoms, chest pain or fever. Acute chest syndrome develops in over 30% of SCD patients in their lifetime, and is a leading cause of death. RBC transfusions can be utilized in the treatment of acute chest syndrome, especially when there is concomitant hypoxemia. Although simple transfusion may be given, rapid clinical decompensation may necessitate an exchange transfusion.

*Acute Multi-organ Failure:* Acute multi-organ failure is a life-threatening complication of SCD, and involves respiratory, hepatic and/or renal failure. It is thought to be a result of multiple tissue infarctions secondary to vaso-occlusion. Treatment includes acute simple or exchange transfusion. In patients with a higher hematocrit (e.g. > 25%) or more severe organ failure, exchange transfusion may be preferable to prevent the increased viscosity and volume overload that may occur with simple transfusion.

*Acute Sequestration:* Splenic or hepatic sequestration occurs when sickled RBCs are trapped within the sinusoids of the organ, leading to engorgement. Sequestration leads to reduced circulating intravascular RBCs, which may lead to symptomatic anemia. Splenic sequestration has a fatality rate of 10%, and can be recurrent. Simple transfusions are the preferred treatment for acute sequestration. RBCs are frequently given in small aliquots (5 ml/kg) over 4 hours, as the release of autologous RBCs from the sequestration may occur during the transfusion. Patients with recurrent splenic sequestration may require chronic transfusion until eligible for splenectomy.

*Acute Stroke:* Approximately 10% of SCD patients will develop a clinical stroke by age 20 years, with a higher percentage experiencing "silent" strokes. The occurrence of ischemic stroke is greatest in children and older patients, whereas the occurrence of hemorrhagic stroke is greatest in patients 20 to 29 years of age. Treatment of an acute ischemic stroke in children includes RBC exchange with a target end hematocrit of 30% and Hgb S level lower than 30%. Patients who have had a stroke, as well as those at high risk for stroke based on transcranial Doppler (TCD) screening, are placed on chronic transfusion therapy. The optimal treatment of adults with stroke is not known at this time, but should follow recommendations for adults with stroke without SCD.

*Acute Symptomatic Anemia:* Patients with SCD are chronically anemic, but are typically asymptomatic. When they do become symptomatic, from blood loss, increased hemolysis, viral suppression of erythropoiesis or RBC sequestration, then an acute simple transfusion is warranted to alleviate cardiac and respiratory symptoms.

*Aplastic Crisis:* Aplastic crisis is defined as a decrease in Hgb of 3g/dl or more with reticulocytopenia, usually resulting from parvovirus B19 infection. Infection results in erythropoiesis suppression for up to 10 days, which, in SCD patients, can lead to a marked anemia and a compensatory expanded plasma volume which can lead to heart failure. Simple transfusion is indicated for severe, symptomatic anemia.

*Intrahepatic Cholestasis:* Intrahepatic cholestasis is a rare yet often fatal complication of SCD where massive sickling and stasis of RBCs in hepatic sinusoids leads to severe hepatocellular and Kupfer cell injury, resulting in a tender, enlarged liver; hemolysis; coagulopathy; and elevated transaminase and bilirubin levels. Intrahepatic cholestasis with bile plugs, dilated sinusoids, sickled RBCs, and Kupffer cell erythrophagocytosis can be demonstrated by liver biopsy. Treatment includes exchange transfusion.

*Preoperative:* General anesthesia increases the risk of intravascular sickling and resulting morbidity and mortality in the perioperative period, which can be reduced by acute simple transfusions with a goal of increasing the hemoglobin to 10 g/dl (SCD patients with HbSC may require RBC exchange transfusions to prevent hyperviscosity). The perioperative treatment should also include hydration, adequate oxygenation, and respiratory therapy. In surgeries that do not require general anesthesia, transfusion is not recommended.

## Indications for Chronic Transfusions:

*Chronic Renal Failure:* Patients with renal failure develop anemia secondary to loss of erythropoietin production. These patients may benefit from chronic simple transfusion to prevent symptomatic anemia.

*Complicated Pregnancy:* Patients who have had previous perinatal fetal mortality, or develop complications of pregnancy such as pre-eclampsia/eclampsia, acute renal failure or acute chest syndrome, may require chronic transfusion therapy. The preferred method of transfusion is determined by the Hgb; if the Hgb is less than 5 g/dl, then a simple transfusion is recommended; if it is 8–10 g/dl, then an exchange transfusion is recommended with a target Hgb of 10 g/dl and Hgb S level reduced to near 30%.

*Prevention of Stroke:* It is estimated that without therapeutic intervention, up to 70% of patients who experience an initial cerebrovascular accident will have a recurrent stroke within 2–3 years. The Stroke Prevention Trial in Sickle Cell Anemia (STOP) compared chronic simple transfusion to transfusions when needed for acute episodes for the prevention of stroke in SCD children with abnormal transcranial Doppler tests, demonstrating a 90% reduction in stroke rate between the chronic transfusion group and the standard of care group. The recommendation is for SCD children at high risk for stroke, as determined by transcranial Doppler, to be given a simple or RBC exchange transfusion every 3–4 weeks to maintain the Hgb S level <30% and hematocrit <30%. Some reports suggest that the Hgb S level can be raised to less than 50% after 3 years of stability. A subsequent study to examine the duration of chronic transfusion (STOP II) demonstrated that once chronic transfusion was discontinued, the high risk of stroke returned or the children had a stroke. Therefore, it is currently recommended that chronic transfusions be maintained indefinitely to prevent stroke in this patient population.

## Controversial Indications for Transfusion:

*Frequent Pain Episodes:* Patients who develop frequent severe pain episodes may benefit from chronic transfusion therapy.

*Priapism:* The initial treatment for priapism is hydration and pain control. If symptoms persist, then simple or exchange transfusion may be indicated. SCD patients who undergo exchange transfusion can develop an unusual syndrome known as ASPEN, the eponym for *A*ssociation between *S*ickle cell disease, *P*riapism, *E*xchange transfusion and *N*eurologic events (headaches, seizures, obtundation). Because of the variable response to exchange transfusion and the risk of ASPEN, priapism is usually treated without transfusion until it persists for more than 24–48 hours.

*Prevention of Pulmonary Hypertension/cor Pulmonale:* Pulmonary hypertension has recently been shown to be a common cause of mortality in SCD patients. One study demonstrated that 32% of adult patients had pulmonary hypertension (defined as a tricuspid regurgitant jet velocity higher than 2.5 m/s on echocardiogram) and 9% had severe pulmonary hypertension (defined as a tricuspid regurgitant jet velocity higher than 3 m/s). Potential treatments for severe pulmonary hypertension are chronic exchange transfusion or inhaled nitrous oxide; ongoing studies are evaluating long-term effects of chronic transfusion therapy on pulmonary hypertension.

*Recurrent Acute Chest Syndrome:* Multiple episodes of acute chest syndrome may lead to the development of restrictive lung disease, severe pulmonary fibrosis, pulmonary hypertension, or cor pulmonale. It is unknown whether chronic transfusion prevents the development of these complications, but it may decrease the frequency of acute chest syndrome.

## Not Indicated for Transfusion:

*Acute Pain Crisis:* The main treatment of acute pain crisis is hydration and pain medication; transfusion is generally not recommended in the treatment of acute pain crisis.

*Leg Ulcer:* Leg ulcers in SCD patients are not a common indication for transfusion.

*Uncomplicated Pregnancy:* Uncomplicated pregnancy is SCD patients is not an indication for transfusion.

## RBC Product Selection:

*Phenotype:* All SCD patients should have extended RBC antigen phenotyping (ABO, Rh, Kell, Kidd, Duffy, Lewis and MNS systems) performed prior to the initiation of transfusion therapy in order to provide partially phenotype blood products (see below) and better manage alloimmunization in the future (i.e. have an extended phenotype performed for future use).

*Provision of Phenotype Matched RBC Products:* Providing partially phenotypically matched RBC products (ABO, C, c, D, E, e, K, and k) for the treatment of SCD patients has been advocated to minimize the likelihood of RBC alloimmunization. Patients who form RBC alloantibodies on this protocol should receive blood with more extensive phenotyping, adding, in addition to the above antigens, $Fy^a$ and $Jk^b$ and the corresponding antigen for which they made an antibody. The disadvantages of phenotype matching of RBC products include increased cost, inventory management, and the inability to procure the desired product at the time of transfusion. In cases of emergency, phenotypic matching may not be possible.

*Leukoreduced Products:* In order to prevent HLA immunization and febrile non-hemolytic transfusion reactions, the use of leukoreduced products is warranted.

*Hgb S Negative:* Sickle cell trait (Hgb SA) RBC products are reported to be safe in SCD patients, but should be avoided if possible when the goal is to decrease Hgb S levels.

## Adverse Effects:

*Autoantibody Formation:* Autoantibody formation has been reported in 8% of transfused SCD patients and in 29% of SCD patients with alloantibodies. These autoantibodies can result in clinically significant hemolysis. The pathogenesis of autoantibody formation, especially in the presence of alloantibody formation, is unknown; theories include alloantibodies possibly causing a conformation change in RBC antigens leading to stimulation of autoantibodies, immune system changes, and genetic predisposition.

*Alloantibody Formation:* Antigenic differences between the transfused RBCs, which are largely obtained from white donors and transfused to AA recipients, is one factor which leads to the increased RBC alloimmunization to RBC antigens observed in transfused SCD patients. Without the implementation of extended phenotypic matching in SCD patients, studies have reported alloimmunization rates in the range of 19–43% in transfused patients with SCD. The most common antibodies found are against K, E, C and $Jk^b$, which are related to the antigenic frequencies in donors versus SCD patients (K: 9 versus 2%, E: 35 versus 24%, C: 68 versus 28%, and $Jk^b$ 72 versus 39% are positive, white versus AA respectively). In addition to a high incidence of anti-erythrocyte alloantibody formation, the number of SCD patients that will make multiple alloantibodies (two to seven or more alloantibodies with separate and differing specificities) with repeated transfusions is high. These events make compatible RBC products difficult to obtain, and also increase the risk of acute and delayed hemolytic transfusion reactions.

*Hyperhemolytic Transfusion Reactions:* A serious complication of RBC transfusion in SCD patients is a hyperhemolytic transfusion reaction (HHTR), in which both donor and recipient RBCs are destroyed and reticulocytopenia may occur. Some patients may have newly detected alloantibodies or autoantibodies, but others have no detectable antibodies to RBC antigens (i.e. negative antibody screen and direct agglutination test). Possible mechanisms include bystandard hemolysis, erythropoiesis suppression, and RBC destruction secondary to contact lysis via activated macrophages. Bystandard hemolysis is an immune-mediated hemolysis, where the RBC does not carry the antigen for which the antibody is directed.

Current recommendations include that future transfusions should be withheld, as they may exacerbate the anemia. Similar reactions may occur after subsequent transfusion even if the RBCs are extensively phenotypically matched. Treatment of patients with HHTRs includes erythropoietin, intravenous immunoglobulin, steroids, and plasma exchange. This serious, life threatening reaction to transfusion has mostly been reported in SCD patients, but HHTRs have also been reported in patients with thalassemia, myelofibrosis, and anemia of chronic disease.

*Iron Overload:* Simple chronic transfusions will result in iron overload, because the body has no mechanism for excreting excess iron. Iron overload can be expected to occur when approximately 120 ml of RBCs/kg of body weight is transfused, or approximately 50 RBC units transfused (see Chapter 65).

β-**Thalassemia:** Thalassemia is a hereditary anemia resulting from mutations in the β-globin gene (β-thalassemia) or α-globin gene (α-thalassemia) that can lead to defective Hgb synthesis and therefore a chronic anemia. The disease is clinically heterogeneous due to genotypically different mutations or compound heterozygozity with other hemoglobinopathies, and to unknown individual patient factors. Patients with thalassemia may require lifelong RBC transfusions to ameliorate the chronic anemia and to suppress the extrameduallary hematopoiesis, which would otherwise lead to severe bone deformities.

**Pathophysiology:** The complications from thalassemia arise from ineffective erythropoiesis and RBC hemolysis, resulting in anemia and iron accumulation. The only definitive treatment is HPC transplantation.

**Indications:** Chronic RBC transfusion is required for patients with homozygous thalassemia and in compound heterozygotes with β-thalassemia major, while patients with β-thalassemia intermedia, β-thalassemia and Hgb E disease, and Hgb H disease (3/4 mutated β-globulin genes) may require periodic simple transfusions. The goals of transfusion therapy in thalassemia major patients are to increase oxygen-carrying capacity, prevent progressive hypersplenism, suppress erythropoiesis which prevents extramedullary hematopoiesis and the resulting pathology fractures from osteopenia, and reduce gastrointestinal absorption of iron. The indications for transfusion are growth retardation, failure to thrive, symptomatic anemia, and for prevention of progressive hypersplenism and of facial and skull deformities.

**Management:** Since these patients are transfused from birth, RBC phenotyping should be performed on the initial pretransfusion sample, or genotyping must be utilized to adequately predict the patients' RBC antigen type. Simple transfusion of leukoreduced RBCs can be used to maintain a Hgb level greater than 9.5 g/dl. Erythrocytapheresis has also been used to treat patients with thalassemia, decreasing the transfusion requirement and iron overload while increasing the transfusion intervals.

**Adverse Effects:**

*Alloantibody Formation:* RBC alloimmunization occurs at a rate of 5–33%, depending on the homogeneity of the population. Alloimmunization rates were lowered by phenotype matching for Rh and Kell from 33 to 2.8% at the Children's Hospital Oakland, and 23.5 to 3.7% at the Aghia Sophia Children's Hospital. In addition, the treatment of these patients can be complicated by the presence of RBC autoantibodies.

*Iron Overload:* See section above, and Chapter 65.

**Recommended Reading**

Adams RJ, McKie VC, Hsu L. (1998). Prevention of a first stroke by transfusions in children with sickle cell anemia and abnormal results on transcranial Doppler ultrasonography. *N Engl J Med* **339**, 5–11.

Aygun B, Padmanabhan S, Paley C, Chandrasekaran V. (2002). Clinical significance of RBC alloantibodies and autoantibodies in sickle cell patients who received transfusions. *Transfusion* **42**, 37–43.

Ho HK, Ha SY, Lam CK et al. (2001). Alloimmunization in Hong Kong southern Chinese transfusion-dependent thalassemia patients. *Blood* **97**, 3999–4000.

Josephson CD, Su LL, Hillyer KL, Hillyer CD. (2007). Transfusion in the patient with sickle cell disease: a critical review of the literature and transfusion guidelines. *Transfus Med Rev* **21**, 118–133.

National Institutes of Health, National Heart, Lung, and Blood Institute. (2002). *Management of Sickle Cell Disease*, 4th edition. Washington, DC: National Institutes of Health  NIH publication 02-2117 (available at: http://www.nhlbi.nih.gov/health/prof/blood/sickle/sc_mngt.pdf).

Rund D, Rachmilewitz E. (2005). Medical progress: beta-thalassemia. *N Engl J Med* **353**, 1135–1146.

Singer ST, Wu V, Mignacca R *et al.* (2000). Alloimmunization and erythrocyte autoimmunization in transfusion-dependent thalassemia patients of predominantly Asian descent. *Blood* **96**, 3369–3373.

The Optimizing Primary Stroke Prevention in Sickle Cell Anemia (STOP II) Trial Investigators. (2005). Discontinuing prophylactic transfusions used to prevent stroke in sickle cell disease. *N Engl J Med* **353**, 2769–2778.

Vichinsky EP, Luban NL, Wright E *et al.* (2001). Prospective RBC phenotype matching in a stroke-prevention trial in sickle cell anemia: a multicenter transfusion trial. *Transfusion* **41**, 1086–1092.

Win N, Doughty H, Telfer P *et al.* (2001). Hyperhemolytic transfusion reaction in sickle cell disease. *Transfusion* **41**, 323–328.

# CHAPTER 46

# Transfusion of Patients Undergoing HPC and Solid-organ Transplantation

Jeanne E. Hendrickson, MD and John D. Roback, MD, PhD

Patients who undergo either solid-organ or hematopoietic progenitor cell (HPC) transplantation typically require transfusion of blood and specialized blood products. Blood product selection needs to take into account the ABO type of both the patient and the recipient, and the special risks of transfusion in these patients (e.g. increased risk of transfusion associated graft versus host disease [TA-GVHD] in HPC recipients).

## Transfusion of Patients Undergoing HPC Transplantation:

**Description:** A number of factors must be taken into consideration when ordering blood products for the HPC transplant recipient. These factors include type of transplant and expected time to engraftment, presence of ABO mismatch between donor and recipient, transfusion triggers for red blood cell (RBC) and platelet transfusions, platelet refractoriness, utility of granulocyte transfusion, and potential transfusion-related complications (including risk of viral transmission and TA-GVHD).

*HPC Source and Recipient Conditioning:* Bone marrow, peripheral blood stem cells (PBSC), and umbilical cord blood are all potential sources of hematopoietic stem cells. Patients receiving PBSCs tend to have more rapid engraftment and therefore decreased transfusion requirements as compared to patients receiving bone marrow; adults receiving umbilical cord blood transplants tend to have the slowest hematologic recovery rates and thus the highest transfusion needs. Rates of engraftment have been attributed to CD34+ cell dose (CD34+ cell count/kg), which is typically highest in PBSCs and lowest in cord blood units. Transfusion requirements of HPC transplant recipients also vary depending on the preparative conditioning regimen. Recipients of non-myeloablative transplants have decreased transfusion requirements as compared to recipients of fully myeloablative transplants.

**Red Blood Cell Transfusion and Product Selection:** Transfusion triggers for RBCs vary from institution to institution, with little evidence-based guidance. The majority of HPC patients are transfused at a threshold Hct of 24–30%. However, clinical status must be taken into consideration to determine individual transfusion thresholds.

*Leukoreduction:* Leukocyte reduction of RBCs and platelets has been proven to decrease febrile non-hemolytic transfusion reactions, CMV transmission, and HLA alloimmunization that may lead to patients becoming refractory to future platelet transfusions. Other as yet unproven considerations for leukocyte reduction include decreased risks of immunomodulation, as well as infections with emerging pathogens. For these reasons, HPC transplant recipients should receive leukoreduced products.

*CMV-safe Products:* HPC transplant recipients are susceptible to multiple viruses, including cytomegalovirus (CMV). Due to their immunocompromised status, primary CMV infection can lead to pneumonitis, gastroenteritis, hepatitis, encephalitis, and retinitis. If donor and recipient are both CMV negative, consideration should be given to providing blood from CMV-negative donors. This can be logistically difficult for the blood bank, given the paucity of CMV-negative donors. Additionally, due to the pre-seroconversion "window period," the risk of acquiring CMV from CMV-seronegative donors is not zero. Leukoreduced products are also considered "CMV safe," although the risk of CMV transmission may be slightly higher than that from CMV-seronegative products (see Chapter 38).

*Irradiation:* TA-GVHD occurs when immunocompetent lymphocytes in a transfused cellular blood component (i.e. RBC, platelet or granulocyte product) proliferate in the transfusion recipient. Proliferating donor lymphocytes invade bone marrow, skin, liver and intestines, leading to symptoms often within 10 days of transfusion. Unlike classical GVHD, TA-GVHD is nearly uniformly fatal due to bone marrow involvement producing pancytopenia. All HPC transplant recipients are considered at-risk for TA-GVHD, and as such should receive irradiated cellular blood products. Blood products must receive a minimum irradiation dose of 25 Gy to the center of the container to prevent TA-GVHD.

*ABO/D Compatibility:* Nearly half of all allogeneic HPC transplants are ABO mismatched, requiring careful selection of compatible blood components. Chapter 76 discusses more fully the transfusion considerations for major, minor and bidirectional incompatible HPCs. Table 46.1 includes recommendations for blood product selection in instances of ABO-mismatched HPC transplants.

Platelet Transfusion: In the peritransplant pre-engraftment phase, the majority of platelet transfusions are administered to non-bleeding patients. The platelet count that triggers a transfusion varies by country and institution, with the majority of patients being transfused prophylactically when the platelet count decreases below 10,000–20,000/μL. Few large studies have examined platelet transfusion triggers specifically in the setting of HPC transplantation; clinical status and other bleeding risk factors must be taken into consideration on an individual basis. Platelet refractoriness can complicate the use of transfusion triggers (see Chapter 49).

*Platelet Product Selection:* Platelet selection is in alignment with RBC selection as stated above, such that all platelet products are leukoreduced, irradiated, and CMV safe.

*ABO/D Compatibility:* Careful selection of platelet products is required, based on recipient and donor blood types. An increased risk of mortality has been associated with the transfusion of ABO-incompatible plasma (which is contained in the platelet product), and therefore some authorities recommend that platelet products should be serologically compatible with both donor and recipient ABO type (Table 46.1), or volume reduced/washed to remove the incompatible plasma.

Should transfusion of D-positive platelets to D-negative individuals be necessary in instances of blood shortages, RhIg should be considered to prevent alloimmunization.

**TABLE 46.1 Transfusion Support for Patients Undergoing ABO-mismatched Allogeneic HPC Transplantation**

| | | PHASE I | | PHASE II | | | PHASE III |
|---|---|---|---|---|---|---|---|
| Recipient | Donor | All Components | RBCs | First Choice Platelets | Next Choice Platelets* | Plasma | All Components |
| O | A | Recipient | O | A | AB; B; O | A, AB | Donor |
| O | B | Recipient | O | B | AB; A; O | B, AB | Donor |
| O | AB | Recipient | O | AB | A; B; O | AB | Donor |
| A | O | Recipient | O | A | AB; B; O | A, AB | Donor |
| A | B | Recipient | O | AB | B; A; O | AB | Donor |
| A | AB | Recipient | A | AB | A; B; O | AB | Donor |
| B | O | Recipient | O | B | AB; A; O | B, AB | Donor |
| B | A | Recipient | O | AB | A; B; O | AB | Donor |
| B | AB | Recipient | B | AB | B; A; O | AB | Donor |
| AB | O | Recipient | O | AB | A; B; O | AB | Donor |
| AB | A | Recipient | A | AB | A; B; O | AB | Donor |
| AB | B | Recipient | B | AB | B; A; O | AB | Donor |

Explanation:

Phase I – from the time when the patient/recipient is prepared for bone marrow transplant/hematopoietic stem cell transplantation

Phase II – from the initiation of chemotherapy until:

for RBCs, DAT is negative and anti-donor ischemagglutinin is no longer detectable (i.e. the back typing is donor type)

for platelets and plasma, recipient's erythrocytes are no longer detectable (i.e. the front typing is consistent with the donor's ABO type)

Phase III – after the front and back type of the patient is consistent with the donor's ABO type.

Beginning from Phase I, all cellular components should be irradiated and leukoreduced.

Modified and adapted from RC Friedberg, University of Alabama, and JD Roback (ed.) (2008) *Technical Manual*, 16th edition, Bethesda, MD: AABB, with permission.

*Platelets should be selected in the order presented. Volume-reduction or washing may be considered if first choice platelets are not available.

However, the degree of immunosuppression in these patients may limit the ability of the transplant recipient to form anti-D antibodies.

**Plasma Transfusion:** Careful selection of plasma products is required, based on recipient and donor ABO blood types. An increased risk of mortality has been associated with the transfusion of ABO-incompatible plasma, and therefore plasma products should be serologically compatible with both donor and recipient ABO types (Table 46.1).

**Granulocyte Transfusion:** Granulocyte transfusions are rarely administered in children, and even less frequently in adults. Granulocytes have a shelf-life of 24 hours; transfusion within 12 hours is suggested for optimal function. Dedicated granulocyte donors are primed with steroids and/or G-CSF. Indications for granulocytes typically include an absolute neutrophil count (ANC) below 500/µl, bacterial or fungal infection non-responsive to antibiotic therapy, myeloid hypoplasia, and a chance for recovery of marrow function and survival (see Chapter 32).

*Granulocyte Product Selection:* Granulocytes should not be transfused through a leukoreduction filter. Therefore, CMV-negative recipients should receive CMV-negative granulocyte transfusions when possible. All granulocyte products transfused to HPC recipients must be irradiated. Due to the large number of contaminating RBCs, granulocytes must be ABO/D compatible and crossmatch compatible with the recipient.

## Transfusion of Patients Undergoing Solid-organ Transplantation:

**Description:** The transfusion of patients undergoing solid-organ transplantation should take into consideration the blood type of both the patient and the donor. Solid-organ transplant recipients are at increased risk for viral transmission, especially CMV. In addition, HLA alloimmunization affects the ability to obtain a compatible organ, and increases the risk of antibody-mediated rejection. For these reasons, solid-organ transplant recipients (or potential recipients) should receive leukoreduced products.

Liver and heart transplantation may require significant amounts of blood products. End-stage liver disease is associated with coagulopathy and thrombocytopenia. Therefore, liver transplantation will often require platelet, plasma, and cryoprecipitate transfusions in addition to RBC products. Heart transplantation is performed using cardiopulmonary bypass, which often leads to transfusion of RBC, platelet, and plasma products. Other solid-organ transplantation surgeries do not require significant blood product support (e.g. kidney transplantation uses 0–2 RBC products on average).

**ABO-incompatible Transplantation:** Historically, due to the expression of ABO antigens on endothelial cells, ABO-incompatible solid-organ transplantation survival rates have been lower than for those that are ABO compatible. However, given the shortage of ABO-compatible organs, a number of strategies have been utilized to improve the survival of ABO-incompatible solid-organ transplants. These strategies include the selection of recipients with low titers of relevant isohemagglutinins (anti-A and/or anti-B), the utilization of $A_2$ donors with lower levels of antigen expression than $A_1$ donors, plasma exchange to lower titers of isohemagglutinins in the peritransplant period, potent immunosuppression, and splenectomy. Even with these strategies, ABO-incompatible solid-organ transplants are reserved for patients without an ABO-compatible organ, in whom death is imminent without a transplant.

**RBC Product Selection:**

*Leukoreduction:* Transplant recipients should receive leukoreduced blood products to decrease HLA alloimmunization as well as CMV and other viral transmission rates. Of note, while contaminating WBCs in RBC transfusions were shown in the 1970s to improve the survival of renal allografts, this practice of intentional WBC exposure has largely been abandoned with the advancement of modern immunosuppressive regimens (see Chapter 64).

*CMV-safe Products:* CMV-safe products should be provided to solid-organ transplant recipients if both the donor and recipient are CMV-negative (see Chapter 38).

*Irradiation:* The provision of irradiated blood products for solid-organ transplant recipients varies from center to center. Although these patients are heavily immunosuppressed, there have been very few reported cases of TA-GVHD. Thus, some centers do not routinely provide irradiated products to solid-organ transplant recipients (see Chapter 62).

*ABO Compatibility:* The transfusion of ABO-identical RBC products may not always be possible for group B or AB patients. If the supply is limited, some centers switch group AB recipients to group A RBCs and group B recipients to group O RBCs. The residual plasma in the RBC products may result in the presence of passive anti-B and possibly a positive direct antiglobulin test (DAT) (and, rarely, acute hemolysis).

Transfusion support for recipients of ABO-mismatched organ transplants requires consideration of a number of factors, including the risk of immediate hemolysis if transfused RBCs are incompatible with the recipient's existing antibodies, as well as the risk of delayed hemolysis (1–2 weeks after transplantation) due to passenger lymphocytes in the transplanted organ (termed *passenger lymphocyte syndrome*). Organs from O donors have the highest risk; lymphocytes contained in the graft are capable of making anti-A/anti-B for up to a month following the transplant, which may lead to a positive DAT, as well as hemolysis, which is rarely severe. The passive antibodies may be missed on immediate spin crossmatch; more sensitive testing includes crossmatch using the antiglobulin phase or DAT. Because of the risk of passenger lymphocyte syndrome, especially with group O organs, some centers will transfuse group O RBC products early in the transplant period. The risk must be balanced with the supply of group O RBC products required to support the patient. Beyond this first month, transfusion of RBCs compatible with the recipient blood type is generally acceptable.

*D Compatibility:* The ability to supply adequate numbers of D-negative RBC products may be limited. If the center is unable to supply adequate D-negative products, one option for males as well as post-menopausal females is to start with D-negative RBCs, switch to D-positive RBCs in the middle of the liver transplantation procedure, and return to D-negative RBCs towards the end of the case. The risk of D sensitization in liver transplantation is likely lower than is seen in the general hospital population.

*Support in a Patient with RBC Alloantibodies:* Patients who have RBC alloantibodies require dialogue with the transplantation team to determine the management of the patient. Many transplantation surgeries are performed at short notice, and therefore the transplantation team should communicate the timeframe of transplantation in order for the transfusion service and their blood supplier to plan appropriately. The transfusion service should determine the strength and identification of the antibodies prior to surgery. Even with an adequate supply of antigen-negative RBC products, the patient may require substantially more than was planned for. This may necessitate the use of antigen-untested RBC products. A similar option, as stated above, is to use antigen-negative products initially, then switch to antigen-untested products in the middle of the transplantation procedure and then return to antigen-negative products at the end of the surgery. Repeat antibody screens may be performed during the surgery to determine the continued presence of the antibody.

**Platelet Product Selection:** Leukoreduction, CMV selection, and irradiation are performed as stated for RBC product selection.

*ABO Compatibility:* When possible, plasma that is ABO compatible with both the donor and recipient should be provided for solid-organ transplant recipients. Volume reduction or washing of the platelets may be considered for ABO-incompatible products. This is especially true for ABO-incompatible transplantations where there is a desire to minimize the anti-A/anti-B titer which is directed against the grafted organ.

*D Compatibility:* D-positive platelets may be administered to D-negative solid-organ transplant recipients if D-negative products are not available. If they are being given in isolation (i.e. without D-positive RBCs also), then RhIg can be administered to prevent the patient from forming an anti-D. Preventing anti-D formation should be considered in the pre-transplant period, as supporting a patient with an anti-D through a transplant requiring large amounts of RBC products (e.g. liver transplantation) can be challenging.

**Plasma Product Selection:** Plasma that is ABO compatible with both the recipient and donor type should be provided for solid-organ transplant recipients. This is especially true for ABO-incompatible transplantations where there is a desire to minimize the anti-A/anti-B titer which is directed against the grafted organ.

There may not be an adequate supply of AB plasma, especially in liver transplantation cases. This requires careful planning and a dialogue with the transplantation team in order to determine the best course of action.

## Recommended Reading

Casanueva M, Valdes MD, Ribera MC. (1994). Lack of alloimmunization to D antigen in D-negative immunosuppressed liver transplant recipients. *Transfusion* **34**, 570–572.

Diedrich B, Remberger M, Shanwell A *et al.* (2005). A prospective randomized trial of a prophylactic platelet transfusion trigger of $10 \times 10(9)$ per L versus $30 \times 10(9)$ per L in allogeneic hematopoietic progenitor cell transplant recipients. *Transfusion* **45**, 1064–1072.

Dominietto A, Raiola AM, van Lint MT *et al.* (2001). Factors influencing haematological recovery after allogeneic haemopoietic stem cell transplants: graft-versus-host disease, donor type, cytomegalovirus infections and cell dose. *Br J Haematol* **112**, 219–227.

Roback JD (ed.). (2007). *Technical Manual*, 16th edition. Bethesda, MD: AABB Press.

Schiffer CA, Anderson KC, Bennett CL *et al.* (2001). Platelet transfusion for patients with cancer: clinical practice guidelines of the American Society of Clinical Oncology. *J Clin Oncol* **19**, 1519–1538.

Stanworth SJ, Massey E, Hyde C *et al.* (2005). Granulocyte transfusions for treating infections in patients with neutropenia or neutrophil dysfunction. *Cochrane Database Syst Rev* (3): CD005339.

# CHAPTER 47

# Transfusion of HIV-positive Patients

Jeanne E. Hendrickson, MD and John D. Roback, MD, PhD

Infection with human immunodeficiency virus (HIV), the virus that causes Acquired Immunodeficiency Syndrome (AIDS), is associated with multiple hematologic abnormalities. These include lymphopenia (noted in 70% of AIDS patients), anemia (70% of AIDS patients), neutropenia (50% of AIDS patients) and thrombocytopenia (40% of AIDS patients). Advanced stage AIDS is associated with pancytopenia. Ineffective hematopoiesis, infection, destruction of hematopoietic cells, and medications all contribute to hematologic abnormalities.

## Pathophysiology:

**Anemia in HIV-positive Patients:** Anemia in HIV-positive patients is multifactorial, from malnutrition, medications, infection, malignancy and HIV itself. Vitamin B12 and/or folate deficiencies can occur from a combination of anorexia and malabsorption. Multiple medications can lead to anemia. Zidovudine (AZT) can lead to bone marrow suppression and anemia; other medications used in HIV-positive patients, including ganciclovir, trimethoprim-sulfamethoxazole, amphotericin B, primaquine, dapsone, and ribavirin can also contribute to anemia. Infections commonly implicated as causes of anemia in HIV-positive patients include parvovirus, EBV, CMV, mycobacteria, and fungi. Malignancies that contribute to anemia in the HIV-positive patient include non-Hodgkin's lymphoma, other lymphomas, and Kaposi's sarcoma. Although a positive DAT is not uncommon in these patients, clinical evidence of autoimmune hemolytic anemia is quite rare.

A complete work-up is warranted for HIV-positive patients with anemia; this includes a complete blood count, blood smear, reticulocyte count, vitamin B12, red blood cell (RBC) folate level, iron studies, and erythropoietin level. A work-up for infections (especially parvovirus, CMV, EBV, mycobacteria and fungi) as well as a bone marrow evaluation may be considered.

Treatment of anemia in HIV-positive patients includes treating the underlying HIV infection, treating vitamin B12/folate/iron deficiency if present, treating any secondary infection, supplementing with recombinant erythropoietin (if serum erythropoietin levels < 500 IU/L), and eliminating contributing medications when clinically feasible. Transfusion of RBCs may be required in cases of severe symptomatic anemia.

**Thrombocytopenia in the HIV-positive Patient:** Primary HIV-associated thrombocytopenia (PHAT) is the most common cause of thrombocytopenia in the HIV-positive patient. PHAT is secondary to platelet autoantibody formation, which results in increased peripheral destruction, ineffective platelet production, and splenomegaly. Despite an increased thrombopoietin (TPO) concentration and increased TPO receptors, patients with PHAT are unable to maintain an appropriate platelet count.

It has been postulated that molecular mimicry between HIV proteins and platelet-specific glycoproteins may be involved in the pathogenesis of PHAT.

The primary treatment of PHAT includes antiretroviral medications. AZT has historically been the primary treatment for PHAT, although early studies show other antiretroviral medications are also effective. IVIG therapy (1000 mg/kg) leads to a predictable (though transient) rise in platelet count in the majority of patients with PHAT. Alternatively, anti-D immunoglobin (RhIg) can also lead to a transient rise in platelet count in D-positive individuals. RhIg should be used with caution in HIV-positive patients with underlying anemia due to the hemolysis that often occurs following this treatment. Other potential treatments for PHAT are similar to those for ITP, and include steroids, androgens, chemotherapy, and splenectomy.

Secondary causes of thrombocytopenia must also be considered in the HIV-positive patient. These include medication effects, underlying infection, and malignancy. Medications known to cause thrombocytopenia in HIV-positive patients include trimethoprim-sulfamethoxazole, ketoconazole, ganciclovir, acyclovir, pentamidine, rifampin, and interferon. Infections implicated in thrombocytopenia of the HIV-positive patient include hepatitis B and C viruses (which may lead to hypersplenism), CMV, EBV, toxoplasmosis, disseminated mycobacteria, and fungal infections. Malignancies leading to thrombocytopenia in the HIV-positive patient include non-Hodgkin's lymphoma, other lymphomas, and Kaposi's sarcoma.

HIV-associated thrombotic microangiopathy (TMA) has also been reported to cause thrombocytopenia in patients with HIV, and should be considered in the differential diagnosis when patients present with microangiopathic hemolytic anemia and thrombocytopenia. HIV patients most at risk of HIV-associated TMA include those with more advanced disease. Transfusion of fresh frozen plasma may be just as effective as plasma exchange in HIV-associated TMA.

## RBC Transfusion:

**Indication:** The indication for RBC transfusion is severe symptomatic anemia unresponsive to (or too urgent to wait for) other treatments including vitamin B12/folate/iron supplementation and/or erythropoietin.

**RBC Product Selection:** Standard blood bank approaches for RBC selection are acceptable.

**Leukoreduction:** Although leukoreduced blood products are recommended by most sources for HIV-positive patients, a randomized, double-blind trial (VATS [Viral Activation Transfusion Study]) comparing HIV activity, immune function, and survival in 531 HIV-positive, CMV-positive patients receiving leukoreduced versus non-leukoreduced RBCs failed to find a difference in HIV or CMV viral load, CD4 counts or other infections (including HBV, HCV, HHV-8, HTLV-1 or HTLV-II) between the two groups.

**Irradiation:** Although HIV-positive patients are clearly immunocompromised, there has only been one reported case of transfusion associated graft versus host disease (TA-GVHD) in a child with HIV, which was transient and atypical. It has been postulated that HIV infects donor T lymphocytes, thus making them incapable of initiating

TA-GVHD. Alternatively, recipient CD8+ T lymphocytes, which are spared until late in the course of AIDS, may prevent the development of TA-GVHD. Many transfusion medicine physicians therefore do not consider irradiation of blood products as warranted for HIV-positive individuals. However, given the nearly uniformly fatal prognosis of TA-GVHD, and the ease of irradiation, others opt to irradiate blood products for HIV-positive individuals to eliminate any potential TA-GVHD risk.

## Platelet Transfusion:

**Indication:** Platelet transfusion is indicated for severe thrombocytopenia, with or without bleeding, unresponsive to other treatments.

**Platelet Product Selection:** Standard blood bank approaches for platelet selection are acceptable. If the patient demonstrates clinical refractoriness, then the appropriate work-up and transfusion of suitable platelet products are warranted (see Chapter 49).

**Leukoreduction and Irradiation:** See the above discussion concerning leukoreduction and irradiation of RBC products.

**Adverse Effects:** Considerations for transfusion of the HIV-positive patient include the potential of transfusion-transmitted viruses and other infections, and the potential immunomodulatory role (if any) the transfusion itself may play in the progression of the HIV disease.

**Transfusion-transmitted Diseases:** The immunocompromised status of patients with HIV increases the risk for morbidity following primary viral infections, including those that may be transfusion-transmitted. In addition to viruses that are primarily found in plasma and for which screening has been uniformly implemented (e.g. HBV, HCV, WNV), there are other viruses of potential concern which are transmitted by white blood cells and that are rarely if ever screened for, such as cytomegalovirus (CMV), Epstein Barr virus (EBV), and human herpesvirus 8 (HHV-8).

CMV is a highly white blood cell-associated, DNA-containing herpes virus that can remain latent indefinitely following primary infection. Among the general population, 60–80% are CMV positive, while approximately 90% of HIV-positive individuals have serological evidence of CMV exposure. CMV viremia is common in late-stage AIDS, with clinical manifestations including retinitis, esophagitis, pneumonitis, hepatitis, biliary disease, and neurological disease. Most CMV infection in HIV-positive individuals is due to reactivation of latent infection. However, for the 10% of HIV-positive patients that are CMV negative, avoiding primary CMV infection is desirable.

The association of CMV infection with blood transfusion is well known, with transfusion-transmitted CMV reported in recipients of RBCs, platelet concentrates, and granulocytes. Frozen plasma products and cryoprecipitate, being largely acellular products, have not been reported to transmit CMV. Leukoreduction of RBC and platelet components significantly reduces the risk of transfusion-transmitted CMV by decreasing the number of monocytes capable of transmitting latent virus (granulocyte components cannot be leukoreduced). Transfusion of components from CMV-seronegative donors ("CMV negative") also reduces the risks of CMV transmission. However, provision of CMV-negative products is often difficult owing to the low number of seronegative donors (Chapter 38).

While it is unclear whether leukoreduced products or CMV-negative products have the lowest rates of transfusion-transmitted CMV, consideration may be given to providing CMV-negative blood to the 10% of HIV-positive patients that are CMV negative, if possible. At a minimum, provision of leukoreduced blood products is recommended by most sources for all HIV-positive patients. This not only decreases the risk of CMV transmission, but also decreases the potential risk for transmission of other strains of human herpesvirus (many of which establish latent infection in lymphocytes).

EBV and HHV-8, which plays a causative role in AIDS-related Kaposi Sarcoma, are present in leukocytes, and epidemiologic data support the possibility of transfusion transmission. There are currently no EBV or HHV-8 serologic tests used in the blood banking industry. Leukoreduction decreases the theoretical risk of transfusion-transmitted EBV and HHV-8 to HIV-positive individuals.

*Toxoplasmosis gondii* can lead to encephalitis, pneumonitis and retinochoroiditis in HIV-positive patients. This agent has been proven to be transfusion-transmitted, but screening tests are not widely available. It is possible that leukoreduction of blood components may reduce the risk of transfusion-transmitted toxoplasmosis.

## Recommended Reading

Collier AC, Kalish LA, Busch MP *et al.* (2001). Leukocyte-reduced red blood cell transfusions in patients with anemia and human immunodeficiency virus infection: the Viral Activation Transfusion Study: a randomized controlled trial. *J Am Med Assoc* **285**, 1592–1601.

Karpatkin S, Nardi M, Green D. (2002). Platelet and coagulation defects associated with HIV-1-infection. *Thrombosis & Haemostasis* **88**, 389–401.

Kruskall MS, Lee TH, Assmann SF *et al.* (2001). Survival of transfused donor white blood cells in HIV-infected recipients. *Blood* **98**, 272–279.

Marti-Carvajal AJ, Sola I. (2007). Treatment for anemia in people with AIDS. *Cochrane Database Syst Rev* CD004776.

Novitzky N, Thomson J, Abrahams L *et al.* (2005). Thrombotic thrombocytopenic purpura in patients with retroviral infection is highly responsive to plasma infusion therapy. *Br J Haematol* **128**, 373–379.

# CHAPTER 48

# Management of Patients Who Refuse Blood Transfusion

Jeanne E. Hendrickson, MD and John D. Roback, MD, PhD

A minority of patients refuse blood transfusion, usually based on religious beliefs (e.g. Jehovah's Witnesses). A doctrine introduced in 1945 by Jehovah's Witnesses teaches that the Bible prohibits the consumption, storage, and transfusion of human blood. The Watch Tower Bible and Tract Society has issued a number of doctrines since that time, citing that "blood is sacred to God," "when a Christian abstains from blood, he or she is in effect expressing faith that only the shed blood of Jesus Christ can truly redeem him or her and save his or her life," and "even in the case of an emergency, it is not permissible to sustain life with transfused blood." These beliefs stem from the interpretation of Biblical scripture, particularly Genesis, Leviticus, Samuel, and Acts.

Although there is some division among Jehovah's Witnesses regarding the Watch Tower doctrines, the majority observe the prohibition against transfusion of red blood cells (RBCs), white blood cells, platelets and plasma. Many Jehovah's Witnesses carry Medical Directive cards, issued by the Watch Tower Society, stating that blood transfusions are unacceptable. The use of blood derivatives, however, is not specifically prohibited, but is "not promoted, in that they may carry on such a life-sustaining role in the body as to be objectionable to Christians." In 2000, the Watch Tower encouraged members to personally decide if accepting these component fractions would violate the doctrine on blood. Examples of such products include albumin, immunoglobulin (including RhIg), clotting factor concentrates prepared from blood, and interleukins (Figure 48.1). Recombinant proteins (such as clotting factors) are generally accepted by Jehovah's Witnesses.

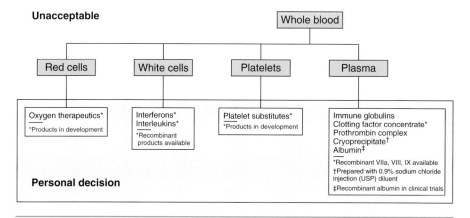

FIGURE 48.1   Blood components and fractions. Reproduced from ZM Bodnaruk, CH Wong, MJ Thomas (2004). Meeting the clinical challenge of care for Jehovah's Witnesses. *Transfus Med Rev* **18**, 105–116, with permission from Elsevier.

While standard transfusions are unacceptable, there are some related procedures that are not prohibited but not promoted, including plasma exchange, dialysis, intraoperative blood salvage, hemodilution, blood donation strictly for the purpose of further fractionation of components, and transfusion of autologous blood "as part of current therapy." Transfusion of preoperatively donated autologous blood is, however, prohibited, due to the belief that blood should not be taken out of the body and stored. Neither blood substitutes nor recombinant erythropoietin are expressly prohibited. Table 48.1 shows acceptable and unacceptable treatments, as well as those left to personal decisions by Witnesses.

From the physician perspective, the right of a competent adult to refuse consent for medical treatment is accepted, and documentation of refusal for transfusion should be placed in the medical record. Worst-case scenario discussions should be held, and documentation to this effect should also be included in the medical record. Some clinicians opt to have Witnesses sign the notes stating these discussions were held. Forcing a Jehovah's Witness to receive a transfusion unwillingly can be viewed as battery. The Witness who accepts a transfusion can be disfellowshipped and spiritually cut off from a community of family and friends.

**Specific Circumstances:** Some specific circumstances require additional consideration:

**Trauma:** Situations of trauma are difficult, in that Medical Directive cards may not be immediately available. If there is any doubt in a clinician's mind as to the wishes of the patient or what is legally appropriate, it is recommended that the clinician treat according to the accepted standards of care until legal documentation is available.

**Pregnant Women and Children:** The treatment of pregnant women and children deserves special attention. Given that minor children are not considered capable of informed consent, it is recommended that the clinician seek legal intervention in cases where the child is placed at risk by parental refusal for transfusion. The laws governing the locality in which the patient is treated determine whether a treating physician can emergently transfuse the child of a Jehovah's Witness. Consultation with legal experts can help clarify these laws.

**Blood Management:** Conversations with the patient regarding treatment options should be held in private, to ensure confidentiality. If the Witness does not allow transfusion, the treating physician should attempt to find the optimal therapy within the boundaries of the patient's religious beliefs. Such options include both blood conservation as well as the use of non-blood adjunctive therapies. Blood conservation can be achieved by reducing blood loss through decreased phlebotomy, intraoperative blood salvage, and/or acute normovolemic hemodilution. Potential adjunctive therapies include iron, folate, vitamin B12, erythropoietin, antifibrinolytic drugs, alternative oxygen carriers, hyperbaric oxygen, and recombinant Factor VIIa. The potential complications of each of these treatments must be weighed carefully against their potential beneficial effects.

The Hospital Information Services department of the Watch Tower Society has established Hospital Liaison Committees. These committees consist of ministers that are available to support physicians, patients, and social workers, and can be reached

| TABLE 48.1  Jehovah's Witness Religious Position on Medical Therapy |
| --- |

**A: Acceptable treatment**

Most surgical and anesthesiological blood conservation measures (e.g. hemostatic surgical instruments, controlled hypotension regional anesthesia, minimally invasive surgery, meticulous surgical hemostasis)

Most diagnostic and therapeutic procedures (e.g. phlebotomy for laboratory testing, angiographic embolization)

Pharmocologic agents that do not contain blood components or fractions such as:

Drugs to enhance hemostasis (e.g. tranexamic acid, epsilon-aminocaproic acid, aprotinin, desmopressin, recombinant factor VIIa, conjugated estrogens)

Hematopoietic growth factors and hematinics (e.g. albumin-free erythropoietin, iron)

Recombinant products (e.g. albumin-free coagulation factors)

Synthetic oxygen therapeutics (e.g. perfluorochemicals)

Non-blood volume expanders (e.g. saline, lactated Ringer's, hydroxyethyl starches)

**B: Personal decision (acceptable to some, declined by others)**

Blood cell salvage* (intraoperative or postoperative autotransfusion)

Acute normovolemic hemodilution*

Intraoperative autologous blood component sequestration* (including intraoperative plateletpheresis, preparation of fibrin gel, platelet gel, platelet-rich plasma)

Cardiopulmonary bypass[†]

Apheresis[†]

Hemodialysis[†]

Plasma-derived fractions (e.g. immune globulins, vaccines, albumin, cryoprecipitate[‡])

Hemostatic products containing blood fractions (e.g. coagulation factor concentrates, prothrombin complex concentrate, fibrin glue/sealant, hemostatic bandages containing plasma fractions, thrombin sealants)

Products containing plasma-derived blood fractions such as human serum albumin (e.g. some formulations of erythropoietin, streptokinase, G-CSF, vaccines, recombinant clotting factors, nuclear imaging products)

Products containing a blood cell-derived fraction

Epidural blood patch

Blood cell scintigraphy (e.g. radionuclide tagging for localization of bleeding)

Peripheral blood stem cell transplantation (autologous or allogeneic)

Other transplants (organ, marrow, bone)

**C: Unacceptable Treatment**

Transfusion of allogeneic whole blood, red blood cells, white cells, platelets, or plasma

Preoperative autologous blood donation (PAD or predeposit)

*Patients might request that continuity is maintained with their vascular system.

†Circuits not primed with allogeneic blood.

‡Cryoprecipitate suspended in 0.9% sodium chloride injection (USP) diluent.

Reproduced from ZM Bodnaruk, CH Wong, MJ Thomas (2004). Meeting the clinical challenge of care for Jehovah's Witnesses. *Transfus Med Rev* **18**, 105–116, with permission from Elsevier.

at 718-560-4300, his@jw.org. The Association of Jehovah's Witnesses for Reform in Blood (AJWRB) is another resource for clinicians and patients (www.ajwrb.org), as are The Society for the Advancement of Blood Management (www.sabm.org) and the Network for Advancement of Transfusion Alternatives (www.nataonline.com).

**Documentation of Consent or Non-consent:** Most (if not all) institutions have policies and forms pertaining to consent to receive blood products. Few facilities, however, have policies and forms specific to the refusal of consent for blood and blood product administration. As above, refusal to provide consent should be included in the patient's chart. A defined refusal to provide consent policy is recommended, as it allows documentation of the patient's refusal, a description of the blood products refused, and potential consequences of this refusal.

## Recommended Reading

Bodnaruk ZM, Wong CJ, Thomas MJ. (2004). Meeting the clinical challenge of care for Jehovah's Witnesses. *Transfus Med Rev* **18**, 105–116.

Jabbour N (ed.). (2005). *Transfusion-free Medicine and Surgery*. Malden: Blackwell Publishing.

Marsh JC, Bevan DH. (2002). Haematological care of the Jehovah's Witness patient. *Br J Haematol* **119**, 25–37.

Remmers PA, Speer AJ. (2006). Clinical strategies in the medical care of Jehovah's Witnesses. *Am J Med* **119**, 1013–1018.

Rogers DM, Crookston KP. (2006). The approach to the patient who refuses blood transfusion. *Transfusion* **46**, 1471–1477.

# CHAPTER 49

# Platelet Transfusion Refractory Patients

Jeanne E. Hendrickson, MD and John D. Roback, MD, PhD

The management of patients who are refractory to platelet transfusions can be difficult, and requires clear communication between the blood bank and clinical team. Platelet refractoriness is defined as an inappropriately low platelet count increment, usually determined by the corrected count increment (CCI), following repeated platelet transfusions.

There are both non-immune and immune causes for platelet refractoriness (Table 49.1), and often the etiology of platelet refractoriness is multifactorial. The primary immune cause of platelet refractoriness is human leukocyte antigen (HLA) alloimmunization, which is present in 30–40% of platelet-refractory patients.

| TABLE 49.1 Etiologies of Platelet Refractoriness | |
| --- | --- |
| **Non-immune Causes** | **Immune Causes** |
| Fever | Anti-HLA antibodies |
| Infection | Anti-HPA antibodies |
| Splenomegaly | Drug-induced antibodies |
| DIC | Plasma protein antibodies |
| Medications | ABO incompatibility |
| Bleeding | |
| HPC transplant | |
| GVHD | |
| VOD | |

Evaluation of the refractory patient includes determining the etiology of the underlying thrombocytopenia, reviewing the patient's underlying illness and current medications, and ascertaining future platelet transfusion requirements and the sites and severity of active bleeding. The hospital transfusion service physician or a designee should communicate the evaluation steps for platelet-refractory patients to the ordering physician, including the need for a 1-hour post-transfusion platelet count, determination of HLA or human platelet antigen (HPA) antibodies, and the timeframe for product availability.

**Non-immune Refractoriness:** Non-immune causes of platelet refractoriness include fever, infection, splenomegaly, disseminated intravascular coagulopathy (DIC), medications, bleeding, hematopoietic progenitor cell (HPC) transplant, graft-versus-host disease (GVHD) and veno-occlusive disease (VOD). Medications most commonly implicated in platelet refractoriness include amphotericin, heparin and vancomycin; other drugs implicated include antithymocyte globulin, granulocyte-macrophage colony-stimulating factor, granulocyte colony-stimulating factor, and interferons. Splenomegaly reduces both the post-transfusion platelet count increment and the time to next transfusion in comparison to individuals with normal spleens. The difference is even greater in comparison with splenectomized individuals. In addition, platelets stored longer than 48 hours have decreased viability following transfusion, though this does not typically affect the CCI.

**Immune Refractoriness:** The most common immune cause of platelet refractoriness is antibodies directed against HLA class I antibodies. These antibodies typically form after exposure to the corresponding HLA class I antigens on either platelets or contaminating white blood cells in transfused blood components, or from exposure during pregnancy. The TRAP (Trial to Reduce Alloimmunization to Platelets) study showed the benefits of leukoreduced products in decreasing HLA class I alloimmunization: a 17% HLA class 1 alloimmunization rate was seen in the leukoreduced study arm, whereas a 45% alloimmunization rate was seen in the non-leukoreduced arm. Inactivation of leukocytes by UVB irradiation also decreased alloimmunization (from 45 to 21%) in that trial, though this treatment modality is not typically available.

Less commonly, platelet refractoriness occurs as a result of antibodies against human platelet antigens. Most studies report the HPA alloimmunization rate to be between 2 and 10% in multiply transfused patients. HPA alloimmunization occurs primarily to HPA-1b and HPA-5b antigens. Patients with Bernard-Soulier syndrome and Glanzmann thrombasthenia may become broadly immunized to the platelet glycoproteins GPIb/IX/V and GPIIb/IIIa, respectively. Of note, The HPA alloimmunization rate in the TRAP trial (8%) was unchanged by leukoreduction.

Drug-induced antibodies, antibodies to plasma proteins, and ABO-incompatible transfusions can also lead to antibody-mediated platelet refractoriness. In general, a 20% decrease in platelet recovery is seen following transfusion with ABO-incompatible units.

An initial step to address cases of platelet refractoriness, while collecting CCI data and other test information (see below), is to determine whether transfusion of ABO-matched platelets of less than 48 hours' storage improves the post-transfusion platelet increment.

**Calculation of Refractoriness:** The precise definition of platelet refractoriness varies, but one widely accepted definition (as applied in the TRAP trial) is two 1-hour CCI values on consecutive days of less than 5000. Others view a CCI less than 7500 measured 10–60 minutes after a transfusion, and less than 4500 measured 18–24 hours after a transfusion, as refractory. The CCI is calculated by the formula:

$$CCI = \frac{\text{Body surface area } (m^2) \times \text{Platelet count increment (PLT/mL)} \times 10^{11}}{\text{Number of platelets transfused}}$$

The BSA can be estimated by the formula:

$$BSA \ (m^2) = 0.00718 \times \text{height } (cm)^{0.725} \times \text{weight } (kg)^{0.425}$$

**Testing:** Once a patient is deemed platelet refractory, a variety of tests are available to guide therapy. Two types of assays are generally most useful:

**HLA Antibody Detection:** Testing for HLA antibodies can be performed by lymphocytotoxicity testing (known as panel-reactive antibody [PRA]), in which serum is reacted with a panel of HLA-typed lymphocytes. HLA antibodies can also be detected by ELISA (enzyme-linked immunosorbent assay) or flow cytometry. If anti-HLA antibodies are identified, selection of patient-specific platelet components by HLA antigen based methods can produce improved posttransfusion platelet increments. If this approach is followed, the patient should be typed for HLA-A and HLA-B antigens. This can be performed by serologic or DNA methods. Crossmatched platelets may also be effective in these patients (see below).

**HPA Antibody Detection:** Testing for HPA antibodies should also be considered in platelet-refractory patients. Multiple platforms for testing are available, including ELISA, the indirect platelet immunofluorescence test, and solid phase red cell agglutination. Refractory patients with anti-HPA antibodies can be treated with crossmatched platelets.

**Treatment of the Alloimmunized Patient:** There are two main methods for selecting matched platelets: 1) HLA antigen based selection, and 2) Platelet crossmatching.

It is recommended that all platelets selected on the basis of HLA or crossmatching results be irradiated to prevent transfusion-associated graft-versus-host disease.

**HLA Antigen Based Selection:** With this strategy, products are selected based on the absence or compatibility of donor HLA-A and HLA-B antigens that correspond to recipient antibodies. Many of the original methods of treating refractory patients were based on selecting platelets with similar HLA phenotypes to the patient (HLA matching). Depending on the patient's phenotype, however, there may be difficulty finding an antigen-matched product. HLA-based matches are graded from A to D, based on the degree of matching (Table 49.2). Certain antigen mismatches are known to fare

TABLE 49.2  Grading of HLA Matching

| Grade | Description |
| --- | --- |
| A | HLA identical |
| BU | 3/3 detected antigens identical; 1 antigen not identified |
| B2U | 2/2 detected antigens identical; 2 antigens not identified |
| BX | 3/4 detected antigens identical; 1/4 cross-reactive |
| BUX | 2/3 detected antigens identical; 1/3 cross-reactive; 1 antigen not identified |
| B2X | 2/4 detected antigens identical; 2/4 cross-reactive |
| C | 1 antigen mismatch, out of cross-reactive group |
| D | ≥ 2 antigen mismatches |

poorly; HLAMatchmaker software (http://tpis.upmc.edu/tpis/HLAMatchmaker/) can help predict HLA compatibility. HLA-matched transfusions have a 20–50% failure rate. As an alternative method, if the HLA antibody specificity can be determined, platelet products that do not express the corresponding HLA antigen(s) can be transfused. This approach, called "antibody specificity prediction," provides a larger number of potentially compatible products than the HLA-matching approach, and is as effective as HLA matching or crossmatching.

Platelet Crossmatching: Crossmatching can be done by solid-phase red cell adherence (SPRCA) assay. In this assay, the serum from the alloimmunized patient is mixed with the apheresis platelet sample. Crossmatching can detect compatible platelets without requiring knowledge of HLA or HPA type or antibody identification. It can be used for detecting compatible products in a recipient with HLA and/or HPA antibodies. Note, however, that a large number of products may need to be crossmatched to locate compatible products for broadly alloimmunized patients.

Sibling Donors: Platelet products from siblings may be beneficial. However, siblings should not be used if there is a possibility the sibling is, or potentially will be, an HPC donor for transplantation.

Bleeding, Platelet-refractory Patient: Transfusion of large numbers of platelets is often attempted in bleeding, platelet-refractory patients, but is usually without benefit. Alternative potential therapies include IVIG, plasma exchange, rituximab, splenectomy, corticosteroids, other immunosuppressive medications, antifibrinolytic agents, and recombinant Factor VIIa. With the exception of IVIG and antifibrinolytic agents, these therapies have generally not proven beneficial.

Prevention: Leukoreduction of transfused blood components decreases alloimmunization to HLA class I antigens. This is hypothesized to occur because HLA antigens on transfused leukocytes initiate a primary immune response that leads to the production of donor-specific anti-HLA antibodies. Some centers advocate prospective HLA antibody screening for all patients with a history of prior transfusions or pregnancies requiring ongoing platelet support.

Recommended Reading

Dzik S. (2007). How I do it: platelet support for refractory patients. *Transfusion* **47**, 374–378.

Petz LD, Garratty G, Calhoun L *et al*. (2000). Selecting donors of platelets for refractory patients on the basis of HLA antibody specificity. *Transfusion* **40**, 1446–1456.

Schiffer CA, Anderson KC, Bennett CL *et al*. (2001). Platelet transfusion for patients with cancer: clinical practice guidelines of the American Society of Clinical Oncology. *J Clin Oncol* **19**, 1519–1538.

Slichter SJ, Davis K, Enright H *et al*. (2005). Factors affecting post-transfusion platelet increments, platelet refractoriness, and platelet transfusion intervals in thrombocytopenic patients. *Blood* **105**, 4106–4114.

The Trial to Reduce Alloimmunizaton to Platelets (TRAP) Study Group (1997). Leukocyte reduction and ultraviolet B irradiation of platelets to prevent alloimmunization and refractoriness to platelet transfusions. *N Engl J Med* **337**, 1861–1869.

# CHAPTER 50

# Massive Transfusion

Beth H. Shaz, MD and Christopher D. Hillyer, MD

Massive transfusion is commonly defined as transfusion of 10 or more red blood cell (RBC) products within 24 hours, which approximates the total blood volume of an adult recipient; other definitions exist and are quite valuable, including RBC replacement of 50% of total blood volume within 3 hours, or blood loss exceeding 150 ml/min. Massive transfusion can occur in a variety of clinical settings, including cardiovascular, spinal and liver surgery; trauma; gastrointestinal bleeds; and obstetrics. Massive transfusion protocols (MTPs) are currently changing in the US and worldwide due to recent data showing that earlier and more aggressive transfusion intervention and resuscitation with blood components that approximate recapitulated whole blood significantly decrease mortality. In this context, MTPs are a key element of "damage control resuscitation."

**Clinical Significance:** The physiologic response to blood loss is to preferentially maintain tissue oxygenation to the brain and heart by shunting blood from other organs, shifting fluid from intracellular to extracellular space and from interstitial to intravascular space, and conserving water and electrolytes. Loss of up to 10% of blood volume results in few symptoms; loss of up to 20% does not usually cause signs or symptoms when the patient is at rest, but will result in tachycardia with exercise; and loss of up to 30% results in hypotension and tachycardia, especially with exercise. Once blood loss exceeds 30%, serious signs and symptoms of cardiovascular compromise occur, including tachycardia with weak pulse, hyperpnea, hypotension, decreased central venous pressure and cardiac output, and cold clammy skin. At approximately 50% blood loss, severe shock and death occur.

In the severely injured trauma patient, uncontrolled hemorrhage with more than 50% blood loss is one of the most common causes of mortality. Over the past three decades, hemorrhage has remained the ultimate cause of mortality in approximately 30% of trauma fatalities, second only to traumatic brain injury. Indeed, in a recent large retrospective review, mortality of patients requiring more than 50 RBC products was 57%.

Severe hemorrhage accompanying obstetric complications, including placenta previa, uterine atony or rupture, and disseminated intravascular coagulopathy (DIC), can lead to hysterectomy and loss of future reproductive capacity, and/or loss of the mother, child, or both. Except in trauma and some obstetrical circumstances, massive transfusion for cardiac and liver transplantation, and other circumstances as mentioned above, follows the general considerations below; however, few clinical trials exist to guide therapy in these and more stratified situations.

**Resuscitation Approaches:** Maintenance of adequate blood flow and blood pressure by infusing a sufficient volume of crystalloid (defined as a substance that can pass through a semipermeable membrane) such as normal saline or lactated Ringer's solution, colloid (defined as a substance that cannot pass through a semipermeable membrane) such as albumin, and/or blood products is paramount to maintaining tissue oxygenation and helping to ensure survival. RBC transfusion is critical to ensure maximal or near maximal arterial oxygen content and thus oxygen delivery to tissues, which is dependent on cardiac output and both the RBC mass and hemoglobin. The ability of transfused RBCs to release oxygen optimally is dependent at least in part on the metabolic status of the patient and the length of storage of the RBC product.

Crystalloid versus Colloid Replacement: Crystalloids distribute quickly into total body water and can cause peripheral and pulmonary edema, but are less expensive than colloid solutions. Colloid solutions primarily remain (at least initially) intravascular, but are more expensive and can cause allergic reactions. Practice has changed to earlier use of colloids (especially plasma) and RBC transfusion while simultaneously decreasing the amount of crystalloids administered due to increasing evidence that large-volume crystalloid administration is associated with abdominal compartment syndrome as well as cardiac, pulmonary, gastrointestinal, coagulation, as well as other complications. In addition, the current goal of volume resuscitation is euvolemia and to avoid supra-normal resuscitation. Euvolemia entails moderate volume resuscitation with the possible use of vasopressor agents to support hemodynamics.

RBC Transfusion: When blood loss is excessive and there is not adequate time for pretransfusion testing, group O RBC and AB plasma products should be issued until the recipient's blood type is known. Each institution should have well-developed policies for:
- emergency release, issuance and delivery of blood products;
- switching blood types (e.g. when to give D-positive RBC products to a D-negative individual;
- issuing antigen positive or antigen untested in a patient with the corresponding alloantibody; and
- issuing ABO incompatible plasma (e.g. an AB patient requiring large amounts of plasma).

Women of child-bearing potential should receive D-negative RBCs, and an inventory of type O, Rh-negative products should be kept for these individuals. As described in detail in Chapter 28, these products are important to prevent formation of alloanti-D, which can lead to future hemolytic disease of the fetus and newborn. Men and older women may receive D-positive RBCs if D-negative RBC products are not available or if the inventory is low and there is a need to reserve these products for women of child-bearing potential.

A patient sample for blood typing and antibody screen should be obtained as soon as possible after patient arrival to ensure that the patient receives type-specific products when available, thus preserving the often limited group O RBC and AB plasma supply in the facility. Transfusion of type-specific products also avoids obfuscation of the patient's true blood type due to the infusion of large amounts of group O RBCs and AB plasma. In addition, patients who receive large amounts of "out of group" components may be receiving ABO-incompatible plasma (such as a group A patient receiving group O RBCs), which can result in a positive direct antiglobulin test and/or hemolysis.

**Component Therapy Based Approaches:** In the recent past, resuscitation and transfusion protocols would start with significant crystalloid or non-plasma (albumin) colloid infusion and many RBC products. This was followed by a component therapy type approach to guide blood product choices, volumes and timing. A component therapy type approach uses clinical findings and laboratory results, and requires that laboratory tests are both timely and reflective enough of the coagulation system to aid in guiding therapy, which is often not the case. An example of using component therapy is to have transfusion thresholds based on laboratory values, such as: hemoglobin $<8\,g/dl$, prothrombin time $>1.5$ times normal, platelet count $<50,000/\mu l$, and fibrinogen $<100\,g/dl$. Using this type of approach, there typically was and still is no set administration ratio of RBC products to plasma, platelets and cryoprecipitate.

Mortality rates using component therapy type strategies were unacceptably high, and the value of earlier and more aggressive blood product administration was considered. Thus MTPs were designed which attempted to "recapitulate" or "reconstitute" whole blood ("recapitulated" whole blood means matching the RBC, plasma, platelets and cryoprecipitate *ratio* of whole blood, while "reconstituted" whole blood refers to premixing components to reassemble whole blood into a single product), or to use fresh whole blood. Early reports in the military setting showed significant reduction in mortality (65% reduced to ~20%; see Figure 50.1), with an optimal plasma to RBC product ratio of 1.4.

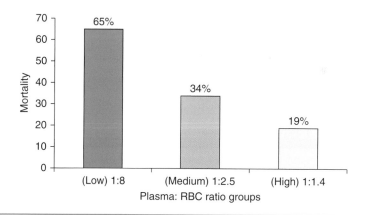

**FIGURE 50.1** Percentage mortality associated with low, medium and high plasma-to-RBC ratios transfused at admission. Ratios are median ratios per group and include units of fresh whole blood counted both as plasma and as RBCs. From Borgman MA, Spinella PC, Perkins JG, *et al.* (2007). The ratio of blood products transfused affects morality in patients receiving massive transfusions at a combat support hospital. *J Trauma* **62**, 805–813.

**Massive Transfusion Protocols:** MTPs define the following: 1. notification of the transfusion service and laboratory, 2. laboratory testing algorithms (e.g. PT, PTT, platelet count, fibrinogen and hemoglobin), 3. blood product preparation (amount of plasma, RBCs, platelets and cryoprecipitate to prepare and issue at set time intervals), and 4. other patient care needs (e.g. blood warmers).

MTPs are designed to ensure optimal transfusion therapy to prevent and treat the multifactorial coagulopathy which can occur early after injury. Indeed, published reports show that a significantly abnormal PT and PTT upon arrival at the trauma center is predictive of a high incidence of mortality.

In 2005, a symposium of surgeons anesthesiologists, hematologists, transfusion medicine specialists, epidemiologists and others, held at the US Army Institute of Surgical Research, led to the recommendation of a 1:1:1 ratio of RBC:plasma:platelets during massive transfusion. A 2007 retrospective study demonstrated improved survival with more plasma transfused per each RBC product (65% mortality with 1:8 and 19% mortality with 1:1.4 plasma:RBC ratio; Figure 50.1) in damage control resuscitation.

Based on the dramatic success of MTPs used in combat, studies are underway to show equivalent performance and patient benefit in the non-military setting. Initial data from the Grady Memorial Hospital, Atlanta, Georgia, support the clinical benefit in both 24-hour and 30-day mortality. The current MTP regimen for blood product administration at the above institution is shown in the Table 50.1 as an example.

### TABLE 50.1 An Example of a Massive Transfusion Protocol

| Package | RBCs | Plasma | Platelets | Cryoprecipitate |
|---|---|---|---|---|
| Initial | 6 units | 6 units | | |
| 1 | 6 units | 6 units | 1 apheresis | |
| 2 | 6 units | 6 units | | 20 units |
| 3 | 6 units | 6 units | 1 apheresis | |
| 4 | 6 units | 6 units | | 10 units |
| 5 | 6 units | 6 units | 1 apheresis | |
| 6 | 6 units | 6 units | | 10 units |

**Multidisciplinary Communication:** The management of patients with acute blood loss requires concise, effective and frequent communication between the trauma team and the hospital transfusion service (HTS). The HTS needs to be able to rapidly prepare blood products for issue, assess RBC component inventory and participate in reconciling laboratory values. Institutions which frequently take care of patients with acute blood loss usually have well-defined MTPs, but this may not always be the case.

MTPs require adequate laboratory support to evaluate the patient's hemoglobin, platelet count, PT, PTT, fibrinogen level, ionized calcium and acidosis in order to address and correct these values. Some institutions use thrombelastograph (TEG) technology, which provides a dynamic and global assessment of the coagulation process, including platelet function, coagulation cascade and fibrinolysis, to guide transfusion management of these patients. Currently, sufficient data are lacking to support its use.

**Complications of Massive RBC Transfusion:** Massive transfusion of RBC products is often complicated by coagulopathy secondary to dilution (termed dilutional coagulopathy) because RBC products are virtually devoid of clotting factors. Once one blood volume has been replaced with RBC products, only ~35% of coagulation factor levels remain as a result of the dilution. Platelets are initially mobilized from the spleen but decrease to ~50,000/μl after replacement of approximately two blood volumes. Fibrinogen decreases to 100 g/dl after approximately two blood volumes are replaced, which significantly limits clot formation.

In addition, the consumption of platelets and coagulation factors, and hypothermia and acidosis which result in decreased function of coagulation factor enzymes, complicate and worsen the coagulopathy. This multifactorial *trauma induced coagulopathy* (TIC) may occur early (ETIC) or as damage control resuscitation is underway, and may lead to ongoing blood loss, decompensated shock and death.

Other complications of massive RBC transfusion are secondary to the storage of RBC products, and are more significant in the patient who receives a large volume of products within a short period of time. Storage of RBCs can lead to the following: 1. Hypothermia (as the products have been recently refrigerated), which can increase tissue oxygen requirements and induce ventricular arrhythmias; 2. Impaired metabolism of citrate and lactate, which increased acidosis; 3. Hyperkalemia; and 4. Decreased oxygen delivery due to decreased 2,3-DPG levels.

As RBCs are stored in citrate containing anticoagulant solutions, administration of large volumes of RBCs leads to citrate binding of free calcium and thus hypocalcemia, which can cause paresthesias, nausea, hyperventilation and depressed cardiac function. In addition, citrate is an acid and thus contributes to the acidosis.

Lastly, complications of massive RBC transfusion arise from the transfusion of older RBC products which have undergone changes during storage (termed the *storage lesion*). The storage lesion includes decreased 2,3-diphosphoglycerate (2,3- DPG), decreased pH, decreased RBC deformability, increased hemolysis, and increased potassium, phosphate and ammonia concentrations, and is further described in Chapter 28. All of these changes may adversely affect the recipient during massive transfusion.

## Recommended Reading

Borgman MA, Spinella PC, Perkins JG *et al.* (2007). The ratio of blood products transfused affects morality in patients receiving massive transfusions at a combat support hospital. *J Trauma* **62**, 805–813.

Holcomb J, Hess J. (2006). Early massive trauma transfusion: state of the art. *J Trauma* **60**, S1–2.

Johanasson PI, Hansen MB, Sorensen H. (2005). Transfusion practice in massively bleeding patients: time for a change? *Vox Sang* **89**, 92–96.

Malone DL, Hess JR, Fingerhut A. (2006). Massive transfusion practices around the globe and a suggestion for a common massive transfusion protocol. *J Trauma* **60**, S91–S96.

# CHAPTER 51

# Perioperative Blood Management

Beth H. Shaz, MD

Blood management is the term given to the practice of minimizing allogeneic blood use, while maximizing patient outcome. Blood management has four main tenets: 1. a focus on guideline-driven proper use of banked blood and minimizing its inappropriate use, 2. pharmaceutical preparations that prevent, minimize or control blood loss, usually in the operative setting, 3. blood conservation methods, and 4. a multidisciplinary approach.

Blood conservation includes preoperative (preoperative autologous donation), intraoperative (hemodilution and blood salvage) and postoperative (blood salvage) techniques. The multidisciplinary approach uses the entire team of healthcare providers, to be uniform in the goal of minimizing transfusion in the treatment of patients. This approach incorporates treating preoperative anemia, using meticulous anesthetic and surgical techniques intraoperatively, and avoiding over phlebotomy postoperatively.

## Factors Influencing the Risk of Transfusion:

**Preoperative Factors:** The strongest predictor of a patient requiring blood during an elective surgery is his or her baseline hematocrit, with other significant contributing factors being the patient's total blood volume (TBV) and red blood cell (RBC) loss during the procedure. Optimization of the patient's hemoglobin (Hgb) prior to surgery will decrease the chance of transfusion. Iron or erythropoietin may be indicated, depending on the cause of the patient's anemia. Preoperative anemia additionally increases the risk of perioperative infection and mortality, which may be secondary to the associated increased risk of allogeneic blood transfusion.

*Discontinuation of Anti-platelet and Anticoagulant Agents:* The correction of impaired hemostasis should decrease blood loss, and therefore decrease the need for transfusion. Impaired hemostasis can be corrected by discontinuing anticoagulants and anti-platelet medications (platelet function returns to normal 72 hours after the last dose of aspirin) in a timely manner, or by treating the cause of the impaired hemostasis. In emergency situations, there may not be adequate time to discontinue the anti-platelet or anticoagulant medications. The effects of anti-platelet medications can be reversed with the transfusion of platelet products, if necessary.

**Unfractionated Heparin:** Heparin can be reversed by protamine sulfate, dosed at 1 mg per 100 U of unfractionated heparin administered over the previous 4 hours. Protamine must be slowly administered (up to 5 mg/min) to reduce the risk of hypotension and bradycardia. The maximum dose of protamine is 100 mg over two hours. Response to protamine can be assessed by measuring the PTT or anti-Xa activity.

**Low Molecular Weight Heparin:** Heparin can be reversed by protamine sulfate, dosed at 1 mg per 1 mg of low molecular weight heparin administered over the previous 4 hours. Protamine must be administered slowly (up to 5 mg/min) to reduce the risk of hypotension and bradycardia. The maximum dose of protamine is 100 mg over two hours. Response to protamine sulfate can be assessed by measuring the PTT or anti-Xa activity.

**Warfarin:** Warfarin can be reversed with vitamin K (dose 2.5–5 mg orally or IV over 30 minutes [risk of acute anaphylactoid reaction with IV use]), which takes 2–24 hours to improve the INR; or, for an immediate effect, plasma products (dose 5–15 ml/kg depending on INR), prothrombin complex concentrate (dose 25–50 U/kg), or recombinant factor VIIa (rFVIIa) (dose 10–90 μg/kg) can be administered. The post-infusion INR can be used to assess the adequacy of warfarin reversal. Prothrombin complex concentrate and rFVIIa do not contain all of the factors that are deficient as a result of warfarin therapy, and therefore may require the supplemental use of plasma products or vitamin K. The half-life of some of the factors, such as 3–6 hours for Factor VII, may require additional doses of rFVIIa, prothrombin complex concentrate or plasma to maintain the target INR.

**Factor Xa Inhibitors:** rFVIIa has been used at a dose of 90 μg/kg to reverse Factor Xa inhibitors with an immediate effect which lasts 2–6 hours.

**Direct Thrombin Inhibitors:** Direct thrombin inhibitors can be reversed by DDAVP (dose 0.3 μg/kg IV over 15 minutes), cryoprecipitate (dose 10 units in an adult), antifibrinolytics (EACA 0.1–0.15 g/kg IV over 30 minutes followed by 0.5–1 g/hour until bleeding subsides or tranexamic acid (TXA) 10 mg/kg IV every 6–8 hours until bleeding subsides), or plasma products (initial dose of 2 units).

*Erythropoietin:* Erythropoietin, a 165 amino acid glycoprotein hormone, corrects anemia caused by renal failure, cancer, cancer therapy and human immunodeficiency virus. Erythropoietin can be used preoperatively, with or without preoperative blood donation or acute normovolemic hemodilution, in elective surgery. Two to four weeks are necessary for adequate erythropoietin-stimulated erythropoiesis to occur. Depending on the estimated blood loss, patient's TBV, and adequate time prior to surgery, erythropoietin may be useful. Concomitant iron supplementation maximizes the benefit of erythropoietin use. Causes of hyporesponsiveness to erythropoietin include folic acid, vitamin B12 and vitamin C deficiency; infection; inflammatory states; and chronic blood loss.

**Adverse Events:** Erythropoietin is associated with thrombotic events, hypertension, seizures and rare cases of pure red cell aplasia, and is contraindicated in patients with uncontrolled hypertension.

*Iron Supplementation:* Correcting iron deficiency anemia prior to surgery decreases the likelihood of need for allogeneic transfusion.

**Intraoperative Factors:** Intraoperative factors to reduce allogeneic transfusion include preventing hypothermia, optimizing surgical technique, and controlling hemostasis. Surgical techniques that reduce bleeding include laparoscopic, robotic or endovascular approaches. Surgical instruments that maximize coagulation include the ultrasonic scalpel and argon beam coagulator.

**Postoperative Factors:** Patients in the intensive care unit have a mean blood draw of approximately 40 ml a day, with a total volume of 760 ml during the ICU stay, which increases to 900 ml in patients with arterial lines. These large losses from phlebotomy contribute to increased transfusion requirements.

**Blood Utilization Guidelines:** Although the blood supply in developed countries has become increasingly safe, there continues to be mortality and morbidity associated with transfusions. An association between transfusion versus non-transfusion, and transfusion of RBC products with longer versus shorter storage time, has been reported to increase length of hospital stay, number of infections, multiple organ failure, and/or mortality in primarily cardiac surgery, trauma and intensive care unit patients. Acknowledging that these studies are either retrospective or prospective cohort studies, making them observational and often difficult to interpret due to concerns about co-morbidities, unrecognized confounding variables and the difficulty in assembling a non-transfused control arm with equal disease severity scores, they have demonstrated that transfusion in general, and transfusion of older stored RBC products in particular, has been repeatedly implicated in increasing morbidity.

These studies must be weighed against the high risk of mortality in anemic (Hgb <8 g/dl) patients who refuse transfusion. Therefore, for each patient, depending on his or her co-morbidities, a transfusion threshold should exist. Creating restrictive transfusion criteria reduces the likelihood of transfusion, with an average savings of a single unit per transfused patient. In addition, in the critically ill population restrictive transfusion criteria (Hgb 7 g/dl) may decrease mortality in some patients (e.g. patients younger than 55 years old and/or with an APACHE II score less than 20) or be equivalent to a liberal transfusion threshold (Hgb 10 g/dl).

Prior to the creation of guidelines, an audit should be performed to understand the established transfusion practice. Transfusion guidelines should be created by a multidisciplinary team and based on evidence in the literature. Prior to implementation of the guidelines, education of the ordering physicians and other members of the healthcare team is required to ensure the guidelines are understood and are followed. After implementing guidelines, a repeat audit must be performed to guarantee they are being followed. This can also identify areas that need continued improvement.

## Pharmaceutical Preparations:

**Desmopressin:** Desmopressin (DDAVP: 1-deamino-8-D-arginine-vasopressin) is a vasopressin analogue that increases the circulating levels of Factor VIII and von Willebrand factor, and has been used to manage bleeding in patients with uremic-induced platelet dysfunction, acquired and hereditary platelet disorders, and type I von Willebrand disease. Studies in adult elective surgery patients (most undergoing cardiac surgery) without underlying bleeding disorders have demonstrated that prophylactic administration of DDAVP does not reduce blood loss, risk of allogeneic transfusion, volume of RBCs transfused, or mortality.

*Dose:* The usual dose is 0.3–0.4 µg/kg over 30 minutes. Doses can be repeated at 8–12 hours. Serial doses are associated with tachyphylaxis, hyponatremia and seizures, especially in children under 2 years of age.

*Adverse Effects:* Side effects of DDAVP include hypotension, hyponatremia, headache, and decreased urine output.

Antifibrinolytics: Antifibrinolytic drugs reduce blood loss and the need for RBC transfusion. The lysine analogues (TXA and EACA) are likely as effective as aprotinin, and are cheaper. In high-risk cardiac surgery, where there is substantial risk of large blood loss, aprotinin may be preferable. Currently aprotinin is not available due to its association with increased mortality, especially secondary to cardiac complications, in comparison to patients who received other antifibrinolytics in a large randomized controlled trial in high-risk cardiac surgery patients.

*Aprotinin:* Aprotinin, a proteinase inhibitor obtained from bovine lung, inhibits multiple mediators of inflammatory responses, fibrinolysis, and thrombin generation. Aprotinin inhibits kallikrein at higher doses and plasmin at lower doses. Aprotinin reduces the need for allogeneic blood transfusion, as well as reducing bleeding and the need for re-exploration in cardiac bypass surgery patients.

**Adverse Effects:** Anaphylactic reactions are possible, especially when patients are re-exposed to aprotinin; therefore, a test dose is required.

*ε-aminocaproic Acid and Tranexamic Acid:* TXA and EACA are synthetic lysine analogs that inhibit fibrinolysis. These drugs block the lysine binding site on the plasminogen molecule, which inhibits the formation of plasmin and, therefore, of fibrinolysis. TXA or EACA is used in a variety of surgeries to decrease allogeneic RBC transfusion.

**Adverse Effects:** There have been cases of thrombotic complications with its use in liver transplantation surgeries.

Recombinant Factor VIIa: Factor VIIa binds to tissue factor to activate Factors IX and X, thereby activating Factor V, which converts prothrombin to thrombin. Thrombin activates platelets and Factors VII, V and XI. This creates a thrombin burst which converts fibrinogen to fibrin for clot formation. rFVIIa is licensed for use in hemophilics with inhibitors, but there are case series demonstrating its success in controlling severe or refractory bleeding. rFVIIa may decrease blood loss and decrease transfusion needs in cases of large blood loss, but has a risk of thrombosis with a resulting increase in mortality. The off-label use of rFVIIa may be useful in certain settings, such as rescue therapy for life-threatening bleeding that is unresponsive to standard hemostatic therapies (i.e. platelet, plasma and cryoprecipitate transfusions) and with no identifiable surgical source of bleeding. Currently, its effectiveness remains uncertain. rFVIIa should be used with caution in patients at increased risk for thrombosis, such as those with a history of thrombotic complications or thrombophilic disorders, or who have disseminated intravascular coagulopathy (DIC).

*Dose:* Dosing for bleeding has ranged from 15 to 180 μg/kg, with lower doses used in Factor VII deficiency secondary to liver failure and higher doses used for life-threatening hemorrhage. Alternatively, lower doses (10–20 μg/kg) can be repeated every 15–30 minutes for a total of 90–180 μg/kg in patients with life-threatening hemorrhage.

*Adverse Effects:* The use of rFVIIa is associated with the risk of thrombotic complications, especially in patients with known hypercoagulability.

## Fibrin Sealant:

Fibrin sealant is the combination of thrombin and fibrinogen mixed with calcium to form fibrin, which is used as a topical hemostatic agent. Products may contain antifibrinolytics (aprotinin) to reduce fibrinolysis, or Factor XIII to increase the strength of the clot. A variety of commercially and individually produced fibrin sealants are available. Currently there are fibrin sealant products consisting of human fibrinogen, human thrombin, calcium chloride and aprotinin (bovine) which are commercially available in the US. In addition, there is recent FDA approval of a product containing human-derived fibrinogen and thrombin without aprotinin. Alternatively, automated devices also exist to produce fibrin sealant from autologous plasma. Another option is autologous fibrin glue prepared from the cryoprecipitated portion of autologous plasma; after thawing, this material is mixed with bovine thrombin immediately before application to the surgical field site. A disadvantage of fibrin sealant is that it takes time to prepare, especially in autologous products, and also time for clot formation. Fibrin sealants reduce allogeneic transfusion and decrease intraoperative and postoperative blood loss for a variety of surgical procedures.

*Adverse Effects:* Bovine thrombin containing products carry the risk of allergic reactions or antibody formation. Antibodies to the bovine thrombin crossreact with human Factor V, leading to Factor V deficiency and risk of hemorrhage. In contrast, human-derived proteins, though virally inactivated, carry the potential risk of transfusion transmitted disease.

## Autologous Blood:

Autologous blood can be collected from a patient in advance of anticipated blood loss (preoperative donation) or at the start of the procedure (acute normovolemic hemodilution); in addition, shed blood can be salvaged for reinfusion both during surgery and in the postoperative period. As the risk of transfusion has decreased and the understanding of the risks and costs of autologous donation has increased, there has been a decline in the utilization of preoperative blood donation.

*Advantages:* The use of autologous blood decreases the risk of adverse events from allogeneic blood, such as transfusion-transmitted diseases, transfusion-related acute lung injury, and other non-infectious hazards of transfusion.

*Disadvantages:* The disadvantages of autologous blood donation include the risk of clerical errors resulting in the transfusion of the incorrect product, and the potential infectious risk of bacterial contamination of the product – especially with *Yersinia enterocolitica*, a common cause of community-acquired diarrhea.

*Costs:* As the allogeneic blood supply has become safer, more attention has been focused on the costs associated with autologous transfusion techniques. Costs arise from unused autologous collections (for example, when a patient has donated enough blood to match the mean number of components used by others undergoing the procedure but requires less). This problem is magnified by over-collection and unnecessary utilization, and by the extra work involved in deviation from routine large-scale allogeneic collection practices.

**Preoperative Autologous Donation:** Provided that the donor has satisfactory iron stores and that bone marrow erythropoiesis can occur in a timely fashion, blood may be comfortably collected from an autologous donor on a weekly schedule (see Chapter 7). The shelf-life of refrigerated whole blood is limited to 42 days with current formulations of anticoagulant-preservative solutions, and a schedule for multiple donations is usually fitted into this 6-week window. Alternatively, some or all of the products can be frozen at −65°C for up to 10 years. Although frozen products allow collections to occur over a longer period, the flexibility of utility at the time of surgery is affected: thawing and deglycerolizing takes a few hours, and the thawed products have a 24-hour outdate.

Autologous blood donations are well tolerated by a variety of high-risk donors, including the elderly, children, pregnant women, and patients with atherosclerotic coronary artery disease. A weekly phlebotomy schedule fosters some degree of RBC regeneration before surgery. However, the most important medical problem associated with autologous donation is anemia developing during the collection interval. When this occurs, it is typically as a result of marginal iron stores and insufficient erythropoietic response (with little or no increase in serum erythropoietin levels), probably because the hematocrit of most donors is not allowed to fall below 30%.

*Indications:* The decision to use preoperative autologous blood donations should be predicated on the type of surgery, the amount of time available for donation and hematopoietic reconstitution, the patient's hematocrit, and the predicted vigor of the erythropoietic response to donation. When large blood loss is expected and multiple autologous blood products are needed, erythropoietin can be used to increase the number of preoperative autologous blood products able to be donated. Patients planning to undergo elective orthopedic surgery are ideal candidates for autologous transfusion, because they require moderate amounts of blood during and immediately after surgery, and typically have sufficient time prior to surgery to make multiple donations. Other surgeries where preoperative autologous blood donation is commonly utilized include cardiac surgery, vascular surgery, radical prostatectomies, hysterectomies and other gynecologic procedures, gastrointestinal surgery, neurosurgery, and bone marrow harvest. As surgical techniques change, the need for blood transfusions may be affected; in such instances, the role of autologous blood should be re-examined.

Long-term frozen storage of autologous RBCs in the absence of an anticipated transfusion episode is both ineffective and expensive. An exception is the storage of blood by individuals with high-frequency or complex alloantibodies, for whom stockpiling of rare autologous products may be beneficial. Even here, however, the likelihood that such blood would be helpful is slim. To be of value, sufficient autologous blood would have to be available to meet the needs of an unexpected emergency; furthermore, delays in sending the blood expeditiously to the hospital where it is needed and in preparing products (thawed and washed free of the glycerol cryoprotectant) would make its use unwieldy.

**Autologous Platelet-rich Plasma:** Autologous platelet-rich plasma can be prepared at the start of cardiac surgery with cardiopulmonary bypass, using apheresis equipment before bypass, to be returned to the patient after heparin reversal at the end of the surgical procedure. Because thrombocytopenia or an acquired platelet defect can occur after blood passes through the membrane oxygenator, the theoretical advantages of transfusing platelet-rich plasma should include an improvement in hemostasis and reduced transfusion requirements. Platelet-rich plasma has been shown to decrease allogeneic transfusion.

**Acute Normovolemic Hemodilution:** The collection of autologous blood at the start of a surgery, for return to the patient at the end of the procedure, had its origins in cardiac surgery. The original goal was prevention of postoperative coagulopathies through *ex vivo* maintenance of a supply of platelets undamaged by exposure to the membrane oxygenator. However, additional advantages regarding the intentionally created anemia were also identified. Hemodilution can contribute to a reduction in RBC loss. In simplest terms, a patient with a hematocrit of 45% and a 2 l blood loss during surgery loses roughly 900 ml of RBCs, whereas a similar patient with a hematocrit of 20% loses only 400 ml of RBCs.

Hemodilution is probably less expensive to accomplish than preoperative autologous blood donation, and it may be the only option available when surgery is performed in non-elective settings. The technique involves removal of blood into standard collection bags with citrate anticoagulation (unless the patient is already heparinized), and replacement of lost volume with either crystalloids or colloids. Close monitoring of the patient's cardiovascular status is necessary during the hemodilution process. Products are stored in the operating room during surgery and reinfused as needed, in reverse order of collection, reserving the bags with the highest concentration of RBCs for the end of the procedure, after blood loss has been controlled. Erythropoietin can also be used preoperatively in conjunction with acute normovolemic hemodilution, especially in patients with small blood volumes and mild anemia.

*Adverse Effects:* The severity of the anemia could affect oxygen transport, although the concomitant drop in blood viscosity, and compensatory cardiac output increases, could restore oxygen delivery.

**Intraoperative Autologous Transfusion:** A number of techniques have been developed for the salvage and reinfusion of blood lost during an operative procedure. The simplest approach, direct reinfusion without washing, involves collection of blood under low vacuum pressure into a plastic bag seated within a hard outer canister where an anticoagulant (usually citrate) is added, and as soon as the bag is full, or within 4 hours after the start of the collection (to prevent bacterial growth), the contents of the bag are reinfused through a standard blood filter to the patient. RBCs shed into a surgical field are accompanied by activated coagulation factors and platelets, cellular debris and soluble factors released from injured tissue cells, pharmaceuticals applied to the field, and irrigant solutions. Alternatively, the contents in the bag can be washed with saline. Devices that include a reservoir for collecting the salvaged blood and a centrifuge for washing are available to collect and process large volumes quickly. With these techniques, intraoperative blood salvage has become practical in situations in which blood loss may be extremely rapid, such as trauma or liver transplantation. Approximately one-half of the blood lost during a surgical procedure can be recovered.

*Adverse Effects:* Complications of intraoperative salvage are infrequent. The hematocrit of salvaged unprocessed blood is typically low because of a combination of

dilution from irrigation fluids and some degree of mechanical hemolysis potentially resulting in dilutional coagulopathy, hemoglobinemia and hemoglobinuria.
In addition, bacterial contamination can occur secondary to environmental organisms such as coagulase-positive and -negative staphylococci, propionibacteria, and
*Corynebacterium* species. Collection of blood from a contaminated site, such as that
associated with spilled intestinal contents, is contraindicated. Many consider cancer
surgery another contraindication for salvaged blood, due to the possibility of tumor
cells within the salvaged blood. Finally, although it is uncommon, the collection process can be associated with fatal air embolism.

**Postoperative Autologous Transfusion:** Both canister systems and RBC processors can be used to collect postoperative blood drainage, such as that from the medi-
astinum after heart surgery, from the peritoneal cavity after hepatic injury, or from the
knee or hip site after orthopedic procedures. Blood salvaged from a serosal cavity has
little residual fibrinogen or platelets, and clotting is usually not a problem; therefore
the addition of anticoagulants to the collection is usually not necessary. The salvaged
blood contains free hemoglobin, and may be contaminated with tissue exudate, bone,
bone marrow and other biologic and surgical materials; nevertheless, most patients
tolerate the infusions well.

*Adverse Effects:* Adverse effects include respiratory distress, hypotension with anaphylaxis, and fever – the latter more likely to occur when the product is collected over
a long time interval (6–12 hours).

**Controlled Hypotension:** Controlled hypotension may decrease blood loss and
transfusion in orthopedic surgery. The concern with controlled hypotension is that
it may produce end-organ ischemia, especially in patients with atherosclerosis. The
decrease in blood loss as a result of controlled hypotension may not outweigh its risk
of ischemia.

## Recommended Reading

Brecher ME, Goodnough LT. (2001). The rise and fall of preoperative autologous
    blood donation. *Transfusion* **41**, 1459–1462.
Carless PA, Henry DA, Anthony DM. (2003). Fibrin sealant use for minimizing peri-
    operative allogeneic blood transfusion. *Cochrane Database Syst Rev* (2): CD004171.
Carless PA, Henry DA, Moxey AJ *et al.* (2004). Desmopressin for minimising periop-
    erative allogeneic blood transfusion. *Cochrane Database Syst Rev* (1): CD001884.
Crowther MA, Warkentin TE. (2008). Bleeding risk and the management of bleeding
    complications in pateints undergoing anticoagulant therapy: focus on new antico-
    agulant agents. *Blood* **111**, 4871–4879.
Despotis G, Eby C, Lublin DM. (2008). A review of transfusion risks and opti-
    mal management of perioperative bleeding with cardiac surgery. *Transfusion* **48**,
    2S–30S.
Dunne JR, Malone D, Tracy JK *et al.* (2002). Perioperative anemia: an independent
    risk factor for infection, mortality, and resource utilization in surgery. *J Surg Res*
    **102**, 237–244.
Fergusson DA, Hebert PC, Mazer CD *et al.* (2008). A comparison of aprotinin and
    lysine analogues in high-risk cardiac surgery. *N Engl J Med* **358**, 2319–2331.

Henry DA, Carless PA, Moxey AJ *et al.* (2008). Anti-fibrinolytic use for minimising perioperative allogeneic blood transfusion. *Cochrane Database Syst Rev* (4): CD001886.

Hill SR, Carless PA, Henry DA *et al.* (2002). Transfusion thresholds and other strategies for guiding allogeneic red blood cell transfusion. *Cochrane Database Syst Rev* (2): CD002042.

Levy JH. (2008). Pharmacologic methods to reduce perioperative bleeding. *Transfusion* **48**, 31S–38S.

Stanworth SJ, Birchall J, Doree CJ, Hyde C. (2008). Recombinant factor VIIa for the prevention and treatment of bleeding in patients without haemophilia. *Cochrane Database Syst Rev* (2): CD005011.

# CHAPTER 52

# Adverse Events and Outcomes Following Transfusion: an Overview

Cassandra D. Josephson, MD and Christopher D. Hillyer, MD

Transfusion of blood products can lead to a number of adverse events and outcomes in recipients, ranging from subclinical infection with a virus that can remain undiagnosed for decades, to acute immune hemolysis with hypotension, shock and death commencing within minutes. A large number of classification schemes exist to categorize adverse events and outcomes in transfusion recipients. These including groupings by pathogenesis (immune versus non-immune; infectious versus non-infectious), reaction type (febrile or allergic), and time to development (acute versus delayed), amongst others.

In this textbook, the classification scheme for adverse events will use two categories: 1) Transfusion Reactions, and 2) Posttransfusion Complications. An additional third grouping is presented, termed the *non-infectious serious hazards of transfusion*, or NiSHOTs.

**Transfusion Reactions:** Transfusion reactions are adverse events temporally associated to the transfusion of blood products (also termed *acute transfusion reactions*), and can range from mild to life-threatening. A list of transfusion reactions and their temporal relationship and severity is presented in Table 52.1.

| TABLE 52.1 Transfusion Reactions | | |
| --- | --- | --- |
| **Name** | **Temporal Relationship** | **Severity** |
| Acute hemolytic | 0–1 hour | Mild–severe |
| Anaphylactic | 0–1 hour | Severe |
| Febrile | 1–6 hours | Mild |
| Hypotensive | 1–3 hours | Mild–moderate |
| Metabolic | 0–4 hours | Mild–moderate |
| Septic | 0–6 hours | Mild–severe |
| TACO | 1–4 hours | Mild–severe |
| TRALI | 0–6 hours | Mild–severe |
| Urticarial/allergic | 0–6 hours | Mild–moderate |

TACO, transfusion-associated circulatory overload; TRALI, transfusion-related acute lung injury.

**Posttransfusion Complications:** Posttransfusion complications are adverse events that are a result of transfusion but do not occur at or around the time of transfusion (also termed *delayed reactions*). These, too, can range from mild to life-threatening. A list of posttransfusion complications and their temporal relationship and severity is presented in Table 52.2.

| TABLE 52.2  Posttransfusion Complications | | |
|---|---|---|
| **Name** | **Time Following Transfusion** | **Severity** |
| Alloimmunization | Days–months | None–severe |
| Delayed hemolytic | Days | Mild–severe |
| Iron overload | Years | Mild–severe |
| PTP | Week–weeks | Moderate–severe |
| TA-GVHD | Week–weeks | Severe |
| TA-MC | Months–years | Unknown |
| TRIM | Week–weeks | Mild–moderate |
| TTD | Days–years | None–severe |

PTP, posttransfusion purpura; TA-MC, transfusion associated microchimerism; TRIM, transfusion related immunomodulation; TTD, transfusion transmitted disease.

**Serious Hazards of Transfusion:**  As the serious infectious hazards of transfusion, including HIV, HCV and HBV, are now rare following transfusion, attention has turned to the non-infectious serious hazards.

NiSHOTs include a variable number of important entities. Universally, mistransfusion, TRALI and bacterial contamination are not uncommon NiSHOTs. Indeed, these three entities are the leading causes of transfusion-associated death in the US and on other continents as well. Data (2004) from the UK hemovigilance scheme regarding morbidity from serious hazards of transfusion (SHOT) showed that incorrect blood component transfused (IBCT; equals "mistransfusion") and acute transfusion reaction ranked highest, followed by TRALI (Figure 52.1).

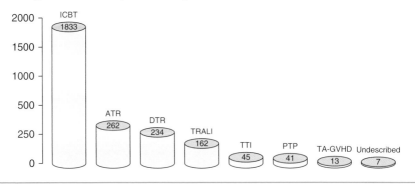

**FIGURE 52.1**  Transfusion recipient morbidity by complication classification according to the 2004 SHOT Annual Report.
ICBT, incorrect blood transfused; ATR, acute transfusion reaction; DTR, delayed transfusion reaction; TRALI, transfusion-related acute lung injury; TTI, transfusion-transmitted infections; PTP, post transfusion purpura; TA-GVHD, transfusion associated graft versus host disease.
Hillyer CD, Blumberg N, Glynn SA *et al.* (2008). Transfusion recipient epidemiology and outcomes research: Possibilities for the future. *Transfusion* **48**, 1530–1537.

## Efforts to Mitigate the Leading Causes of Transfusion-associated Death:

As stated above, the three leading causes of transfusion-associated death are ABO-incompatible transfusion or mistransfusion, TRALI, and bacterial contamination. Significant advances have been made to address these NiSHOTs.

*Mistransfusion:* Mistransfusion, a term generally used to describe an episode where "the wrong patient receives the wrong blood," occurs for a variety of reasons, including improper identification of the intended recipient during initial sample collection for blood typing (so called "wrong blood in tube" [WBIT]), improper typing or pretransfusion testing of the blood component or recipient in the blood bank, or misidentification of the patient recipient and/or the blood product at the time of initiation of the blood transfusion. Mistransfusion, considered by most authorities to be the leading cause of transfusion-related mortality, has a reported estimated incidence approaching 1:14,000 units in the US, with fatality rates approaching 1:500,000 units transfused. A number of technologies exist that are thought to be able to reduce mistransfusion, including bar-coding of patient and blood unit identifiers, radio frequency identification (RFID), combination-locked pouches that require a special bedside code, and bedside ABO tests that would allow an operator to "retype" the patient and "match" the patient blood type with the blood unit label. None of these methods has enjoyed widespread implementation to date, and each involves human actions, interpretation or decisions, thus allowing human error. As yet, a widely employed solution to mistransfusion has not been discovered.

*Transfusion Related Acute Lung Injury:* TRALI was the leading cause of transfusion-related deaths reported to the US FDA in 2003. TRALI is defined as "a new episode of ALI [acute lung injury] that occurs during or within 6 hours of a completed transfusion which is not temporally related to a competing etiology for ALI." The diagnosis of TRALI is considered a clinical and radiographic diagnosis, and not a laboratory-defined one. The pathogenesis of TRALI is not fully understood, but likely requires several "hits"; including donor HLA antibodies. Indeed, TRALI mitigation schemes in the UK, US and elsewhere have focused on limiting exposure to donor HLA-antibodies primarily by using male-only plasma. UK SHOT data from 2007 (http://www.shotuk.org/home.htm) suggest a significant decrease in TRALI following institution of this policy. US mitigation strategies are currently being implemented, but may strain platelet inventories. Close evaluation of these strategies and better understanding of recipients at risk and the mechanism of TRALI are important in reducing the incidence of TRALI.

*Bacterial Contamination:* Both subclinical bacteremia in the donor and inadequate skin decontamination can lead to the collection of bacteria in blood components. These bacteria can multiply and lead to high concentrations of endotoxin, especially in platelet units as they are stored in plasma at room temperature in gas oxygen-permeable plastic bags. The incidence of bacterially contaminated platelets in the 1990s (US) approximated 1:3000 platelet units. Early in this decade the AABB released Standard 5.1.5.1, which requires that bacterial contamination be limited and detected. Several methods to comply with this Standard have led to a decrease in bacterial contamination, septic transfusion reactions and death, by approximately 50%. The true residual risk of bacterial contamination in a culture-tested and released platelet unit is unknown with certainty.

Developing NiSHOTs: A number of significant concerns remain on the horizon and may become NiSHOTs. These are listed in Table 52.3 as "developing." Most notable on this list is "transfusion and mortality."

| TABLE 52.3 NiSHOTs: Developing |
| --- |

*Blood transfusion and mortality: adults and pediatrics*
- **Age of blood and FIE intensive care unit issue**
- **Transfusion Triggers in Critical Care (TRICC). Transfusion Requirement in FIE Pediatric Intensive Care Unit (TRIPICU)**
- **Acute respiratory failure, acute respiratory distress syndrome, myocardial infarction, multiorgan failure**
- **Nitric oxide and S-nitrosohemoglobin**
- **WBC-mediated, TRIM, leukoreduced**

*TRALI: adults*

*Risks or challenges in specialized circumstances: adults and pediatrics*
- Extracorporeal membrane oxygenation
- African Americans
- Anemia of the elderly

*Necrotizing enterocolitis: pediatrics*

*K+ toxicity after irradiation: adults and pediatrics*

*Inflammation and alloimmunization: adults and pediatrics*

*Microchimerism and autoimmune disorders: adults and pediatrics*

*Heavy metals and plasticizers: pediatrics*

Modified from Hillyer CD, Blumberg N, Glynn SA *et al.* (2008). Transfusion recipient epidemiology and outcomes research: Possibilities for the future. *Transfusion* **48**, 1530–1537.

*Transfusion and Mortality:* A significant body of data is developing that associates blood transfusion itself with increased morbidity and mortality in the intensive care unit and after surgery. Some studies have reported an association between transfusions and an increase in length of hospital stay, postoperative infections, lung injury, tissue hypoxia, bleeding or thrombosis, and/or multiple organ failure. These reports are largely observational and often difficult to interpret due to concerns about unrecognized confounding variables inherent in non-randomized studies. At present, the credibility of the association of transfusion and death in some clinical situations and patients has not been firmly established, and, if true, the magnitude and mechanism are not known. Further study is needed and expected in the near term.

## Recommended Reading

Eder AF, Kennedy JM, Dy BA *et al.* (2007). Bacterial screening of apheresis platelets and the residual risk of septic transfusion reactions: the American Red Cross experience (2004–2006). *Transfusion* **47**, 1134–1142.

Hébert PC, Wells G, Blajchman MA *et al.* (1999). A multicenter, randomized, controlled clinical trial of transfusion requirements in critical care. *N Engl J Med* **340**, 409–417.

Hillyer CD, Blumberg N, Glynn SA *et al.* (2008). Transfusion recipient epidemiology and outcomes research: Possibilities for the future. *Transfusion* **48**, 1530–1537.

Kleinman S. (2006). A perspective on transfusion-related acute lung injury two years after the Canadian Consensus Conference. *Transfusion* **46**, 1465–1468.

The Serious Hazards of Transfusion (SHOT) Steering Group (2005). *Serious hazards of transfusion. Annual Report 2004.* Manchester: SHOT Office, Manchester Blood Centre. Available from: http://www.shotuk.org/SHOTREPORT2004.pdf (accessed March 10, 2008).

Tinmouth A, Fergusson D, Yee IC *et al.* (2006). Clinical consequences of red cell storage in the critically ill. *Transfusion* **46**, 2014–2027.

# CHAPTER 53
# Febrile Non-hemolytic Transfusion Reactions

Cassandra D. Josephson, MD

Febrile non-hemolytic transfusion reactions (FNHTRs) are defined as (1) a temperature increase of greater than 1°C (1.8°F) associated with a transfusion, that (2) cannot be attributed to other etiologies, and which (3) may be accompanied by chills, rigors, cold or discomfort. This established definition is becoming recognized as being too narrow, as reactions classified as FNHTR may not have fever, but rather the patients may note chills, a cold sensation or discomfort. Indeed, fever occurring with FNHTRs can be suppressed (or abrogated) by antipyretic medications. Rarely, the reaction can be accompanied by nausea, vomiting and headache. FNHTRs usually occur during or within 2 hours of transfusion. As there are no diagnostic tests to identify FNHTRs, it is a diagnosis of exclusion.

**Incidence:** Multiple factors influence the frequency and reporting of FNHTRs, including type of blood product transfused, patient population, and variability in recording of signs and symptoms. The incidence of FNHTRs due to RBC transfusion has been reported from ~0.1–7.0 reactions per 100 products transfused, though pre-storage leukoreduction has reduced this incidence by more than 50%. FNHTRs occur in ~0.2% to greater than 30% of platelet transfusions, which has been reduced by approximately 90% with the use of pre-storage leukoreduction. Before the general adoption of universal pre-storage leukoreduction, FNHTRs occurred most commonly in older, stored products, especially platelets, due to the relationship between dose and concentration of white blood cells (WBC) and/or cytokines infused (now mitigated by leukoreduction). FNHTRs typically occur later in the course of the transfusion itself. If fever occurs very early in the transfusion event, bacterial contamination should be considered with high suspicion.

**Diagnosis, Differential Diagnosis and Evaluation:** FNHTR is a diagnosis of exclusion, thus all other reasons for fever (e.g. coexisting conditions such as infection, sepsis, or medications and therapies [chemotherapy]) or the accompanying symptoms must be ruled out. A clear temporal association of fever with blood product administration and the lack of any other inciting event usually leads to the conclusion and classification of a FNHTR.

The recognition of fever temporally associated with transfusion is important, not simply because of the possible diagnosis of FNHTR, but rather because fever can accompany much more serious transfusion complications. Fever accompanies acute hemolytic transfusion reactions, TRALI, and the transfusion of bacterially contaminated products, which all can be life-threatening. As recognition of fever is a driver in terminating the presence of a transfusion reaction, stopping the transfusion may be life-saving. In all cases of fever, the reaction should be reported to the transfusion service; evaluation should include clerical and serological evaluation for mistransfusion, and visual inspection of the remaining product, and may include bacterial culture and gram staining of the remaining products in the bag if bacterial contamination is suspected.

**Pathophysiology:** There are two potential mechanisms of FHNTRs in transfused products, which are associated with the presence of contaminating leukocytes. The first potential mechanism involves leukocyte-derived cytokines, and is most commonly associated with platelet products. Leukocyte-derived cytokines (e.g. interleukins [IL-1, IL-6, IL-8] and tumor necrosis factor $\alpha$ [TNF-$\alpha$]) accumulate in the supernatant during room-temperature storage of platelet products. In addition, platelet products must be stored in oxygen-permeable containers, and require the presence of plasma for glucose metabolism. Because the cytokines have already been released during storage, bedside (or post-storage) leukoreduction does not result in decreased incidence of FNHTRs from platelet products. The second potential mechanism involves WBC antibodies, and is most commonly associated with RBC products. Recipient HLA and HNA antibodies, which are against the transfused WBCs, lead to an antigen–antibody complex, resulting in release of endotoxin, which leads to fever and other symptoms via the hypothalamus. Despite implementation of pre-storage leukoreduction, FNHTRs still occur (approximately 0.2% for RBC and 1.0% for platelet transfusions), implying that there may be other mechanisms causing these types of FNHTRs.

**Management:** The transfusion must be stopped immediately. Symptomatic care should be given, including consideration of antipyretics; meperidine may be helpful if chills or rigors are prominent symptoms. Fever and chills are self-limited, and will resolve without treatment.

**Prevention:** Pre-storage leukoreduction of RBCs and platelets decreases the incidence of FNHTRs. If reactions continue to occur, especially with platelets, administering volume-reduced or washed products may help, and, pretransfusion antipyretic administration can be considered, though this may mask fever caused by other serious reactions.

## Recommended Reading

Heddle NM, Blajchman MA, Meyer RM *et al.* (2002). A randomized controlled trial comparing the frequency of acute reactions to plasma-removed platelets and prestorage WBC-reduced platelets. *Transfusion* **42**, 556–566.

Paglino JC, Pomper GJ, Fisch GS *et al.* (2004). Reduction of febrile but not allergic reactions to RBCs and platelets after conversion to universal prestorage leukoreduction. *Transfusion* **44**, 16–24.

# CHAPTER 54

# Allergic, Anaphylactoid and Anaphylactic Reactions

Beth H. Shaz, MD

Allergic reactions can occur following blood transfusion, and are the result of an interaction between an allergen and a preformed antibody – usually IgE. There is a broad spectrum of severity of allergic reactions, ranging from focal mild urticarial reactions to systemic, life-threatening anaphylactic reactions with intractable hypotension or shock and loss of consciousness. Urticarial or mild allergic reactions consist of pruritic urticarial lesions. *Anaphylactoid* is a term that is used to describe reactions less severe than anaphylaxis that are characterized by hypotension, dyspnea, stridor, wheezing and/or diarrhea. In addition, the term anaphylactoid has been used to describe "anaphylactic" reactions not mediated by IgE, due to anti-IgA of limited serologic reactivity, and in recipients with normal IgA levels.

Pathophysiology: Allergic reactions are usually type I hypersensitivity responses, which are mediated by IgE. Antigen-specific Th2 T cells secrete IL-4 and IL-13, which stimulate activated B cells to change their production antibody isotype from IgM to IgE. IgE binds to mast cells and basophils by means of IgE Fc receptors. When an allergen, which is usually a protein in the plasma of the transfused blood component, binds the cell-associated IgE, the mast cell is activated and releases mediators (such as histamine, heparin, leukotrienes, platelet-activating factor, cytokines and chemokines). These mediators, also known as anaphylatoxins, instigate the influx of cells and fluid into the tissues, resulting in the allergic reaction symptoms of the skin, respiratory tract, and cardiovascular and gastrointestinal systems. In addition, IgG may also mediate allergic reactions by leading to complement fixation, one of the results of which is the release of C3a and C5a anaphylatoxins. Much of the pathophysiology of allergic reactions is still poorly understood.

Anti-IgA: Individuals with congenital complete IgA deficiency may develop a class-specific antibody to IgA (anti-IgA). Exposure to IgA will generate an immediate anaphylactic reaction. Patients with severe allergic, anaphylactoid or anaphylactic reactions should have their pretransfusion sample tested for the presence of IgA by nephelometry, a simple and rapid test. Though this method has only modest sensitivity, the presence of IgA will allow the physician to discount the diagnosis of IgA deficiency with anaphylactic transfusion reaction, as these patients cannot make an anti-IgA antibody. If this test does not show demonstrable IgA then more sensitive tests are required, as is an assay for the presence of anti-IgA antibodies (as below). In the future, these patients need to receive blood components, including immunoglobulin products, which lack IgA. Products can either be from an IgA-deficient donor, which is required for plasma components, or be washed to remove as much plasma as possible.

**Antibodies to Other Normal Serum Proteins:** Anaphylactic reactions have been reported in individuals who lack allotypes of other normal serum proteins, such as haptoglobin, C3 and C4, who form IgG or IgE antibodies against these proteins. Chido (Ch) and Rogers (Rg) antigens are on the C4d component of complement, and are bound onto the RBC membrane when complement is activated. Severe anaphylactic reactions have been reported in patients with anti-Ch or anti-Rg who are receiving plasma components.

**Components Within the Blood Product:** The allergen for which the patient has an antibody, or the antibody for which the patient has an allergen, may be in the blood component. For example, the donor may be taking penicillin and the recipient may have a penicillin allergy, or the donor may be allergic to penicillin and the recipient may be on penicillin – both situations possibly resulting in an allergic reaction in the recipient. In addition, anaphylatoxins or other mediators may also be produced in the blood component during storage, or be present in the blood component from the donor.

**Latex Allergy:** Latex may be found in the blood collection or blood transfusion set. If a patient has a latex allergy, the blood supplier or manufacturer may need to be contacted to determine whether latex was used in the manufacturing process, and specific latex-free products may need to be used.

**Clinical Features:** Allergic reactions occur in 0.03–0.61% of RBC transfusions, 0.3–6% of platelet transfusions and 1–3% of plasma transfusions; the majority of these reactions are mild. Anaphylactic reactions occur in approximately 1 in 20,000 to 47,000 components transfused. Symptom onset of anaphylactic reactions is within seconds to 45 minutes after the start of transfusion, while symptom onset for allergic reactions is up to 2–3 hours after completing transfusion. In general, the shorter time to onset of symptoms, the more severe the reaction.

Urticarial reactions consist of pruritic, localized, well-circumscribed, discrete wheals with erythematous, raised, serpiginous borders and blanched centers. Generalized pruritus may precede the eruption, and there may be generalized erythema (flushing) of the skin or angioedema, which is localized, non-pitting and deep edema of the skin.

Symptoms of more severe reactions include upper and/or lower airway obstruction. Upper airway obstruction results in hoarseness, stridor and/or the complaint of a "lump" in the throat. Lower airway obstruction results in audible wheezing, a feeling of chest tightness, substernal pain, dyspnea, cyanosis, anxiety and/or a feeling of impending doom. In addition, there may be profound hypotension, possibly leading to loss of consciousness and shock, and cardiac symptoms of tachycardia, cardiac arrhythmias and cardiac arrest. Severe gastrointestinal symptoms (abdominal cramps/pain, nausea, vomiting, diarrhea) may also be present.

**Diagnosis:** The differential diagnosis of allergic reactions includes immediate hemolytic transfusion reactions, which usually have the presence of fever and absence of cutaneous manifestations. Transfusion related acute lung injury (TRALI) can be differentiated from allergic reactions by the presence of fever, chest X-ray findings of pulmonary edema, and absence of cutaneous and gastrointestinal symptoms. Bacterial contamination is associated with hypotension and shock, but the presence of rigors and fever differentiate it from allergic reaction. Flushing and hypotension may be seen in patients taking angiotensin converting enzyme (ACE) inhibitors, but can be differentiated from allergic reactions by the absence of hypoxia or other pulmonary findings. Circulatory overload can be differentiated from an allergic reaction by the presence of pulmonary edema and hypervolemia with hypertension. Other causes of the patient's allergic reaction should be ruled out, such as coincidental administration of a drug or food.

Patients with a severe allergic reaction should be tested for IgA deficiency and the presence of anti-IgA. Screening for IgA deficiency by measuring levels of IgA can be performed in the majority of large hospital laboratories. Reference laboratories can perform more sensitive tests, which have a lower limit of detection. Anti-IgA testing is currently only performed in reference laboratories, which use a variety of methods, including immunoradiometric assay, passive hemagglutination and flow cytometry. In one study, only 18% of sera from patients with anaphylactic or anaphylactoid transfusion reactions contained anti-IgA; therefore, anti-IgA is only one cause of anaphylactic or anaphylactoid reactions.

## Treatment:

**Mild Allergic Reactions:** Treatment of mild reactions consists of temporarily stopping the transfusion and administering diphenhydramine or another anti-histamine. If the symptoms resolve promptly, then the transfusion can be continued. If the symptoms recur or do not subside, then the transfusion should be discontinued.

**Severe Allergic Reactions:** Appropriate medications and equipment should be available for all transfusions in case of severe immediate transfusion reactions. First, the blood transfusion must be discontinued, and intravenous access maintained with normal saline (0.9% sodium chloride). Oxygen and possible intubation may need to be administered for hypoxia. Epinephrine is the most effective medication for treating anaphylaxis. Vasopressors may be required. In addition, H1-receptor antagonist and intravenous steroids have been used. Observation of the patient for up to 24 hours may be required for the resolution of symptoms, and to ensure that they do not recur.

**Risks of Diphenydramine Administration:** Diphenydramine, which has anticholinergic effects and crosses the blood–brain barrier, can cause dry mouth, urinary retention, drowsiness, decreased alertness (impairing driving abilities), delirium (especially in elderly hospitalized patients), and impaired cognitive performance, memory and attention. In contrast, it can alternatively cause restlessness and nervousness. It has also been associated with cardiotoxicity and arrhythmias.

**Prevention:** The majority of clinical trials demonstrate that the routine use of premedication does not decrease the incidence of allergic reactions.

**Prophylactic Premedication in Patients without Prior History:** Although premedication is a common practice, reported in up to 80% of transfusions, there are no data to demonstrate that routine premedication prior to transfusions decreases the incidence of allergic reactions.

**Prophylactic Premedication in Patients with a Prior History of Allergic Reactions:** In patients who have had previous allergic reactions, it may be beneficial to premedicate with diphenhydramine or another anti-histamine, H1-receptor antagonist, or intravenous steroids, depending on the severity of the previous reaction. In patients with moderate allergic reactions, premedication may be more warranted as a method to decrease the risks of transfusion. In addition, allergic reactions tend to become more severe with repeated plasma/platelet transfusions. In patients with moderate–severe reactions, washing or plasma-reducing red blood cell or platelet products may be warranted to remove the plasma and thereby decrease or eliminate the risk of allergic reactions. If plasma-containing products must be administered, then the patient should be given appropriate premedication and closely monitored during the transfusion, and the product infused slowly.

**IgA Deficiency:** IgA deficiency occurs in approximately 1 in 900 blood donors in the US and UK, 1 in 500 in Finland, and 1 in 18,500 in Japan. Of IgA-deficient individuals who are otherwise healthy, 20–30% form anti-IgA, while approximately 80% of IgA-deficient individuals with autoimmune diseases (such as rheumatoid arthritis, juvenile rheumatoid arthritis, systemic lupus erythematosus or ataxia-telangiectasia) form anti-IgA. Anti-IgA is formed irrespective of a history of pregnancy or previous blood transfusion, but can be stimulated by IgA exposure from a previous transfusion. Anti-IgA is also detected in normal human sera (2–59% of individuals). Given the above, it is thought that anti-IgA may be an alloantibody and/or an autoantibody. Anti-IgA is usually IgG but may be IgM or IgE, and it can be of broad or limited specificity.

In a study of 32,376 blood donors, the frequency of IgA deficiency (<0.05 mg/dl) and class-specific anti-IgA was 1 in 1200, which is greater than the observed frequency of anaphylactic reactions of 1 in 20,000–47,000 transfusions. Therefore, not all individuals with IgA deficiency and anti-IgA have anaphylactic reactions. In addition, reactions have been seen in IgA-deficient patients without detectable anti-IgA. Thus, it may be prudent to monitor closely IgA-deficient individuals for severe allergic reactions, especially those with concomitant anti-IgA, but it is likely unnecessary to restrict them to washed or IgA-deficient products without a trial of unmodified blood products.

**Prophylaxis for Patients with IgA Deficiency and History of Anaphylactic Reactions:** Patients with IgA deficiency and anti-IgA who have a history of anaphylactic reactions may need to receive IgA-deficient products obtained from IgA-deficient donors, or washed products. Red blood cell and platelet products can be washed to remove plasma; some methods of washing remove more plasma than others, and patients with severe reactions likely require a method to remove the majority of plasma. Plasma and cryoprecipitate must be collected from an IgA-deficient donor; these products must be obtained specially from the blood supplier, and take time to acquire. Plasma derivatives contain variable amounts of IgA, and therefore an IgA-deficient product should be carefully selected for these patients. In addition, premedication with steroids and antihistamines may be warranted. Lastly, autologous donation can be considered if the patient is eligible to donate and has specific future needs, such as an impending surgical procedure.

## Recommended Reading

Geiger TL, Howard SC. (2007). Acetaminophen and diphenhydramine premedication for allergic and febrile nonhemolytic transfusion reactions: good prophylaxis or bad practice? *Transfus Med Rev* **21**, 1–12.

Perrotta PL, Snyder EL. (2001). Non-infectious complications of transfusion therapy. *Blood Rev* **15**, 69–83.

Sanders RP, Maddirala SD, Geiger TL *et al.* (2005). Premedication with acetaminophen or diphenhydramine for transfusion with leucoreduced blood products in children. *Br J Haematol* **130**, 781–787.

Sandler SG, Zantek ND. (2004). Review: IgA anaphylactic transfusion reactions. Part II. Clinical diagnosis and bedside management. *Immunohematology* **20**, 234–238.

Thibault L, Beauséjour A, de Grandmont MJ *et al.* (2006). Establishment of an immunoglobulin A-deficient blood donor registry with a simple in-house screening enzyme-linked immunosorbent assay. *Transfusion* **46**, 2115–2121.

Tobian AAR, King KE, Ness PM. (2007). Transfusion premedications: a growing practice not based on evidence. *Transfusion* **47**, 1089–1096.

Vassallo RR. (2004). Review: IgA anaphylactic transfusion reactions. Part I. Laboratory diagnosis, incidence, and supply of IgA-deficient products. *Immunohematology* **20**, 226–233.

# CHAPTER 55

# Acute Hemolytic Transfusion Reactions

Cassandra D. Josephson, MD and Christopher D. Hillyer, MD

Transfusion of incompatible red blood cells (RBCs), or significant amounts of incompatible plasma, can lead to efficient antibody–antigen binding in the recipient, and if the antigen–antibody complex efficiently binds C1q then a cascade of complement protein activation can occur which can rapidly lead to intravascular hemolysis. This adverse event during or following transfusion is termed an *acute hemolytic transfusion reaction* (AHTR) or an *acute immunohemolytic transfusion reaction*. AHTRs can lead to minimal hemolysis with no clinical sequelae, or can result in brisk hemolysis, induction of disseminated intravascular coagulopathy (DIC), hypotension, and shock, followed by renal failure and/or death. Indeed, in a 10-year period ending in 1985, incompatible RBC transfusion accounted for 51% of transfusion-related deaths reported to the FDA.

The text below is divided into transfusion of incompatible RBCs and transfusion of significant amounts of incompatible plasma.

**Transfusion of Incompatible RBCs:** Transfusion of incompatible RBCs is most associated with a mistransfusion event. *Mistransfusion* is the term applied to a transfusion event in which "the wrong blood is transfused to the wrong patient." A more comprehensive definition of mistransfusion is:

> "*transfusion of a unit of blood that has been incorrectly typed, labeled, crossmatched, issued or transfused to a patient, or the transfusion of correctly manufactured and issued unit of blood that is administered to the incorrect patient.*"

Thus, transfusion service errors (typing, labeling, crossmatching and issuing) and patient identification errors (patient sampling for blood typing, or patient and unit identification at the time of transfusion [at the beside]) can lead to mistransfusion.

In extremely rare circumstances, non-ABO incompatible blood is administered knowingly. These include emergent clinical situations where the best RBC product available is incompatible with the recipient's known antibody panel, and when auto or allo-antibodies are present that cannot be either identified or respected.

**Incidence:** AHTR can occur from transfusion of ABO-incompatible blood as well as blood incompatible to other recipient RBC antibodies. The true incidence of ABO-incompatible transfusions is not known; up to 47% of ABO-incompatible mistransfusion events had no adverse effect, according to a notable study by Linden and colleagues. In that study, erroneous administration of blood products yielded an incidence of 1:12,000 and a fatality rate of 1:600,000–800,000. Another estimate suggests that 1 in every 33,000 units is ABO incompatible due to errors made during the transfusion process. In a 2008 report by Janatpour and colleagues, ABO-incompatible transfusions were estimated at 1:38,000–1:100,000 transfusions, and the risk of death from AHTR was 1:1.5 million. Death from mistransfusion has been reported to occur in 2–7% of ABO-incompatible mistransfusion events, and approximately 10–30 recipients die in the US each year from AHTR.

The FDA reported that mistransfusion-related deaths totaled more than twice those due to all infectious hazards combined, and the UK surveillance system reported a mistransfusion rate that was 10 times higher than the rate attributed to infectious disease transmission.

**Clinical Manifestations:** The most common signs and symptoms that occur during AHTR are presented in Table 55.1. Fever is an important early manifestation of AHTR, and it is for this that transfusion recipients are monitored during the initial minutes of blood product administration.

| TABLE 55.1 Signs and Symptoms of Acute Hemolytic Transfusion Reactions | |
| --- | --- |
| Fever | Nausea and vomiting |
| Chills | Dyspnea |
| Rigors | Hyper and hypotension |
| Anxiety; feeling of doom | Hemoglobinuria |
| Facial flushing | Oliguria/anuria |
| Chest and abdominal pain | Pain at infusion site |
| Flank and back pain | Diffuse bleeding |

**Pathophysiology:**

*Intravascular Hemolysis:* The transfusion of incompatible RBC antigens into recipients who have the corresponding antibody can lead to hemolysis and AHTR. Whether hemolysis occurs, and its degree and effects, are dependent on several elements. These include the class and subclass of the recipient antibody, the thermal range over which this antibody can fix complement, the number of RBCs transfused, the density of the corresponding antigen on the RBCs, the efficiency of the complement system, and the complement control proteins in the circulation.

Most AHTRs are due to transfusion of ABO-incompatible RBCs into patients with the pre-existing, naturally-occurring anti-A and/or anti-B antibodies. These antibodies are typically a mixture of IgM and IgG. IgM can efficiently bind C1q and initiate the complement cascade, leading to the formation of the membrane attack complex and erythrocyte lysis. Some IgG antibodies can fix complement efficiently as well, and are usually of the IgG1 and IgG3 subclasses; these can cause mild to fatal AHTRs. Thus, mistransfusion of minor RBC antigens typically in the Kell, Duffy and Kidd systems can cause AHTRs, and these can be fatal as well.

**DIC, Blood Pressure Fluctuation, Shock and Renal Failure:** The binding of complement and the activation of downstream complement proteins leads not only to hemolysis but also to the products of these events, including C3a and C5b. Complement fragments C3a and C5a induce the release of histamines, vasoactive amines, bradykinins, and other inflammatory cytokines which can cause fever, hypo-tension, wheezing, chest pain, nausea and emesis. Intravascular hemolysis produces three products: free hemoglobin, RBC stroma, and no-stroma proteins. All of these are released into the circulation during an AHTR.

Free hemoglobin binds nitric oxide (NO) at a rate 1000× that of the RBC. This hemoglobin scavenging leads to decreased bioavailability of NO, and thus to vasocon-striction and alterations in capillary response to hypoxia. RBC stroma can initiate acti-vation of platelets and the coagulation cascade leading to DIC. RBC stroma has also been shown to increase blood pressure and is toxic to both the glomerulus and the renal tubule, and thus can cause acute renal failure. RBC stroma can therefore cause hypertension, DIC and renal failure.

Ultimately, the increased cytokines and hypotension stimulate a compensatory sympathetic nervous system response contributing to renal, splanchnic and cutane-ous vasoconstriction that, in combination with the pathophysiology described above, leads to shock and circulatory collapse.

**Diagnosis and Evaluation:** Patients with severe AHTR complain of lower back and flank pain and a feeling of dread, as per Table 55.1. These may be early manifesta-tions of changes in renal, neural and splanchnic blood flow. Signs include fever, and hyper- and/or hypotension. Blood specimens may reveal pink to red plasma, and thus evidence of intravascular hemolysis. Together, these clinical findings are suggestive of an AHTR, and require that the transfusion be immediately aborted and an evaluation by the transfusion service for an AHTR started.

*Evaluation by the Transfusion Service:* The transfusion service should be imme-diately informed when an AHTR occurs or is being considered. A clerical check of the patient and product must be performed to determine if the incorrect blood type, product type or patient was transfused, and the blood product with all attached tub-ing returned to the transfusion service. Blood samples are drawn from the recipient in order to visually inspect the specimen for hemolysis, perform a direct antiglobulin test (DAT), and confirm the patient's ABO typing. In addition, a sample from the product is also retyped. The patient's posttransfusion DAT testing may be positive for IgG and C3 (when IgM binds to RBCs it fixes complement, which remains on the RBC and is detected in the DAT). If the DAT is positive for IgG, an eluate may reveal IgG anti-A or anti-B. Repeat ABO testing of the pretransfusion specimen should be performed.

The clinical laboratory evaluation includes an analysis of the recipient's com-plete blood count, serum bilirubin, creatinine and coagulation profile, and urine analysis for hemoglobinuria. The posttransfusion sample will show positive plasma free hemoglobin and decreased serum haptoglobin. Indeed, in anesthetized patients in whom an AHTR occurs, hemoglobinuria may be the only finding suggestive of the AHTR.

*Other Causes of Acute Hemolysis:* Acute hemolysis can occur from transfusion of an improperly stored RBC product resulting in thermal (e.g. exposed directly to ice or placed on a heater), mechanical (e.g. constricted access lines or pressurized infu-sions) or osmolar (e.g. hypotonic solutions or medications infused with the product) injury, and other conditions not related to the transfusion. Patients with underlying hemolytic anemia, such as those with sickle cell disease, thrombotic thrombocyto-penic purpura or an implanted circulatory device, may have acute hemolysis which is not related to the transfusion.

**Management:** Once the possibility of an AHTR is recognized and the transfusion discontinued, supportive care is immediately instituted to manage hypotension and maintain adequate renal perfusion. These measures may include intravenous colloid or crystalloid (10–20 ml/kg of normal saline); a diuretic to maintain urine output between 30 and 100 ml/h or greater (> 1 ml/kg per hour); administration of low-dose dopamine (1–5 µg/kg per minute) for hypotension; and, for active bleeding, the trans-fusion of plasma, platelets and/or cryoprecipitate for DIC. Most textbooks suggest the use of mannitol to induce an osmotic diuresis, but the value of this intervention is not yet well studied.

**Sources of Mistransfusion:** Many opportunities for human error exist in the chain of events leading up to transfusion: the decision to transfuse; the component request; patient sampling; pretransfusion testing; collection of the component from stor-age; and administration to the patient. Dzik described major "zones of error" in the transfusion process. In the first "zone," which encompasses patient identification and pretransfusion testing, there are two broad categories of error: *mislabeled* samples and *miscollected* samples. Mislabeled samples do not meet the criteria for acceptance by the laboratory; miscollected samples contain the wrong patient's blood (wrong blood in tube, or WBIT), but the samples appear to be labeled correctly and are accepted by the lab. WBIT occurs with a median frequency of 0.1–0.5 per 1000 samples.

Laboratory error accounts for approximately 30% of errors resulting in mistrans-fusion. These errors, which occur in the "second zone," include testing the wrong sam-ple, transposing compatibility labels, errors in manual ABO grouping, and retrieving the wrong unit from the blood bank refrigerator. Laboratory errors are more likely to occur at night or on weekends, when there are fewer, often less experienced, staff.

The final bedside check to ensure that correct blood has been issued to the correct patient is the last "zone of error." Administration of the wrong blood unit to the recip-ient is the most common error leading to ABO-incompatible transfusions, accounting for 38% of incidents reported by Linden and colleagues in 2000, and 59% of SHOT-reported incidents between 1999 and 2003. The bedside check is the last opportunity to intercept an erroneously issued unit of blood before it is transfused; however, fail-ure to perform this check properly is common.

Transfusion errors are a sentinel event (which is an unexpected occurrence result-ing in death or a serious adverse event or the risk thereof) according to The Joint Commission. A sentinel event requires a root cause analysis to identify what hap-pened, why it happened, and what factors (human, equipment, and/or environmen-tal) contributed to its occurrence. In addition, an action plan must be developed in order to reduce the risk of the event reoccurring.

**Prevention:** Strict adherence to preransfusion bedside patient identification pro-cedures to assure proper specimen collection will aid in preventing clerical errors and misidentification. Other prevention strategies include technological innovations aimed at reducing human error, including bar coding of blood components, patient identification systems, and barrier systems. In addition, hemovigilance and/or quality assurance programs to monitor transfusion practices, investigate and analyze the root cause of adverse events, and reinforce physician and nurse transfusion practices improve transfusion safety.

**Transfusion of Significant Amounts of Incompatible Plasma:** Occurring less frequently, the passive transfer of donor high-titer anti-A and/or anti-B in plasma-containing products (especially anti-A in group O products) results in hemolysis of the ABO-incompatible recipient's RBCs (termed minor ABO mismatch), which can cause severe morbidity and mortality. The potential for this complication exists for every transfusion of an out-of-group plasma-containing product, owing to the passive transfer of anti-A, anti-B and/or anti-A,B.

Transfusion of incompatible plasma most often occurs in the administration of out-of-group platelets, most commonly group O platelets to a group A recipient, which not infrequently results in AHTR. Passive ABO antibodies can also occur in the administration of non-ABO identical whole blood product or sufficient volumes of group O RBC products (which contain a small volume of plasma) to a non-group O patient. These later situations most often result in a positive DAT and, rarely, a severe AHTR.

*Out-of-group Platelet Transfusion:* This situation invariably occurs with the transfusion of a group O platelet pheresis unit with high-titer anti-A to a group A recipient. The anti-A binds the A antigen on the RBC, starting the complement reaction and downstream events as above. These antibodies are only detectable in recipients when the donor antibody titers are high, and a hemolytic transfusion reaction may ensue when the plasma contains a high-titer antibody, and/or when large volumes of plasma are transfused in relation to the patient's size (such as in neonatal or pediatric transfusions). The increased use of single-donor apheresis platelets in the US, coupled with the low availability of type-specific apheresis platelets, has likely led many adult hospital transfusion services to issue ABO-mismatched platelets that contain large plasma volumes and potentially high titers of isohemagglutinins. In the UK, however, all blood donations undergo anti-A IgM titer testing, and those with titers greater than 1:100 dilution are labeled accordingly. Hospital transfusion services are then aware that these products must be transfused only to group O recipients. Most US hospital transfusion services do not routinely measure isohemagglutinin titers (IgM or IgG) before transfusing ABO-mismatched platelets, although this practice has been recommended by some. This is likely due to the lack of agreement of a "critical titer" that will predict *in vivo* hemolysis. The literature would suggest that titers between 16 and 600 (IgM and IgG) may be clinically relevant. Case reports also support that over 60 deaths have occurred from the transfusion of ABO out-of-group platelets.

The clinical manifestations, pathophysiology, diagnosis, and laboratory work-up and management are virtually identical to those described above for administration of incompatible RBCs, and include in addition determination of the antibody titer in the platelet product.

**Prevention:** Administration of out-of-group platelets would not be necessary if adequate supplies of type-specific platelets were regularly available. Thus, an increase in efforts to maintain adequate inventories is needed by some blood centers and transfusion services. If the situation does occur where only out-of-group platelets are available, the products can be screened for anti-A and those with low titers chosen for administration to a non-group O individual. Washed platelets will have very low residual titers of anti-A, yet washing results in ~33% loss of platelets. In the EU, platelet additive solutions dilute the antibodies as there is minimal remaining plasma, and pathogen-reduction technologies also have wash steps or additive solutions that lead to little residual anti-A.

## Recommended Reading

AABB. (2001). Noninfectious serious hazards of transfusion. *Association Bulletin* #01–4, Bethesda, MD: American Association of Blood Banks, June 14.

Ahrens N, Pruss A, Kieseweter H, Salama A. (2005). Failure of bedside ABO testing is still the most common cause of incorrect blood transfusion in the barcode era. *Transf Apheresis Sci* **33**, 25–29.

Dzik WH. (2002). Emily Cooley Lecture 2002: Transfusion safety in the hospital. *Transfusion* **43**(9), 1190–1199.

Janatpour KA, Kalmin ND, Jensen HM, Holland PV. (2008). Clinical outcomes of ABO-incompatible RBC transfusions. *Am J Clin Pathol* **129**, 276–281.

Josephson CD, Mullis N, Van Demark C, Hillyer CD. (2004). Significant number of apheresis-derived group O platelets units have "high titer" anti-A/AB: Implications for transfusion policy. *Transfusion* **44**, 805–808.

Linden JV, Wagner K, Voytovich AE, Sheehan J. (2000). Transfusion errors in New York State: an analysis of 10 year's experience. *Transfusion* **40**, 1207–1214.

McClelland DBL, Philips P. (1994). Errors in blood transfusion in Britain: survey of hospital hematology departments. *Br Med J* **308**, 1205–1206.

Sazama K. (1990). Reports of 355 transfusion-associated deaths: 1976 through 1985. *Transfusion* **30**, 583–590.

# CHAPTER 56
# Delayed Hemolytic Transfusion Reactions

Cassandra D. Josephson, MD

Delayed hemolytic transfusion reactions (DHTRs) occur 3–10 days after the transfusion of RBC products that appear to be serologically compatible. These reactions occur in patients who have been alloimmunized to minor RBC antigens during previous transfusions and/or pregnancies; pretransfusion testing fails to detect these alloantibodies due to their low titer. Following re-exposure to antigen-positive RBCs an anamnestic response occurs, with a rapid rise in antibody titer. Decreased survival of the transfused RBCs may result, primarily due to extravascular hemolysis. In the majority of cases, however, anamnestic antibody production does not cause detectable hemolysis. The term delayed *serologic* transfusion reaction (DSTR) defines reactions in which an anamnestic antibody is identified serologically, in the absence of clinical evidence of accelerated RBC destruction. Antigens implicated most often in DHTRs and DSTRs are in the Kidd, Duffy, Kell and MNS systems, in order of decreasing frequency.

**Incidence:** In combination, DHTRs and DSTRs occur in approximately 1 in 1500 transfusions, with DSTRs occurring up to four times more often than DHTRs.

**Clinical Manifestations:** DHTRs are characterized clinically by an unexpected drop in hemoglobin or a less than expected posttransfusion increment in hemoglobin following transfusion. This diagnosis should be considered days to weeks after transfusion, although hemolysis may be more prolonged. Symptoms of extravascular hemolysis may include fever, chills, jaundice, malaise, back pain and, uncommonly, renal failure. Additionally, patients with sickle cell disease may experience vaso-occlusive pain as a result of a DHTR.

**Pathophysiology:** A minority of patients exposed to foreign RBCs, either during previous transfusions or during pregnancies, may become sensitized to minor RBC antigens. The primary immune response, which occurs over weeks to months, is typically not clinically significant. Alloantibody titers to these RBC antigens may decrease over time in the patient's plasma without continued exposure to the inciting antigen(s), resulting in negative pretransfusion antibody screening. However, transfusion of offending antigen-positive RBCs may lead to an amnesic response, resulting in rapid production of IgG antibody. This antibody, in turn, may lead to primarily extravascular hemolysis. Thermal range, antibody specificity and IgG subclass are three factors that influence the severity of the DHTR. Those antibodies that react at 37°C *in vitro* are more likely to result in hemolysis.

**Diagnosis:** To investigate a suspected DHTR, posttransfusion specimens should be evaluated for antibody identification and direct antiglobulin (DAT) studies. A positive antibody screen with a newly identified alloantibody and/or a positive DAT will help confirm the reaction; the DAT may show a mixed field reaction, with transfused cells but not autologous cells being coated with antibody. A DAT should be performed with both anti-IgG and anti-C3 reagents. If the DAT is positive for IgG, then eluate testing should be pursued to identify the specificity of the RBC coating antibody. Supplemental serologic testing may be necessary to confirm the diagnosis. This testing includes repeating the antibody screen and a DAT on the pretransfusion specimen (retained in the blood bank), to ensure the previous results were not erroneous; establishing the phenotype on a pretransfusion specimen may also be helpful. Additionally, the RBCs from the transfused product (stored within a "segment" retained in the blood bank) should be phenotyped for the antigen corresponding to the newly identified antibody.

Other laboratory findings suggestive of a DHTR include reticulocytosis, unconjugated hyperbilirubinemia, and urine urobilinogen. In patients with sickle cell disease, an unexpected decrease in the relative proportion of transfused (Hgb A) RBCs provides additional support of a potential DHTR. Although DHTRs primarily involve extravascular hemolysis, some antibodies may fix complement and cause intravascular hemolysis. In such instances hemoglobinuria may also be present, along with elevated serum LDH and decreased haptoglobin.

Delayed *serologic* transfusion reactions are often first identified in the blood bank, when a posttransfusion antibody screen performed three or more days following transfusion is determined to be positive. Close communication with the clinical team is indicated in such instances, to confirm the reaction is solely serological and not hemolytic.

**Differential Diagnosis:** Newly formed autoantibodies must be included in the differential diagnosis when a positive antibody screen and a positive DAT are present in a recently transfused patient. An additional consideration is the fact that alloantibody formation may lead to the development of autoantibodies; the detection of autoantibodies following a DHTR is not uncommon. These RBC autoantibodies are often transient, but may result in hemolytic anemia. DAT positivity beyond 3–4 months following transfusion (and thus beyond the lifespan of the transfused RBCs) supports the diagnosis of the presence of autoantibodies.

In patients with sickle cell disease, other causes of increased RBC destruction must also be included in the differential diagnosis of DHTR. These causes include fever, underlying disease, or hypersplenism. The serologic evaluation in such patients, including antibody screen and DAT, would be expected to be negative.

**Timeline of DHTR:** Table 56.1 details the timeline of a DHTR. Following such a reaction, a transfusion report should be generated by the medical director of the blood bank that includes the testing results and the final classification of the suspected adverse event.

**Management:** Specific treatment is usually not necessary unless symptomatic anemia is present. In those cases, additional RBC transfusions with RBC products that lack the identified antigen corresponding to the newly developed alloantibody may be necessary. However, if the transfusion is ordered prior to specificity identification of the newly formed antibody, the risk of hemolysis must be weighed with the benefits to be gained by transfusion. The clinician and the blood bank/transfusion service physician

| TABLE 56.1 Time Line of DHTR | | |
| --- | --- | --- |
| **Time(days)** | **Event** | **Explanation** |
| 0 | Pretransfusion antibody screening negative | Antibody titer below detectable levels |
| 1 | RBC transfusion | |
| 3–10 | Clinical signs of hemolysis may appear | Accelerated destruction of transfused donor RBCs |
| 10–21 | Posttransfusion sample: positive DAT and positive antibody screen due to newly detected antibody | Antibody titer increases |
| > 21 | DAT may become negative | Antibody-sensitized donor RBCs removed from circulation |
| 21–300 | DAT may persist as positive; eluates may reveal alloantibody specificity or panagglutination | Alloantibody binding non-specifically to autologous RBC, or development of a warm autoantibody |

Modified from Hillyer CD, Silberstein LE, Ness PM, *et al.* (eds). (2007). *Blood Banking and Transfusion Medicine: Basic Principles & Practice*, 2nd edition. San Diego, CA: Elsevier.

should be in close communication regarding this situation. If jointly they decide to transfuse in this situation, sufficient hydration and close monitoring should be performed in light of the risk of ongoing hemolysis.

**Prevention:** To prevent DHTRs and DSTRs, the AABB *Standards for Blood Banks and Transfusion Services* mandates permanent preservation of all records of potentially clinically significant antibodies, as well as a review of previous records prior to RBCs being issued for transfusion. Once a clinically significant antibody has been identified, the patient should receive offending antigen negative units for all future RBC transfusions. The 3-day interval requirement for RBC type and screening of hospitalized patients is based on the finding that anamnestic antibody responses may occur within 3 days of a transfusion.

Some patients (such as those with sickle cell disease) may require additional antigen matching to prevent further alloantibody formation once RBC alloimmunization has occurred. By prospectively avoiding incompatibility to C, E and K antigens, the alloimmunization rate among chronically transfused sickle cell disease patients was reduced from 3% to 0.5% per transfused RBC unit in the Stroke Prevention Trial in Sickle Cell Anemia, and hemolytic transfusion reactions were reduced by 90%.

**Hyperhemolytic Transfusion Reaction:** Hyperhemolytic transfusion reaction (HHTR) have been reported in patients with sickle cell disease, thalassemia and other diseases. In these serious reactions, both donor and recipient RBCs are destroyed, leading to more severe anemia than was present prior to transfusion. In addition, such patients may have reticulocytopenia. Some patients may have alloantibodies or autoantibodies, but others have no detectable antibodies to RBC antigens. Similar reactions may occur after subsequent transfusion even if the RBCs are extensively phenotypically matched. Possible mechanisms include bystandard hemolysis, erythropoiesis suppression, and RBC destruction secondary to contact lysis via activated macrophages. Bystandard hemolysis is thought to be an immune mediated hemolysis, where the RBC does not carry the antigen for which the antibody is directed. Treatments for patients with HHTRs have included erythropoietin, intravenous immunoglobulin, steroids, and plasma exchange.

## Recommended Reading

Garratty G. (2004). Autoantibodies induced by blood transfusion. *Transfusion* **44**, 5–9.

Vamvakas EC, Pineda AA, Reisner R *et al.* (1995). The differentiation of delayed hemolytic and delayed serologic transfusion reactions: incidence and predictors of hemolysis. *Transfusion* **35**, 16–32.

Vichinsky EP, Luban NL, Wright E *et al.* (2001). Prospective RBC phenotype matching in a stroke-prevention trial in sickle cell anemia: a multicenter transfusion trial. *Transfusion* **41**, 1086–1092.

Win N, Doughty H, Telfer P *et al.* (2001). Hyperhemolytic transfusion reaction in sickle cell disease. *Transfusion* **41**, 323–328.

Zimring JC, Spitalnik SL, Roback JD, Hillyer CD. (2007). Transfusion-induced autoantibodies and differential immunogenicity of blood group antigens: a novel hypothesis. *Transfusion* **47**, 2189–2196.

# CHAPTER 57

# Transfusion-Associated Circulatory Overload

Jeanne E. Hendrickson, MD and Christopher D. Hillyer, MD

Traditionally referred to as "volume overload," transfusion-induced cardiogenic pulmonary edema has more recently been called transfusion-associated circulatory overload (TACO). As the name implies, TACO results from circulatory overload following transfusion of blood products, and is due to the inability of the recipient to compensate for the volume of the transfused product. It is likely the most common transfusion reaction, and can be severe.

**Incidence:** TACO is estimated to occur in up to 1% of transfusions, with rates of up to 8% reported in certain elderly orthopedic surgery patients. Unlike transfusion-related acute lung injury (TRALI), TACO is not associated with an immune mechanism and is not any more likely to be associated with the transfusion of plasma-containing products. Epidemiologic data suggests that TACO may be more prevalent in patients that are very young, elderly and/or female, with underlying cardiac dysfunction or a positive fluid balance.

**Clinical Manifestations:** Symptoms of TACO include dyspnea, orthopnea, cough, chest tightness, cyanosis, widened pulse pressure, hypertension, congestive heart failure, and headache. TACO generally occurs towards the end of a transfusion, but may occur up to 6 hours afterwards. Patients at highest risk for TACO include those with diminished cardiac reserve or renal failure, and those receiving large volumes of blood products in relationship to their blood volume within a short period of time.

**Pathophysiology:** TACO is caused by the inability of the cardiopulmonary system in the recipient to tolerate the volume or rate of the transfusion. Thus, TACO represents cardiogenic pulmonary edema.

**Diagnosis:** Physical examination of patients with TACO reveals lung crackles and rales, elevated jugular venous pressure, and possibly an S3 gallop as a result of volume overload. Chest X-ray may show alveolar and interstitial edema, Kerley B-lines, pleural effusions, or cardiomegaly. Brain natriuretic peptide (BNP), a peptide secreted from the ventricles in response to increased filling pressures (and used to aid in the diagnosis of congestive heart failure), may also be used to aid in the diagnosis of TACO. In a study by Zhou and colleagues, BNP was shown to have a sensitivity of 81% and a specificity of 89% if the posttransfusion to pretransfusion BNP ratio was at least 1.5 and the posttransfusion BNP level was at least 100 pg/ml. In a recent case control series by Tobian and colleagues (2008), N-terminal pro-brain natriuretic peptide (NT-proBNP) was also shown to be a sensitive (93.8%) and specific (83.8%) marker for TACO. Elevated posttransfusion NT-proBNP (a more stable BNP analyte) was the only independent variable for the diagnosis of TACO after multivariate logistic regression analysis in this study.

**Differential Diagnosis:** The diagnoses of both TRALI and TACO should be considered in any patient who develops respiratory difficulty during or soon after a transfusion (see Chapter 58 for further details on TRALI; Table 57.1 compares and contrasts TRALI and TACO). Patients with TRALI typically do not have evidence of volume overload, nor do they respond to diuretics or have elevated an BNP. Bacterial contamination, anaphylaxis, and acute intravascular hemolysis should also be in the differential diagnosis of respiratory distress occurs following transfusion.

**TABLE 57.1 Comparison between TRALI and TACO**

| | TRALI | TACO |
|---|---|---|
| Clinical manifestations | Dyspnea, respiratory distress, hypoxia, pulmonary edema, fever, tachycardia, hypotension (within 6 hours of a transfusion). | Dyspnea, respiratory distress, hypoxia, orthopnea, hypertension, jugular venous distention, congestive heart failure (during or soon after a transfusion). |
| Pathophysiology | Antibody mediated (anti-HLA or anti-HNA antibodies, typically against recipient antigens), lipid mediated, or other. | Volume overload. |
| Chest X-ray and laboratory findings | Chest X-ray may show bilateral infiltrates in interstitial and alveolar spaces, lack of cardiomegaly. No elevation in BNP. | Chest X-ray may show alveolar and interstitial edema, Kerley B-lines, pleural effusions, or cardiomegaly. Elevated BNP or post/pre BNP ratio. |
| Treatment | Stop transfusion. Supportive care, including oxygen and possibly ventilatory support. Diuretics typically aren't effective. | Stop transfusion. Supportive care, including diuretics. Sit patient upright. Consider phlebotomy in severe cases. |
| Reporting | Report to transfusion service for transfusion reaction evaluation. Transfusion service will report reaction to donor center for further evaluation (including donor screening for anti-HLA and anti-HNA antibodies). | Report to transfusion service for transfusion reaction evaluation. |
| Future transfusion considerations | Avoid further transfusions from the implicated donor. | Transfuse future blood products more slowly (possibly even in split units, each to be transfused over 3–4 hours). Consider pre-emptive diuretic therapy. |

**Management:** It is critical that the transfusion be stopped if the patient develops respiratory distress. A transfusion reaction should be reported to the blood bank, where an initial investigation will be undertaken. Diuretic therapy is the main treatment for TACO. Other supportive care measures include supplemental oxygen as necessary, and sitting the patient upright. In rare instances, phlebotomy may be necessary.

**Prevention:** Consideration should be given to transfusing future blood products at reduced rates. It may be necessary to split future blood products into two sterile aliquots, and infuse each split over 4 hours.

## Recommended Reading

Tobian AARSL, Tisch DJ, Ness PM, Shan H. (2008). N-terminal pro-brain natriuretic peptide is a useful diagnostic marker for transfusion-associated circulatory overload. *Transfusion* **48**, 1143–1150.

Zhou L, Giacherio D, Cooling L, Davenport RD. (2005). Use of B-natriuretic peptide as a diagnostic marker in the differential diagnosis of transfusion-associated circulatory overload. *Transfusion* **45**, 1056–1063.

# CHAPTER 58

# Transfusion-Related Acute Lung Injury

Jeanne E. Hendrickson, MD and Christopher D. Hillyer, MD

Transfusion-related acute lung injury (TRALI) is a clinical syndrome which presents as shortness of breath secondary to allogeneic transfusion, and can be accompanied by non-cardiogenic pulmonary edema, fever, and hypotension. Although TRALI has been reported following transfusion of virtually all blood products, it occurs most frequently with products containing more than 60 ml of plasma (i.e. plasma and platelet products). The incidence of TRALI is estimated between 0.014% and 0.08% per allogeneic blood product transfused. There are no laboratory tests that define the diagnosis of TRALI, and it is considered to be over-diagnosed (due to its non-specific findings) in some institutions and under-diagnosed (due to lack of suspicion) in others. In the US, TRALI is currently the leading cause of transfusion-related mortality reported to the FDA; this is changing due to AABB guidance documents released in 2006 with requirements for implementation of TRALI mitigation strategies by November 2008.

**Definition:** The diagnosis of TRALI is based on clinical and radiographic findings. To allow improved surveillance and clinical practice, a Canadian Consensus Panel defined criteria for "TRALI" and "possible TRALI" (Table 58.1). TRALI is a *new* episode of acute lung injury (ALI) that *occurs during or within 6 hours* of transfusion, which is *not related to a competing etiology* for ALI. Competing etiologies for ALI include aspiration, pneumonia, toxic inhalation, severe sepsis, shock, trauma, burn injury, cardiopulmonary bypass, drug overdose, and others.

**Clinical Manifestations:** Signs and symptoms of TRALI include sudden onset of respiratory distress, with dyspnea and tachypnea and typically acute hypoxemia, fever, tachycardia, and/or hypotension. Physical examination may reveal rales and diminished breath sounds, without other evidence of fluid overload.

**Pathophysiology:** The pathophysiologic mechanisms of TRALI are not fully understood, but it is generally agreed upon that the end result is an increased permeability of the pulmonary microcirculation such that high-protein fluid enters the interstitium and alveolar air spaces. It is likely that both immunologic and non-immunologic mechanisms play a role, and that two or more "hits" must occur to induce full-blown TRALI.

The primary proposed *immunologic* mechanism is the anti-HLA (or HNA) antibodies in donor plasma reacting with recipient HLA (or HNA) antigens on recipient leukocytes (which may need to be primed by a coexisting process). This may lead to leukoagglutination or collection in the pulmonary vasculature, activation of the complement cascade, and/or mobilization of cytokines affecting lung injury and fluid accumulation in the alveoli. It is also possible that the donor anti-HLA antibodies are directed against HLA antigens on recipient pulmonary interstitial cells. Antibodies to

TABLE 58.1  Criteria for TRALI and Possible TRALI

1. TRALI criteria
   a. ALI
      i. Acute onset
      ii. Hypoxemia
         Research setting:
         $PaO_2/FiO_2 \leq 300$,
         or $SpO_2 < 90\%$ on room air
         Non-research setting:
         $PaO_2/FiO_2 \leq 300$
         or $SpO_2 < 90\%$ on room air
         or other clinical evidence of hypoxemia
      iii. Bilateral infiltrates on frontal chest radiograph
      iv. No evidence of left atrial hypertension (i.e. circulatory overload)
   b. No pre-existing ALI before transfusion
   c. During or within 6 h of transfusion
   d. No temporal relationship to an alternative risk factor for ALI
2. Possible TRALI
   a. ALI
   b. No pre-existing ALI before transfusion
   c. During or within 6 h of transfusion
   d. A clear temporal relationship to an alternative risk factor for ALI

From Kleinman S, Caulfield T, Chan P *et al.* (2004). Toward an understanding of transfusion-related acute lung injury: Statement of a consensus panel. *Transfusion* **44**, 1774–1789.

HLA class I and class II antigens have been reported in the majority of TRALI cases, with blood donated from multiparous women being most likely to contain these antibodies. In about 15% of cases antibodies are not detected. This may be secondary to assay limitations or to a *non-immunologic* mechanism of TRALI, which may be due to co-transfusion of other elements including biologically active lipids and/or cytokines that accumulate during storage.

Both the immune and non-immune mechanisms likely work via a proposed "two-hit" mechanism: the first hit primes lung endothelium and/or leukocytes and is related to the patient's underlying illness (e.g. sepsis), and the second hit is from antibody/antigen binding and/or transfusion of granulocyte-activating lipids and/or cytokines.

**Diagnosis:** A high index of suspicion is necessary when considering the diagnosis of TRALI, which should be considered in all cases of respiratory distress occurring during or soon after transfusion of blood products. Chest X-ray (CXR) often shows bilateral infiltrates involving both alveolar and interstitial spaces, without cardiomegaly. If intubation is required, frothy pink secretions (consistent with acute respiratory distress syndrome, ARDS) may be seen. There are no classic laboratory findings associated with TRALI, but transient leukopenia, neutropenia, monocytopenia, and hypocomplementemia have been reported in some patients. It can, however, be difficult to distinguish TRALI from other causes of respiratory compromise that occur around the time of transfusion, including transfusion-associated circulatory overload (TACO; see Chapter 57, Table 57.1).

**Differential Diagnosis:** TACO, which is caused by volume overload, can be difficult to differentiate from TRALI. Signs of volume overload, including elevated jugular venous pressure, elevated systolic blood pressure, and tachycardia may be present with TACO. Additionally, brain natriuretic peptide (BNP), a peptide secreted from the ventricles in response to increased filling pressures, may be elevated in TACO. TACO may respond to treatment with a diuretic, while TRALI does not usually respond to diuretics.

**Management:** It is critical that the transfusion be stopped if TRALI is suspected, and a transfusion reaction reported to the blood bank where an initial investigation will be undertaken. Management of TRALI is primarily supportive, with severe cases requiring mechanical ventilation. Intravenous steroids have been utilized, but no prospective studies exist analyzing their efficacy. Although TRALI is fatal in 5–10% of cases, the majority of TRALI patients improve clinically within 48–96 hours. Rarely, hypoxemia may persist for longer than 7 days.

**Future Transfusion Considerations:** Patients with a history of TRALI should receive no future blood products from the implicated donor. In TRALI cases where an antibody of donor origin is implicated, the patient is not thought to be at increased risk for future TRALI episodes following the transfusion of products from other donors. However, in the minority of TRALI cases where the implicated antibody is of recipient origin, the possibility exists that the patient may be at increased risk for future TRALI episodes upon exposure to subsequent transfusions. Although leukoreduction of cellular products has been suggested in this subset of patients to decrease TRALI risk, no evidence exists at this time.

**Donor/Recipient Investigation:** Following a potential case of TRALI, the transfusion service should notify the blood collection facility, which investigates the pregnancy and/or transfusion history of the donor, along with determining if the donor has antibodies to HLA class I or II, or HNA antigens. In cases where donor antibodies are discovered, the patient is typically antigen tested to see whether an antibody/antigen interaction could possibly have occurred. AABB *Standards* (5.4.2.1) require that donors implicated in TRALI are evaluated regarding their continued eligibility to donate.

**Prevention:** Eder and colleagues analyzed American Red Cross (ARC) surveillance data on TRALI fatalities from 2003 to 2005. Of the 38 probable TRALI cases, 63% were following plasma transfusion (odds ratio 12.5, 95% CI 5.4–28.9 as compared to RBCs), and a female, antibody-positive donor was significantly more likely to be associated with probable TRALI than with unrelated cases (OR 9.5, 95% CI 2.9–31.1). In 2003, the UK began preferentially transfusing plasma from male donors in an attempt to decrease the transfusion of anti-HLA antibodies from multiparous women and thus decrease the incidence of TRALI. In 2006, the ARC started implementation of a similar plan for male-only plasma.

Also in 2006, the TRALI Working Group (AABB) recommended that hospital transfusion services and blood collection facilities implement TRALI mitigation strategies. Specifically:

- it was recommended minimizing the preparation of high plasma-volume components from donors known to be leukocyte-alloimmunized or at increased risk for leukocyte alloimmunization
- high plasma-volume components were defined as plasma (FFP, FP24, thawed plasma, and cryoprecipitate reduced plasma) obtained from whole blood or apheresis,

apheresis platelets, buffy-coat derived platelets resuspended in plasma from one of the donors in the pool, and whole blood, and

- measures related to plasma components and whole blood should be undertaken by blood collection facilities by November 2007, with implementation of measures related to platelet components by November 2008.

A number of approaches were offered to accomplish these TRALI mitigation strategies, including the use of only male donors in preparation of these components (e.g. "male-only plasma"), with female plasma being preferentially diverted for further manufacturing. Many blood centers have adopted and implemented this strategy. If it were determined that the use of male-only donated products would compromise component availability, then high plasma-volume components from female donors could be used if they were selected in such a way as to minimize their risk of HLA or HNA alloimmunization, including nulliparous donors as well as those female donors with negative HLA/HNA antibody testing.

At the time of this writing, optimal methods for HLA/HNA antibody testing and appropriate nominal background values to define a positive result are not known, and automated platforms are not yet available. It is known, however, that using only male donors for production of apheresis-derived platelets would significantly reduce platelet component availability, and that the prevalence of HLA alloimmunization increases with parity which, if used as a selection criterion, would lead to the deferral of 25–35% of female donors. Thus, strategies for TRALI prevention must carefully weigh supply/demand issues with safety issues, taking into account that not all TRALI cases are associated with the transfusion of products from female donors or the transfusion of HLA or HNA antibody positive products, and also that not all male donors are HLA or HNA antibody negative.

## Recommended Reading

Eder AF, Herron R, Strupp A *et al.* (2007). Transfusion-related acute lung injury surveillance (2003–2005) and the potential impact of the selective use of plasma from male donors in the American Red Cross. *Transfusion* **47**, 599–607.

Kleinman S, Caulfield T, Chan P *et al.* (2004). Toward an understanding of transfusion-related acute lung injury: Statement of a consensus panel. *Transfusion* **44**, 1774–1789.

# CHAPTER 59

# Septic Transfusion Reactions

Cassandra D. Josephson, MD

Bacterial contamination of blood products may result in bacterial infection, sepsis and death in the recipient; as a group, these can be referred to as *septic transfusion reactions*. The highest-risk blood products are platelets, because they are stored at room temperature in a protein- and oxygen-rich environment. In 2004, implementation of techniques to limit and detect bacterial contamination of platelet products was mandated. This has resulted in a significant decrease of these reactions, though the exact residual risk is not known.

Septic reactions usually occur in products stored for longer periods of time secondary to a lag phase followed by an exponential phase of bacterial growth. For platelet products, more severe reactions occur in transfused products which have been stored for 4–5 days after collection. For RBC products, more severe reactions occur after 3–4 weeks of storage. Septic reactions secondary to frozen products usually result from contamination of the water bath, and are rare.

## Incidence:

**RBC Products:** The rate of bacterial contamination is ~2.6 in 100,000 RBC products transfused, and the rate of septic transfusion reactions is ~1 in 250,000 products transfused. The incidence is low, secondary to poor viability of most bacteria during storage at 1–6°C, yet these reactions are associated with a high mortality rate.

**Platelet Products:** The incidence of bacterial contamination is largely based on passive reporting systems using a variety of bacterial testing methodologies. Prior to implementation of the requirement by the AABB for limiting and detecting bacterial contamination, contamination was estimated at 1:1000 to 1:3000 platelet components; the risk of septic reactions was 1:40,000 and 1:240,000 for fatalities. Recently, Eder and colleagues (2007) reported ARC bacterial screening and septic transfusion data from the American Red Cross (ARC); the estimated the residual risk (after implementation of bacterial culture) of septic reactions is ~1:75,000, and of a fatality is ~1:500,000. Therefore, the residual risk of septic transfusion reactions has decreased by approximately 50%. In this same study, the use of a diversion pouch (to collect the first flow of blood for testing samples) decreased the rate of septic transfusion within one collection center from 24 to 5 per 1,000,000 apheresis products, and all 3 fatalities reported in the 27 months were from products not utilizing a diversion pouch. In addition, septic transfusion rates were higher in products which were stored longer; 13 of the 20 septic reactions occurred in individuals who received platelets transfused on day 5 after collection, including all 3 fatalities.

**Plasma and Cryoprecipitate:** Because plasma products and cryoprecipitate are stored frozen they are unlikely to contain viable bacteria, but there are cases of bacterial contamination resulting from water baths used during thawing.

**Clinical Manifestations:** Bacterial contamination may result in none to mild clinical symptoms (e.g. fever or chills), or septic shock and even death. The symptoms occur close to the onset of transfusion. Septic transfusion reactions are characterized by a fever $> 38.5°C$, rigors, hypotension (less often hypertension) and tachycardia. The patient may also experience nausea, vomiting, diarrhea and dyspnea, which may progress quickly to shock, disseminated intravascular coagulopathy (DIC), and renal failure with oliguria.

**Diagnosis:** Possible septic transfusion reactions are assessed by the transfusion service; both the blood product and the patient should be evaluated for the presence of bacteria. Evaluation of the transfused blood product for bacterial contamination includes visual examination, gram staining, and both aerobic and anaerobic microbial culture. In addition, blood culture should be performed on the patient. Definitive diagnosis of transfusion-transmitted bacterial infection requires the isolation in culture of the same organism from both the blood product and the recipient. Septic transfusion reaction can be presumed when the patient has clinical signs of a septic reaction and the blood product culture is positive for bacteria yet the patient is culture-negative, or the patient is culture-positive but the product is culture-negative after the exclusion of other sources of contamination. The blood bank should contact the blood supplier to report the adverse event so the blood supplier can recall and culture other components from the same donation. The blood center may wish to notify a donor implicated in a contaminated unit with a non-skin flora organism, because of diseases associated with asymptomatic bacteremia.

**Differential Diagnosis:** The differential diagnosis of a septic transfusion reaction includes any transfusion reaction that results in fever, including febrile non-hemolytic transfusion reactions, hemolytic transfusion reactions and transfusion-related acute lung injury. In addition, other causes of sepsis in the recipient should be considered.

**Treatment:**
- Stop the transfusion immediately and keep IV access open.
- Resuscitate the patient as needed (shock).
- Begin empiric broad-spectrum antibiotics (pending organism identification).
- Collect blood cultures (opposite arm from transfusion site).
- Send blood product bag for gram stain and culture.

**Contamination Source:** Bacterial contamination may result from skin contamination, asymptomatic bacteremic donor, collection packs, or blood processing.

**Organisms:**

*RBC Products: Yersinia enterocolitica* is responsible for ~60% of the reported cases of RBC-associated sepsis, of which ~70% are fatal. *Yersinia enterocolitica* is contracted from contaminated food, and causes abdominal pain and diarrhea. Other organisms associated with RBC septic reactions include *Enterobacter* spp., *Campylobacter* spp., *Serratia* spp., *Pseudomonas* spp. and *Escherichia coli*.

*Platelet Products:* Bacterial contamination is secondary to skin flora in approximately 75% of cases, especially *Staphylococcus* spp., and asymptomic bacteremia in the donor in the remaining cases. In the ARC study cited above the majority of cases were caused by *Staphylococcus* spp. (especially coagulase-negative *Staphylococcus*), but other causative organisms included *Enterococcus fecalies*, *Enterobacter aerogenes* and *Psudomonas fluorescens*. The majority of fatalities reported to the FDA were mostly from gram-negative organisms (e.g. *Enterobacteriacea*).

*Frozen Blood Products:* *Burkholderia cepacia* and *Pseudomonas aerognosa* have been isolated from products thawed in a contaminated water bath.

**Prevention:** AABB Standard 5.1.5.1. requires limiting and detecting bacterial contamination in platelet products. Methods to fulfill this requirement include donor screening, improved skin disinfection and diversion of the initial aliquot of blood collected, along with screening tests such as automated culture systems, chemiluminescent probes for bacterial nucleic acids and antigens (i.e. lipoteichonic acid for gram-positive bacteria and lipopolysaccharide for gram-negative bacteria), immunoassays, gram staining, pH and glucose detection, and swirling; these tests have varying sensitivity. Pathogen reduction and inactivation technologies, which nearly eliminate the risk, are available for platelet and plasma products outside of the US (Chapter 17).

Culture methods are performed in the blood center prior to release of the product, which may delay its release for up to 48 hours. This delay of release shortens the usable, in-hospital shelf-life of the platelet product (which has a 5-day total shelf-life). Interestingly, the majority of septic transfusion reactions from platelets, especially fatal reactions, occur on day 4 or 5 of storage. The Post Approval Surveillance Study of Platelet Outcomes, Release Tested (PASSPORT) study cultured apheresis platelet products at 24 hours and performed surveillance cultures at 7 days (http://www.passportstudy.com/). Data from other published reports reported up to 50% of bacterially contaminated platelets not being detected through culture at 24 hours after collection; because these undetected contaminated platelets may result in increased risk after increased length of storage time, the decision was made to halt the study.

Culture methods are able to detect less than $10^{2-3}$ CFU/ml. Assays performed at the point of issue, and usually within the transfusion service, have lower sensitivity, but are more rapid and can be performed within approximately 20 minutes. Point-of-issue tests such as metabolic assays (e.g. swirling, pH, glucose and oxygen) detect $10^{6-7}$ CFU/ml, Gram stain detects $10^6$ CFU/ml, and one chemiluscent test detects $10^{3-5}$ CFU/ml.

Until recently in the US, whole blood derived platelets could not be pooled and stored prior to issue (i.e. at the blood center prior to distribution to the transfusion service). A new system allows for pre-storage pooling, and bacterial detection by culture method, of leukoreduced whole blood derived platelet products. Prior to this system, point-of-issue testing methods were used in the majority of whole blood derived platelet products in the US. Outside of the US, alternate and automated systems are available for pre-storage pooling and bacterial testing of whole blood derived platelets. The ability to pre-storage pool and test for bacterial contamination eases the use and safety of this product, because culture methods are more sensitive and pre-storage pooling can be performed at the blood center rather than pooling at the time of issue.

## Recommended Reading

Blajchman MA, Beckers EA, Dickmeiss E *et al.* (2005). Bacterial detection of platelets: current problems and possible resolutions. *Transfus Med Rev* **19**, 259–272.

Brecher ME, Hay SN. (2005). Bacterial contamination of blood components. *Clin Microbiol Rev* **18**, 195–204.

Eder AF, Kennedy JM, Dy BA *et al.* (2007). Bacterial screening of apheresis platelets and the residual risk of septic transfusion reactions: the American Red Cross experience (2004–2006). *Transfusion* **47**, 1134–1142.

Hillyer CD, Josephson CD, Blajchman MA *et al.* (2003). Bacterial contamination of blood components: Risks, strategies, and regulation. Joint ASH and AABB educational session in transfusion medicine. *Hematology Am Soc Hematol Educ Program*, 575–589.

Prowse C. (2007). Zero tolerance. *Transfusion* **47**, 1106–1109.

Yomtovian R, Tomasulo P, Jacobs MR. (2007). Platelet bacterial contamination: assessing progress and identifying quandaries in a rapidly evolving field. *Transfusion* **47**, 1340–1346.

# CHAPTER 60

# Metabolic, Hypotensive and Other Acute Reactions and Complications

Beth H. Shaz, MD

Metabolic complications occur quite frequently, especially when large volumes of blood products are transfused. Hypotensive reactions, back pain and red eye syndrome occur infrequently, and have been associated with the use of leukoreduction filters.

**Metabolic Complications of Transfusion:** Metabolic complications of blood transfusion are most often seen in neonates or in circumstances in which large volumes of blood products are transfused, such as massive transfusion (see Chapter 50). The category of "metabolic complications" typically includes acidosis, citrate toxicity, hyperkalemia, hyperammonemia and hypothermia.

*Pathophysiology:* Metabolic complications are secondary to blood product storage in the cold (1–6°C), citrate within the anticoagulant/preservative solution, and the RBC storage lesion; together these three elements may result in hypothermia, hypocalcemia, acidosis, hyperkalemia and hyperammonemia. Changes in RBCs during storage include decrease in ATP and 2,3-DPG, and increase in hemolysis, lactate and ammonia, while, in the supernatant fluid, increases in potassium and decreases in pH, sodium and glucose are observed. In addition, irradiation of RBC products increases the amount of potassium in the supernatant over time.

*Hypothermia:* Hypothermia has a range of effects, including the following:
1. Decreases in tissue oxygenation due to increases in hemoglobin's affinity for oxygen, which can result in a metabolic acidosis;
2. Increases in the metabolic rate, resulting in increased oxygen consumption;
3. Impairment of the metabolism of citrate and some medications;
4. Inhibition of coagulation factor enzymatic reactions and disruption of platelet function leading to a bleeding diathesis; and
5. Induction of ventricular arrhythmias when large volumes of cold blood are infused through a central catheter in close proximity to the cardiac conducting system. This risk is exacerbated by coexisting hypocalcemia and hyperkalemia.

*Hypocalcemia:* Hypocalcemia is a result of citrate (i.e. citric acid) in the anticoagulant solution, as citrate chelates calcium to prevent blood from clotting. Citrate (an acid) is metabolized primarily in the liver through the Krebs cycle, which results in the release of bicarbonate (a base), which is subsequently excreted by the kidneys. Recipients are at increased risk of hypocalcemia if the amount of citrate is large, such as in massive transfusion or apheresis, or if the recipient is unable adequately to metabolize the citrate secondary to hypothermia or liver failure. In addition, metabolic alkalosis may result if the bicarbonate cannot be excreted, such as in renal failure.

**Clinical Features:** Hypothermia causes increased oxygen requirements, impaired metabolism of citrate and lactate, release of potassium, increased affinity of hemoglobin for oxygen, and ventricular arrhythmias. Citrate toxicity results in hypocalcemia, which causes paresthesias, nausea, hyperventilation and depressed cardiac function.

**Diagnosis:** Laboratory testing of ionized calcium, potassium, ammonia and pH can be used for diagnosis of hypocalcemia, hyperkalemia, hyperammonemia and acidosis.

**Treatment:** Hypothermia can be treated with a blood warmer and other methods to keep the recipient warm. Symptomatic hypocalcemia can be treated with exogenous calcium replacement.

**Prevention:** Use of blood warmers and other patient-warming devices can minimize hypothermia. Neonates and other at-risk patient populations for hyperkalemia, especially patients receiving large RBC volumes, can receive RBC products less than 7 days old or washed older RBC products, and RBC products that are irradiated should be transfused within 24 hours or washed if greater than 24 hours has lapsed since irradiation. Slower infusion rates will decrease the risk of citrate toxicity.

**Hypotensive Reactions:** Hypotensive reactions are sudden decreases in systolic or diastolic blood pressure of at least 30 mmHg occurring usually within 10 minutes of starting the transfusion. The FDA issued a letter in May 1999 to alert the public about hypotensive reactions and bedside leukocyte reduction filters after receiving over 80 reports of these events. Hypotensive reactions are most commonly seen with the use of negatively-charged bedside leukoreduction filters for platelet transfusions in patients receiving ACE inhibitors, but not exclusively so. In addition, hypotensive reactions are seen in patients undergoing extracorporeal blood processing procedures such as hemodialysis and therapeutic apheresis. Notably, most transfusion recipients taking ACE inhibitors do not have hypotensive reactions.

**Pathophysiology:** Hypotensive reactions are thought to be due to bradykinin and des-Arg-bradykinin, which are two vasoactive kinins that are generated by activation of the contact system. Activated Factor XII (Hageman Factor) converts prekallikrein to kallikrein, which cleaves kininogen and releases bradykinin, which causes hypotension and edema by activating B2-kinin receptors on the vascular endothelium through the release of endothelium-derived nitric oxide, prostacyclin, and endothelium-derived hyperpolarizing factor. Bradykinin is rapidly inactivated by angiotensin-converting enzyme (ACE); however, ACE inhibitors hinder this process. Originally, hypotensive reactions were reported with the use of negatively-charged bedside leukoreduction filters, which causes kinin activation. Subsequently, hypotensive reactions have been reported with positively-charged bedside leukoreduction filters. In addition, other causes postulated are blood warmers and other patient-specific factors. Lastly, cases have been reported without ACE inhibitor therapy.

**Clinical Features:** Hypotensive reactions are a sudden decrease of systolic or diastolic blood pressure of at least 30 mmHg occurring within 1 hour (usually within 10 minutes) of starting the transfusion. In addition, patients may have facial flushing, abdominal pain and nausea, loss of consciousness, respiratory distress and shock. In most cases, the hypotension resolves when the transfusion is discontinued.

**Diagnosis:** The diagnosis is based on the clinical symptoms. Hypotensive reactions should be differentiated from other transfusion reactions where hypotension is present, such as transfusion-related acute lung injury (TRALI), which would have respiratory findings consistent with pulmonary edema and fever; and allergic reaction, which would have respiratory findings, cutaneous findings, and possibly gastrointestinal symptoms.

**Treatment:** Hypotensive reactions require immediate intervention, including stopping the infusion and providing supportive care. Once the transfusion is discontinued, the symptoms typically promptly disappear. Supportive care may be necessary until the patient recovers fully.

**Prevention:** The majority of hypotensive reactions can be prevented by avoiding the use of bedside leukoreduction filters. As the majority of blood components within the US is pre-storage leukoreduced, this complication is currently seen infrequently.

**Red Eye Syndrome:** The American Red Cross detected 159 "red eye" reactions in 117 patients receiving RBC products leukoreduced from a specific leukoreduction filter between January 1997 and 1998. These patients exhibited bilateral conjunctival injection or hemorrhage, usually within 24 hours of transfusion, and the symptoms lasted approximately 5 days.

**Pathophysiology:** The mechanism of these reactions is unknown, but most of the reactions have been reported with specific lots of a specific type of leukoreduction filter which contains cellulose acetate membranes. Possible explanations include allergic responses to a component in the system, or direct effect of a chemical, material or product in the system.

**Clinical Features:** These reactions occur usually within 24 hours of RBC transfusion, and are characterized by bilateral erythema of the conjunctiva and eyelids, in addition to possible conjunctival hemorrhage, eye pain, photophobia, decreased visual acuity, headache, periorbital edema, arthralgias, nausea, dyspnea and rash. Symptoms usually resolve within 3 weeks, with a median duration of 5 days. These reactions are seen more commonly in patients who have received more than three leukoreduced RBC products.

**Diagnosis:** The diagnosis is based on the clinical symptoms.

**Treatment:** Treatments have included topical eye drops, including steroids, antimicrobials and artificial tears. No long-term sequelae have been reported.

**Prevention:** The implicated leukoreduction filter has been removed from the market.

**Back Pain:** In 2000, the American Red Cross received reports of 29 reactions of acute back pain during RBC transfusion in 18 patients. These episodes were associated with a specific leukoreduction filter, premedication, and outpatient transfusion.

**Pathophysiology:** The mechanism of these reactions is unknown, but patients who received outpatient transfusions, premedication or a specific fiberglass-containing leukoreduction-filtered RBC units were at increased risk.

**Clinical Features:** These reactions were characterized by severe back pain with additional symptoms of chest pain, dyspnea, headache or flushing, usually occurring within 30 minutes of starting the transfusion. Of the 18 patients, 3 required hospitalization; 2 were admitted to the intensive care unit.

**Diagnosis:** The diagnosis is based on the clinical symptoms. Other causes of acute back pain during transfusion include hemolytic transfusion reactions, which can be differentiated by finding evidence of hemolysis and RBC incompatibility.

**Treatment:** Immediate discontinuation of transfusion is recommended, as well as supportive care.

**Prevention:** The likelihood of these reactions has been decreased by manufacturing changes.

### Recommended Reading

Alonso-Echanove J, Sippy BD, Chin AE *et al.* (2006). Nationwide outbreak of red eye syndrome associated with transfusion of leukocyte-reduced red blood cell units. *Infect Control Hosp Epidemiol* **27**, 1146–1152.

Alvarado-Ramy F, Kuehnert MJ, Alonso-Echanove J *et al.* (2006). A multistate cluster of red blood cell transfusion reactions associated with use of a leucocyte reduction filter. *Transfus Med* **16**, 41–48.

Cyr M, Eastlund T, Blais C *et al.* (2001). Bradykinin metabolism and hypotensive transfusion reactions. *Transfusion* **41**, 136–150.

Perrotta PL, Snyder EL. (2001). Non-infectious complications of transfusion therapy. *Blood Rev* **15**, 69–83.

# CHAPTER 61

# Posttransfusion Purpura

Alfred J. Grindon, MD

Posttransfusion purpura (PTP) is a rare complication of transfusion (reported in 0.3 out of 100,000 components transfused in the UK) and typically occurs 2–14 days after RBC transfusion, resulting in acute, profound thrombocytopenia (platelet count < 10,000/µl). The thrombocytopenia is usually caused by antiplatelet antibodies, which destroy both transfused and autologous platelets.

**Pathophysiology:** PTP is an immune thrombocytopenia resulting from antiplatelet alloantibodies, most often anti-HPA-1a (formerly PL$^{A1}$) alone or in combination with antibodies to other platelet antigens invariably on the GPIIb/IIIa receptor complex. The transfused product (usually RBCs, but platelet and plasma products have been implicated) contains the immunogenic platelet glycoprotein, which induces an anamnesic response. PTP occurs most frequently in women who were previously sensitized during pregnancy, but sensitization from transfusion does occur. About 3% of the white population is homozygous for HPA-1b, yet the incidence of PTP is low, perhaps because of the association of certain HLA class II types to the formation of anti-HPA-1a as seen in neonatal alloimmune thrombocytopenia (NAIT). Interestingly, an increased risk of PTP in women who have previously had children with NAIT has not been reported. Women with a history of PTP may be at increased risk for having a fetus/neonate affected by NAIT. The antibody destroys both transfused and autologous platelets. The mechanism of destruction of autologous platelets is unknown; theories include: (1) the adsorption of donor-derived, soluble platelet glycoprotein on to the autologous platelets; (2) the immune response includes an autoimmune component; and (3) the antibody produced cross-reacts with autologous platelets.

**Clinical Manifestations:** The PTP patient presents with unexplained purpuric rash, bruising or mucosal bleeding 2–14 days after transfusion. There is a mortality rate of approximately 10% secondary to hemorrhage, and approximately 30% have major hemorrhage. The diagnosis is often delayed because of the interval between transfusion and disease onset. A febrile non-hemolytic transfusion reaction often occurs with the implicated transfusion. If untreated the disease is self-limited, with platelet count recovery occurring usually within 21 days.

**Diagnosis:** The differential diagnosis includes other diseases with rapid onset of severe thrombocytopenia, such as idiopathic thrombocytopenic purpura, drug-induced thrombocytopenia, disseminated intravascular coagulopathy, and thrombotic thrombocytopenic purpura. The diagnosis is confirmed by the clinical presentation and the detection of platelet-specific alloantibodies. The majority of cases have HPA-1a antibodies, but antibodies against many other HPA antigens have been reported, and occasionally multiple antibodies are present. Demonstration of platelet antibodies can be determined by ELISA, monoclonal antibody immobilization of platelet antigens assay, and other assays. Documenting the patient's platelet phenotype can be difficult, because of the presence of antibodies reacting with the endogenous platelets and the often severe thrombocytopenia. Genotyping for platelet antigens may be helpful, if available.

**Treatment:** The primary treatment is high-dose intravenous immunoglobulin (IVIG) (typically 2000 mg/kg over 2–5 days), resulting in a platelet count > 100,000/µl in 4–5 days, with an 85% response rate. The mechanism of action of IVIG may be Fc receptor blockade, non-specific binding of Ig to platelet surface, and/or acceleration of IgG catabolism. Plasma exchange was used before discovering the efficacy of IVIG. For those who do not respond, splenectomy has been performed. Corticosteroids appear to be ineffective. If the antibody specificity has been determined, platelets lacking the offending antigen (usually HPA-1a negative) have been used in bleeding patients with modest clinical improvement, though these platelets also have decreased survival. Co-administration of IVIG may improve the survival of the transfused platelets.

**Prevention:** Recurrence of PTP after a future RBC transfusion is uncommon, but has been reported. Some authorities have advocated the use of washed RBC products, or RBC and platelet products from corresponding antigen-negative donors, but the value of this practice is unclear. A report from the United Kingdom described a decrease in cases of PTP after implementation of universal leukoreduction, likely secondary to a decrease in platelets and platelet microvesicles in the leukoreduced RBC product.

### Recommended Reading

Perrotta PL, Snyder EL. (2001). Non-infectious complications of transfusion therapy. *Blood Rev* **15**, 69–83.

Vogelsang G, Kickler TS, Bell WR. (1993). Posttransfusion purpura: a report of five patients and a review of the pathogenesis and management. *Am J Hemat* **21**, 259–267.

Williamson LM, Stainsby D, Jones H *et al.* (2007). The impact of universal leukodepletion of the blood supply on hemovigilance reports of posttransfusion purpura and transfusion-associated graft-versus-host disease. *Transfusion* **47**, 1455–1467.

# CHAPTER 62

# Transfusion Associated Graft versus Host Disease

Beth H. Shaz, MD and Christopher D. Hillyer, MD

Transfusion associated graft versus host disease (TA-GVHD), a rare and almost universally fatal complication of blood product transfusion, is due to the co-transfusion of viable lymphocytes in cellular blood products, such as whole blood, red blood cells, platelets, granulocytes and fresh plasma (not frozen plasma). If the immune system of the recipient cannot recognize and destroy the co-transfused lymphocytes, they can engraft and mount an immunologic response against the host, thus the term *graft-versus-host disease*. There are two reasons that the recipient's immune system can be limited in recognizing the co-transfused lymphocytes: (1) significant immunosuppression, and (2) planned or inadvertent HLA matching of the transfused blood product with the recipient. A number of technologies can abrogate the risk of TA-GVHD in these two circumstances, including irradiation of the product, and pathogen-reduction (also called pathogen-inactivation) technologies. Both of these methods render the DNA in the co-transfused lymphocytes incapable of participating in cell division. Thus, the co-transfused lymphocytes in cellular blood products treated with these modalities cannot engraft. It is critical to identify recipients who are at risk for TA-GVHD and ensure products are so treated in order to prevent this lethal transfusion complication.

**Graft versus Host Disease:** GVHD requires the infusion of viable lymphocytes that engraft in the recipient host and mount an immune reaction themselves. Viable lymphocytes are co-transfused in two situations: (1) the infusion of hematopoietic progenitor cells (HPC) harvested from bone marrow, peripheral blood or umbilical cord blood, for HPC transplantation; and (2) administration of cellular blood products for transfusion. HPC transplant-associated GVHD occurs in up to one-third of patients receiving allogeneic HPCs, and can be fatal. The HPCs cannot be irradiated as this would render the progenitor cells incapable of engrafting and thus defeat the purpose of the transplant. TA-GVHD occurs much less frequently, as the products can be irradiated without compromising the potency of the transfused product and because the dose of lymphocytes in transfused products is significantly smaller, especially in leukoreduced components.

Pathophysiology of TA-GVHD: The mechanism of TA-GVHD is similar to that of acute GVHD after HPC transplantation. There are three separate phases of development: phase one is conditioning, phase two is the afferent phase, and phase three is the efferent phase. The conditioning regimen results in tissue damage and activation of host tissues, which leads to production of inflammatory cytokines. The afferent phase results in donor T-cell activation through antigen presentation, followed by proliferation, and then differentiation of activated T cells. The efferent phase results in the release of inflammatory cytokines that attack host tissues and result in cell death and host tissue destruction. A study characterizing T-cell clones from a patient with TA-GVHD demonstrated three types of clones; type I were CD8+ and specifically lysed cells that express HLA B52 (present on recipient and not donor), type II were CD4+ and specifically lysed cells that express HLA DR15 (present on recipient and not donor), and type III were CD4+ and had no cytotoxic activity yet proliferated in response to stimulation with cells that express HLA DR15.

**HPC Associated versus TA-GVHD:** Once co-administered viable lymphocytes have been infused with HPCs of blood products, they can engraft and mount an immune response against HLA-rich tissues and organs in the recipient. These HLA-rich tissues include the skin, liver and gastrointestinal (GI) tract, and the response is accompanied by fever due to release of inflammatory cytokines. Thus, symptoms of GVHD include the classic triad of erythema, liver dysfunction and GI symptoms, which occur between 4 and 30 days after HPC or blood product administration. TA-GVHD, in contrast to HPC-associated GVHD, also recognizes the HLA-rich bone marrow cells of the recipient as foreign; thus, graft-driven destruction of the bone marrow can ensue, causing devastating pancytopenia. It is the pancytopenia of TA-GVHD that differentiates it from HPC-associated GVHD and which makes it have a near 100% mortality.

TA-GVHD in adults results in death within 3 weeks from symptom onset in over 95% of cases. In neonates the clinical manifestations are similar, but the time between transfusion and symptom onset is longer than for adults: fever occurs at a median of 28 days, rash at a median of 30 days, and death at a median of 51 days. In both groups, fever is usually the presenting symptom followed by an erythematous maculopapular rash, which typically begins on the face and trunk and spreads to the extremities. Liver dysfunction usually manifests as an obstructive jaundice or an acute hepatitis. GI complications include nausea, anorexia or massive diarrhea. Leukopenia and pancytopenia develop later and progressively become more severe, leading to sepsis and candidiasis, which can result in multi-organ failure and death.

**Post-operative Erythroderma:** An unusual variant of TA-GVHD is post-operative erythroderma (POE). POE was first recognized in Japan and Israel following cardiac operations in patients who had received blood products. POE results from the inadvertent and unknown HLA matching of donor and recipient, and occurred particularly in these countries because in both Israel and Japan there is a high level of HLA type similarity within their own populations. As these products by chance are usually only partially matched, and the recipient does have an intact and functioning immune system, the early phases of engraftment that lead to the erythema are aborted by the strengthening host response. Thus, the erythema (POE) does not progress to full GVHD. It was the observation of POE and chance HLA matching that led the Japanese to irradiate all blood products for transfusion (universal irradiation).

## Diagnosis, Treatment and Prevention of TA-GVHD:

**Diagnosis:** The diagnosis of GVHD in any circumstance is based on both clinical findings and laboratory and biopsy results. However, in situations where GVHD is not expected, and TA-GVHD is often not suspected by clinicians, the symptoms, findings and laboratory tests can easily be interpreted as being due to severe viral infection or an adverse reaction to administered medication in ill patients. Thus, TA-GVHD is considered to be significantly under-diagnosed. The discovery of donor lymphocytes or DNA in the patient's peripheral blood or tissue biopsy with the appropriate clinical picture confirms the diagnosis of TA-GVHD. Donor-derived DNA is usually detected using polymerase chain reaction (PCR) based HLA typing, but other methods include the use of restriction fragment length polymorphisms, variable number tandem repeat analysis, microsatellite markers, and cytogenetics.

**Treatment:** Treatment of TA-GVHD is largely palliative and aimed at attempting to improve the function or render recovery of the recipient's immune system and bone marrow. This is largely unsuccessful. Approaches include corticosteroids, antithymocyte globulin and cyclosporin employed with hematopoietic growth factors. There are a few reports of spontaneous resolution, and of successful treatment with a combination of cyclosporin, steroids and OKT3 (anti-CD3 monoclonal antibody) or antithymocyte globulin, or with autologous or allogeneic HPC transplantation. Transient improvement has been seen with nafmostat mesilate, a serine protease inhibitor that inhibits cytotoxic T lymphocytes.

**Prevention:** Since treatment options for TA-GVHD are mostly unsuccessful, patients at increased risk must be identified and transfused with lymphocyte-inactivated products, usually through gamma-irradiation or pathogen-inactivation methods. Gamma radiation is derived from decay of radioactive isotopes, such as cesium-137 or cobalt-60. Some pathogen-reduction technologies have been shown, in human clinical trials, mouse models and other lymphocyte proliferation assays, to inactivate T lymphocytes. Gamma irradiation is the most common method used to prevent TA-GVHD; outside the US, the use of pathogen-inactivated platelets is growing. Chapter 36 describes the irradiation of blood components in significant detail.

## Blood Product Factors Contributing to the Risk of TA-GVHD:

**Age of Blood:** Fresh blood increases the risk of TA-GVHD since, over time, lymphocytes undergo apoptosis and fail to stimulate a MLC response during storage. A Japanese series of cases of TA-GVHD in immunocompetent patients found that 62% of patients had received blood less than 72 hours old, and a US series found that about 90% of cases had received blood less than 4 days old.

**Leukocyte Dose:** Leukocyte reduction of blood products may decrease the risk of TA-GVHD, but does not eliminate it. The SHOT data reported a decrease in the number of TA-GVHD cases following universal leukocyte reduction of blood components in the UK in 1999.

**Blood Products:** All cellular blood products, including RBCs, platelets, granulocytes, whole blood and fresh plasma (not frozen plasma), contain viable T lymphocytes that are capable of causing TA-GVHD. Granulocyte transfusions are the highest-risk product because they are given fresh, have a high lymphocyte count, and are administered to neutropenic and immunosuppressed patients. Therefore, it is recommended that all granulocyte transfusions undergo irradiation prior to transfusion and the remaining cellular blood products be irradiated for patients at increased risk.

**Patients at Increased Risk for TA-GVHD:** There are certain patient populations who are at increased risk for TA-GVHD (Table 62.1). It is difficult to quantify the risk of TA-GVHD for any patient population, because of the unknowns – the number of patients with the disease, the number of patients with the disease who are transfused, and the number of transfusions or type of products each patient received. The risk is therefore derived from case reports or hemovigilance data, which are biased by under-recognition, misdiagnosis, and under- and passive reporting.

---

**TABLE 62.1 Indications for Irradiated Cellular Blood Products to Prevent TA-GVHD**

*Indications for which irradiation is considered to be required*

Congenital immunodeficiency syndromes (suspected or known)

Allogeneic and autologous HPC transplantation

Transfusions from blood relatives

HLA-matched or partially HLA-matched products (platelet transfusions)

Granulocyte transfusions

Hodgkin's disease

Treatment with purine analogue drugs (fludarabine, cladribine and deoxycoformycin)

Treatment with Campath (anti-CD52) and other drugs/antibodies that affect T-lymphocyte number or function

Intrauterine transfusions

*Indications for which irradiation is deemed appropriate by most authorities*

Neonatal exchange transfusions

Pre-term infants/low birth-weight infants

Infant/child with congenital heart disease (secondary to possible DiGeorge syndrome)

Acute leukemia

Non-Hodgkin's lymphoma and other hematologic malignancies

Aplastic anemia

Solid tumors receiving intensive chemotherapy and/or radiotherapy

Recipient and donor pair from a genetically homogeneous population

*Indications for which irradiation is considered unwarranted by most authorities*

Solid-organ transplantation

Healthy newborns/term infants

HIV/AIDS

From Shaz BH and Hillyer CD (2009). Transfusion-associated graft-versus-host disease. In: Murphy M, Pamphilon D (eds). *Practical Transfusion Medicine*, 3rd edition. Oxford: Blackwell Synergy.

---

**Congenital Immunodeficiency Patients:** The first reported cases of TA-GVHD occurred in the 1960s in children with T-lymphocyte congenital immunodeficiency syndromes. TA-GVHD has occurred in children with severe combined immunodeficiency syndromes (SCID) and with variable immunodeficiency syndromes, such as

Wiskott-Aldrich and DiGeorge syndromes. Children may be transfused prior to the recognition of congenital immunodeficiency syndromes. Because of the possibility of pediatric patients having an undiagnosed immunodeficiency, it is prudent to irradiate all blood components for children under a certain age (such as 1 year). This is particularly true with infants undergoing cardiac surgery, who may have unrecognized DiGeorge syndrome. It is recommended that all patients with suspected or confirmed congenital immunodeficiency receive irradiated products.

**Allogeneic and Autologous HPC Recipients:** Both allogeneic and autologous HPC transplant recipients are at increased risk of TA-GVHD. Patients who undergo allogeneic HPC transplantation have received irradiated blood products routinely for over 40 years. Multiple organizations, including the European School of Haematology (ESH), the European Group for Blood and Marrow Transplantation (EBMT) and the Foundation for the Accreditation of Cellular Therapy (FACT), recommend irradiated blood products for allogeneic and autologous HPC recipients, but it is unclear for how long before and after transplantation these patients require irradiated products.

**Patients with Hematologic Malignancies:** Patients with hematologic malignancies are at increased risk for TA-GVHD, especially patients with (or a history of) Hodgkin's disease (HD). It is recommended that patients with hematologic malignancies receive irradiated products; however, it is less clear if this requirement should be only during active treatment, and for how long it should be continued following treatment (it is generally accepted that patients with a history of HD receive irradiated products for life).

**Recipients of Fludarabine and Other Purine Analogues as well as Other Drugs/antibodies that Affect T-lymphocyte Number or Function:** TA-GVHD was initially reported in patients with chronic lymphocytic lymphoma (CLL) receiving fludarabine, which is a purine analogue and results in profound lymphopenia. There are nine reported cases of TA-GVHD in CLL, acute myeloid leukemia (AML) and non-Hodgkin's lymphoma (NHL) patients who received fludarabine up to 11 months prior to transfusion. In addition, TA-GVHD occurred in a patient who received fludarabine for treatment of autoimmune disease. Other purine analogues, including xycoformycin (pentostatin) and chlorodeoxyadenosine (cladribine), have been associated with the development of TA-GVHD. It is recommended that all patients who have received fludarabine or other purine analogues as well as Campath (anti-CD52) or other drugs/antibodies that affect T-lymphocyte function or number be transfused with irradiated products; however, it is unclear for how long these patients should receive irradiated products. The current recommendation is for at least a year, and until recovery from the resulting lymphopenia following the administration of these drugs.

**Fetuses and Neonates:** Fetuses and neonates have immature immune systems, and may be at increased risk of TA-GVHD. In neonates, most cases of TA-GVHD reported are in those with congenital immunodeficiency or those who received products from related donors. At least ten cases were reported after neonatal exchange transfusions; four occurred in infants who had previously received IUT. Seven cases were in preterm infants (excluding those who received a product from a relative). A single case report involved a full-term infant receiving extracorporeal membrane oxygenation (ECMO). The use of irradiated products for fetal and neonatal transfusions is therefore recommended for exchange transfusions and IUT, preterm infants, infants with congenital immunodeficiency, and those receiving products from relatives; its need is less clear for other neonatal transfusions.

**Patients with Aplastic Anemia:** Since patients with aplastic anemia are usually treated with intensive chemotherapy regimens and possible HPC transplantation, some authorities recommend that they receive irradiated products, especially during myelosuppressive therapy.

**Patients Receiving Chemotherapy and Immunotherapy:** TA-GVHD has occurred in patients with solid tumors, including neuroblastoma, rhabdomyosarcoma, urothelial carcinoma and small cell lung cancer, during intensive myeloablative therapy. Therefore, it is recommended that patients with solid tumors receive irradiated products, especially during myelosuppressive therapy.

**Solid-organ Transplantation Recipients:** GVHD is a rare complication of solid-organ transplantation, which usually results from the passenger lymphocytes contained within the solid organ and not from transfusion, even though these individuals are highly immunosuppressed and transfused. There have been four cases of TA-GVHD in solid organ transplant recipients; one was a liver transplant recipient with pre-existing pancytopenia, one was a heart transplant recipient, and two were inconclusive cases in kidney recipients. The risk of TA-GVHD in solid-organ transplant recipients appears to be low, and the use of irradiated products is generally considered to be unwarranted.

**Human Immunodeficiency Virus and Adult Immunodeficiency Syndrome Patients:** HIV/AIDS is not considered a risk factor for TA-GVHD as there is only a single case report of a child with AIDS developing transient TA-GVHD. It is postulated that HIV infects the transfused T lymphocytes, preventing the development of TA-GVHD. The use of irradiated blood products in HIV/AIDS patients is not warranted, but approximately one-fourth of institutions in the US choose to take this precaution, likely because of the immunosuppressive nature of HIV/AIDS, the high degree of fatality of TA-GVHD, and the ability of HARRT to decrease the burden of circulating HIV particles, thus theoretically mitigating the virus from infecting the co-transfused lymphocytes.

**Cardiovascular Surgery Patients:** Prior to the Japanese changing their irradiation policies, the reported incidence of TA-GVHD following cardiovascular surgery was 0.15–0.47%. Of 122 cases of TA-GVHD reported in Japan from 1985 to 1993, 56 were patients after cardiovascular surgery; 28% used blood from a relative and 72% used blood less than 72 hours old. They also reported a lower risk for women than men, possibly secondary to women having previous exposure to leukocytes during pregnancy and childbirth. There are five cases of TA-GVHD after cardiovascular surgery reported in the US and the UK. Possible reasons for this increased risk are that the RBC products transfused are usually less than 72 hours old, and cardiac surgery may result in reduced cell-mediated immunity. The recommendation for irradiation of products for cardiac surgery patients is warranted in Japan, but not in the US at this time.

**Immunocompetent Patients:** TA-GVHD is reported in immunocompetent patients; the majority of cases have occurred with the use of fresh whole blood from a close relative. In a review of 122 cases of TA-GVHD in immunocompetent Japanese patients, 67% had not received products from a related donor and, of the 66 non-cardiovascular surgery patients, 39 had solid tumors and 27 had other conditions. The risk of receiving a blood product from a homozygous donor is greatest in populations with limited HLA haplotype polymorphisms, such as in Japan. The frequency of reported cases is substantially lower than these estimates (1 : 7174 in the US), which may be a result of unrecognized and/or unreported cases, lymphocytes in blood products that are either non-viable or insufficient to cause disease, and/or recipients being able to destroy the donor lymphocytes because of minor HLA differences between donor and recipient. Irradiation of products from close relatives and HLA-matched products are recommended for immunocompetent patients. The risk is otherwise minimal for other immunocompetent patients, and irradiation is not warranted.

## Recommended Reading

BCSH Blood Transfusion Task Force. (1996). Guidelines on gamma irradiation of blood components for the prevention of transfusion-associated graft-versus-host disease. *Transfus Med* **6**, 261–271.

Hume HA, Preiksaitis JB. (1999). Transfusion associated graft-versus-host disease, cytomegalovirus infection and HLA alloimmunization in neonatal and pediatric patients. *Transfus Sci* **21**, 73–95.

Leitman SF, Tisdale JF, Bolan CD *et al.* (2003). Transfusion-associated GVHD after fludarabine therapy in a patient with systemic lupus erythematosus. *Transfusion* **43**, 1667–1671.

Maung ZT, Wood AC, Jackson GH *et al.* (1994). Transfusion-associated graft-versus-host disease in fludarabine-treated B-chronic lymphocytic leukaemia. *Br J Haematol* **88**, 649–652.

Ohto H, Anderson KC. (1996). Survey of transfusion-associated graft-versus-host disease in immunocompetent recipients. *Transfusion Med Rev* **10**, 31–43.

Petz LD, Calhoun L, Yam P *et al.* (1993). Transfusion-associated graft-versus-host disease in immunocompetent patients: report of a fatal case associated with transfusion of blood froma second-degree relative, and a survey of predisposing factors. *Transfusion* **33**, 742–750.

Przepiorka D, LeParc GF, Stovall MA *et al.* (1996). Use of irradiated blood components: practice parameter. *Am J Clin Pathol* **106**, 6–11.

Rososhansky S, Badonnel M-CH, Hiestand LL *et al.* (1999). Transfusion-associated graft-versus-host disease in an immunocompetent patient following cardiac surgery. *Vox Sang* **76**, 59–63.

Schroeder ML. (2002). Transfusion-associated graft-versus-host disease. *Br J Haematol* **117**, 275–287.

Shaz BH, Hillyer CD. (2009). Transfusion-associated graft-versus-host disease. In: Murphy M, Pamphilon D (eds). *Practical Transfusion Medicine*, 3rd edition. Oxford: Blackwell Synergy.

Triulzi DJ, Nalesnik MA. (2001). Microchimerism, GVHD, and tolerance in solid organ transplantation. *Transfusion* **41**, 419–426.

Triulzi D, Duquesnoy R, Nichols L *et al.* (2006). Fatal transfusion-associated graft-versus-host disease in an immunocompetent recipient of a volunteer unit of red cells. *Transfusion* **46**, 885–888.

Williamson LM, Wimperis JZ, Wood ME, Woodcock B. (1996). Fludarabine treatment and transfusion-associated graft-versus-host disease. *Lancet* **348**, 472–473.

Williamson LM, Stainsby D, Jones H *et al.* (2007). The impact of universal leukode-pletion of the blood supply on hemovigilance reports of posttransfusion purpura and transfusion-associated graft-versus-host disease. *Transfusion* **47**, 1455–1467.

# CHAPTER 63

# Microchimerism

Beth H. Shaz, MD and Christopher D. Hillyer, MD

Chimerism is defined as the presence of more than one genetically distinct population of cells in a single organism that originated from more than one zygote. Microchimerism (MC) occurs when the non-host cells represent only <5% of the cells of an individual, and can be a consequence of pregnancy, organ transplantation or transfusion. Blood transfusion can result in a stable persistent minor population of allogeneic cells within the recipient. This situation is called transfusion-associated microchimerism (TA-MC). With increasingly sensitive methods, TA-MC can be detected not infrequently after transfusion, but the conditions that facilitate and the consequences of this phenomenon are unknown.

## Pathophysiology of TA-MC:

**Normal Clearance of Transfused Lymphocytes:** In a study investigating the clearance of co-transfused lymphocytes in immunocompetent recipients, three phases were found: first, 99.9% of lymphocytes were cleared over the first 2 days; second, there was a 1-log *increase* in the number of circulating donor lymphocytes on days 3–5; and third, there was a second clearance event leading to the small number and percent of chimeric cells. It was postulated that the transient increase in donor lymphocytes represents one arm of an *in vivo* mixed lymphocyte reaction with activated donor T lymphocytes proliferating in reaction to HLA-incompatible recipient cells. The second clearance step results from the recipient's immune system mounting an augmented response against the donor cells. Irradiation of the blood component eliminates the expansion of donor cells, and thus abrogates TA-MC.

**Transfusion-associated Microchimerism:** TA-MC is defined as the presence of transfused donor leukocytes constituting up to 5% of the recipient's peripheral blood leukocytes, which can remain for long periods of time (maximum length unknown). A proposed hypothesis for this phenomenon is that there are genetic polymorphisms between the donor and recipient linked to cytokine production, where some of the polymorphisms lead to immune suppression in the recipient that allows for tolerance to the transfused allogeneic cells and the development of TA-MC.

## Clinical Elements of TA-MC:
Irradiation of products prevents TA-MC, while leukocyte-reduction of blood products does not appear to affect its incidence. This suggests that even a low number of residual lymphocytes can cause TA-MC, but those residual lymphocytes are not capable of engraftment (which would result in TA-GVHD). TA-MC has been reported most extensively in trauma patients, where it occurs in up to 25–50% of transfused trauma patients. It has also been reported in sickle cell disease and thalassemia patients, but has not been shown as a sustained event in HIV-positive individuals. In trauma patients, age, sex, injury severity score,

splenectomy and number of products transfused do not appear to correlate with the establishment of TA-MC.

Some authors have referred to TA-MC as an aborted GVH response. The molecular and cellular events that arrest TA-MC as microchimerism and do not allow progression to TA-GVHD have not been identified. It is clear that those patients with TA-MC have good immune function, at least in part. The relationship between TA-MC and postoperative erythroderma is also unclear (see Chapter 62). When patients with TA-MC were studied and observed for signs and symptoms suggestive of chronic GVHD (skin rashes, jaundice, xerostomia, xerophthalmia, diarrhea, or new allergic reactions) several months after transfusion, TA-MC did not appear to be, or to correlate with, chronic GVHD. One study reported a decrease in donor-specific lymphocyte responses in trauma patients with demonstrable TA-MC versus non-TA-MC trauma patients, but the clinical consequences of this observation are unknown. In addition, TA-MC can be sustained for decades after transfusion.

In summary, no clear relationship of TA-MC to clinical outcomes has been elucidated, though speculation has been voiced that those with TA-MC may be at risk for autoimmune disorders at some future point. To determine the association between TA-MC and autoimmune or other diseases, important confounding factors need to be accounted for and long-term follow up is needed, as these diseases can take years to develop and may result in a wide range of subjective symptoms.

**Diagnosis:** The ability to detect TA-MC requires the selection of the optimal genetic difference between donor and recipient DNA, and the capability to detect the small amount of donor DNA amongst a large amount of host DNA. One technique used to detect TA-MC is real-time PCR. Initially, this technique was used to look for the presence of the Y chromosome in women who had received red blood cell transfusions from at least one male donor. The technique was expanded to include a panel of 12 HLA-DR polymorphisms, and a third improvement was the addition of a panel of 12 insertion/deletion (InDel) polymorphisms.

**Testing Limitations:** The ability to detect TA-MC is limited by sample volume and sampling error. Large sample volume will create too much DNA and result in difficulties in testing. Sampling error is likely when an extremely low level of TA-MC exists, because a donor lymphocyte may not be present in the recipient sample. When using techniques that determine the presence of Y chromosomal DNA in women transfused with male blood products, previous pregnancies with male infants are a source of positive results that do not reflect third-party DNA from the donor of the transfused product.

**Prevention:** As TA-MC has no known clinical correlation or adverse effect, there are no recommendations to modify transfusion guidelines or products. Microchimerism can be prevented by irradiation, but this is not an indication for irradiation.

## Recommended Reading

Gill RM, Lee T, Utter GH et al. (2008). The TNF (-308A) polymorphism is associated with microchimerism in transfused trauma patients. *Blood* **111**, 3880–3883.

Lee TH, Donegan E, Slichter S, Busch MP. (1995). Transient increase in circulating donor leukocytes after allogeneic transfusions in immunocompetent recipients compatible with donor cell proliferation. *Blood* **85**, 1207–1214.

Utter GH, Owings JT, Lee T *et al.* (2004). Blood transfusion is associated with donor leukocyte microchimerism in trauma patients. *J Trauma* **57**, 702–708.

Utter GH, Owings JT, Lee T *et al.* (2005). Microchimerism in transfused trauma patients is associated with diminished donor-specific lymphocyte response. *J Trauma* **58**, 925–932.

Utter GH, Nathens AB, Lee T *et al.* (2006). Leukoreduction of blood transfusions does not diminish transfusion-associated microchimerism in trauma patients. *Transfusion* **46**, 1863–1869.

Utter GH, Reed WF, Lee TH, Busch MP. (2007). Transfusion-associated microchimerism. *Vox Sang* **93**, 188–195.

# CHAPTER 64

# Transfusion Related Immunomodulation

James C. Zimring, MD, PhD

Blood products can have profound effects on the recipient's immune system. Most evident is alloimmunization to red blood cell (RBC) blood group and human leukocyte antigens (HLA), the management of which comprises a large part of transfusion medicine. However, transfused blood products have additional effects on recipient immunity other than generation of alloantibodies. Broadly, these effects have been termed *transfusion related immunomodulation* (TRIM). It is reasonable to note that some TRIM effects are well-established and generally accepted, while others are a matter of debate. TRIM effects can be categorized as beneficial or deleterious, and secondary to either humoral or cellular immunity (Table 64.1).

| TABLE 64.1 Established and Proposed TRIM-associated Risks of Transfusion |
| --- |
| **Humoral immunity** |
| • Increased RBC alloimmunization |
| **Cellular immunity** |
| • Decreased renal allograft rejection |
| • Improvement in autoimmune diseases (e.g. Crohn's disease) |
| • Decreased repetitive spontaneous abortions |
| • Increased tumor recurrence |
| • Increased post-operative infection |

## Accepted TRIM Effects:

**Solid Organ Transplantation:** Since the 1960s, it has been appreciated that transfusion of whole blood prior to organ transplantation results in significantly improved allograft survival in both humans and experimental animals. This effect was most notable for transplanted cadaveric kidneys. This phenomenon, or "transfusion effect," occurs to the greatest extent with the transfusion of whole blood, to a lesser extent with packed RBCs, and to an even smaller extent with washed or frozen/thawed RBC products, which are ~80–90% leukocyte reduced. There is little or no effect when giving stringently leukodepleted products (99.9% leukoreduced). The transfusion effect appears to be antigen specific, leading to tolerance to HLA antigens. There is a strong correlation between improved allograft survival and coincidence of HLA antigens on the donor organ and on the transfused WBCs. This is especially true for the HLA-DR locus; a mismatch results in increased sensitization, and a partial mismatch results in the transfusion effect. The risk of transfusion is sensitization to the HLA antigens. The formation of HLA antibodies can cause acute humoral rejection of the allograft, and therefore the majority of transplant centers do not use crossmatch-incompatible allografts (i.e. HLA antigens present on the allograft for which the recipient has HLA antibodies). Current immunosuppressive medications have greatly reduced the utility of the transfusion effect.

**Reduction in the Likelihood of Spontaneous Abortion:** It has been observed that transfusion of allogeneic blood, from paternal or other sources, has a beneficial effect in preventing recurrent spontaneous abortion, possibly by decreasing the T-cell response and generating suppressor T cells. Although the effect is mild, it is reproducible.

## Debated TRIM Effects:

**Cancer, Infection and Autoimmunity:** Subtle TRIM effects have been reported in situations in which mild impairment of cellular immunity may be predicted to result in alterations in pathology. For example, published data suggest that transfusion may increase cancer recurrence and/or metastasis, presumably due to decreased anti-tumor immunity. Likewise, published data demonstrate that increased post-operative infections correlate with transfusion, suggesting TRIM inhibition of anti-microbial immunity. Also, studies suggest that transfusion may have a therapeutic benefit in certain autoimmune states (such as Crohn's disease). In each of these situations, an equally large number of reports show no statistically significant effects. Thus, at this point, it is reasonable to conclude that clinically significant TRIM effects in cancer, post-operative wound infections and autoimmune disorders *may* or *may not* occur, and thus remain disputed.

**Proposed Mechanisms of TRIM:** At first glance, the combination of alloimmunization (which is a positive immune response) and other TRIM effects (which are generally immunosuppressive) may seem contradictory. However, humoral and cellular immunity represent distinct response pathways, which can be mutually antagonistic (e.g. Th1 versus Th2 type responses). Thus, leukocytes in transfused RBCs may simultaneously enhance humoral immunity whilst suppressing cellular immunity. In this way, enhanced antibody responses and cellular immunosuppression may simultaneously occur. The clinical trials referred to above seldom actually measured immunity *per se*, but rather relied on secondary outcomes assumed to be affected by immunologic alterations. The majority of the mechanistic data has been obtained using animal models, and thus may or may not hold true in the human setting. Nevertheless, the observed TRIM correlations are consistent with a hypothesis of immunomodulation being causative.

### Recommended Reading

Blajchman MA. (2002). Immunomodulation and blood transfusion. *Am J Therapeutics* **9**, 389–395.

Blajchman MA. (2005). Transfusion immunomodulation or TRIM: what does it mean clinically? *Hematology* **1**, 208–214.

Vamvakas EC, Blajchman MA. (2007). Transfusion-related immunomodulation (TRIM): an update. *Blood Rev* **21**, 327–348.

# CHAPTER 65

# Iron Overload

Cassandra D. Josephson, MD

Chronic RBC transfusions will result in iron overload, because the body has no mechanism for excreting excess iron. Clinical signs and symptoms of iron overload can begin when total body iron reaches 10–25 g, depending on recipient body weight. As each RBC product contains 200–250 mg of iron, it requires approximately 50–100 transfused RBC products to reach clinically significant iron overload. Patients who require this degree of chronic transfusion typically have sickle cell disease, β-thalassemia major, or a myelodysplastic syndrome. It is important to predict and diagnose iron overload, as, without ongoing treatment, it becomes a progressive and eventually fatal disorder.

**Pathophysiology:** The cause of iron overload in the chronically transfused patient is attributable to the patient's inability to actively excrete the excess iron contained within the erythrocytes that are infused with each RBC product. Each RBC product can be expected to contain approximately 200–250 mg of elemental iron, as 1 ml of RBCs contains about 1 mg of elemental iron. After saturating transferrin and macrophages in the reticuloendothelial system, excess iron accumulates in the cytoplasm of liver, myocardial, and endocrine cells. There it catalyzes the production of free-radicals from hydrogen peroxide which damage cell membranes, proteins, and DNA.

**Clinical Manifestations:** As in idiopathic hemachromatosis, the organs involved in iron overload due to transfusion are predominantly the liver, heart, endocrine glands and skin. The liver is the major physiologic site for normal iron storage, and is the first system to develop pathologic sequelae associated with transfusional iron overload. Hepatomegaly may progress quickly, leading to cirrhosis. Cardiac symptoms may develop without warning. Cardiac toxicity manifests as congestive heart failure or as a restrictive cardiomyopathy and angina (without coronary occlusion). Pancreatic and pituitary dysfunction, insulin-dependent diabetes, delayed growth, shortened stature and delayed sexual maturation are all consequences of excess iron accumulation in the endocrine system. Clinical manifestations depend not only on the amount of excess iron, but also on the rate of iron accumulation, the ascorbate level, alcohol use, and viral hepatitis infection.

**Diagnosis:** The diagnosis of iron overload requires clinical correlation of the signs and symptoms noted above, with an estimation of total body iron accumulation/stores. Estimating total body iron stores is imprecise, and can be accomplished via direct and indirect methods. Indirect methods include serum ferritin measurement, which is simple and inexpensive but also the most inaccurate, and via magnetic resonance imaging (MRI) and a supraconducting quantum interface device (SQUID), which adds accuracy but is costly and not widely available. Cardiac iron can also be indirectly estimated using MRI. Direct measurements can be accomplished via liver biopsy, which maybe more accurate but is invasive and not without risk. It has been shown that patients with liver iron levels greater than 15 mg of iron per gram of liver are at increased risk for hepatic fibrosis, as well as other complications of iron overload. Patients with iron overload should have cardiac, hepatic and endocrine gland function assessed.

**Management:** Without treatment with iron chelating agents, chronically transfused patients will eventually succumb to the complications of transfusional iron overload. Therefore, iron chelation therapy is extremely important, and should be instituted when liver iron concentrations are $\sim 7$ mg Fe/g dry weight or serum ferritin levels are $\sim 1000\,\mu g/l$, which occurs when approximately 120 ml of RBCs/kg of body weight are transfused. Iron chelating agents directly bind iron, and the Fe:chelator complex is excreted in the urine or feces; this is not efficient but is effective over long periods of time. Iron chelation therapy, when accomplished correctly, can and will significantly reduce the complications of iron overload, and can lead to an improved quality of life. The therapeutic goal is to maintain liver iron concentration at $\sim 5$ mg/g liver dry weight or the serum ferritin level at $1000$–$1500\,\mu g/l$.

Both subcutaneous (deferoxamine) and oral (deferipone; deferasirox) iron chelating agents are now available. Deferasirox is preferred because it has a longer half-life and can be taken orally in water or juice, thus increasing compliance with this long-term, chronic therapy. Side effects of deferasirox are minimal, and can include transient gastrointestinal symptoms and skin rash, and mild and reversible increases in serum creatinine and alanine aminotransferase.

### Recommended Reading

Cappellini MD, Cohen A, Piga A et al. (2006). A phase 3 study of deferasirox (ICL670), a once-daily oral iron chelator, in patients with β-thalassemia. *Blood* **107**, 3455–3462.

Vichinsky E, Onyekwere O, Porter J et al. (2006). A randomized comparison of deferasirox versus dereroxamine for the treatment of transfusional iron overload in sickle cell disease. *Br J Haematol* **136**, 501–508.

# CHAPTER 66

# Transfusion Transmitted Diseases

Beth H. Shaz, MD

Transfusion transmitted diseases (TTD) can be caused by viruses, protozoa and prions. A comprehensive, but not all-inclusive, list of potentially TTDs can be found in Table 66.1. While bacteria are transmitted by transfusion, it is usually the products of the bacteria in the donor product (endotoxin) rather than the transfer of infectious bacteria that are the cause of posttransfusion bacteria-related complications. For this reason, bacterial contamination of blood products and methods to limit and detect bacteria in platelet products are discussed in Chapters 59 and 17 of this book, respectively.

Mitigation of transfusion transmission of infectious agents is largely based on donor selection and donor testing in the US, and around the world. Donor selection and testing also occupy individual chapters in this text, including Chapter 5, and Chapters 11–16. Indeed, these strategies have been greatly effective in lowering the residual risk of TTDs in the US, as shown in Figure 66.1. Finally, pathogen-reduction

| TABLE 66.1 Potential Transfusion Transmitted Agents | |
|---|---|
| **Viruses** | Simian foamy virus (SFV) |
| Hepatitis A virus | Chikungunya |
| Hepatitis B virus | Anaplasma |
| Hepatitis C virus | **Protozoa** |
| Hepatitis D virus | *Plasmodium* spp. (malaria) |
| Hepatitis G virus | *Trypanosoma cruzi* (Chagas' disease) |
| TTV and SEN-V | *Toxoplasma gondii* (toxoplasmosis) |
| Human Immunodeficiency Virus (HIV) | *Babesia microti/divergens* (babesiosis) |
| Human T-cell lymphotropic virus (HTLV) 1 and 2 | *Leishmania* spp. (leishmaniasis) |
| Cytomegalovirus (CMV) | **Prions** |
| Epstein-Barr virus (EBV) | Transmissible spongiform encephalopathies (Creutzfeldt-Jakob disease and others) |
| Human herpes virus (HHV)-8 | |
| Lymphocytic Choriomeningitis (LCMV) | |
| Severe acute respiratory syndrome (SARS) | |
| Parvovirus B19 | |
| West Nile Virus | |
| Monkeypox virus | |
| Viral hemorrhagic fevers | |

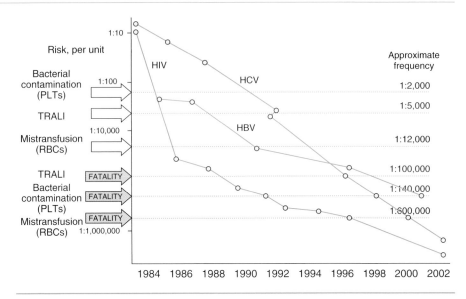

**FIGURE 66.1** Comparison of transfusion risks and their evolution over time. HBV hepatitis B virus; HCV, hepatitis C virus; HIV, human immunodeficiency virus; PLTs, platelets; RBCs, red blood cells; TRALI = transfusion related acute lung injury; From AuBuchon JP. (2004). Emily Cooley Memorial Award. Managing change to improve transfusion safety. *Transfusion* **44**, 1377–1383.

or -inactivation technologies have the potential to eliminate virtually all TTD, though these technologies are only approved for use for platelets and plasma, and then only outside of the US.

**Hepatitis Viruses:** The hepatitis viruses include hepadnaviruses, flaviviruses and picornaviruses, which have all been reported to be transmitted through transfusion, and result in inflammation of the liver as the primary clinical manifestation (other viruses and agents result in hepatitis but it is not the primary disease), with symptoms of jaundice, dark urine, hepatomegaly, anorexia, malaise, fever, nausea, abdominal pain and vomiting. Hepatitis B and hepatitis C, which are transmitted parenterally (i.e. through contact with blood and other body fluids), are the typical hepatitis viruses transmitted through transfusion, but hepatitis A (HAV), which is traditionally transmitted through the fecal–oral route, has rarely been transmitted when sufficient amounts of virus are present.

**Hepatitis A:** HAV in the *Picornaviridae* family, usually results in a mild, self-limited disease with a mortality rate of less than 0.2%. Hepatitis A does not result in chronic infection. The virus is typically transmitted through the fecal–oral route, resulting in foodborne outbreaks. Prophylaxis can be provided through immune serum globulin, and prevention through vaccination is available. The seropositivity rate in the US is approximately 33%.

As most individuals are quite ill during the viremic phase of HAV infection, they most often will not be feeling well enough to donate blood, or will be deferred due to fever or sickness. The risk of transfusion transmitted hepatitis A is less than 1 case per million transfused products. Solvent-detergent processes used to make pathogen-reduced plasma from plasma pooled products fail fully to inactivate the non-enveloped HAV, and as a result transmission has occurred. This has resulted in additional inactivation steps and testing for HAV in products manufactured from large pools of plasma. In addition, recipients of pooled plasma products (e.g. hemophilia patients) should receive the hepatitis A vaccine.

Hepatitis B: Hepatitis B virus (HBV) is a DNA virus and member of the *Hepadnaviridae* family, which is transmitted parenterally, sexually and perinatally. In the US (prevalence 5.6%) and other low-prevalence countries, most infections are horizontal from adult to adult; in countries with a high prevalence of HBV, both horizontal and vertical (i.e. perinatal) infections occur. Widespread vaccination of infants and high-risk adults has resulted in a decreased incidence of the disease (from 260,000 new infections/year to 60,000/year in the US). In addition, hepatitis B immune globulin is available for post-exposure prophylaxis, or for infants born to infected mothers. TT-HBV may result in either an acute infection with subsequent clearance of the virus and immunity, or chronic infection with persistence of viral replication. Chronic infection may resolve with the development of immunity or may reactivate, resulting in further acute disease.

*Acute Infection:* The incubation period is usually 60–150 days. Most individuals are asymptomatic, but 30–50% of infected persons $\geq$ 5 years of age will experience jaundice, fever, loss of appetite, nausea, vomiting and abdominal pain, and 0.5–1% of infected individuals, usually $\geq$ 60 years of age, will have a fulminant acute infection resulting in death. Serologic results in acutely infected adults demonstrate HBsAg positive, anti-HBc positive, IgM anti-HBc positive, and anti-HBs negative. The formation of anti-HBs and the disappearance of HBsAg indicate development of immunity without further clinical disease.

*Chronic Infection:* The likelihood of chronic infection is related to the age of the patient at the time of infection, such that 90% of persons who were infected in infancy, yet only 6% of persons who were infected after age 5 develop chronic disease. Chronic infection may be asymptomatic, while some individuals will develop severe and fatal disease from cirrhosis and hepatocellular carcinoma. Extrahepatic manifestations, which most often result from cryoglobulinemia, include rashes, arthritis, vasculitis and glomerulonephritis. Serologic results in patients with chronic active hepatitis demonstrate HBsAg positive, anti-HBc positive, IgM anti-HBc negative, and anti-HBs negative. Treatment for chronic infection includes interferon, lamivudine, adefovir, and additional antiviral medications.

*Risk of Transfusion Transmission:* In the US, with the current use of testing for HBsAg and anti-HBc, the estimated residual risk of HBV transmission is 1 : 205,000 among *repeat* donors, or 1 : 144,000 among *all* donors, in the absence of nucleic acid testing (NAT). These estimates depend on assumptions, extrapolations and reporting of posttransfusion infection. Transfusion transmitted HBV may either be under-reported or over-reported, as many cases of posttransfusion HBV infection may have not been contracted through transfusion (one report found only 1 in 59 cases could be unequivocally linked to transfusion).

NAT testing for HBV has been implemented in some countries where the prevalence is especially high. In the US, the minipool approach to NAT has the same sensitivity as the most recent generation of serologic tests for HBsAg, as HBsAg rises to high titers very quickly and at approximately the same time as HBV DNA (i.e. minipool NAT testing does not significantly decrease the window period).

**Hepatitis D:** The infectious form of hepatitis D is coated with HBsAg, and therefore hepatitis D is only infectious in the presence of active HBV infection. Co-infection with hepatitis D and B results in more serious disease than infection with hepatitis B alone. Current measures to detect infectivity for HBV also target the elimination of hepatitis D from transfusion transmission.

**Hepatitis C:** Hepatitis C is an RNA virus in the *Flaviviridae* family and is parenterally transmitted, especially through blood transfusions (prior to testing) and intravenous drug use. In the 1970s, 10% of blood transfusion recipients had evidence of hepatitis C infection. With the use of the current NAT testing, risk is estimated at 1 : 1.4 million products or less. The incidence of new infections in the US has declined from 240,000 per year in the 1980s to 19,000 per year in 2006 (http://www.cdc.gov/hepatitis/HCV.htm). The current seroprevalence in the US is 1.8% overall and 79% in intravenous drug users.

*Acute Infection:* Acute infection with hepatitis C virus (HCV) is asymptomatic in 80% of patients. Approximately 20% will have symptoms (fever, jaundice, loss of appetite, fatigue and nausea) of acute infection after an incubation period of 7–8 weeks.

*Chronic Infection:* Chronic infection develops in 75–85% of infected individuals, with cirrhosis developing in 20–30% (after an average of 20 years) and hepatocellular carcinoma (after an average of 30 years). HCV is the leading indication for liver transplantation in the US. Current treatment for chronic infection includes interferon and ribavirin.

*Risk of Transfusion Transmission:* With the use of current enzyme immunoassay tests and NAT, the current risk is less frequently than 1 : 1.4 million donations. A positive test result for HCV antibodies in a donor requires a "lookback" per the FDA, to locate, notify, test and, if appropriate, treat recipients of products donated previously by the same donor, at a time when they did not test positive for HCV either because of having not yet acquired the infection or having markers of infection below the limits of test detection (http://www.fda.gov/cber/gdlns/hcvlkbk.htm). The yield of lookback has been very low since the implementation of NAT testing *circa* 2000.

**Hepatitis E:** Hepatitis E results in a self-limited hepatitis, and is typically transmitted by the fecal–oral route. The hepatitis E virus is found predominantly in tropical countries. Rare cases of transfusion-transmitted hepatitis E have recently been reported.

**Hepatitis G:** Hepatitis G is common in the normal population; 3–15% of normal individuals have antibodies to hepatitis G, and viral RNA is detected in 1–3%. The virus is transfusion transmitted, demonstrated by a high prevalence of infection in multiply transfused individuals. Currently, no disease is associated with this virus, and thus no testing is required.

**TTV and SEN-V:** TTV and SEN-V are widely distributed and are transmissible by transfusion, but currently there is little evidence that they cause disease in recipients. Therefore, testing strategies to mitigate transfusion transmission of these viruses is not warranted.

**Retroviruses:** Retroviruses are RNA viruses with the presence of viral particle-associated reverse transcriptase and a unique replication cycle. The virus particles attach to the cell membrane and subsequently enter the host cell, then the reverse transcriptase enzyme copies viral RNA into cDNA (complementary double-stranded DNA) and the cDNA is then integrated into the host cell's genome. Subsequent transcription, processing and translation of viral genes are mediated by the host cell enzymes. Particles then bud from the cell membrane and infect other cells. In addition, the virus can spread by fusion of infected and uninfected cells, or by replication of the integrated viral DNA during mitosis or meiosis.

**Human Immunodeficiency Virus:** Human immunodeficiency virus (HIV) is a lentivirus, which is a subgroup of the retrovirus family. HIV-1 was discovered in 1983, and the first test was licensed for donor screening in 1985. HIV transmission by intravenous administration of infected blood products is highly efficient. Of hemophiliacs treated with Factor VIII in the early 1980s, 50% were infected; 100% of those who received more than 500,000 units of Factor VIII were infected. In the San Francisco area in 1982, the risk of HIV was ~1 : 100 transfused products.

HIV infection can be transmitted through sexual contact, childbirth, breast-feeding and parenteral exposure to blood. The CDC reported that in 2006 in the US the highest incidence of new disease was in male African Americans, aged 25–44, who had male-to-male sex; of new infections, 50% were from male-to-male sex, 33% from high-risk heterosexual contact, and 13% from intravenous drug use (http://www.cdc.gov/hiv/).

HIV-1 and HIV-2 infection both cause AIDS. The HIV-1 family is divided into main (M), outlier (O) and non-M, non-O (N) groups. Group M has 11 distinct subtypes or clades (A–K). In the US, clade B is almost exclusively prevalent. The greatest genetic diversity is in central Africa. Group O is most common in Cameroon and surrounding West African countries (where it represents 1–2% of HIV infections). Group O infection in the US is very rare, and is usually found in immigrants. Previous generations of HIV antibody assays did not reliably detect group O, but current assays have increased sensitivity to group O and other unusual variants. Still, the FDA continues to recommend permanent deferral of blood donors who were born, resided or traveled in West Africa since 1977, or who has had sexual contact with someone who fulfills these criteria. HIV-2 is also rare in the US (one infected donor identified out of 7.2 million donations), with no reported cases of HIV-2 transfusion transmission in the US.

*Infection:* Approximately 60% of acute HIV infections result in a non-specific flu-like illness with an incubation period of 2–4 weeks. The acute infection resolves in weeks to a few months, resulting in an asymptomatic period that may last years. During this period the HIV viremia persists, and the number of CD4+ lymphocytes (the primary target of HIV) gradually declines. This loss of CD4+ lymphocytes results in opportunistic infections, and in addition there are direct viral effects on multiple organs; these together result in death after, on average, 8–10 years. The course of the disease has changed dramatically with the advent of potent antiretroviral therapy, which has greatly prolonged survival. However, these medications do not eradicate HIV, and have multiple side-effects. Moreover, resistant viral strains have developed which add to the difficulty in treating HIV-infected patients.

*Risk of Transfusion Transmission:* Current testing for HIV infection in blood donors includes serologic assays for antibodies to HIV-1/HIV-2 and by minipool NAT. The current estimates of HIV transmission are less frequently 1 : 2 million products tested, with expected (but unmeasurable) frequencies approximating 1 : 5 to 1 : 8 million products transfused (US). Identification of persons who have received blood products from donors who are later found to test positive for HIV (referred to as "lookback") is mandated by the FDA (CFR 610.46).

**Human T-cell Lymphotropic Virus:** Human T-cell lymphotropic virus (HTLV) is transmitted by transfusion. HTLV-1 predominately infects CD4+ lymphocytes while HTLV-2 infects preferentially CD8+ lymphocytes, and to a lesser extent infects CD4+ lymphocytes, B lymphocytes and macrophages. The seroprevalence in blood donors in the US is approximately 10–20 per 100,000 donors. The primary modes of transfusion are vertical transmission from mother to child, secondary to breast feeding (the infection rate declined from 30% to 3% with the discontinuation of infected mother breast-feeding), sexual transmission and parenteral exposure (intravenous drug use).

*Infection:* Most HTLV infections are asymptomatic, but there is a 2–4% risk of disease that may develop up to 40 years after infection. HTLV-1 is associated with a CD4+ lymphoma, adult T-cell leukemia/lymphoma (ATLL). The risk of ATLL in individuals infected at birth is 4% in their lifetime, with a lower risk in those infected during adulthood (i.e. those who acquire HTLV from transfusion). ATLL has a high mortality rate within 1 year of disease onset. HTLV-1 and HTLV-2 are associated with tropical spastic paraparesis (also known as HTLV-1 associated myelopathy [HAM]; TSP), which is a slowly progressive myelopathy characterized by spastic paraparesis of the lower extremities, hyperreflexia, and bowel and bladder symptomatology. The risk of TSP is about 2% in HTLV-1 positive individuals, with a similar or lower risk in HTLV-2 positive individuals. Other diseases associated with HTLV-1 or HTLV-2 infection include lymphocytic pneumonitis, uveitis, polymyositis, arthritis, bronchitis, dermatitis, and other infectious syndromes.

*Risk of Transfusion Transmission:* RBCs, platelets and whole blood, but not fresh frozen plasma, have resulted in seroconversion of transfusion recipients. Products stored for greater than 7 days before transfusion are less likely to transmit the virus. In the US, seroconversion rates in individuals who receive seropositive cellular components have been reported from 14% to 30%. In a lookback study of recipients from 1999 to 2005 from the ARC, only 38 donors seroconverted in that timeframe; these individuals donated 31 cellular components. None of the four alive recipients who agreed to testing for HTLV was seropositive. The risk of transfusion transmission is estimated to be 1 per 3 million products transfused.

**Herpesviruses:** The herpesviruses have double-stranded DNA, and express viral enzymes that participate in DNA synthesis and nucleic acid metabolism; the viral

DNA synthesis and packaging is confined to the host cell nucleus, the infected cell is destroyed during active viral replication, and the virus is capable of latency indefinitely. CMV is the herpesvirus that is most relevant to transfusion medicine, but other leukocytotropic herpesviruses (EBV, HHV-6, HHV-7 and HHV-8) may also contaminate blood products.

**Cytomegalovirus:** Cytomegalovirus (CMV) is transmitted through transfusion and HPC and solid-organ transplantation. CMV transmission in the community is usually through close contact with a person shedding CMV. The seroconversion rate in blood donors is approximately 1%/year (rates of up to 13%/year in adolescents have been reported). Prevalence ranges from 40 to 90%; the rate increases with age, and is higher in lower socioeconomic groups, urban areas and developing countries. Approximately 50% of US blood donors are CMV seropositive.

*Infection:* Immunocompetent individuals have a mild self-limited disease course, with fever, malaise, hepatosplenomegaly and rash. The immune response does not eliminate the virus, but the virus becomes latent in the peripheral blood leukocytes. Transplacental infection in 5–15% of infected infants results in intrauterine growth retardation, deafness, mental retardation, blindness and thrombocytopenic bleeding. Infection in immunocompromised patients (including premature infants, recipients of solid-organ or HPC transplantation, and AIDS patients) can lead to pneumonitis, hepatitis, retinitis and multisystem organ failure, which may result in death. Treatment of CMV infection includes antivirals such as ganciclovir, cidofovir and foscarnet.

CMV infection can be detected through anti-CMV antibodies, CMV antigenemia assay (which uses immunostaining to identify and quantitate peripheral blood leukocytes containing CMV proteins), and CMV PCR. PCR has allowed earlier detection of CMV, and has largely replaced the need for the CMV antigenemia assay.

*Risk of Transfusion Transmission:* Leukocytes are the primary mode of transfusion transmission of CMV, and therefore leukoreduction greatly decreases the risk of CMV transmission. Transfusion can lead to active CMV infection in a recipient by transfusion transmitted CMV where a seronegative recipient is transfused with a CMV-infected product; reactivated CMV infection where a CMV-seropositive recipient experiences reactivation of their latent infection after transfusion of a CMV-negative product; and CMV superinfection, when a seropositive recipient contracts a new strain of CMV from a CMV-positive product. Transfusion transmitted CMV is an important cause of morbidity and mortality in immunocompromised patients: 13–37% of immunocompromised patients, including low birth-weight neonates, will contract CMV from transfusion of unscreened and unleukoreduced blood products. Up to 4% of recipients who receive seronegative blood products have acquired CMV, and up to 3% of recipients who receive leukoreduced blood products have acquired CMV, based on a number of different reported studies. Indications for CMV-safe products, defined as leukoreduced and/or anti-CMV negative, are reviewed in Chapter 38.

**Epstein Barr Virus:** Epstein Barr virus (EBV) is associated with a variety of diseases, including infectious mononucleosis, Burkitt's lymphoma and nasopharyngeal carcinoma. EBV infection is usually through infected saliva. Acute infection in children is asymptomatic, or characterized by a sore throat and enlarged lymph nodes. Acute infection in adults results in infectious mononucleosis with fever, tonsillar infection, enlarged lymph nodes, hematologic and immunologic abnormalities, hepatitis or other organ involvement. EBV infects B lymphocytes, where it remains latent. Occasional cases of posttransfusion EBV infection have been reported, but screening for EBV is not performed because of the high prevalence of seropositivity (90%) and because donors with active infection (infectious mononucleosis) are usually symptomatic. Leukoreduction may prevent or decrease the risk of transfusion transmission.

**HHV-8:** HHV-8 (also known as Kaposi's Sarcoma Herpesvirus) is associated with Kaposi's sarcoma, primary effusion lymphoma and multicentric Castleman's disease. HHV-8 is primarily transmitted through sexual contact. The incidence in US blood donors is 2.4%. One study in the US demonstrated a 0.082% risk of seroconversion per transfused component. A study in Uganda, where the HHV-8 seroprevalence rate is approximately 40%, showed the seroconversion rate after transfusion with a seropositive product to be 2.8%. Proposed methods to reduce the risk of transfusion transmission include leukoreduction, donor testing and pathogen reduction.

## Other Viruses:

**Parvovirus B19:** Infection with parvovirus B19 usually results in asymptomatic or mild symptoms of rash, vomiting, aching joints and limbs, fatigue and malaise (erythema infectiosum or fifth disease). Infection in sickle cell disease and thalassemia patients may result in aplastic crisis. In immunocompromised individuals (especially HIV-positive individuals), parvovirus infection may result in aplastic anemia. Infection during pregnancy may result in severe fetal anemia or malformation in the infants. The seropositivity rate in blood donors is 30–60%. It is transmitted either through release of virus from the upper respiratory tract, or parenterally. Viremia appears within the first week and persists for 1–2 weeks; chronic infection does not occur. Rare cases of transfusion transmission through blood products have been reported. Parvovirus is not destroyed by solvent-detergent treatment or heat inactivation. Because of seroconversion in recipients of solvent-detergent treated plasma, additional viral inactivation steps as well as NAT testing were implemented for plasma that would be pooled and used to manufacture plasma derivatives.

**West Nile Virus:** West Nile virus (WNV) is a flavivirus that is primarily transmitted through mosquitoes, with birds as the intermediate host. WNV first appeared in the US in New York in 1999, and rapidly expanded its geographic area within 3 years. Transfusion transmission resulted in 23 infections in 2002 during an outbreak in the US where 4156 individuals were infected and there were 284 fatalities.

*Infection:* Infection is often asymptomatic, and does not result in chronic infection with viremia persisting less than 28 days. The incubation period is approximately 3–14 days followed by a range of symptoms including mild fever, headache, rash, eye pain, vomiting, lymphocytopenia, muscle weakness, flaccid paralysis, poliomyelitis and peripheral demyelination. Approximately 1 in 150 infections results in severe neurologic disease, especially in those over the age of 50 years. Prevention is through avoiding mosquitoes and mosquito bites.

*Risk of Transfusion Transmission:* After the documentation of transfusion transmission in 2002, WNV NAT testing was implemented in 2003. Minipool NAT testing only resulted in seven transfusion-transmission cases in 2003–2004. These breakthrough cases resulted in a combination of minipool NAT testing during the non-season coupled with more sensitive individual donor NAT testing in epidemic locations during epidemic times. This testing strategy has resulted in no documented WNV transfusion since the change in testing.

## Protozoa:

*Plasmodium* spp. (Malaria): There are four known *Plasmodium* species that result in malaria in humans: *P. falciparum, P. malariae, P. vivax* and *P. ovale.* The incubation periods range from 12–30 days, depending on the species. *P. falciparum* results in the most serious infection, which can be fatal, compared to the other species. Transmission to humans is through the mosquito, with the lifecycle split between the mosquito and humans. The merozoite form infects the RBCs, where it replicates, resulting in the RBC bursting and releasing organisms into the blood to infect other RBCs. Infection usually lasts for 1–2 years, but can last for up to 30 years in the case of *P. malariae.*

*Infection:* Malarial infection results in fever, chills, headache, hemolytic anemia and splenomegaly. Diagnosis is through examination of thick- and thin-blood smears. A variety of medications are used for prophylaxis or treatment of malaria.

*Risk of Transfusion Transmission:* The risk of malaria is 0.25 cases per million transfusions. Symptoms of infection include fever, chills, headache and hemolysis, and occur a week to several months after transfusion. Transfusion transmitted infection is rarely fatal, and usually results from transmission of *P. falciparum.* Prevention of malarial transmission is through donor deferral, which requires that persons who have had malaria in the preceding 3 years are deferred, travelers to endemic areas are deferred a year, and immigrants of endemic areas are deferred for 3 years after leaving the area. Malarial risk areas are available through the CDC website (http://www.cdc.gov/malaria/risk_map/).

*Trypanosoma Cruzi* (Chagas' Disease): Chagas' disease is confined mainly to Mexico and South and Central America, with increased prevalence in the southern US from immigrants from endemic areas. It is transmitted to humans through the bite of the reduvid bugs, and the lifecycle is split between the two hosts. Acute infections are asymptomatic or mild. Rarely, the site of entry evolves into an erythematous nodule called a chagoma; this may be accompanied by fever and hepatosplenomegaly. Younger children may develop acute myocarditis or meningioencephalitis. Acute infection resolves without treatment, but lifelong low-level parasitemia persists. Of chronically infected individuals, 20–40% develop cardiac or gastrointestinal symptoms years to decades later. Transmission through transfusion is the second major source of human infection, especially in endemic areas. Seven cases of transfusion transmission in the

US and Canada have been reported. The FDA licensed a screening test for antibodies to *T. cruzi* in blood donors in 2006. Testing of about 65% of the US blood supply has demonstrated that 1 : 25,000–35,000 donors are confirmed positive, with the majority of infected donors in Florida and California.

*Toxoplasma Gondii* (Toxoplasmosis): Infection with *T. gondii* occurs in up to 95% of adults in some countries. Cats and mice are intermediate hosts. Acute infection in healthy individuals is usually asymptomatic, but infection in immunocompromised individuals can result in severe disease with CNS involvement, myocarditis and pneumonia. Congenital infection can give rise to serious complications, including liver, CNS disease, abortion or stillbirth. Acute infection resolves with antibody formation, but the organism remains latent in leukocytes. Transmission by transfusion has been documented in immunocompromised individuals. Prevention of transfusion transmission appears possible by leukoreduction, but this has not been evaluated.

*Babesia Microti/Divergens* (Babesiosis): *Babesia microti* in North America and *B. divergens* in Europe are transmitted by tick bite. Symptoms of babesiosis range from mild to severe illness with a hemolysis and fever, and infection lasting more than a year can occur. RBCs are the site of replication. More than 50 cases of transfusion transmission of babesiosis have occurred in the US, with an estimated risk of up to 1 in 1000 in parts of Connecticut. Furthermore, PCR studies reveal that ~1 in 1800 donors in Connecticut are parasitemic. Transfusion transmitted disease occurs with fever developing 1–4 weeks after transfusion; this may be associated with chills, headache, hemolysis, hemoglobinuria and, rarely, life-threatening hemolytic anemia, renal failure and coagulopathy. Asplenic, elderly or severely immunocompromised patients are at risk for more severe infection. Donor selection is currently used to prevent transfusion transmission, including not collecting blood in areas where the disease vectors are endemic during the spring and summer months. Future strategies to prevent disease transmission, especially in high-risk areas, potentially include serologic or NAT testing.

*Leishmania* spp. (Leishmaniasis): There are three forms of leishmaniasis: cutaneous, mucocutaneous and visceral (kala-azar). The organism is transmitted through the bite of sandflies, with each *Leishmania* species restricted to a particular *Phlebotomus* species. The reservoir for the organism includes rodents and small wild mammals. The lifecycle is split between the sandfly and the mammal. In the human the organism invades the reticuloendothelial system, where it replicates and then is released back into the blood. Parasitemia is generally transient and of low levels, and therefore the risk of transfusion transmission is low. Outside of the US, transfusion transmission of *L. donovani*, which causes visceral leishmaniasis, has been reported. Veterans who served in the Persian Gulf are deferred from donating for 1 year upon leaving.

**Prions:** Prion disease results from the benign form of the prion protein (PrP) changing to an insoluble protease-resistant form (PrP$^{Sc}$), which leads to the formation of plaques in the brain.

**Transmissible Spongiform Encephalopathies (Creutzfeldt-Jakob Disease):**
Bovine spongiform encephalopathy (BSE) was initially described in cattle in the UK in 1986. BSE was transmitted through the food chain through meat or bone meal. A ban on ruminant protein in cattle feed has resulted in decreased incidence of BSE. In 1995, BSE was transmitted to humans through the food chain and resulted in variant Creutzfeldt-Jakob disease (vCJD) in the UK. vCJD differs from classical CJD by an earlier age of onset, slower disease progression, and higher levels of $PrP^{Sc}$ in the brain. The disease presents with behavioral changes and dysasthesia, and progresses to cerebellar ataxia, dementia and death in 7–38 months. The majority of cases of vCJD have appeared in the UK, but there have also been reports in other European countries. At least four cases of transfusion transmission of vCJD have occurred in the UK. It is estimated that transfusion transmission by donors who develop vCJD within several years of donation is about 14% for recipients who survive longer than 5 years posttransfusion. Prevention includes deferral for donors who have resided in the UK or Europe for over 6 months, and the use of filters to remove prions; these remove approximately half the infectivity.

**Other Emerging Infections:** There are numerous potentially emerging infections (Dengue fever, SARS, influenza, LCMV) which require continuous surveillance of the blood supply and evaluations of interventions.

## Recommended Reading

Dodd RY. (2007). Current risk for transfusion transmitted infections. *Curr Opin Hematol* **14**, 671–676.

Hladik W, Dollard SC, Mermin J *et al.* (2006). Transmission of human herpesvirus 8 by blood transfusion. *N Engl Med* **355**, 1331–1338.

Stramer SL. (2007). Current risks of transfusion-transmitted agents. *Arch Pathol Lab Med* **131**, 702–707.

Stramer SL, Foster GA, Dodd RY. (2006). Effectiveness of human T-lymphotropic virus (HTLV) recipient tracing (lookback) and the current HTLV-I and -II confirmatory algorithm, 1999 to 2004. *Transfusion* **46**, 703–707.

Zou S, Fang CT, Schonber LB. (2008). Transfusion transmission of human prion diseases. *Transfus Med Rev* **22**, 58–69.

# CHAPTER 67

# Overview of Therapeutic Apheresis

Beth H. Shaz, MD

Therapeutic apheresis (TA) refers to the removal of whole blood from a patient, separation of the whole blood into one or more fractions, followed by the removal of the indicated fraction which will be discarded, and the infusion of a fluid to replace the lost volume. When TA is used to remove plasma and the pathologic antibodies or substances within, this is commonly referred to as *therapeutic plasma exchange* or TPE. When a cellular fraction is removed and discarded, the procedure is referred to as *therapeutic cytapheresis*, which is discussed in Chapters 69–71.

TA can be performed manually, but is much more commonly accomplished via a number of FDA-approved continuous or semi-continuous automated cell separators or apheresis machines. In addition, several specialized apheresis machines remove (or remove, treat and return) targeted blood elements, including lipids in the treatment of familial hyperlipidemia, immunoglobulins in the treatment of rheumatoid arthritis, and neoplasic monocytes in cutaneous T-cell lymphoma.

TA is used in the treatment of a wide variety of diseases. The American Society of Apheresis (ASFA) has created evidence-based guidelines for the use of TA in over 50 disorders (Table 67.1).

This chapter focuses on the basic principles of TA, including methodology, volumes processed, choice of anticoagulant, vascular access, adverse effects, and the approach to a TA request.

Methods: Current apheresis medical devices allow the separation of whole blood into component fractions with removal of the desired fraction, and return of the remaining components by in-line automated technology. These systems require the use of an anticoagulant solution in order to maintain flow through plastic tubing and the blood-containing centrifuge elements of the devices. Heparin and/or citrate may be used; citrate in the form of ACD is much more common. Sterile plastic disposable sets are made for each of the available cell separators which can accommodate flow rates of up to 150 ml/min using either peripheral or central venous access.

Separation of blood components can occur through either centrifugation or filtration, and operation is with either continuous or intermittent flow. Centrifugation allows separation of the component fraction base on differential density, with RBCs having the greatest relative density, followed by neocytes, granulocytes, mononuclear cells, platelets and, lastly, plasma.

Filtration devices separate component fractions based on the size – platelets 3 μm, RBCs 7 μm, lymphocytes 10 μm, and granulocytes 13 μm – while plasma flows freely through the filter/membrane. Cascade filtration uses an arrangement of two or more columns in series, with progressively smaller pore sizes allowing separation and removal of some components from the plasma fraction. This technology has been primarily applied to the removal of plasma lipids and cryoglobulins.

## TABLE 67.1  Indications Categories for Therapeutic Apheresis

|  | TA Modality | Category |
|---|---|---|
| *ABO-incompatible hematopoietic progenitor cell transplantation* | | |
| | Plasma exchange | II |
| *ABO-incompatible solid organ transplantation* | | |
| Kidney | Plasma exchange | II |
| Heart (infants) | Plasma exchange | II |
| Liver | Plasma exchange | II |
| *Acute disseminated encephalomyelitis* | | |
| | Plasma exchange | III |
| *Acute inflammatory demyelinating polyneuropathy (Guillain-Barré syndrome)* | | |
| | Plasma exchange | I |
| *Acute liver failure* | | |
| | Plasma exchange | III |
| *Amyloidosis, systemic* | | |
| | Plasma exchange | IV |
| *Amyotrophic lateral sclerosis* | | |
| | Plasma exchange | IV |
| *ANCA-associated rapidly progressive glomerulonephritis (Wegener's granulomatosis)* | | |
| | Plasma exchange | II |
| *Anti-glomerular basement membrane disease (Goodpasture's syndrome)* | | |
| | Plasma exchange | I |
| *Aplastic anemia; pure red cell aplasia* | | |
| | Plasma exchange | III |
| *Autoimmune hemolytic anemia: warm autoimmune hemolytic anemia; cold agglutinin disease* | | |
| | Plasma exchange | III |
| *Babesiosis* | | |
| Severe | Erythrocytapheresis | II |
| *Catastrophic antiphospholipid syndrome* | | |
| | Plasma exchange | III |
| *Coagulation factor inhibitors* | | |
| | Immunoadsorption | III |
| | Plasma exchange | III |

**TABLE 67.1** *(Continued)*

| | | |
|---|---|---|
| *Chronic inflammatory demyelinating polyradiculoneuropathy* | | |
| | Plasma exchange | I |
| *Cutaneous T-cell lymphoma; mycosis fungoides* | | |
| Erythrodermic | Extracorporeal photopheresis | I |
| Non-erythrodermic | Extracorporeal photopheresis | IV |
| *Cryoglobulinemia* | | |
| | Plasma exchange | I |
| *Dermatomyositis or polymyositis* | | |
| | Plasma exchange | IV |
| | Leukocytapheresis | IV |
| *Erythrocytosis; polycythemia vera* | | |
| Symptomatic | Erythrocytapheresis | II |
| *Familial hypercholesterolemia* | | |
| Homozygotes | Selective removal | I |
| Heterozygotes | Selective removal | II |
| | Plasma exchange | II |
| *Focal segmental glomerulosclerosis* | | |
| Primary | Plasma exchange | III |
| Secondary | Plasma exchange | III |
| *Graft-versus-host disease* | | |
| Skin | Extracorporeal photopheresis | II |
| Non-skin | Extracorporeal photopheresis | III |
| *Heart transplant rejection* | | |
| Prophylaxis | Extracorporeal photopheresis | I |
| Treatment | Extracorporeal photopheresis | II |
| | Plasma exchange | III |
| *Hemolytic uremic syndrome; thrombotic microangiopathy; transplant associated microangiopathy* | | |
| Idiopathic HUS | Plasma exchange | III |
| Transplant-associated microangiopathy | Plasma exchange | III |
| Diarrhea associated pediatric | Plasma exchange | IV |
| Other | Plasma exchange | III |
| *Hyperleukocytosis* | | |
| Leukostasis | Leukocytapheresis | I |
| Prophylaxis | Leukocytapheresis | III |

*(Continued)*

**TABLE 67.1** *(Continued)*

|  | TA modality | Category |
|---|---|---|
| *Hypertriglyceridemic pancreatitis* |  |  |
|  | Plasma exchange | III |
| *Hyperviscosity in monoclonal gammopathies* |  |  |
|  | Plasma exchange | III |
| *Idiopathic thrombocytopenic purpura* |  |  |
| Refractory | Immunoadsorption | II |
| Refractory or non-refractory | Plasma exchange | IV |
| *Inclusion body myositis* |  |  |
|  | Plasma exchange | IV |
|  | Leukocytapheresis | IV |
| *Lambert-Eaton myasthenic syndrome* |  |  |
|  | Plasma exchange | II |
| *Lung transplant* |  |  |
|  | Extracorporeal photopheresis | III |
| *Malaria* |  |  |
| Severe | Erythrocytapheresis | II |
| *Multiple sclerosis* |  |  |
| Acute CNS inflammatory demyelinating disease | Plasma exchange | II |
| Chronic progressive | Plasma exchange | III |
| Devic's syndrome | Plasma exchange | III |
| *Myasthenia gravis* |  |  |
|  | Plasma exchange | I |
| *Myeloma and acute renal failure* |  |  |
|  | Plasma exchange | III |
| *Overdose and poisoning* |  |  |
| Mushroom poisoning | Plasma exchange | II |
| Other compounds | Plasma exchange | III |
| *Paraneoplastic neurologic syndromes* |  |  |
|  | Plasma exchange | III |
|  | Immunoadsorption | III |
| *Paraproteinemic polyneuropathies* |  |  |
| IgG/IgA | Plasma exchange | I |

TABLE 67.1 *(Continued)*

| IgM | Plasma exchange | II |
|---|---|---|
| Multiple myeloma | Plasma exchange | III |
| IgG/IgA/IgM | Immunoadsorption | III |

*Pediatric autoimmune neuropsychiatric disorders associated with streptococcal infections; Sydenham's chorea*

| Severe | Plasma exchange | I |
|---|---|---|

*Pemphigus vulgaris*

| | Plasma exchange | III |
|---|---|---|
| | Extracorporeal photopheresis | III |

*Phytanic acid storage disease (Refsum's disease)*

| | Plasma exchange | II |
|---|---|---|

*POEMS syndrome*

| | Plasma exchange | IV |
|---|---|---|

*Posttransfusion purpura*

| | Plasma exchange | III |
|---|---|---|

*Psoriasis*

| | Plasma exchange | IV |
|---|---|---|

*Rapidly progressive glomerulonephritis*

| | Plasma exchange | III |
|---|---|---|

*Rassmussen's encephalitis*

| | Plasma exchange | II |
|---|---|---|

*Red cell alloimmunization in pregnancy*

| | Plasma exchange | II |
|---|---|---|

*Renal transplantation: antibody-mediated rejection; HLA desensitization*

| | Plasma exchange | II |
|---|---|---|

*Rheumatoid arthritis, refractory*

| | Plasma exchange | IV |
|---|---|---|
| | Immunoadsorption | II |

*Schizophrenia*

| | Plasma exchange | IV |
|---|---|---|

*Scleroderma (progressive systemic sclerosis)*

| | Plasma exchange | III |
|---|---|---|
| | Extracorporeal photopheresis | IV |

*(Continued)*

| TABLE 67.1  *(Continued)* | | |
|---|---|---|
|  | **TA modality** | **Category** |
| *Sepsis* | | |
|  | Plasma exchange | III |
| *Sickle cell disease* | | |
| Life and organ threatening | Erythrocytapheresis | I |
| Stroke prophylaxis | Erythrocytapheresis | II |
| Prevention of iron overload | Erythrocytapheresis | II |
| *Stiff-person syndrome* | | |
|  | Plasma exchange | III |
| *Systemic amyloidosis* | | |
|  | Plasma exchange | IV |
| *Systemic lupus erythematosus* | | |
| Manifestations other than nephritis | Plasma exchange | III |
| Nephritis | Plasma exchange | IV |
| *Thrombocytosis* | | |
| Symptomatic | Thrombocytapheresis | II |
| Prophylactic or secondary | Thrombocytapheresis | III |
| *Thrombotic thrombocytopenic purpura* | | |
|  | Plasma exchange | I |
| *Thyrotoxicosis* | | |
|  | Plasma exchange | III |

Modified from Szczepiorkowski ZM, Shaz BH, Bandarenko N, Winters JL (2007). The new approach to assignment of ASFA categories – Introduction to the fourth special issue: Clinical applications of therapeutic apheresis. *J Clin Apher* **22**, 96–105.

**Volumes Exchanged:** The decision regarding the volume of blood to be exchanged is based on a kinetic model of an isolated one-compartment intravascular space. This model assumes that the component is neither synthesized nor degraded during the procedure, and that it remains within the intravascular compartment. This model works well for components located predominantly in the intravascular compartment, such as IgM and RBCs, and less well for IgG (which is 45% extravascular). When one blood volume is exchanged, then 63.2% of the initial component is removed; at 1.5 volumes 77.7% is removed and at 2.0 volumes 86.5% is removed (Figure 67.1).

**Calculation of Total Blood Volume, Total RBC Volume, and Total Plasma Volume:** All approaches in calculating total blood volumes overestimate the total blood volume (TBV) in obese patients and underestimate it in muscular patients, but they provide a reasonable approximation; two methods to calculate TBV are

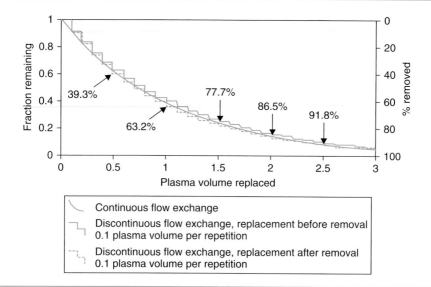

**FIGURE 67.1** Theoretic fraction of solute remaining after plasma exchange for continuous and discontinuous flow separation. From Hillyer CD, Silberstein LE, Ness PM *et al.* (eds). (2007). *Blood Banking and Transfusion Medicine: Basic Principles & Practice*, 2nd edition. Philadelphia: Churchill Livingstone Elsevier, San Diego.

presented in Table 67.2. The total plasma volume is the TBV multiplied by 1 minus the hematocrit (total plasma volume = TBV × [1 − Hct]). The total RBC volume is the TBV multiplied by the hematocrit (total RBC volume = TBV × Hct).

| TABLE 67.2 Calculation of Total Blood Volume | | | | |
|---|---|---|---|---|
| **Rules of Five** | | **Blood Volume (ml/kg of Body Weight)** | | |
| Patient habitus | Obese | Thin | Normal | Muscular |
| Male | 60 | 65 | 70 | 75 |
| Female | 55 | 60 | 65 | 70 |
| Infant/Child | – | – | 80/70 | – |
| **Nadler's formula (for adults)** | | | | |
| Male | (0.00612 × height in inches)/(14.6 × weight in pounds) + 604 | | | |
| Female | (0.005835 × height in inches)/(15 × weight in pounds) + 183 | | | |

**Total Extracorporeal Volume and Total RBC Extracorporeal Volume:** The extracorporeal volume is the amount of blood that is needed outside of the patient to fill the apheresis set and other tubing. This volume varies depending on the system, type of procedure, and ancillary equipment (e.g. blood warmer). The total RBC extracorporeal volume is the RBC volume required to fill the bowl or channel and all the tubing. The total extracorporeal volume should not exceed 15% of the total blood volume, and the total RBC extracorporeal volume should not exceed 15% of the total RBC volume.

**Replacement Solutions:** Replacement fluids are used when the amount removed from the patient will exceed 15% of the patient's total blood volume. This is especially true for plasma exchange, where either 5% albumin and/or plasma products are used in combination with crystalloids (e.g. normal saline) or other colloids (hydroxyethyl starch [HES]). The choice between plasma products versus albumin is based on the disorder the patient is being treated for (e.g. in thrombotic thrombocytopenic purpura plasma is used as a replacement fluid) and the underlying coagulopathic state of the patient (e.g. coagulopathic patients may require replacement of coagulation factors with the use of plasma). For erythrocytapheresis, RBC products are typically used as replacement solutions.

**Vascular Access:** Apheresis procedures require high blood-flow rates that can be achieved with peripheral venous access with one to two large-bore needles (16–18 gauge). In the absence of the ability to place these large-bore needles or the ability of the patient to augment venous return by fist-clenching, then central venous access catheters must be used. These central catheters are designed for positive pressure, and therefore are more-rigid and specifically designed for apheresis or dialysis. The placement of central catheters is not without risk, and is associated with infection, bleeding, pneumothorax, hemothorax and air embolism. Femoral catheters are relatively easy to place, and although they have increased risk of infection they may be appropriate for the patient who requires use of the catheter for only a brief period. In contrast, patients who need prolonged apheresis treatment may require a tunneled catheter or other long-term access device.

**Anticoagulation:** Citrate and/or heparin can be used to prevent coagulation in the extracorporeal circuit. The type of anticoagulant used depends on patient diagnosis, type and length of procedure, type and volume of replacement fluid, type of vascular access, and inlet blood flow.

Citrate: Citrate, usually administered as ACD-A although other forms are available (e.g. ACD-B and trisodium citrate), prevents coagulation by binding ionized calcium, which is required in clot formation. Citrate can result in hypocalcemia, depending on the rate of infusion and the ability of the liver to metabolize the citrate. Mild hypocalcemia may result in mild paresthesias, but more severe hypocalcemia can result in tetany, gastrointestinal symptoms, hypotension, cardiac dysrhythmias and seizures. Symptoms of hypocalcemia can be managed by decreasing the citrate infusion rate and/or administering calcium.

Heparin: Heparin has a half-life of 90 minutes, and results in systemic anticoagulation. Heparin can be used in combination with citrate to decrease the amount of citrate and heparin required, thereby decreasing the citrate toxicity.

**Adverse Events:** In one study which excluded mild vasovagal and transient paresthesias, 4.75% of procedures had complications: 1.2% were citrate related, 1% were vasovagal, and 1% were hypotensive reactions. Another study reported a similar

procedural complication rate with the most common complications being paresthesias (2.7%), hematoma at the puncture site (2.4%), clotting (1.7%), mild to moderate
allergic reactions (urticaria 1.6%) and bleeding (0.06%), and an incidence of severe,
potentially life-threatening adverse reactions of 0.12%. The incidence of hypotensive
reactions increases with the use of more normal saline compared to albumin (8% of
procedures).

**Citrate Toxicity:** Mild hypocalcemia (tingling, oral paresthesia or chest discomfort) is the most common adverse reaction to TA. Citrate reactions can be managed by
slowing the citrate infusion rate, adjusting the whole blood : ACD ratio, or administering calcium.

**Allergic Reactions:** Allergic reactions (ranging from mild urticaria to anaphylaxis)
are mostly associated with the use of plasma as a replacement fluid, but can occur with
the use of albumin and HES. Future procedures may require premedication with antihistames. Atypical allergic reactions are associated with the ethylene oxide gas sterilization of the tubing sets; these are characterized by periorbital edema with chemosis
and tearing. The recurrence of this reaction can be minimized by double priming of
the tubing.

**ASFA Indication Categories:** The indication categories for therapeutic apheresis
are as follows.

**Category I:** TA is standard and acceptable treatment for these diseases, either as a
primary therapy or as a valuable first-line adjunct therapy. The efficacy of TA in these
disorders is usually based on well-designed randomized controlled trials, or on a
broad and non-controversial base of published experience.

**Category II:** TA is generally accepted treatment for these disorders, but is considered to be supportive or adjunctive to other, more definitive treatments, rather than
a primary first-line therapy. Randomized controlled studies are available for some of
these disorders, but in others the literature contains only small series or informative
case studies.

**Category III:** There is a suggestion of benefit for which existing evidence is insufficient, either to establish the efficacy of TA or to clarify the risk/benefit (or sometimes
the cost/benefit) ratio associated with TA use in the treatment of these disorders.
Included are disorders in which controlled trials have produced conflicting results, or
for which anecdotal reports are too few or too variable to support an adequate consensus. TA may reasonably be used in such patients when conventional therapies do
not produce an adequate response, or as part of an Institutional Review Board (IRB)-
approved research protocol.

**Category IV:** Controlled trials have not shown benefit or anecdotal reports have been
discouraging for the use of TA in these disorders. TA for these disorders is discouraged,
and should be carried out only in the context of an IRB-approved research protocol.

**Evaluation of a New Patient for the Initiation of TA:** When considering initiating TA on a patient, the following general issues should be considered.

**Rationale:** What is the rationale for use of TA in the treatment of the patient's disorder? Is there a clear rationale for the use of TA, does the use of TA correct the abnormality, and is there clinical evidence that TA benefits the patient?

**Impact:** What effect does TA have on the patient's comorbidities and medications?

**Technical Issues:** What anticoagulation, replacement fluid, vascular access and volume of whole blood processed should be used?

**Therapeutic Plan:** How many times in total, and at what frequency, should TA be performed?

**Clinical and/or Laboratory End-points:** What clinical and/or laboratory end-points should be used to determine the efficacy of TA, and what are the criteria for discontinuation?

**Timing and Location:** When and where should TA be initiated?

### Recommended Reading

Basic-Jukic N, Kes P, Glavas-Boras S *et al.* (2005). Complications of therapeutic plasma exchange: experience with 4857 treatments. *Ther Apher Dial* **9**, 391–395.

McLeod BC, Price TH, Owen H *et al.* (1999). Frequency of immediate adverse effects associated with therapeutic apheresis. *J Clin Apher* **22**, 96–105.

Shaz BH, Linenberger ML, Bandarenko N *et al.* (2007). Category IV indications for therapeutic apheresis – ASFA fourth special issue. *J Clin Apher* **22**, 176–180.

Shaz BH, Winters JL, Bandarenko N, Szczepiorkowski ZM. (2007). How we approach an apheresis request for a category III, category IV, or non-categorized indication. *Transfusion* **47**, 1963–1971.

Szczepiorkowski ZM, Bandarenko N, Kim HC *et al.* (2007). Guidelines on the use of therapeutic apheresis in clinical practice – evidence-based approach from the Apheresis Applications Committee of the American Society of Apheresis. *J Clin Apher* **22**, 106–175.

Szczepiorkowski ZM, Shaz BH, Bandarenko N, Winters JL. (2007). The new approach to assignment of ASFA categories – introduction to the fourth special issue: clinical applications of therapeutic apheresis. *J Clin Apher* **22**, 96–105.

# CHAPTER 68

# Therapeutic Plasma Exchange

Beth H. Shaz, MD

Therapeutic plasma exchange (TPE, also known as plasmapheresis) is a therapeutic procedure where the blood of the patient is removed and passed through a medical device that separates out and removes the plasma from the other blood components, which are returned to the patient.

**Pathophysiology:** TPE is mostly used for antibody (and, rarely, immune complex) removal from circulation. It can be used to remove other molecules, such as drugs and low-density lipoproteins (LDL), but other apheresis devices are more efficient at LDL removal. Lastly, it can be used to correct deficiencies of a plasma clotting factors when large volumes of plasma infusion (without removal) would result in severe fluid overload in the patient.

**Methods:** Medical devices can remove plasma by membrane filtration devices or by centrifugation in a continuous or intermittent fashion.

**Volume Exchanged:** The decision regarding what volume of blood to be exchanged is based on a kinetic model of an isolated one-compartment intravascular space. This model assumes that the component is neither synthesized nor degraded during the procedure, and that it remains within the intravascular compartment. This model works well for elements located predominantly in the intravascular compartment, such as IgM (80% in the intravascular space) and less well for IgG (45% in the intravascular space). When one blood volume is exchanged, ~65% of the initial component is removed from the intravascular space; at 1.5 volumes ~75% is removed, and at 2.0 volumes ~85% is removed.

**Timing and Frequency:** The time interval between exchanges is generally chosen based on the need to allow the component of interest (e.g. IgG) to re-equilibrate into the intravascular space and the need to minimize the risk of bleeding as a result of dilutional coagulopathy. Re-equilibration of the intravascular IgG with extravascular IgG typically occurs within 2 days; 5–6 one-volume plasma exchanges over 14 days when combined with immunosuppression (to decrease synthesis) achieve a 70–85% reduction in intravascular IgG (Figure 68.1). Unless otherwise specified in the disorders described below, usually 1.0–1.5 plasma volumes are exchanged every other day.

**Replacement Fluids:** Colloids must be used for replacement when >15% of the TBV is removed. The majority of TPE procedures use 5% albumin as replacement fluid. Plasma products are used in the treatment of thrombotic microangiopathies and in patients with underlying coagulopathy. Albumin rarely results in adverse reactions, is free of transmitted transfusion diseases, but does not replace coagulation factors and most other plasma proteins. Plasma products replace all plasma proteins but have

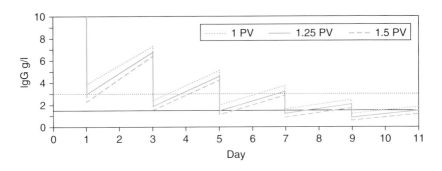

FIGURE 68.1   Apheresis kinetics. Theoretic reduction of immunoglobulin G (IgG) after plasma exchange of 1, 1.25 and 1.5 plasma volumes and after re-equilibration of total body IgG. The *solid line* indicates an 85% reduction, and the *dashed line*, a 70% reduction. The absolute reduction in IgG is reduced with each subsequent exchange. Calculations assume no degradation or synthesis of IgG and re-equilibration of IgG at 2 days. From Hillyer CD, Silberstein LE, Ness PM *et al.* (eds). (2007). *Blood Banking and Transfusion Medicine: Basic Principles & Practice*, 2nd edition. Philadelphia: Churchill Livingstone Elsevier, San Diego.

higher risk of blood product-related reactions (transfusion-transmitted diseases, allergic reactions and transfusion-associated lung injury) and of citrate reactions. Unless otherwise specified, 5% albumin is generally used as the replacement fluid.

Indications:   ASFA recently published guidelines on the use of TPE in clinical practice. Further information on the use of TPE in each of the disorders can be found in the *Journal of Clinical Apheresis Special Issue, Clinical Applications of Therapeutic Apheresis: An Evidence Based Approach*, 4th edition. Specific indications are described below in alphabetical order. After each title, the ASFA category is presented.

ABO-incompatible Hematopoietic Progenitor Cell Transplantation (Category II):   TPE can reduce ABO antibodies, which are responsible for hemolysis. In major ABO incompatibility (the recipient having ABO antibodies against the donor RBCs), if the HPC product cannot be RBC-depleted then TPE should be performed daily before infusion of HPCs with significant RBC contamination to reduce the IgM or IgG antibody titers to <1:16 immediately before HPC transplantation. The replacement fluid is a combination of albumin and plasma (50:50) compatible with both donor and recipient. In minor ABO incompatibility (the donor has ABO antibodies against the recipient RBCs), passenger lymphocytes may produce ABO antibodies 7–12 days after HPC product infusion. In order to prevent hemolysis of the incompatible recipient RBCs, prophylactic RBC exchange with group O RBCs can be performed to deplete the recipient's incompatible RBCs.

ABO-incompatible Solid-organ Transplantation (Category II for Kidney and Heart [Infants Only] and Category III for Liver):   For ABO-mismatched solid-organ transplant (the recipient having ABO antibodies against the solid organ), the antibody titer (the titer of both IgM and IgG ABO antibodies) can be lowered by peri-transplant daily TPE, thus preventing hyperacute rejection in order to improve graft survival. This should be performed in conjunction with immunosuppression and/or intravenous immunoglobulin (IVIG). The replacement fluid for TPE is 5%

albumin with or without plasma (compatible with both the recipient and donor RBCs or group AB), depending upon the presence or absence of coagulopathy. To minimize citrate toxicity, an ACD-A/whole blood ratio of 1:25–50 can be used with calcium infusion in liver transplantation cases. Most protocols use TPE to maintain a low ABO titer for the first 2 weeks after transplantation.

**Acute Disseminated Encephalomyelitis (Category III):** Corticosteroids are the mainstay of therapy; they hasten recovery and result in clinical improvement in up to 60% of patients. IVIG is reserved for patients who do not respond to corticosteroids. TPE has occasionally been used in the treatment of this disease. Most commonly, five every-other-day procedures are used; however, a range of three to six every-other-day procedures has been reported. If improvement is not observed early in treatment, then it is unlikely a response will occur.

**Acute Liver Failure (Category III):** In fulminant hepatic failure, TPE can remove albumin-bound and large molecular weight toxins and restore hemostasis, but the literature on the use of TPE in this setting is minimal and conflicting. If performed, TPE is carried out daily until transplantation or self-regeneration occurs. Basing improvement in liver function on liver enzymes and coagulopathy should be used cautiously, as TPE improves both of these parameters. In order to minimize citrate toxicity, the ACD-A/whole blood ratio can be set at 1:25–50 with the simultaneous use of calcium infusion. Plasma is the preferable replacement fluid, with or without the use of albumin, in order to treat the coagulopathy.

**Acute Inflammatory Demyelinating Polyneuropathy (AIDP; Guillain-Barré Syndrome) (Category I):** TPE and IVIG have demonstrated equal clinical efficacy in the treatment of this disease. The typical course of TPE is five to six every-other-day procedures with albumin replacement. The clinical effects of disease may progress despite the initiation of therapy. Since autonomic dysfunction is present, affected patients may be more susceptible to volume shifts, blood pressure and heart rate changes during extracorporeal treatment. Relapses requiring additional therapy may occur in approximately 10% of patients 2–3 weeks following either treatment. In AIDP patients with axonal involvement, TPE has been reported to be of greater potential benefit than IVIG.

**Amyloidosis, Systemic (Category IV):** No positive data exist to support the use of TPE for neuropathy or other complications associated with amyloidosis.

**Amyotrophic Lateral Sclerosis (ALS) (Category IV):** Multiple small series and a small controlled trial have failed to show any benefit for TPE alone or in combination with immunosuppressive therapy for patients with ALS.

**ANCA-associated Rapidly Progressive Glomerulonephritis (Wegener's Granulomatosis) (Category II):** The current standard approach to management of ANCA small vessel vasculitis is combination therapy consisting of high-dose corticosteroids and cytotoxic immunosuppressive drugs. In severe cases and those with diffuse alveolar hemorrhage, TPE has been added. The European Vasculitis Study Group (EVASC) demonstrated a decrease in the number of patients requiring dialysis after the use of TPE (seven one-volume TPE procedures over 10–14 days) compared to methylprenisolone, both in conjunction with cytoxan and prednisone in the patients with ANCA-positive RPGN and creatinine above 5.7 mg/dl. In patients with diffuse alveolar hemorrhage, replacement with plasma is recommended to avoid dilutional coagulopathy.

**Anti-glomerular Basement Membrane Glomerulonephritis (Anti-GBM GN; Goodpasture's Syndrome) (Category I):** In anti-GBM GN, the current treatment is the combination of TPE, cyclophosphamide and corticosteroids. In general the disease does not relapse, and therefore patients do not need chronic immunosuppression, except for ANCA-positive patients who require long-term immunosuppression because they respond rapidly to treatment but can relapse. It is critical that TPE be implemented early in the course. Several series have demonstrated that most patients with creatinine levels less than 6.6 mg/dl recover renal function, while it is rare for those with an initial creatinine above 6.6 mg/dl to do so. Patients who are dialysis dependent at presentation do not benefit from TPE, and it should not be performed unless diffuse alveolar hemorrhage is present, which can be rapidly fatal and responds to TPE in 90% of affected patients. TPE should be performed daily or every other day for a minimum of 14 days, and if diffuse alveolar hemorrhage is present then plasma should be used as the replacement fluid.

**Aplastic Anemia (AA); Pure Red Cell Aplasia (PRCA) (Category III):** For both AA and PRCA, underlying etiologies, such as medications, malignancies or infections, should be sought and treated. TPE has been successful in the treatment of AA in patients with autoimmune diseases. TPE may be considered in patients with severe AA who do not have a HPC transplant option and have failed to respond to conventional immunosuppressive therapy. TPE may also improve PRCA developing after major ABO-mismatched HPC transplant, or in the setting of erythropoietin therapy with anti-erythropoietin antibodies. TPE should be performed daily or every other day until recovery of hematopoiesis, which can take at least 2–3 weeks.

**Autoimmune Hemolytic Anemia (AIHA): Warm Autoimmune Hemolytic Anemia (WAIHA); Cold Agglutinin Disease (CAD) (Category III):** TPE is most useful in AIHA in severe situations when there is life-threatening anemia which does not respond to transfusion. TPE is more effective in cases of CAD, but for either disease the improvement after TPE is usually temporary. In CAD, RBC agglutination may occur within the cell separator and tubing, which requires that the procedure be performed with both the extracorporeal circuit and the room at 37°C. In addition, the severe anemia may require priming of the extracorporeal circuit with RBCs to safely perform the procedure. TPE is performed daily or every other day until hemolysis is controlled.

**Catastrophic Antiphospholipid Syndrome (CAPS) (Category III):** Full anticoagulation with heparin, high-dose corticosteroids, and TPE or IVIG are the most commonly employed therapies. If CAPS is associated with a flare of systemic lupus erythematosus, cyclosphosphamide is also used. TPE is performed daily with plasma as replacement fluid, with discontinuation based on the patient's clinical response.

## Chronic Inflammatory Demyelinating Polyradiculoneuropathy (CIDP) (Category I):

Corticosteroids, TPE and IVIG all have similar treatment outcomes based on controlled trials. The decision to use one over another is based on cost, availability and side-effects. Individuals may differ in response to any one of these agents. Therapeutic response is measured by improvement or stabilization, at which point treatment can be tapered or discontinued. Of the patients, 60–80% respond to initial therapy; however, long-term prognosis varies. Maintenance therapy, including continuing steroids, periodic TPE, or infusion of IVIG, is usually required because discontinuation of therapy may be followed by relapse. Maintenance therapy is dictated based on the patient's symptoms and clinical exam. TPE is performed two or three times per week until there is improvement, and then the frequency of procedures is tapered as tolerated.

## Coagulation Factor Inhibitors (Category III):

In patients with factor inhibitors (secondary to autoantibodies or alloantibodies), the therapy should be individualized, depending on the clinical setting, the presence or absence of bleeding, and the inhibitor titer. The goals of therapy include cessation of bleeding and suppression of inhibitor production. The current treatment option for bleeding preoperatively is to replace the factor or bypass it. The treatment options for inhibitor suppression include corticosteroids, cyclophosphamide, cyclosporine, rituximab and IVIG. Antibody removal by immunoadsorption, which currently is not routinely available in the US, is more effective than TPE. TPE is performed daily until antibody titer decreases and bleeding can be controlled with other therapeutic modalities.

## Cryoglobulinemia (Category I):

Cryoglobinemia is associated with a wide variety of diseases, including lymphoproliferative disorders, autoimmune disorders and viral infections (e.g. hepatitis B and C). Management is based on treating the underlying disorder and the severity of symptoms. TPE removes cryoglobulins efficiently, and has been most used in active moderate to severe cryoglobulinemia with renal impairment (membranoproliferative glomerulonephritis), neuropathy, vasculitis and/or ulcerating purpura both for short- and long-term management. TPE may be used alone or in conjunction with immunosuppressive agents, and reportedly results in improvement in 70–80% of patients. Cryoglobulin may precipitate in the extracorporeal circuit, and therefore, it is prudent to warm the room, use blood warmers to warm the draw and return lines, and/or warm the replacement fluid. A variety of number of treatments and frequencies (usually every other day) has been reported. For acute symptoms, five or six procedures are performed before re-evaluating for clinical benefit. TPE allows for resolution of acute symptoms prior to treating the underlying disease, and suppression of immunoglobulin production by immunosuppressive drugs. Weekly to monthly maintenance treatments may be necessary in patients who initially responded, in order to prevent recurrent symptoms.

## Dermatomyositis or Polymyositis (Category IV):

TPE did not improve clinical symptoms in a randomized controlled trial compared to sham apheresis.

## Familial Hypercholesterolemia (FH) (Category II):

TPE lowers serum cholesterol in patients with FH who are unresponsive to or intolerant of medical management, but current apheresis devices have been developed to selectively remove apo-B containing lipoproteins. TPE can be effective, but because of the availability of the selective removal systems and their enhanced efficiency of cholesterol removal, the use of TPE to treat FH is uncommon. It may, however, be the only option in small children. TPE is performed once every 2–3 weeks to maintain target time-averaged lipoprotein levels, and is continued indefinitely (see Chapter 73).

**Focal Segmental Glomerulosclerosis (FSGS) (Category III for Primary and Recurrent Disease):** TPE is used in the management of patients with recurrent FSGS in the renal allograft. Although there is no standardized treatment for recurrent FSGS post-transplant, the majority of regimens use a combination of an immunosuppression, angiotensin converting enzyme inhibitors, indomethacin and TPE. TPE should be initiated soon after the diagnosis of recurrence in order to induce remission and improve graft survival. However, the number of treatments needed to control proteinuria, a surrogate marker of FSGS, is quite variable, and treatment may take several weeks to months. TPE has also been used prophylactically immediately prior to and following the transplant, to prevent recurrent FSGS. TPE is performed daily or every other day. One approach is to begin with three daily exchanges followed by at least six more TPEs in the subsequent 2 weeks, for a minimum of nine procedures. Tapering of the frequency of TPE is individually guided by the degree of proteinuria. Some patients require long-term exchanges as maintenance therapy.

**Heart Transplant Rejection (Category III):** TPE has been reportedly used successfully to treat episodes of acute humoral rejection by removing antibodies and/or inflammatory mediators. There are no defined treatment protocols or criteria for the duration or discontinuation of TPE.

**Hemolytic Uremic Syndrome (HUS), Thrombotic Microangiopathy (TMA), and Transplanted Associated Microangiopathy (Category III for Familial HUS, TMA, and TAM and Category IV for Diarrhea-positive Pediatric HUS):**

*Hemolytic Uremic Syndrome:* HUS is characterized by the triad of microangiopathic hemolytic anemia, thrombocytopenia, and acute renal failure. Two types of HUS have been described; diarrhea-associated (also known as typical HUS) and non-diarrhea-associated HUS (also known as atypical HUS [aHUS]).

**Diarrhea-associated HUS (d+ HUS):** In children with d+ HUS, supportive care is the mainstay of therapy without the use of TPE. In adults with HUS, regardless of the presence of diarrhea, clinical manifestations are often severe and the mortality rate is high; therefore, treatment with TPE is common.

**Familial HUS:** Familial HUS is associated with frequent relapses, end-stage renal disease (ESRD) and a mortality rate of 54%. Treatment of familial HUS includes TPE with plasma as the replacement fluid.

*Non-idiopathic Thrombotic Thrombocytopenic Purpura (TTP) or Thrombotic Microangiopathy (TMA):* TMA is often associated with drugs (e.g. cyclosporine, quinine, mitomycin C, ticlopidine, and oral contraceptives), pregnancy, autoimmune diseases, infections (including HIV), or allogeneic HPC or solid-organ transplantation. Clinical and laboratory features are often indistinguishable from those of TTP, and therefore this syndrome has also been described as TTP-HUS. In patients with TMA the role of TPE is uncertain, but this treatment may be appropriate under certain circumstances and with a defined therapeutic trial endpoint, because of the high mortality with idiopathic TTP (see TTP section below) and the inability to differentiate between TTP or TMA. When TPE is used in the treatment of TMA, TPE should be performed daily with plasma as replacement fluid until clinical remission is achieved. In some patients, a therapeutic trial using a limited number of TPE procedures is appropriate.

*Transplant Associated Microangiopathy (TAM):* The effectiveness of TPE for TMA following HPC transplant (TAM) has not been proven, and therefore is not considered a standard of care.

**Hypertriglyceridemic Pancreatitis (Category III):** TPE has been used to treat acute pancreatitis due to hypertriglyceridemia because it lowers triglyceride levels by 70–80%, with improvement in symptoms of pancreatitis following one to two TPE procedures, but a single trial found no difference with regard to mortality, systemic complications and local complications in patients with severe disease. Treatment should be implemented early in the course of disease.

**Hyperviscosity in Monoclonal Gammopathies (Category I):** TPE has been successfully used in the treatment of hyperviscosity. Removal of IgM from the plasma will have a logarithmic viscosity-lowering effect, and therefore TPE is both rapid and efficient in relieving symptoms of hyperviscosity. Patients can be treated daily until acute symptoms abate (generally one to three procedures). The symptomatic serum viscosity threshold for each patient should be determined, and an empirical maintenance schedule may be employed to prevent symptomatic recurrence (generally every 1–4 weeks, depending on additional therapies used).

**Idiopathic Thrombocytopenic Purpura (ITP) (Category IV):** No therapeutic benefit with TPE in the treatment of ITP has been demonstrated in clinical trials.

**Inclusion Body Myositis (IBM) (Category IV):** TPE has not shown to be therapeutically beneficial in the treatment of patients with IBM.

**Lambert-Eaton Myasthenic Syndrome (LEMS) (Category II):** TPE may be a useful adjunct to management of patients with LEMS whose neurological deficit is severe or rapidly developing, or in the case of patients who are unable to wait for immunosuppressive or other medications to take effect. TPE is usually performed daily to every other day, and improvement may not be seen for the first 2 weeks or more after initiation. Treatment should continue until a clinical and electromyography response is obtained, or at least to a full therapeutic trial of a 2- to 3-week course. Repeated courses may be necessary, because the effect lasts only 2–4 weeks in the absence of immunosuppressive drugs.

**Multiple Sclerosis (Category II for Acute CNS Inflammatory Demyelinating Disease and Category III for Devic's Syndrome and Chronic Progressive MS):**

*Relapsing–remitting MS:* TPE has not been specifically studied in relapsing–remitting MS.

*Acute Central Nervous System Inflammatory Demyelinating Disease:* Acute inflammatory demyelinating disease is usually secondary to multiple sclerosis, but can be associated with acute transverse myelitis and neuromyelitis optica (Devic's syndrome). In acute, severe attacks in patients who fail initial treatment with high-dose steroids, TPE may be of benefit. Five to seven TPE procedures are performed over 14 days, with a response rate of 50%.

*Chronic Progressive MS:* The data demonstrate a small or no benefit with weekly TPE in conjunction with other immunosuppressive drugs in multiple randomized controlled trials. It is not clear whether the cost and potential adverse effects of TPE outweigh the small benefit.

**Myasthenia Gravis (MG) (Category I):** The four MG treatment modalities include cholesterase inhibitors, thymectomy, immunosuppression, and TPE or IVIG. TPE is especially used in myasthenic crisis, perioperatively for thymectomy, or as an adjunct to other therapies to maintain optimal clinical status. TPE works rapidly, with a clinical effect within 24 hours, but it may take a week. The benefits will likely subside in 2–4 weeks, if immunosuppressive therapies are not initiated. Usually a series of five procedures is performed; as few as two procedures can be beneficial, but the number and frequency of procedures depends upon the clinical scenario. Some patients may require long-term maintenance TPE. Multiple clinical trials have compared IVIG and TPE; TPE may improve ventilatory status and outcome more rapidly than IVIG, but it has more frequent adverse events. Therefore, the decision to use TPE or IVIG depends on the clinical situation.

**Myeloma and Acute Renal Failure (Category III):** TPE has been used to acutely decrease the delivery of light chains delivered to the renal glomerulus for filtration. Peritoneal dialysis (but not hemodialysis) can also remove light chains, but with lower efficiency than TPE. A recently reported study from the Canadian Apheresis Group, in which almost half of the study subjects received the VAD chemotherapy regimen, failed to demonstrate conclusive evidence that five to seven plasma exchanges over 10 days substantially reduces a composite outcome of death, dialysis dependence or severe renal insufficiency at 6 months. Initial management in non-oliguric patients should focus on fluid resuscitation (2.5–4l/d), alkalinization of the urine and initiation of chemotherapy. If serum creatinine continues to rise, or remains elevated after several days, consideration should be given to adding TPE to the patient's management. For patients who are oliguric, who excrete $\geq 10$ g of light chains per 24 hours, or whose serum creatinine is $\geq 6$ mg/dl, TPE may be considered in initial management. Controlled trials have employed daily to every-other-day TPE as a short-term adjunct to chemotherapy and fluid resuscitation over the period of 2–4 weeks.

**Overdose and Poisoning (Category II for Mushroom Poisoning and Category III for Other Compounds):** TPE is effective in removing highly protein-bound toxins or those with delayed metabolic effects from the intravascular components that are not lipid-soluble or bound to tissue, and do not have a large volume of distribution. Yet there is no correlation between protein binding and a volume of distribution among substances which were successfully treated with TPE, and there have been reports of the failure of TPE to remove substances bound to proteins and lipids. The clinical benefit can be achieved only if toxin levels can be reduced to concentrations below the threshold for tissue damage. Indications for TPE include progressive clinical deterioration, coma, and compromised excretory functions. TPE has frequently and successfully been used in Amanita poisoning. Usually albumin is sufficient as a replacement fluid; however, some toxic substances (e.g. dipyridamole, quinidine, imipramine, propranolol and chlorpromazine) bind to other plasma constituents preferentially over albumin, and therefore plasma may be a more appropriate choice. TPE is usually performed daily until the clinical symptoms have abated and the risk of delayed release of toxin from tissues is minimized.

**Paraneoplastic Neurologic Syndromes (Category III):** TPE cannot be considered standard therapy for autoimmune paraneoplastic neurologic syndromes. If a patient presents prior to development of severe neurological impairment but with a rapidly developing syndrome, aggressive immunosuppression, including daily or every-other-day TPE, may be reasonable in an attempt to halt the process. Five to six TPE procedures over 2 weeks are recommended, and if there is no response then the procedures should be discontinued.

**Pediatric Autoimmune Neuropsychiatric Disorders Associated with Streptococcal Infections (PANDAS) and Sydenham's Chorea (SC) (Category I for Severe Exacerbation):** In severely symptomatic or refractory patients with PANDAS or SC, IVIG or TPE has been shown to reduce symptom severity or shorten the course. A randomized controlled study in patients with SC showed that the mean chorea severity scores decreased by 72%, 50% and 29% in the IVIG, TPE and prednisone groups, respectively. Another randomized placebo-controlled trial of IVIG and TPE in children with PANDAS showed that both therapies at 1 month after treatment produced striking improvements in obsessive compulsive disorder (OCD), with mean improvements of 45% and 58%, respectively. More than 80% of the patients who received IVIG or TPE remained much or very much improved at 1 year. The TPE group appeared to have greater OCD symptom relief than did the IVIG group. Five to six procedures over 7–14 days were utilized in the trials without the benefit of repeat treatment.

**Pemphigus Vulgaris (Category III):** TPE has been utilized in patients with severe symptoms who have either received high doses of conventional agents and/or have an aggressive and rapidly progressive disease. All reported patients have received high-dose systemic corticosteroids and immunosuppressive agents which either produced life-threatening adverse effects or failed to control the disease. The goal of TPE was to reduce the level of autoantibodies with subsequent improvement in clinical symptoms. The decline in autoantibody titers, anti-keratinocyte cell surface and anti-desmoglein 3 antibodies, correlate with clinical response, and the levels of autoantibody may rebound within 1–2 weeks after discontinuation of treatment, which necessitates the use of immunosuppression. The use of TPE should be adjusted with autoantibody titer levels and clinical symptoms. The lack of clinical response after a trial period with concomitant adequate immunosuppression should be sufficient to discontinue treatment.

Phytanic Acid Storage Disease (Refsum's Disease) (Category II): TPE rapidly reduces plasma phytanic acid (PA) from elevated levels. This can be useful in the setting of acute attacks or exacerbation of the disease, as well as for maintenance therapy. Symptomatic levels of PA in Refsum's disease range from 700–8000 μmol/l (normal <33 μmol/l). TPE (usually one or two procedures per week over months) in conjunction with dietary control results in clinical improvements. Response can take weeks to months of treatment. Depending on the clinical response and PA levels, a tapering procedure can be performed.

Paraproteinemic Polyneuropathies (Category 1 for Demyelinating Polyneuropathy with IgG/IgA, Category II for Polyneuropathy with IgM (±Waldenström's), and Category III for Multiple Myeloma with Polyneuropathy): The typical course is five to six treatments over the course of 10–14 days. Long-term TPE or slow tapering of TPE should be considered to prevent relapse. In addition, improvement may be seen weeks following cessation of TPE. If the level of paraprotein is correlative to the polyneuropathy then it can be monitored to evaluate the frequency of treatment, but this correlation does not always exist.

Posttransfusion Purpura (Category III): TPE should be considered as the urgent treatment of hemorrhage and severe thrombocytopenia if IVIG therapy is not effective. Due to severe thrombocytopenia, the ACD-A/whole blood ratio should be adjusted to 1:25–50. Typically the replacement fluid is albumin to avoid further exposure to platelet antigen, but in a bleeding patient plasma supplement may be given. TPE can be discontinued when the platelet count starts increasing (>20,000/μ/l) and bleeding ceases.

Polyneuropathy, Organomegaly, Endocrinopathy, M Protein and Skin Changes (POEMS) Syndrome (Category IV): Among patients with POEMS treated with TPE, no responses were seen among those who received TPE alone, and the response rate among those who received TPE with concurrent corticosteroids was similar to the response rate with steroid therapy alone. Therefore, TPE is likely to be ineffective for this disorder.

Psoriasis (Category IV): Clinical trials have demonstrated that TPE provides no benefit in the treatment of psoriasis.

Rapidly Progressive Glomerulonephritis (RPGN) (Category III): Multiple disorders can produce RPGN, and therefore the appropriate indications for TPE are difficult to determine. TPE appears not to be beneficial in most immune-complex GN cases. However, there are reports of TPE efficacy in RPGN due to IgA nephropathy; these include short-term improvement in renal function and delay in dialysis dependency. Randomized trials of TPE in lupus nephritis have shown no benefit. TPE in cryoglobulinemia has proven successful in several series. TPE may be beneficial in dialysis-dependent patients at presentation. The treatment course is usually daily or every other day for 1–2 weeks, followed by tapering with less frequent treatments. TPE may be discontinued if there is no response after 4 weeks of treatment.

**Rassmussen's Encephalitis (Category II):** Treatment consists of an initial course of five to six TPE procedures over 10–12 days, followed by IVIG with subsequent courses of TPE (with or without IVIG) at 2- to 3-month intervals, as empirically needed, depending on the clinical disease course (seizures and unilateral neurologic deficits). Immunosuppressive medications may increase the interval between courses.

**RBC Alloimmunization in Pregnancy (Category II):** TPE removes the maternal RBC alloantibody that causes hemolytic disease of the fetus and newborn (HDFN). Therefore, it is thought that TPE will decrease the maternal antibody titer and, in turn, the amount transferred to the fetus, thereby protecting it from HDFN. Survival in severe cases of HDFN with the use of TPE and/or IVIG prior to IUT is about 70%. Category II is assigned for patients where there is a previous history of a severely affected pregnancy and the fetus is less than 20 weeks' gestational age. TPE should be considered early in pregnancy (from the seventh to the twentieth weeks) and continued until IUT can safely be administered (from about the twentieth week of gestation). Close monitoring of the fetus for signs of hydrops will aid in guiding treatment. One approach is to use TPE for 1 week (three procedures), followed by weekly IVIG (1 g/kg). In the second or third trimester, it is preferable to place the patient on her left side to avoid compression of the inferior vena cava by the gravid uterus. Hypotension should be avoided, as it may result in decreased perfusion to the fetus.

**Renal Transplantation/Antibody-mediated Rejection (AMR) and HLA Desensitization (Category II for AMR and Desensitization):**

*HLA Desensitization:* TPE with IVIG can be used prior to transplant to remove HLA antibodies (donor-specific antibodies [DSA]) in combination with immuno-suppressive drugs until negative crossmatch is achieved. The ability to obtain a negative crossmatch depends on the DSA titer – for example, usually five TPE treatments preoperatively will allow the titer of <32 to become negative. TPE is usually continued postoperatively and reinitiated in cases where AMR occurs. The risk of AMR is approximately 30%, with a small number of graft losses. TPE is performed daily or every other day, per center-specific protocol, until crossmatch becomes negative; TPE is then also performed postoperatively for a minimum of three procedures. Further treatment is determined by the risk of AMR, by DSA titers or by the occurrence of AMR.

*AMR:* In AMR, DSA can be removed with TPE (with or without IVIG), which is used in combination with other immunosuppressive drugs. Current regimens which include TPE have a graft survival rate of 70–80%. For AMR, some centers use a set number of procedures (usually five to six), daily or every other day. Other centers guide the number of treatments based on improvement in renal function and decrease in DSA titers.

**Rheumatoid Arthritis (RA) (Category IV):** Controlled trials have not shown benefit with TPE in patients with refractory RA, and therefore this disorder is a category IV indication for TPE.

**Scleroderma/Progressive Systemic Sclerosis (Category III):** There are limited and conflicting data regarding the use of TPE in the treatment of this disorder. A treatment course of six procedures over 2–3 weeks should constitute a sufficient therapeutic trial.

**Schizophrenia (Category IV):** A double-blind, randomized trial of TPE versus sham-pheresis demonstrated no benefit.

**Sepsis (Category III):** TPE, due to its non-selective nature, has the potential to remove multiple toxic mediators of the syndrome, and may therefore be more effective than blocking single components of the process. There is limited and conflicting literature regarding the use of TPE in sepsis. The randomized trials have limited treatment to one to two TPEs, while the case series have treated patients daily until improvement. Plasma should be used as a component of the replacement fluid when coagulopathy is present.

**Stiff-person Syndrome (Category III):** TPE can effectively deplete the autoantibodies reactive to a 65-kDa glutamic acid decarboxylase (GAD65, the enzyme responsible for the synthesis of GABA) associated with this disorder. There are few data to support the use of TPE, but TPE may be an alternative if IVIG is not available or the patient does not respond to IVIG. A series of four or five procedures over 8–14 days should effectively deplete the antibody. Repeat series of TPE can be employed empirically if there is an objective clinical improvement that is followed by a relapse of symptoms.

**Systemic Lupus Erythematosus (SLE) (Category III Except Lupus Nephritis, which is Category IV):** There are conflicting data regarding the use of TPE in the treatment of SLE. TPE is beneficial in the treatment of SLE-associated TTP, pulmonary hemorrhage, myasthenia gravis, hyperviscosity and cryoglobulinemia. In addition, TPE has resulted in clinical improvement in patients with CNS involvement, and may result in improvement of other critically ill SLE patients. In contrast, multiple controlled trials have demonstrated lack of efficacy with TPE in lupus nephritis. The typical treatment course is three to six procedures, which is sufficient to assess the treatment response in the patients with lupus cerebritis. The other related diseases are treated in the same fashion as the primary disease (e.g. TTP, myasthenia, hyperviscosity and cryoglobulinemia).

**Thrombotic Thrombocytopenic Purpura (TTP) (Category I):** TPE is a life-saving therapy for TTP which has decreased the overall mortality from >90% to <10%. TPE should be initiated emergently once TTP is recognized. If TPE is not immediately available, plasma infusions should be started at approximately 30–40 ml/kg per day, with care not to induce volume overload, until TPE can be initiated. Both plasma and plasma cryoprecipitate-reduced plasma products have been used as replacement fluid for TPE, with similar results in patient outcome. Corticosteroids are often used as an adjunct at 1 mg/kg per day; however, no definitive trials to prove their efficacy have been performed. Other adjuncts include rituximab, vincristine and splenectomy. Because congenital TTP is characterized by constitutive deficiency of ADAMTS13 activity without an inhibitor, simple infusions of plasma (10–15 ml/kg) or cryoprecipitate (which contains ADAMTS13) are used. Clinical response usually correlates with recovery of the platelet count and normalization of LDH with a clearing of mental status. The median number of TPE procedures to establish hematologic recovery is 7–8 days. The pattern of platelet response is variable, and the platelet count may fluctuate during treatment. Fibrinogen levels may decrease following serial TPE procedures with cryoprecipitate-reduced plasma as replacement fluid. TPE is generally performed daily until the platelet count is >150,000/µl, and the LDH is near normal for 2–3 consecutive days. LDH and bilirubin are removed during TPE; therefore, normalization of these values may be seen post-TPE, but these values will rise post-procedure if the disease is active (see Chapter 93).

Thyrotoxicosis (Category III): $T_3$ and $T_4$ are 99% plasma bound with a half-life of 5–7 days for $T_4$ and 1 day for $T_3$, and $T_3$ is 4 times as active as $T_4$, TPE should, in theory, efficiently reduce their circulating pool, but because of the short half-life the more active $T_3$ it may not be efficacious once other therapies are initiated. The literature exhibits mixed results, but most reported cases note a decrease in the concentration of the total levels of these hormones. In patients with amiodarone-associated thyrotoxicosis, TPE has also been used to reduce the plasma concentration of the drug, which has a half-life of months in patients on chronic therapy. Plasma can be used as a replacement fluid, with the added benefit of having thyroglobulin to improve binding of free hormones. This benefit should be weighed against the increased risks of the procedure with the use of plasma. TPE should be reserved for life-threatening situations when rapid amelioration of symptoms resistant to drug therapy is mandatory. Reported cases suggest that clinical improvement may precede changes in measured hormone levels. TPE should be discontinued once improvement is noted.

## Adverse Effects:

Drug Removal During Apheresis: TPE has been used to treat acute drug toxicity when other modalities, such as gastric lavage, dialysis, hemoperfusion and forced diuresis, are ineffective. Drugs that are lipophilic and highly protein-bound with long half-lives and small volumes of distribution are most effectively removed. More commonly, the issue of therapeutic drug clearance by TPE is a concern. The drugs that are most affected have small volumes of distribution and are protein-bound. Data suggest that prednisone, digoxin, cyclosporine, ceftriaxone, ceftazidime, valproic acid and phenobarbital are not removed by plasma exchange. Salicylates and tobramycin should be supplemented after TPE, and phenytoin should be monitored. The drug is most likely to be removed during the distribution phase, and therefore it is prudent to dose drugs after TPE and not immediately before.

Dilutional Coagulopathy: TPE results in the removal of coagulation factors when plasma is not used as a replacement fluid. The decrease in clotting factor activity ranges from 40% to 70% from the pre-procedure level, which may result in prolongation of PT and PTT. Fibrinogen is most affected, secondary to it being solely intravascular. Clotting factor activity usually returns to baseline within 1–2 days. In the absence of underlying coagulopathy or liver disease, not using plasma as a replacement fluid is appropriate for the prevention of dilutional coagulopathy.

In addition to coagulation factor removal, platelet counts decrease after TPE (reported as 9–53%). Platelet removal is greater when a larger TBV is processed at a slower centrifugal force. In addition, different instruments have varying affects on platelet loss. In general, platelet counts recover within 48 hours.

In patients receiving daily plasma exchanges or who are at risk for coagulopathy, monitoring of coagulation parameters is warranted, and replacement with plasma or platelets may be clinically indicated.

Citrate Toxicity: Mild hypocalcemia (tingling, oral paresthesia or chest discomfort) is the most common adverse reaction to TA. Citrate reactions can be managed by slowing the citrate infusion rate, adjusting the whole blood:ACD ratio, or giving calcium.

Allergic Reactions: Allergic reactions (mild urticaria to anaphylaxis) are mostly associated with the use of plasma as a replacement fluid, but can occur with the use of albumin and HES (hydroxethyl starch). Future procedures may require premedication with antihistames. Atypical allergic reactions are associated with the ethylene oxide gas sterilization of the tubing sets, and are characterized by periorbital edema with chemosis and tearing. The recurrence of this reaction can be minimized by double priming of the tubing.

ACE Inhibitor Reactions: Patients on ACE inhibitors who receive albumin may have facial flushing, hypotension and a feeling of doom. These reactions are secondary to the prekallikrein activator in the albumin which is activated to bradykinin, which is vasoactive. Bradykinin metabolism is inhibited by the ACE inhibitor, which leads to the accumulation of bradykinin. ACE inhibitor therapy should be discontinued 24–48 hours prior to the start of plasma exchange, if possible. Because of the risk of hypotensive reactions with the use of albumin in patients who have recently received ACE inhibitors (i.e. within 24–48 hours), some authorities will use plasma or a colloid starch (e.g. HES) as replacement fluid in to order to decrease the risk, if the patient must have the procedure performed. Another option is to initiate the procedure with albumin, and if a reaction is observed then delay treatment or use an alternative replacement fluid (plasma or a colloid starch). The decision regarding how to proceed is based on the emergent nature of the procedure and the risk/benefits for the individual patient. If albumin replacement is used, slow infusion rates are recommended.

### Recommended Reading

Shaz BH, Linenberger ML, Bandarenko N et al. (2007). Category IV indications for therapeutic apheresis – ASFA fourth special issue. J Clin Apher 22, 176–180.

Shaz BH, Winters JL, Bandarenko N, Szczepiorkowski ZM. (2007). How we approach an apheresis request for a category III, category IV, or non-categorized indication. Transfusion 47, 1963–1971.

Szczepiorkowski ZM, Bandarenko N, Kim HC *et al.* (2007). Guidelines on the use of therapeutic apheresis in clinical practice – evidence-based approach from the apheresis applications committee of the American Society of Apheresis. *J Clin Apher* **22**, 106–175.

Szczepiorkowski ZM, Shaz BH, Bandarenko N, Winters JL. (2007). The new approach to assignment of ASFA categories – introduction to the fourth special issue: clinical applications of therapeutic apheresis. *J Clin Apher* **22**, 96–105.

# CHAPTER 69

# Therapeutic Erythrocytapheresis

Beth H. Shaz, MD

Therapeutic erythrocytapheresis (also known as red blood cell exchange) is a therapeutic procedure where whole blood is removed and passed through a medical device which separates red blood cells (RBC) from the other blood components. The RBCs are discarded, the other blood components are returned to the patient, and the removed RBC mass and/or volume is replaced by banked RBCs and/or saline/albumin. Therapeutic erythrocytapheresis is performed most commonly for the exchange of dysfunctional RBCs (such in sickle cell disease, [SCD]) or infected RBCs (such as in malarial infection) with donor RBCs, or less commonly for the removal of RBCs (such as in polycythemia or hemachromatosis).

## Indications:

**Sickle Cell Disease (ASFA Category I for Life- and Organ-threatening Complications, Category II for Primary and Secondary Stroke Prophylaxis and Prevention of Iron Overload):** In acute life- or organ-threatening SCD complications (such as multiorgan failure and stroke), erythrocytapheresis is preferred over simple transfusion since the HbS concentration is reduced rapidly by removing HbS RBCs and replacing HbS RBCs with HbA RBCs without increasing blood viscosity and volume overload. One procedure is typically sufficient to treat the acute complications of SCD resulting in a HbS of <30%. For patients receiving chronic transfusion therapy (e.g. for stroke prevention), long-term erythrocytapheresis is advantageous over simple transfusion in preventing or markedly reducing transfusion-associated iron accumulation. For chronic transfusion therapy, erythrocytapheresis is typically performed at patient-tailored intervals to maintain the HbS level at <30–50%. The transfusion management of patients with SCD, including simple transfusion versus erythrocytopheresis, acute versus chronic (i.e. prophylactic) transfusions, blood product selection, and indications for transfusion are discussed in Chapter 45.

**Malaria (ASFA Category II for Severe Disease):** Malaria is a vector-borne protozoal infection caused by *Plasmodium vivax*, *P. ovale*, *P. malariae* or *P. falciparum*, which is transmitted by mosquito bites or transfusion. Parasitemia leads to hemolysis and release of inflammatory cytokines, resulting in fever, malaise, chills, headache, myalgia, nausea, vomiting, anemia, jaundice, hepatosplenomegaly and thrombocytopenia. The most severe cases of malaria are a result of *P. falciparum* infection, which can cause high-grade parasitemia, coma, seizures, pulmonary edema, shock, DIC and renal failure. Erythrocytapheresis is used in the treatment of severely ill patients with parasitemia >10%, and may result in rapid clinical improvement.

**Babesiosis (ASFA Category II for Severe Cases):** Babesiosis is a protozal disease transmitted from an animal reservoir to humans by tick bites or transfusion. Most cases are subclinical, or result in fever, anorexia, shaking chills, headaches, myalgia, vomiting and abdominal pain. Asplenic, immunocompromised (HIV-infected patients in particular) and elderly patients may have a much more serious clinical course, where symptoms may include hemolytic anemia, acute renal failure, disseminated intravascular coagulopathy (DIC), congestive heart failure and pulmonary disease. Usually 1–10% of the RBCs are parasitized in normal hosts, but in at-risk individuals parasitemia can reach 85%. The primary treatment is antibiotics. Erythrocytapheresis is used to rapidly lower the level of parasitemia, because the babesia organisms are only present in RBCs, and ameliorate the acute hemolytic process. Erythrocytapheresis is most commonly performed when parasitemia is > 10%, depending on the patient's symptoms and comorbidities, and continued until parasitemia is < 5%.

**Erythrocytosis and Polycythemia Vera (ASFA Category II for Symptomatic Erythrocytosis):** Erythrocytosis (polycythemia) is characterized by an increase in RBC mass and can be primary as the result of a myeloproliferative disorder (MPD) such as polycythemia vera (PV), or secondary as a result of defective congenital hemoglobin, chronic hypoxia related to a respiratory or cardiac disorder, or ectopic (e.g. erythropoietin producing malignancy) or dysregulated (e.g. postrenal transplantation) erythropoietin production. Whole blood viscosity increases significantly as the hematocrit (Hct) level exceeds 50%. Hyperviscosity complications include headache, dizziness, slow mentation, confusion, fatigue, myalgia, angina, dyspnea and thrombosis. Treatment of secondary erythrocytosis focuses on treating the primary disorder. When the primary disorder cannot be reversed, or in patients with MPD, symptomatic hyperviscosity can be treated by therapeutic phlebotomy. The therapeutic endpoint for phlebotomy varies according to the underlying etiology, but the usual goal is a Hct <45% in patients with MPD, a Hct of 50–52% in patients with pulmonary hypoxia or high oxygen affinity hemoglobins, and a Hct of 55–60% in patients with cyanotic congenital heart disease. Erythrocytapheresis may be preferred to phlebotomy in hemodynamically unstable patients as a result of less fluid shifts, or in patients requiring large-volume RBC removal in a short period.

**Volume Exchanged:** When one blood volume is exchanged, then 63.2% of the initial RBCs are removed; at 1.5 volumes 77.7% are removed and at 2.0 volumes 86.5% are removed. Most apheresis devices request the fraction of cells remaining (FCR) to be defined, which is the percentage of the patient's total RBC volume that should be replaced during the procedure. Typically in SCD patients the goal is to have HbS lower than 30% at the completion of the procedure and, therefore, if the patient's initial HbS is 100%, the FCR would be set at 30% for a final HbS level of 30%.

**Replacement Fluids:** For RBC exchange procedures, the patient's RBCs are replaced with leukoreduced donor RBCs. Depending on which anticoagulant-preservative solution is selected, the Hct of the RBC product will differ. CPDA-1, CPD and CP2D RBC products have a Hct of approximately 75%, while additive solution RBC products have a Hct of approximately 60%. In children, the Hct of each product may need to be determined in order to reach the correct end-procedure Hct of the patient. SCD patients should receive phenotype-matched and HbS-negative RBC products. In patients with polycythemia, RBC reduction procedures use normal saline and albumin (if the removal volume is greater than 15% the patient's total blood volume) for replacement fluid.

**End Hematocrit:** Most apheresis devices require inputting a final desired Hct for the patient. For SCD patients, this value is usually 30% in order to prevent hyperviscosity.

**Adverse Effects:** Patients receiving erythrocytapheresis are at risk for RBC transfusion reactions, such as hemolytic transfusion reactions, febrile non-hemolytic transfusion reactions, transfusion-associated lung injury and allergic reactions. These reactions are described in the transfusion reactions section of this book.

## Recommended Reading

Josephson CD, Su LL, Hillyer KL, Hillyer CD. (2007). Transfusion in the patient with sickle cell disease: a critical review of the literature and transfusion guidelines. *Transfus Med Rev* **21**, 118–133.

Sarode R, Altuntas F. (2006). Blood bank issues associated with red cell exchanges in sickle cell disease. *J Clin Apher* **21**, 271–273.

Szczepiorkowski ZM, Bandarenko N, Kim HC *et al.* (2007). Guidelines on the use of therapeutic apheresis in clinical practice – evidence-based approach from the apheresis applications committee of the American Society of Apheresis. *J Clin Apher* **22**, 106–175.

Szczepiorkowski ZM, Shaz BH, Bandarenko N, Winters JL. (2007). The new approach to assignment of ASFA categories – introduction to the fourth special issue: clinical applications of therapeutic apheresis. *J Clin Apher* **22**, 96–105.

# CHAPTER 70

# Therapeutic Thrombocytapheresis

Beth H. Shaz, MD

Thrombocytapheresis is most often employed to remove platelets from patients with thrombocytosis, defined as a platelet count >500,000/μl, which can result in either hemorrhage or thrombosis.

## Indications:

**Thrombocytosis (ASFA Category II for Symptomatic Thrombocytosis and Category III for Prophylactic or Secondary Thrombocytosis):** Thrombocytosis can be a primary disorder occurring as a result of a myeloproliferative disorder, including polycythemia vera [PV], essential thrombocythemia [ET] and chronic myelogenous leukemia [CML]), or as a secondary phenomenon such as acute bleeding, hemolysis, infection, inflammation, asplenia, cancer, or iron deficiency.

Thrombocytapheresis is usually not indicated in secondary disorders, as these are functionally normal, but can be indicated in MPD because the platelets are functionally abnormal, leading to thromboembolic events or, less commonly, bleeding. The majority of thromboembolic events occur when the count is >600,000/μl, although the risk is not directly correlated to the circulating platelet number. Hemorrhagic risk is greatest with platelet counts >1,500,000/μl with associated acquired von Willebrand syndrome (AVWS) secondary to a decrease in VWF multimers, which is directly correlated to the platelet count.

Thrombocytapheresis should be performed daily or as needed to prevent recurrent or progressive thromboembolism or hemorrhage in a patient with a MPD and severe thrombocytosis. In addition, thrombocytapheresis may be appropriate for selected high-risk patients when platelet-lowering agents are contraindicated or intolerable, or when the onset of pharmacologic therapy would be too slow (e.g. before urgent surgery). Usually the goal is a platelet count of <600,000/μl. Platelet-lowering agents must also be given to prevent rapid reaccumulation of circulating platelets.

<u>Volume Exchanged:</u> Usually two total blood volumes are processed, resulting in a 30–60% decrease in platelet count.

## Recommended Reading

Szczepiorkowski ZM, Bandarenko N, Kim HC *et al.* (2007). Guidelines on the use of therapeutic apheresis in clinical practice – Evidence-based approach from the apheresis applications committee of the American Society of Apheresis. *J Clin Apher* **22**, 106–175.

# CHAPTER 71

# Therapeutic Leukapheresis

Beth H. Shaz, MD

Leukapheresis, also called *leukocytapheresis*, is mostly used to remove white blood cells (WBCs) from patients with hyperleukocytosis, defined as a circulating WBC or leukemic blast cell count > 100,000/μl, which is usually secondary to acute leukemia. Hyperleukocytosis results in hyperviscosity, leading to cerebrovascular insufficiency (confusion, coma, and hemorrhage) and pulmonary complications (dyspnea and hypoxemia). It can also lead to disseminated intravascular coagulopathy (DIC).

## Indications:

**Hyperleukocytosis (ASFA Category I for Symptomatic Leukostasis and Category III for Prophylaxis):** Hyperleukocytosis may be symptomatic or prophylactic.

*Symptomatic Leukostasis:* Typically, symptomatic leukostasis is observed in acute myeloid leukemia (AML) when the WBC is >100,000/μl, and in acute lymphoblastic leukemia (ALL) when the WBC is >400,000/μl. Myeloid blasts are larger and more rigid than lymphoid blasts, and myeloid blasts secrete cytokines, which upregulate endothelial cell adhesion molecule expression and activate inflammation, and therefore result in symptoms of leukostasis at lower cell counts. Symptoms of hyperleukocytosis can also occur in patients with chronic myelogenous leukemia or chronic myelomonocytic leukemia, usually in association with an increase in circulating immature myeloid cells and WBC counts >100,000–200,000/μl, and in patients with chronic lymphocytic leukemia and WBC counts >400,000/μl. Cytoreduction with leukapheresis can rapidly reverse the pulmonary and central nervous system (CNS) manifestations of leukostasis. Leukapheresis should be repeated in persistently symptomatic patients until clinical symptoms resolve (typically, WBC or blast counts of <100,000/μl in patients with AML and <400,000/μl in patients with ALL). Concurrent chemotherapy is required in order to prevent rapid re-accumulation of circulating blasts.

Few studies exist directly determining the efficacy of leukapheresis in symptomatic leukemia patients. Bug and colleagues performed a cohort controlled retrospective study comparing 25 leukemic patients with WBC >100,000/μl who received chemotherapy only with leukemic patients with WBC >100,000/μl who received immediate leukapheresis with chemotherapy. Immediate leukapheresis had a significant impact on early death (16% versus 32%; $P < 0.02$). Dyspnea and elevated creatinine and LDH measurements were each independent risk factors for early death.

*Prophylactic:* Prophylactic leukocytapheresis can be performed prior to symptom onset when increased WBC counts are observed. Prophylactic leukocytapheresis reduces early mortality, but does not improve long-term survival in patients with AML and hyperleukocytosis (blast count >100,000/μl). Studies suggest that prophylactic leukocytapheresis offers no advantage over aggressive induction chemotherapy and supportive care in adults with ALL and a WBC count >400,000/μl. In children with ALL and a WBC count >400,000/μl, however, pulmonary and CNS complications develop in over 50% of children. Therefore, prophylactic leukocytapheresis should be considered.

In AML patients, leukapheresis should be continued until the blast cell count is <100,000/μl; in ALL patients, leukapheresis should be continued until the blast cell count is <400,000/μl. Concurrent chemotherapy is required in order to prevent rapid re-accumulation of circulating blasts.

**Volume Exchanged:** Usually two total blood volumes (TBV) are processed, resulting in a WBC count decrease of 30–60%. Efficacy of WBC removal depends on the use of an erythroid sedimenting agent, such as 6% hydroxyethyl starch (HES), which can be used to enhance the removal of immature and mature myeloid cells and thus mobilization from extramedullary disease sites into the intravascular space during the procedure. If removing mature myeloid cells, then HES should be used.

**Timing:** Leukapheresis should be performed immediately in highly symptomatic patients, and then daily as needed to maintain the appropriate WBC count.

**Replacement Fluids:** The majority of leukapheresis procedures result in <15% TBV removal, and therefore normal saline is sufficient. In those procedures where the volume removed will be greater than 15% of the patient's TBV, then replacement with a colloid solution, such as 5% albumin, is recommended.

## Recommended Reading

Blum W, Porcu P. (2007). Therapeutic apheresis in hyperleukocytosis and hyperviscosity syndrome. *Semin Thromb Hemost* **33**, 350–354.

Bug G, Anargyrou K, Tonn T *et al.* (2007). Impact of leukapheresis on early death rate in adult acute myeloid leukemia presenting with hyperleukocytosis. *Transfusion* **47**, 1843–1850.

Szczepiorkowski ZM, Bandarenko N, Kim HC *et al.* (2007). Guidelines on the use of therapeutic apheresis in clinical practice – evidence-based approach from the apheresis applications committee of the American Society of Apheresis. *J Clin Apher* **22**, 106–175.

Szczepiorkowski ZM, Shaz BH, Bandarenko N, Winters JL. (2007). The new approach to assignment of ASFA categories – introduction to the fourth special issue: clinical applications of therapeutic apheresis. *J Clin Apher* **22**, 96–105.

# CHAPTER 72

# Extracorporeal Photopheresis

Mary Darrow, MD, Jeanne Hendrickson, MD and
John Roback, MD, PhD

Extracorporeal photopheresis (ECP) involves the *ex vivo* exposure of peripheral blood mononuclear cells (MNCs), including pathogenic or autoreactive T lymphocytes, to photoreactive 8-methoxypsoralen (8-MOP) and ultraviolet A (UVA) light, followed by reinfusion of the MNCs to the patient. ECP was first successfully applied to the treatment of cutaneous T-cell lymphoma (CTCL). In recent years, ECP has also been applied with variable success to the treatment of alloimmune disorders of cell-mediated immunity, including chronic graft versus host disease (GVHD) and solid-organ graft rejection, as well as to autoimmune diseases such as refractory rheumatoid arthritis, systemic sclerosis and systemic lupus erythematosus.

Pathophysiology: ECP is an immunotherapeutic procedure that results in an antigen-specific immune response directed to autoreactive or pathogenic T cells without causing generalized immunosuppression. The precise mechanisms underlying the efficacy of ECP are currently being investigated, but the desired therapeutic effects are believed to result from multiple synergistic actions. Psoralens used for ECP, such as 8-MOP, intercalate into cellular DNA and upon exposure to UV-A light form DNA adducts that trigger photo-oxidation reactions. Treated cells show increased susceptibility to apoptosis, the extent of which depends on both UV-A dose and psoralen concentration. However, photodestruction of cells most likely underlies only part of the ECP mode of action, as only 2–5% of the entire *in vivo* leukocyte pool is affected. ECP also stimulates the release of cytokines that contribute to the demise of pathogenic lymphocytes. Furthermore, the cell environment alterations caused by ECP are postulated to stimulate monocyte activation and dendritic cell differentiation. When activated dendritic cells re-enter the circulation, they may induce a cytotoxic T-cell response to antigens in the apoptotic, photomodified T cells.

Methods: ECP is an automated procedure that uses Latham bowl-based apheresis via multiple cycles to collect MNCs in plasma and saline for a 270-ml final MNC product. 8-MOP is added, and the MNC-psoralen product is exposed to UV-A light which activates the psoralen, triggering photo-oxidation and cell death. The treated MNC preparation is reinfused into the patient. ECP typically requires heparin as the anticoagulant; ACD may be useful if heparin is contraindicated. Patients with elevated triglyceride or bilirubin levels may be difficult to treat, since photosensors in the ECP instrumentation cannot accurately detect the MNC interface. The procedure requires ~4 hours to complete, and is typically performed on 2 consecutive days, every 2–4 weeks.

Indications: ECP has FDA-approved indications and off-label uses. ASFA also categorizes indications (noted below).

## Cutaneous T-cell Lymphoma (CTCL) (FDA Approved; ASFA Category I for Erythrodermic CTCL; Category IV for Non-erythrodermic CTCL):

CTCL refers to a group of lymphoproliferative disorders caused by pathogenic, clonally-derived CD4+ T lymphocytes that involve the skin. These manifest clinically as erythematous patches, plaques, or tumors, and include mycosis fungoides and Sézary syndrome. Mycosis fungoides is an indolent disease that progresses from an erythematous phase, the classical plaque stage, to skin tumors or erythroderma. Sézary syndrome is an advanced, leukemic form of CTCL that is characterized by lymphadenopathy and excessive numbers of atypical T lymphocytes (Sézary cells) in the circulation. Prognosis and treatment depend on the type and extent of skin or extracutaneous disease. The mean survival in advanced stage CTCL or Sézary syndrome is 2–3 years; the most common cause of death is infection resulting in sepsis. Treatment can be topical, including psoralen photochemotherapy [PUVA]), or systemic, including interferon alpha, retinoids and ECP.

Clinical studies of ECP (2 consecutive days every 4 weeks in CTCL) have demonstrated complete response in 25–50% of patients in 4–12 months, and partial response rates of up to 60% in 4–9 months. A positive response to ECP is more likely in those patients with short duration of disease, absence of bulky lymphadenopathy or internal organ involvement, WBC $<20,000/\mu l$, Sézary cells $\sim$10–20% of MNCs, near normal NK cell activity, CD8+ cells $>15\%$, lack of prior intensive chemotherapy, and plaque-stage disease involving $<10$–15% of the skin. Approximately 75% of patients with Sézary syndrome will respond to ECP, and the addition of interferons or retinoids to ECP improves the response in the remaining 25%.

Some patients require long-term ECP for disease control, while others may tolerate a less frequent procedure schedule. Sézary cells $>20\%$ have been associated with the need for chronic photopheresis. Improvement of the skin lesions is not always achieved with ECP, and the decision to discontinue ECP has to be made in conjunction with the patient and healthcare team after consideration of the degree and duration of response, tolerability, and availability, as well as patient prognosis.

## Graft versus host Disease (GVHD) (Off-label Use; ASFA Category II for Acute and Chronic Skin GVHD; Category III for Non-skin Acute and Chronic GVHD):

GVHD results from activation of donor T cells following allogeneic HPC transplantation (HPCT) and may be acute (aGVHD), typically occurring within 3 months of transplantation, or chronic (cGVHD), typically occurring between 50 and 200 days post-HPCT. Standard therapy for this condition consists of high-dose corticosteroids as well as immunosuppressive agents such as mycophenolate mofetil, tacrolimus, and methotrexate. Mortality in patients with severe GVHD remains high despite aggressive immunosuppressive treatment.

ECP can achieve response rates of up to 70% with demonstrable improvement in joint contractures, lichenified skin changes, xerophthalmia, and xerostomia in patients whose cGVHD is resistant or refractory to immunosuppressive medications. ECP may also be effective in improving bronchiolitis obliterans and the gastrointestinal manifestations of cGVHD. In aGVHD, response rates of up to 60% have been reported with ECP. Some aGVHD patients treated immediately after HPC engraftment have developed pancytopenia associated with ECP. The typical procedure regimen for GVHD is

2 consecutive days every 1–2 weeks until there is a disease response, and then tapered as tolerated.

**Heart Transplant Rejection (Off-label Use; ASFA Category I for Prophylaxis for Rejection; Category II for Treatment of Rejection):** Complications of heart transplantation include cellular rejection, which is mediated by T cells, and allograft vasculopathy, which is an accelerated form of atherosclerosis that occurs in up to 60% of transplanted recipients within 5 years. Prevention and treatment of rejection employ immunosuppressants such as cyclosporine, mycophenolate mofetil, corticosteroids, and antilymphocyte antibodies. Patients treated with ECP prophylactically after cardiac transplantation may exhibit fewer rejection and infection episodes and better survival. In addition, a decrease in coronary artery intimal thickening has been demonstrated in patients treated with ECP versus those treated with immunosuppressants alone. ECP is performed on 2 consecutive days per week, initially, then every 2–8 weeks for several months. The largest randomized clinical trial treated patients with 24 series (2 days) of ECP during the first 6 months following transplantation and demonstrated decreased rejection, while the second largest study showed significant reduction of vasculopathy with one ECP series every 4–8 weeks for 2 years.

**Lung Transplant Rejection (Off-label Use; ASFA Category III):** Acute lung allograft rejection, characterized by bronchiolitis obliterans syndrome (BOS), occurs in 60–80% of patients 5–10 years after the transplant. Lung transplant recipients are maintained on immunosuppressive therapy. The initial treatment of BOS is usually pulsed high-dose methylprednisolone; if the patient does not respond, alternative immunosuppressive therapies have been used. ECP was initially used in the context of refractory BOS, and demonstrated a beneficial effect. In addition, ECP may be effective in patients with persistent acute rejection and early BOS, thus preventing further loss of pulmonary function. A common regimen includes one series (2 consecutive days) every 2 weeks for the first 2 months, followed by once monthly for the next 2 months for a total of six series. Another option is one series weekly for 5 weeks, then every 2 weeks for 2 months, and monthly thereafter.

**Pemphigus Vulgaris (Off-label Use; ASFA Category III):** Pemphigus vulgaris is a rare, potentially fatal, autoimmune mucocutaneous blistering disease with an associated antibody to desmoglein 3. Treatment with corticosteroids or other immunosuppressant agents reduces the mortality rate from 70–100% to ~30%. Clinical response to ECP has been reported.

**Scleroderma/progressive Systemic Sclerosis (Off-label Use; ASFA Category IV):** Systemic sclerosis (or scleroderma), is characterized by the accumulation of connective tissue in skin and viscera. Three randomized trials of ECP therapy for scleroderma have shown conflicting results. An early multicenter study with recent onset and progressive scleroderma patients treated with ECP versus D-penicillamine (n = 79) showed improvement in skin and joint parameters at 6 months. A crossover trial in patients with progressive systemic sclerosis of <5 years' duration (n = 19) showed no benefit of ECP. The latest multicenter trial of scleroderma patients (<2 years' duration) randomized 64 individuals to active or sham ECP; no significant difference was observed between the groups.

**Adverse Effects:** ECP is a well-tolerated procedure with limited side-effects due to volume shifts. Occasionally, transient hypotension may occur. Low-grade fever may occur within 2–12 hours after reinfusion MNCs, likely due to the release of cytokines. A temporary increase in pruritis or erythema may occur in patients with CTCL. Psoralen compounds are contraindicated in patients with aphakia and in those who have exhibited reactions to psoralen compounds or have a history of photosensitive disease (e.g. porphyria cutanea tarda). Patients should avoid sun exposure for 24 hours following ECP treatment.

## Recommended Reading

Bladon J, Taylor PC. (2006). Extracorporeal photopheresis: a focus on apoptosis and cytokines. *J Dermatol Sci.* **43**, 85–94.

Di Renzo M, Sbano P, De Aloe G *et al.* (2008). Extracorporeal photopheresis affects co-stimulatory molecule expression and interleukin-10 production by dendritic cells in graft versus host disease patients. *Clin Exp Immunol.* **151**, 407–413.

Gasová Z, Spísek R, Dolezalová L *et al.* (2007). Extracorporeal photochemotherapy (ECP) in treatment of patients with c-GVHD and CTCL. *Transfus Apher Sci.* **36**, 149–158.

Heshmati F. (2003). Mechanisms of action of extracorporeal photochemotherapy. *Transfus Apher Sci* **29**, 61–70.

Shaz BH, Linenberger ML, Bandarenko N *et al.* (2007). Category IV indications for therapeutic apheresis- ASFA fourth special issue. *J Clin Apher* **22**, 176–180.

Szczepiorkowski ZM, Bandarenko N, Kim HC *et al.* (2007). Guidelines on the use of therapeutic apheresis in clinical practice – evidence-based approach from the apheresis applications committee of the American Society of Apheresis. *J Clin Apher* **22**, 106–175.

# CHAPTER 73
# LDL Pheresis

Beth H. Shaz, MD

LDL pheresis removes apo-B-containing lipoproteins (low density lipoprotein [LDL] and lipoprotein(a) [Lp(a)]). LDL pheresis is indicated in the treatment of familial hypercholesterolemia (FH), which is an autosomal dominant disorder associated with mutations of hepatocyte apolipoprotein-B (apo-B) receptors resulting in decreased LDL removal by the liver. FH homozygotes have cholesterol in the range of 700–1200 mg/dl and heterozygotes have cholesterol in the range of 350–500 mg/dl, resulting in premature coronary heart disease and death. Approximately 1 in 500 Americans is heterozygous FH.

## Indications:

**Familial hypercholesterolemia (ASFA Category I for Homozygotes and Category II for Heterozygotes):** FDA approved indications for LDL pheresis are patients with hypercholesterolemia who are unresponsive to dietary and pharmacologic management, or unable to tolerate medications, with an LDL cholesterol >300 mg/dl, or with known coronary heart disease and LDL cholesterol >200 mg/dl. A single procedure reduces LDL cholesterol levels by 50–60%. Secondary to the slow rise of LDL and Lp(a), treatment intervals every 2–3 weeks maintain an ~50% reduction in cholesterol levels. LDL pheresis is performed indefinitely to maintain target lipoprotein levels. Long-term outcome studies have demonstrated a significant reduction in the number of cardiovascular events.

**Methods:** Mutliple selective removal systems are available within and outside of the US. Each appears to be equally efficatious in the treatment of hypercholesterolemia. All systems require anticoagulation with heparin.

**Adverse Effects:** The use of angiotensin converting enzyme (ACE) inhibitors is contraindicated in patients undergoing LDL apheresis.

## Recommended Reading

Szczepiorkowski ZM, Bandarenko N, Kim HC *et al.* (2007). Guidelines on the use of therapeutic apheresis in clinical practice – evidence-based approach from the apheresis applications committee of the American Society of Apheresis. *J Clin Apher* **22**, 106–175.

# CHAPTER 74

# Immunoadsorption

Beth H. Shaz, MD

Immunoadsorption (IA) utilizes columns which adsorb out immunoglobulin (Ig) by their binding to Staphylococcal protein A. Staphyococcal protein A has a high affinity for the Fc portion of IgG, and for aggregated IgG and IgG containing immune complexes. The effect of IA may not be secondary to antibody removal but by immunomodulation by the release of protein A into the patient, which can subsequently induce B-cell depletion. Currently, Staphylococcal protein A columns are no longer available in the US.

## Indications:

**Coagulation Factor Inhibitors (ASFA Category III):** Patients with congenital factor deficiencies can develop alloantibodies against the factor, known as inhibitors. In addition patients can develop autoantibodies to coagulation factors, leading to an acquired deficiency. Some of these deficiencies place patients at risk for bleeding, while others place patients at risk for thrombosis. In patients with factor inhibitors, the therapy should be individualized, depending on the clinical setting, the presence or absence of bleeding, and the inhibitor titer. The goals of therapy include cessation of bleeding and suppression of inhibitor production. The current treatment options for bleeding, or perioperatively, are either to replace or bypass the factor (e.g. the use of activated Factor VII in patients with acquired Factor VIII deficiency), or to remove/suppress the inhibitor. Inhibitor suppression therapy is usually through immunosuppressants and inhibitor removal through either IA or plasma exchange; IA is more effective than plasma exchange. IA is performed daily until antibody titer decreases and bleeding can be easily controlled with other therapeutic modalities.

**Idiopathic Thrombocytopenic Purpura (ITP) (ASFA Category II for Refractory Disease):** ITP is an autoimmune disease where antibodies or immune complexes bind to platelet surface antigens, primarily GPIIb/IIIa and/or GPIb/IX, resulting in platelet destruction. IA may be considered in patients with refractory ITP, with life-threatening bleeding or in whom splenectomy is contraindicated. Previous studies of IA have demonstrated a large range of outcomes, ranging from no improvement to long-term remission with response usually seen within 2 weeks. IA is generally discontinued when either the patient shows improvement (platelet count $>50,000/\mu l$) or no improvement after six treatments.

**Rheumatoid Arthritis (RA), Refractory (ASFA Category II):** RA is a chronic multisystem autoimmune disease, most commonly associated with inflammatory synovitis. IA has been used in patients who have failed primary treatments and multiple second-line treatments. The usual treatment course is 12 weekly procedures. Clinical improvement occurred at weeks 8–12 in most studies.

**Methods:** Once plasma is separated, the plasma is perfused through the column and then reinfused slowly (less than 20 ml/min) to the patient. The plasma can be separated through whole blood phlebotomy (off-line) or via continuous flow cell separation (on-line); 1000–2000 ml of plasma are treated using on-line method, and 250–500 ml of plasma are treated using the off-line method.

**Adverse Effects:** Use of this column is contraindicated in patients taking angiotensin converting enzyme (ACE) inhibitors, or who have a history of hypercoagulability or thromboembolic events. Common adverse effects include chills, low-grade fever, musculoskeletal pain, hypotension, nausea, vomiting and short-term flare in joint pain and swelling; these appear within 1 hour of the procedure and last for up to 2 hours. Severe respiratory and cardiovascular toxicities can occur, rarely resulting in fatality.

## Recommended Reading

Seror R, Pagnoux C, Guillevin L. (2007). Plasma exchange for rheumatoid arthritis. *Transfus Apher Sci* **36**, 195–199.

Szczepiorkowski ZM, Bandarenko N, Kim HC *et al.* (2007). Guidelines on the use of therapeutic apheresis in clinical practice – evidence-based approach from the apheresis applications committee of the American Society of Apheresis. *J Clin Apher* **22**, 106–175.

# CHAPTER 75

# Therapeutic Phlebotomy

Lawrence B. Fialkow, DO and Beth H. Shaz, MD

Therapeutic phlebotomy, defined as the therapeutic removal of whole blood, is used to decrease abnormally elevated iron stores or red blood cell (RBC) mass. In patients with erythrocytosis, phlebotomy decreases the blood viscosity and improves cardiac hemodynamics, which improves oxygen delivery and decreases the risk of thrombosis. The removal of circulating erythroytes leads to decreased total body iron and mitigates or abrogates the adverse effects of iron toxicity, such as liver, cardiac and endocrine gland dysfunction.

## Indications:

**Hereditary Hemochromatosis:** In the US, the most common cause of iron overload is hereditary hemochromatosis (0.26% of the US population is homozygous), which results from increased iron absorption in the gastrointestinal tract. Other hereditary diseases resulting in iron overload include juvenile hemochromatosis, autosomal dominant hemochromatosis, atransferrinemia, and aceruloplasminemia. Symptoms include liver disease, diabetes, and gonadal and cardiac dysfunction. The severity of liver disease closely reflects the magnitude of hepatic iron deposition. Serum transferrin saturation and ferritin levels are used for screening, as is testing for the presence of *HFE* gene mutations. The presence of parenchymal hemosiderin deposits on liver biopsy establishes a definitive diagnosis. The goal of therapy is to reduce iron stores to normal.

Therapeutic phlebotomy is the treatment of choice, and is usually initiated with ferritin levels >200 ng/ml in children and women, >300 ng/ml in men and >500 ng/ml in pregnant women. The typical regimen is once per week until iron stores are depleted, and then every 3–4 months to maintain normal ferritin levels (typically, levels <50 ng/ml, with transferrin saturation below 50%). Heart and liver function and control of diabetes improve with treatment.

**Secondary Hemochromatosis:** Iron overload can result from chronic transfusion, such as in patients with sickle cell disease, thalassemia major and acquired refractory anemias. These patients typically have anemia, may not tolerate therapeutic phlebotomy and are candidates for iron chelation therapy (see Chapter 65). Other diseases associated with acquired iron overload include chronic liver disease, porphyria cutanea tarda, insulin resistance-associated hepatic iron overload, African dietary iron overload, and medicinal iron ingestion.

**Polycythemia Vera:** Polycythemia vera (PV) is a myeloproliferative disorder (2.3–2.8 in 100,000 individuals per year are affected) which is characterized by an absolute increase in RBC mass and is often associated with leukocytosis, thrombocytosis and splenomegaly. Diagnosis is based on an elevated RBC mass (>25% predicted) with normal arterial oxygen saturation, splenomegaly, thrombocytosis, bone marrow hypercellularity, and low serum erythropoietin levels. Therapeutic phlebotomy once or twice a week decreases the RBC mass, lowering viscosity and improving blood flow, with a goal of a hematocrit of 42% in women and 45% in men.

Secondary Polycythemia: Polycythemia can be secondary to chronic hypoxia resulting in elevated levels of erythropoietin, such as in lung or cardiac disease, or secondary to increase in erythropoietin, as seen with impaired renal perfusion or as a result of erythropoietin-producing tumors. Polycythemia can also occur after kidney transplantation. The elevated hematocrit leads to an increase in blood volume and viscosity, resulting in headache, hypertension, visual disturbances, lethargy, weakness, thrombosis, and cerebral infarction and hemorrhage. As is the case with PV, the treatment is to lower the RBC mass and the viscosity.

Methods: A volume of 450–500 ml of whole blood is removed, which contains 200–250 mg of iron. Patients with a large body mass may tolerate removal of up to 2 units per week. Units drawn via therapeutic phlebotomy cannot be used for allogeneic transfusion unless they are from individuals with hereditary hemochromatosis, it is performed at no expense to the individual, the individual meets FDA allogeneic blood donor criteria, and the center has a FDA variance for use of these units.

Adverse Effects: Adverse effects of therapeutic phlebotomy are similar to those seen occurring with whole blood donation or erythrocytapheresis for other indications (see Chapters 8 and 69, respectively), and include vasovagal reactions and infection, bruising or hematoma formation at the site of venipuncture. Symptoms of hypovolemia can occur in a minority of patients, and can usually be prevented through adequate hydration and avoidance of strenuous exercise within 24 hours following the procedure. In addition, patients undergoing therapeutic phlebotomy for erythrocytosis are at risk for developing iron deficiency.

Recommended Reading

Franchini M, Gandini G, Veneri D et al. (2004). Efficacy and safety of phlebotomy to reduce transfusional iron overload in adult, long-term survivors of acute leukemia. Transfusion 44, 833–837.
Leitman SF, Browning JN, Yau YY et al. (2003). Hemachromatosis subjects as allogeneic blood donors: a prospective study. Transfusion 43, 1538–1544.

# CHAPTER 76

# HPC Products

John D. Roback, MD, PhD and Eleanor S. Hamilton, MT (ASCP)

In this book, the term *hematopoietic progenitor cell* (HPC) is used to denote both those primitive hematopoietic cells that are capable of self-renewal and differentiation into all hematopoietic lineages (sometimes known as hematopoietic stem cells), as well as those more committed progenitor cells that cannot self-renew. Both populations are found in HPC grafts, and each is required for rapid and sustained engraftment. HPC transplantation is usually used to treat malignant disease, although indications for treatment of non-malignant diseases are expanding. This chapter describes the three currently used sources of HPCs for transplantation: peripheral blood (HPC, Apheresis), bone marrow (HPC, Marrow), and umbilical cord blood (HPC, Cord Blood). All three sources can be used for autologous transplantation, syngeneic transplantation between identical twins (although HPC, Cord Blood use for autologous and syngeneic purposes is rare) or allogeneic transplantation. The selection of HPC source and donor must take into account a number of factors, including the underlying disease, the recipient's medical condition, the pre-transplant preparative regimen, and donor/recipient preferences for HPC harvest.

**HPC, Apheresis:** This type of HPC product is harvested by apheresis from the peripheral blood of donors. Efficient collection of progenitor cells typically requires that the cells be "mobilized" from the bone marrow into the blood of donors by chemotherapy (reserved for autologous donors), growth factors and/or more recently developed agents. With appropriate mobilization, most normal donors can donate sufficient HPCs in a single apheresis session for a transplant. In contrast, autologous donors may be challenging to mobilize and collect, depending on their underlying disease process. Given the relative speed, simplicity and non-traumatic nature of this procedure as compared to bone marrow harvest, this is the most commonly used source for transplantation. Additionally, HPC, Apheresis products tend to engraft more rapidly after transplantation.

**HPC, Marrow:** HPC, Marrow is used less frequently for transplantation than HPC, Apheresis, in large part because of the greater difficulty and invasiveness of the collection procedure. HPC, Marrow collection is generally performed as an outpatient, or as a short-stay surgical procedure under general anesthesia. These products have been shown to engraft more slowly than HPC, Apheresis products. Nonetheless, recent studies have raised questions about the relative advantages and disadvantages of these two graft sources. In some patient populations, such as young (<20 year old) patients with aplastic anemia or acute leukemia, overall survival has been reported to be superior when using HPC, Marrow as a donor source. This may relate to differential rates of graft-versus-host disease (GVHD) with HPC, Apheresis and HPC, Marrow. Randomized trials are currently being conducted to address this issue more definitively.

**HPC, Cord Blood:** Previously routinely discarded after childbirth, umbilical cord and placental blood is now known to be a source of high levels of HPC that can be used for transplantation. In contrast to HPC, Apheresis and HPC, Marrow, which are normally collected for a specific patient, HPC, Cord Blood is collected routinely in a number of centers and stored frozen in repositories until compatible recipients require transplantation. Because this product is banked, the time to obtain HPC, Cord Blood for transplantation is often shorter than for HPC, Apheresis and HPC, Marrow. HPC, Cord Blood is also associated with lower rates of GVHD than the other two products. However, comparisons have shown that HPC, Apheresis and HPC, Marrow have other advantages, including more rapid and stable engraftment as well as the opportunity to return to the donor for additional cells if needed. Additionally, the HPC content of cord-blood products, while sufficient for transplantation of many pediatric patients, is often too low for successful transplantation of average-size adults. In these patients, multiple HPC, Cord Blood products have been used for a single transplant.

**Cellular Content:** All three types of HPC products contain not only hematopoietic stem and progenitor cells, quantified in relative terms by CD34+ cell content, but also variable amounts of red blood cells (RBCs), lymphocytes, other WBCs and plasma. Processing to remove these components from the HPC graft is discussed below.

**Indications:** Malignant and pre-malignant diseases are the most common indications for HPC transplantation. These include acute lymphoblastic leukemia (ALL), acute and chronic myelogenous leukemia (AML, CML), myelodysplastic syndrome, plasma cell disorders such as multiple myeloma, and Hodgkin and non-Hodgkin lymphoma. Non-malignant indications for transplantation include inherited metabolic (Hurler syndrome), immune (severe combined immunodeficiency) and RBC (sickle cell disease) disorders; marrow failure states (aplastic anemia); and some autoimmune diseases (systemic sclerosis).

## Collection:

**Collection of HPC, Apheresis:** Many physicians and donors find HPC collection by apheresis preferable to harvesting of marrow. Through the apheresis procedure, significantly higher levels of CD34+ cells can be collected, which often translates into more rapid engraftment for the recipient. Donors require only peripheral venous access, are not anesthetized, and do not have to remain in the medical facility for a prolonged time after the procedure. Although each procedure requires 2–4 hours to complete, and occasionally up to four procedures are needed to collect sufficient HPCs, the associated morbidity and inconvenience are minimal. Occasionally, if peripheral venous access is inadequate, a temporary central venous catheter may be placed for use during the collection period.

Prior to apheresis, HPCs residing in the bone marrow must be mobilized into the peripheral blood for collection. Autologous donors can be mobilized by chemotherapy, which is an effective method to stimulate egress of HPCs into the peripheral blood, as well as through the administration of hematopoietic growth factors (G-CSF and GM-CSF) and/or, more recently, developed agents (e.g. AMD 3100) that block adhesive interactions between HPCs and the marrow microenvironment. While chemotherapy is too toxic for use in allogeneic donors, the other mobilizing regimens can be employed. Studies have shown that combining chemotherapy and growth factors leads to greater numbers of circulating HPCs than either method alone, suggesting that these mobilization regimens work by different mechanisms. Using the appropriate mobilization approaches, either alone or in combination, sufficient HPCs for transplantation can usually be collected from autologous donors. If the HPC yield is suboptimal, a variety of causes should be considered, including the extent of marrow disease and previous treatment, the patient's age and sex, and genetic factors. Despite the use of more limited mobilization regimens for allogeneic donors, almost all can produce sufficient numbers of HPCs by apheresis. If donors mobilize poorly, a variety of approaches can be used, including remobilization with growth factor at increased doses, use of large volume leukapheresis to increase the volume of blood processed to collect more HPCs, and alternative combinations of mobilizing agents such as growth factors together with AMD3100.

The kinetics of mobilization vary with the regimen. AMD3100 treatment mobilizes HPCs within hours of administration. In contrast, growth factor treatment releases CD34+ cells into the peripheral blood more slowly, and apheresis is typically not initiated until about the fifth day after starting G-CSF. The timing of mobilization with chemotherapy is more variable. In practice, these variations between regimens necessitate frequent monitoring of CD34 counts in the peripheral blood. When the counts reach preselected thresholds (typically 5–20 CD34+ cells/$\mu$L), apheresis is initiated. The CD34+ cell yield from each apheresis session is then determined, and the sessions are continued on a daily basis until the required number of HPCs is collected, typically $5 \times 10^6$ CD34+ cells/kg recipient weight.

*Adverse Events:* The main complications of HPC, Apheresis collections are due to the mobilization regimens. Side effects associated with G-CSF administration include bone pain, headache, myalgia, fatigue, malaise and nausea. Rare sequelae include splenomegaly, splenic rupture, stroke and anaphylaxis. AMD 3100 side-effects appear to be mild, including nausea, diarrhea, and minor numbness in fingers and toes. Chemotherapy has the most severe potential adverse effects, as expected. Depending on the regimen, these may include significant nausea, hair loss, diarrhea, mucositis and cytopenias. If venous catheters are used for apheresis, possible complications include pain, bleeding and infection at the site of insertion, and a small risk of air embolism upon removal of the line.

**Harvesting HPC, Marrow:** Bone marrow harvests are the traditional method for obtaining HPCs for transplantation. They are usually performed under anesthesia (local or general) in an operating room by physicians, and occasionally may require the donor to stay in the hospital overnight. The typical target cell dose is $2-4 \times 10^8$ nucleated cells/kg of recipient. Since the percentage of CD34+ cells in bone marrow is relatively constant ($\sim$1–3%), the probable yield can be estimated by performing nucleated cell counts during the procedure. The actual yield can later be determined by a CD34 count of the product at the end of the procedure. The yield of HPC can be a problem, since only a single harvest is performed except in rare circumstances. Investigators have attempted to improve the HPC yield by administering G-CSF prior to marrow harvesting, but this is not a common practice. After collection, the HPC product is mixed with anticoagulant in a sterile container and filtered to remove bone fragments. An advantage of the resulting bone marrow product, as compared to HPC, Apheresis, is much lower T-lymphocyte contamination, which leads to reduced GVHD complications post-transplant.

*Adverse Events:* For HPC, Marrow collection, multiple aspirations from different sites in the iliac crest are typically performed, which can cause prolonged back pain, site bleeding, and/or infection in the donor. Rare risks include damage to bone integrity, blood vessels or nerves, including the possibility of death. Some physicians prefer to collect autologous blood from the donor in the weeks leading up to this procedure, followed by infusion of the units after the harvest is completed to compensate for anemia that may develop. Alternatively, standard allogeneic RBC transfusions can be given. However, if allogeneic RBCs are transfused prior to harvest, the blood should be irradiated.

HPC, Cord Blood: This component is collected from the umbilical cord soon after childbirth, followed by processing, freezing and storage. There are several methods of collection that vary in terms of timing after birth and procedures used to optimize yield. HPC, Cord Blood is either collected by, or under the direction of, representatives of specialized cord blood banks. The minimum cell dose is usually $1.5 \times 10^7$ nucleated cells/kg of recipient body weight, but cell doses of $1.5–3.0 \times 10^7$ nucleated cells/kg may be associated with delayed engraftment.

Processing: In addition to containing the desired HPCs, marrow and apheresis products also contain RBCs, other leukocyte populations such as T lymphocytes, platelets, and additional cells that are not necessary for effective transplantation. While these additional cell types are not problematic for autologous transplantation, they may cause morbidities in allogeneic recipients. For example, ABO-incompatible RBCs and/or plasma in the HPC component may lead to post-transplant hemolysis, while allogeneic T cells can cause significant GVHD. For this reason, allogeneic grafts are often processed prior to transplantation to reduce the likelihood of recipient adverse effects. Additionally, if either autologous or allogeneic grafts will be cryopreserved, they may be volume reduced to decrease the amount of cryoprotectant needed for freezing as well as the space needed for storage. Given the number of possible options for processing, freezing, storage and infusion of HPC components, detailed consultations between treating physicians and HPC laboratory staff are necessary to assure optimal patient outcomes.

Processing of ABO-incompatible Products: HLA compatibility is of paramount importance in selecting donors for successful HPC transplantation. In contrast, ABO compatibility is of only secondary concern, and in fact donor–recipient ABO-incompatibilities occur in nearly half of allogeneic transplants. Because A and B antigens are not expressed on early HPCs, ABO incompatibility does not adversely affect stem cell engraftment. Nonetheless, ABO incompatibilities can affect RBC production and survival after transplantation, necessitating special processing of the graft prior to transplantation as well as appropriate selection of blood products for transfusion. In addition, practitioners should understand the specific requirements related to ABO-incompatible transplants that have been promulgated by organizations including the AABB, the Foundation for the Accreditation of Cellular Therapy (FACT), and the National Marrow Donor Program (NMDP).

Basic processing to remove RBCs or plasma, or to simply reduce the volume of the product, can be accomplished through standard centrifugation in component bags or centrifuge tubes. Effective centrifugation conditions are dependent on the actual instrumentation being used, and need to be developed and validated at each facility. Alternative approaches include density-gradient separation, which uses centrifugation and density solutions such as ficoll-hypaque to promote more effective separation of different product fractions, and buffy coat enrichment. Apheresis instruments and cell washers can also be programmed to effectively enrich for the mononuclear cell fractions that contain the desired HPCs while depleting RBCs and plasma.

Three types of ABO incompatibilities are considered below: major, minor and bidirectional. Blood product support for recipients of ABO-incompatible transplants is discussed in Chapter 46.

*Major ABO Incompatibility:* The occurrence of recipient antibodies that recognize donor RBC antigens is known as a major incompatibility. Typically, the antibodies of concern recognize A and B antigens, although a major incompatibility can also occur with antibodies against other RBC antigens, such as D. Major incompatibilities can have both immediate and delayed consequences. In the first case, the recipient antibodies can hemolyze donor RBCs at the time the donor products are infused. This problem can usually be prevented by depleting RBCs from the graft (typically to levels less than 10–20 mL of total RBCs, calculated by volume of product × hematocrit of product) prior to infusion. Methods for RBC reduction include sedimentation with hydroxyethyl starch or dextran, and automated cell processing. The method should be validated to make certain that sufficient HPCs are retained for transplantation. Alternatively, recipient antibodies can be depleted by plasma exchange, but this approach is used less commonly because it is more invasive.

In the second case, major incompatibilities can have delayed effects through continued hemolysis of RBCs newly produced from the engrafted donor HPCs. On occasion, the implicated antibodies can be produced by the residual recipient immune system for a significant period post-transplantation (~3–4 months), and sometimes longer. Diagnostic findings include the presence of antibodies directed against donor RBCs, a positive direct antiglobulin test (DAT) in a mixed-field pattern, and an eluate reactive with donor RBCs. Rarely, a pure red-cell aplasia can develop, characterized on bone marrow biopsy by the absence of erythroid precursors in the presence of normal numbers of other precursor populations. The appearance of donor RBCs in the circulation heralds the waning of the implicated recipient antibodies. Major incompatibilities do not adversely affect thrombopoiesis or granulopoiesis.

*Minor ABO Incompatibility:* Minor ABO incompatibilities are characterized by donor antibodies to recipient RBC antigens. Minor incompatibilities can also cause significant immediate and delayed hemolytic effects. Immediate hemolytic effects can usually be prevented by depleting plasma from donor HPC products prior to infusion, which requires only brief centrifugation with subsequent aspiration of the supernatant plasma. Delayed hemolytic effects typically manifest between 1 and 2 weeks after transplantation, and are characterized by clinically evident hemolysis of residual recipient

RBCs. Hemolysis can be brisk, and fatal cases have been described. Delayed effects are mediated by donor ("passenger") lymphocytes that engraft and continue to produce anti-recipient RBC antibodies after transplantation. Because these lymphocytes characterize the recipient's new immune profile, hemolysis does not subside until production of recipient-type RBCs ceases, usually within 2 weeks. Particularly severe cases can be managed by RBC exchange to rapidly replace recipient RBCs with donor-type RBCs.

*Bidirectional (Mixed) ABO Incompatibility:* A bidirectional incompatibility is the coexistence of both major and minor incompatibilities – recipient anti-donor antibodies and donor anti-recipient antibodies. Manifestations are typically a combination of those described above. Both RBC and plasma reduction can be applied to the graft to reduce the clinical consequences.

**Special Processing:** Special processing approaches typically involve either negative selection to remove unwanted cells (T-cell reduction and tumor cell purging), or positive selection to isolate cells of interest for transplantation (CD34+ selection).

*T-cell Reduction:* Because GVHD following allogeneic transplantation is mediated by donor T cells, methods that remove T cells from allogeneic grafts have been used to prevent GVHD. These methods typically use beads coated with anti-T-cell antibodies. After binding to T cells in the grafts, the beads (and T cells) can then be removed by either magnetic or sedimentation approaches. Alternative approaches use a combination of anti-T-cell antibodies and complement to lyse donor T cells in the product. While T-cell reduction does reduce the frequency of GVHD, it increases the occurrence of other complications. These include opportunistic infections and disease recurrence, because donor T cells provide immune surveillance functions in the transplant patient. In addition, graft failure can occur, since donor T cells facilitate engraftment in the allogeneic recipient. Thus, the clinician must carefully weigh the benefits of T-cell reduction against the risks for allogeneic transplant patients.

*Purging of Tumor Cells:* Negative selection approaches can also be used to remove, or purge, tumor cells from autologous HPC grafts. This process has drawn significant interest because tumor recurrence continues to reduce the effectiveness of autologous transplantation, and one potential source for the return of tumor is contamination of the autologous HPC product with malignant cells. Similar methods to those used for T-cell reduction can be applied to tumor purging. However, current methods carry the risk of damage to HPCs in the graft, which can impair engraftment. The results of clinical studies have not shown a consistent benefit of *in vitro* purging, and so this method is not routinely used. Instead, current efforts have focused on *in vivo* purging for autologous transplant patients: administering chemotherapy before and/or during mobilization and HPC collection to reduce the tumor burden in HPC products.

*CD34+ Selection:* As an alternative to depleting unwanted cells from HPC products, the stem and progenitor cells can be positively selected. For example, anti-CD34 antibodies have been used to enrich for the cells that are then transplanted. Recent work suggests that a fluorescent marker for the detection of aldehyde dehydrogenase may also serve as an effective method to isolate the desired cells from HPC products. These methods have not yet become routine, but hold future potential.

**Storage:** In many cases of allogeneic transplantation, the donor is collected just prior to the time of the recipient's transplant, and the product is only briefly stored at room temperature or in the refrigerator before it is infused. In contrast, autologous units are usually collected before or at the start of therapy, cryopreserved and stored in the freezer until myeloablative therapy is complete, and then thawed and infused into the patient.

In order to maximize viability of frozen HPC, the cryoprotectant solution, freezing kinetics and frozen storage conditions must be optimized. Addition of dimethyl sulfoxide (DMSO) to HPCs prior to freezing effectively prevents cellular dehydration as well as the formation of ice crystals. When DMSO (at a final concentration of between 5 and 10%) is used in combination with controlled-rate freezing, the HPCs retain high viability after thawing. Computer-controlled freezers are available for precise regulation of the HPC freezing rate. Once frozen, HPCs can be maintained at $-70°C$ or less in a mechanical freezer, or in a liquid nitrogen freezer. If the second option is chosen, the product can be stored beneath the surface of the liquid nitrogen ($-196°C$) or in the vapor phase above the liquid nitrogen surface ($-150°C$ or less). Liquid-phase storage has the advantage of reducing temperature fluctuations. However, the vapor phase is usually chosen for storage of potentially infectious HPC components to limit the risks of contamination of other components. There is no defined expiration date for frozen HPC components, and those frozen for over 10 years have been used successfully for transplantation.

If a component needs to be transported to another facility for transplantation, precautions should be in place to assure that the appropriate storage temperature is maintained and the HPCs are not exposed to conditions that could impair engraftment (e.g. airport X-ray detectors). Liquid nitrogen dry shippers are available which can maintain appropriate temperatures for frozen HPC components for up to 1 week without the risk of liquid nitrogen spillage.

**Preparation and Infusion:** As with standard transfusions, the identity of the recipient and the HPC component should be confirmed by medical staff before infusion is started. Frozen HPC components are typically thawed at 37°C at or near the patient's bedside, and then immediately infused at a rate of approximately 5–20 mL/min. While microaggregate filters (~170 micron) can be used, HPC products must not under any circumstances be infused through leukocyte-reduction filters, which would remove the HPCs from the product. The recipients are often hydrated and their urine alkalinized to improve the excretion of free hemoglobin resulting from RBC hemolysis.

**Adverse Events:** In most cases, cryopreserved HPC products are not washed to remove DMSO, although infusion of DMSO has been associated with some adverse effects. These include hypertension, hypotension (which may be prevented by antihistamines), nausea, vomiting, abdominal pain, diarrhea and, rarely, cardiac problems. Children may be more susceptible to these symptoms, as may patients who receive large DMSO doses (>10 mL of 10% DMSO/kg recipient weight). A simple approach to reduce the risks associated with large DMSO infusions is to divide the transplant into multiple, separated infusions (e.g. two infusions on successive days). If side-effects are observed, they can often be managed by slowing or briefly halting the HPC infusion.

While HPCs can be washed in the laboratory to remove DMSO prior to transplantation, the general belief is that washing can lead to loss of HPCs and reduced viability of remaining HPCs, which could impair engraftment.

**Quality Assurance:** Samples for quality control testing are typically obtained from each component at the time of processing. In most cases, testing is completed prior to HPC infusion. If the product is not cryopreserved it may be transplanted before the completion of testing, although in this case the physician should acknowledge in writing that testing has not been completed at the time of transplantation. Because HPC components are extremely valuable resources, products with positive infectious disease markers are not automatically discarded but may be used at the discretion of the physician and transplant recipient, provided that proper documentation is obtained and maintained. In these cases, the physician will need to determine if the benefit of giving the components outweigh the risk, given the patient's clinical condition.

**Dose:** The transplant dose of CD34+ cells is dependent on the patient, the disease, the donor, and the physician's preferences. Nonetheless, $1 \times 10^6$ CD34 cells/kg recipient weight is a generally accepted minimum dose for successful autologous transplantation, while $2 \times 10^6$ CD34 cells/kg is a minimum for allogeneic transplantation.

**Donor Lymphocyte Infusions:** Following HPC transplantation, donor-derived T cells can both positively and negatively influence outcomes. The positive effects include facilitation of allogeneic HPC engraftment by preventing rejection, providing immunity to opportunistic pathogens, and mediating a graft-versus-leukemia (GVL) effect that reduces the incidence of tumor recurrence. These are, however, balanced by the alloreactive effects of donor T cells that lead to potentially lethal GVHD in the allogeneic transplant recipient. Donor lymphocyte infusion (DLI) describes the post-transplant infusion of unselected donor mononuclear cell fractions (primarily donor T cells), typically in the setting of relapsed disease.

**Indications:** Primary indications for DLI are transplant recipients with either leukemia recurrence (CML, AML) or viral infection (usually cytomegalovirus [CMV] disease or Epstein-Barr virus post-transplant lymphoproliferative disease [EBV-PTLD]). DLI usually induces complete remission in the majority of CML patients with early stage relapse, and the resulting remissions are durable. Similar results can be expected when DLI is administered for viral infections. In contrast, patients suffering from relapsed acute leukemia or other malignancies (myelodysplastic syndrome, multiple myeloma) achieve remission less than half the time following DLI administration, and the remissions are typically only transient in these diseases. Outcomes are usually better if the patient is not experiencing active GVHD at the time of DLI, in large part because DLI will produce some degree of GVHD. In fact, in patients with cancer recurrence, there is continuing debate as to what extent the GVL and GVHD effects are separable, and whether DLI can be effective in the absence of GVHD.

More advanced methods for T-cell immunotherapy are under study, which involve selection, manipulation or other processing of the lymphocytes prior to infusion. These approaches are collectively termed *adoptive immunotherapy*. Some of the earliest work was directed at developing methods to prevent or treat CMV and EBV disease after allogeneic transplantation with reduced GVHD risk. These approaches involved co-culture of donor T cells with cells expressing viral antigens under conditions that favored the selection and expansion of antiviral T cells. After many weeks of *in vitro* selection, the expanded virus-specific donor T cells were shown to rapidly reconstitute anti-viral immunity without causing GVHD after allogeneic transplantation. Despite these successes, the labor-intensive nature of these methods restricted their use to all but a few transplant centers. Ongoing work is directed at identifying additional antigens that can be used for selection of donor T cells to target tumor cells. Antigens under study include major and minor histocompatibility antigens, protooncogenes, and other tumor antigens (PR1, NY-ESO-1). Other efforts are focused at finding more rapid and less expensive methods to select donor T cells targeted at the antigen of interest while eliminating those that cause GVHD. These approaches include depletion of alloreactive cells with immunotoxins, selective *in vivo* expansion of desired T cells by vaccination (veDLI), and transduction of T cells with a suicide vector so that donor cells can be eliminated *in vivo* if GvHD occurs after infusion.

**Dosage:** An escalating dose regimen is often employed for DLI. Patients are first treated with $1 \times 10^7$ T cells/kg recipient body weight. If remission is not achieved, the dose can then be increased to $5 \times 10^7$/kg, and once again to $1 \times 10^8$/kg if remission still does not occur.

**Collection:** Aliquots of HPC, Apheresis can be separated and frozen individually for future use as DLI. Alternatively, DLI can be specifically collected by apheresis, using the same techniques as HPC collections without the requirement of HPC molibilization.

**Processing and Storage:** See HPC, Apheresis (above).

**Preparation and Infusion:** See HPC, Apheresis (above).

**Quality Assurance:** T-cell counts (e.g. CD3) can be performed on collected products for quality assurance purposes.

**Adverse Events:** GVHD is the most common adverse effect of DLI. Below $10^7$ T cells/kg, GVHD is typically not seen; however, these lower doses also do not show significant efficacy. The clinical effectiveness of DLI increases with increasing doses, but so does the likelihood of GVHD. Over 50% of patients infused with $10^8$ cells/kg develop GVHD. Transient mild pancytopenia and an increased predilection to infections are also associated with DLI.

## Recommended Reading

Baron F, Beguin Y. (2000). Adoptive immunotherapy with donor lymphocyte infusions after allogeneic HPC transplantation. *Transfusion* **40**, 468–476.

Dazzi F, Fozza C. (2007). Disease relapse after haematopoietic stem cell transplantation: risk factors and treatment. *Best Pract Res Clin Haematol* **20**, 311–327.

Gajewski JL, Johnson VV, Sandler SG *et al.* (2008). A review of transfusion practice before, during, and after hematopoietic progenitor cell transplantation. *Blood* **112**, 3036–3047.

Montgomery M, Cottler-Fox M. (2007). Mobilization and collection of autologous hematopoietic progenitor/stem cells. *Clin Adv Hematol Oncol* **5**, 127–136.

# CHAPTER 77

# Tissue Banking in the Hospital Setting

Cassandra D. Josephson, MD

A recent paradigm shift has occurred in many of the nation's hospitals in which the management of human tissue, and specifically tissue allografts, has been centralized within hospital transfusion services.

Transfusion services have: (1) evolved to centrally coordinate the supply, dispensing and accounting of blood products amongst numerous entities such as blood donor centers, multiple hospital departments that include purchasing and receiving, intensive care units, surgical and recovery suites, clinics, and emergency rooms; and (2) developed an infrastructure to effectively and efficiently receive, process, store, and manage and issue blood products. Moreover, blood bankers possess the unique knowledge, specialized skills and training, as well as expertise, to handle the varying aspects of transfusion services for their hospitals. Therefore, the hospital transfusion service appears to be a logical place for centralization and accountability for a hospital's tissue services.

**Tissue Suppliers and Hospital Tissue Services:** Tissue suppliers (also known as tissue vendors), unlike hospital tissue services, are the organizations that recover tissue for transplantation from deceased donors (e.g. skin, corneas, cartilage, bones, vessels and valves) after aseptic recovery and processing, which may include freezing. Some of the vendors have patented processing methods to further refine the tissue (e.g. bone and connective tissue) through lyophilization and washing/sterilization methods. Hospital tissue services are the entity in a "centralized model" or entities in a "decentralized model" within a hospital that take ultimate responsibility for tissue, from the request of the transplanting surgeon, to product selection from a qualified vendor, to storage and issuance and, finally, implantation into a recipient. Currently, no specific model for managing tissues has been mandated by any regulatory agency. Hospital tissue services act as the regulator within a hospital, with a major emphasis being placed on traceability and trackability of tissue from donor to recipient and back again. If this aspect of tissue transplantation is more rigorously performed, the safety of tissue transplantation should be markedly improved.

**Tissues Transplanted at Hospitals:** Tissues can be of allogeneic or autologous types. Allogeneic tissues are those recovered from an individual other than the recipient, whereas autologous tissue refers to tissue removed from a patient which is slated to be reimplanted in that same patient at a later date. Most tissue that is transplanted is allogeneic and does not require HLA compatibility, in contrast to organs (e.g. kidney, heart) for transplantation. These types of tissues include bones, skin, tendons, corneas and heart valves. Allogeneic bone and corneas are the most frequently transplanted tissue type in the US. Bony allografts are most commonly used for orthopedic surgeries in order to decrease infection risks, prevent prolongation of surgery, provide structural support and promote bony healing. Corneal allografts are used to correct opacities of diseased corneas and restore clear vision, whereas tendons and ligaments are used for ligament reconstruction and skin allografts are useful to treat deep burn wounds. Table 77.1 describes other types of specific human allogeneic tissue and numbers of transplants performed annually in the US.

| TABLE 77.1 Human Tissue Transplantation in the US Annually | |
|---|---|
| **Human Tissue Allograft Type** | **Number of Transplants/year** |
| Bone | 1,200,000 |
| Cornea | 50,868 |
| Tendon | 32,441 |
| Skin (sq ft) | 19,189 |
| Ligament | 20,293 |
| Fascia lata | 16,754 |
| Pericardium | 11,304 |
| Heart valve (estimate) | 7000 |
| Vessels | 1093 |

Roback JD, Combs MR, Grossman B, Hillyer C, ed. (2008). *Technical Manual,* 16th ed, Bethesda, MD. AABB Press.

Autologous tissues are less frequently recovered and stored than are allogeneic tissues. The temporary removal of tissue from a patient and later reimplantation of the tissue may be critical to a patient's care. Multiple tissue types, such as endocrine (e.g. parathyroid gland, pancreatic tissue) as well as connective tissue, skin and bone (e.g. skull flaps and iliac crests) are reimplanted. The storage conditions for these types of tissue are dictated by the tissue type. Skull flaps recovered from traumatic brain injury craniotomy surgeries are the most common autologous tissue stored in hospital tissue services.

**Regulatory Agencies and Tissue Banking:** In the past decade, the transmission of infectious organisms such as HIV, hepatitis C, *Clostridium spp.*, Group A streptococcus, *Candida albicans* and Creutzfeldt-Jakob disease has been found to have occurred following tissue transplantation. The spread of contagions in transplanted tissue would reach the general public, and thus became the impetus for regulatory agencies that provide direct oversight and accreditation of healthcare organizations to issue new regulations and stringent guidelines for the procurement and use of a safe supply of tissue. More recently, lack of adequate accounting of, and a lapse in, appropriate tracking and tracing of donor procurement by tissue suppliers and hospital tissue services resulted in the distribution and use of tainted products (FN-BTS Recall). Specific agencies now responsible for the administration and oversight of tissue banks and hospital tissue services include the American Association of Tissue Banks (AATB) and the Food and Drug Administration (FDA), namely the division of Good Tissue Practices (cGTP), The Joint Commission (TJC), and AABB.

In 2005, TJC released tissue storage and issuance standards, specifying that written procedures, traceability methods and investigative plans that examine adverse events be required for hospitals in order to achieve compliance. The FDA also generated written regulations specifically to guide tissue suppliers and tissue processors. In 2006, the AABB issued specific standards by which hospitals would have to manage and oversee the use of tissue in their facilities, entitled *Guidelines for Managing Tissue Allografts in Hospitals*. While these agencies provided standards for hospitals on tissue management, specific guidance about which entity should be responsible for tissue services within a hospital was not clearly defined, though TJC emphasized that a single group should be in charge of transplanted tissue in TJC-accredited facilities. Similarly, the FDA provides regulations for tissue banks and processors handling tissue, yet offers few guidelines for healthcare facilities regarding how to handle and control (i.e. order, receive and transplant) tissues. Finally, while the AABB *Standards* provide more governance for tissue services in hospitals, these guiding principles do not explicitly assign responsibility to a definitive department within a hospital for its tissue banking.

**Decentralized Tissue Services in Hospitals:** The decentralized hospital tissue services model is what has historically existed at most hospitals, and is still the infrastructure at many institutions around the country. The transplanting surgeon and/or surgical service in this model are ultimately responsible for the proper handling of tissues, from receipt at the hospital to implantation in the patient. This responsibility encompasses ordering, receipt, storage and tracking, in addition to adherence to all of the regulations set forth by the agencies explained above. Over time this has been efficient for surgeons, eliminating administrative bureaucracy and allowing for specific tailored ordering of tissues for individual surgeons when necessary. In this system, often the salesperson for the vendor delivers the tissue and takes responsibility for proper handling and storage, including passage into the operating room. Regulator agencies such as TJC have recognized the non-standard practices of handling tissues in the decentralized model within hospitals, and acknowledge the difficulties that have been reported for tissue tracking of tissues delivered to the hospital and implanted into patients. Currently, TJC does not prohibit decentralization of hospital tissue banking, but has augmented the process by requiring a single person to keep track of each of the individual practices (e.g. ophthalmologic, orthopedic and cardiovascular separate tissue services).

**Centralized Tissue Services in Hospitals:** The strongest catalyst for hospital administrators to centralize tissue management in their transfusion service stemmed from TJC's standards released in 2005. Drawing upon the vast experience of blood bankers and transfusion medicine specialists, the centralization of tissue management is a logical progression and natural extension of the duties of the transfusion service. Nonetheless, it must be recognized that this transition has not been and will not be a seamless process. It is vital that hospital administrators, transfusion service managers and transfusion medicine specialists evaluate the institution's particular tissue needs and attain a comprehensive understanding of its transplantation services. With the establishment of "tissue services" within the hospital transfusion service, the specific needs of transplantation teams can be evaluated and addressed, and provisions made for a safe tissue supply.

Centralized tissue services can provide hospitals with:

- guidance about the type and amount of equipment,
- recommendations for and supervision of staffing and personnel resources,
- direction regarding physical space and location of tissue services vis-à-vis surgical suites,
- evaluation of the specific tissue needs of physicians in varying specialties,
- leadership in specialized training of medical technologists on diverse aspects of tissue services, and
- supportive assistance to the laboratory's infrastructure on the necessary upgrades to support tissue services within, while sustaining optimal operations of, transfusion service.

As part of the integration of tissue services within the transfusion service, hospital administrators will be required to allocate resources into new strategies, such as billing and revenues derived from surgical services.

As blood banks acquire and maintain accurate tissue product inventory, revenue losses can be expected to diminish. Moreover, expenses will likely diminish with a reduction in product wastage and improved management of ordered products that may lead to revenue savings, if not gains. As revenues accrued from tissue transplantation services expand and support the expenses for tissue services, hospitals will reap the benefit of having invested in this centralization as well as coming into conformance with the newly mandated tissue regulations and guidelines. In the long run, hospitals will garner more savings as well as benefits from having expanded on the established foundation, skills and technology of their transfusion service.

## Recommended Reading

Hillyer CD, Josephson CD. (2007). Tissue oversight in hospitals: the role of the transfusion services. *Transfusion* **47**, 185–187.

Kainer MA, Linden JV, Whaley DN *et al.* (2004). Clostridium infections associated with musculoskeletal-tissue allografts. *N Engl J Med* **350**, 2564–2571.

Kuehnert MJ, Yorita KL, Holman RC *et al.* (2007). Human tissue oversight in hospitals: an AABB survey. *Transfusion* **47**, 194–200.

# PART II
Coagulation

# CHAPTER 78

# Overview of the Coagulation System

Thomas C. Abshire, MD and Shawn M. Jobe, MD, PhD

Hemostasis is an orderly process. When a blood vessel is damaged, von Willebrand Factor (VWF) binds to both exposed connective tissue (e.g. collagen) and circulating platelets, resulting in platelet accumulation at the site of injury. These platelets become activated and both recruit additional platelets and trigger the coagulation system, ultimately leading to the generation of fibrin that overlays the platelet plug. This system is tightly regulated by substantial negative feedback mechanisms, including a series of coagulation factor inhibitors as well as the fibrinolytic system. Finally, repair of the damaged endothelium and remodeling occurs to ensure a smooth vessel surface. This process will be reviewed, focusing first on the platelet and vessel wall interactions and then on the formation of thrombin, which cleaves fibrinogen to fibrin, and the subsequent formation of a fibrin clot. Description of the fibrinogen molecule, the formation of fibrin and the role of Factor XIII, as well as the fibrinolytic pathway, can be found in Chapters 98, 99, and 129–131.

## Platelet/Blood Vessel Interaction:

**Role of the Blood Vessel:** Once the blood vessel is damaged, tissue factor (TF) is released and collagen exposed. The supporting structures of the blood vessel wall (media and adventicia) allow for constriction of the vessel upon injury. There are numerous adhesive proteins in the sub-endothelial basement membrane that provide binding sites for platelets. Besides collagen, these include microfibrils, laminin, thrombospondin, fibronectin, elastin and vitronectin. To keep these adhesive proteins in balance, certain matrix metalloproteinases (MMP), such as collagenase, degrade these adhesive proteins. These MMP are activated largely by the contact system of the coagulation cascade (kallikrein), as well as by the fibrinolytic system (plasmin) and specific enzymes (trypsin).

**Endothelial Cells:** Endothelial cells regulate hemostasis by several mechanisms. In the healthy state, they keep the vessel open by releasing vasodilators such as nitric oxide and prostacycline. Heparin and dermatan sulfates secreted by the endothelium accelerate the action of the anticoagulants anti-thrombin and heparin co-factor II. The conversion of protein C to activated protein C by thrombin occurs on the endothelial surface. Generation of this potent anticoagulant is facilitated by the endothelial membrane proteins, thrombomodulin and the endothelial protein C receptor (EPCR). Endothelial cells also activate the fibrinolytic system by releasing tissue plasminogen activator (TPA) as well as facilitating the release of the inhibitor to TPA, plasminogen activator inhibitor-1 (PAI-1).

Upon injury, the endothelium participates in both hemostasis and inflammation. Endothelial cells can slow bleeding by promoting vasoconstriction of the vessel wall and by secreting certain vasoconstricting proteins, such as renin, endothelin and platelet activating factor. Endothelial cells can also release tissue factor which binds to Factor VII and initiates the coagulation cascade. Finally, the endothelium can focus the inflammatory response to the site of injury through expression of cell adhesion molecules and secreted proteins (e.g. platelet activating factor).

## The Role of Platelets in Blood Vessel Interaction:

**Platelet Structure**: An overview of platelet structure can be found in Figure 78.1. Platelets are anucleate. A network of interconnected channels, the open canalicular system, extends from the inside of the platelet to the outside environment, and may function to allow the rapid release of the constituents of platelet granules. Mitochondria produce ATP, and may also participate in the regulation of the platelet activation response.

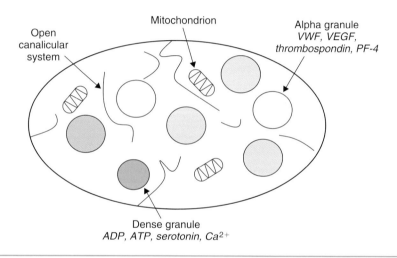

FIGURE 78.1 Overview of platelet structure.

Platelets contain two major types of granules, which release their contents upon activation. The most numerous are the alpha granules, which contain factors that regulate clot formation, new blood vessel formation and inflammation. VWF, fibrinogen and Factor V are procoagulant proteins contained within the alpha granule that may contribute to thrombus formation and stabilization. Alpha granules contain both factors that promote new blood vessel growth (angiogenic factors), and factors that inhibit new blood vessel growth and stabilize established vessels (angiogenesis inhibitors). Angiogenic factors in alpha granules include vascular endothelial growth factor (VEGF), epidermal growth factor (EGF) and platelet derived growth factor (PDGF). Some of the angiogenesis inhibitors include angiostatin, thrombospondin and endostatin. Regulators of inflammation, or cytokines, are also present in alpha granules. Examples include platelet factor-4 (PF-4), CCL5 (RANTES) and interleukin-8 (IL-8). Recent research indicates that despite their similar appearance, there are distinct subtypes of alpha granules. The physiologic significance of this finding is a focus of ongoing research. Alpha granule and dense granule release typically are coordinated. Contents of the dense granule include ATP, ADP, serotonin and calcium. These factors contribute to platelet recruitment and thrombus stabilization at a site of injury.

**Platelet Adhesion:** At various shear rates, different mechanisms mediate the interaction of the platelet with the injured vessel wall. In low shear conditions, the integrin $\alpha_{IIb}\beta_3$ can function alone to mediate the interaction of the platelet with the vessel wall. However, under higher shear conditions present within the arteriole, this mechanism is not sufficient, and a two-step process must occur to allow stable platelet-vessel wall interactions (Figure 78.2). In these high shear conditions, platelets initially attach to exposed collagen in a process mediated by the binding of the A1 domain of von Willebrand Factor (VWF) to collagen. Binding of the platelet glycoprotein GPIb-V-IX to VWF tethers the platelet at the site of injury, and the platelet translocates across the thrombus surface. Platelet stimulation by soluble agonists during translocation activates $\alpha_{IIb}\beta_3$ and promotes the firm adherence of the platelet to VWF and fibronectin within the vessel wall. Under conditions of pathologic shear stress, as might occur at areas of stenosis, the interaction of GPIb-IX-V and VWF can mediate platelet aggregation independent of $\alpha_{IIb}\beta_3$.

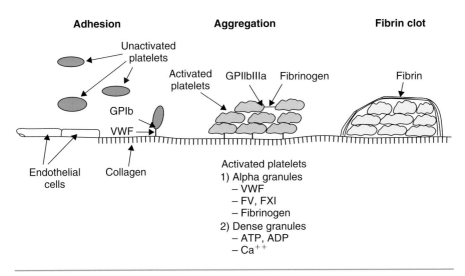

FIGURE 78.2 Overview of platelet–vessel wall interaction.

**Shape Change and Secretion:** Once platelets become activated, they change shape and release the contents of the alpha and dense granules. Events that take place in the cytoplasm include activation of the eicosanoid pathway (thromboxane A2), regulation of cAMP and calcium, activation of kinase signaling pathways, and additional signaling which activates $\alpha_{IIb}\beta_3$.

**Platelet Aggregation:** The major receptor on the platelet surface is $\alpha_{IIb}\beta_3$. It is found on the surface of circulating platelets, but is in an inactive conformation, unable to bind to its primary ligands, fibrinogen and VWF. Stimulation of the platelets by soluble agonists causes $\alpha_{IIb}\beta_3$ to change to an active conformation and bind its ligands. This process is called *inside-out activation*.

**Platelet's Role in Coagulation:** In circulating platelets, negatively charged lipids, such as phosphatidylserine, are limited to the inside surface of the bilayer platelet membrane. After platelet activation, a subpopulation of platelets externalizes phosphatidylserine. At the same time, a membrane rearrangement occurs that results in the release of microvesicles from the platelet. Phosphatidylserine externalized on the platelet and microvesicle surface binds the coagulation factors FVa and FXa, activating the prothrombinase complex, and increasing thrombin generation at the site of injury.

**Clot Retraction and Remodeling:** Once a stable fibrin clot has been formed, the platelets' contractile properties (actin and the cytoskeleton interacting with fibrinogen and platelet $\alpha_{IIb}\beta_3$) aid in remodeling of the clot.

**Formation of Thrombin:** The understanding of how thrombin is formed has evolved since the initial description of the coagulation "cascade" (Figure 78.3). This cascade of serine proteases and other factors interacting in a stepwise fashion was broadly defined as the extrinsic pathway (tissue factor [TF]/Factor VII [FVII]) and the intrinsic pathway (surface-contact factors). In a test tube, the coagulation cascade fairly accurately describes the interactions of clotting factors. However, physiologic thrombin generation is best described by "stages" of protein interaction rather than "pathways."

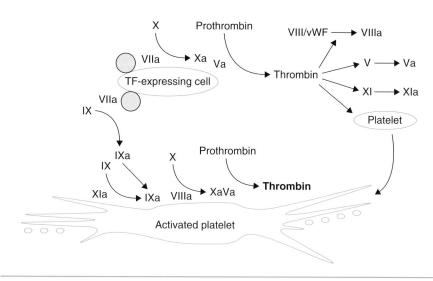

FIGURE 78.3  Overview of thrombin formation.

In the initiation stage, tissue factor located in the vessel wall, from a monocyte or an activated endothelial cell, binds activated FVII (FVIIa), which circulates in small amounts (1–2% of FVII) in the plasma. The enzymatic activity of the TF–FVIIa complex cleaves small amounts of both FX and FIX, thus producing FXa and FIXa. Small amounts of thrombin formed in this initiation stage cleave FVIII and FV, and the propagation stage of clot formation begins.

In the propagation stage, a large amount of thrombin is formed through the actions of the "tenase" (Factors IXa and VIIIa) and "prothrombinase" (Factors Xa and Va) complexes. The tenase complex converts FX to FXa, and the prothrombinase complex converts prothrombin to thrombin. In some, but not all individuals, cleavage of FIX by FXIa contributes substantially to thrombin generation in this propagation phase, perhaps explaining the variable bleeding phenotype in patients with FXI deficiency. Essential cofactors throughout this process include calcium and phospholipids from platelets or the endothelial wall.

Initial production of small amounts of thrombin is essential if a stable thrombin clot is to be generated. Thrombin is a key component to several crucial steps in the formation of fibrin:

1. Release of FVIII from VWF and activation of FVIII
2. Activation of Factors V and XI
3. Activation of FXIII and crosslinking of fibrin
4. Platelet activation (which provides another source of phospholipids)
5. Down-regulation of fibrinolysis by activating the inhibitor, thrombin activatable fibrinolytic inhibitor (TAFI).

In some individuals, activation of FXI by thrombin is an important feedback loop necessary for the generation of additional thrombin. FXI can also be generated by FXIIa (produced by the interaction of high molecular-weight kininogen (HK) with prekallikrein (PK) to form a complex on endothelial cells), but the physiologic contribution of this so-called "contact pathway" to hemostasis is not required for hemostasis, given the lack of clinical bleeding in patients with deficiency of HK, PK or FXII. Thrombin is also important in the regulation of coagulation by releasing tissue plasminogen activator (TPA) from the endothelial cell and in forming a complex with thrombomodulin and activating protein C.

Understanding this process of thrombin formation helps to explain how persons with bleeding disorders such as hemophilia and FXIII deficiency manifest delayed bleeding. Since these disorders do not have a defect in the TF/FVIIa pathway, small amounts of thrombin are generated in the initiation stage, and a fibrin clot is formed. However, the formation of additional thrombin is dependent upon the tenase complex. In patients with hemophilia, either FVIII or FIX is absent and the tenase complex cannot be formed. Thrombin formation is further limited because the TF/FVIIa pathway is down-regulated by an inhibitor to the TF/FVIIa complex known as tissue factor pathway inhibitor (TFPI). Since production of thrombin in patients with hemophilia is dependent upon the TF/FVIIa pathway, this action of TFPI accentuates the defect in thrombin formation and fibrin generation in patients with hemophilia. For those with Factor XIII deficiency, low levels of the protein (usually $< 1–2\%$) contribute to delayed bleeding due to lack of cross-linking of the fibrin clot. This lack of a stable fibrin clot formation then allows for greater fibrinolysis.

## Recommended Reading

Flaumenhaft R. (2003). Molecular basis of platelet granule secretion. *Arterioscler Thromb Vasc Biol* **23**, 1152–1160.

Furie B, Furie BC. (2007). In vivo thrombus formation. *J Thromb Haemost* **5**(Suppl. 1), 12–17.

Heemskerk JW, Bevers EM, Lindhout T. (2002). Platelet activation and blood coagulation. *Thromb Haemost* **88**, 186–193.

Italiano JE, Richardson JL, Patel-Hett S *et al.* (2008). Angiogenesis is regulated by a novel mechanism: pro- and antiangiogenic proteins are organized into separate platelet α granules and differentially released. *Blood* **111**, 1227–1233.

Jackson SP. (2007). The growing complexity of platelet aggregation. *Blood* **109**, 5087–5095.

# CHAPTER 79
# Approach to the Bleeding Patient
Thomas C. Abshire, MD

Patients who present with a bleeding disorder are common in both the inpatient and outpatient settings. It is essential to obtain a thorough history and focused physical examination in order to render a correct diagnosis. This information, along with laboratory screening, can guide the clinician towards a diagnosis. Neither specific nor extensive laboratory testing should be undertaken until these components have been gathered. This chapter will be divided into the history, physical examination and laboratory screening used for rendering a diagnosis.

**History:** A focused hematologic history is necessary to determine whether a patient has a bleeding disorder. Detailed questioning should be carried out on the following topics, for both the patient and family members.

**Mucosal Bleeding:** Skin findings such as purpura and petechiae can be important determinants of a potential bleeding disorder. The size of the bruises, as well as their location and whether they are associated with trauma, should be ascertained. For example, it is common for small children to have one or two small bruises below the knees, but it is rare to have more than five bruises/area, even if these are small (less than 1–2 cm in size). A larger bruise (greater than 2 cm in size), particularly if located on the trunk or upper extremity, and bruises associated with an underlying hematoma are not common in an individual with a normal hemostatic system.

Location of petechiae is also important. For example, a few petechiae underneath the eyes are probably normal in a crying child, but other facial or extremity petechiae should not be considered as normal. The presence of epistaxis should be discerned. Its frequency, duration, location to one nostril and seasonal variation are all important. Nosebleeds lasting greater than 15 minutes are probably clinically important. Any medical intervention (cauterization, iron therapy or other medication, transfusions, etc.) should be noted. Questioning should also take place regarding oral-related bleeding, including bleeding when tooth-brushing or flossing, gum bleeding and blood loss from tooth eruptions or from tooth loss, particularly after wisdom teeth have been extracted.

**Menorrhagia:** The presence of menorrhagia and treatment for menstrual bleeding should be assessed carefully. It is important to record the length of the menses, as well as the number of pads or tampons utilized each day. Menses of 7 days' duration is thought to be at the upper limit of normal. The clinician should ask if the pads or tampons utilized are partially or totally soaked with blood, and whether there are clots associated with the menstrual bleeding. It may also be helpful to question whether 1 or 2 days of menses are particularly "heavy," requiring frequent changing of pads/tampons that are completely saturated with blood, even if the total duration of menses is in the normal range. There are now well-defined scoring systems which aid in assessing the severity of menorrhagia (e.g. the PBAC score).

**Other Bleeding History:**  The patient should be assessed for a history of excessive bleeding post-childbirth, either by vaginal delivery or via cesarean section. The presence of surgically related bleeding or bleeding after minor cuts or lacerations (duration more than 5–10 minutes), especially those requiring sutures, should be tabulated. Poor wound healing and excessive scarring should be documented. Questions should be asked regarding symptoms commonly associated with hemarthrosis and/or muscle bleeding, as patients might be unaware that what they consider as a severe "muscle strain" or "sprained ankle" might actually be a muscle or joint bleed. Lastly, history of bleeding from the gastrointestinal or urinary tract should be ascertained.

**Bleeding in Infancy:**  A bleeding history during infancy should be obtained. This includes bleeding in the scalp at birth, bleeding from circumcision or delayed bleeding from the umbilical cord, bleeding from laboratory sampling obtained in the nursery, and subcutaneous tissue and/or muscle swelling greater than 3–4 cm associated with immunizations administered throughout the first year of life.

**Medications:**  Patients should be questioned regarding any prior transfusion history of plasma, platelets or red blood cell products. Additionally, the use of medications, particularly drugs that affect platelet function – such as aspirin, other anti-inflammatory medication and oral anticoagulants – should be ascertained. Many drugs might affect platelet function, and it is prudent for the clinician to verify all medications and herbal remedies.

**Factor Deficiency versus Platelet/Vessel Interaction Bleeding:**  Careful tabulation of a patient's bleeding symptoms can aid the clinician in differentiating between two broad categories of bleeding disorders: factor deficiency bleeding, or a platelet/vessel interaction defect. In general, factor deficiency bleeding is usually delayed, is often deeper in its tissue penetration, and may more commonly involve a joint or muscle. In contrast, platelet/vessel interaction bleeding is more immediate, can often be superficial, and frequently involves mucosal surfaces such as the nose and mouth, and the skin and the genitourinary tract.

**Physical Examination:** The physical examination can aid in the diagnosis of a bleeding disorder. Initial focus should be on the external examination, with careful attention to mucous membrane bleeding (nose, mouth and conjunctiva) and skin manifestations of purpura and petechiae. In the hospitalized patient, bleeding from IV sites, after blood draws, or in association with a chest tube or endotracheal tube can be particularly instructive. Swelling of the joint or muscle should also be carefully evaluated.

**Newborn Bleeding:**  Special attention needs to be focused on the newborn, who might present with a unique bleeding diathesis compared to older children and adults. The scalp should be assessed carefully for a cephalohematoma, the umbilical cord inspected for bleeding, and any circumcision-related bleeding assessed; the extremities should also be inspected for inappropriate leg swelling after intramuscular medication. Persistent oozing from a heel-stick blood sample might also be a clue to a bleeding disorder. Finally, a patient with an intracranial bleed should be assessed for possible bleeding conditions.

**Surgical versus Systemic Bleeding:** On the consultative service, patients frequently present who might have an underlying bleeding disorder or whose bleeding symptoms might be related to a treatment for which they are hospitalized. It is helpful to realize that local non-systemic bleeding which is confined to a single site (such as from a chest tube) is often surgical in nature, and not indicative of a coagulopathy. This bleeding is often quite brisk and laboratory assessment is normal. In contrast, systemic bleeding, such as might occur with an underlying or acquired bleeding disorder, involves multiple sites of bleeding, and bleeding which is often prolonged and/or delayed. The coagulation laboratory screening tests are often abnormal in this setting, further supporting a defect in hemostasis.

**Summary Comments:** A thorough bleeding history, both for the patient and the family, as well as a focused physical examination are essential in the clinical evaluation of congenital or acquired bleeding disorders. However, it is important to remember that family members might consider an abnormal bleeding and bruising pattern as being "normal," since other family members may present with a similar picture. In addition, patients (and children in particular) with a mild bleeding disorder such as VWD or a platelet function defect may have no symptoms of bleeding, since they have had little opportunity to be hemostatically stressed. If the patient's bleeding history is positive, specific laboratory testing needs to be performed. If the bleeding history and physical examination are completely negative, it is often prudent not to pursue additional laboratory testing unless there is a high clinical index of suspicion or a concerning family history. Moreover, lab testing should be more aggressive in the context of potentially higher-risk surgical bleeding, such as with spine or prostate surgery. In this case, screening or more specific laboratory testing should be considered.

**Laboratory Evaluation:** An overall assessment of the coagulation system can be provided by five screening tests: complete blood count (CBC) including a platelet count; prothrombin time (PT); activated partial thromboplastin time (PTT); thrombin time; and platelet function analyzer (PFA). The laboratory aspects of these tests are discussed in Chapters 113, 114 and 115. Each of these global coagulation tests is essential to focus the clinician on potentially serious causes of bleeding. The platelet count is important to discern the presence of thrombocytopenia, the PFA can sometimes be helpful in evaluating platelet dysfunction, and the PTT is essential to discern the presence of hemophilia. Elevation of the PT alone likely points to Factor VII deficiency. The fibrinogen level is essential in discerning hypofibrinogenemia, and a thrombin time is important to help diagnose a dysfibrinogenemia or heparin contamination.

**PT, PTT and Mixing Studies:** An elevated PT might be indicative of multiple factor deficiencies in the common pathway or underlying liver disease. When a prolonged PTT is discovered, it suggests a defect in the intrinsic pathway; this could be due to a factor deficiency or an inhibitor.

The presence of an inhibitor can be broadly assessed by performing a 1:1 mixing study, but normally a 1:1 mixing study should not be considered unless the initial PTT is prolonged greater than 5 seconds. Some laboratories perform this testing in a

4 : 1 ratio (to help differentiate a less robust inhibitor). If a mixing study is performed, correction of the PTT to the normal range suggests a factor deficiency such as FXII, FXI, FIX or FVIII (see Chapter 127).

Factor deficiencies can be further assessed by measuring specific factor levels. Levels below 40% (especially for FVIII and FIX) might contribute to bleeding. Lack of correction of the PTT suggests an inhibitor. Such inhibitors may promote bleeding (e.g. anti-FVIII), but laboratory inhibitors may also contribute to thrombosis *in vivo* (e.g. lupus anticoagulant); these inhibitors, including the lupus anticoagulant, will be discussed in more detail later in this book. A specific factor inhibitor (e.g. FVIII inhibitor) is suggested by partial correction of the immediate PTT and then prolongation to near its original baseline value at 1 hour (see Chapters 126 and 128).

**PFA and Thrombin Time:** Prolongation of the PFA or bleeding time suggests a defect in platelet/vessel interaction, with the most common condition being VWD or an intrinsic platelet function defect. A long thrombin time is suggestive of heparin contamination, and should be confirmed by treating the sample with heparinase and repeating it, or performing a heparin assay or a reptilase time. Excessive fibrin/fibrinogen degradation products (including the d-dimer), can also prolong the thrombin time. Additionally, a hypo- or dysfibrinogenemia, either on a congenital basis or related to liver disease, may also prolong a thrombin time. Finally, antibodies to thrombin, which can occur after exposing patients to thrombin glue, might also prolong this screening test.

Nonetheless, the most common culprit to prolong both the thrombin time and the PTT is heparin. Its ubiquitous presence in hospitals makes contamination of these screening tests problematic. Accordingly, most laboratories will address this concern by almost uniformly first adding a heparin neutralizing enzyme to a sample with an elevated PTT. Correction of the PTT to normal after heparinase confirms the diagnosis of heparin contamination. Of note, a sample can have both heparin contamination and an underlying coagulation defect; thus, specimens that partially correct after heparinase use should be further analyzed as above.

**Vascular Bleeding:** If the patient has a history of bleeding and all the screening tests previously mentioned are normal, then vascular causes of bleeding should be considered. These can include defects in collagen, such as Ehlers-Danlos or Marfan syndromes. Many viral infections and rickettsial diseases may also cause mucosal bleeding. Henoch-Schonlein purpura is a more common condition in pediatrics which should not be confused with a bleeding disorder. The typical appearance of reddish-looking lesions on the buttocks, as well as other clinical constellations, should focus the clinician on this diagnosis. Certain nutritional deficiencies, such as scurvy (vitamin C deficiency), should be considered. Corticosteroids may also cause bruising, and it must be remembered that senile purpura is relatively common in the elderly.

**Pre-analytical Errors and Pitfalls to Coagulation Testing:** The importance of heparin contamination to samples has already been discussed. Other specimen collection difficulties include traumatic venipuncture, slow blood draw producing an activated specimen, too little or too much blood in a citrated tube, and the presence of polycythemia (newborns and those that suffer from congenital heart disease are two examples who have less plasma and actually need less anti-coagulant in the blood tube).

Additionally, sample validity may be impacted if there is difficulty in not promptly centrifuging the specimen after collection and not separating the platelets from the sample, or if the sample is kept at room temperature for too long before processing. It is important that the local hospital/clinic ensures that the sample is drawn and processed properly before freezing it and sending it to the reference laboratory.

**Overview of Coagulation Testing:** Figure 79.1 summarizes utilization of screening laboratory tests in determining the cause of bleeding in an individual. In the setting

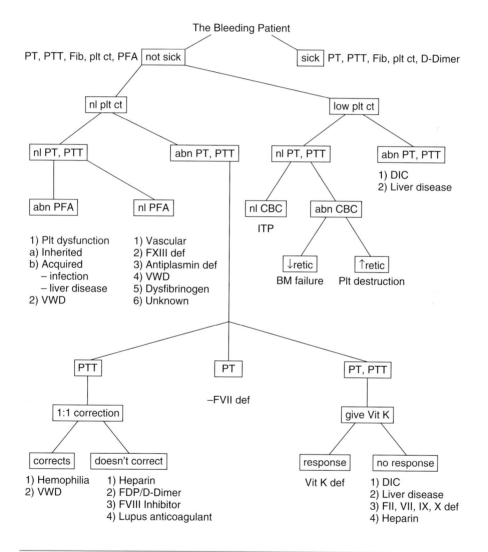

FIGURE 79.1 Overview of the approach to a bleeding patient. abn = abnormal, BM = bone marrow, def = deficiency, DIC = disseminated intravascular coagulopathy, FDP = fibrin degradation products, Fib = fibrinogen level, nl = normal, PFA = platelet function analyzer, plt ct = platelet count, plt = platelet, PT = prothrombin time, PTT = partial thromboplastin time, retic = reticulocyte count, Vit K = vitamin K, VWD = von Willebrand disease.

where laboratory testing is normal, obtaining a history as to the type of bleeding is helpful in discerning the disorder. Despite a normal screen, patients who present with more platelet-/vessel-related bleeding might still have either VWD or platelet dysfunction, or possibly a more rare disorder, such as plasminogen activator inhibitor deficiency (PAI-1). If the bleeding history is more similar to factor deficiency-related bleeding, this presentation might be secondary to FXIII deficiency, antiplasmin deficiency or mild hemophilia (FVIII, FIX or FXI). Suspicion for mild hemophilia is important, since the screening PTT may be normal in individuals with mildly decreased FIX levels.

## Recommended Reading

Hedlund-Treutiger I, Revel-Vilk S, Blanchette VS *et al.* (2004). Reliability and reproducibility of classification of children as "bleeders" versus "non-bleeders" using a questionnaire for significant mucocutaneous bleeding. *J Pediatr Hematol Oncol* **26**, 488–491.

Lillicrap D, Nair SC, Srivastava A *et al.* (2006). Laboratory issues in bleeding disorders. *Haemophilia* **12**(Suppl. 3), 68–75.

Rodeghiero F, Castaman G, Tosetto A *et al.* (2005). The discriminant power of bleeding history for the diagnosis of type 1 von Willebrand disease: an international, multicenter study. *J Thromb Haemost* **12**, 2619–2626.

Rodeghiero F, Tosetto A, Castaman G. (2007). How to estimate bleeding risk in mild bleeding disorders. *J Thromb Haemost* **5**(Suppl. 1), 157–166.

Tosetto A, Castaman G, Rodeghiero F. (2008). Bleeding scores in inherited bleeding disorders: clinical or research tools? *Haemophilia* **14**, 415–422.

Watson HG, Greaves M. (2008). Can we predict bleeding? *Semin Thromb Hemost* **34**, 97–103.

# CHAPTER 80

# Congenital Thrombocytopenia

Shawn M. Jobe, MD, PhD

This chapter will discuss a variety of disorders associated with congenital thrombocytopenia. Although these disorders are most frequently diagnosed in children, diagnosis as an adult is not uncommon. Patients with congenital thrombocytopenia may present with mild mucocutaneous bleeding or petechiae. Many congenital platelet disorders are associated with unique physical characteristics, specific findings on the peripheral blood smear, or other associated pathologies (Table 80.1). A thorough history, physical examination, and evaluation of the platelet size and smear can help differentiate congenital and acquired thrombocytopenias.

Congenital thrombocytopenias have been characterized in a number of different ways. Some causes of thrombocytopenia are associated with a specific pattern of inheritance (Table 80.2). Assessment of the mean platelet volume (MPV) can further narrow the diagnostic possibilities. The normal mean platelet volume (MPV) is 7–11 fl. Platelets smaller and larger than this, as well as giant platelets, are unique to particular congenital thrombocytopenias. In this chapter, the congenital thrombocytopenias are grouped by platelet size. Diagnosis of a specific congenital thrombocytopenia syndrome is most often based on the presence of associated clinical and laboratory abnormalities. In cases where the genetic mutation has been identified, DNA sequencing may aid in confirming the diagnosis.

Differential Diagnosis: Several elements of the history provide important clues that help distinguish congenital from acquired thrombocytopenia, especially immune thrombocytopenic purpura (ITP). A number of medications are associated with thrombocytopenia, including heparin, antibiotics and antipsychotic medications. Family and past medical history may also provide clues about whether the thrombocytopenia is congenital or recently acquired. Close evaluation of the peripheral blood smear and blood count, as well as evaluation for hepatosplenomegaly and lymphadenopathy, will help discern whether myeloproliferative, myelophthistic or aplastic causes might account for the thrombocytopenia.

Management: Bleeding manifestations of congenital thrombocytopenias are typically mild. Supportive hemostatic agents such as DDAVP and anti-fibrinolytic agents will often control bleeding in these patients. Platelet transfusion may be necessary in some cases. Special considerations for the management of specific congenital thrombocytopenias are described below.

Congenital Thrombocytopenias:

Thrombocytopenia with Small Platelets:

Wiskott-Aldrich Syndrome/X-linked Thrombocytopenia: Wiskott-Aldrich syndrome/X-linked thrombocytopenia is a moderate to severe X-linked thrombocytopenia

**TABLE 80.1** Clinical and Laboratory Features Associated with Congenital Thrombocytopenias

**Skin**
    Eczema      WAS gene mutation (Wiskott-Aldrich)

**Skeletal**
*Upper extremities*
    Absent radii with thumbs present/phocomelia      Thrombocytopenia with absent radii (TAR)

    Restricted pronation of the forearm (proximal radio-ulnar synostosis)      Amegakaryocytic thrombocytopenia with radio-ulnar synostosis

    Hand abnormalities (syndactyl, absent digits)      11q terminal deletion disorder

*Lower extremities*
    Various anomalies      Thrombocytopenia with absent radii (TAR)
    Hip dysplasia      Amegakaryocytic thrombocytopenia with radio-ulnar synostosis

**Cardiac**      GPIb/IX/V deficiency (DiGeorge, velocardiofacial syndrome)
     11q terminal deletion disorder
     Thrombocytopenia with absent radii (TAR)

**Renal**
    Hematuria, proteinuria      MYH9-related thrombocytopenia
    Congenital malformations      Thrombocytopenia with absent radii (TAR)

**Gastrointestinal**
    Milk-protein allergy      Thrombocytopenia with absent radii (TAR)

**Auditory**
    High tone sensorineural hearing loss      MYH9-related thrombocytopenia

**Ocular**
    Cataracts      MYH9-related thrombocytopenia

**Neurologic**
    Mental retardation      11q terminal deletion disorder

**Immunologic**
    Autoimmunity      WAS/XLT
    Immunodeficiency      WAS/XLT

**Hematologic**
    Thalassemia/anemia      GATA-1 mutation
    Stomatocytosis      Mediterranean stomatocytosis/thrombocytopenia

**TABLE 80.2** Inheritance Patterns of Congenital Thrombocytopenias, Other than Autosomal Recessive

| | |
|---|---|
| X-linked | WAS gene mutation |
| | Wiskott-Aldrich syndrome |
| | X-linked thrombocytopenia |
| | GATA-1 gene mutation |
| | X-linked thrombocytopenia and dyserythropoiesis with or without anemia |
| | X-linked thrombocytopenia/thalassemia |
| Autosomal dominant | MYH9-related thrombocytopenia |
| | Familiar platelet disorder/acute myeloid leukemia (FPD/AML) |
| | Thrombocytopenia with absent radii (TAR) |
| | Amegakaryocytic thrombocytopenia with radio-ulnar synostosis |
| | Heterozygous GPIb-IX-V deficiency/dysfunction |
| | Mediterranean macrothrombocytopenia |
| | Velocardiofacial syndrome/DiGeorge |
| | Platelet type VWD/type IIB VWD |
| | 11q terminal deletion disorder |

with small platelets (MPV 3.5–5.0 fl) that is often associated with eczema and immunodeficiency. WAS/XLT is caused by mutations in the *WAS* gene, which encodes WASp (Wiskott-Aldrich syndrome protein). The clinical presentation is variable. Some individuals have moderate thrombocytopenia (X-linked thrombocytopenia or XLT), while others have additional findings, including mild to severe eczema, cellular and humoral immunodeficiency, autoimmune diseases, and increased risk of lymphoma.

WASp plays a critical role in the regulation of actin polymerization and intracellular signaling in the activated platelet. Lower levels of WASp are present in patients with more severe phenotypes. In addition to the correlation of WASp levels with disease, genotype–phenotype correlation of disease phenotype with WASp mutations has also been identified. Missense mutations are associated with milder disease, while nonsense and larger gene deletions are associated with severe WAS.

Although splenectomy has been reported to improve the platelet count in individuals with XLT, it should be avoided, since asplenia may aggravate an underlying immunodeficiency. Treatment of individuals with XLT/WAS is primarily focused on the care of acute hemorrhages, and the prevention and treatment of infections and autoimmunity.

## Thrombocytopenia with Normal Platelet Size:

*Familial Platelet Disorder/Acute Myeloid Leukemia (FPD/AML):* Familial platelet disorder/acute myeloid leukemia (FPD/AML) is an autosomal dominant syndrome characterized by thrombocytopenia and a strong predisposition for the development of leukemia. Patients with FPD/AML are thrombocytopenic in the first decade of life, and have a mild bleeding diathesis. Abnormalities in platelet aggregation, particularly to arachidonic acid, have been reported. Up to 30% of individuals with FPD/AML develop myelodysplasia or acute myeloid leukemia by the sixth decade of life. Multiple hematologic malignancies have occurred in affected individuals, including myelodysplasia, acute myeloid leukemia (particularly of the M0 subtype) and lymphosarcoma. Bleeding symptoms are usually mild. Affected individuals should be counseled regarding the potential risk of leukemia development to them and affected family members.

FPD/AML is caused by a loss of function mutation in a single copy of the *AML1* gene (also known as *CBFA2* or *RUNX1*). AML1 is a transcription factor important in thrombopoiesis. Decreased expression of the AML1 protein resulting from the loss of a single allele (gene dosage effect) or a dominant negative effect of the mutated protein can result in the autosomal dominant presentation of FPD/AML.

*Thrombocytopenia with Absent Radii:* In its classical presentation, thrombocytopenia with absent radii (TAR) presents as neonatal thrombocytopenia in association with absent radii. However, severity of the upper limb abnormalities in patients with TAR can vary markedly, ranging from subtle abnormalities of the shoulder girdle to phocomelia, a complete absence of a portion of the upper limb. In addition to the thrombocytopenia and upper limb abnormalities, patients with TAR have a high incidence of renal and cardiac abnormalities and milk-protein allergy (50%).

The gene responsible for TAR remains unknown, but several candidate genes have been excluded, including Hox genes (involved in skeletal development) and the thrombopoietin receptor (c-mpl). Cellular studies suggest that a blockade of megakaryocyte maturation and a decreased responsiveness of megakaryocyte precursors to various cytokines may underlie the observed thrombocytopenia. The inheritance pattern of TAR is autosomal-recessive.

TAR is a diagnosis of exclusion. Prior to making the diagnosis of TAR, a chromosomal breakage syndrome should be excluded (distinguished by the absence of thumbs, and chromosomal breakage testing). Chromosomal analysis for the 22q11 microdeletion should also be performed, as this syndrome can also result in thrombocytopenia (see GPIb-IX-V receptor defects), and limb abnormalities, including absent radii, have been reported.

Although the thrombocytopenia observed in neonates with TAR can be quite severe, gradual remission of the thrombocytopenia frequently occurs. In older individuals with TAR, platelet counts often increase to low normal levels. Removal of milk-containing products has been reported to result in improvement of the thrombocytopenia in some cases.

*Amegakaryocytic Thrombocytopenia with Radio-ulnar Synostosis:* Individuals with this autosomal-recessive syndrome have neonatal thrombocytopenia, proximal radio-ulnar synostosis resulting in limitation of pronation, digital abnormalities (syndactly and clinodactly) and, commonly, hip dysplasia. Amegakaryocytic thrombocytopenia with radio-ulnar synostosis is caused by a mutation of the *HOXA11* gene, a homeobox gene involved in skeletal development. This disorder should be differentiated from TAR because unlike TAR, where thrombocytopenia frequently improves, worsening thrombocytopenia and involvement of other hematologic cell lines often necessitates allogeneic HPC transplant in affected individuals.

*Congenital Amegakaryocytic Thrombocytopenia:* Congenital amegakaryocytic thrombocytopenia (CAMT) is an autosomal-recessive neonatal thrombocytopenia caused by mutations to the *c-MPL* gene, encoding the TPO receptor c-mpl, which results in the absence in signaling through the mutated TPO receptor and therefore megakaryocyte development is aborted. Patients with CAMT have almost complete absence of megakaryocytes at bone marrow biopsy, and serum thrombopoietin (TPO) levels are markedly elevated. These patients have a gradual decline in other hematologic cell lines, progressing to bone marrow aplasia, which may neccesate allogeneic HPC transplantation.

*11q Terminal Deletion Disorder (Paris-Trousseau/Jacobsen Syndrome):* Patients with thrombocytopenia due to an 11q terminal deletion disorder (11q−) have platelets with abnormally large alpha granules, characteristic dysmorphic facies, abnormalities of the upper extremities, cardiovascular defects (52 of 93 patients), and mild to moderate mental retardation. 11q− associated thrombocytopenia is due to decreased levels of the transcription factor Fli-1, which is in the deleted region of chromosome 11. These patients have abnormally large alpha granules in a fraction of the platelet population, and numerous micromegakaryocytes on bone marrow evaluation. Although the thrombocytopenia due to 11q− terminal deletion disorder can be quite severe at a young age, it frequently resolves by adolescence. Bleeding manifestations of 11q− thrombocytopenia are variable, and platelet abnormalities and clinical symptoms of bleeding have been reported to persist after normalization of the platelet count.

## Thrombocytopenia with Large/Giant Platelets:

*GPIb-IX-V Receptor Defects (Bernard-Soulier, Mediterranean Macrothrombocytopenia, Velocardiofacial (DiGeorge) Syndrome):* GPIb-IX-V is the primary platelet receptor for VWF, and is composed of the products of four separate genes (Ibα, Ibβ, IX and V). Mutations of components of this receptor have been implicated as the cause of several thrombocytopenic syndromes, including Bernard-Soulier syndrome (BSS), Mediterranean macrothrombocytopenia, and the thrombocytopenia of velocardiofacial (22q11 microdeletion) syndrome. Manifestations range from asymptomatic thrombocytopenia to severe bleeding associated with BSS. GPIb-IX-V receptor defects are explored in more detail in Chapter 83.

*MYH-9-related Thrombocytopenia (May-Hegglin, Sebastian, Fechtner, Epstein):* *MYH-9*-related thrombocytopenia refers to a spectrum of autosomal dominant clinical syndromes characterized by macrothrombocytopenia and varying degrees of nephritis, hearing loss and cataracts. Bleeding symptoms are usually mild, and care is typically focused on screening and management of renal and auditory complications. Macrothrombocytopenia has been reported in all families with a mutation of the *MYH-9* gene, but penetrance of the other clinical features is variable. One study of families with 21 different causative mutations of *MYH-9* related disease demonstrated high-tone sensorineural hearing loss in 83%, some degree of nephritis (ranging from microscopic hematuria to renal failure) in 62%, and ocular cataracts in 23% of patients.

Patients with *MYH-9* related disorders were previously classified as having one of several different syndromes (May-Hegglin anomaly, Sebastian, Fechtner, and Epstein syndromes) based on their clinical presentation. Recently, identification of the causative gene of the May-Hegglin anomaly resulted in the recognition that these disorders are all the result of mutations within the same gene, *MYH-9*. The protein product of the *MYH-9* gene is non-muscle myosin heavy-chain IIA, a component of non-muscle myosin IIA.

The precise biological function of myosin IIA within the platelet is unknown, but it has been suggested to serve as a molecular motor facilitating changes in cell shape, phagocytosis, and organelle trafficking. Individual mutations of the gene encoding myosin IIA are associated with specific phenotypes. In one study, mutation of amino acid R702 was closely correlated with the presence of hearing loss and nephritis. In contrast, no hearing loss or nephritis was found in individuals with a mutation of amino acid E1841.

Diagnosis of the *MYH-9* related disorders is dependent on the observation of macrothrombocytopenia in association with the characteristic leukocyte inclusions known as Dohle-like bodies. On close observation, Dohle-like bodies can be identified in almost all patients with the syndrome. Platelets from these patients have been reported to have absence of platelet shape-change in platelet aggregation studies. The platelet aggregation pattern is normal, with the exception of a defective response to epinephrine.

*GATA-1 Related Thrombocytopenia:* GATA-1 related thrombocytopenia is a X-linked disease characterized by macrothrombocytopenia and varying degrees of dyserythropoiesis or thalassemia. The severity of erythroid abnormalities is highly variable, and includes thalassemia, mild marrow dyserythropoiesis, and transfusion-dependent anemia with marked anisoctyosis and poikilocytosis. This disorder is associated with mutations of the gene *GATA1*, which encodes the transcription factor GATA-1. GATA-1 is critical for normal megakaryocytopoiesis and erythropoiesis. These patients typically have a moderate bleeding diathesis with a greater than expected amount of bleeding relative to the observed platelet count. The platelets of patients with a GATA-1 mutation are large and have a paucity of alpha granules, and cytoplasmic clusters of smooth endoplasmic reticulum are often noted. Platelet function studies demonstrate abnormalities in platelet shape change following platelet activation and impaired platelet activation by collagen.

*Mediterranean Stomatocytosis/Macrothrombocytopenia:* Mediterranean stomatocytosis/macrothrombocytopenia is an autosomal recessive syndrome caused by mutation of either the *ABCG5* or the *ABCG8* gene. These patients have macrothrombocytopenia and stomatocytic hemolytic anemia with high levels of phytosterols, or plant sterols. Dietary and pharmacologic manipulation of phytosterol levels may be of some benefit in these patients, and prevent cardiac sequelae associated with phytosterolemia.

## Recommended Reading

Crispino JD. (2005). GATA1 in normal and malignant hematopoiesis. *Semin Cell Dev Biol* **16**, 137–147.

Greenhalgh KL, Howell RT, Bottani A *et al.* (2002). Thrombocytopenia-absent radius syndrome: a clinical genetic study. *J Med Genet* **39**, 876–881.

Grossfeld PD, Mattina T, Lai Z *et al.* (2004). The 11q terminal deletion disorder: a prospective study of 110 cases. *Am J Med Genet A* **129**, 51–61.

Ochs HD, Thrasher AJ. (2006). The Wiskott-Aldrich syndrome. *J Allergy Clin Immunol* **117**, 725–738.

Rees DC, Iolascon A, Carella M *et al.* (2005). Stomatocytic haemolysis and macrothrombocytopenia (Mediterranean stomatocytosis/macrothrombocytopenia) is the haematological presentation of phytosterolaemia. *Br J Haematol* **130**, 297–309.

Savoia A, Dufour C, Locatelli F *et al.* (2007). Congenital amegakaryocytic thrombocytopenia: clinical and biological consequences of five novel mutations. *Haematologica* **92**, 1186–1193.

Seri M, Pecci A, Di Bari F *et al.* (2003). MYH9-related disease: May-Hegglin anomaly, Sebastian syndrome, Fechtner syndrome, and Epstein syndrome are not distinct entities but represent a variable expression of a single illness. *Medicine (Baltimore)* **82**, 203–215.

Song WJ, Sullivan MG, Legare RD *et al.* (1999). Haploinsufficiency of CBFA2 causes familial thrombocytopenia with propensity to develop acute myelogenous leukaemia. *Nat Genet* **23**, 166–175.

Thompson AA, Nguyen LT. (2000). Amegakaryocytic thrombocytopenia and radio-ulnar synostosis are associated with HOXA11 mutation. *Nat Genet* **26**, 397–398.

# CHAPTER 81
# Neonatal Alloimmune Thrombocytopenia

Carolyn M. Bennett, MD

Neonatal alloimmune thrombocytopenia (NAIT) is secondary to maternal allo-immunization to fetal platelet antigens and subsequent fetal thrombocytopenia (akin to maternal red blood cell alloimmunization and hemolytic disease of the fetus and newborn [HDFN]). NAIT is the most frequent cause of thrombocytopenia in the first few days of life in otherwise healthy term infants, with an estimated incidence of 1 in 1000–2000 live births. The thrombocytopenia associated with NAIT is often severe and can result in serious bleeding, including intracranial hemorrhage (ICH). Rapid diagnosis and management is essential to prevent the consequences of life-threatening bleeding.

**Pathophysiology:** Exposure to antigen-positive fetal platelets results in mater-nal alloimmunization, with subsequent production of immunoglobulin G (IgG) alloantibodies. These antibodies pass transplacentally into the fetal circulation, bind to fetal platelets, induce clearance by macrophages, and result in fetal and neonatal thrombocytopenia.

Antibodies to the HPA-1a antigen is responsible for more than 75% of NAIT cases, with antibodies to the HPA-5b and other platelet antigens being implicated in a minority of cases. Although 2% of women are HPA-1a negative, only 10% of these individuals form anti-HPA-1a antibodies. Thus, HPA-1a fetomaternal incompatibility alone is not sufficient for antibody production, or NAIT. The risk of HPA-1a alloim-munization is highest in women who are HLA class II DRB3*0101.

**Clinical Manifestations:** NAIT typically occurs in full-term infants. Unlike HDFN, NAIT occurs in the first pregnancy in almost 50% of cases. The thrombocy-topenia, which can occur as early as 20 weeks gestation, is often severe (platelet count $< 20,000/\mu l$). Major bleeding, particularly ICH, occurs in 10–20% of untreated NAIT cases. In other cases, affected infants may present with asymptomatic severe thrombo-cytopenia, or with mild bleeding symptoms such as petechiae or purpura.

**Diagnosis:** In many cases of NAIT, there is no previous history of an affected sibling and the thrombocytopenia is unanticipated. Other known causes of thrombocytope-nia must be rapidly excluded so that appropriate therapy can be initiated. Examination of the peripheral blood smear for other abnormalities and confirmation of throm-bocytopenia is important. An accurate maternal medical history is critical, including medication history and history of maternal hypertension, idiopathic thrombocyto-penic purpura (ITP), systemic lupus erythematosus, other autoimmune disorders and inherited thrombocytopenias.

Laboratory diagnosis of NAIT is based on the presence of maternal platelet alloan-tibodies that bind paternal platelets. In order to determine maternal platelet alloim-munization, rapid platelet antibody testing is typically performed by ELISA on maternal serum; alternative platelet antibody testing techniques include the mono-clonal antibody-specific immobilization of platelet antigens (MAIPA) assay. In order to determine HPA incompatibility between the mother and neonate, usually maternal and paternal platelets are phenotyped and/or genotyped (to prevent drawing blood samples from the neonate). Because the results of these tests may require days to complete, treatment should be initiated based on clinical findings.

**Differential Diagnosis:** The differential diagnosis of infants with thrombocyto-penia includes infection, disseminated intravascular coagulopathy, and necrotizing enterocolitis. Unlike infants with NAIT, however, infants with these conditions may be born with normal platelet counts, which then fall as they become increasingly ill. Chronic hypoxia, intrauterine growth retardation, birth asphyxia and congenital infec-tions should also be included in the differential diagnosis for sick and/or premature infants with thrombocytopenia. Inherited platelet disorders should likewise be consid-ered, such as Bernard-Soulier syndrome, Wiskott-Aldrich syndrome (usually in boys), Fanconi anemia, thrombocytopenia absent radii, amegakaryocytic thrombocytopenia with radio-ulnar synostosis and congenital amegakaryocytic thrombocytopenia.

The differential diagnosis should also include transplacental transfer of maternal platelet autoantibodies. However, the vast majority of infants born to mothers with ITP have platelet counts above 50,000/µl, with ICH being less common in these infants than in those with NAIT.

**Management:** The management of NAIT should be focused on the prevention of severe bleeding, particularly ICH.

**Postnatal Management:** Once the diagnosis of NAIT is considered likely, treatment should be initiated until an alternative diagnosis has been confirmed. Thrombocytopenic infants with platelets under 30,000/µl and/or evidence of bleed-ing should be transfused platelets; the transfusion threshold for ill or preterm infants is lower. The platelet count should be checked within an hour after the transfusion to document response.

Ideally, antigen-negative platelet products (usually HPA-1a and HPA-5b negative are appropriate, as these are the offending antigens in ~90% cases) should be transfused, if they are available at the local blood center. The most reliable source of compatible, anti-gen-negative platelets is the mother, but these may be difficult to obtain in a timely fash-ion. If maternal platelets are transfused, they must be washed to remove the offending alloantibodies from the plasma, and irradiated to prevent transfusion-associated graft-versus-host disease. Importantly, if antigen-negative platelet transfusions are not available, non-antigen negative or antigen untested platelets should not be withheld. Non-antigen negative platelets may result in a lower platelet increment than antigen-negative platelets, but often achieve an adequate response.

Additional treatment approaches, which are often combined with platelet transfu-sion, include intravenous immunoglobulin (IVIG) at a dose up to 1 g/kg per day for 1–2 days. Steroids have not been shown to be effective in NAIT.

Evaluation of the patient for ICH by cranial ultrasound is critical, given the high morbidity/mortality associated with this complication. The risk of ICH is highest dur-ing the first 96 hours of life.

**Antenatal Management:** A history of a previously affected infant with NAIT, a family history of NAIT or the presence of maternal platelet alloantibodies during pregnancy are risk factors for NAIT. Fetuses are at high risk for an adverse outcome if they are antigen positive and have a previously affected sibling, because more severe disease presents earlier in subsequent pregnancies. Evidence of *in utero* ICH has been reported as early as the sixteenth week of gestation. For these reasons, at-risk pregnancies must be managed carefully to avoid the devastating consequences of fetal or neonatal ICH.

Paternal platelet antigen genotyping should be done to determine the risk of the fetus being affected with NAIT. If the father is heterozygous for the antigen in question, fetal HPA genotyping should be performed. Samples for fetal genotyping have historically been obtained via amniocentesis.

Although the optimal strategies for preventing NAIT are controversial, most include the use of frequent fetal monitoring and weekly maternal IVIG (typical dose of 1 g/kg per week). IVIG therapy may be started as early as 12 weeks gestation in high-risk pregnancies. More typically, IVIG is started at 20 weeks gestation, with or without the addition of steroids.

Fetal blood sampling may be used to directly assess fetal platelet counts beginning at 20 weeks gestation. The risks and benefits of this procedure must be carefully weighed. Offending antigen-negative, irradiated platelets must be available during the procedure, with transfusion typically indicated if the fetal platelet count is < 50,000/µl. Weekly intrauterine platelet transfusions may be necessary in severe cases of fetal NAIT.

**Delivery:** Most experts advocate delivery by cesarean section for fetuses with NAIT and platelet counts < 50,000/µl, with a trend towards deliveries earlier in gestation (after fetal lung maturity) for severely affected infants.

**The Future:** Routine screening of women during pregnancy for the presence of platelet alloimmunization, especially anti-HPA-1a, is under investigation in some countries.

## Recommended Reading

Allen D, Verjee S, Rees S et al. (2007). Platelet transfusion in neonatal alloimmune thrombocytopenia. *Blood* **109**, 388–389.

Bassler D, Greinacher A, Okascharoen C et al. (2008). A systematic review and survey of the management of unexpected neonatal alloimmune thrombocytopenia. *Transfusion* **48**, 92–98.

Bussel JB, Primiani A. (2008). Fetal and neonatal alloimmune thrombocytopenia: progress and ongoing debates. *Blood Rev* **22**, 33–52.

Ghevaert C, Campbell K, Walton J et al. (2007). Management and outcome of 200 cases of fetomaternal alloimmune thrombocytopenia. *Transfusion* **47**, 901–910.

Kiefel V, Bassler D, Kroll H et al. (2006). Antigen-positive platelet transfusion in neonatal alloimmune thrombocytopenia (NAIT). *Blood* **107**, 3761–3763.

Overton TG, Duncan KR, Jolly M et al. (2002). Serial aggressive platelet transfusion for fetal alloimmune thrombocytopenia: platelet dynamics and perinatal outcome. *Am J Obstet Gynecol* **186**, 826–831.

Roberts I, Stanworth S, Murray NA. (2008). Thrombocytopenia in the neonate. *Blood Rev* **22**, 173–186.

van den Akker ES, Oepkes D, Lopriore E *et al.* (2007). Noninvasive antenatal management of fetal and neonatal alloimmune thrombocytopenia: safe and effective. *Br J Obstet Gynaecol* **114**, 469–473.

# CHAPTER 82
# Acquired Neonatal Thrombocytopenia
Carolyn M. Bennett, MD

Thrombocytopenia is one of the most common hematologic abnormalities in newborns. By the second trimester of pregnancy, a normal healthy fetus has a platelet count greater than 150,000/μl; therefore, a platelet count below this value is abnormal and warrants further investigation. Fortunately, clinically significant thrombocytopenia (under 50,000/μl) is unusual. Thrombocytopenia can be seen in healthy full-term infants, but is more frequently encountered in sick or premature infants who are admitted to the neonatal intensive care units of tertiary care centers. For many infants the thrombocytopenia is mild and resolves without intervention, or with treatment of the underlying illness. However, in some cases thrombocytopenia can lead to serious bleeding, including intracranial hemorrhage (ICH). Thus, rapid recognition and treatment of thrombocytopenia is necessary. In this chapter we will review the acquired causes of neonatal thrombocytopenia, except for neonatal alloimmune thrombocytopenia (NAIT) which is reviewed in Chapter 81. The congenital thrombocytopenias are reviewed in Chapter 80.

**Pathophysiology:** It is important to determine the underlying cause of thrombocytopenia, because the treatment is determined by the mechanism of underlying pathology. In general, neonatal thrombocytopenia results from decreased platelet production (usually in sick or premature infants), increased platelet destruction (immune or consumptive), or platelet pooling secondary to splenomegaly (rare). These mechanisms are not mutually exclusive, and some infants (particularly sick or premature infants) present with a combination of these mechanisms.

**Decreased Production:** Thrombocytopenia in sick neonates is thought to be secondary to decreased megakaryopoiesis as a result of neonatal asphyxia or infection. In very sick neonates with disseminated intravascular coagulopathy (DIC), there may be platelet destruction as well. Common causes of thrombocytopenia in sick neonates are bacterial or viral sepsis with or without DIC, necrotizing enterocolitis (NEC), perinatal asphyxia or placental insufficiency, and respiratory distress syndrome. Infants with congenital infections, including rubella, herpes, cytomegalovirus, enterovirus and HIV, can also present with thrombocytopenia due to poor platelet production.

**Increased Destruction:** Infants with immune-mediated thrombocytopenia (either alloimmune or autoimmune) have increased platelet destruction, usually without decreased production. A compensatory increase in megakaryopoiesis is often observed in the bone marrow. Neonatal autoimmune thrombocytopenia can occur in infants born to women with ITP. Maternal antiplatelet IgG autoantibodies cross the placenta and bind to neonatal (and fetal) platelets, and cause platelet destruction by macrophages in the reticuloendothelial system. Thrombocytopenia occurs in about 10% of infants whose mothers have ITP.

Kasabach-Merritt syndrome (KMS) is neonatal thrombocytopenia and consumptive coagulopathy caused by giant kaposiform hemagioendothelioma (KHE). Infants with KMS usually have severe thrombocytopenia, and often have evidence of DIC. While the pathophysiology has not been fully established, the thrombocytopenia and coagulopathy is presumed to be due to platelet trapping on the endothelium of the KHE, but can also be the result of DIC.

Neonatal thrombocytopenia may also occur secondary to thrombosis of a major vessel. The thrombocytopenia is thought to result from platelet consumption at the site of thrombosis.

**Clinical Manifestations:** The presenting features of neonatal thrombocytopenia can range from the subtle to the obvious, and often do not correlate predictably with platelet count. Minor bleeding manifestations include petechiae, ecchymoses, purpura, mucocutaneous bleeding, bleeding from venipuncture or circumcision sites, blood-tinged secretions or heme positive stools. More serious bleeding, such as pulmonary, GI or ICH, is less common but does occur, particularly in sick or premature infants with severe thrombocytopenia or in infants with NAIT.

**Diagnosis:** The rapid identification of thrombocytopenia and its cause is imperative in both sick and well neonates, because delays in diagnosis and treatment can have serious consequences. In order to make the diagnosis, maternal history (including medications), perinatal history, physical examination, laboratory testing and sometimes radiographic evaluations are necessary.

*Thrombocytopenia in the Sick or Premature Infant:* The most common cause of early onset thrombocytopenia is chronic fetal hypoxia resulting from pregnancy-induced hypertension, diabetes, maternal hemolysis, elevated liver enzymes and low platelets (HELLP) syndrome or another maternal factor. In these infants the thrombocytopenia is often mild or moderate, resolving within 10 days in the majority of cases. Often these infants have associated findings, such as neutropenia, increased circulating nucleated red cells and evidence of hyposplenism (spherocytes, target cells and Howell-Jolly bodies). Other early causes of thrombocytopenia are prenatal viral infections (TORCH infections, particularly cytomegalovirus), perinatal bacterial infections (group B *Streptococcus*, *Escherichia coli* and *Haemophilus infuenzae*) and perinatal asphyxia. The thrombocytopenia is usually mild to moderate, but can be severe, particularly if DIC is present. Late onset thrombocytopenia, which presents after the first 3 days of life, is caused by bacterial sepsis or NEC in more than 80% of cases. Thrombocytopenia usually develops rapidly over 1–2 days, and is often very severe (under 30,000/µl) with an increased risk of bleeding.

*Autoimmune Neonatal Thrombocytopenia:* This diagnosis should be considered in infants born to mothers with a history of ITP and in mothers with thrombocytopenia. Approximately 10% of infants born to such mothers have clinically apparent thrombocytopenia. Usually these infants are otherwise healthy and have no physical findings besides those resulting from thrombocytopenia. The thrombocytopenia is typically mild, and severe life-threatening bleeding is rare. The risk of ICH in neonates with autoimmune thrombocytopenia is low, but is still higher than older children with ITP.

*Kasabach-Merritt Syndrome:* KMS presents with thrombocytopenia and consumptive coagulopathy in the setting of a large KHE. KMS should be considered in infants who present with unexplained thrombocytopenia. Patients present with thrombocytopenia usually within the first few weeks of life. The KHE is generally solitary, and usually involves the extremities, neck or trunk, although sometimes affects the viscera. Only a small percentage of infants with KHE develop thrombocytopenia and KMS. KHE are cutaneous and visible on physical examination, but radiographic imaging of the belly or retroperitoneum may be necessary to diagnose visceral KHE. The KHE of KMS usually have a period of rapid growth, followed by spontaneous resolution over several weeks to months. During the period of rapid growth, severe thrombocytopenia can occur and require supportive therapy. Infants with KMS often present with evidence of DIC with platelets under $50,000/\mu l$, hypofibrinogenemia and increased fibrin degradation products. DIC can be severe and the mortality high, therefore aggressive treatment is often necessary.

*Thrombosis:* Neonatal thrombocytopenia may occur as a secondary event after thrombosis of a major vessel. Thrombosis is common in sick neonates, especially in infants with indwelling arterial and/or venous catheters. The diagnosis of neonatal thrombosis is made by demonstration of the clot by an imaging study such as ultrasound with Doppler or MRI.

**Differential Diagnosis:** Many conditions are associated with neonatal thrombocytopenia, and therefore the differential diagnosis is broad. In neonates presenting early in life with thrombocytopenia, inherited platelet disorders should be considered including Bernard-Soulier syndrome and other giant platelet syndromes, Wiskott-Aldrich syndrome (usually in boys), Fanconi anemia, thrombocytopenia absent radii, amegakaryocytic thrombocytopenia with radio-ulnar synostosis and congenital amegakaryocytic thrombocytopenia. Sick and/or premature infants with chronic hypoxia, intrauterine growth retardation, birth asphyxia, perinatal infection and sepsis, and congenital infection also present with thrombocytopenia in the first days of life. Infants initially presenting with thrombocytopenia after the first days of life usually have bacterial infections or NEC. Other disorders which should be considered include aneuploidies (trisomy 13, 18 and 21), Noonan syndrome, KMS, metabolic disease and thrombosis. Well infants with moderate to severe thrombocytopenia should be evaluated for neonatal alloimmune and autoimmune thrombocytopenia.

**Management:** The risk of bleeding in neonates with thrombocytopenia varies greatly depending on the underlying cause. Infants with thrombocytopenia due to IUGR have a relatively low risk, and those with sepsis, NEC and autoimmune thrombocytopenia are at intermediate risk. There is very little evidence-based medicine to guide the management of neonates with thrombocytopenia. While platelet transfusions are routinely given to asymptomatic thrombocytopenic neonates prophylactically to prevent bleeding, there are no clinical trials to indicate whether transfusion reduces hemorrhage or improves outcome.

*Thrombocytopenia in the Sick or Premature Infant:* Recognition of the disorders causing thrombocytopenia in sick neonates is based on associated clinical manifestations. Treatment should be focused on the underlying disorder. Infants who have evidence of bleeding or who are seriously ill should be managed with platelet transfusions to keep the platelet count over 50,000/μl, particularly in the first week of life when the risk of ICH is increased. Infants who are clinically stable and who are without additional risk factors for bleeding may tolerate platelet counts of 20,000–30,000/μl, particularly after the first week of life. Sick infants with DIC and coagulopathy should be supported with plasma or cryoprecipitate to keep the fibrinogen over 100 mg/dl. Other supportive measures, such as antibiotics, fluids, vasopressors, mechanical ventilation and red cell transfusions, should be administered as necessary. In stable infants without bleeding with platelets over 20,000–50,000/μl, close observation may be the only treatment necessary until the platelet count recovers.

*Autoimmune Thrombocytopenia:* Infants with NAIT are often healthy term infants with few other presenting signs or symptoms other than thrombocytopenia and mucocutaneous bleeding. Often no treatment is necessary if the platelets are over 50,000/μl, but close observation for bleeding symptoms and serial platelet counts is necessary until the platelet count is stable or rising. The risk of ICH in neonates with autoimmune thrombocytopenia is lower than NAIT, but is higher than that of older children with ITP, particularly in the first week of life; therefore, intervention is often required. In children whose platelet count falls below 50,000/μl or in whom there is overt bleeding, treatment to increase the platelet count is indicated. The most commonly used treatments for neonatal ITP are intravenous immunoglobulin (IVIG, 1 g/kg for 1–2 days) and/or corticosteroids (methylprednisolone 2 mg/kg per day). Both therapies are usually effective in raising the platelets to safe levels within 24–72 hours of initiation. If serious, life-threatening bleeding occurs, IVIG, corticosteroids and platelet transfusions may be necessary to stop the bleeding.

*Kasabach-Merritt Syndrome:* Supportive care with blood products is usually necessary to manage hemorrhagic complications while treatment is initiated. Platelet transfusions should be given for thrombocytopenia and bleeding. Infants presenting with DIC may require plasma infusions and/or cryoprecipitate to correct ongoing coagulopathy.

Several treatment options have been used effectively to manage the KHE of KMS. The ultimate treatment strategy depends on the clinical setting and response. When possible, surgical resection of the KHE is an effective treatment; however, the lesions are often not amenable to surgical intervention. Compression therapy with pressure dressings or pneumatic compression devices can be effective as adjuvant therapy in patients with limb involvement. Some lesions can be treated with vascular ligation or embolization, primarily those with a single vascular supply. Corticosteroids are effective in some patients, and are the most commonly used first-line medical therapy, but up to one-third will not respond. Alternative therapies include chemotherapy (vincristine, cyclophosphamide) and α-interferon (α-IFN). However, irreversible spastic diplegia has been reported in up to 20% of patients treated with α-IFN, therefore α-IFN should be reserved for life-threatening cases. Anticoagulants (heparin and antithrombin III), antiplatelet agents (ticlopidine) and antfibrinolytic drugs (ε-aminocaproic acid and tranexamic acid) have been used to manage the consumptive coagulopathy, but there is little evidence regarding efficacy and safety. Despite aggressive therapy, the mortality from uncontrollable hemorrhage remains as high as 20–30%.

*Thrombosis:* Treatment should be directed at treating the thrombosis, and may involve removal of a catheter, directed thrombolysis and/or systemic anticoagulation. Platelet infusions should be given to keep the platelets over 50,000/µl, especially in patients receiving thrombolytic or anticoagulant therapy.

## Recommended Reading

Christensen RD, Henry E, Wiedmeier SE *et al.* (2006). Thrombocytopenia among extremely low birth weight neonates: data from a multihospital healthcare system. *J Perinatol* **26**, 348–353.

Hall GW. (2001). Kasabach-Merritt syndrome: pathogenesis and management. *Br J Haematol* **112**, 851–862.

Kenton AB, Hegemier S, Smith EO *et al.* (2005). Platelet transfusions in infants with necrotizing enterocolitis do not lower mortality but may increase morbidity. *J Perinatol* **25**, 173–177.

Marks SD, Massicotte MP, Steele BT *et al.* (2005). Neonatal renal venous thrombosis: clinical outcomes and prevalence of prothrombotic disorders. *J Pediatr* **146**, 811–816.

Roberts I, Stanworth S, Murray NA. (2008). Thrombocytopenia in the neonate. *Blood Rev* in press.

Webert KE, Mittal R, Sigouin C *et al.* (2003). A retrospective 11-year analysis of obstetric patients with idiopathic thrombocytopenic purpura. *Blood* **102**, 4306–4311.

# CHAPTER 83

# Bernard-Soulier Syndrome and Other GPIb-IX-V Related Receptor Defects

Shawn M. Jobe, MD, PhD

Mutations in the major platelet receptor for von Willebrand factor (VWF), GPIb-IX-V, result in a number of different syndromes. Clinical manifestations of these syndromes range from a severe platelet function defect in patients with a homozygous deficiency, to mild thrombocytopenia in patients with heterozygous deficiency. Platelet-type VWD, a syndrome similar to type 2B VWD, also is caused by mutations in the GPIb-IX-V receptor.

**Bernard-Soulier Syndrome:** Bernard-Soulier syndrome (BSS) is an autosomal recessive platelet disorder characterized by thrombocytopenia, the presence of giant platelets, and defective ristocetin-induced platelet agglutination.

BSS occurs when a mutation is present in both alleles of one of the components of the platelet receptor GPIb-IX-V. GPIb-IX-V is the receptor that is predominantly responsible for the von Willebrand factor (VWF)-dependent adhesion of platelets in damaged arteries. The absence of GPIb-IX-V results in greatly diminished platelet accumulation at sites of vessel injury, and bleeding. GPIb-IX-V is a multimeric receptor complex composed of the products of four separate genes (Ibα, Ibβ, IX and V). Homozygous and compound heterozygous mutations responsible for BSS have been identified in GPIbα, GPIbβ and GPIX components. Absence of any one of these components results in markedly decreased cell-surface expression of the entire GPIb-IX-V complex, since intracellular association of the GPIbα, GPIbβ and GPIX subunits is necessary for the appropriate trafficking of GPIb-IX-V. BSS caused by mutations in GPV have not been identified.

Giant platelets are a characteristic feature of BSS, and it is not uncommon to see platelets as large as 20µm in diameter (RBC are ~8µm). Platelet counts in individuals with BSS range from 20,000/µl to near normal. Platelet aggregation studies in patients with BSS demonstrate a poor agglutination response to ristocetin and botrocetin. Both ristocetin and botrocetin bind and activate the GPIb-IX-V receptor. Activation of the GPIb-IX-V receptor allows the binding of VWF and platelet agglutination results. Continued absence of agglutination when tested in the presence of added normal platelet-poor plasma indicates that the lack of agglutination is caused by a defect in the patient's GPIb-IX-V receptor, and not by absent or dysfunctional WVF. GPIb-IX-V levels determined by flow cytometry typically demonstrate the absence of the GPIb-IX-V receptor complex from the platelet surface. However, variant forms of BSS have been identified that have expression of a dysfunctional GPIb-IX-V complex on the platelet surface.

BSS patients frequently have a clinically significant bleeding disorder. Mild to moderate bleeds can be adequately treated using DDAVP and anti-fibrinolytic agents. Platelet transfusion is recommended for more severe bleeding manifestations.

**Benign Mediterranean Macrothrombocytopenia:** Mediterranean macrothrombo-cytopenia is an autosomal dominant disorder characterized by moderate thrombocyto-penia, large platelets and a mild bleeding diathesis. The bleeding symptoms can usually be adequately treated using DDAVP and anti-fibrinolytic agents. A causative mutation (Ala156Val) in the gene encoding GPIbα is present in approximately 50% of affected individuals.

Mediterranean macrothrombocytopenia was first described in 1975 in a study that compared 145 asymptomatic Mediterranean and 200 Northern European subjects. In this study, an increased incidence of individuals with thrombocytopenia and large platelets was noted in the Mediterranean population. Recently, a heterozygous mis-sense mutation (Ala156Val) of GPIbα was identified as a potentially important cause of this disorder. Sequencing of consecutive patients who presented to an Italian clinic with symptoms of Mediterranean macrothrombocytopenia identified the Ala156Val mutation in 6 of the 12 patients. This relatively high incidence suggests that Ala156Val mutations in GPIbα may account for a large proportion of cases of macrothrombocy-topenia in Mediterranean populations. The remaining gene mutations responsible for Mediterranean macrothrombocytopenia are unknown.

**Thrombocytopenia and Velocardiofacial (DiGeorge) Syndrome:** Macro-thrombocytopenia often occurs in patients with hemizygous deletion of the chromo-somal region 22q11. This deletion results in a constellation of syndromes called the velocardiofacial syndrome (VCF). Characteristics of patients with VCF include a typi-cal facies, hypocalcemia, thymic aplasia, cardiac anomalies and learning disabilities.

More than half of the 22q11 deletions in patients with VCF include the region encoding GPIbβ. Heterozygous absence of the GPIbβ component of the GPIb-IX-V complex accounts for the mild thrombocytopenia observed in approximately 40% of patients with VCF. Clinically significant bleeding is rarely reported. However, deficien-cies of other genes within the deleted region may result in immune dysregulation, and patients with VCF have an increased tendency for autoimmune thrombocytopenia.

**Platelet-type von Willebrand Disease (Gain of Function Mutation of GPIb-IX-V):** Patients with platelet-type von Willebrand disease (PT-VWD) present clini-cally with thrombocytopenia and low von Willebrand factor (VWF) levels. Significant mucocutaneous bleeding can occur in patients with PT-VWD. Platelet transfusion is typically used as therapy when other supportive measures have been attempted. DDAVP administration or transfusion with plasma-derived VWF concentrates may provide some benefit, but should be used with caution as they may aggravate the thrombocytopenia.

PT-VWD is caused by a unique set of mutations in the GPIbα subunit of the GPIb-IX-V complex. Causative mutations for platelet-type VWD have been identi-fied in the VWF binding domain of GPIbα and a region of GPIbα called the macro-glycopeptide region. These mutations cause GPIbα to bind with VWF with increased affinity. Spontaneous binding of circulating VWF to the platelet GPIb-IX-V recep-tor results in a low plasma VWF level. Thrombocytopenia might result either from clearance of VWF-bound platelets, or from impaired thrombopoiesis due to increased megakaryocyte GPIb-IX-V receptor activation.

Laboratory studies in PT-VWD and type 2B VWD are similar, and distinguishing these two clinical entities can be difficult. Both PT-VWD and type 2B VWD result in mildly decreased levels of plasma VWF and a disproportionate decrease in high molecular weight VWF multimers. Both result in thrombocytopenia, and both show an enhanced platelet agglutination response to ristocetin that is particularly observable at low ristocetin concentrations. In PT-VWD, the enhanced ristocetin agglutination response is due to GPIbα mutations that increase the binding affinity of the platelet GPIb-IX-V receptor to VWF. In type 2B VWD the enhanced agglutination response is due to specific mutations in VWF in its GPIbα binding domain (A1 domain) that increase its binding affinity to the platelet GPIb-IX-V receptor. The two can be distinguished by testing of VWF binding to platelets or by DNA sequencing (Chapter 123).

## Recommended Reading

Kato T, Kosaka K, Kimura M *et al.* (2003). Thrombocytopenia in patients with 22q11.2 deletion syndrome and its association with glycoprotein Ib-beta. *Genet Med* **5**, 113–119.

Lopez JA, Andrews RK, Afshar-Kharghan V, Berndt MC. (1998). Bernard-Soulier syndrome. *Blood* **91**, 4397–4418.

Savoia A, Balduini CL, Savino M *et al.* (2001). Autosomal dominant macrothrombocytopenia in Italy is most frequently a type of heterozygous Bernard-Soulier syndrome. *Blood* **97**, 1330–1335.

# CHAPTER 84

# Glanzmann Thrombasthenia

Shawn M. Jobe, MD, PhD

Glanzmann thrombasthenia (GT) is an autosomal-recessive platelet function disorder where platelet appearance and platelet number are unaffected. GT patients present with platelet-type bleeding, which may be severe, such as purpura, epistaxis, oral mucosal bleeding, menorrhagia or gastrointestinal bleeding.

**Pathophysiology:** Mutations in both alleles of one of the genes that encode for the $\alpha_{IIb}$ or $\beta_3$ polypeptide of the platelet integrin receptor $\alpha_{IIb}\beta_3$ result in GT. $\alpha_{IIb}\beta_3$ is the integrin receptor that mediates platelet adhesion to fibrinogen and VWF, and is normally expressed at high levels on the platelet surface. In the absence of functional $\alpha_{IIb}\beta_3$, both platelet adhesion and platelet recruitment, or aggregation are markedly impaired.

The majority of causative mutations in patients with GT result in severely decreased platelet surface expression of $\alpha_{IIb}\beta_3$. Mutations in the $\alpha_{IIb}$ or $\beta_3$ component result in a similar phenotype. Reported mutations include point mutations, splice defects, and small deletions and gene inversion.

**Variant GT:** Variant forms of GT have been identified in which normal or modestly decreased levels of $\alpha_{IIb}\beta_3$ are expressed on the platelet surface. The phenotype in patients with variant GT is due to expression of a dysfunctional $\alpha_{IIb}\beta_3$ protein on the platelet surface. Mutations in both the extracellular and intracellular domains of $\alpha_{IIb}\beta_3$ can result in variant GT. Events that occur at the $\alpha_{IIb}\beta_3$ intracellular cytoplasmic tail upon platelet activation are required for normal platelet function. These platelet activation events result in a change of $\alpha_{IIb}\beta_3$ structure from a "bent" inactivated form to an activated and extended ligand binding conformation, a process known as inside-out signaling. Mutations in $\alpha_{IIb}\beta_3$ that interfere with this inside-out signaling process have been identified. Another group of patients with variant GT have mutations in the MIDAS (metal-ion-dependent adhesion site) extracellular domain of $\beta_3$. MIDAS domain mutations result in defective receptor activation, impaired ligand binding, or instability of the receptor complex on the platelet surface. Constitutive activation of $\alpha_{IIb}\beta_3$ can also result in variant GT. Platelets from a GT patient with an extracellular mutation (Cys560Arg) in $\beta_3$ spontaneously bound fibrinogen, perhaps preventing appropriate engagement of the platelet receptor at sites of injury. Another variant form of GT, due to a mutation in the integrin binding protein kindlin-3 occurs together with a leukocyte adhesion defect. Integrins present on leukocytes and platelets, including $\beta_1$, $\beta_2$ and $\beta_3$, are all dysfunctional.

**Diagnosis:** Platelet aggregation studies in GT patients demonstrate absent or greatly diminished responses to all agonists. Response to the agglutinating agent, ristocetin, is unaffected, since platelet agglutination that occurs in response to ristocetin is mediated by the GPIb-IX-V receptor complex and is not $\alpha_{IIb}\beta_3$ dependent. Flow cytometry studies for CD41 ($\alpha_{IIb}$) or CD61 ($\beta_3$) can be used to distinguish patients with absent $\alpha_{IIb}\beta_3$ expression from patients with variant GT. $\alpha_{IIb}\beta_3$ can be identified on the surface of platelets of patients with variant GT. Most GT patients will have greatly decreased or absent surface levels of $\alpha_{IIb}\beta_3$.

**Management:** Patients with GT frequently require platelet transfusions for bleeding episodes, often starting from infancy. In patients with absent $\alpha_{IIb}\beta_3$ on the surface of the transfused platelet may be recognized as a foreign protein. Alloimmunization to $\alpha_{IIb}\beta_3$ or HLA antigens may occur, and result in the development of a platelet refractory state. Recombinant Factor VIIa has been used with some effectiveness in patients with GT who are unresponsive to platelet transfusions. HPC transplantation may also be an effective therapeutic option in severe cases.

### Recommended Reading

Bellucci S, Caen J. (2002). Molecular basis of Glanzmann's thrombasthenia and current strategies in treatment. *Blood Rev* **16**, 193–202.

Poon MC, D'Oiron R, von Depka M *et al.* (2004). Prophylactic and therapeutic recombinant factor VIIa administration to patients with Glanzmann's thrombasthenia: results of an international survey. *J Thromb Haemost* **2**, 1096–1103.

# CHAPTER 85

# Platelet Storage-granule Defects

Shawn M. Jobe, MD, PhD

A number of clinical syndromes are associated with storage granule defects. The release of platelet cytoplasmic granules following platelet activation contributes to propagation and stability of the forming thrombus. These storage granule defects typically result in a mild bleeding disorder.

There are two types of platelet cytoplasmic granules:

1. *Platelet dense granules*, sometimes called δ granules, contain calcium, serotonin, adenosine diphosphate (ADP) and adenosine triphosphate (ATP) and have a dark electron dense appearance by electron microscopy.

2. *Platelet α-granules* contain proteins with a number of different functions. Procoagulant molecules, pro- and anti-angiogenic proteins, inflammatory cytokines, platelet receptors, and bactericidal proteins have been identified within the α-granule. Recent data suggest that numerous subtypes of α-granules are present within the platelet. Release of these different α-granule subpopulations may be differentially regulated depending upon the agonist used.

**Diagnosis:** Dense-granule content and release can be evaluated using luminometry. Luminometry is an assay, often performed in concert with aggregometry, that measures the release of dense-granule ATP into the plasma. Electron microscopy can be utilized to distinguish defects in dense-granule release and formation. Platelets from patients with no α-granules appear gray when evaluated by light microscopy, due to the absence of the basophilic α-granule. Expression of the P-selectin adhesive protein on the platelet surface is frequently used as a marker of α-granule membrane fusion. Fusion of the α-granule membrane with the external platelet membrane occurs upon platelet activation. As a result of this membrane fusion, P-selectin is expressed on the platelet surface. The presence of additional systemic manifestations or hematologic findings may also aid in diagnosis.

**Management:** Bleeding manifestations of platelet storage defects are typically mild and can be treated with supportive agents, such as DDAVP and anti-fibrinolytic agents. Platelet transfusion may be necessary in cases of severe bleeding. Care of these patients is often more focused on the accompanying systemic manifestations.

**Hermansky-Pudlak Syndrome:** Hermansky-Pudlak syndrome (HPS) is an autosomal recessive disease characterized by an absence of platelet dense granules and oculocutaneous albinism. Ophthalmologic complications include nystagmus and cataracts. In addition, pulmonary fibrosis, typically noted clinically in the fourth decade of life, occurs in 60% of patients, and granulomatous colitis affects approximately 15% with HPS1.

Mutations in seven different genes have been reported to result in HPS. All mutations result in defective lysosome packaging and formation. HPS1 is the most common of the seven HPS subtypes (HPS1-7), and is responsible for approximately 90% of the cases. Most cases of HPS1 occur in members of a pedigree residing on the Caribbean island of Puerto Rico, secondary to a founder effect.

**Chediak-Higashi Syndrome:** Chediak-Higashi syndrome (CHS), like HPS, is characterized by a paucity of dense granules and variable degrees of oculocutaneous albinism. The gene responsible for CHS, called *Lyst*, is presumed to play a role in vesicular trafficking. The blood smear of patients with CHS is remarkable for the presence of granulocytes with huge cytoplasmic granules. Neutropenia and pronounced immunodeficiency occur in individuals with CHS. Patients with CHS often develop a potentially life-threatening complication in the first decades of life, a non-malignant lymphohistiocytic infiltration referred to as the accelerated phase of CHS. HPC transplantation cures the hematologic manifestations of the illness, but neurologic manifestations of the illness, such as ataxia and decreased cognitive abilities, will continue to progress and may present later in life.

**Gray Platelet Syndrome:** Patients with gray platelet syndrome (GPS) have a gray appearance when evaluated on a blood smear, due to a paucity of basophilic α-granules. In some pedigrees, GPS is associated with a mild to severe myelofibrosis and extramedullary hematopoiesis. In another pedigree, the granule deficiency affects other hematopoietic cells, and gray-appearing neutrophils with absent granules are observed. Macrothrombocytopenia has been reported in some pedigrees.

The gene defect(s) responsible for gray platelet syndrome is unknown. Electron microscopy reveals the presence of some α-granule-like structures, but those few that remain are empty. Platelet protein analysis shows markedly decreased levels of several α-granule proteins, including fibrinogen, thrombospondin and Factor V. An X-linked form of GPS has been identified that is due to a mutation in the DNA-binding region of the megakaryocyte and erythroid transcription factor GATA-1, but autosomal recessive and dominant forms of GPS have also been reported.

**White Platelet Syndrome:** White platelet syndrome is an autosomal dominant disorder characterized by macrothrombocytopenia and decreased α-granule content. This syndrome was recently described in a large multigenerational pedigree from Minnesota. The α-granule content of the platelets is decreased, giving some of the platelets a gray appearance. The unique characteristic of platelets in patients with White platelet syndrome is the presence of large, fully developed Golgi complexes. Typically, Golgi bodies are only present during thrombopoiesis. The gene defect(s) responsible are unknown.

**Other Granule Defects:** Quebec platelet syndrome is an autosomal dominant bleeding disorder. α-granules are present, but α-granule protein levels are decreased. Degradation of α-granule proteins is due to ectopic expression of the protease urokinase in the platelet α-granule. Release of the ectopically-expressed urokinase in the formed clot may also result in accelerated clot lysis and delayed bleeding manifestations. Anti-fibrinolytic agents may substantially benefit these patients.

Patients with absence of both α- and dense granules have also been reported in the literature. These patients typically had a mild to moderate bleeding diathesis.

## Recommended Reading

Gunay-Aygun M, Huizing M, Gahl WA. (2004). Molecular defects that affect platelet dense granules. *Semin Thromb Hemost* **30**, 537–547.

Introne W, Boissy RE, Gahl WA. (1999). Clinical, molecular, and cell biological aspects of Chediak-Higashi syndrome. *Mol Genet Metab* **68**, 283–303.

McKay H, Derome F, Haq MA *et al.* (2004). Bleeding risks associated with inheritance of the Quebec platelet disorder. *Blood* **104**, 159–165.

Tubman VN, Levine JE, Campagna DR *et al.* (2007). X-linked gray platelet syndrome due to a GATA1 Arg216Gln mutation. *Blood* **109**, 3297–3299.

White JG, Key NS, King RA, Vercellotti GM. (2004). The White platelet syndrome: a new autosomal dominant platelet disorder. *Platelets* **15**, 173–184.

# CHAPTER 86

# Failure to Release and Aspirin-like Defects

Shawn M. Jobe, MD, PhD

Patients with platelet granule release and aspirin-like defects typically have mild mucocutaneous bleeding.

**Pathophysiology:** In these disorders, components of platelet signaling and synthetic pathways necessary for the optimal propagation and stabilization of the forming thrombus are missing or dysfunctional. Although platelet adhesion by the platelet receptors GPIb-IX-V and $\alpha_{IIb}\beta_3$ is required for efficient platelet accumulation and thrombus formation at a site of injury, additional steps facilitated by pathways affected in these disorders enhance platelet recruitment and stabilize the formed thrombus. Mutations in platelet receptors, signaling proteins and synthetic pathways have been identified in patients with granule release and aspirin-like defects.

## Examples of Failure of Granule Release and Aspirin-like Defects:

**Thromboxane Pathway Defects:** Thromboxane (TxA2) is an agonist released by activated platelets that supports the activation of adjacent platelets. The importance of this pathway in thrombus formation is illustrated by the clinical effectiveness of aspirin. Aspirin's anti-thrombotic effects are mediated through irreversible inhibition of cyclooxygenase, a key enzyme in the TxA2 synthetic pathway. Aspirin-like platelet defects are observed in patients with cyclooxygenase deficiency, and a similar phenotype is observed in patients with a deficiency of another enzyme necessary for TxA2 synthesis, thromboxane synthase. TxA2's effectiveness depends on its ability to bind and activate the G-protein coupled TxA2 receptor, and mutations in the TxA2 receptor have been found in kindred with a mild-platelet function disorder.

**ADP/ATP Receptor Defects:** ADP and ATP are soluble platelet agonists released from the dense granules of activated platelets. ADP binding to the receptor P2Y12 on adjacent platelets stabilizes the thrombus. Two clinically important anti-thrombotics, clopidogrel and ticlopidine, function by inactivating P2Y12. Patients with dysfunctional or absent P2Y12 receptor present clinically with a mild bleeding diathesis and have decreased aggregation in response to ADP. Dominant negative mutation of another ADP/ATP receptor P2X1 results in a similar clinical phenotype.

**Defects in Platelet Intracellular Signaling Pathways:** The soluble platelet agonists ADP, TxA2 and thrombin bind G-protein coupled receptors. These receptors depend on the actions of several G-proteins for activation of downstream signaling pathways. A mutation in the G-protein, $G\alpha_q$, has been found in a patient with a mild bleeding diathesis and decreased aggregation to multiple agonists. Alterations in other intracellular signaling pathways, including calcium mobilization, tyrosine kinase activity and phospholipase activity, have also been identified in patients with bleeding disorders and abnormal platelet aggregation responses. The genes responsible for many of these alterations remain unidentified.

**Defects in Platelet Procoagulant Activity:** Disorders of platelet procoagulant activity have been demonstrated in several bleeding disorder patients with normal platelet aggregation activity and normal platelet count. One such disorder, Scott syndrome, is characterized by normal platelet aggregation and platelet counts, and *decreased* platelet procoagulant activity. Decreased procoagulant activity occurs due to impaired phosphatidylserine externalization, a critical factor accelerating the activity of the procoagulant tenase and prothrombinase enzyme complexes on the platelet surface. Interestingly, another bleeding disorder, Stormorken syndrome, is characterized by *increased* platelet phosphatidylserine externalization and platelet procoagulant activity on circulating platelets. Defects in platelet adhesiveness have been suggested to account for some of the bleeding manifestations in these patients. The gene(s) responsible for either of these disorders is unknown.

**Diagnosis:** Typically, aggregation studies indicate an impaired second wave of aggregation and decreased ATP-mediated luminescence when luminometry is performed. Aggregation studies will not identify a patient with a procoagulant defect, and specialized testing is required to examine for a defect in platelet phosphatidylserine externalization.

**Management:** Adequate hemostasis can usually be achieved using DDAVP or an anti-fibrinolytic agent. More severe bleeding may require platelet transfusion.

### Recommended Reading

Rao AK. (2003). Inherited defects in platelet signaling mechanisms. *J Thromb Haemost* **1**, 671–681.

Stormorken H, Holmsen H, Sund R *et al.* (1995). Studies on the haemostatic defect in a complicated syndrome. An inverse Scott syndrome platelet membrane abnormality? *Thromb Haemost* **74**, 1244–1251.

Weiss HJ, Lages B. (1997). Platelet prothrombinase activity and intracellular calcium responses in patients with storage pool deficiency, glycoprotein IIb-IIIa deficiency, or impaired platelet coagulant activity – a comparison with Scott syndrome. *Blood* **89**, 1599–1611.

# CHAPTER 87

# Acute (Childhood) Immune Thrombocytopenic Purpura

Carolyn M. Bennett, MD

Immune thrombocytopenic purpura (ITP) is a bleeding disorder characterized by immune-mediated platelet destruction and resultant thrombocytopenia. Two forms of ITP have been described: acute and chronic. Acute ITP occurs almost exclusively in children and is one of the most common acquired bleeding disorders in the pediatric population, with an estimated prevalence of 4–8 per 100,000 children per year. Children typically present with severe thrombocytopenia, but rarely have serious bleeding. In the majority of children, the disease resolves spontaneously over a period of weeks, irrespective of treatment. In contrast, most adults who present with ITP develop chronic disease with thrombocytopenia persisting beyond 6 months, and spontaneous recovery occurs in less than 10% of cases. Only 10–20% of children with ITP go on to have chronic disease. Acute and chronic ITP are approached differently, and therefore are discussed in two separate chapters; acute childhood ITP will be reviewed in this chapter, while chronic ITP in both adults and children will be presented in Chapter 88.

Pathophysiology: While ITP is an autoimmune disease, the precise pathophysiologic mechanisms responsible for the disorder are poorly understood. A major feature of ITP is $F_c$-receptor-mediated clearance of IgG autoantibody-coated platelets by macrophages in the reticuloendothelial system. Most affected children produce both IgG and IgM autoantibodies reactive to multiple platelet antigens, usually platelet surface glycoproteins such as GPIIb/IIIa and GPIb/IX. However, not all patients have measurable antibodies, and the mechanisms that trigger autoantibody production are not clear. Children with acute ITP show increased expression of gamma-interferon dependent genes in the early stages of the disease, indicating the presence of a global pro-inflammatory state. T-cell abnormalities are also involved in the pathophysiology of acute ITP. Imbalances in the mutually inhibitory Th1 and Th2 patterns of cytokine response in T-helper lymphocytes have been measured in patients with ITP. Platelet autoreactive T-cell clones have been identified in the peripheral blood of children with ITP, and cytotoxic T cells from ITP patients can destroy platelets with a high effector cell to target ratio. In addition, abnormalities in the numbers and function of regulatory T cells (CD4+, CD25+ T cells) have been shown to be important for the development of many autoimmune diseases, including ITP. Ultimately, the normal mechanisms of self-tolerance and autoantibody suppression are re-established in the majority of children with ITP, and the thrombocytopenia resolves.

Clinical Manifestations: The physical findings of ITP are usually limited to those associated with thrombocytopenia, and include petechiae, bruising and mucocutaneous bleeding. Patients with acute ITP usually have a rapid onset of bleeding symptoms, which appear over a few hours or days. Typically, children with ITP have surprisingly few other complaints. There may be a history of a recent upper respiratory infection, febrile illness or immunization, but usually the history is unremarkable. In the majority of children with ITP there is no history of fever, recurrent infection, weight loss, bone pain, fatigue or other constitutional symptoms. Despite severe thrombocytopenia in the majority of cases, there is a surprisingly low incidence of overt or severe bleeding. Occasionally children with ITP develop nasal or oral bleeding, but more serious or life-threatening bleeding is unusual. The most serious complication of acute ITP is intracranial hemorrhage (ICH). While ICH (and other serious bleeding) does occur in childhood ITP, the risk is low (approximately 1:600). Some children with ITP will have other findings, such as fever, lymphadenopathy or splenomegaly, but these findings are atypical, and deserve further scrutiny for a more serious condition such as leukemia, myelodysplasia or aplastic anemia.

**Diagnosis:** ITP is a clinical diagnosis that is characterized by the acute onset of thrombocytopenia and associated symptoms in an otherwise well child, the absence of other physical or laboratory findings, and the exclusion of other causes of thrombocytopenia. Careful clinical and laboratory assessments are necessary to exclude other causes of thrombocytopenia, particularly in patients with atypical features.

*Clinical Evaluation:* The clinical evaluation of a patient presenting with possible ITP must include a detailed medical history (including past medical history, family history, medication history, review of systems, growth and development, etc.) and physical examination. A history of arthritis, diabetes, inflammatory bowel disease or other autoimmune disease should be investigated further. A history of poor growth, frequent infections, family history of blood disease or other unusual findings are also not consistent with ITP, and further investigation may be warranted. Full physical examination is extremely important for diagnosis. Children with ITP usually have few presenting physical signs or symptoms other than those associated with thrombocytopenia. Physical findings such as fever, weight loss, pain, limp, jaundice, lymphadenopathy and hepatosplenomegaly may be present, but are atypical of patients with acute ITP and should be investigated further to rule out a more serious malignant, genetic or autoimmune disorder.

*Laboratory Evaluation:* The most important laboratory testing for the diagnosis of ITP is the complete blood count and the examination of peripheral blood smear. Usually, the platelet count is severely low in children with ITP, often under 20,000/μl, but the white blood count, differential, hemoglobin, hematocrit and red blood cell indices should be normal. The presence of large platelets is characteristic of ITP, but giant platelets are more typical of platelet disorders, such as Bernard-Soulier syndrome or other MYH-9 mutation related disorders. Irregularities in the red blood cell or white cell number or morphology are not characteristic of ITP, and should trigger further investigation. The presence of white cell blasts is indicative of leukemia, and bone marrow evaluation is necessary.

Other laboratory testing may be performed as indicated, but no additional tests are part of the routine evaluation for the diagnosis of ITP. Bone marrow evaluation is usually unnecessary in the majority of typical cases, and should be reserved for children presenting with atypical features. Additional laboratory studies, such as coagulation tests (PT, PTT and fibrinogen), platelet survival and/or function studies, rheumatologic studies (ANA, lupus anticoagulant, complement, etc.), blood chemistries, serum immunoglobulins or platelet-associated antibody testing, are unnecessary for diagnosis unless specifically indicated. Assays for platelet antibodies have poor specificity and sensitivity for ITP, and are therefore not recommended for routine evaluation.

**Differential Diagnosis:** The differential diagnosis in children presenting with thrombocytopenia is broad, and patients presenting with thrombocytopenia and atypical features require more intensive evaluation. In infants presenting with thrombocytopenia, congenital thrombocytopenic disorders such as amegakaryocytic thrombocytopenia and thrombocytopenia-absent radius (TAR) should be considered. Wiskott-Aldrich syndrome should be ruled out in infant boys presenting with low platelets in the setting of eczema, failure to thrive, chronic infections and small platelets on peripheral smear. Patients with skeletal anomalies should be evaluated for Fanconi anemia or other bone marrow failure disorders. It is important to distinguish between the primary acute presentation and ITP secondary to other illnesses such as HIV, hepatitis C, systemic lupus erythematosus, autoimmune lymphoproliferative syndrome or other autoimmune disease, as treatment in these cases is different. Leukemia, myelodysplasia and aplastic anemia may present with thrombocytopenia, but rarely as the only laboratory finding. These children usually have other presenting signs or symptoms, such as hepatosplenomegaly, limb pain, lymphadenopathy, weight loss, persistent fever, anemia, leukocytosis or neutropenia. Bone marrow evaluation should be performed in all patients with atypical features.

**Management:** There is an ongoing debate among hematologists regarding treatment for children with uncomplicated acute ITP. Recent evidence suggests that serious, life-threatening bleeding is exceedingly rare in children with acute ITP, even when the platelet count is under 20,000/µl, and may not be prevented by therapy directed at raising the platelet count. The most serious complication of acute ITP is intracranial hemorrhage (ICH). Prevention of ICH is the major motivation for treating children with acute ITP presenting with a severely low platelet count (under 20,000/µl). A recent prospective, international study of children with acute ITP showed the incidence of ICH in children with ITP to be very low (about 0.17%). All of the most commonly used treatments for ITP have side-effects. Therefore, some children with typical acute ITP with no or mild bleeding symptoms may be managed with close observation alone without specific pharmacologic therapy, providing that close follow-up and frequent re-evaluations can be provided. Patients must be followed closely for bleeding symptoms, or for the development of a more serious condition such as aplastic anemia. For children who present with bleeding or who are unable to comply with close follow-up for geographic or other reasons, or in whom the risk of ICH is increased, treatment to raise the platelet count to over 20,000/µl may be indicated.

The most common and most effective pharmacologic therapies for acute ITP include intravenous immunoglobulin (IVIG), Rh immune globulin (RhIg) and corticosteroids (Table 87.1). There is no evidence that any one of these treatments is more effective than the other, and all have side-effects, therefore treatment choice is somewhat arbitrary. Most patients (>70%) will respond to any of the therapies by 72 hours.

TABLE 87.1 Treatment Options for Acute (Childhood) ITP

| Treatment | Dose | Course | Side-effects | Response (platelets >20,000/$\mu$l) |
|---|---|---|---|---|
| Corticosteroid<br>• prednisone<br>• prenisolone<br>• solumedrol | 1–4 mg/kg per day | 1–14 days | Hyperglycemia, irritability, weight gain, hypertension, anxiety, dysphoria, | 60% by 72 hours |
| IVIG | 1–2 g/kg | 1–2 days | Headache, chills, fever, rarely aseptic meningitis, anaphylaxis | 24–48 hours |
| RhIg | 75 $\mu$g/kg | 1 dose | Headache, chills fever, fatigue, decrease in hemoglobin, vomiting, hemolysis, anemia | 24–48 hours |

IVIG and RhIg are costly and must be given via intravenous infusion, but are associated with a more rapid platelet response (platelets over 20,000/$\mu$l often within 24 hours). RhIg is less expensive then IVIG and has a shorter infusion time (few minutes), but is only effective in patients who are D positive with an intact spleen. RhIg also causes hemolysis, and should not be given to patients with anemia (hemoglobin < 10 g/dl), as the hemoglobin decreases 0.5–2 g/dl in most patients. Corticosteroids may be given orally and are relatively inexpensive, but may take 48–72 hours to have a similar effect.

In the majority of children ITP resolves spontaneously in weeks to months, with or without treatment. Occasionally children will have recurrent episodes of thrombocytopenia in the months following presentation, particularly in the setting of infection. Ultimately, approximately 80% of patients will have complete resolution of their thrombocytopenia by 6 months. While the risk of late relapse is unknown, it is thought to be quite low, and in general children with ITP can expect complete resolution of symptoms without evidence of an underlying autoimmune or other hematologic or malignant process.

In about 20% of children and adolescents with ITP, the thrombocytopenia persists beyond the 6 months and becomes chronic. Chronic ITP in both children and adults is presented in detail in Chapter 88.

## Recommended Reading

Blanchette V, Bolton-Maggs P. (2008). Childhood immune thrombocytopenic purpura: diagnosis and management. *Pediatr Clin North Am* **55**, 393–420.

Bolton-Maggs P, Tarantino MD, Buchanan GR *et al.* (2004). The child with immune thrombocytopenic purpura: is pharmacotherapy or watchful waiting the best initial management? A panel discussion from the 2002 meeting of the American Society of Pediatric Hematology/Oncology. *J Pediatr Hematol Oncol* **26**, 146–151.

Cooper N, Bussel J. (2006). The pathogenesis of immune thrombocytopaenic purpura. *Br J Haematol* **133**, 364–374.

George JN, Woolf SH, Raskob GE *et al.* (1996). Idiopathic thrombocytopenic purpura: a practice guideline developed by explicit methods for the American Society of Hematology. *Blood* **88**, 3–40.

Kuhne T, Buchanan GR, Zimmerman S *et al.* (2003). Intercontinental Childhood ITPSG. A prospective comparative study of 2540 infants and children with newly diagnosed idiopathic thrombocytopenic purpura (ITP) from the Intercontinental Childhood ITP Study Group. *J Pediatr* **143**, 605–608.

Nugent DJ. (2006). Immune thrombocytopenic purpura of childhood. *Hematology Am Soc Hematol Educ Program*, 97–103.

Tarantino MD, Young G, Bertolone SJ *et al.* (2006). Single dose of anti-D immune globulin at 75 microg/kg is as effective as intravenous immune globulin at rapidly raising the platelet count in newly diagnosed immune thrombocytopenic purpura in children. *J Pediatr* **148**, 489–494.

# CHAPTER 88

# Chronic Immune Thrombocytopenic Purpura

Carolyn M. Bennett, MD

Immune thrombocytopenic purpura (ITP) is a bleeding disorder characterized by immune-mediated platelet destruction and resultant thrombocytopenia and muco-cutaneous bleeding. Chronic ITP is defined by the persistence of immune thrombo-cytopenia beyond 6 months, with spontaneous recovery occurring in less than 10% of adults with the disease. The estimated incidence of ITP is approximately 100 cases per 1 million persons per year, with about half of these cases presenting in adults. Approximately twice as many women are affected as men. Patients with chronic ITP often have problematic bleeding that requires ongoing therapy, and hemorrhagic deaths are not uncommon. Chronic ITP in both adults and children is presented in this chapter. Acute childhood ITP is reviewed in Chapter 87.

Pathophysiology: ITP is an autoimmune disorder characterized by immune-mediated destruction of platelets and subsequent thrombocytopenia. However, the precise pathophysiologic mechanisms responsible for the platelet destruction are not clearly understood. A major feature of ITP appears to be the $F_c$-receptor-mediated clearance of IgG autoantibody-coated platelets by macrophages in the reticuloendothe-lial system. Platelet-associated IgG autoantibodies can be measured in about 50–60% of patients with ITP and are reactive to multiple platelet antigens, usually platelet surface glycoproteins such as GPIIb-IIIa, GPIa-IIa and GPIb-IX. Autoreactive B cells producing anti-platelet antibodies play a role in the pathogenesis of ITP, but the mech-anisms that trigger autoantibody production are not clear. Approximately 40–50% of patients do not have measurable antibodies, and therapies aimed at reducing autoanti-body production and autoantibody-mediated platelet destruction (anti-B-cell therapy and splenectomy, respectively) are effective in only 50–70% of patients. Alternative mechanisms for platelet destruction must be present in most, if not all, patients. T cells and their secretory factors are important for the stimulation of antibody-producing B-cell clones, and may be important in the development of antiplatelet antibody pro-duction. T lymphocytes may be directly responsible for platelet destruction, and this may be an important cause of thrombocytopenia in some patients with ITP. Recent studies have shown that cytotoxic T cells from ITP patients can destroy platelets with a high effector cell to target ratio. Platelet autoreactive T-cell clones have been identi-fied in patients with chronic ITP, and genes involved in T-cell mediated cytotoxicity are unregulated in many ITP patients.

The pathophysiology of ITP may also involve decreased or inadequate production of platelets in addition to increased destruction. Several recent studies suggest that antiplatelet glycoprotein antibodies and possibly antiplatelet T cells have effects on bone marrow megakaryocytes. In addition, megakaryocyte production appears to be inadequate and thrombopoietin levels inappropriately low in many thrombocytopenic patients with ITP. Treatment with thrombopoietic agents results in dramatic platelet increases in some patients.

ITP is a heterogeneous disease and the initiating immunologic event is not obvious, but a complex immunologic process involving platelets, megakaryocytes, B cells, T cells and other components of the immune system is likely for the development of disease.

**Clinical Manifestations:** The physical findings associated with ITP are usually limited to those associated with thrombocytopenia, and include petechiae, ecchymoses and purpura. Symptoms at presentation may be abrupt in onset, but in some patients may have been present for months or years. Mucocutaneous bleeding can be widespread, and may include epistaxis, gingival bleeding, hematuria and menorrhagia. Rarely, patients present with signs of more severe bleeding, such as intracranial hemorrhage. Patients with ITP are usually healthy and have few other physical manifestations of disease. Additional findings, such as fever, malaise, night sweats, joint pain, lymphadenopathy or hepatosplenomegaly, are atypical and may suggest alternative diagnoses. Patients who present with ITP secondary to an underlying illness such as HIV, hepatitis and systemic lupus erythematosus may present with symptoms of the primary illness. Further evaluation is necessary for patients presenting with atypical features.

**Diagnosis:** ITP is a diagnosis of exclusion, and other causes of thrombocytopenia must be considered before making this diagnosis. Careful clinical and laboratory assessments are necessary to exclude other causes of thrombocytopenia, particularly in patients with atypical features.

*Clinical Evaluation:* The clinical evaluation of a patient presenting with ITP must include a detailed past medical and family history, medication history and review of systems. In general, patients with ITP have a surprisingly unremarkable medical history. A history of arthritis, diabetes, inflammatory bowel disease or other autoimmune disease may indicate secondary ITP, and should be investigated further. A history of frequent infections, family history of blood disease or other unusual findings are also not consistent with ITP, and further investigation may be warranted. A complete physical examination is important for making the diagnosis. Patients with ITP usually have few presenting physical signs or symptoms other than those associated with thrombocytopenia. Physical findings such as fever, weight loss, pain, jaundice, lymphadenopathy and hepatosplenomegaly can occur in ITP, but are unusual, and may be indicative of a more serious disorder.

*Laboratory Evaluation:* The most important laboratory testing for the diagnosis of ITP is the complete blood count and the examination of peripheral blood smear. Platelet count is low in patients presenting with ITP, often under 30,000/µl, but the white blood count, differential, hemoglobin, hematocrit and red blood cell indices are normal. The presence of large platelets is a distinguishing feature of the peripheral smear of patients with ITP, but irregularities in the red blood cell or white blood cell morphology are atypical and should trigger further investigation. The presence of white blood cell blasts is indicative of leukemia, and bone marrow evaluation is necessary.

Other laboratory testing may be performed as indicated, but no additional tests are necessary in the routine evaluation for the diagnosis of ITP. Bone marrow evaluation is not needed in the majority of typical cases, but should be reserved for patients presenting with atypical features and has also been advocated in patients over 60 years of age. Additional laboratory or radiographic studies, such as coagulation tests (PT, PTT and fibrinogen), platelet survival or function studies, rheumatologic studies (ANA, lupus anticoagulant testing, complement assessment, etc.), blood chemistries, serum and serum immunoglobulins, may be helpful in eliminating alternative diagnoses, but are usually unnecessary in typical cases. Assays for antigen-specific platelet autoantibodies have poor specificity and sensitivity for ITP, and are therefore not recommended as routine testing. Leukemia and other bone marrow infiltrative or failure disorders rarely present without other cellular or physical findings; therefore, careful attention to the history, physical exam and laboratory findings is important so that subtle findings are not overlooked.

**Differential Diagnosis:** ITP is the most common cause of acquired severe thrombocytopenia in an otherwise healthy individual, but the diagnosis is one of exclusion, and other causes of thrombocytopenia must be excluded. Bone marrow failure or infiltrative disorders such as Fanconi anemia, aplastic anemia, leukemia and myelodysplasia should be considered in patients with atypical physical or laboratory findings. Patients presenting with a splenomegaly should be evaluated for diseases of abnormal platelet distribution, such as congestive splenomegaly from portal hypertension. Many drugs are responsible for thrombocytopenia, and careful drug history is important. Other diseases that result in thrombocytopenia, such sepsis, thrombotic thrombocytopenic purpura and DIC, rarely present without associated presenting features, but should be considered in the differential diagnosis.

## Management:

**Initial Management:** The initial treatment must be tailored to the presenting platelet count and bleeding symptoms. Patients presenting with platelets >50,000/μl are often discovered incidentally, have no or few symptoms and frequently do not require therapy. Patients with moderate thrombocytopenia (platelets 20,000–50,000/μl) can usually be followed closely on an outpatient basis and treated only to relieve symptomatic bleeding. Patients with severe thrombocytopenia (platelets <20,000/μl) may have overt bleeding and require therapy. Management decisions should not be based exclusively on the platelet count, but also must depend on age, lifestyle and other medical conditions, since these factors contribute to the overall risk of serious bleeding.

The most common first-line therapy for adults with ITP is corticosteroid, usually prednisone, 1–2 mg/kg per day for 2–4 weeks, with a taper if there is a platelet response; however, other regimens have been utilized. Pulsed high-dose dexamethasone at 40 mg/d for 4 days without taper is an effective alternative oral regimen, and may be associated with a higher sustained response rate. Approximately 50–75% of patients will respond to corticosteroids, irrespective of dose or duration. Intravenous immunoglobulin (IVIG, 1 g/kg per day for 1–3 days) and Rh immune globulin (RhIg, 50–75 μg/kg) are equally effective as first-line therapies for adults with ITP. However, both require IV administration and are considerably more expensive. IVIG is usually reserved for adults presenting with severe thrombocytopenia with serious bleeding after corticosteroids have been utilized and found ineffective. RhIg is less expensive then IVIG and has a shorter infusion time (few minutes), but is effective only in patients who are D positive with an intact spleen. RhIg also causes hemolysis and should not be given to patients with anemia (hemoglobin <10 g/dl), as the hemoglobin decreases 0.5–2 g/dl in most patients.

## Relapse:

*Children:* In the majority of children ITP resolves spontaneously in weeks to months, with or without treatment. Occasionally children will have recurrent episodes of thrombocytopenia in the months following presentation, particularly in the setting of infection. Ultimately, approximately 80% of patients will have complete resolution of their thrombocytopenia by 6 months. For approximately 20% of children and adolescents with ITP, the thrombocytopenia persists beyond 6 months. The therapeutic goal in these children is to prevent and/or manage bleeding symptoms, as up to one-third of these children will ultimately have spontaneous remission by 1 year. Bleeding in children with ITP within the first 6–12 months of diagnosis is managed with first-line therapy (i.e. corticosteroids, IVIG and RhIg) unless patients become refractory (see below). Splenectomy is avoided in young children, due to the risk of overwhelming bacterial sepsis, unless there is life-threatening bleeding uncontrolled by other interventions.

*Adults:* In more than 70% of adults, ITP persists beyond the 6-month period and becomes chronic. Many of these patients have moderate to severe thrombocytopenia with bleeding symptoms, and require alternative therapies. While splenectomy is associated with significant morbidity and an increased risk of *Streptococcus pneumonia* sepsis, it is the most effective and durable treatment for adults with chronic ITP who have not had a sustained response to initial therapy. Approximately 70% of patients will achieve sustained remission after splenectomy. Laparoscopic splenectomy appears to be preferable to open splenectomy, as the recovery time is shorter and the outcome is similar. Splenectomy is an appropriate therapeutic option in adults and older children with refractory ITP and ongoing symptomatic bleeding.

**Refractory ITP:** Refractory ITP is defined as the persistence of thrombocytopenia despite initial treatment with corticosteroids, IVIG, RhIg and, in adults and older children, splenectomy. Approximately 10% of patients (adults and children) will be refractory to all standard therapy, including splenectomy. While the overall prognosis of ITP is good, the mortality in this refractory group of patients is high – between 6% and 15%. Since the likelihood of durable and complete remission in these patients is much lower, the goal is to balance the bleeding risk with lifestyle issues and toxicity of therapy. The American Society of Hematology guidelines suggest that maintaining a platelet count of 30,000–50,000/μl in chronic ITP patients without other risk factors and without overt bleeding is a generally acceptable level for reducing the risk of spontaneous hemorrhage. An individualized approach is important for disease management, as no single protocol is appropriate for all patients. No randomized clinical trials have been performed comparing the numerous therapies for chronic ITP, and no trials have assessed the outcomes of bleeding and death. Therefore, no evidence-based algorithm can be proposed as the standard of care for patients with chronic refractory ITP. However, several agents have been shown to be useful for the treatment of chronic ITP and, while none offers a cure, some result in durable remission and are useful in alleviating bleeding symptoms.

Rituximab, an anti CD-20 monoclonal antibody, is probably the most effective, relatively non-toxic drug for the treatment of severe, refractory ITP. The most common dosing regimen is $375\,mg/m^2$ weeks for 4 consecutive weeks, but other dosing regimens have been utilized. Up to 50% of patients achieve sustained platelet response with rituximab. Rituximab is considered a "splenectomy-sparing" therapeutic approach by some experts, particularly in children, or in patients who are high surgical risks who cannot undergo splenectomy. Azathioprine ($1–4\,mg/kg$ per day) is the most commonly used chemotherapeutic agent for ITP, and is generally well tolerated. Azathioprine may take up to 3–6 months to be effective, and the long-term efficacy and toxicity is not known. Mycophenolate mofetil (MMF) is a purine nucleotide synthesis inhibitor that was initially developed to prevent solid-organ transplant rejection, but has been used more recently for the treatment of autoimmune disease such as systemic lupus erythematosus. Some patients with refractory ITP treated with MMF have shown a favorable response with minimal toxicity, but whether sustained remission can be achieved after discontinuation of the drug is not known. Other treatments that have been used with minimal toxicity and varying efficacy include danazol, dapsone and 6-mercaptopurine.

Cytotoxic or other immunosuppressive agents such as cyclophosphamide, cyclosporin, vinca alkaloids, combination therapies and hematopoietic progenitor cell transplantation have more significant toxicities, and are usually reserved for highly refractory patients with severe thrombocytopenia and bleeding.

Clinical trials with newer investigational agents, such as thrombopoietin receptor agonists and monoclonal immunosuppressive agents, are currently under investigation. Thrombopoietin agents are particularly promising. In a recent randomized, placebo-controlled study of eltrombopag, an oral thrombopoietin receptor agonist, 80% of adult patients with chronic ITP had platelet responses (Bussel et al., 2007). Severe side-effects were not observed, but more long-term studies of efficacy and safety are necessary prior to widespread use. Romiplostim, a thrombopoietin mimetic peptibody, and eltrombopag have recently been FDA approved for the treatment of chronic ITP in patients who have failed other treatments.

## Recommended Reading

British Committee for Standards in Haematology General Haematology Task Force. (2003). Guidelines for the investigation and management of idiopathic thrombocytopenic purpura in adults, children and in pregnancy. Br J Haematol 120, 574–596.

Bussel JB, Cheng G, Saleh MN et al. (2007). Eltrombopag for the treatment of chronic idiopathic thrombocytopenic purpura. N Engl J Med 357(22), 2237–2247.

Cines DB, McMillan R. (2007). Pathogenesis of chronic immune thrombocytopenic purpura. Curr Opin Hematol 14, 511–514.

Cooper N, Bussel J. (2006). The pathogenesis of immune thrombocytopaenic purpura. Br J Haematol 133, 364–374.

George JN, Woolf SH, Raskob GE *et al.* (1996). Idiopathic thrombocytopenic purpura: a practice guideline developed by explicit methods for the American Society of Hematology. *Blood* **88**, 3–40.

Godeau B, Provan D, Bussel J. (2007). Immune thrombocytopenic purpura in adults. *Curr Opin Hematol* **14**, 535–556.

Olsson B, Andersson PO, Jernas M *et al.* (2003). T-cell-mediated cytotoxicity toward platelets in chronic idiopathic thrombocytopenic purpura. *Nat Med* **9**, 1123–1124.

# CHAPTER 89

# Drug-induced Thrombocytopenia

Carolyn M. Bennett, MD

Drug-induced thrombocytopenia (DIT) is a fairly common clinical problem, and numerous drugs have been implicated in the development of thrombocytopenia. The risk of thrombocytopenia for any drug is low, and fortunately only a small number of patients taking a suspected medication will develop thrombocytopenia. However, many patients who develop drug-induced thrombocytopenia are taking multiple medications and are critically ill. Rapid recognition of thrombocytopenia in affected patients and identification and removal of the offending agent before clinically significant bleeding occurs is imperative. Unfortunately, there are no laboratory tests with sufficient specificity or sensitivity to be useful for making the diagnosis. DIT is a clinical diagnosis that can be supported only by resolution of thrombocytopenia after cessation of drug. Certain drugs have a well-documented association with thrombocytopenia (see Table 89.1). Heparin-induced thrombocytopenia is discussed in Chapter 90.

**TABLE 89.1** Drugs Commonly Implicated as Triggers of Drug-induced Thrombocytopenia*

| Drug category | Drugs implicated in five or more reports | Other drugs |
|---|---|---|
| Heparins | Unfractionated heparin, low molecular weight heparin | |
| Cinchona alkaloids | Quinine, quinidine | |
| Platelet inhibitors | Abciximab, eptifibatide, tirofiban | |
| Antirheumatic agents | Gold salts | D-penicillamine |
| Antimicrobial agents | Linezolid, rifampin, sulfonamides, vancomycin | |
| Sedatives and anticonvulsant agents | Carbamazepine, phenytoin, valproic acid | Diazepam |
| Histamine-receptor antagonists | Cimetidine | Ranitidine |
| Analgesic agents | Acetaminophen, diclofenac, naproxen | Ibuprofen |
| Diuretic agents | Chlorothiazide | Hydrochlorothiazide |
| Chemotherapeutic and immunosuppressant agents | Fludarabine, oxaliplatin | Cyclosporine, rituximab |

*For a more extensive list, see the University of Oklahoma website (http://moon.ouhsc.edu/jgeorge/DITP.html).
From Aster RH, Bougie DW (2007). Drug-induced immune thrombocytopenia. *N Engl J Med* **357**, 580–587.

**Pathology:** Hundreds of drugs have been implicated in DIT. Drugs cause thrombocytopenia by two basic mechanisms: decreased platelet production or increased platelet destruction.

**Decreased Production:** DIT resulting from decreased production is usually the result of generalized dose-dependent myelosuppression. Many chemotherapeutic agents suppress the bone marrow and cause leukopenia, anemia and often severe thrombocytopenia. The cytopenias resolve with cessation of chemotherapy. Other non-chemotherapeutic agents, such as the anti-epileptic drug valproate, can cause myleosuppression. The dose-dependent association between valproate and neutropenia and thrombocytopenia is well documented. Selective depression of megakaryocyte production has been associated with drugs such as thiazide diuretics, ethanol and tolbutamide, and can lead to isolated thrombocytopenia.

**Increased Destruction:** Platelet destruction in DIT is usually immune-mediated, and the mechanism involves the drug-induced production of antiplatelet antibodies which bind to platelets and trigger their consumption by macrophages in the reticuloendothelial system.

*Hapten-induced Antibody:* Certain small-molecule drugs called haptens can become covalently linked to platelet surface antigens and form hapten–protein complexes. Antibodies directed at these cell surface protein complexes are produced, and result in platelet destruction and thrombocytopenia. The phenomenon is thought to be responsible for the thrombocytopenia observed with penicillin and penicillin derivatives. Hapten-induced immune hemolytic anemia occurs much more commonly with penicillins. DIT by this mechanism has been reported, but is less uncommon.

*Drug-dependent (Quinine Type) Antibody Formation:* This form of drug-induced thrombocytopenia is characterized by the formation of an antibody that binds to platelets only in the presence of the drug in the soluble state. These antibodies are usually directed at glycoprotein GPIIb/IIIa or GPIb/V/IX complexes, and bind non-covalently to produce "compound" epitopes or induce conformation changes to which antibodies are specific. Quinine, sulfonamide antibiotics and non-steroidal anti-inflammatory drugs trigger thrombocytopenia by this mechanism.

*GPIIb/IIIa Inhibitors:* Specific antiplatelet medications, GPIIb/IIIa inhibitors such as tirofiban, eptifibatide and abciximab, cause DIT by antibody-dependent mechanisms. Acute thrombocytopenia often occurs within hours of drug exposure, and may suggest a non-immune mechanism. However, in the presence of tirofiban and eptifibatide, conformational changes in the GPIIb/IIIa molecule occur, resulting in a neo-epitope. Antibodies are produced that are specific to these neo-epitopes. Antibody mediated platelet destruction can lead to severe thrombocytopenia and bleeding. In the case of abciximab, antibody formation is against the murine component of the chimeric Fab fragment specific for GPIIIa.

*Drug-induced Antibody:* In rare instances, drugs induce the production of true antiplatelet autoantibodies that cause platelet destruction even in the absence of the sensitizing agents. In these cases the autoantibodies may persist indefinitely, leading to chronic DIT. Gold salts and procainamide can induce this form of DIT.

*Immune Complex:* In this form of DIT, antibodies causing thrombocytopenia are produced as a result of drug-antibody immune complex formation leading to platelet destruction and consumption. Heparin-induced thrombocytopenia is triggered by this mechanism.

**Clinical Manifestations:** Patients with DIT usually present with moderate to severe thrombocytopenia (platelets < 50,000/μl) and evidence of bleeding. Usually the thrombocytopenia becomes clinically apparent 1–2 weeks after starting the drug, but it can occur in patients who have been taking a drug for months or years. Thrombocytopenia may occur more quickly if the drug has been administered previously, or with drugs such as quinine or abciximab. Bleeding symptoms are variable, and range from mild cutaneous findings (such as petechiae and ecchymoses) to more significant overt hemorrhage. When the thrombocytopenia is severe, patients can present with severe bleeding from the nose, gums, GI tract or uterus. Life-threatening spontaneous intracranial hemorrhage, while rare, can occur in patients with severe thrombocytopenia.

**Diagnosis:** The diagnosis of drug-induced immune thrombocytopenia must often be made on clinical findings alone, as the laboratory assays do not often have sufficient sensitivity to detect drug-induced antibodies.

*Clinical Evaluation:* Drug-induced thrombocytopenia should be considered in any patient taking medication who presents with sudden, unexplained thrombocytopenia. It is important to obtain a complete and accurate medical history and list of medications, including all prescription and non-prescription medications, non-conventional therapies and herbal remedies. Specific attention should be given to higher-risk medications such as quinines and sulfonamides. Often, the diagnosis is confirmed only after platelet recovery with discontinuation of a suspected, sensitizing drug.

*Laboratory Evaluation:* Laboratory testing may not be particularly helpful in making the diagnosis of DIT. However, various *in vitro* methods exist to detect drug-induced binding of IgG to platelets. Radiolabeled or fluorescein-labeled anti-IgG can be used to detect platelet-bound immunoglobulin, but these tests may not help distinguish between primary immune thrombocytopenia and DIT. Enzyme-linked immunosorbent assay (ELISA), flow cytometry and immuno-precipitation Western blotting have more specificity for detecting platelet-reactive antibodies induced by drugs. Unfortunately, these tests are not widely available, may take several days to complete, and have low sensitivity; therefore, they are not generally useful for the acute care of patients with suspected DIT.

**Differential Diagnosis:** Other causes of thrombocytopenia should be considered, particularly in hospitalized patients taking multiple medications. Thrombocytopenia from sepsis, disseminated intravascular coagulopathy (DIC) or thrombotic thrombocytopenic purpura (TTP) may be difficult to distinguish from DIT, particularly in critically ill patients. Heparin-induced thrombocytopenia should be considered in patients on heparin. Patients with immune thrombocytopenic purpura (non-drug-induced), either primary or secondary, can present with acute thrombocytopenia. Patients with HIV infection can present with thrombocytopenia which may be immune-mediated or infection-induced. Patients who present with other physical or laboratory findings should be evaluated for bone marrow failure or infiltrative disorder. Patients with thrombocytopenia and hepatosplenomagaly should be evaluated for congestive disorders, such as portal hypertension, malignancy or hepatitis.

**Management:** Discontinuation of the sensitizing drug is the necessary "therapy." As many drugs as possible should be discontinued in patients with suspected DIT. If a particular therapy is medically necessary, a non-immunologic, non-cross-reacting alternative should be considered. Platelet transfusions should be used for life-threatening bleeding, but may not be effective due to the presence of antibody. In patients with severe drug-induced thrombocytopenic who are acutely ill, intravenous immunoglobulin (IVIG), plasma exchange, and corticosteroids have been used, but the clinical benefit is uncertain. Usually, the thrombocytopenia improves within the first few days after discontinuing the drug. Sensitivity to the drug persists indefinitely, and further use of the suspected causative agent should be avoided.

## Recommended Reading

Aster RH, Bougie DW. (2007). Drug-induced immune thrombocytopenia. *N Engl J Med* **357**, 580–587.

George JN, Raskob GE, Shah SR *et al.* (1998). Drug-induced thrombocytopenia: a systematic review of published case reports. *Ann Intern Med* **129**, 886–890.

Li X, Swisher KK, Vesely SK, George JN. (2007). Drug-induced thrombocytopenia: an updated systematic review, 2006. *Drug Safety* **30**, 185–186.

Visentin GP, Liu CY. (2007). Drug-induced thrombocytopenia. *Hematol Oncol Clin North Am* **21**, 685–696, vi.

# CHAPTER 90

# Heparin-induced Thrombocytopenia

Christine L. Kempton, MD

Heparin-induced thrombocytopenia (HIT) occurs in 1–3% of patients treated with unfractionated heparin for greater than 5 days. Although thrombocytopenia occurs, bleeding is rare; in contrast, both arterial and venous thrombosis are more common sequelae (odds ratio 36.9).

Pathophysiology: HIT is an immune-mediated disorder caused by IgG antibodies directed against the complex of heparin and platelet factor 4 (PF4). Although anti-heparin/PF4 antibodies are detected in nearly all patients with HIT, they are also detected in many patients without HIT. Both the frequency of asymptomatic anti-heparin/PF4 antibodies and incidence of HIT vary by clinical situation (Table 90.1). HIT has no known HLA association, but is 1.5–3.0 times more likely to occur in women.

| TABLE 90.1  Incidence of Heparin/PF4 Antibodies and HIT | | |
|---|---|---|
| Clinical population | Incidence of anti-heparin/PF4 antibodies (%) | Incidence of HIT (%) |
| Orthopedic surgery | 14 | 3–5 |
| Cardiac surgery | 25–50 | 1–2 |
| General medical patient | 8–20 | 0.8–3.0 |
| Chronic hemodialysis | 0–2.3 | 0–0.1 |

Adapted from Arepally GM (2006). Clinical practice. *N Engl J Med* **355**, 809–817.

The propensity for thrombosis in patients with HIT is thought to occur as a result of platelet aggregation and activation of coagulation. There are several potential mechanisms by which coagulation is activated: platelet activation via binding of the IgG Fc regions leading to cross-linking of FcγIIa receptors on the platelet surface; PF4 binding heparin sulfates (a naturally occurring glycosaminoglycan) on the surface of the endothelial cell, leading to endothelial cell injury and tissue factor expression; and monocyte binding of heparin/PF4-IgG immune complexes that may lead to monocyte activation and tissue factor expression.

## Clinical Manifestations:

The major manifestations of HIT are thrombocytopenia, thrombosis and allergic reactions.

**Thrombocytopenia:** The majority of patients ($>90\%$) with HIT have a greater than 50% decrease in their platelet count. The remainder of patients, who have a decrement that is less than 50%, have typically had clinical events such as thrombosis or skin lesions at the site of heparin injection. The median platelet count nadir is approximately 50,000/μl, and the nadir is rarely less than 20,000/μl. In patients who have had an initial drop in platelet count associated with surgery, the highest postoperative platelet value should be used as the baseline platelet count for comparison. If the patient has an elevated platelet count when heparin is started, a greater than 50% drop in platelet count is consistent with HIT, even though the patient may not be thrombocytopenic by absolute numbers of platelets.

**Thrombosis and Other Sequelae:** Approximately 25–50% of patients with HIT will develop a thrombosis. The majority of these will occur at the time of diagnosis or within the first week; however, the prothrombotic risk remains for up to 30 days. Risk factors for thrombosis include orthopedic or trauma surgery, a large platelet count decrease, and a higher optical density on ELISA testing for anti-heparin/PF4 antibody. Venous thrombosis occurs more frequently than arterial thrombosis, especially in patients undergoing orthopedic surgery. However, arterial thrombosis is as frequent as venous thrombosis in the post-cardiac surgery setting. Overall, in those with arterial thrombosis, involvement of large lower-limb arteries is more common than stroke or myocardial infarction. In patients without clinical symptoms of thrombosis, venous thrombosis may be detected in up to 50% by duplex ultrasonography. For these reasons, a screening ultrasound of the upper and lower extremities is recommended to further refine the clinical pretest probability and to assist in determining the duration of anticoagulation.

Erythematous plaques or necrosis at the site of subcutaneous heparin injection may be a feature of HIT. The skin lesions are typically painful, and begin five or more days after initiation of subcutaneous heparin.

Up to one-fourth of patients with HIT may develop an allergic reaction within 5–30 minutes of intravenous heparin injection. Signs and symptoms may include fever, chills and respiratory distress. An abrupt drop in platelet count, if related to HIT, will follow the allergic reaction.

**Diagnosis:** HIT is a clinicopathologic diagnosis that requires careful clinical assessment *prior* to any laboratory testing. Parameters to consider when developing a clinical pretest probability are the degree of thrombocytopenia, the timing of the onset of thrombocytopenia, the presence of thrombosis or other sequelae, and the likelihood of other causes of thrombocytopenia (Table 90.2).

In patients who have not been recently exposed to heparin (within the prior 100 days), antibodies require 5–10 days to develop. Thus, the drop in the platelet count does not begin until 5–10 days after starting heparin therapy. The nadir of platelet counts is usually reached several days later. In those who have had heparin within the past 100 days, anti-heparin/PF4 antibodies may persist in circulation and lead to a drop in platelet count within the first 24 hours. Unlike normal humoral responses, anti-heparin/PF4 antibodies are not thought to demonstrate true amnestic responses; nevertheless, a patient who has a history of HIT may theoretically have immunological memory and demonstrate HIT faster than five days after exposure to heparin. Rarely,

### TABLE 90.2 "4T"s HIT Clinical Pre-test Probability

|  | 2 points | 1 point | 0 points |
|---|---|---|---|
| Thrombocytopenia | Platelet count fall >50% and nadir ≥20,000/μl | Platelet count fall 30–50% or nadir 10–19,000/μl | Platelet count fall <30% or nadir <10,000/μl |
| Timing | Clear onset between 5 and 10 days or platelet count fall ≤1 day (prior heparin exposure within 30 days) | Consistent with onset between 5–10 days, but not clear (e.g. missing platelet counts); onset after day 10 or fall ≤1 day (prior heparin exposure 30–100 days ago) | Platelet count fall <4 days without recent exposure |
| Thrombosis or other sequelae | New thrombosis; skin necrosis; acute systemic reaction post IV unfractionated heparin | Progressive or recurrent thrombosis; non-necrotizing (erythematous) skin lesions; suspected thrombosis (not proven) | None |
| Other causes of thrombocytopenia | None apparent | Possible | Definite |

≤ 3 points: Low clinical probability; 4–5 points: Intermediate clinical probability; 6–8 points: High clinical probability. Adapted from Lo GK *et al.* (2006). Evaluation of pretest clinical scores (4 T's) for the diagnosis of heparin-induced thrombocytopenia in two clinical settings. *J Thromb Haemost* **4**, 759–765.

the onset of thrombocytopenia and/or thrombosis occurs five or more days following the discontinuation of heparin, and is considered delayed-onset HIT.

*Laboratory Testing:* Testing for HIT includes both the demonstration of anti-heparin/PF4 antibodies by an ELISA assay and the determination that those antibodies induce platelet activation in the setting of heparin using a functional platelet assay (see Chapter 121). Positive testing (i.e. antibodies are both present and induce platelet activation in the presence of heparin) occurs in < 1%, 11–28% and 30–100% of patients with a low, intermediate and high clinical pretest probability, respectively. Accordingly, testing is warranted in those with an intermediate and high clinical probability (Figure 90.1). In those with an indeterminate diagnosis, alternative causes of thrombocytopenia should be re-evaluated. If no alternative cause for thrombocytopenia is found, repeat testing can be considered in several days. Any positive ELISA that follows an initial negative test should be confirmed with a functional assay in this setting. If the ELISA remains negative, HIT is excluded. However, as treatment of HIT requires immediate discontinuation of all heparin-containing products, changing to alternate anticoagulants (see below) is recommended while awaiting laboratory results.

**Differential Diagnosis:** When evaluating a patient for potential HIT, other causes of thrombocytopenia must be considered. Disseminated intravascular coagulation and infection are not uncommon in the hospitalized patient, and should be considered in the differential diagnosis of thrombocytopenia. Careful review of the patient's other drug exposures, baseline platelet count prior to admission, and peripheral blood

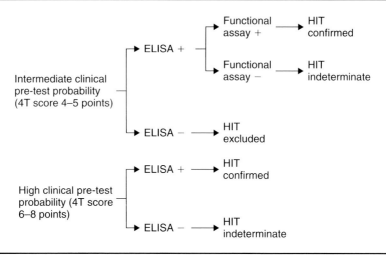

FIGURE 90.1  Diagnostic algorithm.

smear are important parts of the diagnostic work-up of a patient suspected of having HIT. In 5–10% of patients receiving heparin, there may be a non-immune-mediated reduction in platelet count within the first 1–2 days. The nadir is typically not less than 100,000/µl, and the platelet count improves despite continuation of heparin. This phenomenon, previously called Type 1 HIT, is of no known clinical consequence.

**Management:** In patients with an intermediate or high clinical probability of HIT, all heparin should be discontinued. Because risk of thrombosis persists even with cessation of heparin, an alternative non-heparin anticoagulant should be started immediately, regardless of the presence or absence of thrombosis. In a prospective study, 10.4% of subjects developed a thrombosis between discontinuation of heparin and initiation of an alternative anticoagulant. Great care must be taken to avoid all sources of heparin, such as intravenous (IV) flushes and heparin added to total parenteral nutrition formulas. Although low molecular weight heparin has a much lower rate of inducing anti-heparin/PF4 antibodies, low molecular weight heparin can potentiate HIT once it is present, and thus low molecular weight heparins should also be avoided in a patient suspected of having HIT. Direct thrombin inhibitors (lepirudin, argatroban and bivalirudin) (DTI) are currently approved for use in the United States for treatment of patients with HIT (Table 90.3). Both lepirudin and argatroban directly bind thrombin, and do not require anti-thrombin for their effect. Major bleeding can occur in up to 17.6% of those receiving lepirudin when initiated with a bolus, and 5.3% of patients receiving argatroban.

Initiation of warfarin should be delayed until the platelet count has recovered to at least 100,000/µl or near baseline. If started earlier, while there is persistent thrombocytopenia, there is a risk of venous gangrene once the DTI is discontinued. If warfarin is initiated prior to identification of HIT, vitamin K, 5 mg orally or IV, should be used to reverse the effect of warfarin.

Warfarin should be overlapped with the DTI for at least 5 days, and a therapeutic INR should be present on two consecutive INR readings 24 hours apart prior to

TABLE 90.3 FDA-approved Alternative Anticoagulants

| | Dosing | PTT target ratio/timing first PTT | Effect on PT | Elimination | Half-life | |
|---|---|---|---|---|---|---|
| Lepirudin | *Bolus, 0.4 mg/kg (max 44 mg); infusion, 0.15 mg/kg/h (max. 16.5 mg/h); alternative approach 0.05–0.1 mg/kg/h infusion without a bolus | 1.5–2.5 × patient's initial PTT; check PTT 4 hours after initiation and dose adjustments | Minor | Renal | 80 min | Dose reduce even for mild renal impairment CrCl < 60; anti-lepirudin antibodies occur in up to 40%, this may lead to reduced elimination and an increased anticoagulant effect |
| Argatroban | #2 μg/kg/min; alternative approach 1.5 μg/kg/min | 1.5–3 × patient's initial PTT, NTE 100 seconds; check PTT 2 hours after initiation and 2–4 hours after dose adjustment | Yes | Hepatic | 40–50 min | |
| Bivalirudin | 0.15–2.0 mg/kg/h | Monitored by ACT | | Enzymatic (80%) and renal (20%) | 25 min | Approved for use in PCI only |

*Alternative approach is based on observations that the risk of bleeding with higher doses of lepirudin is substantial without added reduction in thrombotic episodes. At lower doses (alternative approach), similar rates of thrombosis were seen with less bleeding.
#Lower doses (alternative approach) of argatroban may be adequate, particularly in those with multi-organ dysfunction.

discontinuing the DTI. These recommendations are similar to those used when over-lapping warfarin with heparin. They are intended to avoid a period of increased coag-ulation as a result of warfarin-induced protein C and S deficiency, and assure adequate anticoagulation. Since argatroban, and to a lesser degree lepirudin, may affect the baseline prothrombin time prior to initiation of warfarin, determining when a patient is therapeutic with warfarin can be difficult. An approach is outlined in Figure 90.2; Warfarin affects Factors II, VII, IX, and X, whereas argatroban is specific for thrombin. Thus, a chromogenic Factor X assay can be used to distinguish the effects of warfarin as distinct from argatroban.

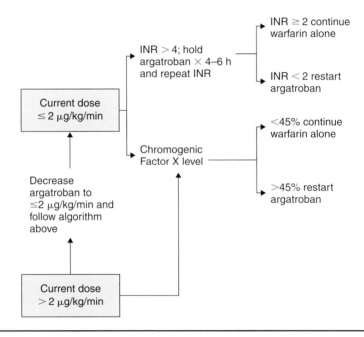

**FIGURE 90.2** Conversion of argatroban to warfarin. Argatroban and warfarin should overlap for at least 5 days, and warfarin should be therapeutic for 24 hours prior to discontinuation of argatroban. The INR or chromogenic Factor X level can be used to determine if warfarin is thereapeutic. If the INR is used, patients need to be receiving <2 μg/kg per minute for accurate assessment of the INR. In patients receiving argatroban at doses <2 μg/kg per minute, the target INR while on the combination of warfarin and argatroban is >4. Once the INR is >4, the argatroban can be discontinued and an INR repeated after 4–6 hours. If the INR is 2–3, then the warfarin is therapeutic. In patients receiving >2 μg/kg per minute of argatroban, the infusion rate should be decreased to a rate of 2 μg/kg per minute and a repeat INR obtained 4–6 hours after the dose adjustment; the algorithm for argatroban at doses <2 μg/kg per minute can then be followed. If a chromogenic Factor X level is used, argatroban can be continued without dose adjustment. A chromogenic Factor X level of 45% is consistent with an INR of 2. The chromogenic Factor X level and INR are inversely proporational. As the chromogenic Factor X level decreases, the corresponding INR increases.

In the setting of a thrombosis, warfarin should be continued for 3–6 months. The shorter duration can be used for those with a catheter-related thrombosis. In the absence of a thrombosis, there are no clinical studies to guide the duration of anticoagulation. In a retrospective study of 62 patients with isolated HIT (though asymptomatic thrombosis was not excluded by ultrasonography), 52.8% developed thrombosis during 30 days of follow-up; however only 10% of those developed after day 8. Accordingly, one month of warfarin anticoagulation is likely adequate in those with HIT without thrombosis where no evidence of thrombosis is discovered by ultrasonography.

**Platelet Transfusions:** Since bleeding is rare and there are anecdotal reports of clinical worsening when platelet transfusions are given to patients with HIT, it is recommended that prophylactic platelet transfusion be avoided

**Cardiac Surgery:** Patients with a history of HIT may require cardiac bypass. If anti-heparin/PF4 antibodies are not detected on ELISA, patients can be re-exposed to heparin for a short period (intraoperatively) without increasing antibody titers. A DTI should be used prior to and after surgery. Overall, this approach of using heparin during cardiac bypass in those with a history of HIT, but who currently do not have anti-heparin/PF4 antibodies detected on ELISA, is preferred over using a DTI. In those with a recent clinical history of HIT and persistent antibodies, a DTI should be used. Bivalirudin is typically the preferred DTI in the setting of cardiac surgery. Its use should be undertaken with an experienced surgeon and anesthesiologist following a published protocol.

**Patient Education:** Patients with HIT need to be clearly informed of their reaction to heparin, and it should be listed as an allergy. A card stating this as an allergy should be provided to the patient. Further heparin exposure should be avoided, except in limited and well-controlled situations in which the benefits outweigh the risks (as in cardiac surgery).

## Recommended Reading

Arepally GM, Ortel TL. (2006). Clinical practice. Heparin-induced thrombocytopenia. *N Engl J Med* **355**, 809–817.

Greinacher A, Farner B, Kroll H *et al.* (2005). Clinical features of heparin-induced thrombocytopenia including risk factors for thrombosis. A retrospective analysis of 408 patients. *Thromb Haemost* **94**, 132–135.

Lo GK, Juhl D, Warkentin TE *et al.* (2006). Evaluation of pretest clinical scores (4 T's) for the diagnosis of heparin-induced thrombocytopenia in two clinical settings. *J Thromb Haemost* **4**, 759–765.

Warkentin TE, Kelton JG. (2001). Delayed-onset heparin-induced thrombocytopenia and thrombosis. *Ann Intern Med* **135**, 502–506.

Warkentin TE, Sheppard JA, Sigouin CS *et al.* (2006). Gender imbalance and risk factor interactions in heparin-induced thrombocytopenia. *Blood* **108**, 2937–2941.

Warkentin TE, Maurer BT, Aster RH. (2007). Heparin-induced thrombocytopenia associated with fondaparinux. *N Engl J Med* **356**, 2653–2655.

Warkentin TE, Greinacher A, Koster A, Lincoff AM. (2008). Treatment and Prevention of Heparin-induced Thrombocytopenia: American College of Chest Physicians Evidence-Based Cinical Practice Guidelines (8th edition). *Chest* **133**, 340s–380s.

Zwicker JI, Uhl L, Huang WY *et al.* (2004). Thrombosis and ELISA optical density values in hospitalized patients with heparin-induced thrombocytopenia. *J Thromb Haemost* **2**, 2133–2137.

# CHAPTER 91
# Autoimmune Lymphoproliferative Syndrome

Michael A. Briones, DO

Autoimmune lymphoproliferative syndrome (ALPS) involves a triad of lympho-proliferative disease, autoimmune cytopenias and increased susceptibility to malignancy. This syndrome is caused by a primary defect in apopotosis, or programmed cell death.

**Pathophysiology:** ALPS is most often associated with heterozygous mutations in the gene encoding the Fas protein, TNFRSF6 (tumor necrosis factor receptor super-family), and other related effector proteins that regulate lymphocyte survival. Most of these mutations are inherited in an autosomal dominant fashion with variable penetrance.

**Clinical Manifestations:** Lymphoproliferation is the major clinical finding in ALPS, presenting as lymphadenopathy and massive splenomegaly. The median age at initial presentation is 24 months. More pronounced in infancy, the lymphadenopathy often regresses during adolescence and may resolve spontaneously in selected patients. Additionally, 75% of patients have hepatomegaly. ALPS is not usually associated with systemic symptoms, such as fever, rigors or night sweats; these symptoms should suggest an alternative diagnosis.

Approximately one-third of ALPS patients have autoimmune hemolytic anemia, immune-mediated thrombocytopenia or autoimmune neutropenia. Anticardiolipin antibodies and rheumatoid factor are also commonly seen. More rare autoimmune phenomena include glomerulonephritis, optic neuritis, Guillian-Barré syndrome, primary biliary cirrhosis, autoimmune hepatitis, arthritis, vasculitis, childhood linear IgA disease and acquired Factor VIII coagulopathy. Skin rashes of probable autoimmune origin and urticaria are also both common. The risk of developing autoimmunity is lifelong, and becomes greater with age.

Increased susceptibility to malignancy is the most worrisome clinical finding associated with ALPS, with lymphoma developing in about 10% of patients. A wide range of other malignancies has also been described.

**Diagnosis:** The diagnosis of ALPS is based upon the findings shown in Table 91.1. The identification of double-negative T cells by flow cytometry is a useful screening tool, especially if combined with other laboratory features, including peripheral lymphocy-tosis, circulating autoantibodies and polyclonal hypergammaglobulinemia. The upper limit of normal for double-negative T cells is 1%. Confirmatory testing of Fas-mediated apoptosis is done by *in vitro* testing. Apoptotic cell death has been assessed by a variety of methods, including hypodiploid quantification, terminal deoxynucleotidyl transferase biotin-dUTP nick end labeling (TUNEL), propidium iodide surface staining, annexin surface staining, and dye exclusion. ALPS patients show wide variation in the degree of impairment to Fas-mediated apoptosis. Molecular studies have revealed that germline

| TABLE 91.1 Diagnostic Criteria for ALPS |
| --- |

**Required**

Chronic non-malignant lymphoproliferation, lymphadenopathy and/or splenomegaly

Defective *in vitro* Fas-mediated lymphocyte apoptosis

Increase in circulating double negative T cells that are CD4 negative CD8 negative and express the alpha/beta

T-cell receptor above the normal range of 0.1–0.9% of lymphocytes (absolute counts range normally from 2 to 17/ml)

**Supporting**

Associated autoimmune disease

Mutations in Fas gene, Fas ligand gene, or Caspase 8 or 10 genes

Family history of autoimmune lymphoproliferative syndrome (ALPS)

Mutations of genes encoding Fas or related apoptosis signaling proteins

heterozygous mutations in the genes encoding apoptosis signaling proteins Fas, Fas ligand (FasL), Caspase 8 or Caspase 10 are present in most patients with ALPS. The current ALPS classification is based upon the mutation present. Type Ia involves mutation of the TNFRSF6 (Fas) gene, type Ib involves mutation of the TNFSF6 (FasL) gene, type II involves mutation of caspase 8 or 10 genes, and type III involves no currently known mutation.

**Differential Diagnosis:** Autoimmune diseases such as systemic lupus erythematosus (SLE) may present with symptoms similar to ALPS. Acute leukemias, lymphomas, and viral infections including EBV, CMV and HIV must be ruled out. Immune disorders such as IgA deficiency and common variable immunodeficiency must also be considered. Paroxysmal nocturnal hemoglobinuria, Rosai Dorfman disease, histiocytic disorders and other causes of hypersplenism must also be considered.

**Management:** Immune suppression with corticosteroids such as prednisone has been the standard initial therapy. High-dose pulse therapy with IV methylprednisolone (5–30 mg/kg per day) has been used, followed by lower-dose oral prednisone (1–2 mg/kg per day) for maintenance, with tapering of the corticosteroids over several months depending on response. Intravenous immunoglobulin (1–2 g/kg) has also been used, and may benefit some patients with severe AIHA by abrogating antibody-mediated red blood cell and platelet destruction. Some ALPS patients with chronic neutropenia and associated infections benefit from three times a week granulocyte colony stimulating factor (G-CSF). Rituximab has been utilized (375 mg/m$^2$ weekly × 4 doses) for treatment of refractory, chronic cytopenias. Patients refractory to standard drug regimens and red blood cell transfusions who require long-term immune-suppression or who display severe refractory cytopenias (especially thrombocytopenia) may respond to mycophenolate mofetil or cyclosporine. Chemotherapeutic agents such as vincristine have been utilized with variable results. Hematopoietic progenitor cell transplantation has been used successfully for those with severe disease that do not respond to other therapy. To avoid the risk of splenic rupture, participation in contact sports is not recommended for ALPS patients with significant splenomegaly.

## Recommended Reading

Alvarado CS, Straus SE, Li S *et al.* (2004). Autoimmune lymphoproliferative syndrome: a cause of chronic splenomegaly, lymphadenopathy, and cytopenias in children – report on diagnosis and management of five patients. *Ped Blood Cancer* **43**, 164–169.

Rao VC, Straus SE. (2006). Causes and consequences of the autoimmune lymphoproliferative syndrome. *Hematology* **11**(1), 15–23.

Rieux-Laucat F, Blachere S, Danielan S *et al.* (1999). Lymphoproliferative syndrome with autoimmunity: A possible genetic basis for dominant expression of the clinical manifestations. *Blood* **94**, 2575–2582.

Sneller MC, Wang J, Dale JK *et al.* (1997). Clinical, immunologic, and genetic features of an autoimmune lymphoproliferative syndrome associated with abnormal lymphocyte apoptosis. *Blood* **89**, 1341–1348.

Worth A, Thrasher AJ, Gaspar HB. (2006). Autoimmune lymphoproliferative syndrome: molecular basis of disease and clinical phenotype. *Br J Haematol* **33**, 124–140.

# CHAPTER 92
# Hemolytic Uremic Syndrome

Michael A. Briones, DO

Hemolytic uremic syndrome (HUS) is a thrombotic microangiopathy (TMA) characterized by thrombocytopenia, acute renal failure and microangiopathic hemolytic anemia. HUS is associated with a range of disorders, but is most commonly associated with bloody diarrhea caused by shiga-like toxin-producing bacteria (~90% of HUS in children). Non-diarrhea associated HUS is commonly termed atypical HUS, diarrhea-negative HUS, or non-shiga-like toxin-associated HUS, which can result from a wide variety of disorders, such as transplantation, neuraminidase from *Streptococcal* infection, cyclosporine, chemotherapy, other medications, pregnancy, and malignancy. The treatment of diarrhea-associated HUS differs from that of atypical HUS, and differentiating between atypical HUS and thrombotic thrombocytopenic purpura (TTP) can be difficult. Atypical HUS is often treated with plasma exchange until TTP can be excluded (see Chapter 93). In contrast, plasma exchange is not utilized in the treatment of diarrhea-associated HUS in children. The outcome of atypical HUS is often worse than diarrhea-associated HUS; in atypical HUS, 50% of patient develop end-stage renal disease and 25% die in the acute phase of the disease.

<u>Pathophysiology:</u> TMA is defined by endothelial cell damage, resulting in platelet-associated thrombosis and vessel obstruction. Histologically, this is manifested by accumulation of platelet thrombi containing von Willebrand factor (VWF) with very little fibrin. Red blood cell (RBC) destruction occurs in the vessels narrowed by these thrombi, leading to schistocyte formation (Figure 92.1).

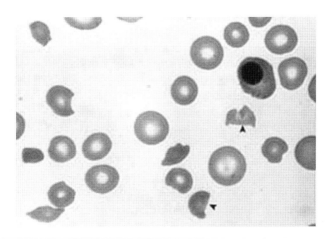

**FIGURE 92.1** Note the absence of platelets and the presence of a nucleated erythrocyte and schistocytes (arrowheads), consistent with a microangiopathic process. From Hoffman R, Benz EJ, Shattil SJ *et al.* (eds) (2005). *Hematology: Basic Principles and Practice*, 4th edition. Philadelphia, PA: Elsevier.

**Diarrhea-associated HUS:** HUS occurs in as many as 15% of children approximately 1 week after an episode of hemorrhagic gastroenteritis caused by shiga-like toxin-producing strains of *Escherichia coli* O157:H7 (STEC). Infections with other *E. coli* serotypes, *Shigella dysentariae* and other microbes can also cause HUS in children and adults. HUS following urinary tract infection with STEC has also been reported.

Cattle are the major vectors for *E. coli* O157:H7, with human infection occurring after ingestion of contaminated undercooked meat, non-pasteurized milk, or fruits/vegetables. The enterohemorrhagic bacteria bind to mucosal epithelial cells of the colon and invade. Besides damage to the underlying tissues and consequent bloody diarrhea, the shiga-like toxin then enters the blood stream, binds to polymorphonuclear leukocytes and is transported to target organs, including the kidneys. Renal glomerular endothelial cells are the primary targets for shiga-like toxins as they are particularly rich in glycolipid Gb3 (globotriaosylceramide), the predominant membrane receptor for the shiga toxin. The endothelial cells become separated from the basement membrane and platelets are activated by contact with exposed collagen and ultra-large VWF (ULVWF) in the subendothelium. The subsequent binding of fibrinogen to activated platelet glycoprotein IIb/IIIa complexes induces aggregation of platelets in the glomerular microcirculation. In addition, shiga-like toxin induces renal endothelial cells to secrete chemokines, cytokines and reactive oxygen species to upregulate the expression of adhesion molecules on cell surfaces. Finally, local exposure of tissue factor may activate the coagulation cascade, leading to thrombin generation and fibrin formation. These events act together to amplify renal microvascular thrombosis in *E. coli*-associated HUS. Other organs, including the brain, pancreas and liver, may also be damaged by circulating microthrombi.

**Atypical HUS:** Atypical HUS may occur secondary to *Streptococcus pneumoniae* infection, genetic causes, medications and pregnancy, and in association with other disorders. *Streptococcus* or other neuramidase producing bacteria accounts for up to 10% of pediatric HUS cases, typically occurring in children <2 years old. These patients present with pneumonia (rarely, skin infection), bacteremia and acute renal failure. The proposed mechanism is that the neuraminidase cleaves sialic acid residues on RBCs, platelets and glomeruli, leading to exposure of the Thomsen-Friedenreich antigen, which is recognized by a naturally occurring IgM antibody leading to complement fixation and cellular damage (T-activation).

A minority of atypical HUS cases are related to an autosomal dominant or recessive inheritance genetic pattern. These patients have low levels of the third component of the complement system (C3), which is most commonly associated with genetic mutation of Factor H gene (*CHF*). Factor H inhibits activation of the alternative pathway, while other mutations are in the membrane cofactor protein (*MCP*; MCP degrades C3b and C4b), or Factor I gene (*IF*; IF inactivates cell bound C3b). The resulting complement dysregulation results in enhanced complement activation and increased cell lysis. Of 156 atypical HUS patients (58 familial and 98 sporadic) whose disease was triggered by various causes (pregnancy, medications, infection), 56% demonstrated no mutations in these three genes. Interestingly, these genetic mutations can influence disease prognosis and response to treatment; with the *MCP* mutations being associated with an improved response compared to the *CHF* mutations.

Drug-induced TMA has been associated with antineoplastic agents (mitomycin, cisplatin, bleomycin, gemcitabine), immunosuppresive agents (cyclosporine, tacrolimus, muromonab-CD3), antiplatelet agents (ticlopidine, clopidogrel) and other medications (quinidine, interferon). Some of the medications produce HUS in a cumulative dose-dependent manner, such as mitomycin and cyclosporin, while others act through an idiosyncratic immune-mediated response, such as quinidine, ticlopiden and clopidogrel. Pregnancy-associated TMA can be a result of HELLP, acute fatty liver associated with DIC, atypical HUS, or TTP (see Chapter 103).

**Clinical Manifestation and Laboratory Features:** Hemolytic uremic syndrome is characterized by thrombocytopenia, microangiopathic hemolysis and acute renal failure. Approximately 90% of typical childhood HUS cases present 1–2 weeks following a prodromal diarrheal illness, which often involves bloody diarrhea. Stool cultures may be negative by the time of presentation.

## Diagnosis:

**CBC and Coagulation Markers:** HUS is defined by the presence of microangiopathic hemolytic anemia, thrombocytopenia and acute renal failure. Schistocytes are seen on peripheral blood smear, which reflect the fragmentation of the RBCs that occur as the RBCs traverse vessels partially occluded by platelet and hyaline microthrombi. In addition to schistocytes, giant platelets may be seen on peripheral smear. Elevation of the reticulocyte count, lactate dehydrogenase and indirect bilirubin (rarely >2–3 mg/dl), together with low haptoglobin levels, reflect intravascular hemolysis. The direct antiglobulin test is typically negative. Thrombocytopenia is mild to moderate in severity (platelet count <60,000/μl), and purpura or active bleeding are rare. The anemia and thrombocytopenia are unrelated to the severity of the renal dysfunction. However, hemolysis, RBC fragmentation and thrombocytopenia, if prolonged for more than 10 days, are associated with long-term renal sequelae. A moderate leukocytosis may also be present. The PT and PTT are within the normal range, differentiating HUS from DIC, but the fibrinogen, VWF, FVIII and D-dimer, as well as other markers of thrombin generation and fibrinolysis, may be elevated.

**Renal Function and Other Complications:** Blood urea nitrogen (BUN) and creatinine are typically markedly elevated. The urinalysis typically shows varying degrees of nephritis with associated protein, RBCs, WBCs and casts. Persistent proteinuria may be associated with an increased risk of progressive renal dysfunction. Complement (C3, C4) levels may be assessed and are usually reduced.

**Differential Diagnosis:** The differential diagnosis of HUS includes other TMAs: thrombotic thrombocytopenic purpura (TTP); disseminated intravascular coagulation (DIC); malignant hypertension; post-HPC (or solid-organ) transplantation TMA; Kasabach-Merritt syndrome; radiation nephritis; hemolysis with prosthetic cardiac devices (Waring Blender hemolysis); drug-induced microangiopathy; antiphospholipid antibody syndrome (APS); HIV-associated TMA; or hemolysis, elevated liver enzymes, low platelets (HELLP) syndrome. No laboratory test is currently able to distinguish between these disorders. The patient's history and underlying medical diseases are most useful in determining the cause of the TMA and resulting therapy. For example, plasma exchange initiated soon after diagnosis for TTP is the standard of care. However, in HPC transplantation-associated TMA, plasma exchange is no longer considered standard of care as effectiveness has not been proven. Plasma exchange in malignant hypertension-induced TMA or secondary to a prosthetic intravascular device is also not typically indicated. For the remainder of associated disorders, the utility of plasma exchange is uncertain, but a therapeutic trial may be indicated, especially if idiopathic TTP cannot be excluded.

## Management/Prognosis:

**Diarrhea-associated HUS:** The mainstay of therapy for diarrhea-associated HUS is supportive care, including dialysis. Early recognition and intravenous volume expansion have been shown to provide renal protection. Thrombocytopenia typically lasts 1–2 weeks, with anemia lasting for a slightly longer time period. Platelet transfusions are not recommended except in exceptional circumstances. Evidence suggests that antibiotics should be avoided unless the patient displays clinical signs of sepsis. Other treatments, including plasma therapy and use of intravenous immunoglobulin (IVIG), fibrinolytic and antiplatelet agents, corticosteroids and antioxidants have been ineffective in controlled clinical trials when utilized during the acute phase of the disease. As previously mentioned, dialysis is required in over 50% of children. However; dialysis is rarely needed for longer than 7 days. Hypertension is a common finding. Renal injury progressing to end-stage renal failure and kidney transplantation is rare in diarrhea-associated HUS (occurring in less than 10% of patients). Gastrointestinal complications, including cholelithiasis, pancreatitis and hepatitis, may occur. Neurological complications (seizures, altered mental status, stroke) may occur in up to 25% of children with HUS. Death rarely occurs (2–4%).

**Atypical HUS:** Neuraminidase-associated HUS is treated with antibiotic therapy and supportive care. Owing to this complication, some clinicians try to avoid plasma transfusion to decrease exposure to naturally occurring antibodies against the Thomsen-Freidenreich antigen. In other causes of atypical HUS, plasma exchange/infusion has been used with an overall drop in the mortality rate from 50% to 25%, but its efficacy is debated. Plasma exchange should be considered if renal or heart failure is present, and should be initiated within 24 hours of the patient's presentation; therapy should also be continued for at least 2 days after complete remission. Renal transplantation is a limited option for atypical HUS because of the 50% recurrence rate and >90% rate of graft failure in patients with recurrence.

## Recommended Reading

Caprioli J, Pengb L, Remuzzi G. (2005). The hemolytic uremic syndromes. *Curr Opin Crit Care* **11**, 487–492.

Caprioli J, Noris M, Brioschi S *et al.* (2006). International Registry of Recurrent and Familial *HUS/TTP*. Genetics of HUS: the impact of MCP, CFH, and IF mutations on clinical presentation, response to treatment, and outcome. *Blood* **108**, 1267–1279.

Noris M, Remuzzi G. (2005). Hemolytic uremic syndrome. *J Am Soc Nephrol* **16**, 103.

Siegler R, Oakes R. (2005). Hemolytic uremic syndrome; pathogenesis, treatment, and outcome. *Curr Opin Pediatrics* **17**, 200–204.

Zheng XL, Sadler JE. (2008). Pathogenesis of thrombotic microangiopathies. *Annu Rev Pathol* **3**, 177–249.

# CHAPTER 93

# Thrombotic Thrombocytopenic Purpura

Christine L. Kempton, MD

Thrombotic thrombocytopenic purpura (TTP), a thrombotic microangiopathy (TMA), is a syndrome consisting of microangiopathic hemolytic anemia (MAHA), thrombocytopenia, and end-organ damage secondary to microvascular thrombi. Anemia, thrombocytopenia, fever, neurological signs and renal abnormalities make up the classic pentad. However, this classic pentad is rarely fully present, and treatment should be initiated in the setting of unexplained MAHA and thrombocytopenia. Treatment of TTP with therapeutic plasma exchange has drastically decreased the mortality rate from greater than 90% to less than 20%.

**Pathophysiology:** The pathophysiology of TTP has been recently clarified as an acute deficiency of von Willebrand factor (VWF)-cleaving metalloprotease (ADAMTS13 [a disintegrin and metalloprotease with thrombospondin type 1 motif, 13]). ADAMTS13 deficiency results in the accumulation of ultra-large VWF multimers, which bind platelets, and leads to both thrombi in the microvasculature and thrombocytopenia. The microthrombi then cause MAHA by shearing of red blood cells (RBCs).

In idiopathic TTP, autoantibodies directed against ADAMTS13 cause enzyme dysfunction by increased clearance or impaired enzyme attachment. Mutations of the *ADMATS13* gene are associated with congenital TTP (Upshaw-Shulman syndrome).

The pathophysiology of secondary TTP is varied, and depends on the underlying cause; some disorders are associated with ADAMTS13 autoantibody formation, while other disorders are associated with endothelial injury without autoantibody formation.

Despite the link between ADAMTS13 and TTP, ADAMTS13 deficiency is not pathognomonic for TTP. Low ADAMTS13 activity can be seen in healthy individuals, as well as in individuals with DIC, ITP, sepsis, hepatic dysfunction, malignancy and pregnancy. Furthermore, patients with TTP may have normal ADAMTS13 activity. Therefore, the diagnosis of TTP remains a clinical diagnosis.

**Epidemiology:** The annual incidence of TTP is 11 per $10^6$ persons, with the incidence of idiopathic TTP being 4 per $10^6$ persons. The incidence of idiopathic TTP is 2.5 times greater in women as compared to men, and 4.9 times greater in African Americans as compared to other races. The median age at presentation is approximately 40 years, but may be younger in those with idiopathic or pregnancy-related TTP.

**Clinical Manifestations:** The classic pentad of anemia, thrombocytopenia, fever, neurological signs and renal failure is infrequently present. Fever is present in 24% of patient, renal abnormalities in 59%, and neurological signs in 63%. Renal involvement may manifest as proteinuria or increasing serum creatinine level. Neurologic involvement may manifest as seizures, focal neurological deficits, or problems with memory or confusion that can be subtle. Patients may also complain of non-specific symptoms, such as abdominal pain, nausea and weakness. Importantly, the absence of fever, renal dysfunction, neurologic findings or other end-organ damage does not exclude the diagnosis of TTP.

Patients with congenital TTP related to a mutation of the *ADAMTS13* gene, and resulting ADAMTS13 deficiency, may present anytime from the neonatal period through young adulthood. In two-thirds of patients the interval between relapses is every 2–3 weeks, and in one-third of patients the interval between relapses may be as long as years.

**Diagnosis:** TTP should be considered in the differential diagnosis of all patients presenting with MAHA and thrombocytopenia in the absence of underlying disease. MAHA is defined as schistocytes on peripheral smear, and hemolytic anemia. Hemolytic anemia is diagnosed by an elevated lactate dehydrogenase (LDH), indirect bilirubin and reticulocyte count. A microangiopathic cause for the hemolytic anemia is supported by the presence of schistocytes on the peripheral smear (>2 per field at 100× magnification).

Given that MAHA and thrombocytopenia are not specific for TTP, other causes should also be evaluated (Table 93.1). PT, PTT, direct antiglobulin test (DAT), renal and liver function tests may help in diagnosis. PT and PTT typically are within normal limits in TTP, and DAT is typically negative. Although low activity levels of ADAMTS13 and the presence of inhibitory antibodies to ADAMTS13 may estimate the risk for future relapse, the diagnostic value of these tests remains unclear (see Chapter 120).

| TABLE 93.1  Differential Diagnosis of Microangiopathic Hemolytic Anemia |
| --- |
| Autoimmune diseases (antiphospholipid antibody syndrome, systemic lupus erythematosus) |
| Disseminated intravascular coagulopathy |
| Hematopoietic progenitor cell transplantation |
| HIV/AIDS |
| Kasabach-Merritt Syndrome |
| Malignancy |
| Malignant hypertension |
| Medications: |
| • Calcineurin inhibitors: cyclosporine and tacrolimus |
| • Cancer chemotherapy: mitomycin C, cisplatin and gemcitabine |
| • Quinine |
| Pregnancy (hemolysis elevated liver enzymes and low platelets [HELLP] syndrome, eclampsia) |
| Prosthetic heart valves |
| Radiation nephritis |
| Scleroderma renal crises |

**Differential Diagnosis:** The patient's history and physical examination are most useful in determining the cause of MAHA and thrombocytopenia. In addition to the difficulty distinguishing between idiopathic and non-idiopathic TTP, the distinction between hemolytic uremic syndrome (HUS) and TTP may be difficult; patients with HUS typically have significantly worse renal function than those with TTP, and lack neurologic symptoms. In the adult patient, atypical HUS is not readily distinguishable from TTP and is managed similarly. Children with diarrhea-associated HUS are not routinely treated with plasma exchange (see Chapter 92).

**Management:** Left untreated, the mortality associated with TTP is greater than 90%. Therapeutic plasma exchange (TPE) is the primary therapy for TTP. TPE is hypothesized to work by removing ultra-large VWF multimers and anti-ADAMTS13 autoantibodies, while concurrently restoring ADAMTS13 protease activity. TPE should be instituted as soon as possible after TTP is suspected, based on the findings of an unexplained MAHA and thrombocytopenia. If TPE is not available, plasma infusion (15–30 ml/kg) should be initiated while arrangements for TPE are made (Chapter 68).

Although some patients may have had signs or symptoms of TTP for weeks, instituting TPE should not be delayed on the basis of apparent clinical stability. There are no clinical prognostic factors that can accurately predict the onset of potentially catastrophic end-organ damage.

There are clinical situations, such as pregnancy and systemic lupus erythematosis, where TPE should be instituted as in idiopathic TTP, because these disorders are associated with ADAMTS13 autoantibody formation (i.e. have pathophysiology identical to idiopathic TTP). In contrast to TTP, TPE in TMA secondary to malignant hypertension, prosthetic intravascular devices or hematopoietic progenitor cell transplantation are typically not indicated. For the remainder of cases of non-idiopathic TTP, the utility of TPE is uncertain; typically, a trial of TPE is initiated.

TPE can be initiated using plasma products (fresh frozen plasma [FFP] or cryoprecipitate reduced plasma [CRP]). Although underpowered to be definitive, a small study comparing CRP and FFP found no difference in patient outcome. The typical TPE regimen is 1.0 total plasma volume exchanged daily until the platelet count is above 150,000/μl. Most authorities will continue TPE for 2 days after the platelet count is >150,000/μl, and some authorities will taper off TPE. The patient should be closely followed for the next few weeks for evidence of recurrence.

During treatment and in the immediate period after discontinuing TPE, daily laboratory monitoring should include hematocrit/hemoglobin, platelet count, reticulocyte count and LDH. If the platelet count decreases or the LDH increases after discontinuing TPE, the patient should be re-evaluated for re-initiating TPE. Infection is associated with worsening or relapses of TTP.

All patients should receive folic acid to facilitate RBC production secondary to the increased RBC destruction. Given the potential for an underlying autoimmune disorder leading to ADAMTS13 inhibitory antibodies, steroids have been used in the treatment of TTP. They should be considered as an adjunct to treatment with TPE in patients that are refractory to TPE (defined below) have an exacerbation when TPE is discontinued, have life-threatening episode or relapse after remission. Steroid regimens include 1–2 mg/kg of prednisone daily until remission is achieved, or 1 g of methylprednisolone per day for 3 days.

Other immunosuppressive agents that have been reported in case series include rituximab, cyclophosphamide and vincristine. The demonstrated presence of an inhibitor to ADAMTS13 theoretically makes the use of such immunosuppressive agents rational. Although there are no historical clinical trials to guide usage, a current NHLBI-funded randomized clinical trial is comparing patient outcome in patients with idiopathic TTP receiving TPE and steroids to those receiving TPE, steroids, and rituximab (Study of TTP and Rituximab [STAR]).

Given the risk of platelet transfusions worsening disease, platelets should be reserved for patients with severe or life-threatening hemorrhage.

**Refractory Disease:** Refractory disease has been defined in various ways; definitions include a lack of improvement in platelet count after seven daily TPE procedures, lack of normalization of platelet count after 3 weeks of treatment, and worsening of disease despite treatment. A variety of treatment approaches have been used for refractory TTP, but none in a randomized clinical trial. Approaches include changing the replacement fluid (e.g. FFP to CRP or conversely), increasing the exchange volume to 1.5–2.0 plasma volumes, performing two exchanges a day with 1.0 plasma volume per each exchange, and adding steroids or other immunosuppressive agents if not already utilized. The approach to the refractory patient depends on the disease severity, response to previous TPE procedures, other medications, and patient's underlying diseases. In life-threatening disease, an increase in TPE frequency or exchange volume as well as immunosuppressive therapy should be considered.

In a small study, those patients that had ADAMTS13 deficiency and an inhibitor to ADAMTS13 had a more prolonged clinical course as compared to those with ADAMTS13 deficiency without an inhibitor (see Chapter 120). Those with an inhibitor had a longer time to platelet recovery (23 days versus 7 days) and a higher proportion of relapses (62% versus 25%).

**Exacerbation:** Recurrence of active TTP prior to 30 days after the last TPE procedure should be considered an exacerbation of the initial episode rather than a relapse (see below), and treated with re-initiation of TPE; 20–45% of patients will experience an exacerbation.

**Relapse:** Relapse is typically defined as disease recurrence (i.e. thrombocytopenia and elevated LDH) after 30 days from the last TPE procedure from the previous episode. Approximately 35% of patients experience a relapse, which can be years later. TPE should be reinitiated. In addition, multiple other therapies, including splenectomy and immunosuppressive medications, may be considered in patients who have recurrent relapsing TTP, although mixed reports exists with respect to the effectiveness of these procedures.

**Familial TTP:** For patients with congenital TTP secondary to a congenital deficiency of ADAMTS13, the treatment of choice is replacement of the missing cleaving protease by prophylactic plasma infusions. Infusions of plasma (10 ml/kg) or CRP may be required as frequently as every 2–3 weeks, depending on the severity of the deficiency.

## Recommended Reading

Coppo P, Wolf M, Veyradier A *et al.* (2006). Prognostic value of inhibitory anti-ADAMTS13 antibodies in adult-acquired thrombotic thrombocytopenic purpura. *Br J Haematol* **132**, 66–74.

George JN. (2006). Clinical practice. Thrombotic thrombocytopenic purpura. *N Engl J Med* **354**, 1927–1935.

Moake JL, Rudy CK, Troll JH *et al.* (1982). Unusually large plasma factor VIII:von Willebrand factor multimers in chronic relapsing thrombotic thrombocytopenic purpura. *N Engl J Med* **307**, 1432–1435.

Rock GA, Shumak KH, Buskard NA *et al.* (1991). Comparison of plasma exchange with plasma infusion in the treatment of thrombotic thrombocytopenic purpura. Canadian Apheresis Study Group. *N Engl J Med* **325**, 393–397.

Terrell DR, Williams LA, Vesely SK *et al.* (2005). The incidence of thrombotic thrombocytopenic purpura-hemolytic uremic syndrome: all patients, idiopathic patients, and patients with severe ADAMTS-13 deficiency. *J Thromb Haemost* **3**, 1432–1436.

Wyllie BF, Garg AX, Macnab J *et al.* (2006). Thrombotic thrombocytopenic purpura/haemolytic uraemic syndrome: a new index predicting response to plasma exchange. *Br J Haematol* **132**, 204–209.

Zeigler ZR, Shadduck RK, Gryn JF *et al.* (2001). Cryoprecipitate poor plasma does not improve early response in primary adult thrombotic thrombocytopenic purpura (TTP). *J Clin Apheresis* **16**, 19–22.

# CHAPTER 94

# Antiphospholipid Antibody Syndrome

Michael A. Briones, DO

Antiphospholipid antibody syndrome (APS) is an autoimmune and multi-system disorder of recurrent arterial and venous thrombosis, pregnancy loss and immune cytopenias associated with the presence of antiphospholipid antibodies (APA), and a persistently positive anticardiolipin (ACL) or lupus anticoagulant testing. It is now well established that APA are heterogenous and bind to various protein targets, among them the plasma protein beta 2 glycoprotein I (β2GPI) and prothrombin. Antiphospholipid antibodies were first described in 1906 in a study by Wassermann and colleagues, among patients with positive serologic testing for syphilis. Antiphospholipid antibodies can be broadly categorized into those antibodies that prolong phospholipid-dependent coagulation assays, known as lupus anticoagulant (LA), or ACL, which targets cardiolipin. APS can either be primary when it is not associated with another disease or secondary when it occurs in association with other conditions, such as systemic lupus erythematosus (SLE).

Pathophysiology: The mechanisms which cause thrombosis in APS are unclear. There is evidence that the antibodies interfere with the protein C pathway by impairing both protein C activation and the function of activated protein C (APC). There are data implicating endothelial cell dysfunction, with these cells expressing significantly higher amounts of the adhesion molecules, intercellular cell adhesion molecule-1 (ICAM-1), vascular cell adhesion-1 (VCAM-1) and E-selectin, when these are incubated with APA and β2GPI *in vitro*. Similarly, the incubation of endothelial cells with antibodies reacting with β2GP1 has been shown to induce endothelial cell activation with upregulation of various adhesion molecules, IL-6 production and alteration in prostaglandin metabolism. Another possible mechanism suggested to explain the prothrombotic and proinflammatory activities of APA is the upregulation of tissue factor (TF). IgG from patients with APS significantly increases TF function and transcription in monocytes and has also shown increased levels of soluble TF in patients with APS and thrombosis. Hence, there is evidence that APA induce TF expression and procoagulant activity *in vitro* and in patients with APS.

Recent studies have suggested that activation of the complement cascade may be necessary for APA-mediated thrombosis. Thrombus formation induced by antibodies to β2GPI requires a priming factor such as bacterial lipopolysaccharide (LPS), and is complement dependent. Hypocomplementemia has been found in a significant proportion of patients with primary APS, and has been associated with thrombosis. The mechanism by which APA mediate disease, however, is only partially understood, and is limited by the apparent polyspecificity of the antibodies, the multiple potential end-target organs and the variability of clinical context. Antiphospholipid antibodies are heterogeneous, and more than one mechanism may be involved in causing thrombosis.

**Clinical Manifestations:** APS should be considered in a patient with arterial or venous thrombosis, recurrent fetal loss, immune-mediated thrombocytopenia, and hemolytic anemia in association with a LA or ACA. Males with primary APS will usually demonstrate high-titer APA in association with thrombocytopenia, livedo reticularis, arterial thrombosis or a cerebrovascular accident (CVA). Children can present with a CVA, manifesting symptoms such as seizures, chorea and paralysis. Adults and children with APS can also present with deep venous thrombosis, Budd-Chiari syndrome, cutaneous vasculopathy (e.g. Henoch-Schonlein purpura), pulmonary embolism, optic neuropathy or renal disease (Table 94.1). Fetal and newborn complications can occur with maternal transfer of APA. Fetal loss/miscarriage can occur in up to 20% of pregnancies who have APS. The episodes of fetal loss usually occur during the late first or early second trimester of pregnancy, and are thought to be secondary to thrombosis in the uteroplacental vasculature.

**TABLE 94.1  Clinical Presentations of Antiphospholipid Antibody Syndrome**

| | |
|---|---|
| Deep vein thrombosis | Budd-Chiari syndrome |
| Pulmonary embolism | Recurrent fetal loss |
| Arterial thrombosis | Nephropathy |
| Myocardial infarction | Livedo reticularis |
| Cerebrovascular accident | Hemolytic anemia |
| Chorea | Thrombocytopenia |
| Seizures | Rarely, hemorrhagic disorders |

**Diagnosis:** The diagnosis of APS relies heavily on persistent laboratory abnormalities. Although multiple etiologies of thrombosis can occur simultaneously, diagnosis of APS often looks for persistent positive APS labs in the absence of other causes of thrombosis. The labs measure both antiphospholipid antibodies by solid phase assay and lupus anticoagulant activity by fluid phase assays. Given the heterogeneous nature of lupus anticoagulants, there is a complex panel of testing involved to maximize sensitivity to lupus anticoagulants with different specificities. Use and interpretation of these diagnostic tests are described in Chapter 137.

**Management:** The treatment of thrombosis in APS involves anticoagulation, initially heparin, then followed by warfarin. Because of the high incidence of recurrence, indefinite anticoagulation therapy is warranted in patients with persistently elevated APA. In patients with recurrent thrombotic events despite adequate anticoagulation, low-dose aspirin (81 mg daily) is often added. Immunosuppressive agents such as cyclophosphamide and corticosteroids are sometimes utilized with variable results in APS, but are helpful in the treatment of patients experiencing exacerbation of concomitant SLE or other autoimmune disease.

When utilizing the oral anticoagulation warfarin, the International Normalized Ratio (INR) should be maintained between 2.0 and 3.0. Higher INR values of 2.5 to 3.5 are recommended by some authors to prevent arterial thromboses. The presence of underlying thrombophilia should be investigated (e.g. Factor V Leiden or prothrombin gene mutation). It should be emphasized to the APS patient that coexistent prothrombotic factors and lifestyle issues, such as hormone replacement therapy containing estrogens, smoking, and increased weight, should be addressed.

For catastrophic APS, plasma exchange or intravenous immunoglobulin combined with heparin, and high-dose corticosteroids are often used, although there are no controlled studies to validate this approach. Rituximab, a monoclonal antibody which selectively depletes CD20-positive B lymphocytes, has been successfully employed in a small number of patients with resistant APS. If thrombocytopenia or hemolytic anemia is the only manifestation of APS, management should be similar to that routinely utilized in patients with idiopathic thrombocytopenia purpura or autoimmune hemolytic anemia. Standard approaches to both hematological problems include intravenous immunoglobulin, corticosteroids and/or splenectomy.

## Recommended Reading

Crowther MA, Ginsberg JS, Julian J et al. (2003). A comparison of two intensities of warfarin for the prevention of recurrent thrombosis in patients with the antiphospholipid antibody syndrome. *N Engl J Med* **349**, 1133–1138.

Finazzi G, Marchioli R, Brancaccio V et al. (2005). A randomized clinical trial of high-intensity warfarin vs conventional antithrombotic therapy for the prevention of recurrent thrombosis in patients with the antiphospholipid syndrome (WAPS). *J Thromb Haemost* **3**, 848–853.

Levine JS, Branch DW, Rauch J. (2002). The antiphospholipid syndrome. *N Engl J Med* **346**, 752–763.

Miyakis S, Lockshin MD, Atsumi T et al. (2006). International consensus statement on an update of the classification criteria for definite antiphospholipid syndrome (APS). *J Thromb Haemost* **4**, 295–306.

Pierangeli SS, Chen PP, Gonzalez EB. (2006). Antiphospholipid antibodies and the antiphospholipid syndrome: an update on treatment and pathogenic mechanisms. *Curr Opin Hematol* **13**, 366–375.

# CHAPTER 95

# von Willebrand Disease

Thomas C. Abshire, MD

von Willebrand disease (VWD) is a disorder characterized by deficiency of the von Willebrand factor (VWF) and is the most common inherited bleeding disorder, affecting 0.1–1% of the general population. VWD was first characterized in 1926 by Erik von Willebrand after describing a group of patients in the Aland archipelago off the coast of Finland. These patients had bleeding symptoms different from classic hemophilia, and an autosomal inheritance pattern. The clinical manifestations are varied, but usually comprise mucosal bleeding symptoms and bleeding immediately after invasive procedures or surgery.

**Pathophysiology:** VWF is a large multimeric protein which is encoded from chromosome 12p. It is secreted from both megakaryocytes and endothelial cells, and stored as a large VWF multimer in the Weibel Palade bodies of endothelial cells and in platelet α granules. VWF is synthesized in the endoplasmic reticulum as a large proVWF and the two C terminal portions are linked tail to tail as a dimer and stored as a 230-kDa pro-VWF protein. Glycosylation of the molecule takes place and the C-terminal dimers are then N-terminal multimerized up to 20,000 kDa in size. The VWF propolypeptide (VWFAgII) has a distinct role in the multimerization and storage of the protein, and is cleaved off prior to storage. The larger multimeric forms of the protein help in mediating platelet adhesion. The VWF protein has two essential functions; the first is to act as a carrier protein for FVIII, and the second is to bridge platelets to the blood vessel wall forming a link between glycoprotein (GP) Ib on the surface of the platelet with collagen and the underlying sub-endothelial matrix.

**Classification:** VWD is classified into three types: type 1, a partial quantitative deficiency of VWF; type 2, a qualitative deficiency of VWF; and type 3, a complete deficiency of VWF.

**Type 1 VWD:** Type 1 VWD accounts for approximately 80% of all VWD with a purported prevalence of 1% of the population, although the true prevalence is more likely between 0.1% and 0.4%. Type 1 VWD is a quantitative defect, where the VWF functions normally but is present in lower than normal amounts. Type 1 is autosomal dominantly inherited. This manifests itself as a decrease in both antigen levels and functional activity. Multimer patterns, detected by Western blot, show normal size distribution but decreased band intensity. In addition, Factor VIII (FVIII) levels are typically low, as VWF stabilizes FVIII. The vast majority of type 1 VWD is due to decreased synthesis, with the exception of type 1C or Vicenza, which is known to have a decreased half-life. Type 1C or Vicenza has lower VWF levels at baseline (single digits to teens) and an increased VWF propeptide level as well as increased peak levels of VWF when measured 15–30 minutes after desmopressin challenge.

**Type 2 VWD:** Type 2 VWD comprises the vast majority of the remaining cases that are not type 1 (15–20%), and in general is characterized as a qualitative defect with normal to low antigen levels but substantially decreased functional activity. Given the complexity of domains and functions of VWF, there are four different specific activities that can be altered in mutated VWF, and each constitutes a separate sub-type of type 2 VWD. Figure 95.1 demonstrates the domain sequence of the molecule, and where the known type 2 defects are located.

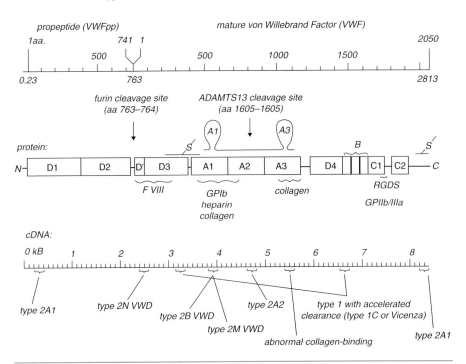

**FIGURE 95.1** Depiction of the protein sequence aligned with the cDNA sequence and the location of various type 2 VWD disorders as they relate to the domain regions of the VWF. Protein structure and its relation to VWD variant subtype mutation location. VWF contains specific domains that have been shown to have discrete functions, including (1) the interaction with Factor VIII (D′ and D3), GPIb (A1), collagen (A1 and A3), ADAMTS13 (A2), platelet GPIIb/IIIa (C1) and furin (D2-D′). The lower panel localizes hotspots for DNA mutations which cause VWF variants that tend to correlate with the functional domains. The variants include type 2A1 (2 A variants that fail to multimerize properly), type 2 N (decreased interaction with Factor VIII), type 2B (increased binding to platelet GPIb), type 2 M (decreased binding to GPIb but with normal VWF multimers), type 2A2 (2 A variants caused by increased spontaneous proteolysis of VWF by ADAMTS13), abnormal binding to collagen, and type 1C (type 1 VWD variant with accelerated VWF clearance). Reproduced courtesy of RR Montgomery.

*Type 2A VWD:* Type 2A VWD accounts for 10–12% of all VWD. It is characterized by absence of both large and intermediate weight VWF multimers. Type 2A VWD is characterized by either decreased multimerization during synthesis (assembly of VWF multimers within the cell) or increased susceptibility to proteolysis from increased cleavage by ADAMTS13 metalloprotease. Deficiencies in multimerization can be due to mutations found in the D2 and D3, A2 and the C terminal portions of the VWF protein. Increased proteolysis has been discovered to be exclusively due to mutations in the A2 region of the molecule, which enhances susceptibility to the effect of the ADAMTS13 protease.

*Type 2B VWD:* In type 2B VWD, a gain of function mutation results in increased affinity for the VWF protein for GP Ib on the platelet surface. This leads to a mild thrombocytopenia and a lack of the large molecular weight multimers due to increased binding to and clearance of platelets. This defect is caused by mutations in the A1 region of the molecule where interaction with the GP 1b is known to occur, and affects approximately 3–5% of the VWD population. A unique laboratory feature of this type of VWD is its increased platelet aggregation in response to low doses of the antibiotic ristocetin. This testing will be discussed in Chapter 123.

*Platelet-type or Pseudo VWD:* From a laboratory perspective, pseudo VWD has a similar profile to type 2B VWD. As in type 2B, there is increased affinity between VWF and platelets, but in pseudo VWD it is due to a gain of function mutation on the platelet GP 1b receptor and not in the VWF. Other laboratory features, such as loss of high molecular weight multimers, mild thrombocytopenia, and increased aggregation to low dose ristocetin, are similar between platelet/pseudo VWD and type 2B. Distinguishing pseudo VWD from type 2B is discussed in Chapter 123.

*Type 2M VWD:* In type 2M VWD, the functional activity of the molecule is approximately one-half of the antigenic activity. The defect is now known to be related to abnormalities in the A1 region of the VWF, and is secondary to decreased affinity to GP 1b – the opposite of type 2B VWD. Type 2M VWD affects 1–2% of individuals with VWD. Multimer patterns are typically normal in type 2M VWD. This might be explained by decreased platelet binding and resultant exposure of the VWF to cleavage by the ADAMTS13 protease.

*Type 2N VWD:* Type 2N VWD (also known as the Normandy variant of type 2 VWD) is characterized by a low FVIII activity (5–15%). Type 2N VWD affects 1–2% of individuals with VWD. A mutation in the FVIII binding domain of VWF in the D' region prevents the normal stabilization of FVIII by VWF. Type 2N has been termed *autosomal hemophilia A*, as it has some symptoms similar to hemophilia A, but is not sex linked. In some cases, type 2N may be misdiagnosed as hemophilia A, but careful attention to the predominant mucosal bleeding symptoms of VWD 2N and the FVIII level in the mild hemophilia range should help to differentiate the two conditions. It is inherited in a recessive or compound heterozygous fashion.

**Type 3 VWD:** Type 3 VWD is also known as the autosomal recessive form of VWD, where the individual is either homozygous or compound heterozygous. Type 3 VWD is characterized by the complete absence of VWF, with no functional activity and no antigenic protein. The FVIII coagulant activity is in the moderate hemophilia range (3–5% FVIII activity). Type 3 VWD occurs in approximately one in $1 \times 10^6$ individuals and accounts for, at most, 1% of all VWD individuals.

<u>**Clinical Manifestations:**</u> The clinical history is the most important criteria utilized in establishing a diagnosis of VWD. Without a history of bleeding in the patient

or within the family, it is difficult to assign a diagnosis. Details of the bleeding history are found in Chapter 79. European investigators and others have recently determined a bleeding score based upon the history. This approach may not be as applicable to children as to adults. First, children have often not been carefully questioned, and nor have they been hemostatically challenged. Second, "milder" bleeding symptoms such as purpura and epistaxis are more prevalent among children, and may be overlooked in the history. Therefore, an underlying bleeding condition, such as VWD, is more likely to be missed in the younger age group.

**Diagnosis:** An important aspect in the diagnosis of VWD is carefully collected bleeding history. If the history is suggestive of VWD it is essential to perform additional diagnostic testing, because the PT, PTT, thrombin time, fibrinogen, PFA/bleeding time, CBC and platelet count are often normal. Therefore, more specific testing for VWD is required. The PTT may be prolonged if the FVIII activity is less than 40%. Similarly, the PFA/bleeding time can sometimes be prolonged if the VWF levels are less than 30%.

The different laboratory tests utilized in making a diagnosis and in differentiating the types of VWD are summarized in Figure 95.2. More detailed assessment of the laboratory testing in VWD is found in Chapter 123.

*VWD Screening Tests:* Screening tests for VWD include the VWF antigen, VWF ristocetin cofactor activity (VWF:RCo), FVIII activity, and VWF multimers, as each of these tests is important in discerning whether a patient has VWD and, if so, the type of VWD. Most hematologists will usually perform the VWF antigen, VWF ristocetin cofactor and the FVIII activity on initial screening, and will perform the multimer testing and blood type once a diagnosis is reached. The abnormal range for VWF antigen, ristocetin cofactor and FVIII activity is usually less than 50% (usually less than two standard deviations below the mean), but may vary based on the normal ranges determined by the local or reference laboratory. There are a number of modulating factors which impact on interpretation of VWF testing, most importantly the ABO blood type. The blood type O is associated with lower VWF levels than other ABO blood types; therefore, assessment of bleeding symptoms is essential to establish true disease. Other modulating factors include physiologic conditions, such as pregnancy, mid-menstrual cycle and exercise; all of which may elevate the VWF level. Advancing age also has a role in raising the VWF. Finally, VWF is an acute phase reactant, which can increase substantially in response to physiological stress. For individuals who are sick with other illnesses, or patients who are stressed or frightened by blood draws (children in particular), the stress of being tested may cause a mild type 1 VWD patient to test in the normal range.

*Making a Diagnosis:* Some investigators suggest that VWF levels must be less than 30% in order to make a diagnosis of VWD. For levels between 30% and 50%, care must be taken in rendering a diagnosis, particularly if the bleeding symptoms are minimal. Conversely, if the patient has definite bleeding symptoms and levels are above 50%, these results may have been confounded by some of the previously described factors which may raise VWF levels, and testing should be repeated.

**Differential Diagnosis:** Mucosal bleeding symptoms are most commonly associated with VWD, but can also be a manifestation of inherited platelet dysfunction,

**von Willebrand disease variants and platelet disorders affecting VWF binding**

| | Normal | Type 1 | Type 3 | Type 2A | Type 2B | Type 2N | Type 2M | PT-VWD* | BSS* |
|---|---|---|---|---|---|---|---|---|---|
| VWF:Ag | N | ↓ | absent | ↓ | ↓ | N or ↓ | ↓ or N | ↓ | N |
| VWF:RCo | N | ↓ | absent | ↓↓↓ | ↓↓ | N or ↓ | ↓↓ | ↓↓ | N |
| FVIII | N | ↓ or N | 1–6% | N or ↓ | N or ↓ | ↓↓ | N | N or ↓ | N |
| RIPA | N | often N | absent | ↓ | often N | N | N or ↓ | often N | absent |
| LD-RIPA | absent | absent | absent | absent | ↑↑↑ | absent | absent | ↑↑↑ | absent |
| PFA | N | N or ↑ | ↑↑↑ | ↑ | ↑ | N | ↑ | ↑ | ↑↑↑ |
| BT | N | N or ↑ | ↑↑↑ | ↑ | ↑ | N | ↑ | ↑ | ↑↑↑ |
| Platelet count | N | N | N | N | ↓ (rarely N) | N | N | ↓ | → |
| Usual Tx | | DDAVP VWF conc | VWF conc | VWF conc (DDAVP) | VWF conc | VWF conc (DDAVP) | VWF conc | platelets | platelets |
| Response to DDAVP | | good | none | poor | decreases platelets | poor | poor | decreases platelets | none or modest |
| Response to VWF conc | | good | good | good | good | good | good | decreases platelets | no response |
| Frequency in general population | | reported as 1–2% | very rare 1:250,000 | rare | rare | rare | rare | rare | rare |
| VWF multimers | N | N but ↓ | absent | abnormal | abnormal | N but ↓ | N but ↓ | abnormal | normal |

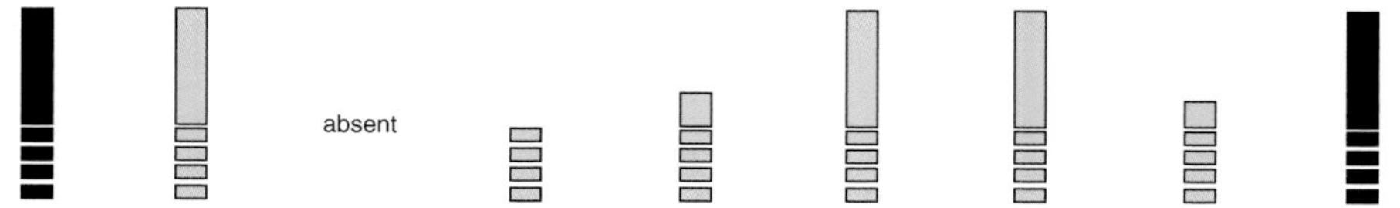

* platelet disorders that affect VWF interaction with platelets.

FIGURE 95.2 Summary of von Willebrand disease (VWD) types (types 1, 3, 2A, 2B, 2M, 2N) and absence of or defective platelet GPIb receptor for VWF, platelet-type VWD (PT-VWF) or Bernard–Soulier syndrome (BSS). The lower portion illustrates the VWF multimers that are identified by SDS-agarose gel electrophoresis of VWF for each of these variants. VWF:Ag, von Willebrand Factor antigen; VWF:RCo, VWF activity by ristocetin cofactor assay; LD-RIPA, ristocetin-induced platelet aggregation to low-dose ristocetin; N, normal; Tx, treatment; Conc, concentrate; PFA, platelet function analyzer; BT, bleeding time. Reproduced courtesy of RR Montgomery.

plasminogen activator deficiency 1 (PAI-1) or a vascular disorder such as Ehlers-Danlos syndrome. If the bleeding manifestations are not lifelong, an acquired platelet dysfunction associated with many medications should be considered, as well as acquired VWD. Most acquired VWD is associated with a lower VWF ristocetin cofactor compared to the VWF antigen. Most cases of acquired VWD usually presents in the elderly, and are characterized by accelerated clearance of the von Willebrand protein in the reticulo-endothelial system, occasionally antibody mediated but most often by increased clearance of the protein. The pathophysiology surrounding acquired VWD is varied, encompassing: autoimmune disorders (e.g. systemic lupus erythematosis), certain monoclonal gammapathies (e.g. multiple myeloma) and certain malignancies, such as lymphoma. This condition is now best known as acquired von Willebrand syndrome (AVWS). Other underlying conditions associated with AVWS should be considered, especially in children. These include: Wilm's tumor, congenital or acquired hypothyroidism (lower VWF levels as a result of decreased protein synthesis), and certain cardiac conditions such as ventricular septal defect (VSD) and valvular defects such as aortic stenosis, pulmonary stenosis and mitral valve prolapse. Prosthetic heart valves can result in AVWS in some settings. Finally, hemangiomas may also cause AVWS.

## Management:

**Minor Bleeding:** Most minor bleeding in VWD is considered mucosal in nature and does not need to be treated. However, persistent epistaxis or oral bleeding can be treated by antifibrinolytics and/or desmopressin (DDAVP).

*Antifibrinolytics:* The most useable antifibrinolytic (aminocaproic acid) may be given at 50–100 mg/kg per dose, every 6–12 hours (maximum 6 g per dose) for a duration of 3–7 days, depending upon the extent of the mucosal bleeding. Occasionally, one dose of an antifibrinolytic may be sufficient to stop epistaxis in a child. Additionally, tranexamic acid has been effective at 25 mg/kg per dose, to a maximum of 1.5 g per dose, given every 8 hours over 3–7 days (this is only available in the liquid form). The topical nasal saline gel can also be helpful in the prevention of epistaxis. Occasionally, other topical agents, such as gel foam, collagen and topical thrombin and fibrin glue, can also be helpful in epistaxis or oral bleeding. Topical vasoconstricting agents have been used in conjunction with cauterization for recalcitrant nosebleeds.

*Desmopressin:* Intravenous and/or nasal desmopressin (DDAVP) has been utilized as primary treatment for VWD types 1 and 2, although it is important to note that DDAVP is generally considered to be contraindicated for type 2B as it may exacerbate thrombocytopenia and potentially promote thrombosis. DDAVP works by inducing release of VWF from endothelial cells and by contact pathway activation. Additionally, it helps with GP IIbIIIa interaction and platelet micro-particle release. Intravenous DDAVP is given as 0.3 µg/kg in 30 cc of normal saline over 15 minutes. The intranasal form is administered as one spray (150 µg) for a patient less than 50 kg and two sprays (300 µg) for those over 50 kg. Both the intranasal and the intravenous forms of DDAVP may be given once a day for up to 2–3 days with careful monitoring of fluid intake. The most concerning and rare side-effects are hyponatremia and seizures.

More common side-effects include facial flushing and headache and, occasionally, alteration in blood pressure and abdominal cramps. If the medication is given too frequently (for more than 3 consecutive days), tachyphylaxis may occur.

**Major or Perioperative Bleeding:** Often in type 1 VWD, DDAVP may be the only treatment needed in more severe bleeding or surgery. A therapeutic trial with DDAVP should be carried out and VWF levels and activity monitored to establish DDAVP responsiveness prior to surgery. However, in most cases of major bleeding or surgery, a VWF replacement therapy is utilized to maintain VWF:RCo and FVIII levels above the hemostatic level (see Chapter 144). In major surgery, the usual goal is to maintain the FVIII/VWF:RCo activity at greater than 50% for 1 week and then greater than 30% for most of the second week to allow for complete healing.

*Acquired von Willebrand Syndrome:* Treatment is aimed at eliminating the underlying disorder along with addressing the underlying bleeding manifestations that accompany acquired von Willebrand syndrome (AVWS). Treatment with VWF/FVIII containing concentrates is the preferred method to alleviate bleeding, although rFVIIa has also been used effectively.

If the etiology of AVWS is antibody mediated, then blockade of the Fc receptors of the reticuloendothelial system is best achieved by intravenous immunoglobulin. Occasionally, plasma exchange or immunosuppressive agents such as corticosteroids or cyclophosphamide are utilized.

## Recommended Reading

Federici AB, Mannucci PM. (2007). Management of inherited von Willebrand disease in 2007. *Annals Med* **39**, 346–358.

Peake I, Goodeve A. (2007). Type 1 von Willebrand disease. *J Thromb Haemost* **5**(Suppl. 1), 7–11.

Sadler JE, Budde U, Eikenboom JC et al. (2006). Update on the pathophysiology and classification of von Willebrand disease: a report of the Subcommittee on von Willebrand Factor. *J Thromb Haemost* **4**, 2103–2114.

Tosetto A, Rodeghiero F, Castaman G et al. (2006). A quantitative analysis of bleeding symptoms in type 1 von Willebrand disease: results from a multicenter European study (MCMDM-1 VWD). *J Thromb Haemost* **4**, 766–773.

Tosetto A, Rodeghiero F, Castaman G et al. (2007). Impact of plasma von Willebrand factor levels in the diagnosis of type 1 von Willebrand disease: results from a multicenter European study (MCMDM-1VWD). *J Thromb Haemost* **54**, 715–721.

# CHAPTER 96

# Hemophilia A

Amy L. Dunn, MD

Hemophilia A (also known as classical hemophilia) results from congenital deficiency of Factor VIII (FVIII). It is an X-linked recessive disorder that results in decreased or absent circulating FVIII activity, leading to lifelong bleeding. Hemophilia A has an incidence of approximately 1 : 5000 male births.

Pathophysiology: FVIII is a plasma glycoprotein consisting of six domains A1–A2–B–A3–C1–C2 (Figure 96.1). The encoding gene is found on the long arm of the X chromosome (Xq28). The mature protein is a heterodimer with a light chain consisting of domains A3–C1–C2 and a heavy chain with the domains A1–A2–B. The majority of FVIII is thought to be synthesized in liver endothelial cells, but may also be produced in endothelial cells in general (e.g. elevated FVIII levels during liver failure). Upon release into the circulation it is non-covalently linked to von Willebrand Factor (VWF), which prevents enzymatic degradation of FVIII until needed during coagulation. During coagulation, tissue factor (TF) combines with FVIIa at the site of injury. This complex activates FX and FIX, leading to conversion of prothrombin to thrombin. The initial thrombin cleavage of the FVIII light chain causes FVIII to be released into the circulation and then activated to FVIIIa by further thrombin-mediated proteolysis. FVIIIa, along with FIXa, then act as cofactors on a phospholipid surface during activation of Factors X, V and, ultimately, thrombin (see Chapter 78). Patients with hemophilia A are unable to generate adequate thrombin due to the lack of FVIII, and become dependent upon the TF pathway. Circulating tissue factor pathway inhibitor (TFPI) efficiently downregulates the TF–FVIIa pathway as well as FXa, and as a result patients with hemophilia bleed. Thrombin-activatable fibrinolysis inhibitor (TAFI) production is also decreased in hemophilia, leading to more rapid dissolution of the fibrin clot. A large number of molecular defects have been described in the pathology of hemophilia A, including gross gene deletions and rearrangements of the DNA sequence, single gene rearrangements, deletions and insertions. A list of mutations leading to hemophilia A can be found at http://europium.csc.mrc.ac.uk.

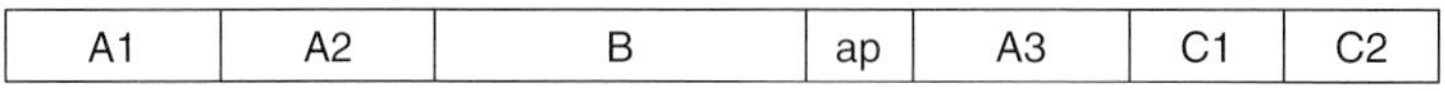

| A1 | A2 | B | ap | A3 | C1 | C2 |

FIGURE 96.1 Domain structure of Factor VIII.

Clinical Manifestations: The hallmark of hemophilia-related bleeding is delayed bleeding along with joint and muscle bleeding. In comparison, patients with VWD more commonly manifest immediate and mucocutaneous bleeding. As hemophilia A is X-linked, the vast majority of affected patients are male. Females can be affected in cases of extreme X chromosome lionization, or gene abnormalities such as Turner's syndrome. There is a high rate of spontaneous mutations within the FVIII gene, and as a result approximately 30% of newly diagnosed patients will have a negative family history.

In general, the severity of bleeding depends upon the percentage of circulating clotting factor activity. Patients with levels of >5–40% are classified as having mild hemophilia, patients with levels of 1–5% as moderate and those with less than 1% activity as having severe disease. Commonly, patients with severe disease will suffer from spontaneous bleeding while those with mild to moderate disease more typically bleed with trauma or surgery. In the newborn period, the most common findings are bleeding from blood draws, heel sticks, immunizations and circumcision. Infants born of known carrier mothers should not be circumcised until testing for FVIII rules out hemophilia. Older children and adults may experience excessive bruising, hematomas, intracranial bleeding, joint bleeding, muscle bleeding and mouth bleeding. The single largest preventable cause of morbidity is degenerative joint disease due to recurrent hemarthrosis. Carrier females may experience bleeding symptoms such as menorrhagia, oral bleeding, surgical and trauma-related bleeding.

**Diagnosis:** An X-linked inheritance pattern, elevated PTT and decreased plasma FVIII assay confirm the diagnosis.

**Differential Diagnosis:** Hemophilia A and B are clinically indistinguishable, and individual factor levels must be used to clarify the diagnosis. Patients with mildly low FVIII levels and an autosomal inheritance pattern may have type 1 von Willebrand disease (VWD). They will exhibit a commensurately decreased VWF antigen and ristocetin cofactor. Patients with type 3 VWD will have moderately low FVIIII, absent multimers, along with essentially absent VWF antigen and ristocetin cofactor activity. VWD type 2 N should also be considered in the setting of mildly low FVIII levels, inheritance in an autosomal recessive fashion, and poor response to recombinant FVIII therapy. In VWD type 2 N, the pathophysiology involves decreased FVIII binding to VWF. This leads to rapid proteolysis of FVIII despite adequate production. This type of VWD can be evaluated via a VWF to FVIII binding assay. Acquired low FVIII levels can also result from autoantibody formation.

**Management:**

**Factor Concentrates:** The mainstay of hemophilia care is FVIII replacement with intravenously delivered FVIII concentrates. Concentrates are either plasma derived, containing varying amounts of VWF, or recombinant products (see Chapter 145), and both currently undergo multiple viral and pathogen attenuation steps. No infectious complications have been reported since these steps were incorporated into the manufacturing process (mid-1980s for human immunodeficiency virus [HIV] and late 1980s for hepatitis B virus [HBV] and hepatitis C virus [HCV]); however, the possibility of contamination with new infectious agents such as prions cannot be excluded with plasma-derived products.

Infusions can be delivered in response to bleeding episodes ("demand therapy") or to prevent bleeding ("prophylaxis"). To prevent or minimize long-term sequelae, demand therapy should be given as soon as possible after a bleeding episode is recognized.

Because the need for urgent treatment is so important, many patients affected by hemophilia are educated in home infusion techniques. During a bleeding episode, factor replacement therapy should never be delayed to perform imaging or laboratory studies.

The dose and frequency of factor delivery is calculated based upon the half-life of the product (typically 12 hours), the intravascular volume of distribution (1 unit of FVIII per kilogram raises the plasma concentration by about 2%) and the desired clotting factor activity. For example, to raise the factor level of a 20-kilogram child with severe hemophilia to 100%, the dose should be $20\,kg \times 50\,IU = 1000\,IU$. Correction to FVIII levels of 40% activity is considered hemostatic in most cases; however, in the setting of surgery or life-/limb-threatening hemorrhage, higher FVIII levels (80–100%) are recommended. Post-surgical hemostasis should maintain FVIII levels above 50–70% for the first week, and above 30% for the second week. Ancillary measures such as compressive dressings, cauterization, packing and splinting should also be implemented when appropriate. Table 96.1 illustrates a suggested approach to factor replacement therapy for commonly encountered bleeding events.

**TABLE 96.1 Suggested Approach to Treatment of Bleeding Episodes**

| Bleed Site | Desired % Activity | Length of Therapy | Ancillary Measures to Consider |
|---|---|---|---|
| Central nervous system | 100% | 7–14 days, then strongly consider prophylactic therapy for a minimum of 6 months | Continuous infusion FVIII; antiepileptic prophylaxis; surgical intervention |
| Persistent oral/mucosal | 30–60% | 3–7 days | Antifibrinolytic therapy; custom mouthpiece; topical thrombin powder |
| Retropharyngeal | 80–100% | 7–14 days | Continuous infusion FVIII; antifibrinolytic therapy |
| Nose | 30–60% | 1–3 days | Packing, cautery; saline nose spray/gel; nasal vasoconstrictor spray; antifibrinolytic therapy |
| Gastrointestinal | 40–80% | 3–7 days | Antifibrinolytic therapy; endoscopy with cautery |
| Persistent gross urinary | 40–60% | 1–3 days | Vigorous hydration; evaluation for stones/UTI; avoid antifibrinolytic therapy; glucocorticoids |
| Muscle | 40–80% | Every other day until pain-free movement | Rest, ice, compression, elevation; physical therapy |
| Iliopsoas | 80–100% | Until radiographic evidence of resolution | Continuous infusion FVIII; bedrest; physical therapy |
| Joint | 40–80% | 1–2 days | Rest, ice, compression, elevation; physical therapy |
| Target joint | 80% day 1, 40% days 2 & 4 | 3–4 days | Rest, ice, compression, elevation; physical therapy |

Prophylaxis: In developed countries, prophylactic therapy delivered one to four times per week is considered the standard of care, and is the only therapy proven to prevent the long-term complication of degenerative joint disease. It is common practice to begin prophylaxis prior to the onset of recurrent joint bleeding.

Desmopressin: Desmopressin, or DDAVP (1-deamino-8-D-arginine vasopressin), is a synthetic form of the hormone vasopressin. The product may be given intravenously, subcutaneously or via nasal delivery. Details on dosing can be found in Chapter 95. DDAVP causes release of FVIII and VWF from storage sites. Patients with mild hemophilia A and some with moderate disease can be tested with DDAVP, and if they show a response by manifesting hemostatic levels or at least a three-fold response of FVIII, then DDAVP is often sufficient to treat mild bleeding symptoms such as nose, mouth and soft tissue bleeding. FVIII storage pools become depleted after multiple doses, so this treatment is not adequate for lengthy therapy, and fluid intake must be monitored closely as hyponatremia may result, particularly in children less than 2 years of age and in the elderly. In most cases, life- or limb-threatening bleeding episodes require FVIII replacement. Antifibrinolytic therapies such as aminocaproic acid or tranexamic acid are utilized to prevent excessive fibrinolysis, and are particularly useful in diminishing bleeding symptoms in locations with prominent fibrinolytic activity, such as the mouth, gastrointestinal tract and uterus.

## Complications:

Infectious complications: In the late 1970s and early 1980s, prior to incorporation of viral attenuation steps (mid-1980s for HIV and late 1980s for HBV and HCV) in factor manufacturing, many patients became infected with HCV and/or HIV from contaminated concentrates. Although many patients succumbed to these infections, there is a large cohort of long-term survivors as a result of highly active anti-retroviral therapies for HIV and the combination of ribavirin and interferon for HCV.

Inhibitors: A common complication of congenital hemophilia A is development of inhibitory antibodies to FVIII. These inhibitory antibodies occur in approximately 30% of patients with severe hemophilia, and to a lesser extent in those with mild and moderate disease (2–3%). They are more likely to occur in the setting of a positive family history of inhibitors, in association with large gene deletions such as introns 1 and 22 inversions, and in non-white patients. Non-genetic risk factors such as factor replacement during inflammatory states (known as the "danger theory") and delivery via continuous infusion may play a role. There is no clear evidence as to whether the rate of antibody formation is influenced by the type of product utilized; however, it has been suggested that plasma-derived products that contain VWF may be less immunogenic.

Antibodies to FVIII in congenital hemophilia are most commonly directed against the A2 or C2 domains and often develop within the first 10–20 exposures to exogenous FVIII, but can develop at any age. They typically neutralize both endogenous and exogenous FVIII. The antibody titer is measured utilizing a Bethesda assay, and is expressed in units (BU). One Bethesda unit is that amount of antibody which lowers the plasma factor level by 50% (see Chapter 128). Autoantibodies can develop in patients without hemophilia, leading to a condition known as acquired hemophilia (described in Chapter 111). Low-titer inhibitors ($< 5$ BU) are often transient but may be persistent and carry clinical significance, while high-titer inhibitors ($> 5$ BU) can significantly impact patient care and quality of life.

Immune tolerance therapy (ITT) with repeated exposure to FVIII concentrate over a period of months to years may eradicate the antibody. ITT has been accomplished utilizing both high-dose 100–200 IU/kg per day and low-dose 50 IU/kg thrice weekly of FVIII. Both recombinant and VWF containing plasma-derived products have been used successfully. Immunosuppressive agents such as cyclosporine and rituximab have been utilized as well, with some success. Bleeding in the setting of a high-titer inhibitor often requires bypassing therapy with either high doses of recombinant FVIIa (Chapter 147) or an activated prothrombin complex concentrate (Chapter 143).

Joint Disease: Degenerative joint disease due to recurrent hemarthrosis is the single largest preventable cause of morbidity for patients with hemophilia A. The pathogenesis of this arthropathy is multi-factorial, with iron being the most likely culprit triggering the degenerative changes. Recurrent hemarthrosis causes a combination of anatomic/mechanical changes along with an inflammatory reaction resulting in a thickened hyperemic synovium. This inflamed synovium participates in a vicious cycle of further bleeding, and ultimately the destruction of articular cartilage. The impact of hemophilic arthropathy is wide-ranging. Unfortunately the physical findings are often subtle in the early stages of joint disease, and investigation with MRI may be required to demonstrate hemosiderin deposition and or hypertrophic synovium. Initial treatment is to reduce or prevent further bleeding with clotting factor. Arthroscopic and radionuclide synovectomy have been employed to address recurrent bleeding in these target joints, but it remains to be seen whether they can halt articular destruction.

## Recommended Reading

Abshire T, Kenet G. (2004). Recombinant factor VIIa: review of efficacy, dosing regimens and safety in patients with congenital and acquired factor VIII or IX inhibitors. *J Thromb Haemost* **2**, 899–909.

DiMichele DM. (2006). Immune tolerance: critical issues of factor dose, purity and treatment complications. *Haemophilia* **12**(Suppl. 6), 81–85; discussion 85–86.

Dunn AL, Abshire TC. (2006). Current issues in prophylactic therapy for persons with hemophilia. *Acta Haematol* **115**, 162–171.

Franchini M. (2007). The use of desmopressin as a hemostatic agent: a concise review. *Am J Hematol* **82**, 731–735.

Gouw SC, van der Bom JG, Marijke van den Berg H. (2007). Treatment-related risk factors of inhibitor development in previously untreated patients with hemophilia A: the CANAL cohort study. *Blood* **109**, 4648–4654.

Hooiveld M, Roosendaal G, Vianen M *et al.* (2003). Blood-induced joint damage: long-term effects *in vitro* and *in vivo*. *J Rheumatology* **30**, 339–344.

Manco-Johnson MJ, Abshire TC, Shapiro AD *et al.* (2007). Prophylaxis versus episodic treatment to prevent joint disease in boys with severe hemophilia. *N Engl J Med* **357**, 535–544.

Oldenburg J, Pavlova A. (2006). Genetic risk factors for inhibitors to factors VIII and IX. *Haemophilia* **12**(Suppl. 6), 15–22.

# CHAPTER 97

# Hemophilia B

Amy L. Dunn, MD

Hemophilia B, also known as Christmas disease, results from a congenital deficiency or absence of Factor IX (FIX), leading to lifelong bleeding risk. It is an X-linked recessive disorder with an incidence of approximately 1 : 25,000 male births and accounts for 20% of hemophilia cases.

**Pathophysiology:** Factor IX is vitamin K dependent protein, and has a molecular weight of 57,000 Daltons. The gene is 33 kb long on the end of the X chromosome (Xq27). FIX is a serine protease believed to be synthesized in the liver and released into the circulation in its inactive form. During coagulation, tissue factor (TF) along with Factor VII activates FIX. Factor IXa in turn activates FVIII. FVIIIa along with FIXa and calcium then act as cofactors on a phospholipid surface (typically a platelet) during activation of Factors X and V and, ultimately, thrombin (see Chapter 78). Patients with hemophilia B are unable to generate adequate thrombin due to the lack of FIX and become dependent upon the TF pathway. Tissue factor pathway inhibitor (TFPI) downregulates the TF pathway and, as a result, patients with hemophilia bleed.

A number of mutations have been described in the FIX gene, including small and large deletions and additions, rearrangements and missense mutations. A list of mutations can be found at http://www.kcl.ac.uk/ip/petergreen/haemBdatabase.html. There is a unique type of hemophilia B, known as the Leyden variant, where point mutations in the promoter region disrupt transcription factor binding sites. During puberty, androgen affects on this promoter region lead to rising FIX levels.

**Clinical Manifestations:** The hallmark of hemophilia-related bleeding symptoms is delayed bleeding along with joint and muscle bleeding. In comparison, patients with von Willebrand disease more commonly manifest immediate and mucocutaneous bleeding. As hemophilia B is X-linked, the vast majority of affected patients are male. Females can be affected in cases of extreme X-chromosome lyonization or gene abnormalities such as Turner's syndrome.

In general, the severity of bleeding depends upon the percentage of circulating clotting factor activity. Levels of >5–40% fall into the mild range, while those of 1–5% are moderate and <1% is severe disease. The lower the factor level, the more likely the patient is to bleed from even minor trauma. Infants typically suffer from induced bleeding, such as delivery-related intracranial hemorrhage, and bleeding from blood draws, heel sticks, immunizations and circumcision. Traumatic delivery and use of forceps and vacuum assist should be avoided. Additionally, infants born of known carrier mothers should not be circumcised until testing has ruled out hemophilia. Older children and adults may experience excessive bruising, hematomas, intracranial bleeding, joint bleeding, muscle bleeding and mouth bleeding. Carrier females can also experience bleeding symptoms such as menorrhagia, mouth bleeding, and surgical and trauma-related bleeding.

**Diagnosis:** An elevated PTT and decreased plasma FIX assay confirm the diagnosis.

**Differential Diagnosis:** Hemophilia A and B often are clinically indistinguishable, and individual factor levels must be measured to clarify the diagnosis. Low FIX levels can be seen in advanced vitamin K deficiency; however, the other vitamin K dependent proteins (FII, FVII, FX) will be affected as well, and the PT will be elevated. Acquired low FIX levels are rarely a result of autoantibody formation.

**Management:** The mainstay of hemophilia B care is FIX replacement with intravenously delivered FIX concentrates. Both plasma-derived and recombinant concentrates are available (see Chapter 146). Due to a history of viral contamination of plasma-derived products, both types of product undergo several viral and pathogen attenuation steps. Since these important steps have been incorporated into the manufacturing process, no infectious complications have been reported. However, the possibility of contamination with new infectious agents such as prions cannot be excluded with plasma-derived products.

An infusion delivered in response to bleeding is known as *demand therapy*. Infusions given to prevent bleeding and its long-term consequences are known as *prophylaxis*. To minimize long-term sequelae, demand therapy should be given as soon as possible after a bleeding episode is recognized, and many patients affected by hemophilia are educated in home infusion techniques. During a bleeding episode, factor replacement should never be delayed to perform imaging or laboratory studies.

The dose and frequency of factor delivery is calculated based upon the half-life of the product (typically 24 hours), the intravascular volume of distribution (1 unit of plasma derived FIX per kilogram raises the plasma concentration by about 1%) and the desired clotting factor activity. For example, to raise the factor level of a 20-kilogram child with severe hemophilia to 100%, the dose should be $20\,kg \times 100\,IU = 2000\,IU$. Recombinant FIX has an increased intravascular volume of distribution in children, and is conventionally dosed at 1.2–1.4 units per kilogram to achieve a 1% rise in activity. Activity of 30–50% is considered hemostatic in most cases; however, in the setting of surgery or life-/limb-threatening hemorrhage, higher levels are recommended. Ancillary measures such as rest, ice, compression and elevation should also be implemented when appropriate. Antifibrinolytic therapy is particularly useful in diminishing bleeding symptoms in locations with prominent fibrinolytic activity, such as the mouth, gastrointestinal tract and uterus. Table 97.1 illustrates a suggested approach to factor replacement therapy of commonly encountered bleeding events. In developed countries, prophylactic therapy delivered one to three times per week is considered standard of care. It is common practice to begin prophylaxis around the age of 2 years, or prior to the onset of recurrent joint bleeding.

**Complications:**

**Anaphylactoid Reactions and Inhibitors:** Hemophilia B is rarely complicated by the development of inhibitory antibodies. While these antibodies occur in only approximately 3% of patients with severe hemophilia B, anaphylactoid reactions to exogenous FIX which can be life-threatening have been reported prior to recognition of a positive antibody titer. This is most common in patients with complete gene mutations or major

**TABLE 97.1  Suggested Approach to Treatment of Bleeding Episodes**

| Bleed site | Desired % activity | Length of therapy | Ancillary measures to consider |
|---|---|---|---|
| Central nervous system | 100% | 7–14 days then strongly consider prophylactic therapy for a minimum of 6 months | Continuous infusion FIX; anti-epileptic prophylaxis; surgical intervention |
| Persistent oral/mucosal | 30–60% | 3–7 days | Antifibrinolytic therapy; custom mouthpiece; topical thrombin powder |
| | 80–100% | 7–14 days | Continuous infusion FIX; antifibrinolytic therapy |
| Nose | 30–60% | 1–3 days | Packing, cautery; saline spray/gel; vasoconstrictor spray; antifibrinolytic therapy |
| Gastrointestinal | 40–80% | 3–7 days | Antifibrinolytic therapy; endoscopy with cautery |
| Persistent gross urinary | 40–60% | 1–3 days | Vigorous hydration; evaluation for stones/UTI; avoid antifibrinolytic therapy; glucocorticoids |
| Muscle | 40–80% | Every third day until pain-free movement | Rest, ice, compression, elevation; physical therapy |
| Iliopsoas | 80–100% | Until radiographic evidence of resolution | Continuous infusion FIX; bedrest; physical therapy |
| Joint | 40–80% | 1–2 days | Rest, ice, compression, elevation; physical therapy |
| Target joint | 80% day 1, 40% day 3 | 3–4 days | Rest, ice, compression, elevation; physical therapy |

derangements. It is thought that the relatively small FIX molecule can diffuse into the extravascular space, causing this reaction. Most patients with an anaphylactoid reaction will subsequently develop an inhibitory antibody, most of which are high-titer in nature. The rate of inhibitor development appears to be similar between plasma-derived and recombinant products. The antibody titer is measured utilizing a Bethesda assay, and is expressed in units (BU). One Bethesda unit is the amount of antibody which lowers the plasma factor level by 50%. Successful desensitization to FIX has been achieved in some patients, and immune tolerance therapy (ITT) with repeated exposure to FIX concentrate may eradicate the antibody.

**Nephrotic Syndrome:** Nephrotic syndrome is a reported complication of ITT with FIX products, and commonly does not occur until several months into therapy. This is due to FIX immune complex deposition in the kidney. Immunosuppressive agents such as cyclosporine and Rituximab have been utilized with some success. Bleeding in the setting of a high-titer inhibitor often requires bypassing therapy with recombinant FVIIa.

**Joint Disease:** Degenerative joint disease due to recurrent hemarthrosis is a common complication for patients with hemophilia B. While the exact mechanism of blood-induced damage is unknown, iron released from lysed red blood cells likely instigates an inflammatory reaction. This inflammation results in a vascular, hypertrophic synovium. Recurrent hemarthrosis leads to hemosiderin deposition, as well as degenerative changes in the articular cartilage and bone. Unfortunately the physical findings are often subtle in the early stages of joint disease, and investigation with MRI may be required to demonstrate hemosiderin deposition and/or hypertrophic synovium. Different techniques, such as arthroscopic and radionuclide synovectomy, have shown promise in reducing bleeding frequency and pain; however, it is unclear whether they can slow progression of bone or cartilage changes.

## Recommended Reading

Cross DC, Van Der Berg HM. (2007). Cyclosporin A can achieve immune tolerance in a patient with severe haemophilia B and refractory inhibitors. *Haemophilia* **13**, 111–114.

Crossley M, Ludwig M, Stowell KM *et al.* (1992). Recovery from hemophilia B Leyden: an androgen-responsive element in the Vactor IX promoter. *Science* **257**, 377–379.

DiMichele D. (2007). Inhibitor development in haemophilia B: an orphan disease in need of attention. *Br J Haematology* **138**, 305–315.

Dunn AL, Abshire TC. (2006). Current issues in prophylactic therapy for persons with hemophilia. *Acta Haematologica* **115**, 162–171.

Manco-Johnson MJ, Abshire TC, Shapiro AD *et al.* (2007). Prophylaxis versus episodic treatment to prevent joint disease in boys with severe hemophilia [see comment]. *N Engl J Med* **357**, 535–544.

Oldenburg J, Pavlova A. (2006). Genetic risk factors for inhibitors to Factors VIII and IX. *Haemophilia* **12**(Suppl. 6), 15–22.

Roosendaal G, Lafeber FP. (2003). Blood-induced joint damage in hemophilia. *Sem Thromb Hemostasis* **29**, 37–42.

# CHAPTER 98

# Congenital Disorders of Fibrinogen

Shannon L. Meeks, MD

Congenital afibrinogenemia, hypofibrinogenemia, dysfibrinogenemia and hypodysfibrinogenemia are a spectrum of inherited fibrinogen disorders which are secondary to mutations in transcription, mRNA processing, translation, polypeptide processing and assembly, or secretion of fibrinogen. These mutations result in quantitative (type I) and/or qualitative (type II) fibrinogen defects: absent fibrinogen (afibrinogenemia), low levels of fibrinogen (<150 mg/dl of fibrinogen; hypofibrinogenemia), dysfunctional fibrinogen (dysfibrinogenemia) or low levels of dysfunctional fibrinogen (hypodysfibrinogenemia). These disorders can result in no clinical manifestations (60%), in hemorrhagic tendencies with varying severity (28%), in thrombosis (20%) or in hemorrhage and thrombotic complications (2%). In addition, these disorders can be associated with spontaneous abortions and impaired wound healing.

<u>Pathophysiology:</u> Fibrinogen is a 340-kDa glycoprotein which is primarily synthesized in the liver and has a normal plasma concentration of 200–400 mg/dl. Fibrinogen is also synthesized in megakaryocytes and stored in the α-granules of platelets. It has a half-life of 4 days, and the minimal hemostatic amount is ~50 mg/dl. The protein is made up of two copies each of three different chains, α, β and γ, which are linked by disulfide bridges. The chains are symmetrically distributed around the central E domain, which is made from the amino terminal end of all six chains. This central E domain also contains fibrinopeptide A (FPA) and fibrinopeptide B (FPB) on the amino terminal ends of the α and β chains respectively (Figure 98.1).

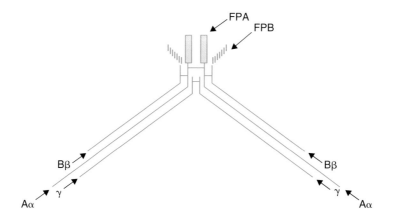

**FIGURE 98.1** Fibrinogen is composed of three chains, the Aα, Bβ and γ chains, arranged as a heterodimer, Aα2Bβ2γ2. The conversion of fibrinogen to fibrin, α2β2γ2, requires the cleavage of peptide bonds to release fibrinopeptide A and fibrinopeptide B. From Hoffman R, Benz EJ, Shattil SJ *et al.* (eds) (2005). *Hematology: Basic Principles and Practice*, 4th edition. Philadelphia, PA: Elsevier.

Carbohydrates are found on the β and γ chains; however, there is heterogeneity in the carbohydrate composition.

**Role in Coagulation:** When initiated by a hemostatic challenge, the coagulation cascade produces thrombin, which cleaves fibrinogen to fibrin, allowing polymerization and formation of a fibrin clot. This clot is further stabilized by Factor XIIIa and the resultant cross-linking of fibrin. Thrombin cleavage of fibrinogen releases FPA and FPB from the central E domain. Cleavage of FPA is faster than cleavage of FPB, and FPA cleavage is the only step necessary to initiate polymerization of fibrin monomers. The release of FPA exposes a site in the central E domain that then binds to a site on the carboxy terminal end of the γ chain in the D domain. Cleavage of FPB then allows for a more compact structure in the fibrin polymers. Fibrinogen release from platelet α-granules serves as a bridge between glycoprotein IIb/IIIa receptors on the platelet surface during platelet aggregation. Fibrinogen also interacts with plasminogen, tissue plasminogen activator and $\alpha_2$-antiplasmin to help regulate the balance between fibrin deposition and fibrinolysis.

**Genetics:** A separate gene is responsible for the synthesis of each of the fibrinogen chains, which are clustered together on chromosome 4q and appear to be regulated as a group. Mutations have been found in the genes for each of the fibrinogen chains; the α gene contains the majority of the associated mutation. A list of mutations can be found in the Fibrinogen Database (www.geht.org/databaseang/fibrinogen). Mutations in the fibrinogen genes may be inherited in an autosomal recessive or dominant fashion. Homozygotes or compound heterozygotes for severe mutations display afibrinogenemia, in which no fibrinogen is produced. Patients with hypofibrinogenemia have a decreased amount of a normal fibrinogen protein, which usually occurs in people who are heterozygous for a single normal fibrinogen allele and a mutated allele. Dysfibrinogenemia or hypodysfibrinogenemia typically results from mutations inherited in an autosomal dominant fashion, or as compound heterozygotes; these mutations alter the structure of the fibrinogen protein, leading dysfunctional properties.

**Clinical Manifestations:** Bleeding symptoms in patients with afibrinogenemia may be similar to those seen in patients with moderate to severe hemophilia. Umbilical cord bleeding and mucosal bleeding are the most commonly reported symptoms, but joint and muscle bleeding, intracranial hemorrhage and bleeding with surgery or trauma occur as well. Patients with hypofibrinogenemia typically have a less severe bleeding phenotype, with the most common symptoms including bleeding after surgery, menorrhagia, and mucocutaneous bleeding. Women with afibrinogenemia and hypofibrinogenemia have an increased risk of miscarriage. Splenic rupture has been reported in patients with afibrinogenemia.

Patients with hypodysfibrinogenemia and dysfibrinogenemia have a diverse clinical phenotype, although there is some correlation between molecular defect and clinical phenotype. Clinical phenotypes include asymptomatic patients diagnosed because of family history or laboratory abnormalities, patients with bleeding symptoms, patients with thrombosis, and patients with both bleeding symptoms and thrombosis. Patients may have bleeding postsurgery or trauma. They also may have umbilical cord bleeding and intracranial hemorrhage, with these more severe symptoms being associated with lower functional fibrinogen levels. Delayed wound healing has also been reported.

**Diagnosis:** Patients with afibrinogenemia have an abnormal prothrombin time (PT), activated partial thromboplastin time (PTT), and increased bleeding time. The thrombin time, which measures thrombin-induced conversion of fibrinogen to fibrin, is also prolonged. The thrombin time can also be prolonged by heparin and in the presence of some fibrin degradation products. The PT and PTT are typically normal until fibrinogen levels are less than 100 mg/dl. Thus, patients with hypofibrinogenemia may or may not have an abnormal PT and PTT. They will have mildly prolonged thrombin times and low levels of fibrinogen detected by a thrombin clotting method or by immunoassays. Immunoassays utilizing antifibrinogen antibodies quantify fibrinogen antigen levels, and a discrepancy between an activity and antigen level suggests a dysfibrinogenemia. Fibrinogen is an acute phase reactant, and significant elevations may occur during periods of stress. Age, gender, race, smoking, obesity and pregnancy are also known to alter fibrinogen levels (see Chapter 129).

**Differential Diagnosis:** Congenital abnormalities in fibrinogen must be separated from the more common acquired disorders. Acquired causes of hypofibrinogenemia are liver disease, ascites, disseminated intravascular coagulation, and L-asparaginase therapy. Abnormalities of thrombin clot formation, such as occurs with elevated paraproteins, heparin and fibrin/fibrinogen degradation products, may elevate the thrombin time, PT and PTT, and interfere with the diagnosis of hypofibrinogenemia.

**Management:** Patients with afibrinogenemia may require fibrinogen replacement for acute bleeding or for prophylaxis; therapy is less often needed in patients with hypofibrinogenemia or dysfibrinogenemia. No fibrinogen concentrates are currently available in the US, although they are available outside of the US. Cryoprecipitate, which contains more fibrinogen ($\sim$250 mg of fibrinogen per cyroprecipitate unit) is most often utilized for treatment and for prophylaxis. The usual dose is 1 unit per 5 kg followed by 1 unit per 15 kg as needed to keep the fibrinogen level above 75 mg/dl. Rarely, fatal thrombosis has been associated with replacement therapy. Treatment with an antifibrinolytic agent such as tranexamic acid or aminocaproic acid is helpful for mucosal bleeding. Antibody development to fibrinogen replacement therapy is rare, but anaphylaxis has been reported; therefore, this replacement therapy should be monitored closely.

## Recommended Reading

Asselta R, Duga S, Tenchini ML. (2006). The molecular basis of quantitative fibrinogen disorders. *J Thromb Haemost* **4**, 2115–2129.

Bolton-Maggs PH, Perry DJ, Chalmers EA *et al.* (2004). The rare coagulation disorders – review with guidelines for management from the United Kingdom Haemophilia Centre Doctors' Organisation. *Haemophilia* **10**, 593–628.

Peyvandi F, Kaufman RJ, Seligsohn U *et al.* (2006). Rare bleeding disorders. *Haemophilia* **12**(Suppl. 3), 137–142.

Vu D, Neerman-Arbez M. (2007). Molecular mechanisms accounting for fibrinogen deficiency: from large deletions to intracellular retention of misfolded proteins. *J Thromb Haemost* **5**(Suppl. 1), 125–131.

# CHAPTER 99

# Factor XIII, $\alpha_2$-Antiplasmin and Plasminogen Activator Inhibitor-1 Deficiencies

Shannon L. Meeks, MD

Factor XIII (FXIII) deficiency is a rare (<1:1,000,000 individuals) coagulation factor deficiency, which is autosomal recessive and results in moderate to severe delayed bleeding as a result of the inability to adequately cross-link fibrin strands. $\alpha_2$-Antiplasmin ($\alpha_2$-AP) deficiency results in excess fibrinolysis due to the inability to inhibit plasmin. Plasminogen activator inhibitor-1 (PAI-1) deficiency results in excess fibrinolysis due to the inability to inhibit plasminogen conversion to plasmin. $\alpha_2$-AP and PAI-1 deficiencies are very rare causes of bleeding.

**Pathophysiology:** FXIII is a transglutaminase that circulates in plasma as a heterotetramer consisting of 2A subunits and 2B subunits. The 2A subunits of FXIII are synthesized in megakaryocytes, placenta, uterus and macrophages; the 2B subunits are synthesized in the liver and often circulate in the plasma in excess of the A subunits. FXIII has a half-life of 9 days, and levels of 2–3% are thought to be adequate for hemostasis.

$\alpha_2$-AP is a glycoprotein that is a serine protease inhibitor (serpin). It is synthesized in the liver, and primarily circulates bound to plasminogen. $\alpha_2$-AP circulates at a concentration of $0.7\,\mu M$, and has a half-life of 2.6 days.

PAI-1 is also a serine protease inhibitor. Synthesized in the liver and endothelial cells, PAI-1 inhibits tissue plasminogen activator (TPA). PAI-1 is typically present in trace amounts in the plasma.

Role in Coagulation: The regulation of hemostasis is the result of a delicate balance between procoagulant and anticoagulant mechanisms. Activation of the clotting cascade results in the formation of thrombin, which then converts fibrinogen to fibrin and activates FXIII to FXIIIa. The A subunits of FXIII are the catalytic components, while the B subunits serve as carrier proteins. Upon thrombin cleavage, the active cysteine site on the A subunit is exposed and the B subunits are released. FXIIIa then catalyzes the cross-linking of the gamma and alpha fibrin chains by formation of a covalent bond between glutamine and lysine residues forming a stable clot. FXIIIa also cross-links $\alpha_2$-AP, fibrinogen, fibronectin and collagen to form a stable clot which is resistant to fibrinolysis. While coagulation is ongoing, the anticoagulant system is also in place to limit the clot extention. One part of this system is the fibrinolytic system, in which plasminogen is activated primarily by tissue plasminogen activator to plasmin, which in turn breaks down both fibrin and fibrinogen. The fibrinolytic system is regulated by $\alpha_2$-AP, which inhibits plasmin. $\alpha_2$-AP also binds to plasminogen and serves as a competitive inhibitor of plasminogen binding to fibrin. $\alpha_2$-AP is cross-linked to the alpha chains of fibrin by FXIIIa, increasing fibrin resistance to plasmin. The fibrinolytic system is also regulated by PAI-1, which inhibits the activation of plasminogen by inhibiting TPA (Figure 99.1).

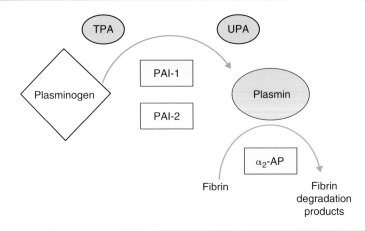

FIGURE 99.1 Schematic representation of the fibrinolytic system. $\alpha_2$-AP, $\alpha_2$-antiplasmin; PAI-1 and PAI-2, plasminogen activator inhibitor-1 and -2; TPA, tissue-type plasminogen activator; UPA, urokinase-type plasminogen activator. ▭ = inhibitors; ⬭ = activators. From Hoffman R, Benz EJ, Shattil SJ *et al.* (eds) (2005). *Hematology: Basic Principles and Practice*, 4th edition. Philadelphia, PA: Elsevier.

**Genetics:** FXIII deficiency is an autosomal recessive disorder with a prevalence of approximately 1 in 2 million people. The gene for the A subunit is found on chromosome 6, while the gene for the B subunit is located on chromosome 1. Most patients have mutations in the A subunit, leading to detectable levels of the B subunit but minimal to no detectable A subunit. These patients have minimal to no functional FXIII activity. Mutations in the B subunit are rare, and result in no detectable A or B subunits.

$\alpha_2$-AP and PAI-1 deficiencies are very rare autosomal recessive disorders. The $\alpha_2$-AP gene is located on chromosome 17p. There are only seven reported mutations in the Human Gene Mutation Database at the Institute of Medical Genetics in Cardiff (www.hgmd.cf.ac.uk), with the majority being type I deficiencies in which the decrease in antigen level is proportional to the decrease in activity level. The PAI-1 gene is located on chromosome 7, and is linked to the cystic fibrosis gene.

## Clinical Manifestations:

**FXIII Deficiency:** Patients with FXIII deficiency can manifest moderate to severe bleeding symptoms. Most present early in life with severe or life-threatening bleeding. Umbilical cord bleeding is seen in 80% of untreated patients, and 30% of patients have intracranial hemorrhage which can be spontaneous or following minimal trauma. Excessive bruising, muscle bleeding, joint bleeding, menorrhagia, and bleeding following surgery or trauma can also occur. Bleeding symptoms are often delayed, as is wound healing. Recurrent miscarriage is seen unless treatment is given during pregnancy.

**$\alpha_2$-AP Deficiency:** Homozygous deficiency results in levels between 0.01 and 0.15 U/ml. Bleeding symptoms in these patients range from moderate to severe, and include soft tissue hematomas, mucocutaneous bleeding, hemarthroses, and bleeding following surgery or trauma. Additionally, intramedullary hematoma formation in long bones has been described in four patients. Heterozygotes are typically asymptomatic, but mild bleeding symptoms have been reported.

**PAI-1 Deficiency:** Both homozygous and heterozygous deficient patients have been rarely reported. Bleeding symptoms in homozygous patients may be similar to those seen in $\alpha_2$-antiplasmin deficiency.

**Diagnosis:** The routine hemostasis screening labs are normal in FXIII, $\alpha_2$-AP and PAI-1 deficiencies, including prothrombin time (PT), activated partial thromboplastin time (PTT), bleeding time and thrombin time. As a qualitative screening test, clot solubility is increased in the presence of 5-M urea, 2% acetic acid or 1% monochloroacetic acid in patients with severe FXIII deficiency. Confirmatory testing with FXIII activity levels can be performed at specialized labs. In $\alpha_2$-AP or PAI-1 deficiency, patients may have a shortened euglobulin lysis time. Both functional and immunological assays can be used to detect $\alpha_2$-AP or PAI-1 activity and antigen levels.

**Differential Diagnosis:** Acquired deficiencies of FXIII can occur secondary to liver disease, leukemia, Henoch-Schönlein purpura (HSP) and a variety of autoimmune disorders. Severe and life-threatening bleeding has been reported with FXIII deficiency due to an acquired FXIII inhibitor. Acquired deficiencies of $\alpha_2$-AP may occur in liver disease, nephrotic syndrome, disseminated intravascular coagulation, cardiopulmonary bypass and metastatic cancer. Acquired deficiency of PAI-1 has also been rarely reported.

**Management:** Patients with FXIII deficiency can be treated with FXIII concentrates as well as plasma products and cryoprecipitate. A plasma-derived FXIII heterotetramer concentrate is approved outside of the US, and is available via Investigational New Drug (IND) in the US. A recombinant FXIII product consisting of only the A subunit is in clinical trials. Patients with mutations in the B subunit must be treated with a product containing both the A and the B subunits. Because the B subunit acts as a carrier protein for the active A subunit, infusion of the recombinant FXIII A subunit in the setting of B subunit deficiency leads to rapid clearance of the recombinant FXIII. Given the high risk of severe or life-threatening bleeding, the long half-life and the low levels of FXIII needed for hemostasis, patients with severe FXIII deficiency are ideal candidates for prophylaxis. Prophylaxis against bleeding can be achieved in most patients by infusing 10–20 U/kg of FXIII concentrate or, if that is unavailable, 15–20 ml/kg of plasma product, or 1 unit per 10 kg of cryoprecipitate every 4–6 weeks. Acute bleeding episodes and surgical procedures may be treated with FXIII concentrate, plasma products or cryoprecipitate, with a goal of maintaining a FXIII level greater than 5% until healing is complete. If not already on prophylaxis, women wishing to become pregnant should begin a prophylactic regimen to maintain FXIII through levels above 3% to prevent pregnancy loss. FXIII inhibitors are rare in patients with congenital FXIII deficiency receiving replacement therapy.

Antifibrinolytic agents prevent the binding of plasminogen to fibrin, thereby inhibiting fibrinolysis and stabilizing the fibrin clot. These products include tranexamic acid or ε-aminocaproic acid, and may be given orally or intravenously to treat patients with FXIII, $\alpha_2$-AP or PAI-1 deficiency who have bleeding symptoms. These agents have also been used for prophylaxis prior to invasive procedures, and in cases of acquired deficiency. Additionally, successful use of plasma transfusion for pre-operative management of $\alpha_2$-AP deficiency has been reported.

## Recommended Reading

Bolton-Maggs PH, Perry DJ, Chalmers EA *et al.* (2004). The rare coagulation disorders – review with guidelines for management from the United Kingdom Haemophilia Centre Doctors' Organisation. *Haemophilia* **10**, 593–628.

Favier R, Aoki N, de Moerloose P. (2001). Congenital alpha(2)-plasmin inhibitor deficiencies: a review. *Br J Haematol* **114**, 4–10.

Ivaskevicius V, Seitz R, Kohler HP *et al.* (2007). International registry on Factor XIII deficiency: a basis formed mostly on European data. *Thromb Haemost* **97**, 914–921.

Lovejoy AE, Reynolds TC, Visich JE *et al.* (2006). Safety and pharmacokinetics of recombinant Factor XIII-A2 administration in patients with congenital Factor XIII deficiency. *Blood* **108**, 57–62.

Peyvandi F, Kaufman RJ, Seligsohn U *et al.* (2006). Rare bleeding disorders. *Haemophilia* **12**(Suppl. 3), 137–142.

# CHAPTER 100

# Factor XI Deficiency

Thomas C. Abshire, MD

Factor XI (FXI) deficiency, also known as hemophilia C, was first described in 1953. The bleeding tendency varies from no to mild bleeding to more severe bleeding (but not to the degree of severity in patients with hemophilia A or B). FXI deficiency can result in homozygotes or compound heterozygotes (<15% FXI levels), who usually have more severe bleeding tendency, or heterozygotes (25–70% FXI levels), who have little or no bleeding. The disorder is mostly found in the Ashkenazi Jewish population, where the gene frequency is 4.3%, but can also be found in other ethnic groups.

## Pathophysiology:

**Biochemistry:** FXI is a 160-kDa glycoprotein which separates into two 80-kDa subunits linked by disulfide bonds. It comprises heavy chains with four repeats that have binding sites for high molecular weight kininogen (HK), thrombin, platelets, FXI and FXII. It also comprises a light chain where the serine protease is located. Originally, FXI was thought predominantly to be activated by FXII or HK, due to the intrinsic pathway and the role of FXI and calcium activating FIX. As described in Chapter 78, the role of contact activation in the formation of FXIa is now thought to be a minor pathway compared to the role of thrombin activation of FXI. Thrombin's role in activation of FXI is crucial to stable clot formation and protection against fibrinolysis by thrombin activatable fibrinolytic inhibitor (TAFI). Decreased production of TAFI may play a role in the predominant mucosal bleeding manifestations which are common in FXI deficiency. FXI is synthesized in both the liver and megakaryocytes. It is converted to the active serine protease form, FXIa. This conversion is accelerated with calcium, platelets (source of phospholipids) and thrombin.

**Genetics:** The gene is located on chromosome 4q34-35, near the gene which also encodes for prekallikrein (PK). Most defects are quantitative, resulting from the inability to make adequate protein; it is unusual to find a qualitative defect due to a dysfunctional protein. Accordingly, in most cases, the amount of functional protein correlates with the protein antigen level. There are three main mutations: type I occurs in the last intron of the gene, which results in a disruption of the mRNA splicing or a premature translation termination; type II in exon 5, resulting in a premature stop codon; and type III in exon 9, resulting in a nucleotide substitution (missense mutatation). Type I is rare amongst the Jewish population (accounts for only 1%). Type II is the earliest founder gene traced to early patriarchal times. Accordingly, it exists not only in the Jewish population, but also among other Middle Eastern peoples. Ashkenazi Jews comprise both types II and III, whereas those of Middle Eastern origin only manifest type II. Homozygous defects of type II have the lowest FXI levels (approximately 1%). Type III FXI deficiency displays the highest FXI levels, approximately 10% in the homozygous state. Doubly heterozygous mutations of type II/III comprise intermediate levels of FXI, in the 2–5% range.

Persons of Arab and Iraqi Jewish decent are usually type II. The condition is also seen amongst whites, Asians and Indians, and these FXI defects can arise from other exon mutations within the gene. A new defect in exon 3 has recently been described amongst whites.

**Clinical Manifestations:** The bleeding manifestations can be quite variable, and assessment must be individualized. It can be helpful to note carefully the bleeding history in the family, as the bleeding manifestations are often similar within families.

**Homozygous or Compound Heterozygous FXI Deficiency:** The bleeding manifestations for those who are homozygous (<15% FXI level) are usually not as severe as in hemophilia A or B or other severe bleeding disorders. Spontaneous bleeding is rare, and usually bleeding occurs with trauma in areas where there is increased fibrinolysis, such as in the nose, mouth and genitourinary tract. The bleeding in these sites is thought to be related to decreased production of TAFI with the concomitant increase in fibrinolysis. Increased menstrual bleeding and bleeding post-partum is also encountered in 25% of homozygous FXI-deficient patients. Bleeding from orthopedic and abdominal surgery is less common than in hemophilia.

**Heterozygous FXI Deficiency:** Similar to the homozygous state, bleeding associated with heterozygous deficiency (25–70% FXI levels) also occurs in areas of increased fibrinolysis (nose, mouth, teeth-related and genitourinary tract), but is usually mild.

**Diagnosis:**

*Laboratory Diagnosis:* There is wide variability in the laboratory detection of heterozygous FXI deficiency. Since these patients may have levels ranging from 25% to 70%, the PTT may be normal and is not a reliable screening tool. Also, since FXI levels can be "normal" (i.e. 50–150%) in the heterozygous patient, it is important to do genotyping in those of Jewish descent or known Middle Eastern ancestry who have a mucosal bleeding history. The FXI levels do not reliably predict who will bleed, or how severe the bleeding might be.

Inhibitory antibodies to FXI are encountered in approximately 5% of patients who are homozygous for the type II mutations. See Chapter 128 for diagnosis of factor inhibitors.

The sample should be freshly drawn and processed properly, as stored samples that are frozen with inadequate platelet removal, or samples exposed to glass, may alter FXI levels.

**Differential Diagnosis:** Since the bleeding manifestations may vary in FXI deficiency, other conditions must be considered, including: platelet function defects, VWD, and combined deficiencies of other factors such as FVII and FIX. Occasionally, a combined defect FXI/FVIII deficiency is encountered. Also, acquired platelet dysfunction from drugs must be considered. Other conditions such as Noonan's syndrome and Gaucher's Disease (more prevalent in Ashkenazi Jews) and their association with FXI deficiency should also be considered.

**Management:** Nose and mouth-related bleeding can mostly be managed with anti-fibrinolytic agents such as aminocaproic acid or tranexemic acid. Invasive surgery such as tonsillectomy or intra-abdominal surgery is often managed with both a FXI containing product and antifibrinolytic therapy. Pure FXI products are not available in the US (although they are available in some countries outside the US), and therefore plasma products (i.e. fresh frozen plasma or plasma frozen within 24 hours of phlebotomy) are used. The goal in surgery is to keep the FXI level above 50% for the first week, and above 30% for the second week. Less invasive surgery can be managed by maintaining lower FXI levels of 30% for a shorter period of time (5 days). FXI has a half-life of approximately 48 hours. Therefore, plasma products can usually be given every 12–24 hours to keep through levels above 50%. Fibrin glue can be used to help with hemostasis. DDAVP has also been utilized for mucosal bleeding or minor surgery. If inhibitors to FXI develop, recombinant FVIIa has been utilized effectively to achieve hemostasis.

## Recommended Reading

Gomez K, Bolton-Maggs P. (2008). Factor XI deficiency. *Haemophilia* **14**, 1183–1189.

O'Connell NM. (2003). Factor XI deficiency – from molecular genetics to clinical management. *Blood Coagul Fibrinolysis* **14**, S59–S64.

Salomon O, Zivelin A, Livnat T *et al.* (2003). Prevalence, causes, and characterization of Factor XI inhibitors in patients with inherited factor XI deficiency. *Blood* **101**, 4783–4788.

Salomon O, Zivelin A, Livnat T, Seligsohn U. (2006). Inhibitors to Factor XI in patients with severe Factor XI deficiency. *Semin Hematol* **43**, S10–S12.

# CHAPTER 101

# Factor VII Deficiency

Shannon L. Meeks, MD

Factor VII (FVII) is a 50-kiloDalton, vitamin K dependent serine protease which is produced in the liver and circulates in the blood at a concentration of 0.5 μg/ml. The half-life of FVII is 3–4 hours. FVII deficiency is quite rare, with a prevalence of approximately 1:500,000. Homozygotes and compound heterozygotes often have FVII levels <0.01–0.03 U/ml and may have a severe bleeding phenotype, while heteozygotes with levels approaching 0.5 U/ml are typically asymptomatic. However, for the individual patient, FVII levels do not predict bleeding severity. In the majority of patients FVII deficiency is associated with a mild bleeding phenotype characterized primarily by increased bruising, epistaxis, menorrhagia, gum bleeding, and bleeding with trauma or surgery. More severe phenotypes are at risk for bleeding similar to classical hemophilia, including hemarthrosis and intracranial hemorrhage.

**Pathophysiology:** Upon vessel injury, tissue factor (TF) is exposed and binds both FVII and activated FVII (FVIIa). The TF–FVII/FVIIa complex activates Factor X to initiate coagulation along with Factor IX. FVII is proteolytically cleaved to FVIIa by several enzymes, including thrombin, Factor Xa, Factor IXa and the TF/FVIIa complex. FVIIa is 40- to 120-fold more active than FVII. Tissue factor pathway inhibitor, a lipoprotein, is the primary inhibitor of FVIIa.

The FVII gene is located on chromosome 13 (13q34), and is inherited in an autosomal recessive fashion. More than 130 mutations can be found in the Medical Research Council's FVII Mutation Database (http://europium.csc.mrc.ac.uk). The majority of these are missense mutations.

**Diagnosis:** Isolated prolongation of the protime (PT) is found in the initial laboratory work-up in the patient with FVII deficiency. In heterozygotes the PT is ~1–3 seconds prolonged, while the PT can be greater than 20 seconds prolonged in a patient with severe disease. The diagnosis is then confirmed with a FVII activity level. It is important to remember that different sources of thromboplastin in this assay can lead to varying results. Samples should not be stored on ice, as this may induce cold activation of coagulation factors and underestimate the amount of FVII in the plasma sample. Similarly, the presence of recombinant FVIIa in the plasma leads to an overestimation of the FVII level. FVII antigen levels can be used to identify patients with dysfunctional proteins. Isolated FVII deficiency is differentiated from a combined deficiency of the vitamin K dependent proteins by normal levels of other vitamin K dependent proteins (Factors II, IX, and X).

**Differential Diagnosis:** An inherited FVII deficiency must be differentiated from the more common acquired deficiency of FVII. Acquired FVII deficiency may be seen in vitamin K deficiency, liver disease and disseminated intravascular coagulation. FVII has the shortest half-life among the coagulation factors, and thus FVII deficiency occurs earlier in the course of a disease process than other acquired factor deficiencies.

**Management:** First-line treatment for bleeding in FVII deficiency is recombinant FVIIa, typically at a dose of 20–30 µg/kg. A single dose may be sufficient for mild to moderate bleeding or for a minor surgical procedure. Most surgical procedures and severe bleeding episodes require dosing at 20–30 µg/kg every 4–6 hours until hemostasis is obtained; the dose and frequency should be individualized. This dosing and interval is quite different from the dosing of patients with hemophilia and inhibitors who are treated with much higher and more frequent doses. Patients with FVII deficiency have an intact intrinsic coagulation pathway and are able to amplify the thrombin burst and generate significantly more thrombin and thus a more stable fibrin clot. The exact level of FVII needed to prevent bleeding is unknown, and clinical response should be carefully followed. If recombinant FVIIa is unavailable, fresh frozen plasma (FFP) and/or prothrombin complex concentrates (PCC) may be utilized. In addition, some countries outside of the US have Factor VII plasma-derived concentrates available. FFP contains approximately 1 U/ml FVII, and 15 ml/kg can achieve a plasma level of approximately 15%. Antifibrinolytics such as tranexamic acid and aminocaproic acid may be beneficial for mucosal bleeding. Prophylactic regimens with recombinant FVIIa or PCC have been used for patients with severe phenotypes. Development of inhibitors to infused FVII has been reported.

## Recommended Reading

Acharya SS, Coughlin A, Dimichele DM. (2004). Rare Bleeding Disorder Registry: deficiencies of Factors II, V, VII, X, XIII, fibrinogen and dysfibrinogenemias. *J Thromb Haemost* **2**, 248–256.

Brenner B, Wiis J. (2007). Experience with recombinant-activated Factor VII in 30 patients with congenital factor VII deficiency. *Hematology* **12**, 55–62.

Huth-Kuhne A, Rott H, Zimmermann R, Halimeh S. (2007). Recombinant factor VIIa for long-term replacement therapy in patients with congenital Factor VII deficiency. *Thromb Haemost* **98**, 912–915.

Mariani G, Herrmann FH, Dolce A *et al.* (2005). Clinical phenotypes and Factor VII genotype in congenital Factor VII deficiency. *Thromb Haemost* **93**, 481–487.

Peyvandi F, Kaufman RJ, Seligsohn U *et al.* (2006). Rare bleeding disorders. *Haemophilia* **12**(Suppl. 3), 137–142.

# CHAPTER 102

# Factor II, Factor V and Factor X Deficiencies

Shannon L. Meeks, MD

Inherited deficiencies of Factors II, V and X (FII, FV, FX) are rare (estimated frequencies are: FII, 1 in 2,000,000; FV, 1 in 1,000,000; and FX, 1 in 500,000). Patients who are homozygous or compound heterozygous for defects in the FII, FV or FX genes can have moderate to severe bleeding symptoms, with patients having FX deficiency more likely to manifest severe symptoms. Given the small number of cases reported in the literature, the genotype/phenotype correlation is difficult to ascertain; however, in general, the lower the levels of factor, the more severe the bleeding. A spectrum of bleeding symptoms have been reported, including easy bruising, mucosal bleeding, surgical bleeding and trauma-related bleeding, while hemarthroses and intracranial hemorrhage are less common. Heterozygotes are typically asymptomatic; however, easy bruising, mucous membrane bleeding and bleeding with surgery have been reported.

**Pathophysiology:** FII (also known as prothrombin) is a 72-kDa, vitamin K dependent glycoprotein with a plasma concentration of approximately $100\,\mu g/ml$ and a half-life of 3 days. It is synthesized in the liver and is the inactive zymogen of thrombin, a serine protease. FV is a 330-kDa non-enzymatic cofactor that is synthesized in both hepatocytes and megakaryocytes, with 80% being secreted and 20% being stored in platelet $\alpha$-granules. The half-life of FV is 36 hours. FX is a vitamin K dependent glycoprotein that is synthesized in the liver as an inactive zymogen and circulates at a concentration of $8–10\,\mu g/mL$. The half-life of FX is 40 hours.

**Role in Coagulation:** In the coagulation cascade FX can be activated by FVIIa from the tissue factor (extrinsic) pathway or FIXa from the contact factor (intrinsic) pathway. FXa, in the presence of calcium and as part of the prothrombinase complex with FVa and phospholipid, sequentially cleaves two peptide bonds in prothrombin to form thrombin. Thrombin in turn proteolytically cleaves fibrinogen to fibrin, and contributes to forming a stable fibrin clot. Thrombin also acts on a variety of other substances in the hemostatic pathway. It promotes platelet activation, activates Factor XIII to crosslink the fibrin clot, enhances clot stability by activating thrombin-activatable fibrinolysis inhibitor (TAFI), and upregulates its own production by activating FIX, Factor VIII (FVIII) and FV. Thrombin's procoagulant activity is downregulated by antithrombin, heparin cofactor II and protease nexin I. Thrombin also binds to thrombomodulin and activates protein C, which in turn cleaves, and thus inactivates, FVIIIa and FVa, the cofactors in the factor Xase and prothrombinase complexes, respectively.

**Genetics:** The prothrombin gene is on chromosome 11p11.2, and the deficiency is inherited in an autosomal recessive fashion. The hemostatic level of prothrombin is thought to be between 20 and 40%, and thus the majority of heterozygous individuals do not manifest a bleeding diathesis. More than 40 mutations leading to prothrombin deficiency have been described. Combined deficiencies of the vitamin K dependent proteins (FII, FVII, FIX and FX) occur when there is an abnormality in the γ-glutamyl carboxylase gene or the vitamin K epoxide reductase complex.

The FV gene is on chromosome 1q24.2, with a 40% sequence homology with FVIII in their A and C domains, and the deficiency is inherited in an autosomal recessive fashion. The hemostatic level for FV is thought to be between 10% and 30%, and thus heterozygotes do not typically manifest bleeding diathesis. Over 50 mutations have been identified to date, and most of these have been unique to individual families. A combined deficiency of Factors V and VIII also occurs at a frequency of 1 in 1 million. This combined deficiency is due to a mutation in the *LMAN1* gene found on chromosome 18, which encodes a transmembrane protein that resides in the endoplasmic reticulum and Golgi and acts as a chaperone in the intracellular transport of both FV and FVIII. Mutations in the *MCFD2* gene and protein which acts as a cofactor for LMAN have also been described.

The FX gene is found on chromosome 13q34, and the deficiency is inherited in an autosomal recessive fashion. The hemostatic level for FX is thought to be 10–20%, and thus heterozygotes do not manifest bleeding manifestations. The majority of mutations found to date have been single-point mutations. FX deficiency, like prothrombin deficiency, can be associated with a combined deficiency of the vitamin K dependent proteins.

**Diagnosis:** FII, FV and FX are essential parts of the common pathway of coagulation, and deficiencies in any of these factors can lead to prolongation of the prothrombin time (PT) and activated partial thromboplastin time (PTT). The finding of a prolonged PT and PTT which correct when mixed with normal plasma suggest a factor deficiency, and individual factor levels should be checked. However, the prolongation of the PT and PTT may be minimal and reagent dependent; thus, if the clinical index of suspicion for factor deficiency is high, individual factor levels should be measured. Most clinical labs use functional assays to test for factor levels. Antigen testing can be performed using an immunoassay, and can be useful in separating type I and type II deficiency. Type I (quantitative) or hypoproteinemia occurs when there is a decreased level of normally functioning protein, characterized by a proportional decrease in protein antigen and activity. Type II (qualitative) or dysproteinemia occurs when there is a normal antigen level but a decreased level of activity.

**Differential Diagnosis:** Both FII and FX are vitamin K dependent proteins, and therefore their levels decrease in vitamin K deficiency and with warfarin therapy. FII, FV and/or FX deficiencies may be seen in liver disease. Acquired prothrombin deficiency may occur in the presence of inhibitors similar to those seen with the lupus anticoagulant.

**Management:** Acute bleeding episodes for these factor deficiencies can be treated with fresh frozen plasma (FFP) at 15–20 ml/kg, which will usually raise the factor level by 25% and should be hemostatic in most cases. For surgical procedures or more severe bleeding episodes, a loading dose of 15–20 ml/kg of FFP followed by 3–6 ml/kg every 12–24 hours is usually adequate for hemostasis. Bleeding symptoms as well as factor levels may be followed. For FII or FX deficiency, another option is a pro-thrombin complex concentrate (PCC) at a dose of 20–30 IU/kg based on FIX units. PCCs contain FII, FVII, FIX and FX, and vary in the amounts of each factor from product to product and lot to lot – for example, Bebulin VH is a PCC with more FX (FX > FII > FIX) and Profilnine SD has more FII (FII > FIX > FX). However, these products are labeled based on FIX units, and the exact amount of FII or X in the products is unknown. Platelets can also be given for FV deficiency, potentially aiding in hemostasis by delivering FV to the site of bleeding. No pure FII, FV or FX con-centrates are available in the US. Antifibrinolytic therapies such as ε-aminocaproic acid or tranexamic acid may be administered orally or intravenously to treat mucosal bleeding with or without factor replacement. Hormonal therapy with estrogens and/or progesterones may help reduce menstrual blood loss in patients with menorrhagia. Prophylaxis with once-weekly home administration of PCCs for FII or FX deficiency or FFP for FII, FV or FX deficiency has been reported for patients with severe disease.

## Recommended Reading

Acharya SS, Coughlin A, Dimichele DM. (2004). Rare Bleeding Disorder Registry: deficiencies of Factors II, V, VII, X, XIII, fibrinogen and dysfibrinogenemias. *J Thromb Haemost* **2**, 248–256.

Asselta R, Tenchini ML, Duga S. (2006). Inherited defects of coagulation Factor V: the hemorrhagic side. *J Thromb Haemost* **4**, 26–34.

Bolton-Maggs PH, Perry DJ, Chalmers EA *et al.* (2004). The rare coagulation disorders – review with guidelines for management from the United Kingdom Haemophilia Centre Doctors' Organisation. *Haemophilia* **10**, 593–628.

Herrmann FH, Auerswald G, Ruiz-Saez A *et al.* (2006). Factor X deficiency: clinical manifestation of 102 subjects from Europe and Latin America with mutations in the Factor 10 gene. *Haemophilia* **12**, 479–489.

Peyvandi F, Kaufman RJ, Seligsohn U *et al.* (2006). Rare bleeding disorders. *Haemophilia* **12**(Suppl. 3), 137–142.

Zhang B, McGee B, Yamaoka JS *et al.* (2006). Combined deficiency of factor V and factor VIII is due to mutations in either LMAN1 or MCFD2. *Blood* **107**, 1903–1907.

# CHAPTER 103
# Bleeding Disorders in Pregnancy

Christine L. Kempton, MD

Post-partum hemorrhage (PPH) is a leading cause of maternal death worldwide. Although the majority of PPH occurs as a result of anatomical defects resulting from obstetrical trauma, the risk is increased when a congenital or acquired hemostatic defect is present. This chapter will review some of the congenital and acquired hemostatic disorders affecting pregnancy.

## Congenital Disorders:

**von Willebrand Disease:** During pregnancy von Willebrand factor (VWF) and Factor VIII (FVIII) levels increase, peaking between the twenty-ninth and thirty-fifth weeks of gestation; they return to baseline within 7 to 10 days following delivery.

In patients with type 1 von Willebrand disease (VWD), FVIII levels >50% are considered adequate for both neuroaxial anesthesia (NA) and delivery. Accordingly, in those with FVIII levels >50% treatment can be delayed until immediately after delivery, when levels will begin to return to baseline (Figure 103.1). In patients with third-trimester FVIII levels <50%, treatment should be initiated prior to NA and delivery. In all patients, treatment should continue to maintain FVIII levels >50% for 3–4 days following vaginal delivery and 4–5 days following cesarean section, or until abnormal bleeding has ceased.

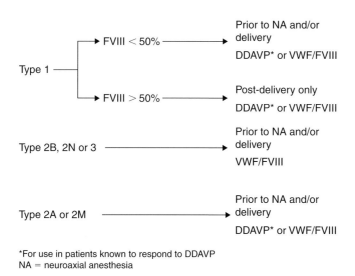

FIGURE 103.1 Treatment of VWD during labor and delivery.

In contrast to management of patients with type 1 VWD, FVIII levels alone cannot be used to predict hemostasis and stratify management strategies in patients with types 2 and 3 VWD. Because of dysfunctional VWF in type 2 A, 2B and 2 M, the VWF activity is more significantly reduced than the VWF and FVIII levels. In patients with types 2 N and 3 VWD, VWF and FVIII levels are low and not expected to appreciably change during pregnancy. Accordingly, both VWF activity and FVIII levels need to be considered when making treatment decisions, and most pregnant women with types 2 or 3 VWD will need treatment prior to NA and delivery.

Options for treatment include 1-deamino-8-arginine-vasopressin (DDAVP) or VWF-FVIII containing concentrates (see Chapter 144). DDAVP can be given intranasally (1.5 mg/ml, total dose equals 300 μg) or intravenously (0.3 μg/kg). DDAVP increases FVIII and VWF in most patients with type 1 VWD. However, not all patients respond adequately, and therefore it is preferable to have confirmed a response to DDAVP in the non-pregnant state prior to its use at the time of a hemostatic challenge. Since hyponatremia is a potential side-effect, fluid intake should be reduced to 75% of normal and drinks with higher NaCl content should replace free water. DDAVP has been used safely throughout pregnancy in patients with diabetes insipidus, albeit at lower doses, and at the time of invasive perinatal testing in women with congenital bleeding disorders. There is concern regarding neonatal hyponatremia if the fetus is exposed to DDAVP during labor and delivery. However, if maternal fluids are managed properly after receiving DDAVP, the risk of neonatal hyponatremia is minimal, since neonates take little oral fluids and no free water during their first 24 hours. However, to be cautious, neonates born to mothers that have received DDAVP prior to delivery should have their sodium checked shortly after delivery and be managed accordingly. More conservative options include foregoing NA, or receiving a VWF-FVIII concentrate prior to NA.

**Carriers of Hemophilia A or B:** In general, no therapy is needed for women with FVIII or Factor IX (FIX) levels >50%. For those with levels <50%, treatment is necessary before NA and delivery. Treatment should maintain FVIII and FIX levels at >50% for 3–4 days post-partum for vaginal delivery and 4–5 days post-partum for cesarean section. Treatment options in women with FVIII deficiency include DDAVP and FVIII products. Treatment of FIX carriers is with FIX products only.

**Factor XI Deficiency:** In a cohort of pregnant women with severe Factor XI (FXI) deficiency only 30% experienced PPH, and PPH was more likely if there was a past history of surgical bleeding. Accordingly, the decision to utilize prophylaxis during labor and delivery should be based on a prior history of bleeding. As with major surgery, the goal of therapy is to maintain a FXI level of >40% using plasma product transfusion (e.g. fresh frozen plasma or plasma frozen 24 hours after phlebotomy), approximately 15 ml/kg. Since the half-life of FXI is >50 hours, additional plasma products can be transfused every 48 hours to maintain FXI levels above 40% for 3–4 days following a vaginal delivery or 4–5 days following a cesarean section. For women without a history of bleeding despite severely reduced FXI levels, plasma product transfusions can be withheld until there is evidence of clinical bleeding as it is unlikely to occur. However, prior to NA, prophylactic plasma product transfusion is recommended in severe FXI deficiency (<15%) even in the absence of a history of bleeding.

**Platelet Function Defects:**  Management of platelet function defects at the time of labor and delivery should be based on the prior history of bleeding, patient preferences regarding NA, and the response to DDAVP during other hemostatic challenges. Decisions regarding management in pregnancy are similar to those during major surgery (see Chapters 83–86). In general, options for treatment include DDAVP, platelet transfusions and recombinant Factor VIIa (rFVIIa). In many cases, DDAVP alone will be sufficient.

**Other Inherited Defects:**  Deficiencies of fibrinogen and Factor XIII (FXIII) may result in recurrent pregnancy loss and peri-partum hemorrhage. Cryoprecipitate to achieve a fibrinogen concentration above 100 mg/dl is required in the setting of fibrinogen abnormalities. Cryoprecipitate or plasma products can be used to treat FXIII deficiency. Additionally, a FXIII concentrate is currently available as part of a clinical investigation (and in some countries outside of the US), and could potentially be obtained for compassionate use. Regular infusions may be required to facilitate carrying the pregnancy to term.

Factor VII (FVII) levels rise during normal pregnancy, and levels of 15–25% are considered adequate for hemostasis. rFVIIa, 20–30 µg/kg, should be used for those women with FVII deficiency who have persistently low FVII levels after reaching the third trimester. A first dose can be given prior to NA and repeated every 2–6 hours, depending on the patient's prior bleeding history and severity of PPH. For many patients, a single dose will suffice.

**Prenatal Counseling and Delivery of Fetus with a Potential Bleeding Disorder:**  Prenatal counseling should be considered for women with congenital bleeding disorders who are contemplating pregnancy. At the time of delivery, if prenatal testing has not excluded a bleeding disorder then the fetus should be handled as though one is present. Specifically, assisted deliveries using forceps and vacuum extraction should be avoided. Circumcision should be delayed until adequate workup can be performed, typically at 1 year of age.

## Acquired Disorders:

**Acquired FVIII Inhibitors:**  FVIII inhibitors occur rarely in pregnancy. They occur most commonly within the first 3 months post-partum, and only approximately 5% ante-partum. Bleeding is variable, ranging from asymptomatic to fatal (mortality 5.6%). Although pregnancy-related FVIII inhibitors often disappear spontaneously, this resolution may take a year or longer. Corticosteroid therapy should be considered in all patients to facilitate inhibitor eradication. Cyclophosphamide and other therapies are reserved for those that fail to respond adequately to corticosteroids (see Chapter 111).

**Thrombocytopenia:**  When evaluating gravid women with thrombocytopenia, the same general approach is utilized as in non-gravid patients (see Chapter 87). However, several additional disorders and treatment options should be considered.

*Gestational Thrombocytopenia:*  Gestational thrombocytopenia is the most common cause of thrombocytopenia in pregnancy. It is typically mild (platelet count

$>70,000/\mu l$), occurs late in pregnancy, resolves following delivery, does not affect the fetus and occurs in women without a history of thrombocytopenia (they may have a prior history of gestational thrombocytopenia). No treatment is needed.

*Immune Thrombocytopenia:* Immune thrombocytopenia (ITP) can lead to more profound thrombocytopenia, and occurs earlier in pregnancy than is seen with gestational thrombocytopenia. The platelet count may progressively decline during pregnancy, reaching a nadir in the third trimester. A platelet count $>50,000/\mu l$ is considered safe for delivery, but a platelet count $>75,000/\mu l$ is recommended for NA.

If treatment is required, options include intravenous immunoglobulin (IVIG) (0.4 g/kg per day for 5 days or 1 g/kg per day for 2 days) or corticosteroids. IVIG is generally preferred to steroid therapy. With longer corticosteroid exposure, the risks of hypertension, hyperglycemia and osteoporosis increase. Prednisone, prednisolone or methylprednisolone are favored over dexamethasone because the former are well metabolized by the placenta, exposing the fetus to only 10% of the maternal dose. Splenectomy has been successfully performed during pregnancy, and can be considered in severe and refractory ITP. The second trimester is the optimal time to perform this intervention.

Maternal antibodies can affect the fetal platelet count, leading to thrombocytopenia in the newborn. Unfortunately, the maternal platelet count does not predict the infant's platelet count. Despite the potential for significant thrombocytopenia in up to 20% of infants, intracranial hemorrhage (ICH) occurs in $<1.5\%$ of infants born to mothers with ITP. The likelihood of ICH or other fetal complications is not influenced by the method of delivery; therefore, obstetric indications alone should determine the method of delivery.

The neonate's platelet count should be measured immediately following delivery and serially in those that are thrombocytopenic, as the nadir may not occur until 2–5 days following delivery. It is generally recommended that infants with a platelet count $<20,000/\mu l$ or those with hemorrhage receive treatment (IVIG, corticosteroids and/or platelet transfusion).

*Disseminated Intravascular Coagulation:* Disseminated intravascular coagulation (DIC) is associated with major obstetrical complications such as placental abruption, amniotic fluid emboli, retained fetus syndrome and eclampsia. The diagnosis and treatment of DIC in pregnancy is similar to that in the non-pregnant patient (see Chapter 110).

*Hemolysis, Elevated Liver Enzymes and Low Platelets:* Hemolysis, elevated liver enzymes and low platelets (HELLP) is a syndrome that is part of a spectrum associated with pre-eclampsia and eclampsia. Mild thrombocytopenia may precede other signs and symptoms, such as right upper quadrant pain and elevated liver transaminases. There are no standard laboratory values to use as diagnostic criteria. Prompt delivery is the treatment of choice.

*Acute Fatty Liver of Pregnancy:* Acute fatty liver of pregnancy is a rare disorder, with liver failure dominating the clinical picture. Thrombocytopenia is typically mild, and hemolysis is usually not present unless DIC has developed.

*Thrombotic Thrombocytopenia Purpura:* Thrombotic thrombocytopenia purpura (TTP) in pregnancy has been reported to occur at any time period during gestation, but most commonly occurs later in pregnancy or after delivery. Signs and symptoms are similar to those of TTP not associated with pregnancy (see Chapter 93). In contrast to the other pregnancy-related disorders listed above, the clinical course of TTP is not altered by delivery. The presence of pre-eclampsia, right upper quadrant pain and elevated liver enzymes may point towards other pregnancy-related microangiopathies, although there may be times that this distinction is difficult and the response to therapy, delivery and/or plasma exchange may be the only means to distinguish between these disorders.

Subsequent pregnancies carry a risk of recurrent TTP. In the Oklahoma TTP-HUS registry, 19 women with a history of TTP became pregnant and approximately 25% of these women developed recurrent TTP. Of those with a history of pregnancy/post-partum-related TTP, 18% had recurrent TTP during a subsequent pregnancy. If the cause of TTP was idiopathic, it was more likely for these women to develop TTP during a subsequent pregnancy (43%). There were no maternal deaths associated with pregnancy. However, the pregnancies were often complicated and fetal death occurred in 40%, although not all fetal deaths appeared related to maternal TTP.

## Recommended Reading

British Committee for Standards in Haematology General Haematology Task Force. (2003). Guidelines for the investigation and management of idiopathic thrombocytopenic purpura in adults, children and in pregnancy. *Br J Haematol* **120**, 574–596.

Asahina T, Kobayashi T, Takeuchi K, Kanayama N. (2007). Congenital blood coagulation Factor XIII deficiency and successful deliveries: a review of the literature. *Obstet Gynecol Survey* **62**, 255–260.

Conti M, Mari D, Conti E *et al.* (1986). Pregnancy in women with different types of von Willebrand disease. *Obstet Gynecol* **68**, 282–285.

Demers C, Derzko C, David M, Douglas J. (2006). Gynaecological and obstetric management of women with inherited bleeding disorders. *Intl J Gynaecol Obstet* **95**, 75–87.

Hauser I, Schneider B, Lechner K. (1995). Post-partum Factor VIII inhibitors. A review of the literature with special reference to the value of steroid and immunosuppressive treatment. *Thromb Haemost* **73**, 1–5.

Kujovich JL. (2005). Von Willebrand disease and pregnancy. *J Thromb Haemost* **3**, 246–253.

Kulkarni AA, Lee CA, Kadir RA. (2006). Pregnancy in women with congenital Factor VII deficiency. *Haemophilia* **12**, 413–416.

Mannucci PM. (2004). Treatment of von Willebrand's disease. *N Engl J Med* **351**, 683–694.

Mannucci PM. (2005). Use of desmopressin (DDAVP) during early pregnancy in Factor VIII-deficient women. *Blood* **105**, 3382.

Salomon O, Steinberg DM, Tamarin I *et al.* (2005). Plasma replacement therapy during labor is not mandatory for women with severe factor XI deficiency. *Blood Coagul Fibrinol* **16**, 37–41.

Vesely SK, Li X, McMinn JR *et al.* (2004). Pregnancy outcomes after recovery from thrombotic thrombocytopenic purpura–hemolytic uremic syndrome. *Transfusion* **44**, 1149–1158.

# CHAPTER 104

# Vascular Bleeding Disorders

Michael A. Briones, MD

The vascular disorders are a heterogeneous group of conditions characterized by easy bruising, petechiae, ecchymosis, mucosal bleeding and spontaneous bleeding from small vessels. The underlying defect is either in the vessels themselves or in the perivascular connective tissue. Platelet count, prothrombin time (PT), partial thromboplastin time (PTT), bleeding time and platelet function analyzer (PFA) are generally normal. Congenital disorders are caused by either vascular malformations or disorders of connective tissue; acquired disorders have multiple causes, such as an increase in vessel fragility, vasculitis, and decreased collagen.

## Hereditary Vascular Malformations:

### Hereditary Hemorrhagic Telangiectasia (Osler-Weber-Rendu Disease):

Hereditary hemorrhagic telangiectasia (HHT), the most common inherited vascular bleeding disorder, is inherited in an autosomal-dominant pattern and affects both sexes equally. Bleeding occurs spontaneously or after minor trauma, from malformations of dilated small vessels lacking smooth muscle in the skin and mucosa (telangiectases). These vessels rupture easily and do not contract effectively, causing prolonged bleeding. Telangiectases are 2–4 mm in diameter, appear dark red, and are frequently localized on the face, the lips, the mucous membranes of the mouth or nose, and the fingertips. Bleeding, including epistaxis and gastrointestinal (GI) hemorrhage, becomes more common in adulthood as the number of lesions increases, and may lead to iron deficiency anemia. Arteriovenous malformations may develop in the lungs. Coagulation studies are normal. Genetic testing is available for two types of HHT (types 1 and 2). Treatment is primarily symptomatic, although recurrent GI or pulmonary bleeding may require surgical treatment. Acute epistaxis may first be treated by local measures (compression, topical hemostatic drugs). In some instances, antifibrinolytic therapy with tranexamic acid or aminocaproic acid may be beneficial.

### Giant Cavernous Hemangiomata and Kasabach-Merritt Syndrome:

Giant cavernous hemangiomas are usually congenital, although they usually do not become clinically relevant until several months to several years of age. Most common in girls, they are categorized into three main types: capillary, cavernous and capillary–cavernous mixture. Cavernous hemangiomas are benign vascular tumors having dilated thin-walled vessels and sinuses lined with abnormal endothelium, and are found in the GI tract, skin, bone, liver and mucous membranes. Thrombi may develop within the capillaries, which can lead to disseminated intravascular coagulopathy (DIC). This later condition is known as Kasabach-Merritt syndrome. The DIC is typically low grade, but rarely can be life-threatening. Therapy for DIC often includes transfusion with platelet (for thrombocytopenia) and plasma products (for elevated PT and PTT), cryoprecipitate (for low fibrinogen), and the use of low-dose heparin infusion or low molecular weight heparin to interrupt the excessive thrombin generation. Specific therapy for the hemangioma may include surgery, radiation, sclerosing agents, embolization of the abnormal vasculature, systemic chemotherapy, corticosteroids and/or vinca-alkaloids.

## Hereditary Connective Tissue Disorders:

**Ehlers-Danlos Syndrome:** Ehlers-Danlos syndrome is a rare autosomal dominant syndrome that results from a mutation of a gene in collagen synthesis, leading to loss of skin elasticity, joint hypermobility and systemic organ tissue fragility. Cutaneous findings include thin skin and a tendency to develop non-palpable purpuric lesions as a result of dermal blood vessel fragility. A correlation has been reported between increased thumb flexibility and unexplained bleeding tendencies, suggesting that many patients with undiagnosed bleeding may have a modified form of Ehlers-Danlos syndrome or other collagen defects.

**Pseudoxanthoma Elasticum:** Pseudoxanthoma elasticum, an autosomal dominant disorder, is characterized by mineralization and fragmentation of elastin in the skin, retina and blood vessels. Cutaneous lesions include small white or yellow papules, or larger confluent areas of purpura or necrosis.

**Mitochondrial Encephalomyopathy with Lactic Acidosis and Stroke-like episodes Syndrome:** Mitochondrial encephalomyopathy with lactic acidosis and stroke-like episodes syndrome (MELAS) is associated with a mutation in the mitochondrial transfer gene, and signs of the disease may include non-palpable purpuric lesions on the palms and soles and hypertrichosis/ichthyosis. Transmission electron microscopy of involved skin reveals evidence of endothelial degeneration and morphologically irregular mitochondria.

## Acquired Vascular Disorders:

**Purpura Simplex (Idiopathic Purpura):** Purpura simplex is defined as spontaneous easy bruising on the trunk and the lower extremities that typically occurs following minor trauma. It is likely secondary to increased fragility of skin vessels, and is one of the most frequent reasons for increased bruising. If the investigation into increased bruising does not reveal other bleeding history from the patient, or a family history of bleeding and screening coagulation laboratory are normal, the patient should be reassured and advised to avoid taking aspirin or other medications which might cause platelet dysfunction.

**Senile Purpura:** Progressive loss of collagen in the dermis and vascular wall may lead to characteristic lesions of old age, or senile purpura. Usually occurring on the dorsal side of the hands and wrists or on the forearms, these lesions may be large and well demarcated. Venipuncture may result in rapidly spreading purpura, and thus prolonged pressure should be applied after venipuncture in elderly patients. No specific treatment exists.

**Purpura Due to Infections:** Infections may cause purpura and mucous membrane bleeding through direct vessel damage by micro-organisms or toxins, vasculitis, thrombocytopenia or DIC. DIC from infections, especially meninogococcal disease, may be associated with renal failure, rapidly spreading purpura with skin necrosis, and circulatory failure. Treatment of the underlying infection is critical, with supportive care for the coagulopathy. Heparin treatment to counteract excessive thrombin generation may be helpful in certain instances.

**Scurvy:** Scurvy is a result of insufficient intake of vitamin C, which is required for collagen synthesis. Clinical manifestations include oral bleeding, mucous membrane bleeding, orbital hemorrhage leading to propotosis, and subperiosteal hemorrhage (especially in infants). Rarely, hematuria, hematochezia and melena are noted; costochondral beading (scorbutic rosary) may also occur. Low-grade fever, anemia, and poor wound healing are typical signs of scurvy. Vitamin C administration is effective in curing infantile scurvy. The symptoms of irritability, fever, tenderness to palpation and hemorrhage generally resolve within 7 days of initiating treatment.

**Henoch-Schönlein Purpura:** The most common vasculitis in children, Henoch-Schönlein purpura (HSP) is characterized by the acute onset of abdominal pain and lower extremity urticarial plaques and palpable purpura. Nephritis, arthritis and scrotal swelling may also occur. Of affected patients, 90% are younger than 10 years of age. In pediatric cases, boys are more commonly affected than girls (2 : 1). The etiology of the syndrome remains unclear, but reported triggers include viral (hepatitis B virus, hepatitis C virus, parvovirus B19, and human immunodeficiency virus) and bacterial (*Streptococcus* species, *S. aureus*, and *Salmonella* species) infections in children, medications (non-steroidal anti-inflammatory drugs [NSAID], angiotensin-converting enzyme inhibitors, and antibiotics), food allergies, and insect bites in adults. A specific genetic risk factor for the disorder is complement factor 4 (C4) deficiency.

Leukocytoclastic vasculitis in HSP is the result of IgA immune complex and complement deposition on vessel walls. Increased serum IgA levels may be found in these patients, along with elevated white blood cell counts, platelet counts and C-reactive protein levels. HSP is self-limited, and treatment is supportive. Steroids may be used for patients with renal involvement, persistent purpura or severe abdominal pain.

## Recommended Reading

Bristow J, Carey W, Egging D *et al.* (2005). Tenascin-X, collagen, elastin, and the Ehlers-Danlos syndrome. *Am J Med Genet C Semin Med Genet* **139**, 24–30.

Fuchizaki U, Miyamori H, Kitagawa S *et al.* (2003). Hereditary haemorrhagic telangiectasia (Rendu-Osler-Weber disease). *Lancet* **362**, 1490–1494.

Mao JR, Bristow J. (2001). The Ehlers-Danlos syndrome: on beyond collagens. *J Clin Invest* **107**, 1063–1069.

Mukhtar IA, Letts M. (2004). Hemangioma of the radius associated with Kasabach-Merritt syndrome: case report and literature review. *J Ped Orthoped* **24**, 87–91.

Trapani S, Micheli A, Grisolia F *et al.* (2005). Henoch Schönlein purpura in childhood: epidemiological and clinical analysis of 150 cases over a 5-year period and review of literature. *Semin Arthritis Rheum* **35**, 143–153.

# CHAPTER 105
# Bleeding Risks with Liver Disease

Thomas C. Abshire, MD

Liver disease can either be acute (e.g. acute hepatitis) or chronic (e.g. malignancy, biliary atresia, hepatitis C cirrhosis), and is usually associated with hemorrhagic and occasionally, thrombotic complications. The hemorrhagic complications of liver disease are related to: decreased hepatic synthesis of coagulation factors and vitamin K deficiency, thrombocytopenia and platelet dysfunction, and hypofibrinogenemia and dysfibrinogenemia. Changes associated with an increased risk of thrombosis include increased levels of Factor VIII (FVIII) and von Willebrand factor (VWF) levels, decreased levels of protein C, protein S and antithrombin III, and decreased clearance of activated coagulation factors. The hemorrhagic complications are more pronounced in the setting of end-stage liver disease, which is often complicated by variceal bleeding or infection/sepsis.

**Pathophysiology:** A summary of the hemostatic defects common in liver disease is found in Table 105.1, and these are described in more detail below.

---

### TABLE 105.1  Hemostatic Defects in Liver Disease

- Decreased synthesis of clotting factors
- Vitamin K deficiency
- Thrombocytopenia (decreased production and hypersplenism)
- Platelet function defect (increased fibrin/fibrinogen degradation products and decreased adhesion to GPIb)
- Disseminated intravascular coagulopathy (DIC)
- Systemic fibrinolysis (decreased $\alpha 2AP$ and decreased clearance of TPA)
- Dysfibrinogen (increased sialic acid production)

---

**Decreased Synthesis of Clotting Factors:** Since the liver is the exclusive source of all clotting factors (except FVIII and VWF), impairment of liver function will affect the production of these factors. The rate of fall in factor level is proportional to the coagulation factor half-life. As FVII has the shortest half-life (approximately 4 hours), FVII deficiency is the first coagulation defect to appear during acute liver failure. The vitamin K dependent factors are disproportionately affected in liver disease, resulting in a greater effect on the PT compared to the PTT. VWF, FVIII and fibrinogen are each acute phase reactants that tend to rise early in liver failure; however, as liver disease progresses fibrinogen will subsequently fall due to decreased hepatic production. FVIII and VWF do not fall, as they are synthesized by endothelial cells and megakaryocytes (VWF).

**Vitamin K Deficiency:** The liver has an essential role in vitamin K absorption, and therefore vitamin K deficiency should be considered in conditions such as biliary cirrhosis and bile duct obstruction, where bile acid metabolism is impaired. Poor nutritional intake as a result of alcoholism may also lead to vitamin K deficiency.

**Thrombocytopenia:** Thrombocytopenia is secondary to multiple factors. First, thrombocytopenia is significantly related to splenomegaly, with the accompanying sequestration and trapping of platelets. Second, platelet survival may be affected by the production of platelet-associated antibodies which may be seen in certain infectious complications (e.g. hepatitis C and autoimmune hepatitis). Third, decreased platelet production from megakaryocytes can be a result of decreased thrombopoietin production by the liver. Fourth, alcohol, which is associated with liver disease, can also affect megakaryocyte production. Lastly, viral and other infections associated with some causes of liver failure may also inhibit the production of megakaryocytes/platelets.

**Platelet Dysfunction:** Liver disease is associated with platelet dysfunction and is multifactorial related to the effect of fibrin degradation products (FDP) which impairs platelet function, and to decreased GPIb on the platelet surface, and secondary to defective signal transduction due to altered platelet content. Also, high levels of high density lipoprotein may alter nitric oxide content and decrease aggregability.

**Disseminated Intravascular Coagulopathy:** Disseminated intravascular coagulopathy (DIC) is not uncommon in severe liver disease where the primary defect is excessive activation of clotting factors and formation of thrombin. Demonstration of excessive thrombin is supported by elevated prothrombin fragment 1.2 and thrombin–antithrombin complexes (TAT), as well as increased D-dimers. The underlying pathophysiology of DIC is not known, but might be related to decreased natural inhibitors to clotting such as anti-thrombin (ATIII) and protein C, inability to clear activated clotting factors, injury to hepatocytes and tissue factor release. Cytokine production is evident in severe liver disease, and concomitant events such as sepsis may also play a role. The laboratory differences between DIC and liver disease may be difficult to differentiate. In both cases, coagulation times are prolonged (due to consumption in DIC and decreased factor synthesis in liver failure). Similarly, D-dimers are also elevated (due to increased synthesis in DIC and decreased clearance in liver failure). The nuances of the laboratory differentiation between DIC and liver failure are discussed below.

**Systemic Fibrinolysis:** In addition to DIC, increased systemic fibrinolysis is also evident in severe liver disease. The euglobulin lysis time (ELT) is short in almost 50% of tested patients. This is a result of decreased synthesis of $\alpha_2$-antiplasmin and decreased clearance of fibrinolytic enzymes such as tissue plasminogen activator (TPA). The presence of excessive fibrinolysis should be considered when bleeding in liver disease is difficult to manage.

**Abnormal Fibrinogen (Dysfibrinogenmia):** As previously mentioned, fibrinogen levels tend to remain normal to only slightly depressed despite severe hepatic impairment. However, the fibrinogen that is produced has increased amounts of sialic acid (similar to fetal fibrinogen). This increased amount of sialic acid renders the fibrinogen abnormal and prolongs the thrombin time. Despite this alteration, it is unlikely that the dysfibrinogen of liver disease contributes to systemic bleeding.

**Clinical Manifestations:** The predominant clinical manifestation from liver disease is bleeding. This bleeding may come from gastrointestinal hemorrhage associated with esophageal and gastric/intestinal varices as well as gastritis and gastric and duodenal ulcers. Also, systemic bleeding may be manifested from intravenous and chest/endotracheal tube sites and from the skin (purpura and petechiae), mouth and nose.

**Diagnosis:** Laboratory testing to differentiate between liver disease and DIC helps guide therapy. In liver disease, typically the PT is much more prolonged in comparison to the PTT, primarily due to the effect of the vitamin K dependent factors on the PT. FVIII is often increased, and FV and FVII are low. Fibrinogen levels are usually greater than 100 mg/dl unless there is significant acute hepatitis or end-stage liver disease. Finally, the platelet count is only mildly decreased and in some cases normal.

**Differential Diagnosis:** The main clinical differential diagnosis in severe liver disease is DIC. In general, in DIC, the PT is not elevated out of proportion to the PTT as in liver disease, and the fibrinogen level and platelet count are lower in DIC than in liver disease. Additionally, the amounts of FDP and D-dimer are more elevated in DIC.

Additional laboratory findings, such as schistocytes on peripheral smear, are typically seen more often in DIC than in liver failure. Importantly, DIC is an inherently unstable state with ongoing consumption of coagulation factors and, in most cases, consumption of platelets. In contrast, liver failure typically progresses more slowly with relatively stable platelet counts. Thus, serial testing over time may help to distinguish between DIC and liver failure.

Some clinicians advocate comparing FVIII levels to FV or FVII. The reasoning is that FVIII is made in endothelial cells. Thus, a decrease in FVIII concomitant with FV and FVII supports consumption, whereas a normal FVIII with decreased FV and FVII suggests liver failure. FVIII is an acute phase reactant, which can be elevated to superphysiological levels and then drop back to normal levels during a consumptive coagulopathy. Thus, a "normal" FVIII level cannot rule out DIC. Importantly, coagulopathy of liver failure and DIC are not mutually exclusive, and can occur simultaneously in some patients.

Hemostatic screening tests may help differentiate certain types of liver failure: acute hepatitis, obstruction of the biliary system and cirrhosis of the liver.

- *Acute severe hepatitis* is associated with marked abnormalities of the screening tests, including markedly elevated PT/PTT and extremely low fibrinogen levels. FV and FVII are moderately decreased accompanied by a significantly elevated FVIII.
- *Biliary obstruction* is accompanied by a disproportionately elevated PT compared to an only slightly elevated PTT, a normal to slightly increased fibrinogen level and markedly low FVII but normal to only slightly depressed FV.
- *End stage liver disease* is similar to biliary obstruction in terms of the markedly elevated PT and decreased FVII and only a normal to slightly increased PTT. It is differentiated from biliary obstruction by a slightly decreased fibrinogen and platelet count, and moderately decreased FV. FVIII also tends to be more elevated in end-stage liver disease.

**Summary of Laboratory Differentiation in Hepatic Failure:** Biliary cirrhosis (such as occurs with congenital biliary atresia) is very similar to end-stage liver disease, except this condition manifests a slightly elevated fibrinogen level and demonstrates only slight depression of both FV and the platelet count. End stage liver disease is differentiated from acute hepatitis by manifesting only a slightly increased PTT and slightly decreased fibrinogen level, compared to the marked discrepancies in both of these screening tests in acute hepatitis.

## Management:

**Conventional Management:** There is usually an element of vitamin K deficiency in any form of liver disease, and therefore vitamin K should be administered daily at least for 3 days. The dose should be 2–3 mg for an infant, 7 mg for a child and 10 mg for an adult, all given intramuscularly. If a patient is acutely ill, the vitamin K should be administered intravenously, by slow push. In the acute setting, if both the PT and PTT are elevated and the intravascular volume is not problematic, plasma products (e.g. fresh frozen plasma and plasma frozen within 24 hours of phlebotomy) should be administered at 15–20 ml/kg. Cryoprecipitate may be given for low fibrinogen levels. Platelet transfusions should be administered if the platelet count is less than 50,000/μl and accompanied by clinical signs of bleeding. Platelet transfusions may also be given if the platelet count is less than 50,000 μl in preparation for a high-risk invasive procedure. Most clinicians would recommend lowering of the PT (usually the goal is an INR of less than 1.5 or 2.0, depending on the clinical situation) and fibrinogen (usually a goal of greater than 100–150 mg/dl) if major surgery is planned. Additionally, some authorities maintain the platelet count above 75,000/μl in the surgical setting, but the response to platelet transfusions may be blunted secondary to splenomegaly and therefore this goal may be difficult to obtain.

**Difficult Bleeding Management:** If bleeding is difficult to manage with the interventions mentioned in the preceding paragraph, several other treatment modalities may be entertained. DDAVP may be utilized to help reverse the platelet dysfunction, which is common in liver disease. If intravascular volume is a limiting factor in treatment options, a plasma exchange may be attempted; however, in a very ill patient, plasma exchange becomes a difficult intervention due to secondary volume shifts which can increase hemodynamic instability. The use of antifibrinolytics, such as aminocaproic acid or tranexemic acid, can also be considered while being circumspect regarding the potential risk of thrombosis with these agents. Finally, rFVIIa and prothrombin complex concentrates can also be utilized if FVII or other vitamin K dependent factors are low (e.g. FII and FX) or if there is concern about volume overload. It is important also to be aware of the potential thrombotic risk when utilizing these agents (due to potential underlying DIC and excessive thrombin activation).

## Recommended Reading

Blonski W, Siropaides T, Reddy KR. (2007). Coagulopathy in liver disease. *Curr Treat Options Gastroenterol* **10**, 464–473.

Caldwell SH, Hoffman M, Lisman T *et al.* (2006). Coagulation in Liver Disease Group. Coagulation disorders and hemostasis in liver disease: pathophysiology and critical assessment of current management. *Hepatology* **44**, 1039–1046.

Kujovich JL. (2005). Hemostatic defects in end stage liver disease. *Crit Care Clin* **21**, 563–587.

Leonis MA, Balistreri WF. (2008). Evaluation and management of end-stage liver disease in children. *Gastroenterology* **134**, 1741–1751.

Trotter JF. (2006). Coagulation abnormalities in patients who have liver disease. *Clin Liver Dis* **10**, 665–678.

# CHAPTER 106

# Bleeding Risks with Vitamin K Deficiency

Thomas C. Abshire, MD

In the early 1930s, the condition known as hemorrhagic disease of the newborn was first discovered to be related to a deficiency of vitamin K. It was later appreciated that deficiency of vitamin K presents with unique bleeding manifestations at various ages. Vitamin K is an essential co-factor for the clotting proteins, FII, FVII, FIX and FX, as well as to the natural inhibitors of the non-vitamin K dependent factors (FV and FVIII), protein C and protein S. The most common deficiency of vitamin K dependent coagulation factors is a result of warfarin anticoagulation.

<u>Pathophysiology:</u> Vitamin K is essential in the post-translational modification of glutamic acid residues in specific clotting factors. These residues are carboxylated via the vitamin K dependent γ-glutamyl carboxylase to a specific γ-carboxyglutamic acid clotting factor. Other essential cofactors to this reaction are reduced vitamin K, carbon dioxide, and oxygen. The essential role of vitamin K in activation of these clotting proteins is illustrated in Figure 106.1. This carboxylation of glutamates occurs near the $NH_2$ terminus. These γ-carboxyglutamic acid clotting factors (Gla) are important in binding calcium to a phospholipid membrane, and in forming the coagulation complexes important to the formation of thrombin (tenase and prothrombinase complexes). Warfarin blocks the formation of vitamin K-hydroquinone (reduced vitamin K) from vitamin K-epoxide.

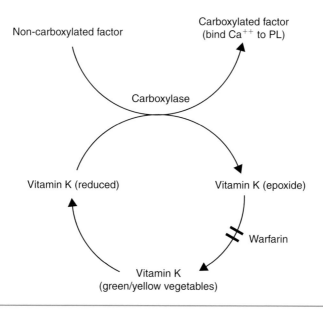

FIGURE 106.1 The vitamin K cycle and its role in production of carboxylated clotting factors.

The major source of dietary vitamin K is vitamin $K_1$ (phylloquinone), which is primarily derived from yellow or green plants. The daily dietary requirement is 1 μg/kg. Vitamin K is lipid soluble, and an intact intestine is required for optimal absorption. Vitamin $K_2$ (menaquinones-n) is another source of vitamin K, which is produced by bacteria and is also found in animal and soy protein products. This form of vitamin K is less readily absorbed. It is important to realize that both human and cow's milk are low in vitamin K. Any condition which might impair absorption of vitamin K, such as malabsorption syndromes and liver dysfunction, will impact the production of active clotting factors.

**Clinical Manifestation:** There must be a high degree of suspicion to entertain a diagnosis of vitamin K deficiency in the neonatal period, infancy, childhood or adult life. In particular, this diagnosis should be considered in the context of unexplained intracranial or gastrointestinal hemorrhage. These two clinical manifestations are more common in the first year of life, whereas skin manifestations, such as purpura and/or nose and mouth-related bleeding, are more common in children and adults. Table 106.1 reviews the age of presentation, the bleeding site, cause and treatment for different clinical manifestations of vitamin K deficiency in children.

**TABLE 106.1  Vitamin K Deficiency in Infants**

| Syndrome | Age at presentation | Bleeding site | Cause | Treatment |
|---|---|---|---|---|
| Early HDN (5%, at-risk mothers) | 1 day | ICH, GI, umbilical, bony abn | Anticonvulsants; anti-TB meds | 10 mg vitamin K daily × 2 weeks |
| Classic HDN (0.01–1%) | 1–7 days | ICH, GI, umbilical, ENT | Idiopathic; maternal drugs | 0.5–1 mg vitamin K at birth |
| Late HDN (0.02%) | 2–24 weeks | ICH, GI, skin, ENG | Idiopathic; malabsorption; liver disease; breastfeeding | Vitamin K at birth; repeat as needed orally/IM |

Vitamin K deficiency in infancy by syndrome: early HDN (hemorrhagic disease of the newborn), classic HDN and late HDN. Listed are: age of presentation; common bleeding sites (intracranial hemorrhage, ICH; gastrointestinal, GI, ear/nose/throat, ENT), cause and treatment.

Accidental or intentional ingestion of warfarin must be considered, as well as ingestion of long-acting rat poison, especially when reversal of presumed warfarin overdose has been attempted, only to have its bleeding manifestations recur within 24 hours. In the adult population, excessive alcohol intake and a combination of prolonged fasting, excessive vomiting and antibiotic therapy may present a triad of severe vitamin K deficiency, particularly in the elderly.

Because vitamin K is lipid soluble, normal absorption of fats is required for vitamin K uptake. Lack of bile salts secondary to cholestatic disease can lead to deficiency. Similarly, deficiency can be observed secondary to use of drugs that decrease cholesterol absorption, such as cholestyramine. Finally, metabolic disorders that involve dysregulated lipid metabolism can also lead to a vitamin K deficient state (e.g. Gaucher disease).

## Diagnosis:

*Laboratory Testing:* As previously discussed, proteins containing $\gamma$-carboxyglutamic acid (Gla) are widespread throughout the body, and deficiency of vitamin K has its effects in several organ systems, in addition to coagulation. Unfortunately, the earliest evidence of vitamin K deficiency occurs in areas that are difficult to assess clinically. Thus, in theory, early vitamin K deficiency could be detected through analysis of liver stores (liver biopsy) or through altered measures of bone metabolism (i.e. increased decarboxylated osteocalcin). Less invasive measuring of vitamin K in plasma can be performed, but precise measurements can be difficult. However, if obtained accurately, adult plasma levels less than 100 pg/ml or cord blood levels less than 50 pg/ml are suggestive of vitamin K deficiency. Other tests not readily performed include decreased urinary excretion of Gla proteins and the more sensitive assessment of increased decarboxylated proteins, the so called "*protein induced in vitamin k absence*" (PIVKA). The PIVKA most commonly assessed in laboratories involves PIVKA II (non-carboxylated prothrombin). From a practical standpoint, measurement of vitamin K dependent coagulation factors, (FII, FVII, FIX, FX) and decreased protein C and protein S are the easiest way to assess vitamin K deficiency. While this approach may miss very early deficiency, it assesses one of the main pathological outcomes of decreased vitamin K. If these vitamin K dependent factors are low (usually less than 30–35%), the screening tests for coagulation (PT, PTT) are both prolonged and bleeding is more likely. Early vitamin K deficiency might not affect the PTT to the same degree as the PT. Also, in isolated vitamin K deficiency, both the platelet count and the fibrinogen level are normal.

*Clinical Features Important in Diagnosis:* In addition to the previously described diagnostic tests, the clinical scenarios which present with vitamin K deficiency must be weighed carefully. Vitamin K deficiency in the immediate neonatal period is most likely due to maternal use of anti-seizure medication or anti-tuberculosis drugs. Classic hemorrhagic disease of the newborn appearing in the first week of life is either idiopathic or related to maternal ingestion of medication. Vitamin K deficiency which appears in the first several months of life is either idiopathic or related to gastrointestinal (GI) malabsorption, liver disease, or breastfeeding in conjunction with decreased oral intake of vitamin K secondary to loss of vitamin K via the GI tract or to the poor vitamin K content found in breast milk. Warfarin poisoning from ingestion of rat poison or accidental overdose of adult medication can also cause vitamin K deficiency in children.

In the adolescent and adult population, prolonged fasting, excessive vomiting, medications (including certain antibiotics such as cephalosporin, overdose of vitamin E, and salicylates), and malabsorption from many different causes (alpha 1 anti-trypsin deficiency, biliary atresia, celiac disease, chronic diarrhea, cystic fibrosis, cholestatic liver disease and thermal burn injuries) all may cause vitamin K deficiency. Finally, alcoholism is associated with vitamin K deficiency. Several herbal remedies may also have the potential to effect vitamin K deficiency (e.g. Etpapain).

*Differential Diagnosis:* When excessive mucosal bleeding is encountered (nose, mouth, GI/GU tract and bruising), screening laboratory testing is essential. The hemostasis screen (complete blood count, PT, PTT, fibrinogen, thrombin time and

PFA) will help focus towards a correct diagnosis. As previously mentioned, an elevated PT and PTT along with normal other screening labs essentially rules out other causes of bleeding and confirms vitamin K deficiency as the likely etiology. Both DIC and liver disease will display abnormalities of all the other laboratory screening tests. Heparin contamination and dysfibrinogenemia are ruled out by a normal thrombin time, and fibrinogen deficiency is unlikely given a normal fibrinogen level. Occasionally, either homozygous or severe heterozygous deficiency of FII, FV and FX may cause an elevation in both the PT and PTT. In this case, FX and FV are the most likely deficiencies to consider. If there were a congenital factor deficiency, a bleeding history should be evident for many years. Acute onset of bleeding could be attributed to antibodies to these factors.

**Management:** Management of vitamin K deficiency varies by the clinical manifestations and underlying cause. Careful attention to maternal anticonvulsants and other medications will help reduce early hemorrhagic disease of the newborn. In this situation, maternal use of vitamin K will be helpful in reversing any possible effects. Classic hemorrhagic disease of the newborn has been effectively eliminated by administration of vitamin K at birth. Parents who choose not to allow their newborn to be given vitamin K at birth may be placing the infant at risk for vitamin K deficiency, particularly if they are breastfed or encounter malabsorption for any reason (including infectious diarrhea) and decreased oral vitamin K intake.

Acute treatment of vitamin K deficiency varies, and is similar to the treatment for warfarin reversal. Treatment is based upon the screening laboratories and the clinical manifestations. If there is not excessive bleeding and the PT/PTT are the only tests elevated, intramuscular vitamin K may be given at 1–5 mg for an infant, 5–10 mg for a child and 10 mg for an adult. However, if there is CNS, GI or extensive mouth and nose bleeding, intravenous vitamin K should be given as a slow infusion and plasma products should be transfused at 15–20 ml/kg, which should raise most factor levels 20–25% above their baseline. If there is concern about volume restriction, rFVIIa or, preferably, prothrombin complex concentrates which include most of the vitamin K dependent factors might be more helpful (see Chapters 29 and 143).

## Recommended Reading

Clarke P, Shearer MJ. (2007). Vitamin K deficiency bleeding: the readiness is all. *Arch Dis Child* **92**, 741–743.

Cranenburg EC, Schurgers LJ, Vermeer C. (2007). Vitamin K: the coagulation vitamin that became omnipotent. *Thromb Haemost* **98**, 120–125.

Doneray H, Tan H, Buyukavci M, Karakelleoglu C. (2007). Late vitamin K deficiency bleeding: 16 cases reviewed. *Blood Coagul Fibrinolysis* **18**, 529–530.

Flood VH, Galderisi FC, Lowas SR et al. (2008). Hemorrhagic disease of the newborn despite vitamin K prophylaxis at birth. *Pediatr Blood Cancer* **50**, 1075–1077.

Laposata M, Van Cott EM, Lev MH. (2007). Case records of the Massachusetts General Hospital. Case 1-2007. A 40-year-old woman with epistaxis, hematemesis, and altered mental status. *N Engl J Med* **356**, 174–182.

Pavlu J, Harrington DJ, Voong K et al. (2005). Superwarfarin poisoning. *Lancet* **365**, 628.

Spahr JE, Maul JS, Rodgers GM. (2007). Superwarfarin poisoning: a report of two cases and review of the literature. *Am J Hematol* **82**, 656–660.

van Hasselt PM, de Koning TJ, Kvist N *et al.* (2008). Netherlands Study Group for Biliary Atresia Registry. Prevention of vitamin K deficiency bleeding in breastfed infants: lessons from the Dutch and Danish biliary atresia registries. *Pediatrics* **121**, e857–e863.

# CHAPTER 107
# Bleeding Risks with Cardiac Disease

Thomas C. Abshire, MD

The main risk of bleeding with cardiac disease occurs when the patient is placed on cardiopulmonary bypass (CPB). Immediate postoperative bleeding, both surgical and secondary to platelet dysfunction, is common in patients coming off cardiopulmonary circuits. It is rarer to have hemostatic defects associated with cyanotic heart disease or valvular heart disease; these will also be discussed.

## Pathophysiology:

**Cardiopulmonary by Pass:** The combination of surgically related bleeding (wound site, chest tubes) and platelet dysfunction can account for almost the entirety of post-operative bleeding after CPB. Platelet dysfunction can be a result of activation of platelets with subsequent α- and dense-granule release. Platelet dysfunction may also occur with hypothermia, which is often induced during surgery and may also be secondary to the pump and/or filters, with subsequent activation of platelets. Bleeding may also be the result of mild thrombocytopenia ($<100,000/\mu l$), from shortened platelet survival secondary to platelet consumption during the procedure or as a result of the circuit.

A mild decrease in all the clotting factors can occur immediately after the patient is placed on the bypass circuit, secondary to hemodilution. The level of dilution of clotting factors is approximately 50%. The liver may also be temporarily affected during bypass surgery, with decreased blood flow causing decreased hepatic production of clotting factors. Fibrinolysis is increased in CPB as a result of increased tissue plasminogen activator (TPA), decreased plasminogen activator inhibitor-1 (PAI-1) and decreased α2 antiplasmin (α2AP).

Activated coagulation can be present during CPB, but is less likely to be encountered if there is adequate anticoagulation. If under-heparinization occurs, increased generation of thrombin–antithrombin complexes (TAT) and prothrombin fragment 1.2 and increased D-dimer production can be observed. When hemostatic parameters have been carefully studied in children and adults undergoing CPB, investigators have demonstrated lower heparin levels in children. This result is probably due to a combination of factors: (1) greater hemodilution in children (due to a smaller blood volume) resulting in decreased clotting factors, especially fibrinogen and antithrombin (contributing to "heparin resistance"), and (2) a greater degree of hypothermia in children, which inappropriately prolongs the ACT, leading the clinician to surmise that adequate heparinization is in place. Additionally, the ACT may also be inappropriately prolonged in adults, leading to suboptimal heparin dosing. Adults also undergo hypothermia, hemodilution of clotting factors/platelets and antifibrinolytic therapy (with resultant kallikrein inhibition). Thus, the whole blood clotting time assessment (ACT) utilized in most centers may not be adequate to optimize anticoagulation in the CPB setting.

**Cyanotic Heart Disease:** The primary hemostatic problem evident in cyanotic heart disease stems from polycythemia. In general, coagulation abnormalities are rare, but occasionally hepatic congestion from an increased hematocrit can lower production of clotting factors. The platelet count can also be slightly lower due to shortened survival of platelets, and there is often a mild platelet function defect for the reasons previously described. In addition, there may be factitious elevation of the screening clotting tests (PT, PTT), as a result of an excessive citrate : plasma ratio in the tube utilized for testing, secondary to the increased hematocrit (see Chapter 113). It is essential to correct for the hematocrit and the resulting decreased plasma volume. Therefore, less anticoagulant is needed in testing the patient with polycythemia. Until correction of the cyanotic heart defect is possible, lowering of the hematocrit will help ameliorate any underlying coagulation defects.

**Non-cyanotic Heart Disease/Valvular Heart Disease:** Certain cardiac conditions, such as a ventricular septal defect (VSD) and aortic stenosis or pulmonic stenosis, may all contribute to acquired von Willebrand syndrome (AVWS) (see Chapter 95) with a laboratory picture similar to type 2 AVWS. In addition, artificial prosthetic cardiac valves may likewise cause AVWS. In this condition, there is loss of the high molecular weight multimers secondary to a sheering effect on the von Willebrand factor (VWF). Additionally, thrombocytopenia may also occur from a sheering effect resulting in increased destruction/removal of the platelets. The VWF findings are comprised of inappropriately low VWF : RCo compared to VWF : Ag, and loss of high molecular weight multimers. Correction of the VSD or abnormal heart valve will correct both the AVWS and thrombocytopenia.

**Clinical Manifestations:** Bleeding related to CPB surgery is usually evident from persistent bleeding from nasogastric tubes and urinary catheters, chest/endotracheal tubes, intravenous sites, sites of blood draws and the surgical site. Bleeding from all these sites helps to differentiate the systemic bleeding sometimes associated with CPB from localized surgical bleeding. The bleeding associated with cyanotic and non-cyanotic heart disease is usually milder, and mucosal in nature (epistaxis, petechiae and purpura). As in CPB-associated bleeding, bleeding solely from the surgical site points towards surgically-related bleeding as the primary cause of hemorrhage.

## Diagnosis and Differential Diagnosis of Bleeding Associated with CPB:

**Initial Bleeding Assessment Associated with CPB:** Screening laboratory tests such as the PT, PTT, fibrinogen and platelet function analyzer (PFA) are not uniformly performed after CPB. Instead, anesthesiologists utilize the whole blood clotting assay (ACT) and the platelet count to guide management after CPB. As previously discussed, because most causes of bleeding with CPB are known the focus of bleeding management after CPB is on chest tube and surgical site/wound drainage, as well as on consideration of a platelet transfusion when a surgical cause for excessive bleeding has been ruled out and significant bleeding continues. Platelet transfusions will correct the thrombocytopenia and the platelet dysfunction. However, prior to a platelet transfusion, heparin should be reversed to allow the hemostatic system to function in the absence of high-dose anticoagulant.

**Heparin Reversal:** After the procedure, the anesthesiologist is careful to reverse the effects of heparin with protamine. Rarely, patients may develop toxicity to protamine, presenting with symptoms of hypotension, hypoxia and disseminated intravascular coagulopathy (DIC). Usually this occurs in patients who have received protamine either from previous CPB, or by use of protamine-containing insulin. The increased use of recombinant insulin makes the latter mechanism unlikely. The etiology of protamine toxicity appears to be either IgE-mediated anaphylaxis or complement anaphylactoid-mediated, secondary to heparin–protamine or protamine–antiprotamine complexes. Protamine contains arginine, which, when converted to nitric oxide, may also contribute to the systemic toxicity of this disorder.

**Screening Laboratory in CPB Bleeding Assessment:** If the above causes of bleeding have been effectively excluded, it is important to order additional screening laboratory tests to determine a specific diagnosis and to guide further management: PT, PTT, fibrinogen, complete blood count and platelet count, PFA, heparin level, D-dimer and euglobulin lysis time (ELT). These screening tests will first confirm the lack of excessive heparin or thrombocytopenia as potential causes for continued bleeding. Second, dilutional coagulopathy may be present, suggested by an elevated PT and PTT, and low fibrinogen. This is best treated with plasma products (e.g. fresh frozen plasma and plasma frozen within 24 hours of phlebotomy) and cryoprecipitate. The presence of DIC would be suggested by an elevated D-dimer, and would necessitate evaluation for an underlying etiology. The baseline D-dimer is expected to be elevated as a result of normal post-surgical hemostasis. Thus, measuring D-dimer levels in this context and monitoring coagulation parameters over time is essential to assess for consumptive coagulopathy. In this setting, plasma products and/or cryoprecipitate would be utilized along with consideration of additional anticoagulation therapy. The ELT may be used to help guide therapy for excessive fibrinolysis. Finally, an elevated PFA may suggest the ongoing presence of a platelet dysfunction, as well as need for additional therapy besides platelet transfusions.

**Management:** A normal hemostasis screening laboratory panel will help to exclude heparin, dilutional coagulopathy (including low fibrinogen levels) and platelet dysfunction as contributing causes to ongoing bleeding. Nonetheless, platelet dysfunction, abnormal fibrinolysis, DIC and possible antithrombin antibodies may still play a role in continued bleeding, despite normal laboratory screening. The D-dimer in particular often lags behind the clinical findings, being slower to rise initially in DIC and lingering for days (due to increased reabsorption from the extracellular space) after DIC has resolved.

**Desmopressin:** Presumed platelet dysfunction may be treated with desmopressin (DDAVP). DDAVP may help with platelet dysfunction by releasing VWF from its endothelial and platelet storage sites, which fosters better bridging of platelets to the vessel wall. DDAVP will also facilitate increased GP IIb/IIIa expression and improved platelet–platelet interaction. Finally, DDAVP causes microparticle release from platelets, with resultant increased platelet aggregation. DDAVP must be monitored carefully due to potential side-effects, including hyponatremia and seizures, especially in infants and the elderly – the two age groups most likely to undergo CBP-related surgery.

**Antifibrinolytic Therapy:** Treatment with an antifibrinolytic, such as aminocaproic acid or tranexemic acid, may be helpful if the ELT is short (a third antifibrinolytic, aprotinin, is currently not available in the US). These agents may be utilized empirically if prompt ELT testing is not possible, or if ongoing bleeding dictates more urgent therapy. The potential side effects of thrombosis must be weighed. Aminocaproic acid or tranexemic acid are similar in structure to lysine, and competitively inhibit the binding of plasminogen and TPA to the lysine binding sites on fibrin. Antifibrinolytic therapy may also reduce platelet dysfunction, since the agents may decrease fibrin degradation products and their effect on platelet–platelet interaction. Since use of an antifibrinolytic agent may also cause thrombosis, careful monitoring is encouraged.

**Anticoagulation Therapy:** If DIC after CPB is a clinical concern, therapy with anticoagulants such as heparin may be reinstituted. For those patients unable to tolerate heparin (e.g. heparin-induced thrombocytopenia), direct thrombin inhibitors such as argatroban or hirudin may be utilized.

**Therapy for Inhibitors to Bovine Thrombin:** Increasingly, fibrin glue with human thrombin is utilized as a local tissue sealant; however, topical bovine thrombin may also still be used. Antibodies can be formed from this bovine source which may cross-react with human thrombin or FV. The source of the anti-FV antibodies may be secondary to bovine FV in topical thrombin, or from exposure to intra-operative plasma products. Antibodies may also be formed against the vitamin K dependent clotting factors (FII, FVII, FIX and FX) but cross-reactivity to FV is the more common. Late bleeding, several weeks after surgery, may occur. Antibodies to thrombin are often detected by an elevated PT and PTT with lack of correction by mixing studies, but can always be discerned by an elevated thrombin time in the face of a normal reptilase time. The latter test rules against a dysfibrinogen. These antibodies are best treated with appropriate blood product support and with immunosuppressive agents, corticosteroids, intravenous immunoglobulin or plasma exchange.

## Recommended Reading

Eisses MJ, Chandler WL. (2008). Cardiopulmonary bypass parameters and hemostatic response to cardiopulmonary bypass in infants versus children. *J Cardiothorac Vasc Anesth* **22**, 53–59.

Fergusson DA, Hébert PC, Mazer CD *et al.* (2008). A comparison of aprotinin and lysine analogues in high-risk cardiac surgery. *N Engl J Med* **358**, 2319.

McEwan A. (2007). Aspects of bleeding after cardiac surgery in children. *Paediatr Anaesth* **17**, 1126–1133.

McLaughlin KE, Dunning J. (2003). In patients post cardiac surgery, do high doses of protamine cause increased bleeding? *Interact Cardiovasc Thorac Surg* **2**, 424–426.

Owings JT, Pollock ME, Gosselin RC *et al.* (2000). Anticoagulation of children undergoing cardiopulmonary bypass is overestimated by current monitoring techniques. *Arch Surg* **135**, 1042–1047.

Yoshida K, Tobe S, Kawata M, Yamaguchi M. (2006). Acquired and reversible von Willebrand disease with high shear stress aortic valve stenosis. *Ann Thorac Surg* **81**, 490–494.

# CHAPTER 108
# Bleeding Risks with Renal Disease

Thomas C. Abshire, MD

Renal disease has long been associated with a bleeding diathesis, primarily mucosal bleeding such as bruising, nasal or oral bleeding, and gastrointestinal bleeding. The association of a prolonged bleeding time, anemia and uremia has been observed since the 1950s. Dialysis and erythropoietin have greatly decreased the bleeding tendency in renal-failure patients. This chapter will focus on the hemostatic disorders associated with chronic renal disease, although acute renal failure can also present with similar bleeding manifestations. The laboratory findings, differential diagnosis and management issues associated with renal disease will also be addressed.

**Pathophysiology:** Uremic patients consistently demonstrate disordered platelet/vessel interaction. Over the years, this has been best characterized by the prolonged bleeding time. Several substances present in uremic patients can contribute to this pathophysiology. Guanidinosuccinic acid (GSA) is known to accumulate in uremic plasma, and affects platelet and vascular function. GSA is a derivative of L-arginine; both are precursors of nitric oxide (NO). NO production is increased in uremic patients, which leads to increased cyclic GMP, producing vascular relaxation and decreased aggregation of platelets.

**Platelet Dysfunction:** In addition to the effect of NO on platelet function, there are other abnormalities seen in the uremic patient that can contribute to poor platelet aggregation and bleeding. First, platelet phospholipids are modified, which affects the release of arachidonic acid, which is subsequently converted to thromboxane $A_2$. Second, thromboxane $A_2$ is decreased, as are the dense granules in platelets and resultant ADP release. Finally, increased platelet aggregation and resultant platelet dysfunction occurs when platelets interact with the surface membrane of the dialysis machine.

**Anemia-associated Bleeding:** Chronic renal disease results in decreased erythropoietin production, leading to decreased red blood cell (RBC) mass. Since the initial discovery in the 1950s of the correlation between prolongation of the bleeding time and anemia, clinicians have utilized RBC transfusions to reduce the bleeding time and resultant bleeding manifestations. This normalization of the bleeding time from an increased RBC mass may be related to increased availability of ADP from the RBCs, but is more likely a rheologic phenomenon. In essence, increasing the RBC mass allows platelets and clotting factors to be effectively guided towards the endothelium, and thus be able to participate in coagulation.

**Thrombocytopenia:** The platelet number can be slightly decreased in chronic renal disease, but is rarely below 100,000/μl. The platelet size may also be smaller than normal.

**Bleeding Secondary to Medications:** Chronic renal disease patients are often exposed to heparin during dialysis. Reversal of heparin at the end of each treatment is standard, but heparin overdose should be considered in the chronic renal disease patient who unexpectedly bleeds after dialysis. Additionally, medications, including various antibiotics and anti-inflammatory drugs, may affect platelet function.

**Hypofibrinolysis:** Chronic renal patients occasionally have decreased fibrinolysis secondary to low tissue plasminogen activator (TPA) and increased plasminogen activator inhibitor-1 (PAI-1). Occasionally, this may contribute to clotting within the shunt placed for dialysis. It may also contribute to more systemic clotting problems.

**Nephrotic Syndrome:** Nephrotic syndrome is described separately due to the myriad of potential coagulation abnormalities seen. Elevated fibrinogen, FVIII and VWF levels are common in nephrotic syndrome and in other renal diseases, primarily as acute phase reactants. Fibrinogen is also abnormal, and the resultant dysfibrinogen is characterized by a prolonged thrombin time and reptilase time. Certain clotting factors, such as FIX and FXII, can also be diminished. The low FIX level rarely cause bleeding. Decreased antithrombin (ATIII) and plasminogen may also be observed; these might contribute to the thrombosis occasionally seen in patients with nephrotic syndrome.

**Clinical Manifestations:** Since chronic renal-disease related bleeding is most often secondary to a platelet dysfunction, the clinical manifestations present as mucosal bleeding such as epistaxis, oral bleeding and bruising. Occasionally, gastrointestinal bleeding may be encountered.

**Diagnosis:** The primary bleeding manifestations of chronic renal disease are platelet dysfunction and occasionally, a mild thrombocytopenia. The PT and PTT are often normal unless FIX and FXII are low. The platelet count may be lower than normal, and the fibrinogen level is often elevated and associated with an increased thrombin/reptilase time. The bleeding time and/or platelet function analyzer (PFA) are often elevated. Despite the association of increased bleeding time or PFA with chronic renal disease, these tests should not be utilized to predict bleeding risk in surgery or during invasive procedures. If a patient has nephrotic syndrome, the laboratory abnormalities associated with this condition should be assessed appropriately. In particular, both a FIX level and an ATIII level should be performed.

**Differential Diagnosis:** The differential diagnosis of the bleeding manifestations and laboratory findings associated with chronic renal disease is fairly straightforward. It must be determined whether the acquired platelet dysfunction associated with renal disease is due to the disorder, or to a congenital platelet dysfunction. A personal and family history of bleeding prior to the onset of renal disease will help in this differentiation.

**Management:** The primary treatment for bleeding associated with chronic (and sometimes acute) renal disease is dialysis. This intervention should be performed prior to any surgery or invasive procedure. The second line of treatment is to maintain a RBC mass in a more normal range, aiming for a hematocrit of approximately 30%. This level can be achieved by RBC transfusions or, more commonly, with the use of erythropoietin. This agent, just as with RBC transfusions, can often correct the bleeding time and/or PFA in uremic patients. If these two treatment modalities do not correct the bleeding manifestations, intravenous or nasal desmopressin (DDAVP) may be used. DDAVP asserts its action by initiating the clotting cascade through contact pathway activation as well as releasing VWF from both platelets and endothelial cells. The agent is also effective in microparticle release from platelets, as well as promoting greater GP IIb/IIIa interaction. The onset of action of DDAVP is between 30 minutes and 4 hours after administration. Finally, estrogen may be used in the form of a patch, via oral administration or IV, to help correct bleeding manifestations. Estrogen is thought to exert its action by blocking the effect of nitric oxide in its multiple influences; both on the blood vessel wall and on platelet activation and aggregation. The estrogen peak effect is 6 hours after administration, and may last as long as 14 days. Cryoprecipitate has been utilized to correct the platelet dysfunction of uremia, but this is less attractive than the previously mentioned modalities due to variability, the potential prolonged duration of response and the risk of transfusion adverse events. Platelet transfusions have also been utilized in the face of acute bleeding but are only transiently effective, as the transfused platelets will become ineffective in the uremic recipient.

## Recommended Reading

Holden RM, Harman GJ, Wang M et al. (2008). Major bleeding in hemodialysis patients. *Clin J Am Soc Nephrol* **3**, 105–110.

Kaw D, Malhotra D. (2006). Platelet dysfunction and end-stage renal disease. *Semin Dial* **19**, 317–322.

Noris M, Remuzzi G. (1999). Uremic bleeding: closing the circle after 30 years of controversies? *Blood* **94**, 2569–2574.

# CHAPTER 109
# Bleeding Risks in Cancer

Michael A. Briones, DO

Thrombosis is a more common, and potentially fatal, complication of cancer than bleeding. Bleeding, which may also be severe and life-threatening, occurs in approximately 10% of patients with cancer. Liver failure, infection, medications (including chemotherapy), radiotherapy, surgery and the malignancy itself may lead to local vessel invasion, disseminated intravascular coagulopathy (DIC), or abnormalities in platelet function and number. Bleeding manifestations include hematemesis, hematochezia, melena, hemoptysis, hematuria, epistaxis, vaginal bleeding, ulcerated skin lesions, echymoses, petechiae and bruising.

## Clinical Manifestations:

### Platelet Disorders and Thrombocytopenia:

*Thrombocytopenia:* Thrombocytopenia results from reduced production, increased destruction, or sequestration (e.g. from splenomegaly) of platelets, and is the major cause of bleeding in untreated cancer patients. Low platelet count in malignancy is most frequently encountered as a result of chemotherapy with cytotoxic medications; other causes include marrow invasion/infiltration or immune mediated thrombocytopenia (ITP). The extent and severity of thrombocytopenia secondary to chemotherapy depends upon the agent administered. Significant replacement of the marrow by tumor cells may occur in leukemia or in advanced metastatic disease. ITP is most commonly seen in association with lymphoid malignancies (especially chronic lymphoid leukemia), or with lung, breast and testicular cancers. The overall frequency of this complication in lymphoma is 0.4–1%, and may either precede the diagnosis or present after therapy is initiated. Of note, fever, infection, mucosal damage, prior radiotherapy and intensive chemotherapy may increase the risk of bleeding.

*Platelet Dysfunction:* Reduced platelet adhesion, aggregation abnormalities, poor clot retraction and storage-pool defects result in platelet dysfunction. Platelet dysfunction is observed in patients with chronic myeloproliferative disorders (e.g. essential thrombocythemia, polycythemia vera, and myelodysplasia), as well as in some patients with acute leukemias. In most malignancies, thrombocytopenia is usually the more significant cause of bleeding compared to platelet dysfunction. However, in myeloproliferative disorders, bleeding rather than thrombosis may occur with an elevated platelet count, due its association with acquired von Willebrand syndrome (AVWS).

Dysproteinemia leads to qualitative platelet abnormalities usually as a result of paraproteins coating the platelets. Platelet dysfunction or thrombocytopenia is observed frequently: in 15% of patients with IgG myeloma, 38% with IgA myeloma and 60% with Waldenstroms macroglobulinemia.

## Coagulation Abnormalities:

*Disseminated Intravascular Coagulopathy:* DIC, a consumptive coagulopathy, is an infrequent cause of bleeding in cancer patients, with the exception of its association with tumor lysis syndrome, certain medications, and acute myeloid leukemia–especially monocytic (M4/M5) and promyelocytic leukemia (APL). DIC can be associated with fibrinolysis, especially with APL, which can increase the risk of bleeding. Subclinical DIC (termed *compensated DIC*) results in laboratory abnormalities of fibrinolysis (elevated FDP and D-dimer) without prolongation of the prothrombin time (PT) or partial thromboplastin time (PTT) or depressed fibrinogen. This finding is seen frequently in untreated cancer patients (~50%), and in ~90% of patients with metastatic disease. Laboratory evidence of intravascular coagulation activation includes elevated levels of fibrinopeptide A, prothrombin $F_{1+2}$, thrombin–antithrombin complexes, fibrin monomers and D-dimers. The most commonly recognized laboratory hallmarks of DIC are thrombocytopenia, hypofibrinogenemia, and a prolonged PTT and PT (see Chapter 140).

*Acquired Inhibitors:* Acquired hemophilia is a potentially severe hemorrhagic disorder secondary to autoantibodies against various clotting factors (usually FVIII). In a recent survey in the UK, approximately 15% of acquired FVIII antibodies were found in cancer patients and the associated bleeding manifestations appear more common and severe in cancer patients than in non-cancer patients. Other autoantibodies include those to Factor V and von Willebrand Factor (VWF), or heparin-like substances released by cancer cells. These inhibitors are rare, but may cause severe hemorrhage. AVWS has been reported in association with myeloproliferative and lymphoproliferative disorders, as well as in Wilms' tumor (see Chapter 89).

*Medications:* Most chemotherapeutic agents, through their myelosuppressive action, cause thrombocytopenia and bleeding. Chemotherapeutic agents may also affect coagulation and fibrinolytic activity.

*L-asparaginase:* This agent is utilized in the treatment of acute lymphoblastic leukemia (ALL) and acts on the amino acid L-asparagine, which is essential for leukemic cell growth. During treatment, there is an increased risk of both bleeding and clotting, although the overall incidence of bleeding complications is low. This prothrombotic tendency increases following the cessation of L-asparaginase therapy, since the recovery of the coagulant proteins (fibrinogen and Factor VII) and the platelet count occurs earlier than that of the anticoagulant proteins. This recovery is not necessarily related to the half-life of the proteins. Additionally, lower antithrombin levels are seen in comparison to the clotting proteins, and this imbalance also tends to make patients more hypercoaguable.

**Diagnosis:** A complete blood count (especially the platelet count), PTT, PT, fibrinogen, and fibrin degradation products or D-dimer tests are important to assess in cancer patients presenting with bleeding. Peripheral blood smear review is imperative in determining whether there is evidence of thrombotic microangiopathy (specifically the presence of schistocytes).

**Management:** When cancer-related bleeding occurs, correction of thrombocytopenia with platelet transfusions, treatment of a prolonged PT and PTT with plasma products, correction of hypofibrinogenemia with cryoprecipitate, and correction of anemia with red blood cells may be required. Reduced response to platelet transfusions may be associated with infection, fever, DIC, splenomegaly, bleeding and medications (including heparin and antibiotics). Treating the underlying malignancy (particularly in DIC caused by acute leukemias) is essential to correction of the DIC. Treatment options for DIC, ITP and other coagulopathies are discussed in alternate chapters in this book.

## Recommended Reading

Kwaan HC, Vicuna B. (2007). Thrombosis and bleeding in cancer patients. *Oncol Rev* **1**, 14–27.

Mannucci PM. (2003). Overview of bleeding in cancer patients. *Pathophysiol Haemost Thromb* **33**, 44–45.

Pereira J, Phan T. (2004). Management of bleeding in patients with advanced cancer. *Oncologist* **9**, 561–570.

Zakarija A, Kwaan HC. (2003). Bleeding and thrombosis in the cancer patient. *Expert Rev Cardiovasc Ther* **1**, 271–281.

# CHAPTER 110
# Disseminated Intravascular Coagulopathy

Amy L. Dunn, MD

Disseminated intravascular coagulopathy (DIC) is "an acquired syndrome characterized by the intravascular activation of coagulation with loss of localization arising from different causes. It can originate from and cause damage to the microvasculature, which if sufficiently severe, can produce organ dysfunction" (International Society of Thrombosis and Hemostasis [ISTH]). The clinical severity of DIC varies such that DIC may result in laboratory abnormalities only, or in hemorrhage or thrombosis. Hemorrhage, which is secondary to the consumption of coagulation factors and fibrinolysis, is a more common manifestation in 70–90% of patients. Thrombosis, which is secondary to activation of coagulation factors and platelet activation resulting in both microvascular thrombosis and large-vessel thrombosis, is less common, occurring in 10–40% of patients. The treatment is focused on treating the underlying disease and on replacement of coagulation factors by blood product transfusion.

Pathophysiology: During normal hemostasis, three major components of the coagulation system – vascular integrity, normal quantity and function of coagulation factors, and normal platelet count and function – are coordinated to result in the delicate balance between bleeding (anticoagulation) and clotting (procoagulation). When this balance is disrupted, DIC can occur. The most common causes of DIC are endotoxin generated (e.g. sepsis); other causes include trauma, burns, obstetrical complications, vascular malformations, envenomations or exposure to malignant cells. The complications of DIC include bleeding and end-organ failure, due primarily to microthrombi (renal, respiratory, hepatic and CNS). During DIC, a *generalized* activation of the coagulation system leads to thrombin generation. Fibrin formation and fibrinolysis can then result in microthrombi and consumption of clotting factors and platelets. Excessive or prolonged tissue factor (TF) exposure leading to thrombin production is the usual triggering mechanism. In normal coagulation, exposure of TF leads to rapid induction of tissue factor pathway inhibitor, which then downregulates thrombin formation by feedback inhibition. In the setting of DIC, however, continued thrombin generation can occur due to new or ongoing TF exposure from an inciting process such as sepsis/trauma or activation of the intrinsic pathway of coagulation, leading to secondary thrombin generation. Exposure of negatively charged phospholipid surfaces, which occurs during cell membrane phosphatidylserine externalization, may also play a role in sustaining thrombin generation. This phospholipid switching occurs during activation or apoptosis of platelets, and allows for a readily available surface for ongoing coagulation. The generation of platelet microparticles from cell damage may also provide a surface for sustaining thrombin generation. Additionally, consumption of endogenous anticoagulant proteins, such as protein C, protein S and antithrombin, can contribute to coagulation disarray.

**Clinical Manifestations:** Patients with DIC manifest a heterogeneous presentation based upon the inciting incident. The most common manifestations include hemorrhage, renal dysfunction, hepatic dysfunction, respiratory dysfunction, shock and CNS dysfunction. Dermatologic signs may include petechiae, purpura and skin necrosis. Overt signs and symptoms of thrombosis may also be present.

**Diagnosis:** The diagnosis of DIC is made by a combination of clinical circumstances and laboratory evaluations. Diagnosis can be aided by utilizing a five-step algorithm developed by the ISTH (Figure 110.1). This scoring system was designed to assist in diagnosis and to be utilized as a reference standard. Since its inception, it has proven both sensitive and specific for the diagnosis of overt DIC, and increasing scores strongly correlate with patient mortality.

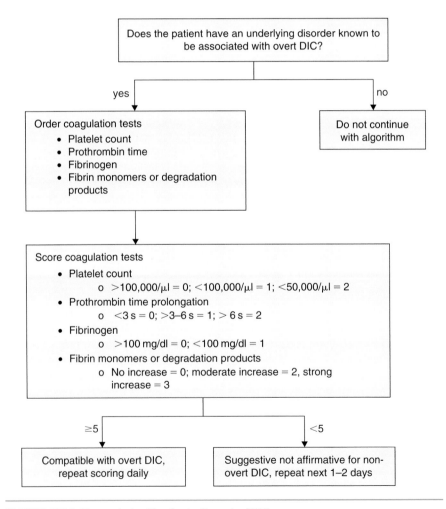

FIGURE 110.1 Diagnostic algorithm for the diagnosis of DIC.

The complete blood count is used to assess platelet number and the degree of anemia associated with the underlying illness. The peripheral smear will show evidence of thrombotic microangiopathy (schistocytes, thrombocytopenia and anemia). The activated partial thromboplastin time (PTT) is an *in vitro* measurement of the intrinsic pathway. It is prolonged in deficiencies of Factors XII, XI, IX, VIII, prekallikrein and high molecular weight kininogen (low levels of Factors XI, IX and VIII are associated with increased bleeding). The prothrombin time (PT) is an *in vitro* measurement of the extrinsic pathway. It is prolonged in DIC due to deficiencies of Factors X, VII, V, II (prothrombin) and I (fibrinogen). Factor levels below 20% and or PT more than 3 seconds over normal are more likely to be associated with bleeding complications. Fibrinogen is the most abundant clotting protein. Levels below 150 mg/dl are typically the result of consumption (DIC) ± decreased production (liver failure). Levels less than 100 mg/dl significantly increase the risk of bleeding. The thrombin time measures the time to conversion of fibrinogen to fibrin. Decreased concentrations of fibrinogen, decreased clearance of fibrin degradation products and the presence of heparin prolong the thrombin time. Fibrin degradation products are elevated during DIC, and are formed when fibrinogen and *non-cross-linked* fibrin are degraded by plasmin. D-dimers are a result of plasmin breakdown of *cross-linked* fibrin (excessive thrombin activation), and are elevated in the circulation during ongoing intravascular clotting. Levels of the endogenous coagulation inhibitor antithrombin are typically decreased in DIC. Levels less than 80% may be associated with thrombosis. The euglobulin clot lysis time screens for excessive fibrinolysis. Shortened lysis times are common in DIC, and may indicate hyperfibrinolysis (see Chapter 140).

**Differential Diagnosis:** Liver failure and severe vitamin K deficiency can mimic DIC. To distinguish between liver disease and vitamin K deficiency, it is helpful to measure Factor V and Factor VII levels. In liver disease both factors will be decreased, but in vitamin K deficiency only Factor VII will be low. As compared to DIC, neither is typically associated with severe reduction of fibrinogen. Thrombotic thrombocytopenic purpura and hemolytic uremic syndrome manifest are also thrombotic microangiopathies, but rarely show evidence of consumption of clotting proteins.

**Management:** The goal of DIC management is identification and appropriate treatment of the inciting process. Thereafter management should be guided by physical findings and laboratory assessment, with the goal of treatment to assist and maximize circulatory support and oxygen delivery while replacing consumed coagulation components. Table 110.1 shows potential treatment options.

**Platelet and Plasma Support:** There are no clinical trials on which to base guidelines for platelet therapy in the setting of DIC. It has even been suggested that platelet transfusions may feed the fire by providing a phospholipid surface for ongoing thrombin generation. In the absence of scientific evidence, however, typically a platelet count greater than 20,000/μl in the presence of mild bleeding and >50,000/μl in the setting of active bleeding is suggested.

The role of plasma replacement in DIC has been studied in one small randomized controlled trial in neonates. In that study, plasma and platelet therapy did not show a survival advantage over no therapy or exchange transfusion. Infusion of plasma

TABLE 110.1 Treatment Options for DIC

| Product | Dose | Indication |
| --- | --- | --- |
| Antithrombin | (desired AT level % − baseline AT level %) × body weight (kg) divided by 1.4% units/kg | Reduced antithrombin levels |
| Aminocaproic acid | 100 mg/kg orally every 6 hours<br><br>33.3 mg/kg per hour IV | Increased fibrinolysis (use cautiously in the setting of hematuria) |
| Cryoprecipitate | Adult: 10 units<br>Pediatric: 1 unit per 10 kg IV | Decreased fibrinogen |
| Heparin | 500 IU/h (no loading dose)<br><br><br><br>5–10 IU/kg per hour | DIC with evidence of ongoing thrombin generation unresponsive to conventional measures. |
| Plasma products | Adult: 4 units<br>Pediatric: 15–20 ml/kg | Decreased fibrinogen and clotting factors |
| Platelet products | Adult: 1 apheresis product equivalent<br>Pediatric: 10–15 ml/kg | Thrombocytopenia |
| Protein C concentrate | 24 mg/kg per hour for 96 hours | Adult patients with severe sepsis at high risk of death |

products or cryoprecipitate (in the setting of severe hypofibrinogenemia) should be considered in the setting of bleeding associated with low fibrinogen levels; however, there is currently no evidence for prophylactic use of plasma support in the absence of bleeding. The goal of fibrinogen replacement is to maintain levels >100 mg/dl to prevent or treat bleeding. The use of recombinant Factor VIIa as a means to enhance thrombin generation to treat refractory bleeding in the setting of DIC has been utilized in small numbers of patients with some success. However, randomized trials will be required prior to recommending or optimizing its use in the setting of DIC.

Anticoagulant Pathway Therapy: Depletion of anticoagulant proteins such as protein C, protein S and antithrombin has been postulated to play a role in sustaining DIC, and therefore replacement of these proteins has been explored as a therapeutic option. Over the past decade evidence has been slowly mounting to support this approach. Activated Protein C concentrates (APC) have been utilized effectively in treating purpura fulminans in neonates with homozygous protein C deficiency. Several pilot studies have suggested efficacy in the setting of DIC as well. Subsequently, in a large randomized controlled trial, adult patients were treated with continuous infusion APC concentrates for 96 hours upon presentation with DIC. This study demonstrated decreased overall mortality at 28 days compared to controls; however, there was a trend toward higher bleeding rates in the treated arm. Patients at highest risk of death showed the most benefit. Additionally, patients in the treatment arm exhibited lower inflammatory markers, thought to be due to the anti-inflammatory activity of APC. Further studies have failed to show a benefit of this therapy in pediatric patients or in adult patients with DIC who were not considered to be at high risk of death. Further studies will help to define the role of this agent.

Antithrombin (AT) concentrates have also been shown, in animal models of sepsis, to improve coagulation and improve survival rates. They have subsequently been utilized in several small human randomized controlled trials. In adult patients with severe sepsis and DIC, two of three trials showed a beneficial effect of antithrombin on 28- to 30-day mortality of all causes. The remaining trial showed no survival difference. There is no clear advantage to achieving supranormal levels of antithrombin. Additional trials are needed to define clearly the indications and dosing strategies.

**Anticoagulants:** If the balance of coagulation in a particular patient is tipped toward thrombosis, with evidence of microvascular or large-vessel thrombosis, low-dose continuous infusion heparin can be considered at a dose of 500 IU/hour in adults or 5–10 U/kg per hour in children if hemorrhage cannot be controlled by factor replacement alone. The theory behind the use of heparin in this setting is that it can interrupt coagulation activation; however, to date no well-designed randomized controlled trial has demonstrated the efficacy of this approach.

**Antifibrinolytic Therapy:** Primary systemic fibrinolysis can occur in patients with DIC, particularly in the setting of acute promyelocytic leukemia. This can be recognized by an abnormally short euglobulin clot lysis time (ELT), and an antifibrinolytic agent such as aminocaproic acid may need to be added to control bleeding. Suggested dosing and indications for all of the above agents are included in Table 110.1.

## Recommended Reading

Bernard GR, Ely EW, Wright TJ et al. (2001). Safety and dose relationship of recombinant human activated protein C for coagulopathy in severe sepsis. *Crit Care Med* **29**, 2051–2059.

Fourrier F. (2004). Recombinant human activated protein C in the treatment of severe sepsis: an evidence-based review. *Care Med* **32**, S534–541.

Taylor FB, Toh CH, Hoots WK et al. (2001). Towards definition, clinical and laboratory criteria, and a scoring system for disseminated intravascular coagulation. *Thromb Haemost* **86**, 1327–1330.

Toh CH, Downey C. (2005). Back to the future: testing in disseminated intravascular coagulation. *Blood Coag Fibrinol* **16**, 535–542.

Toh CH, Hoots WK, ISTH SSCoDICot. (2007). The scoring system of the Scientific and Standardisation Committee on Disseminated Intravascular Coagulation of the International Society on Thrombosis and Haemostasis: a 5-year overview. *J Thromb Haemost* **5**, 604–606.

Wiedermann CJ, Kaneider NC. (2006). A systematic review of antithrombin concentrate use in patients with disseminated intravascular coagulation of severe sepsis. *Blood Coag Fibrinol* **17**, 521–526.

# CHAPTER 111
# Acquired Coagulation Factor Inhibitors

Christine L. Kempton, MD

Acquired coagulation factor inhibitors are autoantibodies directed against native clotting factor in persons without an underlying bleeding disorder. Although rare, these disorders can result in life-threatening hemorrhage which can be difficult to manage. Inhibitors to the following factors will be discussed: Factor VIII (FVIII), von Willebrand Factor (VWF), Factor V (FV), prothrombin and thrombin.

## Inhibitors of FVIII:

**Pathophysiology:** FVIII inhibitors are autoantibodies that bind to native FVIII in a person without congenital hemophilia. Antibody binding leads to apparent FVIII deficiency. The incidence is 1.48 per million person years, although this rate increases with age. In those over the age of 85, the incidence is 14.66 per million person years; in those under the age of 16 it is only 0.0045 per million person years. In approximately 40–50% of cases, the acquisition of the FVIII inhibitor is related to an underlying condition (Table 111.1). The association between FVIII inhibitor and underlying disorder decreases with age. In a large observational study, an acquired inhibitor was secondary to an associated condition in all patients aged less than 40 years but in only 25% of those aged over 85.

| TABLE 111.1 Underlying Disorder in Patients with FVIII Inhibitors | |
|---|---|
| **Disorder** | **Frequency** |
| Autoimmune disorders: SLE, rheumatoid arthritis, polymyalgia rheumatica | 17% |
| Malignancy: lung, GI, prostate | 15% |
| Dermatological: psoriasis, pemphigoid | 3% |
| Pregnancy | 2% |

Pregnancy-associated FVIII inhibitor development is discussed in Chapter 103.

**Clinical Manifestations:** The severity of bleeding is quite variable. Approximately one-third of patients will have only minor bleeding that does not require any treatment. The most common sites of bleeding include the subcutaneous tissue, muscle, gastrointestinal tract and genitourinary tract. In contrast to congenital hemophilia A, hemarthroses are rare.

Approximately one-third of patients with an acquired FVIII inhibitor have spontaneous resolution; however, this may not occur until greater than 1 year (range 10–21 months) after diagnosis. In one report, fatal bleeding occurred in 9–22% of patients at a median of 19 days after diagnosis, but it can occur months later. There are no parameters that are known to be predictive of inhibitor resolution or fatal bleeding. In retrospective series, the inhibitor titer in those with fatal hemorrhage was identical to that in those who did not require any hemostatic therapy.

**Diagnosis:** These patients have a prolonged PTT, with a normal PT and thrombin time. Factor VIII, IX and XI activity assays typically demonstrate a markedly reduced FVIII activity level, and other clotting factor activities will be normal. On rare occasions a lupus inhibitor may coexist with a FVIII inhibitor and lead to alterations in other clotting assays. An inhibitor to FVIII, detected using the Bethesda assay (see Chapter 128), will be present. Because of type 2 inhibitor kinetics, some patients may have residual FVIII activity despite a disproportionately high inhibitor titer.

**Management:** Two major aspects of management of acquired FVIII inhibitors include treatment of acute bleeding, and eradication of the inhibitor.

*Hemostatic Treatment:* Minor bleeding such as ecchymoses or epistaxis that are self-limited do not require active treatment, although close monitoring and follow-up is necessary. For bleeding requiring treatment, several options are available: DDAVP, FVIII infusions and bypassing agents (recombinant Factor VIIa [rFVIIa] and aPCC [such as FEIBA]) (Table 111.2). DDAVP is most effective when FVIII activity is >5%. The effect of DDAVP decreases with subsequent dosing, and thus its use is limited to bleeding that requires a short duration of therapy. Additionally, the half-life of the released FVIII will be shortened secondary to binding of the inhibitory antibody, but the exact duration is variable from patient to patient.

**TABLE 111.2 Treatment Options for Acquired FVIII Inhibitors**

|  | Dose | Patient population | Duration of response | Monitoring | Potential side-effects |
|---|---|---|---|---|---|
| DDAVP | 0.3 µg/kg | FVIII > 5% | Variable | FVIII activity | Hyponatremia, seizures |
| FVIII | 100 U/kg or 20 U/kg for each Bethesda titer + 40 U/kg | Inhibitor titer < 5–10 BU/ml | Variable | FVIII activity | |
| rFVIIa | 70–90 µg/kg | All | 2–3 hours | None | Thrombosis |
| aPCC | 75–100 U/kg | All | 8–12 hours | None | Thrombosis |

Infusion of FVIII can be utilized when the inhibitor titer is <5 BU/ml and perhaps up to 10 BU/ml. Regardless of the initial dose, FVIII activity should be measured 15–30 minutes after the bolus. Since the half-life of the factor will be reduced, levels should be monitored carefully to determine the dosing interval. Alternatively, a continuous infusion could be started at a dose of 8 U/kg per hour and adjusted to achieve the targeted FVIII activity (80–100% for severe, limb- or life-threatening bleeding).

When the inhibitor titer is >5–10 BU/ml a bypassing agent should be used, either rFVIIa or aPCCs. rFVIIa (see Chapter 147) has been FDA approved for use in patients with acquired FVIII inhibitors, and is effective or partially effective in approximately 75% of bleeding episodes. Thrombotic events are uncommon but have been reported, and may be more likely in elderly populations who also have underlying vascular disease. There is no available laboratory testing to monitor efficacy or toxicity. The dose and frequency of dose should be adjusted to the clinical severity and the patient's response.

aPCC is not FDA approved for use in patients with acquired inhibitors, although it is routinely used in patients with congenital hemophilia complicated by an inhibitor. aPCC carries a risk of thrombosis similar to rFVIIa. There are no laboratory tests to monitor efficacy. Laboratory parameters of DIC (see Chapter 140) should be monitored if therapy is given for more than 5 days. The dose and frequency should be adjusted to the clinical severity and the patient's response.

The duration of therapy will be determined by bleeding severity and the risk of toxicity secondary to hemostatic therapy. With severe bleeds, hemostatic therapy should continue for at least several doses beyond the cessation of bleeding.

*Eradication of the Inhibitor:* Because of the inability to predict who may have a fatal hemorrhage or who will spontaneously remit, immune suppression should be considered in all patients. Primary treatment options include corticosteroids, cyclophosphamide and rituximab. Corticosteroids (prednisone 1 mg/kg per day), should be apart of the initial therapy. Cyclophosphamide (oral or intravenous) can be used in combination with coticosteroids initially, or following several weeks of corticosteroid monotherapy if an adequate response is not seen. Pulse intravenous cyclophosphamide (500–750 mg/m$^2$) given every 4 weeks will lead to less bladder exposure to acrolein and therefore a lower risk of hemorrhagic cystitis than with oral cyclophosphamide. In clinical practice, pulse intravenous dosing of cyclophosphamide has been safe and effective. If oral cyclophosphamide is utilized, a daily dose of 1–2 mg/kg is recommended.

Rituximab has been used in the setting of FVIII inhibitor eradication as well. In one series of 10 subjects who received rituximab alone, 8 of 10 responded. In those that responded, the inhibitor became undetectable within 3–12 weeks. Thus, some postulate that rituximab may facilitate a more rapid disappearance of the inhibitor; however, comparative studies are lacking. After 28.5 months of follow-up, 3 of the 8 initial responders had relapsed.

Intravenous immunoglobulin (IVIG) has been considered a part of early therapy; however clinical studies have not demonstrated a level of positive response that supports its up front use.

An overall approach to inhibitor eradication in those with an acquired FVIII inhibitor would include corticosteroids as a first-line approach in all patients. Cyclophosphamide and/or rituximab can be added to corticosteroids as first-line therapy in those with little reserve to tolerate additional bleeding, or who are at high risk for complications related to hemostatic therapy. Additionally, rituximab or cyclophosphamide may be used as second-line therapy in those that have failed to respond to corticosteroids alone after 3–4 weeks.

## Inhibitors of VWF:

Pathophysiology: The exact incidence and prevalence of acquired von Willebrand syndrome (AVWS) is unknown. Kumar and colleagues estimated, based on a retrospective review, that 0.049% of the population has AVWS. Although AVWS can occur by mechanisms other than an antibody-mediated inhibition of VWF activity (Table 111.3), this section will focus on antibody-mediated mechanisms. Lymphoproliferative disorders and monoclonal gammopathy of undetermined significance (MGUS) are associated with nearly half of all AVWS.

| TABLE 111.3   Diseases Associated with AVWS | |
| --- | --- |
| **Mechanisms of AVWS** | **Associated disease(s)** |
| Antibodies | Lymphoproliferative disorders/MGUS |
| Inhibition of VWF function | |
| Non-specific binding leading to clearance of VWF | Myeloproliferative disorders |
| Adsorption on malignant cells | Cardiac disease   (VSD/AS) |
| Proteolysis | |
| Decreased production | Hypothyroidism |

VWF, von Willebrand Factor; VSD, ventricular septal defect; AS, aortic stenosis.

Clinical Manifestations: AVWS typically manifests similarly to congenital von Willenbrand disease (VWD) (see Chapter 95), with mucocutaneous or post-surgical bleeding. However, one-fourth of patients may be asymptomatic. In contrast to congenital VWD, patients with AVWS will have previously normal hemostasis without a personal or family history of abnormal bleeding.

Diagnosis: Testing for AVWS is similar to that for congenital VWD, and includes measurement of von Willebrand Factor (VWF) activity, VWF antigen, FVIII assay, and multimer analysis. A type 2 VWD pattern with discordance of the VWF activity and antigen is most commonly seen (activity : antigen ratio <0.6). Multimer analysis may demonstrate the loss of large molecular weight multimers in 50–80% of cases (see Chapter 124).

Detection of an inhibitor can be attempted by performing a VWF activity assay after mixing the patient's plasma with equal parts of normal pooled plasma (NPP). Unfortunately, this method will only detect antibodies that directly affect the function of the assay used. Because of these limitations, testing for an inhibitor is rarely positive, but is slightly more likely to occur in the setting of lymphoproliferative disorders, neoplasia and immunological disease (approximately 30% of these cases).

Management: As with other autoimmune coagulation defects, the two major components of treatment include hemostatic therapy for acute bleeds and antibody eradication.

*Hemostatic Therapy:* DDAVP and a FVIII concentrate rich with VWF, such as Humate P or Alphanate (see Chapter 144), are the primary treatments. Both DDAVP and VWF-containing products are effective in one-third of patients. However, since the recovery and half-life of individual patient response may be reduced, careful monitoring is required. The choice of agent will depend on baseline VWF activity, anticipated duration of therapy, and ability to tolerate the potential side-effects of DDAVP. DDAVP is preferred over a VWF product in the setting of (1) mildly reduced VWF activity, (2) therapy that is anticipated to be of 3 days' duration or less, or (3) the patient not having uncontrolled hypertension or cardiovascular disease. Antifibrinolytic therapies such as aminocaproic acid or tranexamic acid may also be used as adjunctive therapies in the setting of mucocutaneous bleeding.

*Antibody Eradication:* In the setting of an associated condition for which therapy exists, treatment of the underlying condition is the most appropriate route for antibody eradication. If adequate treatment for the underlying disease is not available, or a quick response to therapy is not anticipated, IVIG or plasma exchange may be considered. IVIG has been utilized in patients with AVWS and has been reported to be effective in approximately one-third of patients, although responses were more frequent in the setting of lymphoproliferative disorders, MGUS, neoplasia and immunological diseases. Plasma exchange has been used infrequently, and is effective in only about 20% of reported cases.

An overall approach to the treatment of AVWS suspected to be secondary to an autoantibody is to treat acute bleeding with DDAVP or a VWF-containing product and monitor response. If the response is inadequate, the addition of IVIG (typically 2 g/kg infused in divided doses daily over 2–5 days) can be considered, since IVIG may improve the recovery and half-life of the VWF-containing product in addition to potentially improving the patient's native VWF antigen level and activity. Treatment should also be directed at eliminating the underlying disease, when applicable.

**Inhibitors of FV:** The development of an anti-FV inhibitor is rare. It may occur in association with the use of topical bovine thrombin, aminoglycosides and malignancy. Patients may develop an immune response to bovine thrombin after use in cardiovascular surgery; these antibodies cross-react to human thrombin as well as FV. Patients may be asymptomatic, or display severe bleeding weeks after the exposure. Anti-FV inhibitory antibodies should be suspected in the setting of abnormal bleeding with a newly prolonged PT and PTT but normal thrombin time. The PT and PTT will fail to correct on mixing with normal pooled plasma, and the FV activity will be low. An inhibitor titer can be measured utilizing the Bethesda assay. Plasma and platelet (FV is contained within platelet α-granules) products can be used to support hemostasis. Since the inhibitor is typically short-lived, immunosuppressive therapy is not warranted.

**Inhibitors of Prothrombin and Thrombin:** Antibodies that bind thrombin may occur after exposure to bovine thrombin. Not all antibodies will inhibit the function of thrombin. Antibodies that neutralize the function of thrombin will lead to prolongation of the PT, PTT and thrombin time. Prolongation of the thrombin time helps to distinguish inhibitors to thrombin from those directed against FV and prothrombin.

Anti-prothrombin antibodies occur most commonly in association with antiphospholipid antibody syndrome (see Chapter 94). Although antiprothrombin antibodies have been purported to be associated with lupus anticoagulant activity and a prothrombotic state, rarely these antibodies lead to increased clearance of prothrombin resulting in a bleeding diathesis. On laboratory testing, both the PT and PTT will be prolonged and Factor II activity will be reduced. The thrombin time will be unaffected.

## Recommended Reading

Collins PW. (2007). Treatment of acquired hemophilia A. *J Thromb Haemost* **5**, 893–900.

Favaloro EJ, Posen J, Ramakrishna R *et al.* (2004). Factor V inhibitors: rare or not so uncommon? A multi-laboratory investigation. *Blood Coag Fibrinol* **15**, 637–647.

Franchini M, Lippi G. (2008). How I treat acquired Factor VIII inhibitors. *Blood* **112**, 250–255.

Lollar P. (2005). Pathogenic antibodies to coagulation factors. Part II. Fibrinogen, prothrombin, thrombin, Factor V, Factor XI, Factor XII, Factor XIII, the protein C system and von Willebrand Factor. *J Thromb Haemost* **3**, 1385–1391.

# CHAPTER 112
# Introduction to Coagulation Testing

James C. Zimring, MD,PhD

The laboratory analysis of *in vitro* coagulation function can be, in itself, a fairly complicated task. Even more daunting is the ultimate goal of predicting the capacities and tendencies of hemostasis and thrombosis in the patient, based upon extrapolation from *in vitro* data. The intrinsic difficulties are due, at least in part, to the fact that regulation of coagulation is a function of the coordinated interaction of blood phase proteins, circulating cells, cells of the vasculature, and extracellular matrix proteins in the vessel wall. The clinical labs are restricted to the analysis of what can be obtained from a blood sample. Such sampling provides a fair measure of soluble circulating coagulation proteins, a reasonable sampling of circulating cells, and very poor or absent sampling of vascular elements.

In addition to the above limitations in sampling, coagulation testing is further compromised by the necessity of a highly artificial testing environment (i.e. *in vitro* conditions). For example, analysis of specimens in a static silicate tube at room temperature and in the absence of the vasculature is a poor approximation of dynamic intravascular blood circulation over living endothelium and exposed extracellular matrix. Coagulation testing is further affected by the intrinsically labile nature of the analytes being studied. Many of the soluble plasma factors are proteases that have evolved specifically to destroy other members of the same process being analyzed. Similarly, platelets are intrinsically unstable cells, which can rapidly activate, degranulate and/or die during storage.

Additional challenges are presented by historical difficulties in standardizing the reagents and methods utilized in coagulation testing. Reagents can differ substantially depending upon source, or even from the same source over time. In the case of the prothrombin time (PT), the international normalization ratio (INR) has been implemented to minimize lab-to-lab variation with regard to warfarin monitoring. However, such stringent normalization has not been applied to many of the other systems used in coagulation laboratory medicine.

Listing the above difficulties in coagulation testing is not to imply in any way that coagulation testing is not medically meaningful. In contrast, it can be essential to the diagnosis and management of a variety of hemophilias, thrombophilias and more complicated coagulopathic states. Moreover, with time, sophistication of laboratory methods and understanding of human coagulation medicine has begun to solve several of the aforementioned problems. For example, while direct sampling of patient endothelium is certainly not appropriate, it is now possible to screen for certain genetic traits that are known to correlate with endothelial dysfunction regarding coagulation and hemostasis. Moreover, increased sophistication of measuring coagulation parameters in whole blood has helped to decrease some of the artificial components of *in vitro* testing. The development of peptide substrates with colormetric readouts has helped to allow direct measurement of factor activity and decreased interference at different points in the coagulation cascade. Interference by confounding variables has been further decreased by the progressive development and implementation of antibody-based assays that measure the amount of protein based upon immunoreactivity. Finally, the identification of genetic polymorphisms that are associated with defects in coagulation has contributed to diagnosis of coagulopathy in assays independent of biological activity.

Although there are still many problems to solve in the development of comprehensive laboratory diagnosis and management of coagulation, the above advances and ongoing innovation along these lines is greatly improving laboratory analysis of coagulation medicine. Of course, great caution must be taken to require evidence-based criteria (whenever possible) when altering medical management based upon novel testing methodologies. Given the complexities of coagulation medicine, perhaps more than in many other fields, what seems "logical" based upon our current understanding can turn out not to have diagnostic validity. For example, the rationale behind the bleeding time test is that if attempting to predict how a patient will respond to surgery, then a controlled incision would be a good test for hemostasis. However, it is now established that the bleeding time test has little to no predictive value in screening for increased risk of intraoperative bleeding, in patients without other indications of bleeding tendencies. Thus, careful attention must be paid to the rapidly evolving literature in this dynamic field, and stringent criteria must be applied to the diagnostic validity of new and existing methodologies.

## Analysis of Proteins Involved in Coagulation:

**Measurement of Activity:**  Traditionally, coagulation-based tests have consisted of biological assays that measure functions associated with clotting. At the most primitive and rudimentary level, coagulation assays simply consist of drawing blood and watching it clot. Chelating out the calcium and initiating the clot under controlled conditions in the lab using re-calcification plus initiators of clotting (e.g. PT and activated partial thromboplastin time [PTT]) allows a more standardized approach, and clotting has been measured in a number of different ways, such as via visualization, light absorption and mechanical impedance. As the technology has developed, more controlled and sophisticated methodologies have been incorporated to measure specific factors and factor inhibitors by using factor-deficient plasma and mixing studies; however, numerous points of interference and obfuscation persist due to the reliance upon the coordinated interaction of multiple clotting factors in addition to the particular substance being studied. This latter difficulty has in some ways been reduced by using chromogenic peptide (amidolytic) substrates, which allow the activity determination of a single factor in an assay that does not depend on all of the other factors being present and functional. However, amidolytic substrates introduce their own potential artifacts. Peptide substrates are much smaller than the natural substrate, resulting in two main changes:

1. The enzyme being studied may require a binding site on the natural substrate, in addition to the enzymatic target. As the peptide may be cleaved without the requirement for such binding sites, the enzymatic activity being studied may occur for the peptide but would not occur for the natural substrate.

2. Antibody inhibitors may bind to an enzyme and block access of the natural substrate through steric hindrance, but a small peptide substrate may be able nevertheless to bypass the inhibitor and slip into the catalytic pocket. In this way the peptide-based assay would not detect the inhibitor.

The frequency of these two occurrences is quite low, and they do not substantially detract from the significant advance made by amidolytic substrates; nevertheless, such potential artifacts must always be considered in the interpretation of peptide substrate based assays.

**Measurement of Protein:** Over the past several decades, laboratory analysis of coagulation has been revolutionized by the cloning/identification of proteins involved in coagulation, along with the advent of monoclonal antibodies and immunoassays. This technology allows the direct measurement of proteins involved in coagulation, regardless of their biological activity. Such tools allow the differentiation of type I (quantitative) versus type II (qualitative) defects for a number of different factors. Moreover, it has allowed the rapid and accurate quantification of factors that have difficult to measure activities, and for products of coagulation that don't have an obvious natural activity (e.g. fibrin degradation products). As with all systems, antibody-based assays have potential pitfalls. For example, false positives can be caused by heterophilic antibodies, and false negatives can occur due to prozone effects (in the case of agglutination-based assays) and the hook effect (in the case of some ELISA-based systems). Such potential problems must always be considered in the interpretation of such assay systems.

**Measurement of Gene Sequence:** The relative explosion in genomics information and analytic capabilities has allowed for diagnostic analysis of genetic traits as opposed to gene products. Although complete genomic sequencing has not yet made its way to the clinical labs, analysis of specific polymorphisms known to encode for defective or altered gene products or gene regulatory elements is now routine practice. Perhaps most ubiquitous is analysis of polymorphisms associated with thrombosis, such as Factor V Leiden and the prothrombin gene mutations. Typically, these traits are diagnosed by PCR-based analysis. When run properly, PCR is a robust and accurate assay system. However, given its extreme sensitivity, it is important always to be mindful of potential false positives due to sample contamination. Also, as it is an amplification-based technology, the presence of amplification inhibitors must be controlled for. However, these caveats notwithstanding, genetic analysis has revolutionized coagulation medicine. The advent of DNA analysis, in combination with an expanding database of known functional polymorphisms, promises to extend our diagnostic capabilities past the activities that can be measured in blood; indeed, to the extent that genetic traits influence the role of endothelium and vascular structure in coagulation, this provides a route to allow diagnostic evaluation of *in vivo* factors not previously amenable to analysis.

**Analysis of Cellular Biology in Coagulation:** Clearly, the platelet has long enjoyed a role as the central cellular player in hemostasis. Diagnostic analysis of platelet function has traditionally been carried out in large part by aggregometry-based assays in which a panel of agonists is added to blood fractions enriched in platelets and aggregation is observed. This approach has been very efficient in diagnosing a series of well-defined defects in platelets, such as Bernard-Soulier syndrome and Glanzmann's Thrombasthenia. Aggregometry has also been very useful in diagnosing von Willebrand disease, in which certain activities and functions of von Willebrand Factor can be measured by platelet clumping in response to ristocetin.

Despite the accuracy of aggregometry-based assays in diagnosing a select group of defined disorders, perhaps more often than not, aggregometry picks up vague platelet dysfunctions that do not fit into a clear pathological category. Defects such as storage pool disease and certain granule secretion defects can be detected by aggregometry, but may or may not correlate to a functional defect in a patient. Likewise, hyperaggregatable platelets can be detected, but they may or may not correlate to increased risk of thrombotic disease.

Flow cytometry based assays have added a higher level of sophistication to the diagnostic arsenal. Moreover, in an attempt to minimize the artificial variables introduced during testing, additional methodologies that use less fractionated or even whole blood (and under flow conditions) have been devised (such as the platelet function analyzer and thromboelastography). Although such approaches may have some advantages, similar to aggregometry, these approaches can detect a large number of non-specific defects with unclear clinical significance. Perhaps more importantly, in the absence of a patient history or family history of bleeding, these tests have poor predictive value regarding the likelihood of a hemostatic defect in a given patient (e.g. in a pre-surgery work-up). This is in no way to suggest that platelet analysis should not be carried out; on the contrary, it can lead to invaluable information in the setting of certain pathologies. However, in many settings, the current methodologies still have poor predictive value. Newer and more powerful assay systems are needed, and, as a number are in development, careful attention to the evolving literature is required to keep abreast of ongoing developments.

**Analysis of Therapeutic Interventions:** There is an ongoing expansion of our understanding of anticoagulation and thrombolysis therapy. We have an increasing understanding of the toxicology of anticoagulants (e.g. heparin-induced thrombocytopenia). We likewise have begun to enter an era of pharmacogenomics, such as linking genetic polymorphisms in cytochrome P450 to increased metabolism and resistance to warfarin. Finally, an expanding number of anticoagulants are increasingly becoming available to the clinician, many of which require distinct approaches to monitoring and balancing therapeutic effects. Each of these topics is covered in subsequent chapters. It is the role of the laboratory to assist in monitoring anticoagulant therapy, in addition to its more traditional role of diagnostics.

**Integration of Coagulation Labs with Patient Care:** Diagnosis and management of coagulation medicine at the laboratory level can be a convoluted and confusing topic. A wide variety of different methodologies is used to generate the data, each of which has its own unique susceptibility to artifact. Moreover, perhaps more than many other laboratory disciplines, interpretation of coagulation tests must take place in an integrated fashion, combining multiple different laboratory parameters with clinical history and physical examination. A patient with increased sodium is hypernatremic, regardless of the etiology. In contrast, an increasing baseline coagulation with escalating fibrinogen and C4b binding protein can be a normal finding during pregnancy, but may indicate a hypercoaguable patient with a substantial risk of thrombosis in other settings.

The landscape of coagulation medicine is further complicated by the fact that evidence-based data are lacking for many assays of coagulation, especially given the rapid pace of development of new systems or modification of existing tests. The laboratory pathologist must be able to integrate multiple analytic systems and communicate these findings to the clinician in the context of the patient presentation. There is as much harm to be done in over-interpreting as in under-interpreting the data. While diagnosing a thrombophilia may save a patient from a thrombotic event, incorrect assignment of a hypercoaguable state may result in a patient unnecessarily receiving lifelong warfarin, which in of itself carries a substantial risk for serious bleeding episodes. The up-to-date coagulation laboratory will pay ongoing attention to evolving knowledge concerning what can and cannot be confidently determined with existing assays, and critically consider new diagnostic tests as they become available.

# CHAPTER 113

# Prothrombin Time and Activated Partial Thromboplastin Time

James C. Zimring, MD, PhD

The prothrombin time (PT) and activated partial thromboplastin time (PTT) are two separate but related *in vitro* laboratory assays that are used to evaluate a wide variety of coagulation factors. Factors involved uniquely in the PTT are traditionally referred to as *intrinsic* pathway factors, whereas factors unique to the PT are traditionally referred to as *extrinsic* pathway factors. Factors involved in both pathways are termed *common* pathway factors (Figure 113.1).

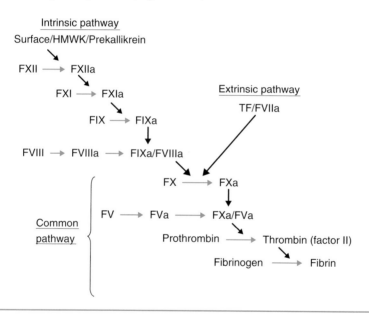

FIGURE 113.1 Coagulation cascade. Diagonal lines indicate activation, vertical lines indicate combination with another factor.

Many but not all of the factors detected by the PT and PTT are essential to maintaining normal hemostasis. The PT and PTT pathways do not directly reflect *in vivo* hemostasis; nevertheless, they represent useful laboratory tests by which the function of factors can be measured. Because multiple factors are simultaneously required for normal clotting times, the PT and PTT are used as general screening tests for coagulation factors.

<u>**Prothrombin Time:**</u>  The PT represents the *in vitro* analysis of coagulation factors in the extrinsic (Factor VII) and common pathways (Factors II, V, X and fibrinogen).

Although the PT will be prolonged by substantial decreases in fibrinogen, it is not sensitive to moderately decreased fibrinogen; thus, a specific fibrinogen assay should be run to evaluate fibrinogen levels and function.

**Indications:** The indications for PT are: 1. Assessment of coagulation factor function in patients being tested for a bleeding disorder, 2. Presurgical assessment of baseline coagulation, 3. Monitoring of coagulation during surgical procedures, and 4. Monitoring of anticoagulation therapy (used extensively to monitor warfarin therapy).

**Method:** Clotting is initiated by the addition of tissue factor (TF). Traditionally, homogenized brain tissue was used as the source of TF; however, recombinant purified TF is used in some modern assay systems. The PT is determined by measuring the amount of time from TF addition to clotting of the sample.

**Sources of Error:**

*Falsely Prolonged PT:* Citrate prevents coagulation by chelating calcium, and is used as the anticoagulant when collecting a specimen for PT. In the controlled setting of the laboratory, a fixed amount of calcium is then added back to the specimen so as to exceed the capacity of citrate to a degree that restores normal calcium levels – so called *recalcified plasma.* In the event that too little sample is drawn into the coagulation tube (i.e. a "short draw") then there will be a relative excess of citrate to plasma calcium, and the calcium added in the lab may not be sufficient to overcome the excess citrate. This can lead to an artificially prolonged PT. Similarly, if the patient is polycythemic (hematocrit over 60%) then the relative amount of plasma in a normal volume draw is decreased. Thus, specimens from polycythemic patients can likewise have a falsely prolonged PT, which may lead to inappropriate administration of procoagulant therapies.

**International Standards:** The PT is the predominant laboratory assay used to monitor warfarin therapy. However, standardization has traditionally been a problem, due to intrinsic variability from lab to lab. To remedy this, the International Normalized Ratio (INR) is used, which calculates a standard value as a function of the normal range in a given laboratory and a correction factor determined for each batch of reagents. Importantly, the INR is used to standardize anticoagulation in warfarin-treated patients, and is not designed to assess other coagulopathic states (e.g. coagulopathy of hepatic failure).

**Activated Partial Thromboplastin Time:** The PPT represents the *in vitro* analysis of coagulation factors in the intrinsic (HMWK, prekallikrien, Factors XII, XI, IX and VIII) and common pathway factors (Factors II, V, X and fibrinogen). Although the PTT, like the PT, will be prolonged by substantial decreases in fibrinogen, it is not sensitive to only moderately decreased fibrinogen; thus, a specific fibrinogen assay should be run to evaluate fibrinogen level and function.

**Indications:** The indications for PTT are: 1. Assessment of coagulation factor function in patients being tested for a bleeding disorder, 2. Presurgical assessment of baseline coagulation, 3. Monitoring of coagulation during surgical procedures, and 4. Monitoring of anticoagulation therapy (used extensively to monitor heparin therapy, although a heparin level is preferable to a PTT in many cases).

Method: Clotting is initiated by the addition of a source of phospholipid, from which tissue factor is absent, and a contact initiator, such as silica. The use of different initiators can lead to different sensitivity to interfering factors or to even factor levels. This will vary based upon the assay components, which should be characterized by the manufacturer.

Sources of Error:

*Falsely Prolonged PTT:*

1. As with the PT, citrate is used to prevent clotting of the collected specimen and recalcified plasma is used to perform the assay. Thus, as with the PT, a short draw of a specimen or a high hematocrit can give an artificially prolonged PTT (see PT above).

2. The PTT is very sensitive to the effects of heparin, which is why it is used to monitor heparin therapy. However, this also provides the opportunity for contaminating heparin to give a falsely prolonged PTT. As heparin is used in flushing intravenous lines, great care must be taken to avoid drawing a specimen from a line containing heparin. Likewise, drawing a specimen from a vein downstream of the infusion of heparin may result in contamination. Additional heparin-like substances (e.g. danaproid, etc.) can have a similar effect on PTT.

3. Antiphospholipid antibodies (lupus anticoagulants) prolong the PTT *in vitro*, but can cause thrombosis *in vivo*. Thus, the presence of an undetected lupus anticoagulant may lead to the opposite therapeutic maneuver from that which should be taken. Lupus anticoagulants should be suspected any time that an isolated prolonged PTT is observed without an alternate explanation.

## Interpretation of PT and PTT Tests:

Prolonged PTT with Normal PT: With the exception of a source of error (see above), an isolated prolonged PTT indicates a defect in the activity of one of the intrinsic pathway factors. A defect in a common pathway factor is essentially excluded due to the normal PT. The defect can be due to decreased production, which is typically a congenital deficiency (e.g. hemophilia). Decreased production can also be acquired (e.g. hepatic failure); however, in such cases an isolated intrinsic factor deficiency is not observed and either both the PTT and PT are prolonged, or the PT can be selectively prolonged in early liver failure (see below). However, acquired decreased activity of a specific intrinsic pathway factor can be observed due to an autoantibody inhibitor. Distinguishing between a deficiency and an inhibitor can be accomplished by using mixing studies. Finally, an isolated prolonged PTT may indicate the presence of a lupus anticoagulant, which is a risk factor for thrombosis.

The clinical significance of an isolated prolonged PTT is highly variable from patient to patient. The reason for this is that whereas defects in Factors VIII, IX or XI can lead to profound bleeding problems, defects in Factor XII, HMWK or prekallikrein do not lead to problems in hemostasis. An isolated prolonged PTT can be caused due to a deficiency in any of these factors. Thus, in the context of an isolated prolonged PTT, specific factor assays should be performed to identify whether a defective component is likely to be clinically significant. In addition, physical examination and clinical history of bleeding are key components in guiding these analyses.

**Prolonged PT with a Normal PTT:** With the exception of a source of error (see above), an isolated prolonged PT suggests a defect in Factor VII. A defect in a common pathway factor is excluded due to the normal PTT. However, it should be noted that Factor VII has a short half-life compared to other coagulation factors. Thus, an isolated prolonged PT may be the early finding in a generalized coagulopathy that leads to decreased amounts of multiple factors. In addition, while Factors II and X are required for the PTT as part of the common pathway, higher quantities of Factors II and X are required for the PT. This explains why the PT is selectively prolonged during warfarin therapy or in vitamin K deficient states. Thus, an isolated prolonged PT may also reflect vitamin K deficiency or early liver disease.

**Prolongation of the PT and PTT:** With the exception of a source of error (see above), dual prolongation of the PT and PTT can indicate a generalized coagulopathy that is affecting both the intrinsic and extrinsic pathways (e.g. liver failure or a consumptive coagulopathy). However, this pattern can also be seen in an isolated defect of a factor in the common pathway (i.e. Factors X, V or II, or fibrinogen). Fibrinogen deficiencies are most common, and can be seen either due to decreased levels (hypofibrinogenemia) or due to abnormal fibrinogen that cannot function as a normal substrate for thrombin (dysfibrinogenemia). These two latter scenarios can be distinguished by performing antigen-based fibrinogen assays in combination with a thrombin time (TT) or reptilase assay. It is important to note that because the PT requires higher levels of Factors II and X than does the PTT, a common factor defect (i.e. Factors II or X) may initially show up as a selective prolonged PT. Massive heparin overdose/contamination may also prolong both the PT and the PTT. Finally, both the PT and PTT can be prolonged due to therapeutic use of a direct thrombin inhibitor (e.g. argatroban).

**Shortening of the PT and PTT:** On occasion, patients with a short PT or PTT will be encountered. The clinical significance of a short PT is unknown, but may occur after use of rFVIIa. However, a short PTT may indicate the presence of elevated levels of Factor VIII. Factor VIII is an acute phase reactant that is induced to high levels by inflammation and acute illness. Thus, an elevated Factor VIII is common in the setting of a hospitalized patient, and a PTT may be transiently shortened during illness. A persistent elevated Factor VIII, especially in a patient not experiencing inflammation, is an independent risk factor for thrombosis. Based upon the clinical setting, a shortened PTT may justify an evaluation of Factor VIII levels and assessment for risk of thrombosis.

# CHAPTER 114

# Platelet Count

James C. Zimring, MD, PhD

The platelet count, also described as the quantitative analysis of platelets, is routinely performed as part of the complete blood count and in the evaluation of bleeding patients or those at risk for bleeding (e.g. preoperatively or in patients undergoing chemotherapy). In addition, the platelet count is used to evaluate patients with myeloproliferative disorders or thrombosis who may have thrombocytosis.

**Methods:** Traditionally, manual counting of platelets by microscopy has been the gold standard for quantitative assessment of circulating platelets. This can be performed by phase contrast using a hemocytometer, but can also be carried out by reviewing the peripheral smear. Although manual counts may still be performed routinely in small labs or for confirmation of automated results, the large volume of specimens in busy laboratories almost uniformly requires automation. A variety of commercially available instruments is available. A general discussion of the methodologies and technical issues for different methodologies is presented below.

**Electronic Impedance Aperture:** *Coulter counters* utilize this particular methodology, which consists of detecting resistance across two electrodes that are conducting a direct current. As a cell passes through the aperture the resistance increases, indicating the detection of a cell, referred to as a *counting event*. The amplitude of the resistance is a function of the size of the cell. As platelets are substantially smaller than other cell types, accurate platelet counts can be determined by counting the number of events that fall within the size-range of platelets. This method was the first to be automated.

**Optical and Flow-cytometry Based Methods:** A variety of systems take advantage of flow cytometry, adding different parameters, and are sometimes used in combination with impedance technology. In its simplest form, individual particles from a blood specimen are pulsed with a laser. Light scatter can be used to give an estimation of event size. This technique has been called *optical platelet counting*. The addition of two-dimensional light scatter also allows measurement of the density of a cell using the refractive index. Addition of a fluorescent dye that binds nucleic acids allows separation of RBCs, reticulocytes and platelets. Finally, additional fine specificity can be provided by staining with fluorescent antibodies against platelet-specific antigens (e.g. anti-CD61 and anti-CD41a). Different instruments utilize distinct combinations of these approaches; however, as the measurement parameters for platelets are known for each of the above variables, gating on the platelet population is achieved by each approach, albeit with different degrees of specificity.

**Sources of Error:** There are three main sources of error in automated platelet counting, as described below.

**Mistakenly Counting a Small Particle that is not a Platelet:** Ultimately, each of the automated counting techniques uses size as a determining factor in identifying a counting event as a platelet. However, there are additional particles in the blood that can fall within the size-range of a normal platelet. In some instruments, highly microcytic RBCs may fall within the platelet gate. In addition, if a patient is sufficiently septicemic, circulating microbes may be counted as platelets. Moreover, during treatment of neoplasia or during tissue injury, products of cell lysis, including apoptotic bodies, may be mistakenly counted as platelets.

A histogram of the events in the platelet gate is typically generated, and an abnormal distribution will result in a "flag" that an abnormality is present. In addition, the more specific the criteria utilized to determine the platelet count, the less likely an instrument is to mistake a non-platelet small body as a platelet (e.g. erroneous small bodies may be the size of a platelet but have different light side scatter, and would not be bound by platelet specific antibodies).

A flag from the instrument that an abnormality exists should result in a detailed analysis of the specimen by the laboratory, including manual analysis by microscopy to assess the presence of platelets and/or other small particles.

Perhaps most importantly, the accuracy of the platelet count should be questioned in any patient who has platelet/vessel type bleeding (i.e. mucocutaneous bleeding) with a normal platelet count by automated instrumentation, and who has no alternate explanation for such bleeding. Manual assessment of platelets in such cases is warranted. Of course, it is possible to have a normal platelet count with abnormal platelet function. However, if a discrepancy is found in automated platelet counting, subsequent specimens from the patient should be analyzed manually, unless a remedy to the automated artifact can be devised.

**Mistakenly Missing a Large Platelet:** There are several scenarios in which platelets are unusually large including Bernard-Soulier syndrome, May-Hegglin anomaly bone marrow proliferative disorders, and in some cases of ITP. The platelet count may be artificially low if the platelets are sufficiently large to fall outside the platelet counting gate. As above, the likelihood of such an error depends upon the specificity of parameters utilized to establish the platelet gate.

**Platelet Clustering or Aggregation (Pseudothrombocytopenia):** Platelets can become activated during blood-drawing or storage of the sample, which can result in aggregation in the specimen tube. This may be due to inadequate sample mixing resulting in an uneven distribution of anticoagulant. In addition, it has been noted that platelets from some patients cluster in an EDTA-dependent fashion, which can often be remedied by substituting a citrate-based anticoagulant. Also, *ex vivo* platelet aggregation due to platelet-specific cold agglutinins has been reported. Because the size of the aggregation/clusters can shift outside of the platelet gate and because multiple platelets are aggregated together, the reported platelet count can be substantially lower than the actual platelet count. A compensatory increase in the leukocyte count can also be observed, as these aggregates may be counted as leukocytes. Some (but not all) automated counters can detect alterations in size distribution and generate an error flag, which should always be investigated for potential pseudothrombocytopenia. Additionally, patients with inordinately low platelet counts without any symptoms of thrombocytopenia may be considered as potentially having pseudothrombocytopenia. Pseudothrombocytopenia can result in both inappropriate transfusion of platelets and additional unnecessary diagnostic evaluations. The presence of psuedothrombocyto-penia can be assessed by a manual smear, which can visualize platelet clusters.

## Recommended Reading

Briggs C, Harrison P, Machin SJ. (2007). Continuing developments with the automated platelet count. *Intl J Lab Hematol* **29**, 77–91.

Segal HC, Briggs C, Kunka S *et al.* (2005). Accuracy of platelet counting haematology analysers in severe thrombocytopenia and potential impact on platelet transfusion. *Br J Haematol* **128**, 520–525.

# CHAPTER 115

# Global Tests of Primary Hemostasis

Connie H. Miller, PhD

The bleeding time (BT) and platelet function analyzer (PFA) are used as screening tests of primary hemostasis. The bleeding time is an *in vivo* screening test of primary hemostasis which measures the time to cessation of bleeding of a uniform skin cut. It is used to assess the integrity of platelet, plasma and vessel-wall components. The platelet function analyzer is an *in vitro* test of platelet plug formation which measures the time to occlusion of a window in a coated membrane through which citrated blood is forced at high shear rate. It is used to test for platelet function defects.

**Bleeding Time:** The BT is performed by applying a blood-pressure cuff inflated to 40 mmHg to the arm, making a uniform cut on the volar surface of the forearm, blotting the top of the blood drop with standard filter paper every 30 seconds, and timing until the cessation of bleeding. Spring-loaded devices with retracting blades are most commonly used.

**Test Performance:** BT performance is highly operator-dependent. A review of evidence in 1998 concluded that:

1. In the absence of a bleeding history, BT is not a useful predictor of the risk of hemorrhage with surgery;
2. A normal BT does not exclude the possibility of excessive hemorrhage with invasive procedures;
3. BT cannot be used to reliably identify patients who may have recently ingested aspirin or non-steroidal anti-inflammatory agents; and
4. The best preoperative screen is a careful clinical history, with performance of screening or diagnostic tests when the history warrants.

A test of primary hemostasis remains an important part of evaluation of a patient with a personal or family history of excessive bleeding, and also for screening under certain circumstances with a high risk of bleeding – such as tonsillectomy and adenoidectomy in young children. Today, the BT has largely been replaced by *in vitro* tests of platelet plug formation, usually the PFA (described below), which is less invasive, easier to perform and more reproducible.

BT may result in scarring or keloid formation. In severe disorders, such as von Willebrand disease (VWD) Type 3, BT cuts may bleed profusely and be difficult to close. BT is prolonged at a platelet count below 50,000–100,000/µl, depending on the cause of thrombocytopenia; in more severe forms of VWD; and in major platelet function disorders. It is often normal in mild platelet function defects and in VWD Type 1. Anti-platelet drugs variably prolong the BT.

**Sources of Error:** False positives can be caused by poor operator technique. In addition, connective tissue disorders can result in a prolonged BT despite normal platelet function. Although such a finding does not reflect on the ability of platelets to participate in hemostasis, it may nevertheless demonstrate bleeding tendency due to connective tissue defects. False negatives can be caused by poor operator technique or a mild disorder.

Quality Assurance: Controls are not usually measured; the reference range should be calculated by each laboratory.

Platelet Function Analyzer: The PFA uses whole blood which is forced at high shear rate through a window in a membrane coated with collagen and epinephrine (CEPI), or collagen and adenosine diphosphate (CADP). Platelets adhere and aggregate to produce closure, measured as closure time (CT).

Test Performance: Coefficient of variation (CV) ranges from 6% to 13% for normal specimens. CT appears to be inversely correlated with functional von Willebrand Factor, and is equivalent to or better than BT as a screening tool for VWD, although both lack sensitivity for platelet function disorders. In vitro tests will fail to detect defects of the vessel wall in connective tissue disorders, such as Ehlers-Danlos syndrome, which may result in prolonged BT. Like the BT, the CT lacks sufficient sensitivity and specificity to be used alone for screening of individuals for platelet disorders, and should be used along with clinical history to determine which patients should undergo diagnostic studies. Few studies have as yet been undertaken to establish a clear role for PFA in predicting clinical outcomes and monitoring therapy.

Sources of Error: False positives are seen in specimens older than 4 hours, those sent through pneumatic tubes, and those with low hematocrit. False negative may be a result of mild disease.

Quality Assurance: No quality control materials are provided. Reference ranges should be calculated by each laboratory.

### Recommended Reading

Harrison P. (2005). The role of PFA-100® testing in the investigation and management of haemostatic defects in children and adults. *Br J Haematol* **130**, 3–10.

Hayward CPM, Harrison P, Cattaneo M *et al.* (2006). Platelet function analyzer (PFA)-100® closure time in the evaluation of platelet disorders and platelet function. *J Thromb Haemost* **4**, 312–319.

Peterson P, Hayes TE, Arkin CF *et al.* (1998). The preoperative bleeding time lacks clinical benefit. *Arch Surg* **133**, 134–139.

# CHAPTER 116

# Platelet Aggregation Studies

Connie H. Miller, PhD

The platelet is a dynamic structure, covered externally with glycoprotein (GP) receptors. Internally, the platelet possesses alpha ($\alpha$) granules, which contain a number of proteins, including von Willebrand Factor (VWF), fibronectin and fibrinogen; and dense ($\delta$) granules, containing adenosine triphosphate (ATP), adenosine diphosphate (ADP), serotonin, pyrophosphate, magnesium and calcium. Primary hemostasis is initiated by platelet adhesion to vessel-wall components at the site of vessel injury. VWF is key for this process at a high shear rate. Platelet activation occurs through exposure to collagen and thrombin, and leads to release of granule contents. Aggregation of additional platelets produces a physical barrier. Platelets participate in clot formation through release of stored factors and provision of a surface for assembly of coagulation factor complexes to complete the platelet plug.

*In vitro* platelet aggregation (PAGG) testing measures the response of platelets to added aggregating agents, or agonists, with measurements that primarily reflect activation, aggregation and, with some instruments, secretion. Although PAGG testing is both time- and labor-intensive, requires use of fresh blood and may be difficult to interpret, it remains the gold standard for assessing platelet function.

## Platelet Aggregation by Light-transmission Aggregometry (LTA):
Traditionally, PAGG has been tested by light-transmission aggregometry (LTA), which measures the change in optical density (OD) of stirred platelet-rich plasma (PRP) when an aggregating agent, or agonist, is added (Figure 116.1). As the platelets aggregate, more light is transmitted due to decreased absorbance as a result of decreased platelet dispersion. The patient's PRP is set as 0% aggregation and autologous platelet-poor plasma (PPP) at 100%. A curve of OD versus time is recorded which may reveal lag phase, shape change, and first and second waves of aggregation. Calculated parameters include the aggregation rate (slope), maximum amplitude, and percent aggregation.

## Platelet Aggregation by Whole Blood Aggregometry (WBA): In LTA, the
centrifugation of plasma to produce PRP and the use of additionally centrifuged PPP or buffer to standardize the platelet count are manipulations that may alter platelet response or result in loss of platelet subpopulations. Measurement of PAGG in whole blood eliminates these steps and maintains the physiologic blood-cell milieu. In addition, WBA requires only one-fourth the volume of blood required for LTA, making it particularly valuable in the pediatric setting.

WBA is measured by the impedance method. Platelets adhere in a monolayer to wires separated by a gap across which a voltage is applied. When an agonist is added, additional platelets aggregate to the layer, increasing the resistance across the gap, measured in ohms. Change in impedance is plotted as a function of time, giving

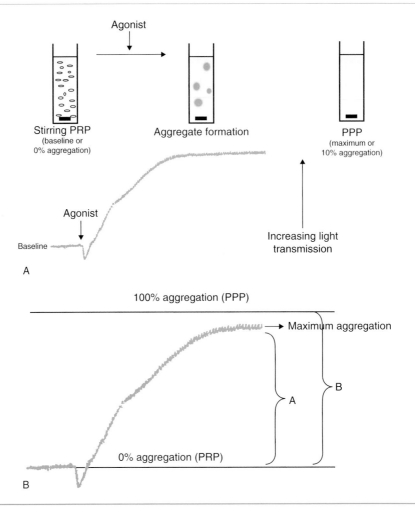

FIGURE 116.1 Process of light-transmission aggregometry (LTA). Stirred platelet-rich plasma (PRP) is set as 0% aggregation. After an agonist is added, aggregates form and light transmission increases. The maximal amount of light transmission is seen in autologous platelet-poor plasma (PPP), which is set as 100% aggregation. % platelet aggregation is calculated as the distance from 0 to maximum aggregation divided by the distance from 0 to 100% aggregation × 100. Adapted from Jennings and White (2007).

similar curves to those produced by OD changes in LTA. WBA is more sensitive to most agonists, and requires lower concentrations to achieve aggregation. The second wave of aggregation is not seen in WBA curves, but such visualization is not necessary if release is measured directly rather than inferred from the curves. Appearance of curves and numerical results differ somewhat between LTA and WBA, requiring method-specific interpretation.

**Platelet Adenosine Triphosphate (ATP) Release:** Neither LTA nor WBA is sensitive to all cases of storage pool and release defects. PAGG and platelet nucleotide release may be measured simultaneously in specialized instruments. The secretion of

ATP is quantitated by its ability to cause the firefly enzyme luciferase to cleave its substrate luciferin and generate luminescence. The amount of luminescence is compared with that generated by a standard amount of ATP added to patient blood or PRP, and is reported in nanomoles (nm) (Figure 116.2).

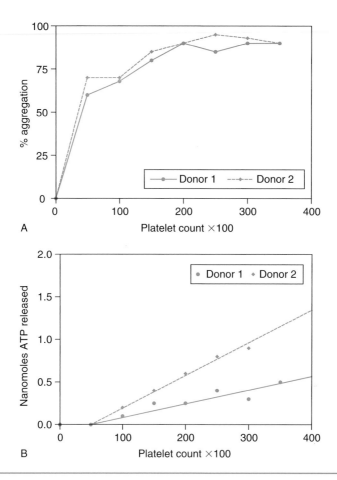

FIGURE 116.2 Platelet aggregation (A) and ATP release (B) with 10-μM ADP as a function of platelet count. Aggregation occurs normally above a platelet count of 50,000. ATP release is linearly related to platelet count. From Jennings and White (1999).

**Interpretation of Results from LTA, WBA and ATP Release:** PAGG and ATP release are evaluated with a panel of agonists. For some agonists, several concentrations may be used to define the threshold concentration at which aggregation occurs, or standard concentrations may be set. The latter facilitates comparison with ATP release data. The curves should be inspected visually, but today's instruments provide quantitative measures, which can be compared to reference ranges. Reference ranges differ significantly among laboratories, and should be established in each laboratory with the instrument and reagent system used.

Certain agonists are commonly used. Platelet function is evaluated by both the responses to a given agonist and the general pattern of responses to multiple agonists. A detailed description of these patterns and corresponding diagnoses can be found in Chapters 117 and 118.

*ADP* may be used at concentrations of 1–10 μM. Low concentrations produce a monophasic or biphasic curve with disaggregation. Higher concentrations show a single wave of aggregation, post-release. Studies of anti-GPIIb-IIIa drugs have used 10–20 μM.

*Epinephrine* is a mild agonist used at 5–10 μM. It produces a slight initial response and a second-wave response that is highly variable. The presence of decreased aggregation with epinephrine alone is quite common in normal individuals. Epinephrine is not used in WBA.

*Collagen*, a strong agonist, is used at concentrations of 1–5 μg/ml. It produces adhesion to collagen fibrils, shape change, and release, followed by aggregation.

*Arachidonic acid*, used at 0.5 mM, reacts with cyclo-oxygenase to produce thromboxane $A_2$. This reaction is inhibited by aspirin and other anti-platelet drugs. The effect of aspirin is not reversible, and continues through the life of the platelet. Effects may still be seen at 2 weeks post-ingestion. WBA is more sensitive to aspirin than LTA.

*Ristocetin* is sometimes used to screen for von Willebrand disease (VWD), but it is not very efficient for that purpose. VWD type 1 patients often have normal platelet aggregation with standard concentrations of ristocetin. Plasma tests for VWF antigen and ristocetin cofactor are more sensitive and specific. Ristocetin is, however, useful for diagnosis of Bernard-Soulier syndrome at concentrations of 1.25 mg/ml and above. Low concentrations, 0.5 mg/ml and below, are used to detect the hyperaggregability of platelets in type 2B VWD and platelet-type VWD. ATP release is not measured with ristocetin.

*Thrombin* at 1 U/ml produces the maximum release of ATP from storage. Since thrombin clots plasma or blood, aggregation cannot be measured unless a thrombin analogue is used. Failure to respond to strong thrombin indicates a defect in the platelet storage pool.

*Thrombin receptor activating peptide* (TRAP), used at 5–10 μM, is a synthetic peptide mimicking the sequence of the thrombin protease-activated receptor (PAR-1) after thrombin hydrolysis. It is sometimes used in place of thrombin to generate activation without clotting, primarily in the measurement of the effects of anti-platelet drugs.

## Recommended Reading

Brass LF, Stalker TJ, Zhu L, Woulfe DS. (2007). Signal transduction during platelet plug formation. In: Michelson AD (ed.). *Platelets*, 2nd edition. San Diego, CA: Academic Press, pp. 319–343.

CLSI. (2007). *Platelet Function Testing by Aggregometry*. CLSI document *H58-P*. Wayne, PA: Clinical and Laboratory Standards Institute.

Gurbel PA, Becker RC, Mann KG *et al.* (2007). Platelet function monitoring in patients with coronary artery disease. *J Am Coll Cardiol* **50**, 1822–1834.

Jennings LK, White MM. (1999). *Platelet Protocols*. San Diego, CA: Academic Press.

Jennings LK, White MM. (2007). Platelet aggregation. In: Michelson AD (ed.). *Platelets*, 2nd edition. San Diego, CA: Academic Press, pp. 495–508.

Moffat KA, Ledford-Kraemer MR, Nichols WI, Hayward CP. (2005). Variability in clinical laboratory practice in testing for disorders of platelet function: results of two surveys of the North American Specialized Coagulation Laboratory Association. *Thromb Haemost* **93**, 549–553.

Zhou L, Schmaier AH. (2005). Platelet aggregation testing in platelet-rich plasma: description of procedures with the aim to develop standards in the field. *Am J Clin Pathol* **123**, 172–183.

# CHAPTER 117

# Laboratory Diagnosis of Genetic Platelet Function Defects

Connie H. Miller, PhD

Genetic platelet function defects (PFD) should be considered in patients with symptoms suggestive of a defect in primary hemostasis, particularly if lifelong or familial. Normal platelet function involves over 100 gene products, defects in many of which could potentially cause platelet function disorders. Disorders of major clinical significance (such as Glanzmann thrombasthenia) or striking clinical presentation (such as Hermansky-Pudlak syndrome) have been well characterized but are rare in most populations. Table 117.1 provides a detailed but not exhaustive list of the known disorders categorized by their mechanisms. A number of the disorders listed have been described in only a few individuals. Samples of the platelet aggregation results expected in these disorders are shown in Table 117.2.

**Platelet Aggregation Studies:** Platelet aggregation studies, as described in Chapter 116, are used to determine the cause of a defect in primary hemostasis. When sensitive platelet aggregation studies are performed in large numbers of individuals with mild bleeding symptoms, such as women with menorrhagia, a wide spectrum of functional defects is seen – most as yet uncharacterized. Control individuals also show a frequency of defects approaching 20%, suggesting a high false-positive rate. Indeed, abnormalities detected in the clinical laboratory are much more likely to have an acquired etiology. Results suggesting a specific genetic diagnosis, therefore, should be confirmed by repetition, with drug effects carefully excluded. Most disorders cannot be diagnosed reliably without confirmatory tests such as visualization of granules, receptor studies, or gene sequencing.

**Platelet Function Analyzer:** The platelet function analyzer (PFA) is an *in vitro* test of platelet plug formation by measuring the time to occlusion of a window in a coated membrane through which citrated blood is forced at high shear rate (see Chapter 115), which is used to test for platelet function defects (PFD). The test has been reported to be not sensitive or specific enough to use for screening to determine who should have platelet aggregation studies. It may exclude more severe PFD, such as Glanzmann thrombasthenia and Bernard-Soulier syndrome, but milder defects may be missed.

**Electron Microscopy:** Electron microscopy allows for visualization of dense granules (whole mount) or α granules (transmission), and is used for the diagnosis of storage pool disorders and granule defects.

**DNA Sequence Analysis:** DNA sequence analysis identifies specific regions of involved genes that are associated with PFD through polymerase chain reaction (PCR)

## TABLE 117.1  Genetic Disorders of Platelet Function

| Type of defect | Disorder | Defect in: |
|---|---|---|
| Adhesion defects | von Willebrand disease (VWD) | Plasma von Willebrand factor |
| | Bernard-Soulier syndrome | GPIbα, GPIbβ, GPIX |
| | Platelet-type VWD | GPIb |
| Aggregation defects | Glanzmann thrombasthenia | GPIIb/IIIa |
| | Afibrinogenemia | Plasma fibrinogen |
| Storage pool disorders | | |
| Dense granules | Primary δ SPD | Heterogeneous |
| | Hermansky-Pudlak syndrome | HPS gene |
| | Chediak-Higashi syndrome | Membrane structure |
| | Wiskott-Aldrich syndrome* | WAS protein |
| α granules | Gray platelet syndrome | Granule packaging |
| | Quebec platelet disorder* | u-tPA |
| α/dense granules | αδ SPD | P-selectin ? |
| Receptor defects | Thromboxane A₂ | Thromboxane A₂ receptor |
| | Collagen | Integrin α2β1, GPVI |
| | Adenosine diphosphate (ADP) | P2Y₁₂ |
| | Epinephrine | α-adrenergic receptor |
| | Platelet activating factor | PAF receptor |
| | Serotonin | Serotonin receptor |
| Release defects | Cyclo oxygenase deficiency | Cyclo oxygenase |
| | Thromboxane synthetase deficiency | Thromboxane synthetase |
| G protein defects | Gαδ deficiency | Gαq |
| Cytoskeletal defects | May-Hegglin anomaly | MYH9, non-muscle myosin |
| | Wiskott-Aldrich syndrome* | WAS protein |
| Procoagulant function | Scott syndrome | Phosphatidylserine transport |
| | Quebec platelet disorder* | Multimerin, FV, u-tPA |

*Entered in two categories.
GP, glycoprotein; SPD, storage pool disorder; WAS, Wiskott-Aldrich syndrome; FV, Factor V; u-tPA, urokinase type tissue plasminogen activator.

TABLE 117.2 Examples of Whole Blood Platelet Aggregation (AGG) and ATP Release (REL) Findings in Patients with Known Platelet Function Defects

| | ADP 20 mM | | Collagen 2 µg/ml | | Arachidonic acid 0.5 mM | | Thrombin 1 U/ml | Ristocetin 1.0 mg/ml | Ristocetin 0.25 mg/ml |
|---|---|---|---|---|---|---|---|---|---|
| | AGG | REL | AGG | REL | AGG | REL | REL | AGG | AGG |
| Glanzmann thrombasthenia | A | N | A | N | A | N | N | N* | N |
| Hermansky-Pudlak | N | D | N | D | N | D | D | N* | N |
| Bernard-Soulier | N | N | N | N | N | N | N | A | N |
| VWD Type 2A | N | N | N | N | N | N | N | A | N |
| VWD Type 2B | N | N | N | N | N | N | N | N | I |
| Collagen receptor defect | N | N | D | D | N | N | N | N | N |
| Scott syndrome | N | N | N | N | N | N | N | N | N |

*Qualitatively abnormal pattern with disaggregation.
N, normal; D, decreased; A, absent; I, increased.

and direct sequencing. This method is used to confirm a diagnosis of Bernard-Soulier syndrome, Glanzmann thrombasthenia, platelet-type VWD, and others.

## Interpretation of Results Based Upon Combined Application of the Above Lab Tests:

**Glanzmann Thrombasthenia:** Glanzmann thrombasthenia (GT) exhibits a primary aggregation defect with all agonists, except ristocetin, due to a defect or deficiency of the receptor GP IIb/IIIa. ATP release usually occurs normally. Curves with ristocetin may be irregular and show disaggregation. Afibrinogenemia also results in a similar primary aggregation defect; however, unlike GT, afibrinogenemia will also result in prolongation in coagulation studies (i.e. PT, PTT, thrombin time).

**Storage Pool Disorders:** Storage pool disorders (SPD) may involve granules other than those in platelets, producing multi-system disease. Hermansky-Pudlak and Chediak-Higashi syndromes also produce oculocutaneous albinism. The latter also involves a severe immunodeficiency and lymphoproliferative syndrome. Wiskott-Aldrich syndrome (WAS) has immune deficiency and eczema, but not albinism. WAS also is classified as a platelet cytoskeletal defect. SPD affecting dense granules without other features may be among the most common platelet function defects, however, the diagnosis may be missed if ATP release is not measured. Use of thrombin at 1 unit/milliliter causes maximum release of dense body contents, documenting their presence or absence, even when the release mechanism is faulty.

**Alpha Granule Disorders:** Alpha granule disorders include the gray platelet syndrome, so named for its colorless platelets lacking granules, and the Quebec platelet disorder, a complex disorder with increased urokinase-type plasminogen activator in platelets.

**Defects Detected using Ristocetin:** Selective decrease in response to ristocetin is most often due to von Willebrand disease (VWD), a plasma deficiency or defect of von Willebrand factor (VWF). Typically, platelets will aggregate normally to other agonists, but have decreased response to ristocetin. The same platelet aggregation results are exhibited in Bernard-Soulier syndrome (BSS), due to a defect in the glycoprotein (GP) Ib-IX-V complex. The aggregation defect in VWD will correct with addition of normal plasma, while that of BSS will not. BSS can also be distinguished by the presence of large platelets and mild thrombocytopenia.

In contrast to most forms of VWD, both type 2B VWD and platelet-type VWD will have increased aggregation in response to ristocetin, which is observed at lower ristocetin doses. Both disorders are gain-of-function mutations (VWF in the case of type 2B VWD, and the VWF ligand GPIb in the case of platelet-type VWD).

**Scott Syndrome:** Scott syndrome is a disorder of phosphotidylserine transport that limits the construction of the prothrombinase complex on the platelet membrane, leading to insufficient thrombin generation. Platelet function studies are normal.

## Recommended Reading

Hayward CPM, Rao AK, Cattaneo M. (2006). Congenital platelet disorders: overview of their mechanisms, diagnosis, and treatment. *Haemophilia* **12**(Suppl. 3), 128–136.

# CHAPTER 118

# Laboratory Diagnosis of Acquired Platelet Function Defects

Connie H. Miller, PhD

Most platelet function defects (PFD) encountered in clinical practice are acquired rather than genetic in both inpatients and outpatients. They may be a cause of excessive bleeding, an incidental finding, or a purely *in vitro* phenomenon. Platelet aggregation and ATP release tests, described in Chapter 116, are sensitive to drugs, such as aspirin, which may cause irreversible effects as long as 2 weeks post-ingestion. Ideally, abnormal studies should be repeated after 10–14 days with no medications. Marked improvement in platelet function tests should cause the drug to be considered not only as an assay confounder but also as a potential cause of the patient's symptoms. Poor recollection of drug use may be responsible for the high frequency of PFD in healthy individuals, which is approximately 20% in several studies. It is more likely, however, that a high false-positive rate is inherent in the methodology. Table 118.1 gives a list of drugs influencing platelet function. It is not exhaustive and, since the impact of many drugs is unknown, exclusion from published lists does not indicate a lack of effect. Moreover, the consumption of herbal products and specific diet can also have an effect on platelet function.

Acquired PFD may result from the disorders discussed below and lead to excessive bleeding, usually mild to moderate; however, it can be life-threatening in the presence of other hemostatic defects, such as thrombocytopenia and coagulation factor deficiencies. There are insufficient data to support use of the bleeding time to predict clinical bleeding in most settings. Likewise, the platelet function analyzer (PFA) has been shown to lack sensitivity and specificity for this purpose. The global status of the patient rather than a specific test should be used to guide therapy. The laboratory diagnosis of PFDs includes platelet aggregation studies, which are used to identify the cause of a defect in primary hemostasis resulting in excessive bleeding, or abnormal screening tests.

**Anemia:** Anemia of any etiology may increase the bleeding risk through reduction of platelet–vessel wall interaction. Uremia results in platelet aggregation and ATP release defects with a variety of agonists, including ADP, collagen, epinephrine and arachidonic acid. Thrombin and ristocetin results are often normal. *In vitro* platelet function does not, however, correlate well with clinical bleeding. Uremic plasma can inhibit platelet aggregation, perhaps by guanidinosuccinic acid (GSA) mediated through nitric oxide. Findings are reported to return to normal after renal transplant.

**Myeloproliferative Disorders:** Myeloproliferative disorders, including essential thrombocythemia, polycythemia vera, chronic myelogenous leukemia and agnogenic myeloid metaplasia, result in PFD that may lead to excessive bleeding. Some effects are asymptomatic, however, and thrombosis is often a greater risk. Abnormalities

| TABLE 118.1 Drugs that Inhibit Platelet Function | |
|---|---|
| Non-steroidal anti-inflammatory drugs | Aspirin, ibuprofen, indomethacin, naproxen, meclofenamic acid, mefenamic acid, diflunisal, phenylbutazone, piroxicam, tolemetin, sulindac, sulfinpyrazone, zomepirac |
| Antibiotics | Penicillins, cephalosporins, nitrofurantoin, hydroxychloroquine, miconazole |
| Thienopyridines | Ticlopidine, clopidogrel |
| GPIIb/IIIa antagonists | Abciximab, tirofiban, eptifibatide |
| Platelet cAMP modifiers | Prostacyclin, iloprost, dipyridamole, cilostazol, caffeine, theophylline, aminophylline, nitric oxide |
| Cardiovascular drugs | Nitroglycerin, propanalol, isosorbide dinitrate, nitroprusside, nifedipine, verapamil, diltiazem, quinidine, furosemide |
| Volume expanders | Dextran, hydroxyethyl starch |
| Psychotropic drugs | Amitriptyline, imipramine, nortriptyline, chlorpromazine, promethazine, flufenazine, trifluoperazine, haloperidol, fluoxetine, paroxetine |
| Anesthetics | Halothane, dibucaine, tetracaine, metycaine, cyclaine, butacaine, nupercaine, procaine, cocaine, plaquenil |
| Chemotherapeutic agents | Mithramycin, daunorubicin, BCNU |
| Anticoagulants, fibrinolytics, and antifibrinolytics | Heparin, streptokinase, tissue plasminogen activator, urokinase, $\varepsilon$-aminocaproic acid |
| Antihistamines | Diphenhydramine, chlorpheniramine, mepyramine |
| Miscellaneous drugs | Clofibrate, halofenate, radiographic contrast media, ethanol |
| Foods and food supplements | Vitamin E, omega-3 fatty acids, fish oil, onions, garlic, ginger, cumin, tumeric, clove, black tree fungus, Ginko biloba, caffeine, alcohol |

in aggregation and in ATP release, and an acquired storage pool disorder, have been described, as well as acquired von Willebrand disease due to loss of high molecular weight multimers.

Acute leukemias and myelodysplastic syndrome produce bleeding primarily through thrombocytopenia and disseminated intravascular coagulation (DIC); however, PFD also play a role. Decreased platelet aggregation with ADP, epinephrine and collagen, as well as ATP release defects, have been reported. Acquired VWD and Bernard-Soulier syndrome also have been described.

**Hepatic Disease:** In liver disease, platelet dysfunction plays a secondary role to coagulation factor deficiencies and DIC as a cause of bleeding. Aggregation and release may be decreased with collagen, thrombin, ADP, epinephrine and ristocetin.

<u>Gammopathies:</u> In paraproteinemias, PFD are thought to be due to non-specific binding of immunoglobulins to the platelet membrane producing primarily aggregation defects, which may occur as part of the bleeding diathesis in multiple myeloma, Waldenstrom macroglobulinemia, IgA myeloma and monoclonal gammopathy of undetermined significance.

<u>Autoimmunity to Platelet Proteins:</u> Antiplatelet antibodies may interfere with normal platelet function by blocking glycoprotein receptors, inducing an acquired Glanzmann thrombasthenia, Bernard-Soulier syndrome or other receptor deficiency in patients with immune thrombocytopenic purpura (ITP). Aggregation may be decreased with ristocetin, ADP, epinephrine and collagen. Antibody-mediated platelet activation may also produce an acquired storage pool deficiency. Interpretation of platelet function tests may be difficult due to low platelet count.

<u>Disseminated Intravascular Coagulopathy:</u> Disseminated intravascular coagulopathy (DIC) may result in activation of platelets by thrombin or other agonists, producing an acquired storage pool deficiency or exhausted platelet syndrome. This effect is less important clinically than the consumption of coagulation factors and platelets.

<u>Extracorporeal Membrane Oxygenation:</u> Cardiopulmonary bypass produces PFD in most patients due to platelet activation and physical damage. Platelet aggregation by most agonists is reduced, and $\alpha$ and dense granules are depleted. The defects usually resolve within 24 hours.

## Recommended Reading

Escolar G, Cases A, Vinas M. (1999). Evaluation of acquired platelet dysfunction in uremia and cirrhotic patient using the platelet function analyzer (PFA-100): influence of hematocrit elevation. *Haematologica* **84**, 614–619.

Rao AK. (2007). Acquired disorders of platelet function. In: Michelson AD (ed.). *Platelets*, 2nd edition. San Diego, CA: Academic Press, pp. 1051–1076.

Shen YMP, Frenkel EP. (2007). Acquired platelet dysfunction. *Hematol Oncol Clin N Am* **21**, 647–661.

# CHAPTER 119

# Laboratory Diagnosis of Immune Thrombocytopenic Purpura

Carolyn M. Bennett, MD

Immune thrombocytopenic purpura (ITP) is the result of humoral anti-platelet autoimmunity. ITP can be considered the platelet equivalent of autoimmune hemolytic anemia (AIHA). However, unlike AIHA, which has specific laboratory criteria (i.e. a positive direct antiglobulin test in the presence of laboratory evidence of hemolysis), ITP is a diagnosis of exclusion based largely on clinical criteria rather than specific laboratory testing. Other causes of thrombocytopenia that must be excluded include:

*Increased removal from circulation.* A number of pathologies must be considered, including sequestration due to hypersplenism. In addition, exclusion of consumptive coagulopathies (such as disseminated intravascular coagulopathy [DIC]) and thrombotic microan giopathies (such as thrombotic thrombocytopenic purpura [TTP]) through analysis of LDH, presence of schistocytes on peripheral smear, and additional coagulation testing is also important.

*Decreased production.* Peripheral blood analysis and bone marrow biopsy are used to assess platelet production. Bone marrow studies are not usually indicated in early stages of evaluation, unless there are atypical features. The presence of leukemic cells in peripheral blood, or bone marrow changes consistent with neoplasia, provides an alternate cause for thrombocytopenia. Likewise, findings consistent with bone marrow failure syndromes provide an alternate etiology. In ITP, bone marrow biopsy typically shows normal or increased numbers of megakaryocytes, many of which may have an immature appearance, presumably do to a compensatory increase in megakaryopoiesis.

*False diagnosis of thrombocytopenia.* Examination of a peripheral smear will help to exclude a false diagnosis of thrombocytopenia (e.g. platelet aggregation, pseudo-thrombocytopenia or cold agglutinins); see Chapter 114.

*Anti-platelet autoantibodies.* Although anti-platelet antibodies are central to the pathophysiology, the presence of anti-platelet antibodies is a somewhat non-specific finding, and thus is not generally considered useful in the diagnosis of ITP. Use of anti-platelet antibodies in ITP diagnosis is not recommended by the American Society of Hematology or a similar panel of experts from the United Kingdom. Nevertheless, some clinicians continue to analyze anti-platelet antibodies when evaluating ITP. However, due to the lack of a "gold standard" test for detection and suboptimal sensitivity, a negative result does not exclude a diagnosis of ITP.

## Methods for Platelet Antibody Testing:

Flow Cytometry: The earliest assays using flow cytometry measured platelet-associated IgG (PAIgG). These assays had sensitivity rates of 90–100%, but were non-specific (specificity of 40%) and failed to discriminate between pathologic and non-pathologic platelet antibodies. More recent flow cytometric assays using microbeads coated with platelet-specific glycoproteins, platelet-specific antibodies, or control platelets incubated with patient serum or platelets are analyzed in the flow cytometer and reported as a percentage or ratio of reactivity.

The use of flow cytometry for the detection of platelet autoantibodies has improved the sensitivity and specificity of diagnosis; however, tests which continue to utilize control platelets are less sensitive as a result of interferences from both HLA and ABO antibodies. Recently, incorporation of glycoprotein-specific mouse monoclonal antibody has increased the sensitivity to 86% with a specificity of 100%. These results have also been compared to other antigen-specific assays and show good concordance rates. In addition, studies have shown the practical use of the newer bead based flow cytometric assays for both initial diagnosis and follow-up of patients with ITP.

**Monoclonal Antibody Specific Immobilization of Platelet Antigen:** In monoclonal antibody specific immobilization of platelet antigen (MAIPA), patient platelets with bound autoantibody or control platelets sensitized with patient serum are incubated with mouse monoclonal antibody, which recognizes a target on the platelet surface. After several steps, including washing, addition of buffer and centrifugation, the supernatant is added to a microtiter plate containing goat-specific anti-mouse IgG. As a result, the mouse monoclonal antibody is captured and the platelet surface glycoprotein with bound antibody is immobilized. The human antibody is then detected with an enzyme-labeled goat anti-human IgG probe.

MAIPA has a sensitivity reported as between 49% and 66% with a specificity of 78–92% and a positive predictive value of ~80%. The discrepancy in sensitivity and specificity typically results from false-negative results due to low platelet counts and low levels of autoantibody, lack of antibody specificity for the major platelet glycoproteins, or other patient factors, including concurrent immunosuppressive treatment. As a result, use of the MAIPA for detection of platelet-specific antibodies is typically reserved for initial diagnosis. False-positive results can also occur when human anti-mouse antibodies are present. Because of the specificity for anti-platelet antibodies, interference from HLA and ABO antibodies is less likely.

**Modified Antigen Capture ELISA and Immunobead Assays:** With similar specificity and sensitivity to MAIPA, modified antigen capture ELISA (MACE) and immunobead assays slightly vary from the methodology of MAIPA. The former two utilize monoclonal antibody-coated microtiter plates or plastic beads, respectively. Patient platelets or control platelets incubated with patient serum are added, which allow capture of the specific platelet glycoprotein and antibody immune complex. Use of control platelets for either of these assays including the MAIPA adds an additional aspect of complexity to the above testing because platelets must be supplied by the user.

**Enzyme-linked Immunosorbent Assays:** Commercially available ELISAs (enzyme-linked immunosorbent assays) detect glycoprotein-specific antibodies in patient serum. These assays are not uncommonly used in thrombocytopenic hospitalized patients. Microtiter plates are coated with platelet glycoproteins, GPIIb/IIIa, GPIb/IX and GPIa/IIa. After addition of patient serum, bound autoantibodies are detected using enzyme-labeled secondary antibodies. False-positive results can occur in patients with human anti-mouse antibodies (HAMA). In addition, patients with low-level antibody may have false-negative results. The sensitivity and specificity of the ELISA has less sensitivity and specificity for ITP than the other methods; one ELISA commercial assay was recently reported to have a sensitivity of 53%, specificity of 72%, positive predictive value of 90% and negative predictive value of 24%.

**Platelet Antibody Testing in Other Diseases:** Platelet antibody testing is used in the diagnosis of diseases associated with platelet alloantibodies (i.e. platelet anti-bodies directed towards non-self platelet antigens), which include neonatal alloim-mune thrombocytopenia (Chapter 81) and posttransfusion purpura (Chapter 61). In neonatal alloimmune thrombocytopenia, the mother is sensitized to platelet antigens on the fetal platelets, which results in thrombocytopenia, which can lead to intracra-nial hemorrhage. In posttransfusion purpura, a patient has severe thrombocytopenia 5–10 days after a transfusion, which is associated with a platelet alloantibody. Platelet antibody testing is indicated in these clinical situations.

## Recommended Reading

Cines DB, Blanchette VS. (2002). Immune thrombocytopenic purpura. *N Engl J Med* **346**, 995–1008.

Cooper N, Bussel J. (2006). The pathogenesis of immune thrombocytopaenic pur-pura. *Br J Haematol* **133**, 364–374.

McMillan R. (2003). Antiplatelet antibodies in chronic adult immune thrombocyto-penic purpura: assays and epitopes. *J Pediatr Hematol Oncol* **25**, S57–S61.

McMillan R. (2005). The role of antiplatelet autoantibody assays in the diagnosis of immune thrombocytopenic purpura. *Curr Hematol Rep* **4**, 160–165.

Tomer A. (2006). Autoimmune thrombocytopenia: determination of platelet-specific autoantibodies by flow cytometry. *Pediatr Blood Cancer* **47**, 697–700.

Tomer A, Koziol J, McMillen R. (2005). Autoimmune thrombocytopenia: flow cyto-metric determination of platelet-associated autoantibodies against platelet-specific receptors. *J Thromb Haemost* **3**, 74–78.

Yildirmak Y, Yanikkaya-Demirel G, Palanduz A, Kayaalp N. (2005). Antiplatelet anti-bodies and their correlation with clinical findings in childhood immune thrombo-cytopenic purpura. *Acta Haematol* **113**, 109–112.

# CHAPTER 120
# ADAMTS13 Testing

Christine L. Kempton, MD

The diagnosis of thrombotic thrombocytopenic purpura (TTP), a thrombotic microangiopathy, is critical, as the rapid initiation of plasma infusion or exchange therapy can be life-saving. To date, the diagnosis of TTP is dependent upon the demonstration of an otherwise unexplained microangiopathic hemolytic anemia and thrombocytopenia and clinical findings, including fever, neurologic changes and renal abnormalities. Laboratory findings consistent with a hemolytic anemia include an elevated LDH, increased reticulocyte count and elevated indirect bilirubin. A microangiopathic cause of the hemolytic anemia is supported by the presence of schistocytes on peripheral blood smear. The presence of end-organ damage, particularly kidney, should be assessed by measuring the patient's creatinine level. Although progress in laboratory diagnosis has been made, TTP is a clinical diagnosis supported by the above laboratory and clinical findings. Currently, treatment should be instituted based on these findings; however, the role of ADAMTS13 analysis is evolving.

The major mechanism that underlies TTP is the accumulation of ultra-large von Willebrand multimers (VWF), secondary to a deficiency of the ADAMTS13 enzyme, that leads to platelet microthrombi in end organs. TTP has been associated with congenital deficiencies of ADAMTS13 or inhibitors directed against ADAMTS13, as seen in idiopathic TTP, both leading to reduced protease activity. This chapter will review both the assays for ADMANTS13 activity and the assessment for the presence of an inhibitor for the ADMANTS13 enzyme.

**ADAMTS13 Activity:** Better diagnostic tests are required for TTP, as early symptoms and laboratory findings can also be found in thrombotic microangiopathies, such as disseminated intravascular coagulopathy. With the discovery of ADAMTS13 and its association with TTP, laboratory assessment of ADAMTS13 enzyme activity is now desirable. Numerous assays have been developed to measure ADAMTS13 activity. To date, none of these assays is sensitive or specific enough to allow reliable diagnostic use in a timely fashion; however, as rapid and accurate ADAMTS13 testing may become an integral component of TTP diagnosis in the future, a synopsis of existing testing is present here.

**Methods:**

*Whole Substrate Methods:* A variety of tests of ADAMTS13 activity have been developed using the natural substrate, von Willebrand Factor (VWF). Such assays require two steps. The first step incubates patient's plasma with purified von Willebrand Factor that has been treated to cause partial unfolding. The second step determines the ADAMTS13 activity in the patient's plasma by measuring proteolysis of VWF. This has been accomplished by several methods, including the following: 1. The disappearance of large VWF multimers via immunoassay, 2. The generation of proteolytic fragments, 3. Residual VWF collagen-binding, and 4. Residual VWF ris-tocetin cofactor activity. These assays are time consuming, cumbersome to perform, and of variable reliability.

*Peptide Substrate Methods:* More recently, the cloning of ADAMTS13 and detailed mapping of its recognition sites on VWF has facilitated the development of modified ADAMTS13 substrates, leading to more rapid and reproducible assays.

*FRET-VWF73 Assay:* A novel ADAMTS13 substrate, VWF73, contains the 73 amino-acid residues that are the minimum amino-acid sequence required for ADAMTS13 cleavage of VWF between Y1605 and M1606 in the A2 domain (Figure 120.1). Using the VWF73 substrate eliminates the need to treat purified plasma or recombinant VWF to cause unfolding. To facilitate detection of ADATMS13 cleavage of VWF73, fluores-cence resonance energy transfer (FRET) is utilized. One side of the peptide has a fluo-rescent molecule while the other has a quenching molecule; thus, cleavage of the peptide dequenches the fluorochrome, resulting in positive signal. The fluorescence signal is detected every 5 minutes for 1 hour and, as more substrate is cleaved, the fluorescence increases. The change in fluorescence over time (reaction rate) is calculated by linear regression analysis. Pooled normal plasma is used as a reference. ADAMTS13 activity is expressed as a percentage of the reaction rate found in the test plasma compared to the reaction rate found in pooled normal plasma. Baseline levels of ADAMTS13 should be acquired prior to plasma exchange or infusion, as ADAMTS13 levels will be increased by therapy.

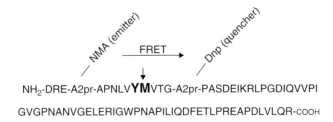

FIGURE 120.1 Structure of FRET-VWF73l. 73 amino acids are required for ADAMTS13 cleavage. P7 and P5′ are replaced with A2pr and Nma and Dnp are attached. Nma emits energy that is transferred to Dnp when the sequence is intact. The arrowhead indicates the ADAMTS13 cleavage site. FRET = fluorescence resonance energy transfer.

**Test Performance:** Two studies have compared the FRET-VWF73 assay with older whole substrate methods. When compared to the proteolytic multimer assay in 79 patients, 100% agreement between the two assays was seen. In another study com-paring FRET-VWF73 with five different assays there was good correlation, with cor-relation coefficients ranging from 0.898–0.971. The assay reproducibility was excellent (coefficient of variation 6%). The FRET-VWF73 assay is now considered to be supe-rior to these other assays because it is a one-step assay that can be completed in 1 hour with excellent precision. However, there are several limitations to the FRET-VWF73 assay.

*False-positive ADAMTS13 Deficiency:* Severe hyperbilirubinemia has been reported to interfere with the assay, leading to a false reduction in the ADAMTS13 activity. However, the degree of reduction is mild, and would not lead a normal level being measured as severely reduced. Free plasma hemoglobin greater than 2 g/l also reduces ADAMTS13 activity by inhibition of the enzymatic activity, although not specific to this assay.

Interpretation: As our understanding of the pathophysiology of TTP evolves, it seems likely that ADAMTS13 is involved in many cases of TTP, but it is unclear to what extent altered ADAMTS13 activity is required for TTP in some cases. Thus, while that newer-generation assays give more rapid and more accurate measurement, the utility of such measurements in acute medical management continues to be a matter of debate. A prospective cohort showed that only 16 of 48 patients with idiopathic TTP had severe ADAMTS13 deficiency.

*Thus, a normal ADAMTS13 activity should not delay initiation of plasma exchange in a patient with a high degree of suspicion for TTP. In prospective studies, severe deficiency of ADAMTS13 (<5%) has a sensitivity for the diagnosis of TTP ranging from 33% to 80%. Accordingly, a severely reduced ADAMTS13 activity is neither necessary nor sufficient for the diagnosis of TTP.*

In a prospective study, those with idiopathic TTP had responses to treatment and mortality rates that were similar regardless of ADAMTS13 activity. However, those with severe ADAMTS13 deficiency were more likely to relapse than those without (43% versus 8%).

The utility of testing patients in follow-up has been investigated in several small studies. In a retrospective study, patients with an undetectable ADAMTS13 activity at first remission had a greater risk of relapse (odds ratio 2.9) than those with detectable ADAMTS13 activity. The risk was even greater in those with an undetectable ADAMTS13 activity and a positive inhibitor titer (odds ratio 3.6). In a prospective study of 35 patients with a first episode of idiopathic TTP and undetectable ADAMTS13 activity at presentation, the presence of undetectable ADAMTS13 activity at the time of remission had a positive and negative predictive value of 28% and 95%, respectively. Among subjects with an undetectable ADAMTS13 activity at the time of remission 38% relapsed, whereas only 5% of those with a detectable level relapsed. Accordingly, a detectable ADAMTS13 activity suggests relapse is unlikely, but an undetectable level is not predictive of relapse over the following 18 months.

ADAMTS13 Inhibitor: Assays have been developed to measure the activity of anti-ADAMTS13 inhibitors. To date, none of these assays is sensitive or specific enough to allow reliable diagnostic use in a timely fashion; however, as rapid and accurate anti-ADAMTS13 activity may become an integral component of TTP diagnosis in the future, a synopsis of existing testing is present here.

Methods: Several tests for anti-ADAMTS13 inhibitory antibodies exist, including mixing studies to look at inhibition of known activity (similar to testing for coagulation factor inhibitors – see Chapter 128) and ELISA-based procedures.

Interpretation: In many cases of TTP, the deficiency of ADAMTS13 activity is due to anti-ADAMTS13 autoantibodies that inhibit activity. However, this is not always the case. Measuring anti-ADAMTS13 antibodies without measuring ADAMTS13 activity will miss some patients with decreased ADAMTS13; this is especially true for patients with Upshaw-Shulman syndrome, who have a congenital deficiency of ADAMTS13. Overall, the role of anti-ADAMTS13 testing is unclear at the current time; however, it may have some power in guiding plasma exchange, as the antibody levels can decrease preceding a return of ADAMTS13 activity. The presence of an inhibitor to ADAMTS13 has been associated with a slower response to plasma exchange and a greater chance of relapse (43–62% versus 5–25%).

## Recommended Reading

Kremer Hovinga JA, Mottini M, Lammle B. (2006). Measurement of ADAMTS-13 activity in plasma by the FRETS-VWF73 assay: comparison with other assay methods. *J Thromb Haemost* **4**, 1146–1148.

Lammle B, Kremer Hovinga JA, George JN. (2008). Acquired thrombotic thrombocytopenic purpura: ADAMTS13 activity, anti-ADAMTS13 autoantibodies and risk of recurrent disease. *Haematologica* **93**, 172–177.

Peyvandi F, Lavoretano S, Palla R *et al.* (2008). ADAMTS13 and anti-ADAMTS13 antibodies as markers for recurrence of acquired thrombotic thrombocytopenic purpura during remission. *Haematologica* **93**, 232–239.

Tripodi A, Chantarangkul V, Bohm M *et al.* (2004). Measurement of von Willebrand Factor cleaving protease (ADAMTS-13): results of an international collaborative study involving 11 methods testing the same set of coded plasmas. *J Thromb Haemost* **2**, 1601–1609.

# CHAPTER 121

# Laboratory Diagnosis of Heparin-induced Thrombocytopenia

Anne M. Winkler, MD and James C. Zimring, MD, PhD

Heparin-induced thrombocytopenia (HIT) is a clinicopathologic syndrome characterized by the presence of a constellation of clinical findings (see Chapter 90) and laboratory results. Classically, after heparin administration, progressive thrombocytopenia is observed. Because the mechanism of thrombocytopenia involves platelet activation, thrombosis can also occur in some cases. The clinical sequelae of thrombocytopenia and thrombosis can be severe and life-threatening, and a decrease in platelet count by 50% after heparin administration is sufficient evidence to replace heparin with an alternate anticoagulant pending laboratory analysis for HIT. As heparin is a safe and often preferred drug for inpatient thromboprophylaxis, laboratory assessment of HIT is valuable in distinguishing true cases of HIT compared to thrombocytopenia from other etiologies. Moreover, once a patient is diagnosed with HIT, future thrombophrophylaxic therapies must be tailored to take this into account.

Developing understanding of the molecular mechanisms involved in HIT has led to a model in which antibodies are formed against a complex between platelet factor 4 (PF4) and heparin. These pathogenic HIT antibodies provoke platelet activation by interaction with the platelet FcγIIa receptor resulting in platelet aggregation, thromboxane generation, granule release, formation of platelet-derived microparticles, and overall stimulation of a procoagulant response. Other mechanisms potentially contributing to hypercoaguability include activation of endothelial cells and monocytes with resultant exposure of tissue factor and neutralization of the anticoagulant effect of heparin by released PF4, which further propagates thrombogenesis.

**Laboratory Testing for HIT Antibodies:** Functional serotonin release assays (SRA) have been and remain the gold standard for laboratory diagnosis of HIT. However, these assays are difficult to perform, requiring specialized equipment, use of radioactivity, well-characterized platelet donors, and experienced operators (see below). Thus, SRAs are not feasible as a general screening test for HIT antibodies. Accordingly, a number of different methodologies have been developed to allow routine screening for HIT antibodies (see below). These tests have high sensitivity, but can result in a substantial rate of false positives due to lower specificity. Currently, the best approach is considered to be identification of these anti-PF4/heparin antibodies through a combination of solid phase antigen assays and functional platelet activation assays (such as SRA).

**Antigen Assays:** In general, most laboratories utilize commercially available enzyme immunoassays (EIAs) for detection of anti-PF4/heparin antibodies. These assays detect HIT antibodies by using PF4 as a substrate. To simulate the *in vivo* epitopes, the PF4 is complexed with either heparin or other polyanion molecules that mimic heparin.

**Solid Phase Enzyme Immunoassays:** In solid phase EIAs, PF4/polyanion complexes are bound to microtiter plates and, after incubation with patient sample and addition of an enzyme-linked anti-human globulin reagent, a colorimetric response can be quantified using an automated microplate reader. The resultant optical density (OD) correlates with a qualitative (positive or negative) result depending on pre-validated reference ranges. However, subtle differences in commercially available kits exist, regarding the source of PF4 (recombinant versus platelet derived) and the use of heparin or other polyanions such as polyvinyl sulfonate. Despite these differences, solid phase EIAs have similar sensitivities of approximately 99%; however, there is variability in clinical specificity.

In the past, commercially available kits incorporated combinations of anti-human IgG, IgA and IgM, and thus detected anti-PF4/heparin antibodies of the IgG, IgM and IgA isotypes. However, it is now understood that the pathogenesis of HIT relies upon antibody binding to the platelet FcγIIa receptor, which is IgG dependent. Although isolated cases of HIT associated with IgA and IgM anti-PF4/heparin antibodies have been described, these antibodies are not typically considered pathogenic and acceptance of positive results without confirmation by a functional assay may be partly responsible for the over-diagnosis of HIT. As a result, elimination of broad spectrum anti-human globulin reagents and use of more recently approved IgG-specific commercial kits have increased the clinical specificity of antigen assays.

More recently, a commercially available assay incorporating addition of interleukin-8 (IL-8) has been FDA approved for detection of anti-IL-8/heparin antibodies. The addition of IL-8 increases sensitivity through identification of rare cases of HIT attributed to platelet activation by antibodies against IL-8/heparin complexes.

**Fluid Phase Enzyme Immunoassays:** As a result of the inherent problem of protein denaturation with solid phase EIAs, a fluid phase EIA in which anti-PF4/heparin IgG antibodies bind PF4/heparin antigens followed by capture with beads was developed. Although not commonly used clinically, this assay has greater sensitivity than the aforementioned solid phase EIAs. In addition, this method avoids non-specific binding by minimizing exposure of cryptic antigens affiliated with denatured PF4.

**Rapid Immunoassays:** For rapid detection of HIT antibodies, particle gel immunoassay (PaGIA) and particle immunofiltration assays have been added to the armamentarium of HIT testing.

In the PaGIA, PF4/heparin complexes are bound to high-density polystyrene beads. If present in a patient specimen, anti-HIT antibodies bind the beads. Bead-bound anti-HIT antibodies are detected through agglutination that occurs as a result of the addition of a secondary anti-human globulin reagent. Similar to gel testing widely used in blood banks, agglutinated beads fail to migrate through the gel, producing a visible band that indicates a positive result. This assay is currently approved for use in Europe and Canada, and is being investigated in the United States. Due to limited data, the sensitivity and specificity of this assay is currently uncertain; however, it is postulated to be intermediate between more commonly used solid phase EIAs and functional platelet assays.

The particle immunofiltrate assay is an alternate but similar approach, which utilizes a reaction well that contains dyed particles coated with PF4 without heparin. The lack of heparin does not adversely impact detection of anti-PF4/heparin antibodies as a result of close approximation of PF4. If the patient specimen contains HIT antibodies, particles agglutinate and fail to migrate through the membrane filter, resulting in lack of detectable color and thus indicating a positive result. This assay has been approved for use in the United States. However, due to the lack of performance characteristics, it is not commonly used.

**Potential Sources of Error:** As with other EIAs and antibody-based tests, presence of interfering immune complexes may result in false-positive results. In addition, inadequate washing during testing may cause false-negative results as a result of failure to remove excess unbound antibody in patients with high-titer antibodies.

**Interpretation of Antigen Assays:** As a general rule, due to the sensitivity of antigen assays, a negative result excludes a diagnosis of HIT in 80–90% of cases. However, results should always be interpreted in conjunction with the pretest probability and in consideration of the clinical picture. If the clinical suspicion indicates intermediate to high probability, all positive EIAs should be confirmed by a functional platelet assay. As an adjunct to guide further testing, a recent publication correlated OD measurements of solid phase EIAs to SRAs. In this study, the probability of pathogenic HIT antibodies as defined by strong positivity (>50% release) in the SRA increased in relation to the magnitude of OD measurements of the EIA. It was determined that most cases of HIT were associated with an EIA OD >1.40 units, while weak positive results (0.40–1.00 OD units) typically excluded the diagnosis of HIT. Due to lower EIA OD cut-offs supplied by manufacturers, the authors recommend reporting the qualitative and quantitative result with an associated probability of presence of clinically significant anti-PF4/heparin antibodies by correlation with SRA results to aid clinician interpretation.

**Platelet Activation (Functional) Assays:** Functional platelet assays are almost exclusively performed in specialized coagulation laboratories, which are capable of performing high-quality platelet activation assays. In accordance with a recent survey of specialized coagulation laboratories, only 14 of 44 laboratories performed functional assays, with the SRA being more commonly performed in the US and Canada compared to the heparin-induced platelet activation (HIPA) test being performed in Europe.

**Platelet Aggregation Test:** Historically, conventional aggregometry using citrated platelet-rich plasma (PRP) was used to assess platelet aggregation; however, in comparison to the functional assays (SRA and HIPA), which incorporate addition of washed platelets to eliminate non-specific aggregation due to heparin, fibrinogen or other acute phase reactants, the platelet aggregation test (PAT) method demonstrated suboptimal sensitivity (50–80%). As a result, the PAT is not advocated for initial detection of anti-PF4/heparin antibodies.

**Serotonin Release Assay:** Since first described in 1986, a radiolabeled $^{14}$C-SRA has been accepted as the gold standard for detection of "pathogenic" anti-PF4/heparin antibodies. The $^{14}$C-SRA is performed by incubating a patient's sample with radiolabeled washed donor platelets, which have been determined to be susceptible to platelet activation by HIT antibodies. After activation, radiolabeled serotonin is released and detected; a positive result requires at least 20% release of serotonin over background reactivity. Specificity is increased through including incubation with high concentrations of heparin (10–100 IU/ml), and addition of a monoclonal antibody to block platelet FcγIIa receptors. The assay can also be used to test whether HIT antibodies cross-react with alternate anticoagulants, such as danaparoid. Positive, negative and "weak positive" controls to ensure platelet reactivity are necessary to further confirm results. The [14]C-SRA is a time-consuming and technically challenging assay that requires strict adherence to established procedures. In addition, special licensure is required for use of radioisotopes, limiting the ability of many labs to perform this test. There can be substantial variability in this assay between laboratories due to the lack of standardization and intrinsic differences in platelet donors. However, when performed correctly, the [14]C-SRA has a sensitivity of 92–100% with a specificity of 98% for detection of pathogenic HIT antibodies.

Other end-points have been analyzed using an SRA, which includes quantification of serotonin release by high-performance liquid chromatography (HPLC), enzyme immunoassays (EIA) and flow cytometry. These detection methods have resulted in similar sensitivity and specificity to the [14]C-SRA.

**Heparin-induced Platelet Activation Test:** In the HIPA assay, patient sample and buffer are added to microtiter wells containing washed donor platelets. A magnetic stirrer agitates the mixture to maintain suspension of unaggregated platelets. At 5-minute time intervals the wells are examined for platelet aggregation using an indirect light source, and change of the mixture from turbid to transparent indicates a positive result. Both pharmacologic and high doses of heparin are used as previously described. This technique is sensitive and specific; however, visual interpretation can result in substantial interobserver variability. On the other hand, an advantage of the HIPA is possibility for repeat testing over time.

**Interpretation of Functional Platelet Assays:** Despite advances in laboratory testing for HIT, ability to detect pathogenic HIT antibodies relies upon their detection using functional platelet assays. EIAs are useful for screening; however, it is important to be aware of the differences of antibody specificity used in coagulation laboratories.

**Other Methodologies:** Quantification of platelet microparticles and annexin V binding assays have also been published as effective methods of anti-PF4/heparin antibody detection; however, these are not in widespread clinical use.

### Recommended Reading

Greinacher A, Juhl D, Strobel U et al. (2007). Heparin-induced thrombocytopenia: a prospective study on the incidence, platelet-activating capacity and clinical significance of antiplatelet factor 4/heparin antibodies of the IgG, IgM, and IgA classes. J Thromb Haemost 5, 1666–1673.

Lo GK, Sigouin CS, Warkentin TE. (2007). What is the potential for overdiagnosis of heparin-induced thrombocytopenia? Am J Hematol 82, 1037–1043.

Price EA, Haywayr CPM, Moffat KA et al. (2007). Laboratory testing for heparin-induced thrombocytopenia is inconsistent in North America: a survey of North American specialized coagulation laboratories. Thromb Haemost 98, 1357–1361.

Warkentin TE. (2007). Heparin-induced thrombocytopenia. *Hematol Oncol Clin N Am* **21**, 589–607.

Warkentin TE, Sheppard JI. (2006). Testing for heparin-induced thrombocytopenia antibodies. *Transfus Med Rev* **20**, 259–272.

Warkentin TE, Sheppard JI, Moore JC *et al.* (2005). Laboratory testing for the antibodies that cause heparin-induced thrombocytopenia: How much class do we need? *J Lab Clin Med* **146**, 341–346.

Warkentin TE, Greinacher A, Koster A, Lincoff AM. (2008). Treatment and prevention of heparin-induced thrombocytopenia: American College of Chest Physicians Evidence-Based Clinical Practice Guidelines (8th edition). *Chest* **133**, 340s–380s.

Warkentin TE, Sheppard JI, Moore JC *et al.* (2008). Quantitative interpretation of optical density measurements using PF4-dependent enzyme-immunoassays. *J Thromb Haemost* **6**, 1304–1312.

# CHAPTER 122

# Molecular Biology of von Willebrand Disease

Connie H. Miller, PhD

von Willebrand Factor (VWF) is a large adhesive glycoprotein required for platelet adhesion to sub-endothelium at the site of vessel injury, platelet aggregation to form the platelet plug, and stabilization of Factor VIII (FVIII) in the circulation. Deficiency or defect of VWF leads to the common disorder of hemostasis von Willebrand disease (VWD). Most cases of VWD, but not all, are due to defects in the structural gene for VWF, located on chromosome 12p. VWF forms a non-covalently bound complex with FVIII, which may lead to a secondary decrease of FVIII coagulant activity (VIII:C) in VWD, although the FVIII gene on the X chromosome is normal.

The VWF gene includes 52 exons, covering 178 kilobases (kb) of DNA (Figure 122.1); it encodes an mRNA of about 9 kb. In addition, there is a pseudogene for VWF located on chromosome 22q, which is a copy of exons 23–34. The VWF gene produces a 2813 amino acid (AA) polypeptide. After intracellular processing, the mature subunit consists of 2050 AA with a molecular weight of >260,000 Da. Processing occurs in the endoplasmic reticulum (ER) and the Golgi. In the ER, a 22-AA signal peptide is cleaved, dimers are formed by disulfide bonds at the carboxyl termini, and N-linked glycosylation occurs. In the Golgi, O-linked glycosylation occurs, large polymers called multimers are formed by disulfide bonds at the amino termini, and a 741-AA propeptide (VWFpp) is cleaved. The protein is secreted or stored as multimers reaching >20 million daltons. After release from cells, the ultra-large multimers are reduced in size by proteolytic cleavage by the metalloproteinase ADAMTS13. Failure of cleavage results in the ultra-large multimers seen in thrombotic thrombocytopenic purpura (TTP).

VWF is made up of 11 domains (Figure 122.1). Four D domains are present at the N-terminus. FVIII binding involves the D' and D3 domains. Three central A domains include the GPIbα and collagen binding sites. The GPIIb/IIIa binding site is in the C domains at the carboxyl-terminus. Dimer formation involves the carboxyl terminus, while multimer formation occurs at the N-terminus. Normal VWF function requires the presence of high molecular weight multimers.

Over 250 mutations in the VWF gene have been shown to cause VWD, including missense, nonsense, frameshift and splice-site mutations, as well as insertions and deletions. They are listed in the ISTH SSC VWF database (http://www.vwf.group.shef. ac.uk/). Many mutations are unique to individual families, while a few recur, such as Y1584C in type 1 VWD. Locations of common mutations for various types of VWD are shown in Figure 122.1.

VWD is classified based on quantitative and qualitative tests of VWF, as described in Chapters 123 and 124. Type 1 VWD, a partial quantitative deficiency of VWF, may result from a wide range of mutations throughout the gene. Two large studies characterizing type 1 mutations have demonstrated that not all patients with type 1 have a

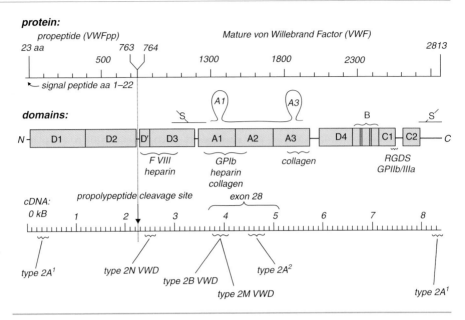

**FIGURE 122.1** The von Willebrand factor (VWF) protein sequence (top), domain structure and binding sites (middle) and cDNA sequence (bottom). Type $2A_1$ is due to abnormal synthesis and $2A_2$ is due to increased proteolysis. Reproduced by permission of RR Montgomery.

mutation in the VWF gene. Those without VWF mutations may have defects in other genes, as yet unidentified.

VWD with a qualitative defect in VWF is defined as type 2, with subtypes. Type 2 A is due to decrease in high molecular weight multimers, through either decreased synthesis or increased proteolysis. Mutations for the former are found in the D1 domain. Mutations causing increased proteolysis are often found in the A2 domain. Type 2B is due to increased binding of VWF multimers to platelets and results from mutations in the GPIb binding region of the A1 domain. Type 2 M VWD results in decreased binding to GPIb due to mutations located primarily in the A1 domain. Most mutations for types 2 A, 2B and 2 M VWD can be identified by sequencing exon 28. In type 2 N VWD, VWF functions normally in its interaction with platelets but fails to bind to and stabilize FVIII. In the absence of VWD binding, FVIII is highly labile, with steadystate levels as low as those seen in hemophilia A. This form, which mimics hemophilia A, is caused primarily by mutations in exons 18–20 or 23–24.

Type 3 is a complete or nearly complete deficiency of VWF, due to homozygosity or compound heterozygosity for VWF mutations. Type 3 mutations are scattered throughout the gene, and nonsense and frameshift mutations are the most common. Large deletions of the VWF gene have been linked to formation of alloantibodies to normal VWF after replacement therapy.

**DNA Sequence Analysis:** Specific regions of involved genes are sequenced to identify mutations associated with subtypes of VWD by polymerase chain reaction (PCR) and direct sequencing. These tests are used to confirm a diagnosis of Type 2A, 2B, 2M, 2N or platelet-type VWD. Sequencing of the entire VWF gene is not usually performed for clinical purposes, due to its large size. Specific areas known to have a high frequency of mutations for certain subtypes may be sequenced relatively easily with automated methods. Mutations occurring outside of the targeted area may be missed.

## Recommended Reading

Lillicrap D. (2004). *The Basic Science, Diagnosis, and Clinical Management of von Willebrand Disease*. Montreal: World Federation of Hemophilia, Monograph No. 35 (available at: http://www.wfh.org/2/docs/Publications/VWD_WomenBleeding Disorders/TOH-35_English_VWD.pdf/).

Sadler JE, Budde U, Eikenboom JCJ *et al.* (2006). Update on the pathophysiology and classification of von Willebrand disease: a report of the Subcommittee on von Willebrand Factor. *J Thromb Haemost* **4**, 2103–2114.

# CHAPTER 123

# Laboratory Diagnosis of Inherited von Willebrand Disease

Connie H. Miller, PhD

von Willebrand disease (VWD) includes both quantitative and qualitative disorders of von Willebrand Factor (VWF). Routine screening tests are not useful in diagnosis of most forms of VWD. The activated partial thromboplastin time (PTT) is prolonged only in the minority of VWD patients who have reduced Factor VIII (FVIII) coagulant activity (VIII:C). Prothrombin time, thrombin time and fibrinogen are normal. The platelet function analyzer (PFA) will detect moderate to severe VWD, but is not sufficiently sensitive to allow exclusion of the diagnosis. Accurate diagnosis and characterization of VWD subtype require a panel of tests measuring different aspects of the Factor VIII/VWF complex, which are described below.

All VWF measures are influenced by ABO blood group, with group O individuals having 20–25% lower VWF and VIII:C than those with non-O groups in the relationship $O < A < B < AB$. African Americans show significantly higher levels of VWF antigen (VWF:Ag) and VIII:C. Use of ABO blood group- and race-specific reference ranges for diagnosis has been proposed, but remains controversial. VWF is an acute phase reactant, and is elevated with pregnancy, inflammation and chronic disease. VWF is lowest in days 1–4 of the menstrual cycle, and testing women during menses has been advocated to minimize variability. If test results are inconclusive, repeat testing is warranted. Comparison to a single standard, preferably measured against the World Health Organization international standard, is important to allow use of ratios of the test results to reach a diagnosis.

**Ristocetin Cofactor:** Ristocetin cofactor (VWF:RCo) is the most useful clinical test for diagnosis of VWD and for monitoring most VWD treatment. The test is an indicator of the ability of VWF to link platelet glycoprotein (GP) IIb/IIIa receptors. Platelet aggregation or agglutination by VWF occurs *in vitro* only when the antibiotic ristocetin induces a conformational change in VWF allowing binding. VWF:RCo is expressed in units per milliliter (U/ml) or in units per deciliter (U/dl), which is equivalent to %. VWF activity is measured by agglutination of washed, lyophilized or formalin-fixed normal platelets by patient plasma in the presence of ristocetin. Agglutination can be measured by aggregometry, automated coagulation analyzer or visual agglutination.

**VWF Antigen:** VWF:Ag is measured immunologically by its reaction with heterologous antibodies with the use of enzyme-linked immunosorbent assay (ELISA) or latex immuno-assay (LIA). Gel techniques have been shown to be less reproducible than ELISA or LIA. VWF:Ag is expressed in units per milliliter (U/ml) or in units per deciliter (U/dl), which is equivalent to %.

**Factor VIII Coagulant Activity:** VIII:C is secondarily reduced in VWD due to lack of sufficient VWF to stabilize it in the circulation, or decreased ability of VWF to bind FVIII. VIII:C is used in the diagnosis of inherited or acquired VWD, in the monitoring of VWD therapy, and in the diagnosis of inherited or acquired hemophilia A. Measurement of VIII:C is by clotting or chromogenic assays. VIII:C is expressed in units per milliliter (U/ml) or in units per deciliter (U/dl), which is equivalent to %.

**Collagen Binding Assay:** Collagen binding assay (VWF:CB) measurement is used in the diagnosis and classification of VWD. VWF:CB is determined by ELISA of VWF binding to collagen on microtiter plates. This test is highly dependent on the type of collagen used. It is sometimes performed as a substitute for VWF:RCo, but measures a different function of VWF: its ability to bind to subendothelial collagen to allow platelet adhesion. VWF:CB is expressed in units per milliliter (U/ml) or in units per deciliter (U/dl), which is equivalent to %. This test is not widely available for clinical use.

**Ristocetin-induced Platelet Aggregation:** Ristocetin-induced platelet aggregation (RIPA) is measured by aggregation of patient's platelets in platelet-rich plasma (PRP) by addition of ristocetin in two concentrations. RIPA is not specific for VWD; it also detects Bernard-Soulier syndrome (BSS). Decreased RIPA from VWD can be corrected by addition of normal plasma, while that from BSS cannot. RIPA is not sensitive to all forms of VWD, and is frequently low in African Americans without VWD. When RIPA is performed with low concentrations of ristocetin, at which platelets from normal individuals fail to aggregate, increased aggregation occurs in two disorders: type 2B VWD and platelet-type VWD. A test measuring binding of VWF to fixed normal platelets can differentiate the two disorders.

**VWF Multimers:** VWF multimer analysis is used in the diagnosis of inherited or acquired VWD and classification of VWD subtype (rarely, it is used in the diagnosis of thrombotic thrombocytopenic purpura). Separation of VWF multimers is by SDS-agarose gel electrophoresis and reaction with polyclonal anti-VWF antibodies, usually by Western blot. VWF structural abnormalities revealed by analysis of VWF multimers are used to distinguish type 1 from type 2 VWD, and to distinguish between subtypes of type 2 VWD. Type 1 has normal sized multimers, but may have decreased intensity of bands due to quantitative defects. In contrast, type 2A has decreased high molecular weight multimers as a result of mutations that prevent multimerization or lead to increased clearance. Type 2B can also have decreased high molecular weight multimers because they are removed from circulation as a result of their increased binding affinity for platelets. Normal multimers are seen in types 2M and 2N. Examples are shown in Figure 123.1.

**VWF Inhibitor:** VWF inhibitor is used in the detection of antibodies interfering with VWF function in patients with type 3 VWD or acquired VWD. VWF:RCo measurement in a mix of patient plasma and normal pool plasma (NPP) is compared to that in a mix of buffer and NPP. The test is analogous to the Bethesda method for FVIII inhibitors (Chapter 128). Decrease in the VWF:RCo in a mix of patient and NPP when compared to a mix of buffer and NPP suggests the presence of an inhibitor. Inhibition of VIII:C in the Bethesda assay may also occur.

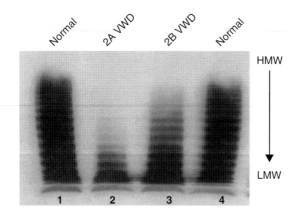

FIGURE 123.1  von Willebrand Factor multimer patterns in types of von Willebrand disease. Modified from Lillicrap (2004).

**VWF Binding to Platelets:**  A monoclonal antibody to VWF is used to detect binding of patient's VWF to formalin-fixed normal platelets in the presence of low concentrations of ristocetin by ELISA or fluid phase binding assay. This assay is used to distinguish between type 2B and platelet-type VWD. Plasma VWF from a type 2B patient will show increased binding to normal platelets. Plasma VWF from a platelet-type VWD patient will bind normally to platelets, since the defect is in the patient's platelets. The test may be less sensitive if the patient's VWF:Ag is less than 10 U/dl.

**Factor VIII Binding to VWF**:  Patient VWF is captured by a monoclonal antibody and quantitated, then reacted with normal FVIII. FVIII bound is measured by chromogenic methods (ELISA or fluid phase binding assay). This test is used to distinguish between type 2 N VWD and mild hemophilia or hemophilia carrier. Results are expressed as the ratio between bound VWF and bound FVIII. Patients with type 2 N VWD show a reduced ratio.

**DNA Sequencing:**  DNA sequencing is described in Chapter 122.

**Diagnosis of VWD Types:**  Expected test results for different types of VWD are shown in Table 123.1.

**Type 1 VWD:**  Type 1 VWD is the most common form. VWF:Ag, VWF:RCo and VWF:CB are usually decreased proportionately to 10–50 U/dl. VIII:C is often normal. Mutation analysis has found 85% of type 1 patients with VWF:Ag $\leq$ 40 IU/dl and 52% of those with VWF:Ag > 40 IU/dl to have a VWF mutation. VWF multimers are normal with the usual methods; however, recent studies suggest that there may be subtle defects detectable only with more sensitive techniques.

**Type 2 VWD:**  Type 2 VWD forms have structurally abnormal VWF, which may be present in normal or reduced quantities. The VWF:RCo to VWF:Ag ratio is usually below 0.6–0.7. In type 2A, high molecular weight multimers are missing due to abnormal synthesis or increased clearance. RIPA is decreased. In type 2B, increased binding of VWF to platelets leads to loss of HMW multimers and mild thrombocytopenia. RIPA is increased with low amounts of ristocetin. In type 2M, the ratio of

| Feature | 1 | 2A | 2B | 2M | 2N | 3 |
|---|---|---|---|---|---|---|
| VWF antigen | D | N/D | N/D | N/D | N | A |
| VWF Ristocetin cofactor | D | D | D | D | N | A |
| FVIII activity | N/D | N/D | N/D | N/D | D | D |
| VWF:RCo/VWF:Ag | N | D | D | D | N | – |
| RIPA | N/D | D | I | D | N | A |
| FVIII binding | N | N | N | N | D | N |
| VWF multimers | N | AB | AB | N | N | A |
| PFA-100 | N/AB | AB | AB | AB | N | AB |
| Inheritance pattern | AD | AD/AR | AD/AR | AD/AR | AR | AR |

**TABLE 123.1 Diagnostic Test Results in Various Types of von Willebrand Disease**

A, absent; AB, abnormal; AD, autosomal dominant; AR, autosomal recessive; D, decreased; I, increased; N, normal.

VWF:RCo to VWF:Ag is reduced but multimers appear normal. This is due to a defect in the binding site for GPIb; clinically, this is a subtle distinction from type 1 and may be missed if tests are not appropriately standardized. Type 2N is caused by a defect in the VWF binding site for FVIII. VIII:C is reduced to 5–20 IU/dl, while VWF appears quantitatively and qualitatively normal. This form of VWD may be misdiagnosed as mild hemophilia A or hemophilia carrier. Type 2N may be seen in compound heterozygosity with other forms of VWD, in which case it is suspected because of a reduced VIII:C to VWF:Ag ratio. It may be confirmed by measurement of VIII:C binding to patient's VWF, or detection of a 2N mutation.

**Type 3 VWD:** Type 3 VWD which is the most severe type, results from two abnormal alleles. Essentially, no VWF is produced, resulting in undetectable levels of VWF: Ag, VWF:RCo and VWF:CB. VIII:C is $\leq 10$ U/dl. Type 3 individuals may develop alloantibodies against VWF, and should be monitored for that event.

**Platelet-type or Pseudo-VWD:** Platelet-type or pseudo-VWD mimics type 2B VWD but is due to a defect in platelet GPIbα not VWF. The two can be distinguished by testing of VWF binding to platelets or by DNA sequencing.

## Recommended Reading

Federici AB. (2006). Diagnosis of inherited von Willebrand disease: a clinical perspective. *Sem Thromb Hemost* **32**, 555–565.

Gill JC. (2004). Diagnosis and treatment of von Willebrand disease. *Hematol Oncol Clin N Am* **18**, 1277–1299.

Lillicrap D. (2004). *The Basic Science, Diagnosis, and Clinical Management of von Willebrand Disease*. Montreal: World Federation of Hemophilia, Monograph No. 35 (available at: http://www.wfh.org/2/docs/Publications/VWD_WomenBleeding Disorders/TOH-35_English_VWD.pdf/).

National Heart, Lung, and Blood Diseases Institute (2007). *The Diagnosis, Evaluation, and Management of von Willebrand Disease*. NIH Publication No. 08-5832 (available at: http://www.nhlbi.nih.gov/guidelines/vwd/vwd.pdf/).

# CHAPTER 124
# Laboratory Diagnosis of Acquired von Willebrand Syndrome

Connie H. Miller, PhD

Acquired von Willebrand syndrome (AVWS) is a rare disorder in which laboratory findings and clinical symptoms mimic various types of von Willebrand disease (VWD). It may be suspected in a patient with abnormal VWD tests and no previous history of excessive bleeding, who has one of the disorders and conditions listed in Table 124.1 or one with a similar mechanism. The International Registry on AVWS (http://www.intreavws.com/) lists over 100 cases, only 2 of which are idiopathic. AVWS may be due to the presence of an autoantibody, adsorption of VWF by cells or surfaces, increased proteolysis, or decreased or aberrant production. VWF may be adsorbed to tumor cells, lymphocytes, plasma cells or activated platelets. In congenital or acquired cardiac defects, AVWS may result from adsorption of HMW multimers to aberrant structures or from increased proteolysis due to high shear stress, and often resolves with surgical correction. Hypothyroidism leads to decreased VWF production, which is restored by treatment.

**TABLE 124.1 Disorders and Conditions Associated with Acquired von Willebrand Syndrome**

| | |
|---|---|
| Lymphoproliferative disorders | Monoclonal gammopathy of undetermined significance, multiple myeloma, non-Hodgkin lymphoma, hairy cell leukemia, chronic lymphocytic leukemia, Waldenstrom macroglobulinemia |
| Myeloproliferative disorders | Polycythemia vera, chronic myeloid leukemia, essential thrombocythemia, myelofibrosis |
| Neoplastic disorders | Wilms tumor (nephroblastoma), peripheral neuroectodermal tumor, adrenocortical carcinoma, gastric carcinoma, acute lymphoblastic leukemia, lung cancer, acute myeloid leukemia |
| Autoimmune disorders | Systemic lupus erythematosus, scleroderma, mixed connective tissue disease, Ehlers Danlos syndrome, autoimmune hemolytic anemia, Felty syndrome |
| Endocrine disorders | Hypothyroidism, diabetes mellitus |
| Cardiovascular diseases | Cardiac defects (VSD, ASD), aortic stenosis, angiodysplasia, mitral valve prolapse |
| Infectious diseases | Epstein-Barr virus, hydatid cyst |
| Drugs | Ciprofloxacin, valproic acid, griseofulvin, hydroxyethyl starch |
| Other | Uremia, hemoglobinopathies, reactive thrombocytosis, pesticide ingestion, glycogen storage disease, sarcoidosis, telangectasis, ulcerative colitis, bone marrow transplant, graft-versus-host disease |

The diagnosis of AVWS uses the assays described in Chapter 123 for inherited VWD: Von Willebrand Factor antigen (VWF:Ag), ristocetin cofactor (VWF:RCo), collagen binding (VWF:CB), VWF multimers and VWF inhibitor. Measurement of VWF:RCo is the most sensitive test. In contrast to inherited VWD, however, more than 80% of AVWS cases show loss of high molecular weight VWF multimers. Inhibitory autoantibodies directed against VWF or the Factor VIII/VWF complex are detected in fewer than 20% of AVWS cases by inhibition of VWF:RCo or VWF:CB in mixing studies. Non-neutralizing antibodies may lead to increased clearance of immune complexes, yet be undetectable *in vitro*. The presence of immune complexes containing VWF has sometimes been demonstrable by gel techniques, such as crossed immunoelectrophoresis, or by their adsorption to staphylococcal protein A. Adsorption of VWF to cells and surfaces may be manifested only as loss of high molecular weight multimers.

## Recommended Reading

Federici AB, Rand JH, Bucciarelli P *et al.* (2000). Acquired von Willebrand syndrome: data from the International Registry. *Thromb Haemost* **84**, 345–349.

Kumar S, Pruthi RK, Nichols WL. (2002). Acquired von Willebrand disease. *Mayo Clin Proc* **77**, 181–187.

# CHAPTER 125

# Laboratory Assessment of Treatment of von Willebrand Disease

Connie H. Miller, PhD

von Willebrand disease (VWD) may be treated by raising the patient's own von Willebrand Factor (VWF), by replacement of VWF, and by use of adjuvant therapies with a global effect on hemostasis. A treatment plan must be based on an accurate diagnosis of the type and severity of VWD present in the patient and the specific clinical situation (i.e. actively bleeding or undergoing minor or major surgical procedure).

Laboratory tests used to monitor treatment of VWD include ristocetin cofactor (VWF:RCo), Factor VIII activity (VIII:C) and platelet function analyzer (PFA). VIII:C level is thought to be the most important determinant for surgical and soft tissue bleeding, while VWF:RCo level is more important for mucous membrane bleeding. The PFA closure time (CT) has also been used to monitor therapy, and has largely replaced the bleeding time for that purpose.

## Treatment Modalities:

**DDAVP:** DDAVP (1-deamino-8-D-arginine vasopressin, desmopressin) has been used for more than 25 years to treat VWD. It acts by causing secretion of stored VWF from endothelial cells. It is most effective in patients with type 1 VWD, who have stores of normal VWF available, and is less effective in patients with type 2 variants, in which the VWF produced is abnormal in structure. It is generally ineffective in type 3 patients. DDAVP is contraindicated in type 2B VWD, because release of more avidly binding VWF may cause or worsen thrombocytopenia. Because there are individual differences in response to DDAVP, a test dose is often given to gauge the patient's response. A three- to five-fold increase over baseline levels of VWF:RCo is expected. VWF:RCo should rise to >30 U/dl. The maximum levels of VIII:C and VWF occur 30–60 minutes after DDAVP administration. Measurements of VWF:RCo and VIII:C should be made at baseline, within 1 hour, and 2–4 hours later. The PFA CT has been shown to correlate with VWF:RCo levels following DDAVP in patients with type 1 and type 2 VWD, and can be used rapidly to measure response.

**VWF Replacement:** VWF replacement requires use of FVIII plasma-derived concentrates specifically labeled to contain VWF. Most recombinant and monoclonally prepared VIII:C concentrates do not contain VWF. Available concentrates differ in their ratios of VWF:RCo to VIII:C, and the presence of high molecular weight multimers. Concentrate use should be monitored with VWF:RCo and VIII:C levels at least every 12 hours. PFA CT does not correct when VWF:RCo rises in type 3 patients. Peak levels of VWF:RCo and VIII:C over 200 U/dl should be avoided because of increased thrombotic risk. Those receiving concentrates for prophylaxis also should have factor recovery measured to determine that they do not have sustained high levels. Type 3 patients treated with VWF-containing products should be monitored for the appearance of inhibitors directed against VWF.

**Adjunctive Therapies:** Adjunctive therapies, such as antifibrinolytic and topical agents, require no laboratory monitoring. Estrogens, such as oral contraceptives, may increase factor levels; however, they have additional effects promoting hemostasis that are not detectable in the laboratory. Some type 3 women respond to hormones with decreased bleeding, although VWF levels do not rise.

### Recommended Reading

Cattaneo M, Federici AB, Lecchi A *et al.* (1999). Evaluation of the PFA-100 system in the diagnosis and therapeutic monitoring of patients with von Willebrand disease. *Thromb Haemost* **82**, 35–39.

Fressinaud E, Veyradier A, Sigaud M *et al.* (1999). Therapeutic monitoring of von Willebrand disease: interest and limits of a platelet function analyser at high shear rates. *Br J Haematol* **106**, 777–783.

Gill JC. (2004). Diagnosis and treatment of von Willebrand disease. *Hematol Oncol Clin N Am* **18**, 1277–1299.

Mannucci PM. (2004). Treatment of von Willebrand disease. *N Engl J Med* **351**, 683–694.

# CHAPTER 126

# Coagulation Factor Testing

Connie H. Miller, PhD

Coagulation factors may be measured by methods assessing both their presence, as antigens, and their ability to function, or activity. Inherited coagulation factor deficiencies may be of two types. Type I defects (quantitative) have decreased absolute amounts of the factor, resulting in both decreased activity and antigen levels. Type I defects result from lack of production or increased clearance of the gene product. Type II defects (qualitative) have a defective gene product, resulting in decreased activity but normal or slightly decreased levels of antigen. Type II defects result from a mutation in the gene product or altered post-translational modification. Acquired defects may mimic either type by decreased or faulty production, increased clearance or inactivation, or the presence of a circulating inhibitor, usually an antibody, which reacts with the factor to mask its activity or remove it from the circulation. Figure 126.1 illustrates the relationship among functional and immunologic tests and how they reveal underlying structure and function. This chapter reviews both the coagulation factor activity (functional) and antigen (immunologic) assays for Factors II (FII), V (FV), VII (FVII), VIII (FVIII), IX (FIX), X (FX), XI (FXI), XII (FXII) and XIII (FXIII), which are indicated in the diagnosis of congenital and acquired coagulation factor deficiencies, characterization of coagulation factor defects and monitoring of coagulation factor therapy.

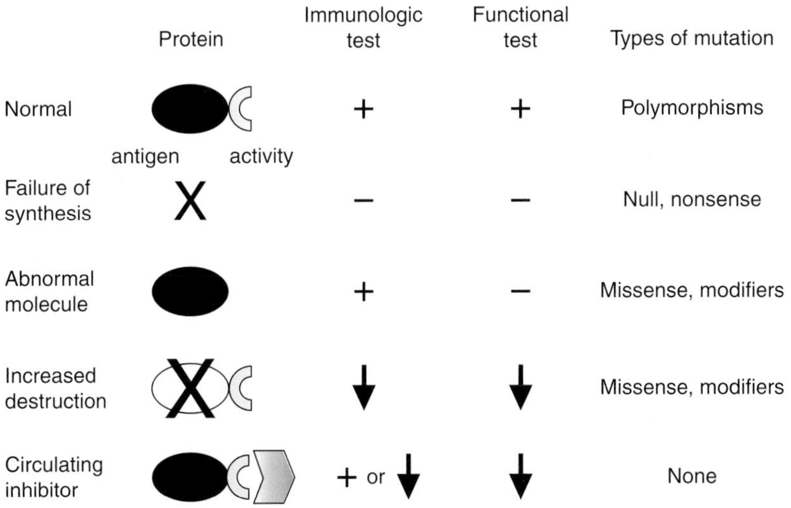

FIGURE 126.1 Immunologic and functional tests of coagulation factors. +, present; −, absent; ↓, decreased. From Miller CH (2006). Laboratory tests for the diagnosis of thrombotic disorders. *Clin Obstet Gynecol* **49**, 844–849.

**Methods:** Activity assays may be clot-based or chromogenic (amidolytic). Clot-based factor assays are modifications of the screening tests to allow a single factor to be rate-limiting in clot formation. Each compares the relative abilities of a test plasma and a control plasma to correct the defect of plasma deficient in a given factor. Correction is measured in an activated partial thromboplastin time (PTT) or prothrombin time (PT) based assay. Chromogenic assays also provide all components needed except for the factor to be measured; the end-point is not a clot but the cleavage of a synthetic substrate through an enzymatic reaction, detected colorimetrically. Immunologic tests for measuring coagulation factors include enzyme-linked immunosorbent assay (ELISA), latex immunoassay (LIA) and other standard methods. Because the measurement of most clinical relevance is activity, immunologic methods are used primarily to characterize the type of defect present as quantitative or qualitative.

**Factor VIII Activity:** Factor VIII activity can be measured by a one-stage assay based on the PTT, a two-stage assay (not widely used in the US), or a chromogenic method. Results are comparable when testing patient plasma, except for a small number of mild hemophilia A patients, who give lower values with the two-stage and chromogenic assays, and those treated with recombinant FVIII lacking the B domain. For quantitation of low levels of VIII:C, a second curve may be generated using VIII:C reference plasma at concentrations of 0.5–20 U/dl. Measurement of VIII:C concentrates requires use of a concentrate standard.

**Factor VIII Antigen:** Factor VIII antigen can be measured using only human or monoclonal antibodies, and does not precipitate. It is not routinely measured in clinical practice. The old term "Factor VIII-related antigen" or VIII:RAg seen in the literature actually refers to von Willebrand Factor antigen (VWF:Ag).

**Factor IX Activity:** Factor IX activity is usually measured by one-stage assay based on the PTT. A low curve may also be used to quantitate more accurately the levels of IX:C seen in hemophilia patients.

**FIX Antigen:** FIX antigen may be detected with heterologous or monoclonal antibodies.

**FXI and FXII Activities:** FXI and FXII activities are usually measured by one-stage PTT-based assay. FXII deficiency is often the cause of a prolonged PTT in a non-bleeding patient.

**FII, FV, FVII and FX Activities:** FII, FV, FVII and FX activities are usually measured in a one-stage PT-based method, although other methods are available. All can also be detected immunologically.

**FXIII:** FXIII is not detected by tests with the endpoint of clot formation or thrombin generation. Most laboratories screen for FXIII function by assessing stability of a clot in either 5 M urea or 1% monochloroacetic acid. This method detects patients with clinically significant FXIII deficiency. Actual quantitation of FXIII activity requires cleavage of a specific substrate. FXIII subunits A and B can be measured immunologically.

<u>Test Performance:</u> For quantitation, clotting time or optical density is plotted against concentration using a curve generated from multiple dilutions of the reference plasma of known concentration. Results are expressed as units per milliliter (U/ml) or, more commonly, as units per deciliter (U/dl), which is equivalent to percent of "normal" (%). A result of 100 U/dl or 1.00 U/ml is equivalent to 100%. Results can be expressed in international units (IU) if they are derived from a standard which is defined by the manufacturer or calibrated in-house against an international standard.

Deficient plasmas may be obtained from individuals with a known factor deficiency; however, because of viral risk, artificially depleted plasmas have been developed. They appear to be satisfactory, except in the case of FVIII-deficient plasma depleted by monoclonal antibodies, which lacks von Willebrand Factor and performs differently in FVIII inhibitor assays.

Reference ranges for coagulation factors vary with the population to be tested and methods used, and should be derived locally whenever possible. Some factors, such as FVIII, do not follow a normal distribution in the population. Their ranges must be calculated in transformed units or as fifth to ninety-fifth percentiles using non-parametric methods. It is not appropriate to remove outlying values to produce normality. Some coagulation factors show differences among populations that affect the reference ranges. FVIII activity levels vary significantly by ABO blood group, with type O individuals having lower levels. FVIII is significantly higher and FXII significantly lower in African Americans.

<u>Sources of Error:</u> Factor assays performed with clotting techniques are subject to interference from heparin. Heparinase may be used to remove heparin prior to assay. Lupus anticoagulants will influence clot-based tests for multiple factors. Chromogenic assays are not influenced by heparin, lupus anticoagulants or fibrin degradation products (FDP). Measurement of antigen alone may give false assurance that a factor is present, since it may be non-functional.

Measurement of factor activity using only one dilution of patient plasma may mask the presence of an inhibitor. Multiple dilutions may show that the patient's plasma has a dilution curve that does not parallel the normal, usually indicating presence of an inhibitor, often a lupus anticoagulant.

<u>Quality Assurance:</u> Factor assays require a reference plasma, used to produce a curve against which the quantity of a factor can be measured. Commercial reference plasmas should provide the source of the quantitation used to set the calibration for each factor. Positive and negative controls should be run with each assay.

<u>International Standards:</u> Use of a national or international standard provides consistency over time and allows results to be compared across laboratories. Such calibrated standards are preferable to use of pooled plasma from a small number of individuals. International standards are available for FII, FV, FVII, FVIII, FIX, FX, FXI and FXIII in plasma (National Institute for Biological Standards and Control, Potter's Bar).

## Recommended Reading

Andrew M, Paes B, Johnston M. (1990). Development of the hemostatic system in the neonate and young infant. *Am J Pediatr Hematol Oncol* **12**, 95–104.

Andrew M, Vegh P, Johnston M *et al.* (1992). Maturation of the hemostatic system during childhood. *Blood* **80**, 1998–2005.

de Alarcón PA, Werner EJ. (2005). Neonatal values and laboratory methods. In: de Alarcón PA, Werner EJ (eds). *Neonatal Hematology.* Cambridge: Cambridge University Press, pp. 406–430.

Miller CH. (2006). Laboratory tests for the diagnosis of thrombotic disorders. *Clin Obstet Gynecol* **49**, 844–849.

# CHAPTER 127
# Mixing Studies

Connie H. Miller, PhD

In mixing studies, patient plasma with a prolonged activated partial thromboplastin time (PTT) or prothrombin time (PT) is mixed with normal pool plasma (NPP). PTT or PT is measured after mixing the two samples. Mixing studies are used to distinguish among potential causes for a prolonged screening test – in particular, to distinguish between a factor deficiency and the presence of an inhibitor.

**Method:** The standard procedure for PTT or PT is performed on a 1:1 mixture of patient plasma and NPP, usually immediately and after 1 hour's incubation at 37°C.

**Interpretation:** Mixing study outcomes are summarized in Table 127.1. Complete correction of a 1:1 mix suggests a factor deficiency, either congenital or acquired. Failure to correct completely suggests an inhibitor interfering with one or more coagulation factors, or a lupus anticoagulant. If repeating the PTT mix after incubation for 1 hour at 37°C gives a longer time, a FVIII inhibitor, which is time-dependent, may be present. Rarely, time-dependent inhibitors may be directed against FV. A PT which corrects most often indicates a multifactor deficiency due to the presence of liver disease, vitamin K deficiency or warfarin; however, it may represent FVII deficiency alone.

TABLE 127.1 Interpretation of Activated Partial Thromboplastin Time and Prothrombin Time Mixing Studies in Specimens without Heparin

| Observation | Possible cause | Tests to perform |
|---|---|---|
| **APTT 1:1 mix of patient and NPP** | | |
| Corrected immediately and at 1 hour | Factor deficiency | Factor VIII, IX, XI, XII activities (Factors II, V, X if PT also prolonged) |
| Corrected immediately and prolonged at 1 hour | Factor VIII inhibitor | Factor VIII activity, if low; FVIII inhibitor |
| | Weak LA | LA tests |
| Incompletely corrected | LA | LA tests |
| | Factor inhibitor | Factor VIII, IX, XI, XII activities |
| **PT 1:1 mix of patient and NPP** | | |
| Corrected immediately | Factor deficiency | Factor II, V, VII, X activities |
| Incompletely corrected | LA | LA tests |
| | Factor inhibitor | Factor II, V, VII, X activities |

NPP, normal pool plasma; LA, lupus anticoagulant.

The definition of "correction" varies from laboratory to laboratory because of differences in the PTT and PT methods used. Some thromboplastin reagents are more sensitive than others. Attempts to standardize across laboratories have been generally unsuccessful. Consultation with the particular laboratory involved or someone experienced in interpreting its results may be helpful, particularly when partial correction occurs.

**Test Performance:** In clinical situations where specimen volume is limited, as in pediatrics, mixing studies may be skipped and appropriate diagnostic tests performed instead. With automated analyzers, performing factor assays may be quicker and use less plasma than mixing studies. A lupus anticoagulant or non-specific inhibitor may be indicated by the reduction of multiple factors.

## Sources of Error:

**Heparin Contamination:** In a patient who is hospitalized or has a central venous access device, the specimen should be treated first with a heparin-neutralizing agent (such as heparinase) and the PTT or PT repeated. The tests should normalize if slight heparin contamination or a therapeutic level of heparin is present. This step is often omitted in outpatients, particularly children unlikely to have heparin exposure, to conserve the specimen.

**Weak Inhibitors:** Mixing studies may be misleading if the PTT or PT is only slightly prolonged, because a weak inhibitor can be diluted sufficiently in a 1:1 mix to disappear.

**Quality Assurance:** This is the same as for standard PTT and PT. NPP should be checked for the presence of all coagulation factors and the absence of inhibitors.

## Recommended Reading

Forte K, Abshire T. (2000). The use of Hepzyme in removing heparin from blood samples drawn from central venous access devices. *J Pediatr Oncol Nurs* **17**, 179–181.

Kaczor DA, Bickford NN, Triplett DA. (1991). Evaluation of different mixing study reagents and dilution effect in lupus anticoagulant testing. *Am J Clin Pathol* **95**, 408–411.

# CHAPTER 128

# Specific Factor Inhibitor Testing

Connie H. Miller, PhD

The coagulation factor inhibitor assay is used to identify inhibitors occurring in individuals with inherited factor deficiencies (alloantibodies) and those who are not congenitally deficient (autoantibodies). Coagulation factor inhibitors are detected in the laboratory by their ability to neutralize specific coagulation factors and must be distinguished from non-specific inhibitors, like the lupus anticoagulant. Antibodies with type 1 or "simple" kinetics show high affinity and can be saturated. In antibody excess, all activity is neutralized. Antibodies with type 2 or "complex" kinetics show lower affinity and can dissociate from antigen. Both free antigen and free antibody may be present. Most hemophilic inhibitors have type1 kinetics; type 2 is more common among autoantibodies. Factor VIII (FVIII) inhibitor quantitation was standardized in 1974, when investigators meeting in Bethesda, Maryland, agreed on a standard method and a unit of measure: the Bethesda unit (BU), defined as the amount of inhibitor which destroys one-half of the FVIII activity (VIII:C) in 1 milliliter (ml) of normal plasma within 2 hours. This method has been extended to quantitation of inhibitors to other coagulation factors.

**Methods:** A specific factor assay is performed on a mixture of patient plasma and normal pool plasma (NPP) to detect an inhibitor (antibody) to a specific coagulation factor.

**Issues Surrounding Laboratory Detection of Inhibitors:** Inhibitors to clotting factors occur as autoantibodies or alloantibodies. These may be suspected when multiple dilutions of patient plasma give different activity levels, suggesting that the patient's dilution curve does not parallel the normal. Acquired inhibitors often have complex kinetics, such that factor activity may be measurable, even in the presence of a significant antibody. Lack of ability to measure *in vitro* inhibition does not preclude the presence of an inhibitor, since immune complexes may be cleared *in vivo*.

**FVIII Inhibitors:** FVIII inhibitors are time dependent. The Bethesda assay is performed by mixing one part of patient plasma with one part of NPP, incubating at 37°C for 2 hours, and measuring the VIII:C remaining in the mixture. This remaining activity is divided by the activity remaining in a control 1:1 mixture of NPP with imidazole buffer and multiplied by 100 to give the % residual activity (RA). The RA is converted to BU by means of a graph plotting the logarithm of RA against BU (Figure 128.1) or by use of the equation:

$$BU = (2 - \log RA)(0.301)^{-1}$$

For RA = 100%, BU = 0. For RA 25–100%, BU of the lowest dilution in that range is reported. If no dilution falls above 25%, patient plasma is tested at higher dilutions,

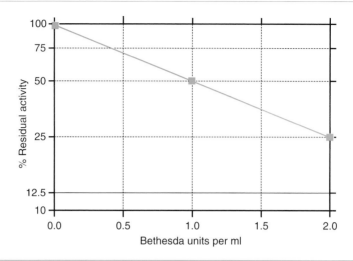

FIGURE 128.1   Calculation of Bethesda units in the Bethesda assay for inhibitors.

with the first dilution falling between 25% and 75% RA multiplied by the dilution factor and reported. For inhibitors with complex kinetics, which do not respond linearly to dilution, the dilution closest to a RA of 50% is reported. For increased sensitivity, the assay may be performed using a higher proportion of patient plasma (e.g. a 3 : 1 mix).

Modifications to increase the sensitivity and reproducibility of the Bethesda method have been widely adopted. Substitution of FVIII-deficient plasma for buffer in patient dilutions and the control mix and buffering of NPP with imidazole to pH 7.4 serve to maintain the protein concentration and the pH of the mixtures during incubation. This has been called the Nijmegen-Bethesda (NB) method, and the results are Nijmegen-Bethesda units (NBU).

**Factor IX Inhibitors:** Factor IX (FIX) inhibitors, which occur infrequently in patients with hemophilia B, may be measured by the Bethesda method without incubation, since they are not time dependent.

**Factor XI Inhibitors:** Factor XI (FXI) inhibitors may occur in up to one-third of congenitally deficient patients homozygous for the Type II mutation (Glu117Stop) and may be time-dependent. These inhibitors are detected by the Bethesda method.

**Factors V, VII, X and XIII Inhibitors:** Inhibitors also are seen rarely in congenital deficiencies of Factors V, VII and X, and may be detected by Bethesda-type assays. FXIII inhibitors may be detected by inhibition of photometric assays of FXIII activity.

**von Willebrand Factor Inhibitors:** Inhibitors directed against von Willebrand Factor (VWF) in type 3 von Willebrand disease may sometimes be detected by mixing patient and NPP and performing a ristocetin cofactor or collagen binding assay. VIII: C may also be inhibited in a Bethesda assay. Many VWF inhibitors, however, are non-neutralizing. VWF inhibitors in non-VWD patients will be detected only rarely in a Bethesda-type assay.

## Sources of Error:

**False Positive:** Lupus anticoagulants and non-specific inhibitors may also be present in hemophilia patients and give false-positive results. Heparin contamination may also give a false-positive inhibitor test.

**False Negative:** Circulating FVIII may complex with antibody, leading to underestimation of the inhibitor titer. Heating plasma to 56°C, to destroy endogenous and infused FVIII without affecting the antibody, may increase the inhibitor titer in some patients.

Diagnosis of both alloantibodies and autoantibodies may be complicated by their presentation as multifactor deficiencies (Table 128.1). At dilutions used in most clinical factor assays, these antibodies interfere with tests of other factors in the intrinsic pathway. When patient plasma is tested in higher dilution, accurate measurement of the non-inhibited factors is obtained. The same effects may be seen with hemophilic inhibitors.

### TABLE 128.1 Dilutional Effect of Antibodies on Factor Levels

|                       | Dilution | Factor VIII | Factor IX | Factor XI | Factor XII |
|-----------------------|----------|-------------|-----------|-----------|------------|
| Factor VIII inhibitor | 1:5      | <1          | 23        | 19        | 40         |
|                       | 1:20     | <1          | 45        | 38        | 55         |
|                       | 1:100    | <1          | 78        | 65        | 82         |
| Factor IX inhibitor   | 1:5      | <1          | <1        | <1        | –          |
|                       | 1:20     | 110         | <1        | 70        | –          |

Current clinical inhibitor assays detect only those antibodies which neutralize factor activity in the fluid phase. *In vivo* factor recovery studies are required for detection of non-neutralizing inhibitors and those with atypical *in vitro* behavior.

## Recommended Reading

Giles AR, Verbruggen B, Rivard GE *et al.* (1998). A detailed comparison of the performance of the standard versus the Nijmegen modification of the Bethesda assay in detecting Factor VIII:C inhibitors in the haemophilia A population of Canada. *Thromb Haemost* **79**, 872–875.

Kasper CK. (1991). Laboratory tests for Factor VIII inhibitors, their variation, significance and interpretation. *Blood Coag Fibrinol* **2**, 7–10.

Key NS. (2004). Inhibitors in congenital coagulation disorders. *Br J Haematol* **127**, 379–391.

Manco-Johnson MJ, Nuss R. (2000). Heparin neutralization is essential for accurate measurement of Factor VIII activity and inhibitor assays in blood samples drawn from implanted venous access devices. *J Lab Clin Med* **136**, 74.

Salomon O, Zivelin A, Livnat T *et al.* (2003). Prevalence, causes, and characterization of Factor XI inhibitors in patients with inherited Factor XI deficiency. *Blood* **101**, 4783–4788.

# CHAPTER 129

# Laboratory Diagnosis of Dysfibrinogenemia and Afibrinogenemia

Thomas C. Abshire, MD

Abnormalities of fibrinogen may be either genetic or acquired, and may lead to bleeding or thrombosis. Fibrinogen is a large glycoprotein present in plasma at a concentration of 200–400 mg/dl. It is composed of three separate polypeptide subunits (alpha ($\alpha$), beta ($\beta$) and gamma ($\gamma$) chains), each produced by a distinct gene of 6500–7500 base pairs located at chromosome 4q32. Expression of the closely linked genes is regulated together.

The structure of fibrinogen and normal assembly of fibrin monomers after activation by thrombin are shown in Figure 129.1. Fibrinopeptide A (FPA) is first cleaved by thrombin, forming fibrin monomer with exposed sites in the A$\alpha$ chain of the E domain of fibrinogen with the carboxy terminal region of the $\gamma$ chain (D region of fibrinogen). This action allows for polymerization of the fibrin monomers into two-stranded, elongating structures known as protofibrils. Generation of soluble fibrin formed after FPA release occurs much more quickly than that achieved by the release of fibrinopeptide B (FPB) from the action of thrombin. This latter thrombin cleavage facilitates the lateral assembly of fibrin and growth of the fibrin sheaths into a thicker formation. Finally, thrombin activates Factor XIII (FXIII), which covalently links the fibrin monomer assembly via lysine and glutamic acid bonds to form gamma–gamma dimers in the D region of fibrin and then, more slowly, in the alpha polymers. The effect of FXIII allows for "tightening" of the D–D region, providing longitudinal strength and alpha

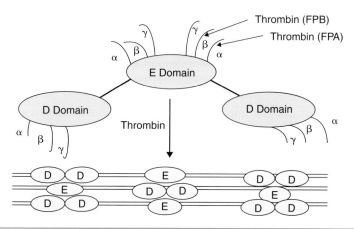

FIGURE 129.1 Fibrinogen structure and fibrin formation. As described in the text, fibrinopeptide A (FPA) and fibrinopeptide B (FPB) are progressively cleaved by thrombin to form soluble fibrin and fibrin protofibrils. Thrombin then activates FXIII to form a stable fibrin clot. The lower part of the figure illustrates the covalently linked stable fibrin.

polymer tightening, promoting greater end-to-end strength. The effect of FXIII on the D–D covalent bond helps to prevent plasmin digestion of fibrin. The role of thrombin and plasmin in fibrinogen/fibrin assembly/degradation has importance in the laboratory decisions regarding assessment of disorders of fibrinogen and fibrinolysis.

**Description, Methods and Indications:** Fibrinogen assays, thrombin and reptilase clotting times are measured in the assessment of fibrinogen abnormalities.

**Fibrinogen Assay Measurements:** Testing for fibrinogen is primarily by two methods. A functional assay using the Clauss or thrombin clotting method (the more common method in the US) measures the clotting activity upon addition of excess thrombin and diluted plasma. Fibrinogen is detected by measuring electromagnetic viscosity. An automated detection system compares the patient level to a standard curve. The second method to assay fibrinogen is an antigenic assay, which is an immunologic measure of the protein, usually by enzyme-linked immunosorbant assay (ELISA). These two assays serve different diagnostic purposes and are used in different clinical situations, as described below.

**Assessment of Hypofibrinogenemia and Afibrinogenemia:** Hypofibrinogenemia and afibrinogenemia have decreased or absent levels of fibrinogen, respectively. The decrease is observed in both functional and antigenic assays; both assays are proportionately reduced. If the fibrinogen is sufficiently low, both the PT and PTT screening tests may be prolonged; however, the PT and PTT are insensitive to moderate decreases in fibrinogen. Decreased or absent fibrinogen may be due to either lack of synthesis (genetic defect or liver failure) or consumption. If fibrinogen becomes sufficiently low that hemostasis is compromised, then factor deficiency type bleeding would be expected (see Chapter 98). Replacement of fibrinogen can be achieved with cryoprecipitate (see Chapter 31).

**Assessment for Dysfibrinogenemia:** In the dysfibrinogenemic state, fibrinogen is synthesized in a defective manner, such that its ability to serve as a substrate for thrombin is compromised. Thus, the fibrinogen molecule is present and antigen assays are typically normal or elevated. In contrast, functional assays show decreased levels. This discordance between the functional and antigenic assays for fibrinogen is the hallmark of diagnosing dysfibrogenemia. Dysfibrinogenemia can be a congenital defect, or acquired. For example, nephrotic syndrome, fetal fibrinogen and liver disease all have increased amounts of sialic acid which impair fibrin formation.

Two tests are typically performed to further assess dysfibrinogenemia: the thrombin time and the reptilase time. The thrombin clotting time is a measurement of the conversion of fibrinogen to fibrin, and is performed by adding thrombin to platelet-poor plasma. It cleaves both FPA and FPB from fibrinogen. The reptilase time is a helpful confirmatory test to perform if the thrombin time is prolonged. Reptilase is an enzyme isolated from the snake venom *Bothrops atrox*. It cleaves FPA from fibrinogen, but does not cleave FPB. The reptilase time is not affected by heparin, but may be affected by the same confounding variables that affect the thrombin time. In typical dysfibrinogenemia, both the thrombin time and the reptilase time are prolonged.

## Sources of Error:

Fibrinogen Detection: Fibrinogen levels that are less than 100 mg/dl or above 400 mg/dl are re-assayed at alternative dilutions. This automated form of fibrinogen assessment is quite reproducible and relatively insensitive to the effect of heparin and fibrinogen degradation products (FDP).

Screening tests such as the PT and PTT are often not prolonged unless the fibrinogen level is below 100 mg/dl. When either antigenic or functional fibrinogen levels are below 100 mg/dl, the thrombin time (thrombin clotting time, TCT) is often elevated.

Dysfibrinogen Detection: The thrombin time is affected by several conditions. It is exquisitely sensitive even to small amounts of heparin. The reptilase time, which is not sensitive to heparin, can be used to assess possible heparin contamination. Additionally, degradation products from either fibrin or fibrinogen and abnormal circulating proteins (e.g. gammopathies in multiple myeloma) may prolong both the thrombin and reptilase times.

Genetic Testing for Fibrinogen Abnormalities: As discussed in the clinical section on hypo/afibrinogenemia and dysfibrinogenemia, genetic testing must be considered to supplement the screening tests discussed here. This is particularly important since different fibrinogen abnormalities are known to present with either bleeding or clotting, and management is quite different. The website describing the details of what is known about the genetics of these conditions has been previously described in Chapter 98.

## Recommended Reading

Abshire TC, Fink LK, Christian J, Hathaway WE. (1995). The prolonged thrombin time of nephrotic syndrome. *J Pediatr Hematol Oncol* **17**, 156–162.

Lefkowitz JB, DeBoom T, Weller A *et al.* (2000). Fibrinogen Longmont: a dysfibrinogenemia that causes prolonged clot-based test results only when using an optical detection method. *Am J Hematol* **63**, 149–155.

Martinez J. (1997). Congenital dysfibrinogenemia. *Curr Opin Hematol* **4**, 357–365.

Okuda M, Uemura Y, Naka K, Tatsumi N. (2003). Preparation of a purified fibrinogen calibration material for Clauss method and turbidimetric immunoassay possessing biological activity and antigenicity. *Clin Lab Haematol* **25**, 167–172.

# CHAPTER 130

# Laboratory Assessment of Fibrinolysis

Thomas C. Abshire, MD

The fibrinolytic pathway is important in the maintenance of blood flow, ensuring that fibrin clot formation is localized to the site of vascular injury and the clot is dissolved once the injury is healed. Fibrinolysis is carefully regulated by a number of pro- and anti-fibrinolytic enzymes (Figure 130.1). Increased fibrinolysis can result in excessive bleeding, but it has been more difficult to prove in animal models and humans that reduced fibrinolysis can result in excessive clotting. Hyperfibrinolysis should be considered in a patient with non-surgical bleeding whose initial laboratory testing reveals no defects within platelets or the coagulation system. The focus of this chapter will be on the laboratory parameters helpful in the assessment of a patient with hyperfibrinolysis and bleeding, with a brief review on the lack of utility of fibrinolytic assays in assessing the evaluation of thrombosis.

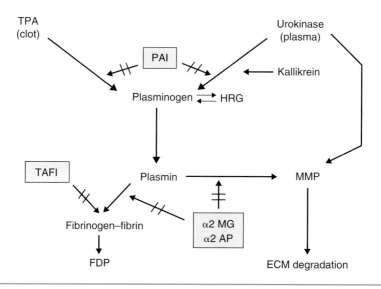

**FIGURE 130.1** The fibrinolytic pathway in coagulation. TPA, tissue plasminogen activator; PAI-1, plasminogen activator inhibitor-1; HRG, histadine-rich glycoprotein; TAFI, thrombin activatable thrombinolysis inhibitor; α2MG, α2 macroglobulin; α2AP, α2 antiplasmin; MMP, matrix metalloproteinases; ECM, extracellular matrix; FDP, fibrin degradation products.

Increased fibrinolysis can either be primary (excessive plasmin and/or tissue plasminogen activator [TPA] production in the absence of proportional antecedent coagulation activation) or secondary (compensatory response to disseminated intravascular coagulopathy [DIC] or excessive thrombin activation).

Primary fibrinolysis can arise spontaneously or as a result of thrombolytic therapy. TPA converts plasminogen to plasmin, mostly within fibrin strands, with subsequent lysis of the clot matrix into fibrin degradation products (FDP). The balance of the effects of thrombin and plasmin on fibrinogen/fibrin lead to a stable fibrin clot (Figure 130.2). Plasmin's initial effect on fibrinogen is to form FDP and fibrin X. Fibrin X is then degraded into fragments Y and D. Fragment Y is subsequently converted into smaller fragments (less than 100 kDa), known as fragments D and E. These can be produced prior to cross-linking of fibrin by FXIII, but may also occur after a stable fibrin clot is formed. However, the predominant FDPs generated from a stable clot with cross-linked fibrin are complexes DD/E, DY and D-dimer.

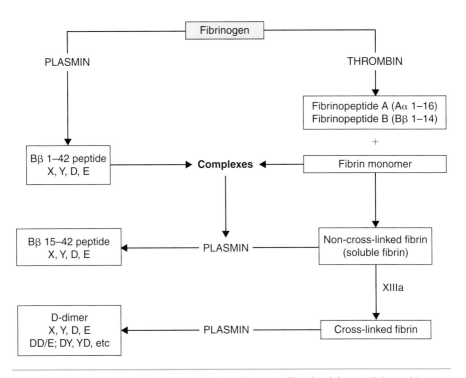

FIGURE 130.2 The role of thrombin and plasmin in fibrinogen/fibrin breakdown and the resulting fibrin degradation products produced based upon this interaction (FDP X, Y, D, E, DD/E, DY, YD); XIIIa: Factor XII. From Goodnight SH, Hathaway NE (2001). *Disorders of Hemostasis and Thrombosis, A Clinical Guide.* New York, NY: The McGraw-Hill Companies. Reproduced with permission.

Under normal conditions, the fibrinolytic process remains strictly controlled and localized to the fibrin clot; however, in secondary fibrinolysis, usually due to DIC, local control is lost and there is often systemic bleeding.

The fibrinolytic pathway is primarily initiated (under physiologic conditions) by release of TPA from endothelial cells after vascular injury and fibrin deposition. TPA is usually localized within the clot, is bound to fibrin, and facilitates the conversion of plasminogen to plasmin. Another activator of plasminogen, urokinase, is predominantly produced in the kidney, with small amounts circulating in the plasma as pro-urokinase or single chain urokinase. These may be converted to the active form of urokinase via the contact pathway system (kallikrein). Plasminogen partially (50%) circulates as a reversible complex with histidine-rich glycoprotein (HRG). Both urokinase and plasmin are essential to activating the matrix metalloproteases (MMP), a family of zinc peptidases which degrade collagen, fibronectin and other subendothelial contact proteins

Inhibitors to the fibrinolytic pathway are present in the plasma and in the fibrin clot. Free (non-clot based) plasmin is neutralized by the inhibitor, $\alpha$2 antiplasmin ($\alpha$2AP) as well as by $\alpha$2 macroglobulin ($\alpha$2MG). Both of these inhibitors are important to keep plasmin in balance and block any unopposed degradation of key clotting factors such as FVIII and fibrinogen. Plasminogen activator inhibitor-1 (PAI-1) has a greater inhibitory effect on TPA in the plasma. PAI-1 is an acute phase reactant which is inactivated by physiologic conditions (e.g. neutral pH). The liver is also instrumental in clearing TPA from plasma. Both PAI-1 and liver clearance contribute to the short half-life of TPA (approximately 5 minutes). PAI-1 is found in platelets, endothelial cells and plasma. Another fibrinolytic inhibitor, thrombin activatable fibrinolysis inhibitor (TAFI), regulates TPA within the fibrin clot. TAFI cleaves lysine from partially degraded fibrin, preventing plasminogen and TPA from binding to fibrin.

The fibrinolytic pathway and inhibitors of fibrinolysis are generally balanced, but occasionally systemic fibrinolysis can occur when excessive plasma TPA is present in conjunction with deficiency of fibrinolytic inhibitors such as PAI-1. Additionally, impaired ability of the liver to remove TPA and plasmin can also produce a fibrinolytic state, as can certain conditions such as malignancy, which may secrete novel fibrinolytic procoagulants.

**Laboratory Testing of Fibrinolysis:** The following tests are important in assessing the patient who is clinically bleeding when routine screening testing has not uncovered a cause, and where antifibrinolytic therapy might be helpful.

**Euglobulin Lysis Time:** The euglobulin lysis time (ELT) is the standard screening test for hyperfibrinolysis. The euglobulin fraction of plasma contains: fibrinogen, plasminogen, plasminogen activators and plasmin, and is formed as a precipitate. The precipitate is lacking fibrinolytic inhibitors. The ELT is performed by isolating the euglobulin fraction from patient plasma, suspending it in buffer to which thrombin is added, allowing a clot to form, and then measuring clot lysis. A positive test is evident with a shortened ELT (usually < 60 minutes). This indicates that increased fibrinolysis is present, although the exact cause of the excessive fibrinolysis is not determined by this methodology. It is important to note that the ELT can be shortened in consumptive coagulopathies (e.g. DIC); thus, primary hyperfibrinolysis would consist of a shortened ELT in the absence of other labs that indicate consumptive processes (e.g. schistocytes on peripheral smear).

**Alpha 2 Antiplasmin:** Low levels of $\alpha$2AP are associated with excessive fibrinolysis due primarily to the low (50%) ratio of $\alpha$2AP compared to plasminogen. Accordingly, excessive plasmin, unopposed by $\alpha$2AP, will contribute to a bleeding state. $\alpha$2AP is measured by its ability to inactivate a known amount of plasmin, with the endpoint detected chromogenically.

**Plasminogen Activator Inhibitor 1:** PAI-1 may be measured by a functional assay involving its reaction with either TPA or urokinase. Several methods have been described, most utilizing a chromogenic endpoint. Available commercial methods have been developed to detect high levels of PAI-1 as a thrombotic risk factor, but, for reasons stated below, these are not helpful in this setting. Additionally, they are not well standardized for detecting low levels of PAI-1, which may cause excessive bleeding. PAI-1 antigen may also be measured by ELISA, which detects both the active and inactive forms.

**Tissue Plasminogen Activator:** TPA may be measured by a functional assay due to its ability to convert plasminogen to plasmin with cleavage of a chromogenic substrate. TPA antigen may also be measured by ELISA. PAI-1 and TPA functional measurements vary inversely, and often are measured together. Patients with decreased PAI-1 may have increased TPA. Most assays are designed to detect low TPA as a risk factor for thrombosis, but, for reasons described below, are quite variable.

**Thrombin Activatable Fibrinolysis Inhibitor:** TAFI may be measured by a functional assay or immunologically. Activated TAFI (TAFIa) can also be measured. Although the clinical significance of these tests is not yet clear, TAFI provides a marker for thrombin generation and the role of thrombin in regulating fibrinolysis.

**Fibrinogen:** Fibrinogen assays are a readily available, and can aid in determining the effectiveness of fibrinolytic therapy. Usually, fibrinogen levels are approximately 50% of normal during fibrinolytic therapy, such as occurs with the use of TPA. However, attempts should be made to maintain the fibrinogen level above 100 mg/dl to minimize the bleeding potential.

**Fibrin Degradation Products:** A variety of tests have been described to monitor different FDPs. In general, these represent more global tests for fibrinolysis and fibrinogen/fibrin breakdown. These assays are most sensitive to the smaller fibrinolytic fragments of fibrin/fibrinogen, fragments D and E. It is performed utilizing a latex agglutination methodology (latex particles coated with antibodies to fragments D and E). D-dimers can be particularly informative because, unlike fragments D and E, D-dimers are only formed after Factor XIIIa has cross-linked fibrin. Thus, D-dimers indicate a more stable clot formation.

**Sources of Error:** There is a variety of pre-analytical errors that can occur in tests of fibrinolysis. It is important that these specimens are carefully drawn from patients in a fasting state, usually in the morning and, if possible, without concomitant inflammation and stress. Lack of attention to these variables may contribute to false elevation, especially for TPA and PAI-1. The variability associated with testing is one reason why it is difficult to correctly attribute lack of fibrinolysis to either an acquired or a congenital thrombotic state, and why many investigators are not convinced that either reduced synthesis of TPA or high levels of PAI-1 actually contribute to thrombosis.

In general, increased circulating plasminogen activators can occur with excessive thrombolytic agents such as TPA, or in conditions such as liver disease, certain malignancies (e.g. acute promyelocytic leukemia) and CPB. Decreased inhibitors to fibrinolysis, such as diminished α2AP and decreased PAI-1, can rarely occur on a genetic basis or in the acquired conditions of liver disease, and occasionally in amyloidosis.

Finally, D-dimers constitute a special case that deserves particular attention. First, D-dimers are cleared by the liver. Thus, in advanced liver failure, D-dimers can be elevated by mechanisms other than increased fibrinolysis. However, plasminogen activators are also cleared more slowly in liver failure and there is decreased production of α2AP, which can contribute to authentic hyperfibrinolysis. Thus, interpretation of elevated D-dimers in liver failure must include these confounding factors. As the FDP assays are often agglutination based (including D-dimers), they are susceptible to the prozone effect. In particular, very high levels of D-dimers may give a false-negative result. If there is a clinical picture that would lead to a high index of suspicion for pathophysiology that would typically result in elevated D-dimers, but the D-dimer levels are normal, the laboratory should be asked to perform a dilution of the specimen to test for the prozone effect. If the prozone effect is present, dilution of the specimen will convert a low or normal level to a high level.

## Recommended Reading

Agren A, Wiman B, Stiller V *et al.* (2006). Evaluation of low PAI-1 activity as a risk factor for hemorrhagic diathesis. *J Thromb Haemost* **4**, 201–208.

Agren A, Wiman B, Schulman S. (2007). Laboratory evidence of hyperfibrinolysis in association with low plasminogen activator inhibitor type 1 activity. *Blood Coag Fibrinol* **18**, 657–660.

Bouma BN, Mosnier LO. (2006). Thrombin activatable fibrinolysis inhibitor (TAFI) – how does thrombin regulate fibrinolysis? *Ann Med* **38**, 378–388.

Kim PY, Foley J, Hsu G *et al.* (2008). An assay for measuring functional activated thrombin-activatable fibrinolysis inhibitor in plasma. *Anal Biochem* **372**, 32–40.

Law RH, Sofian T, Kan WT *et al.* (2008). X-ray crystal structure of the fibrinolysis inhibitor alpha2-antiplasmin. *Blood* **111**, 2049–2052.

Nesheim M. (2003). Thrombin and fibrinolysis. *Chest* **124**, 33S–39S.

Rau JC, Beaulieu LM, Huntington JA, Church FC. (2007). Serpins in thrombosis, hemostasis and fibrinolysis. *J Thromb Haemost* **5**, 102–115.

Santamaría A, Borrell M, Mateo J *et al.* (2007). What is the clinical impact of low plasminogen activator inhibitor-1 (PAI-1) activity? A case report and study of the incidence of low PAI-1 antigen in a healthy population. *J Thromb Haemost* **5**, 1565–1566.

# CHAPTER 131
# General Overview of the Hypercoaguable State

Michael A. Briones, DO

It has long been appreciated that multiple factors can contribute to the evolution of a thrombophilic state. Virchow's triad is a concept delineating the pathogenesis of venous thromboembolism (VTE) proposing that a predisposition to VTE occurs as a result of (1) alterations in blood flow (blood stasis), (2) vascular endothelial injury, and (3) alterations in the constituents of the blood that favor clotting (hypercoaguability). In normal hemostasis, the complex interaction between endothelium, platelets and clotting cascade sets in motion a hemostatic response, which is rapid and localized to the injury site. In contrast, the thrombophilic state ensues when clotting occurs in the absence of injury, or when appropriate clotting extends beyond appropriate boundaries in an unchecked response. Perturbation in each of the components of Virchow's triad can lead to thrombophilia, due either to inappropriately increased activation or to deficits in natural anticoagulant and/or clot removal pathways.

## Pathophysiology:

**Blood Stasis:** Venous stasis represents an important pathogenic factor in the development of thrombosis. The role of venous stasis has been investigated in patients with spinal cord injury and other forms of paralysis. These studies show that the majority of venous thrombi originate in regions of slow blood flow, such as the large venous sinuses of the calf and thigh or in valve cusp pockets or bifurcations of the venous system. This becomes particularly apparent in situations of physical inactivity such as bed rest and during air travel, where the lack of pumping action of the large muscles causes decreased blood flow or stasis. It has been suggested that blood-pooling leads to activation of the coagulation system, thus resulting in a state of local hypercoaguability. In addition, possible endothelial damage from distension of the vessel walls by the pooling blood leads to further activation of the homeostasis system. The activation products of clotting and fibrinolysis can also induce endothelial damage, which, in a feed-forward loop, promotes a local state of hypercoaguability.

**Vessel Wall Injury:** Vessel wall injury can occur due to physical trauma, inflammatory vasculitis from a variety of etiologies, or coagulation activation itself by effects of coagulation factors on the endothelium. Surgical manipulation of vessels is a key source of vessel wall injury and vascular activation. For example, damage to the vascular endothelium is a key predisposing factor to venous thrombosis after major hip or knee surgery.

**Hypercoaguability:** The risk of venous thrombosis is increased when the homeostatic balance between pro- and anti-coagulant forces is shifted in favor of coagulation. When this imbalance is due to an inherited defect, the resulting hypercoaguable state remains a lifelong risk factor for thrombosis. In contrast, hypercoaguability due to a transient factor should be treated only as long as the risk factor is present. In most cases, disturbances in the coagulation cascade arise in inherited thrombophilias. Since each genetic defect represents an independent risk factor for thrombosis, individuals with multiple defects have a significantly increased risk of thrombosis.

**Specific Perturbations Leading to Hypercoaguability:** A variety of changes in the balance of coagulation activation and natural anticoagulation are now understood to play a role in thrombophilic states, as a result of either increased coagulation activation or decreased natural anticoagulation. The list of such factors continues to grow, with an evolving array of potential molecular changes that may promote thrombosis. Such factors include (but are not limited to) increased coagulation factors, decreased antithrombin, decreased protein C and protein S, and decreased tissue factor pathway inhibitor. Moreover, mutations that make factors resistant to their natural inhibitors (e.g. Factor V Leiden) can lead to thrombophilia. Thrombosis can also be promoted due to decreased fibrinolysis from changes in proteins such as plasmin, tissue plasminogen activator (t-PA) and PAI-1. A general overview of these factors is presented here, and more thorough details can be found in individual chapters devoted to these different factors.

**Increased Coagulation Factors:** Increased levels of certain coagulation factors are in themselves an independent risk factor for thrombosis. Most notably, elevated Factor VIII levels increase thrombotic risk. In addition, the G20210A mutation in the prothrombin gene results in increased levels of prothrombin, which can promote clot formation.

**Antithrombin System:** There are two principal mechanisms by which thrombin activity is regulated. *Antithrombin* (AT) is a circulating plasma protease inhibitor. It neutralizes several of the enzymes in the clotting cascade, especially thrombin, Factors Xa and IXa, as well as Factors XIIa and XIa. AT has two active functional sites: the reactive center, Arg393-Ser394; and the *heparin*-binding site located at the amino terminus of the molecule. The binding of endogenous or exogenous heparins to the heparin-binding site on AT produces a conformational change in AT which accelerates the inactivating process 1000- to 4000-fold.

Heparins contain a specific pentasaccharide that is important for binding to AT. The glycosaminoglycan heparan sulfate found on endothelial surfaces also contains this pentasaccharide and may mediate part of the physiologic action of AT. This endothelial system in which the cell surface is coated with activated AT is poised to rapidly inactivate any excess thrombin in the general circulation. The potential of this system is magnified by the surface–volume geometry of the microcirculation, in which 1 ml of blood can be exposed to as much as $5000\,cm^2$ of endothelial surface. Thrombin activity is also inhibited by $\alpha_2$-macroglobulin, heparin cofactor II and $\alpha_1$-antitrypsin.

**Activated Protein C and Protein S System:** As coagulation activation progresses, thrombin activates protein C which, together with its natural cofactor protein S, inactivates Factors Va and VIIIa. In this way, proteins C and S inactivate the prothrombinase and the intrinsic tenase, complexes. Deficiencies in either protein C or S can lead to a hypercoaguable state. In addition, a functional equivalent of a defect in protein C activity can arise from altered Factor V. In particular, the Factor V Leiden trait carries a mutation such that it is a poor substrate for protein C and is therefore inactivated more slowly, resulting in a hypercoaguable state.

**Tissue Factor Pathway Inhibitor:** Tissue factor pathway inhibitor (TFPI) circulates in plasma, but at very low concentrations compared to AT. TFPI inhibits Factor Xa in two ways: it directly inhibits Factor Xa, and it complexes with Factor Xa and the complex inhibits TF/FVIIa, thereby impairing the triggering mechanism of the extrinsic pathway. TFPI is primarily synthesized by the microvascular endothelium. About 20% of TFPI circulates in plasma in association with lipoproteins, while the majority remains associated with the endothelial surface, apparently bound to cell-surface glycosaminoglycans. The plasma concentration of TFPI is greatly increased following intravenous heparin administration; this release of endothelial TFPI may contribute to the anti-thrombotic efficacy of heparin and low-molecular-weight-heparin. Recombinant TFPI is currently being evaluated as one of several new anticoagulants.

**Fibrinolytic System:** The fibrinolytic system is essential for removal of excess fibrin deposits in order to preserve vascular patency. The circulating proenzyme plasminogen is a single-chain glycoprotein with an molecular weight of 90 kDa. Cleavage of the Arg560–Val561 bond converts plasminogen to an active two-chain plasmin molecule. Plasmin digests cross linked fibrin. Conversion of plasminogen to plasmin is achieved by a variety of plasminogen activators that include physiological substances such as urokinase and t-PA, as well as streptokinase. Hereditary or acquired defects in finbrinolysis can lead to hypercoaguability.

The most important circulating plasminogen activator in humans is t-PA. As both t-PA and plasminogen bind to the fibrin gel and are incorporated into the developing thrombus, the stage is set for dissolution of the fibrin clot from its inception. The control of plasmin generation is as important as the previously described control of thrombin. The strongest inhibitor of plasmin is $\alpha$2-antiplasmin, a single-chain protein that forms a 1:1 complex with plasmin so rapidly that, under normal circumstances, free plasmin is never detectable. Other inhibitors of minor significance are $\alpha$2-macroglobulin and $\alpha$1-antitrypsin. The most important inhibitor of t-PA is the PAI-1, which not only inhibits t-PA but also urokinase plasminogen activator (u-PA). PAI-1 circulates in plasma in a free, non-complexed form of 48 kDa and as a complex with t-PA of 110 kDa. In normal human plasma, most of the t-PA is complexed. Plasma-PAI-1 behaves as an acute phase reactant, rising in a variety of pathological conditions and occurring in relatively large quantities in platelets.

## Recommended Reading

Bockenstedt PL. (2006). Management of hereditary hypercoagulable disorders. *Hematology Am Soc Hematol Educ Program*, 444–449.

Kroegel C, Reissig A. (2003). Principle mechanisms underlying venous thromboembolism: epidemiology, risk factors, pathophysiology and pathogenesis. *Respiration* **70**, 7–30.

Wagman LD, Baird MF, Bennett CL. (2006). Venous thromboembolic disease. *J Natl Compr Cancer Netw* **4**, 838–869.

# CHAPTER 132

# Antithrombin Testing

James C. Zimring, MD, PhD

Antithrombin (AT) (previously called antithrombin III) is a natural anticoagulant which, upon activation by heparin, is a potent inactivator of coagulation. The predominant activities of AT are inhibition of distal coagulation factors thrombin (Factor IIa) and Factor Xa. However, AT also has some activity against more proximal factors (Factors IXa, XIa and XIIa), contact factors and, in some cases, Factor VIIa.

AT functions normally in the absence of administered heparin, presumably by activation by heparan sulfate associated with the vascular endothelium. However, therapeutic administration of heparin greatly enhances this activity. Inherited AT deficiency can consist of quantitative (type I) deficiencies in which AT is normal but present at decreased levels, or qualitative (type II) deficiencies in which AT is present but mutations alter AT function. AT deficiency can also be acquired through mechanisms resulting in decreased synthesis (e.g. hepatic disease) or increased loss (e.g. consumptive coagulopathies and nephrotic syndrome). Due to the presence of type I and type II defects, tests of both AT activity and AT protein quantities (antigen test) are required.

Both thrombosis itself and treatment with anticoagulants can substantially alter measurable levels of AT. Thus, when possible, AT testing should be performed after a thrombotic event has resolved and when the patient is not on anticoagulant therapy. The risk of decreasing anticoagulation in order to assess AT deficiency must be evaluated on a case-by-case basis, based upon clinical judgment. In addition, AT levels can be decreased in the neonatal period; during pregnancy; in the presence of liver disease, burn injury, trauma, sepsis, disseminated intravascular coagulation and nephrotic syndrome; and secondary to treatment with L-asparaginase and estrogens.

**AT Activity:** The indication for measuring AT activity is the clinical assessment of congenital or acquired AT deficiency such as in patients with thrombophilia. Several methodologies have been described and used, but in general AT activity is determined by measuring the ability to inhibit a known quantity of natural targets Factors IIa or Xa. Typically, heparin is added to the patient specimen in order to activate all of the AT present. Usually, an amidolytic peptide substrate for Factors II or X is used, which gives a chromogenic readout.

**Test Interpretation:** Decreased AT activity is consistent with either a type I or a type II deficiency. Additional testing for AT antigen is required to distinguish these possibilities.

**Sources of Error:** Although inhibition of either Factor IIa or Factor Xa can provide accurate measurements of AT activity, there is an additional serum factor that can inhibit Factor IIa in a heparin-dependent fashion (i.e. heparin cofactor II

(HCII)). Thus, in some cases, HCII will mask an AT deficiency. Unlike Factor IIa, which is inhibited by both AT and HCII, Factor Xa is inhibited by AT but not HCII. Thus, decreases in AT are not obfuscated by HCII when FXa is used as a substrate. Compared to human thrombin, bovine thrombin is less sensitive to the effects of HCII. Thus, assays using either Factor Xa or bovine thrombin are preferred to human thrombin. HCII deficiency alone is not known to be a risk factor for thrombosis, although it may contribute to thrombosis in combination with other risk factors.

**AT Antigen:** As for AT activity, the indication for measuring the AT antigen level is for clinical assessment of congenital or acquired AT deficiency, such as seen in patients with thrombophila. AT antigen is measured using monoclonal antibodies that recognize AT by several techniques, most commonly by ELISA. AT antigen levels are typically ordered after an abnormal AT activity assay, to distinguish type I and type II defects.

**Test Interpretation:** Decreased AT antigen indicates either a type I inherited defect or an acquired AT decrease. However, a normal AT antigen level alone does not rule out a type II defect, which has normal amounts of AT (antigen levels) but decreased activity on a molecule by molecule basis.

### Recommended Reading

Blajchman MA. (1994). An overview of the mechanism of action of antithrombin and its inherited deficiency states. *Blood Coag Fibrinol* **5**, 5–12.

Perry DJ. (1994). Antithrombin and its inherited deficiencies. *Blood Rev* **8**, 37–55.

# CHAPTER 133

# Proteins C, S and Z Testing

James C. Zimring, MD, PhD

Proteins C, S and Z are vitamin K dependent factors that function as anticoagulant proteins. Congenital or acquired deficiencies in protein C and/or S can lead to clinical thrombophilia, and therefore the measurement of their activity is indicated in the evaluation of thrombophilia. The role of protein Z in the evaluation of thrombophilia is unclear.

**Protein C and Protein S:** Proteins C and S are vitamin K dependent factors that function as anticoagulant proteins, and congenital or acquired deficiencies in these proteins can lead to clinical thrombophilia. Proteins C and S are cofactors, which largely become active as a negative feedback in response to coagulation activation. Activation of coagulation converts prothrombin to thrombin (T) (Figure 133.1). In addition to promoting clot formation by converting fibrinogen to fibrin, thrombin also activates a negative feedback anticoagulation pathway by activating protein C. Although low levels of protein C can be activated directly by free thrombin, the process is greatly enhanced by the formation of thrombin thrombomodulin (TTM) complexes, when thrombin binds to thrombomodulin on the endothelial surface. This

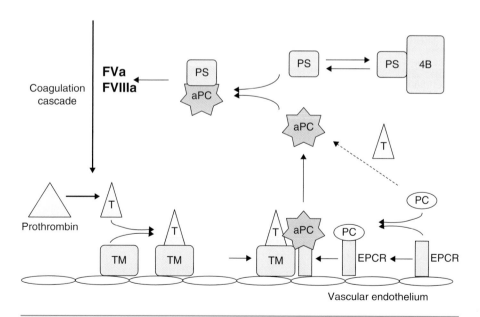

FIGURE 133.1 Activation of coagulation converts prothrombin to thrombin. aPC = activated protein C; EPCR = endothelial protein C receptor; PS = protein S; T = thrombin; TM = thrombomodulin; 4B = C4b binding protein.

process is enhanced further by the binding of protein C to an endothelial protein C receptor (EPCR), which places protein C in close proximity to the TTM complex. Once activated and freed from the EPCR, protein C can inhibit clotting Factors Va and VIIIa, but only does so very weakly unless complexed to protein S as a cofactor. The protein C/S complex provides strong anticoagulant activity and negative feedback to coagulation.

The activity of protein S, and therefore the ability of the protein C/S complex to function, is tightly regulated by another protein named C4b binding protein (C4bBP). Under normal conditions, approximately 60% of protein S is complexed to C4bBP, and is in equilibrium with free protein S. As only free protein S is able to complex with protein C, protein S activity is a function of both total protein S and C4bBP levels.

In general, testing for protein C and/or protein S can be broken down into functional assays (measuring the amount of activity) versus antigen assays (measuring the amount of the protein present). These two assays can be combined to assess qualitative versus quantitative defects. There is a variety of methodologies for both types of assays.

*Protein C Activity (Coagulation-based Assays):* Functional assays measure the activity of protein C. Because much of the protein C in a patient's specimen has not yet been activated, a protein C activation step is required prior to assaying protein C activity. This can be accomplished by addition of the physiological activators (thrombin or TTM). Alternatively, protein C can be activated by copperhead snake venom. After this step, the specimen is added to protein C deficient plasma, and a partial thromboplastin test (PTT) is performed. Prolongation of the PTT is a measure of the protein C activity. Accurate interpolation of protein C levels requires a standard curve of known protein C quantities.

*False Positives and Negatives:* Because the readout of PTT based assays is dependent upon the other factors of the intrinsic pathway functioning normally, it is susceptible to underestimation of protein C activity by changes that shorten the PTT, such as elevated Factor VIII levels during acute phase reactions. Protein C activity can likewise be underestimated if the PTT has decreased susceptibility to protein C activity due to a Factor V Leiden mutation. Conversely, protein C may be overestimated due to other factors that prolong the PTT, such as a lupus anticoagulant. Finally, the functional assay may not be valid in the context of anticoagulant therapy that alters the PTT (e.g. heparin or a direct thrombin inhibitor).

*Protein C Activity using Artificial Substrates:* Instead of measuring the activity of protein C in the context of the entire PTT, a separate strategy is to engineer a small peptide substrate that contains the recognition sequence for protein C activity (amidolytic assays). By linking the peptide to an indicator to measure cleavage (e.g. a para-nitroaniline group), protein C activity can be measured directly. This assay is not susceptible to interfering factors involving other members of the coagulation system.

*False Positives and False Negatives:* Because the peptide substrate is artificial in its cleavage conditions, it does not require that protein C interact with its normal binding partners. For example, a defect in the protein C domain that binds Factor V would prevent protein C from cleaving Factor Va (even if its enzymatic activity were intact), because the Factor V affinity was decreased. Such a defect would be detected in a PTT-based assay where protein C binds its authentic substrate, but would not be detected by a peptide substrate. In addition, in theory, an antibody inhibitor of protein C could prevent protein C from binding to its natural substrates, while a peptide substrate may still access the enzymatic domain. Case reports of patients with decreased protein C activity in a PTT-based but not peptide based assay have been described due to either a specific mutation in protein C or, more commonly, an incompletely characterized inhibitor found in uremic patients that inhibits the PTT but not a peptide-based assay.

*Protein C Antigen Assays:* Determination of protein C antigen can be carried out by a variety of antibody-based detection methodologies (e.g. ELISA). However, the levels of protein C antigen do not necessarily reflect potential protein C activity. Normal antigen with decreased activity can be found in type II qualitative genetic deficiency or, in very rare cases, the presence of an acquired protein C inhibitor. Measuring antigen levels may be helpful in evaluating patients on warfarin, whose protein C is inactive due to lack of vitamin K dependent modification. However, this approach is likely of limited value, as warfarin can also decrease the amount of protein C present.

False Positives and Negatives: Assays for protein C antigen are susceptible to the general factors that cause artifacts in antibody–antigen-based assays. The presence of heterophilic antibodies can lead to false positives, whereas very high concentrations can lead to false negatives in ELISA testing (i.e. the hook effect) or in bead agglutination assays (i.e. the prozone effect).

*Protein S Activity:* Due to the fact that protein S does not have an intrinsic activity of its own but functions by enhancing the activity of protein C, the tests of protein S function depend upon enhancement of protein C activity. Thus, functional protein S assays can be performed by mixing patient specimens with protein S deficient plasma. Increases in coagulation time correspond to levels of protein S. Any of the above assays for protein C activity can be used for the readout; however, in practice, PTT-based tests are most common. Accordingly, measurement of protein S levels is susceptible to each of the above factors that can lead to false positives or negatives in the measurement of protein C activity.

*Assays of Protein S Antigen:* Protein S is only able to act as a cofactor for protein C when it is in its free form (i.e. not bound by C4bBP). Some protein S antigen assays only give a total protein S level. Normally, about 60% of protein S is bound by C4bBP. However, C4bBP is an acute phase reactant that can be induced to high levels by inflammation and illness; C4bBP is also elevated during pregnancy. To distinguish free protein S from C4bBP bound protein S, antibody reagents that distinguish between the two forms can be utilized, or a precipitation step can be performed with polyethylene glycol, which precipitates C4bBP bound, but not free protein S.

False Positives and Negatives: Assays for protein S antigen are susceptible to the general factors that cause artifacts in antibody–antigen-based assays. The presence of heterophilic antibodies can lead to false positives, whereas very high concentrations can lead to false negatives in ELISA testing (i.e. the hook effect) or in bead agglutination assays (i.e. the prozone effect).

*Iatrogenic Alterations in Protein C and Protein S:* Perhaps the most common confusion in protein C and protein S testing comes from patients who have had a thrombosis and have begun warfarin therapy prior to a laboratory work-up that includes protein C and protein S testing. This is not a "false" negative *per se*, as the assays are correctly measuring protein C and S; rather, the levels are actually low due to anticoagulant therapy. Similar to Factors II, VII, IX and X, proteins C and S are vitamin K dependent factors. Thus, patients on warfarin have decreased protein C and protein S activity.

A genetic deficiency in both protein C and S would be expected to present with profound thrombophilia from birth. Thus, any sample showing decrease in both protein C and S in an adult should immediately be suspected of warfarin. Measuring protein C and S antigen levels may address this issue to some extent, as warfarin mainly targets the activity. However, warfarin can also decrease total antigen levels. Some would consider whether it is worth switching patients to different anticoagulation to allow for more accurate protein C and S testing. Implementation of such a maneuver depends upon clinical judgment on a case-by-case basis.

Diagnosis of Clinically Significant Protein C and/or Protein S Defects: Diagnosis of protein C or protein S deficiency is complicated by several issues. The lower levels of these factors in normal individuals overlap with the upper levels of these factors in heterozygous deficient states. Overall, less than 50% activity is typically associated with a deficient state. Often, 60–70% is a borderline finding that leads to retesting. Genetic defects in either protein C or protein S can present as a quantitative defect (type I) that has decreased activity and antigen levels, or a qualitative defect (type II) that has normal antigen levels with decreased activity. As above, some type II protein C patients have normal activity in amidolytic assays and require a PTT-based assay to detect the deficiency, due to a defect in required binding domains other than the enzymatic site. A type III protein S deficiency has been described, which has decreased total protein S with normal free protein S, but recent data challenge whether type III patients have an increased risk of thrombosis.

Both protein C and protein S deficiencies can be acquired. Decreased levels of either can be seen during the neonatal period, sepsis, DIC, acute thrombosis, liver disease, and as a result of warfarin or L-asparaginase treatment. Selective acquired protein S deficiency is usually the result of increased levels of C4bBP, which occur during inflammation, in pregnancy, and as a result of oral contraception (estrogens). Relative elevated levels of C4bBP can also be seen in nephrotic syndrome due to renal loss of free protein S but not C4bBP-bound protein S.

Protein Z: Like proteins C and S, protein Z is a vitamin K dependent factor that participates in anticoagulant pathways. However, the activity of protein Z is a distinct pathway from the protein C/S complex. Protein Z is a cofactor for the Z dependent protease inhibitor, and together these two proteins can inactivate Factor Xa. There are conflicting data as to whether isolated perturbed protein Z levels are of clinical relevance. Evidence exists to suggest that the combination of protein Z deficiency with other thrombophilic factors (e.g. Factor V Leiden) may increase the risk of thrombosis. In contrast, some data indicate an apparently paradoxical protection against stroke by decreased levels of protein Z. In any case, while protein Z may theoretically play a role in predicting thrombosis, measurement of protein Z is not typically used in thrombophilia evaluations and remains a factor of unknown clinical significance. The indication for measuring protein Z level is in a thrombophilia evaluation; however, it is not a routine laboratory measurement. The primary method of evaluation is ELISA.

# CHAPTER 134

# Activated Protein C Resistance and Factor V Leiden Testing

Anne M. Winkler, MD and James C. Zimring, MD, PhD

The Factor V Leiden (FVL) mutation promotes thrombophilia due to decreased efficiency of a natural anticoagulant pathway. Inactivation of Factor Va is one of the main targets by which proteins C and S exert their natural anticoagulant effects. However, several allelic variants of Factor V have been described, which contain amino acids substitutions that decrease the ability of Factor V to be inactivated by activated protein C (APC). The most predominant of these mutations is a single point mutation in the Factor V gene (G1691A), known as Factor V Leiden. This mutation results in a substitution of arginine by glutamine at position 506 (R506Q), the dominant APC cleavage site of Factor Va. There are two main approaches to testing for FVL: APC resistance and direct genomic analysis.

**Activated Protein C Resistance:** APC resistance is a fluid phase functional assay that measures the ability of protein C to inactivate Factor Va. As the mechanism by which FVL promotes thrombosis is a decrease in its ability to be inactivated by APC, measuring this "APC resistance" can be predictive of FVL. The primary indication for this assay is the clinical assessment of thrombophilia.

**Methods:** Most commercially available assays for detection of APC resistance utilize PTT-based or PT-based Factor V assays for detection. The principle of the assay involves measuring a PTT prior to and after the addition of APC and calcium to a plasma sample. Theoretically, addition of APC to a normal plasma sample should prolong the PTT due to inactivation of Factors Va and VIIIa, whereas in an APC-resistant individual the degree of prolongation of the PTT should be decreased. As a result, a calculation known as the APC sensitivity ratio is performed:

$$\text{APC sensitivity ratio} = \frac{\text{PTT in the presence of APC}}{\text{PTT in the absence of APC}}$$

A resultant ratio less than or equal to the cut-off value is consistent with resistance to APC. It has been reported that most patients with ratios less than 0.71 are heterozygous or homozygous for FVL.

**Test Interpretation:** A positive APC resistance may indicate FVL, and should lead to direct molecular testing. It is essential to remember that there are mutations other than FVL that can affect APC resistance (see Table 134.1). FV Liverpool and R485K give a positive APC resistance, and may contribute to thrombosis. In contrast, FV Cambridge gives a positive APC resistance but does not appear to cause thrombosis. Additional

**TABLE 134.1  Variant Factor V Mutations**

| Variant | Nucleotide variation | Amino acid substitution | APC resistance | Increased risk of thrombosis |
|---------|----------------------|-------------------------|----------------|------------------------------|
| FV Cambridge | G1091C | R306T | Yes | No |
| FV Hong Kong | A1090G | R306G | No | No |
| FV Liverpool | T1250C | I359T | Yes | Yes |
| R485K mutation | G1628A | R485K | Yes | Yes |
| R2 haplotype | T1328C | M385T | Yes | Unknown |
|  | A4070G | H1299R |  |  |
|  | A5380G | M1736V |  |  |
|  | A6755G | D2194G |  |  |
| A/G allele | A2391G | S739S | Yes | Unknown |
|  | A2663G | K830R |  |  |
|  | A2684G | H837R |  |  |
|  | A2863G | K897D |  |  |

mutations that give APC resistance have unknown clinical effects. Thus, a positive APC resistance in the absence of the FVL mutation can still be an independent risk factor for thrombosis.

A negative APC resistance has been used to rule out the presence of FVL. While false negatives are very rare, especially in second-generation assays, any coagulation activity based system is susceptible to interference from potentially unanticipated substances. Thus, a direct assay for the FVL mutation may be indicated despite a normal APC resistance if there is a high clinical suspicion of thrombophilia.

**Test Performance:** In first-generation assays, an additional normalization step was required; however, the addition of Factor V-deficient plasma to current assay systems has abrogated this requirement. Current generation assays have a sensitivity and specificity approaching 100%.

**Sources of Error:**

*False Positives:* Using Factor V-deficient plasma has substantially reduced interference from non-Factor V factor deficiencies and use of oral anticoagulants such as warfarin that were previously the causes of false-positive results. However, false positives can and still do occur in the second-generation assays due to baseline alterations of the PTT resulting from the presence of a lupus anticoagulant, other antibody-based factor inhibitors, Factor V deficiency, increased Factor VIII levels, or as a result of newer anticoagulants such as direct thrombin inhibitors and Factor Xa antagonists. Prolongation of the PTT due to heparin can be avoided by the use of heparin-neutralizing agents, and further dilution with normal plasma or addition of exogenous phospholipids can be utilized to neutralize lupus anticoagulants. However, any result consistent with APC resistance should be confirmed with a direct test for FVL.

**Factor V Leiden:** The FVL test directly detects the presence of a polymorphism in the Factor V gene. This polymorphism encodes for the FVL gene product, which is a risk factor for thrombosis. Testing for FVL is typically used in patients who have had an unexplained thrombosis and are being evaluated for thrombophilia, or for evaluation of recurrent pregnancy loss. In some cases, it is used for genetic counseling of patients whose family members have been found to have the FVL mutation.

Methods: A variety of DNA analytic methods can be used to test for FVL, including polymerase chain reaction (PCR) based and non-PCR based methods. As the FVL mutation (G1691A) creates a novel restriction endonuclease recognition site, initial PCR based methods included restriction fragment length polymorphism (RFLP) analysis. Primers that recognize conserved sequences in Factor V were used for PCR amplification of the Factor V gene followed by digestion using Mnl1 endonuclease and agarose gel electrophoresis. Subsequent methods have been developed, including (1) the amplification-refractory mutation system (ARMS), (2) single-strand conformation polymorphism (SSCP), (3) enzyme-linked immunosorbent assays (ELISA) for nucleic acid products, (4) real-time PCR with fluorophore-labeled allele-specific hybridization probes and fluorescence resonance energy transfer (FRET) assays, and (5) direct sequencing of DNA. In addition, newer, non-PCR based assays have also been utilized for the detection of FVL.

Test Interpretation: As a general principle, nucleic-acid based tests result in either the presence or absence of the allele, interpreted as homozygous normal, heterozygous FVL or homozygous FVL. When compared, the various molecular methods for detection of FVL have a high concordance rate. However, it is important to note that the error rate in testing has been estimated as between 3% and 6%. As diagnosis for a genetic risk factor for thrombosis may result in lifelong anticoagulation, it has been recommended that positive tests for FVL be confirmed with repeat testing, preferably by an alternate method.

There are also other variants of Factor V, which are not routinely tested for, that may have clinical significance (see Table 134.1). Such variants will not be detected by methodologies focused on detecting the presence of the G1691A mutation.

Sources of Error: Sources of error are predominantly restricted to the general issues intrinsic to amplification based systems.

*False Positives:* With any PCR- or amplification-based systems, contamination (theoretically as few as 1–10 molecules) from a positive patient or a previous amplification is sufficient to give a false positive.

*False Negatives:* Inhibitors of PCR, such as heparin or unusual contaminating salts, can prevent amplification. This will give a false-positive result in systems geared to amplify the FVL, but not normal gene products by sequence-specific primers. However, in systems that amplify the Factor V gene and then detect FVL through RFLP or sequence-specific probe, the presence of an inhibitor will be easily detected, as no amplification of any product will be observed. In addition, false negatives have been reported after allogeneic HPC transplantation, as peripheral blood was utilized as a source of DNA and the HPC donor did not carry the FVL polymorphism, but the patient carried the mutation.

There are additional polymorphisms that have been described in the Factor V gene other than FVL (e.g. Cambridge, Hong Kong, Liverpool, R485K, R2 haplotypes, and A/G allele – see Table 134.1). Of these, FV Liverpool and FV R485K appear to be of clinical significance, but will not be detected by DNA analytic techniques specific for FVL. Thus, a normal FVL test does not rule out an abnormal Factor V, which may lead to thrombophilia. However, as previously described, APC resistance assays will often detect these variants.

## Recommended Reading

Cooper PC, Rezende SM. (2007). An overview of methods for detection of Factor V Leiden and the prothombin G20210A mutations. *Intl J Lab Hem* **29**, 153–162.

Rosendorff A, Dorfman DM. (2007). Activated protein C resistance and Factor V Leiden: a review. *Arch Pathol Lab Med* **131**, 866–871.

Segers K, Dahlbäck F, Nicolaes GA. (2007). Coagulation Factor V and thrombophilia: background and mechanisms. *Thromb Haemost* **98**, 530–542.

# CHAPTER 135

# Prothrombin Gene Mutation Testing

Anne M. Winkler, MD and James C. Zimring, MD, PhD

The prothrombin (Factor II) gene mutation assay detects the presence of the G20210A mutation in the 3′ untranslated region of the prothrombin gene. This assay is indicated in the assessment of thrombophilia.

Prothrombin is a vitamin K dependent protein that is intimately involved in coagulation. Upon activation, prothrombin is converted to thrombin, which in turn converts fibrinogen into fibrin. Thus, thrombin is a critical component of the coagulation cascade, directly resulting in deposition of fibrin and clot formation. In addition, thrombin also activates compensatory negative-feedback anticoagulant and thrombolytic pathways.

In 1996, a single guanosine to adenosine polymorphism at nucleotide position 20210 of the prothrombin gene located on chromosome 11 was reported as a risk factor for venous thrombosis. Unlike many other mutations that change the amino acid sequence of a protein, G20210A is located in the 3′ untranslated region of the prothrombin gene; hence, the amino acid sequence of the expressed prothrombin is unaltered. However, the G20210A mutation results in both increased mRNA production through enhanced RNA processing, and stabilization of mRNA leading to a longer half-life. Through these two mechanisms, greater amounts of prothrombin mRNA are present, leading to increased prothrombin and thrombin protein levels in plasma. This results in a risk factor for coagulation activation and thrombosis; in particular, increased rates of cerebral and deep vein thrombosis. Coincidence of the G20210A mutation with Factor V Leiden (see Chapter 134) synergistically increases risk of thrombosis.

## Prothrombin Gene Mutation Assay:

**Methods:** A variety of assays have now been described for detecting the G20210A mutation. Direct tests for G20210A rely predominantly on polymerase chain reaction (PCR) based assays. Initial PCR based methods included restriction enzyme digestion to detect the G20210A mutation. Subsequent assays have been described using real-time PCR, the amplification-refractory mutation system (ARMS), single-strand conformation polymorphism (SSCP), enzyme-linked immunosorbent assays (ELISA) for nucleic acid products, and direct sequencing of DNA. A non-PCR based assay has also been described.

Prothrombin protein levels can be measured in the plasma using Factor II activity or antigen assays. However, these assays are not recommended for screening of the prothrombin G20210A mutation due to the insufficient sensitivity and specificity.

**Test Interpretation:** As a general principle, nucleic-acid based tests result in either the presence or absence of the allele, interpreted as homozygous normal, heterozygous G20210A or homozygous G20210A. However, it is important to note that the error rate in testing has been estimated between 3% and 6%. As diagnosis for a genetic risk factor for thrombosis may result in lifelong anticoagulation, it has been recommended that positive tests for G20210A be confirmed with repeat testing, preferably by an alternate method.

It is important to note that there are additional polymorphisms that have been described in the prothrombin gene that will not be detected by G20210A specific assays; polymorphisms have been found at positions 20207, 20209, 20218 and 20221. The clinical significance of these alternate mutations is unclear, although data exist to suggest that the polymorphism C20209T is not associated with an increased risk of thromboembolism. However, as the clinical significance of these mutations is determined, additional characterization of the prothrombin gene sequence may be required for assessment of thrombotic risk.

**Sources of Error:** Sources of error are predominantly restricted to general issues intrinsic to amplification-based systems.

*False Positives:* With any PCR or amplification based systems, contamination (theoretically as little as 1–10 molecules) from a positive patient or a previous amplification is sufficient to give a false-positive result.

*False Negatives:* Inhibitors of PCR, such as heparin or unusual contaminating salts, can prevent amplification. This will give a false-negative result in systems geared to specifically amplify G20210A, but not the normal gene variant. However, in systems that amplify the prothrombin gene and then detect through RFLP or sequence-specific probes, the presence of an inhibitor will be easily detected, as no amplification of any product will be observed. In addition, false negatives for PCR based tests have been reported after allogeneic HPC transplant, as peripheral blood was used as a source of DNA and the HPC donor did not encode the mutation, but the patient carried the mutation.

## Recommended Reading

Cooper PC, Rezende SM. (2007). An overview of methods for detection of Factor V Leiden and the prothombin G20210A mutations. *Intl J Lab Hem* **29**, 153–162.

McGlennen RC, Key NS. (2002). Clinical and laboratory management of the prothrombin G20210A mutation. *Arch Pathol Lab Med* **126**, 1319–1325.

Nguyen A. (2000). Prothrombin G20210A polymorphism and thrombophilia. *Mayo Clin Proc* **75**, 595–604.

Poort SR, Rosendall FR, Reitsma PH, Bertina RM. (1996). A common genetic variation in the 3′-untranslated region of the prothrombin gene is associated with elevated plasma prothrombin levels and an increase in venous thrombosis. *Blood* **88**, 3698–3703.

# CHAPTER 136

# Laboratory Diagnosis of Hyperhomocysteinemia

James C. Zimring, MD, PhD

The presence of hyperhomocysteinemia is associated with increased risk for atherosclerotic vascular disease and thrombophilia. Homocysteine is an amino acid in humans, which is not incorporated into proteins during translation. Rather, it is an intermediate product that is synthesized from methionine, along the path of cysteine synthesis. Once formed, homocysteine can then be converted into cysteine, or recycled back into methionine. The conversion of homocysteine into cysteine or methionine requires vitamins B6 and B12, respectively. Thus, deficiencies in vitamin B6, B12 or folate can result in elevated homocysteine levels (hyperhomocysteinemia). In addition, homocysteine levels can be substantially elevated due to impaired metabolism in renal failure, and consumption of certain drugs (predominantly the fibrate class of lipid-lowering drugs, methotrexate, and some anti-seizure medications). Environmental factors such as smoking and consuming large amounts of coffee have been linked to increased homocysteine levels, especially when combined with low folate. Finally, there are genetic causes of hyperhomocysteinemia, most commonly due to mutations in the methylene tetrahydrofolate reductase (MTHFR) gene. The (T) mutation (C677T) is the best described mutation that leads to a thermolabile form of the enzyme.

Hyperhomocysteinemia can promote both vascular thrombosis and venous thromboembolism. Homocysteine is a reactive molecule that degrades proteins through thiolation of lysine and cysteine residues. Thus, elevated levels of homocysteine can have widespread effects on host proteins. With relevance to thrombosis, there is evidence that highly elevated levels of homocysteine can induce vascular injury and promote atherosclerotic plaques. Additional data indicate direct effects promoting platelet activation and decreasing thrombolytic pathways.

While profound elevations in homocysteine clearly cause substantial pathology, there are widespread conflicting data as to whether moderate elevations are a significant risk factor for thrombotic disease. Thus, it is unclear whether homocysteine levels are an appropriate screening test for evaluating atherosclerosis and thrombosis risk. Currently, it is not included in most screening recommendations. However, given its known potential contribution, evaluation of homocysteine may be meaningful in the evaluation of patients with other risk factors (in particular Factor V Leiden) or in the setting of a thrombotic patient for whom no alternate cause can be found.

**Plasma Homocysteine Level:** Plasma homocysteine levels are used in the assessment of increased risk of atherosclerotic vascular disease and thrombophilia.

**Methods:** High-pressure liquid chromatography (HPLC) is the most common methodology for assaying homocysteine in plasma. The readout can be detected either by electrochemical detection, or by using thiol-specific derivatization agents that contain fluorochromes. Alternate methods can also be employed, including enzyme-linked immunosorbent assay (ELISA), capillary electrophoresis and gas-liquid chromatography.

**Sources of Error:** Pre-analytical issues represent the predominant source of error in determination of homocysteine concentration.

*Specimen Storage:* One source of homocysteine in serum is direct release from red blood cells (RBCs). Because RBCs continue to release homocysteine *ex vivo*, storage of unseparated blood specimens for greater than 1 hour can result in an artificial increase in plasma homocysteine. If necessary, specimens can be stored for up to 4 hours on ice.

*Random versus Fasting Homocysteine Levels:* Plasma homocysteine levels decrease in the first few hours after eating, followed by a progressive increase. It has been suggested that random screening be performed, followed by a fasting homocysteine, if elevated levels are detected; others have suggested fasting homocysteine alone as a screening test. Although less convenient for the patient, fasting homocysteine gives more consistent results.

*Orthostatic Variation:* Plasma homocysteine has orthostatic variation, and can be decreased up to 30% at supine rest compared to standing, suggesting a role in blood pressure regulation and/or vascular tone. This factor should be controlled for, and becomes significant most often with supine hospitalized patients compared to outpatient laboratory sampling.

**MTHFR Mutation:** Genotyping of an individual for the C677T (T) mutation of MTHFR is indicated in the assessment of potential genetic risk factors for atherosclerotic disease or thrombophilia.

**Methods:** MTHFR genotyping is carried out mainly by PCR, with either restriction length polymorphism as a sequence-specific readout or with sequence-specific fluorescent probes in a real-time PCR assay. In addition, a cleavase invader assay is also available.

**Test Interpretation:** The C677T mutation is a common variant that results in decreased MTHFR activity due to lability of the enzyme. The presence of the (T) form is linked to increased levels of plasma homocysteine. However, whether this correlates to increased risk of vascular or thrombotic disease is a matter of dispute. Moreover, the (T) mutation appears to give substantial elevation of homocysteine, particularly when compounded with decreased folate levels.

An additional mutation has been described in MTHFR (A1298C), which is found in *trans* with C677T. Together these mutations further destabilize MTHFR, but the addition of A1298C may not result in further increases in homocysteine. Testing for A1298C is available by PCR in combination with C677T screening.

As a general principle, nucleic-acid based tests are interpreted as homozygous normal, heterozygous, or homozygous for the mutant allele. However, it is important to note that the error rate in testing is not zero. As diagnosis for a genetic risk factor for thrombosis may result in lifelong anticoagulation, it is recommended that positive tests for MTHFR mutations be confirmed with repeat testing, preferably by an alternate method.

## Sources of Error:

*False Positives:*  With any PCR- or amplification-based systems, contamination (theoretically as little as 1–10 molecules) from a positive patient or a previous amplification is sufficient to give a false positive.

*False Negatives:*  Inhibitors of PCR, such as heparin or unusual contaminating salts, can prevent amplification. In addition, false negatives for PCR based tests have been reported after allogeneic HPC transplant, as peripheral blood was used as a source of DNA and the HPC donor did not encode the mutation, but the patient carried the mutation.

## Recommended Reading

Eldibany MM, Caprini JA. (2007). Hyperhomocysteinemia and thrombosis: an overview. *Arch Path Lab Med* **131**, 872–884.

Lievers KJ, Kluijtmans LA, Blom HJ. (2003). Genetics of hyperhomocysteinaemia in cardiovascular disease. *Ann Clin Biochem* **40**, 46–59.

Ubbink JB. (2000). Assay methods for the measurement of total homocyst(e)ine in plasma. *Sem Thromb Hemost* **26**, 233–241.

# CHAPTER 137

# Laboratory Diagnosis of Lupus Anticoagulant and Antiphospholipid Antibodies

Michael A. Briones, DO

The lupus anticoagulant (LA) was first described as a poorly characterized substance that prolonged a partial thromboplastin time (PTT) and was associated with systemic lupus erythematosus (SLE). However, it has subsequently been appreciated that, in contrast to the bleeding tendencies that might be expected as a result of its PTT-prolonging activity, LAs promote thrombophilia *in vivo*, leading to outcomes such as venous and arterial thrombosis, thrombocytopenia and/or recurrent pregnancy loss (see Chapter 94). LAs can be observed in the context of SLE and other collagen vascular diseases, and as a primary syndrome without associated autoimmune diseases. The "lupus anticoagulant" is neither associated exclusively with SLE, nor is it an anticoagulant, and therefore LA is an inaccurate name, which has led to much confusion.

Biochemical analysis has led to the identification of most LAs as antiphospholipid antibodies, and the thrombotic pathophysiology caused by LAs is called the antiphospholipid antibody syndrome (APS). However, these "antiphospholipid antibodies," named due to their original detection using crude cardiolipin as a substrate, are in fact highly heterogeneous in their substrate specificity and can recognize a number of targets, including cardiolipin, β-2-glycoprotein I (β2GPI), prothrombin, annexin V, protein C and phospholipids. Thus, most of the "antiphospholipid" antibodies in APS are not specific for phospholipids themselves, but recognize protein/phospholipid complexes.

Patient evaluation for APS usually occurs in the setting of young patients without other known risk factors who have sustained a venous thromboembolism, stroke, peripheral arterial thrombosis, myocardial infraction or pregnancy complication (e.g. recurrent fetal loss, pre-eclampsia, intrauterine growth retardation), or as part of the evaluation of SLE. The International Society of Thrombosis and Hemostasis (ISTH) has created the Sapporo criteria, and later the modified Sapporo criteria, for the classification of APS. According to these criteria, APS is present in patients with one clinical and one laboratory criterion. Clinical criteria include objectively confirmed arterial, venous or small-vessel thrombosis; or pregnancy complications consisting of recurrent fetal loss before the tenth week of gestation, one or more unexplained fetal deaths at or beyond the tenth week of gestation, or premature birth due to placental insufficiency, eclampsia or pre-eclampsia. Laboratory criteria include either an LA or a medium to high titer antiphospholipid IgG or IgM on two or more occasions at least 6 weeks apart. Thus, repeat testing with ongoing clinical monitoring is required for patients suspected of APS. Moreover, evaluation of other causes of thrombophilia and the proper clinical identification of high-risk patients is required to limit the pre-test probability to a level that makes the testing meaningful.

**Diagnosis:** Given the complexity of the biochemical nature of LAs/antiphospholipid antibodies, diagnosis of the APS does not depend on any one laboratory test. Instead, a panel of tests is usually performed (fluid phase tests for LA activity and solid phase tests for antiphospholipid antibodies). Laboratory support for the diagnosis of APS relies on the demonstration of an LA and/or an antiphospholipid antibody. Diagnosis of APS depends upon persistent positivity of the above tests in the context of clinical findings consistent with APS (see Chapter 94).

**Methods and Interpretation:** LAs are identified as activities in coagulation-based assays, in which they prolong phospholipid-dependent clotting due to a phospholipid specific inhibitor, and not as a result of specific inhibition of any one coagulation factor. In addition to phospholipid neutralization (see below), specific factor assays and inhibitor assays can be employed to rule out inhibition due to a non-LA based specific factor inhibitor.

Because LAs are highly heterogeneous in their epitope specificity, it is necessary to perform more than one assay to have sufficient sensitivity. The more inclusive the diagnostic criteria, the less specific the tests become. Many diagnostic algorithms advocate sensitive screening tests, starting with a mixing-based study to assay for PTT inhibition, followed by more specific diagnostic assays. However, which assay(s) should be used for screening is a matter of debate, given the lack of predictability regarding which LA assay may be positive for a given patient. Moreover, certain coincident factors associated with acutely ill patients (e.g. an elevated Factor VIII [FVIII]) can sufficiently shorten PTT-based tests that a "positive" screening test may be missed due to an abnormally short PTT, which is moved back into the normal range by a LA. In contrast, other approaches advocate using multiple assay systems as part of the screening test, which decreases the likelihood of missing a given patient but can lead to a high rate of false positives.

Like LA activity, detection of antiphospholipid antibodies is difficult secondary to the fact that the epitope specificity of such antibodies varies from patient to patient. Thus, an ELISA restricted to a given specificity may not detect a pathogenic antibody in a given patient. Due to this heterogeneous nature of antiphospholipid antibodies, multiple assays are often used. Some diagnosticians advocate use of anticardiolipin as a screening test, especially $\beta2$ GPI dependent anticardiolipin tests that are more specific. However, others argue that clinically significant antiphospholipid antibodies that do not react with cardiolipin are too frequent to give this approach sufficient sensitivity; thus, multiple screening tests with different epitope specificities are often used.

**Lupus Anticoagulant:** The LA is defined as an activity that prolongs a clotting test in a phospholipid-dependent fashion. Prolongation is measured by assessing the clotting time in a given test platform. Phospholipid dependency is measured either by neutralizing with excess phospholipid or by juxtaposing a phospholipids-dependent and -independent activity. Typically, the source of neutralizing phospholipids is either hexagonal phase extracts, or preparations from platelet membranes.

**PTT Sensitive to LA:** A LA-sensitive PTT requires a PTT reagent containing a low amount of phospholipid, specifically designed as a screening test for an LA. However, in the setting of acute phase reactants (e.g. elevated FVIII levels), the LA-sensitive

PTT may not be prolonged even if a LA is present. Therefore, if an LA is clinically suspected, further tests should be performed even if the LA-sensitive PTT result is normal. An additional problem in the LA-sensitive PTT is mistaking an inhibitor against an intrinsic factor as a LA. Phospholipid neutralization typically addresses this issue, but as some FVIII inhibitors target the phospholipid binding site of FVIII, FVIII inhibitors can potentially be mistaken for an LA in some settings. However, FVIII inhibitors typically lead to clinical bleeding, whereas LAs typically lead to clotting.

**Dilute Russell's Viper Venom Test:** Russell's viper venom activates FX in the patient's plasma in a phospholipid-dependent fashion. Because the dilute Russell's viper venom test (DRVVT) screening reagents contain a low amount of phospholipids, a DRVVT phospholipid neutralization test can be performed to assay phospholipid dependence. A ratio is derived from the screen clotting time divided by the confirmatory clotting time. If the ratio exceeds the established cut-off, then a LA is confirmed. In some laboratories, DRVVT screen and DRVVT confirmation are performed simultaneously for convenience. Because Russell's viper venom directly activates FX, then any inhibitors against intrinsic pathway factors upstream of FX should not interfere with the DRVVT.

**Textarin/Ecarin Time:** Textarin is a snake toxin that directly activates prothrombin in a phospholipids-dependent fashion, whereas ecarin activates prothrombin in a phospholipids-independent fashion. Thus, in the presence of phospholipid, the ratio of activities reflects the phospholipid dependence of clotting, which can be used to test for an LA. Because textarin also requires FV, a FV inhibitor or FV deficiency can give a false positive. In contrast, a false negative can result from heparin, which prolongs the ecarin time through activation of heparin cofactor II.

**Dilute Prothrombin Diagnostic Assay:** A new dilute PT assay has recently become available, which uses a re-lipidated recombinant human tissue factor-based reagent in the presence of calcium to activate the tissue factor pathway. Similar to the above tests, the screening test contains a lower amount of phospholipids and the confirmatory test contains a higher amount. The clotting times of the screening assay are compared with those of the confirmatory assay. A ratio is derived from the screen clotting time divided by the confirmatory clotting time and used to determine if the test is positive.

**Kaolin Clotting Time:** The Kaolin clotting time (KCT) utilizes mineral clay (kaolin) to initiate the PTT. For reasons not entirely clear, PTT-based tests utilizing kaolin are particularly sensitive to LAs. Mixing studies carried out on samples with a prolonged KCT can be very sensitive to detecting LAs. Although a prolonged KCT may indicate a LA, the KCT does not in of itself contain a phospholipid neutralization step, and thus doesn't establish phosopholipid dependence.

**Antiphospholipid Antibodies:** Antiphospholipid antibodies form a heterogeneous group of IgG or IgM antibodies directed against phospholipid or phospholipidprotein components. Epitopes that have been shown to be recognized by antiphospholipid antibodies associated with thrombosis include cardiolipin, $\beta$2GPI, prothrombin, annexin V, protein C and phospholipids. As above, use of multiple substrates gives the highest sensitivity but at the cost of specificity, and different screening approaches are utilized by different labs. In the event that a patient has a high clinical index of suspicion for APS but routine enzyme-linked immunosorbent assays (ELISA) are negative, an extended panel encompassing each of the known substrates might be warranted. A detailed review of each of these substrates is outside of the scope of this chapter. However, discussions of cardiolipin and β2GPI are provided to highlight specific details.

**Anticardiolipin Antibodies:** Anticardiolipin antibodies share a common *in vitro* binding affinity for cardiolipin, and can be detected using enzyme-linked immunosorbent assays. Traditionally, anticardiolipin antibodies were used as a screening test; however, specificity is poor. The immunoglobulin isotype may be IgG, IgM or IgA. It is widely believed that the IgG isotype is most strongly associated with thrombosis. IgM anticardiolipin antibodies can be seen during inflammatory responses, common to ill or hospitalized patients, and are of unknown significance. However, the contribution to thrombosis cannot be unequivocally ruled out in a patient with an IgM antiphospholipid antibody, especially at very high titer. ELISAs for anticardiolipin antibodies are poorly standardized, and anticardiolipin antibody testing has shown poor concordance between laboratories. Anticardiolipin antibodies are reported as a titer specific to the isotype (IgG, IgM or IgA phospholipid antibody titer) but, because the accuracy and reliability of assays are limited, consensus guidelines recommend semiquantitative reporting of results (low, medium or high titer). As a general principle, many groups advocate using anticardiolipin antibody tests that measure antibodies in a β2GPI-dependent fashion to add specificity.

**Anti-β2-glycoprotein I Antibodies:** It has been observed that many of the anticardiolipin antibodies that correlate with thrombosis are directed to an epitope on β2GPI contained in the cardiolipin. This led to the development of the anti-β2GPI immunoassay. Anti-β2GPI is strongly associated with thrombosis and other features of the APS. Although rarely the case, anti-β2GPI has been found to be the sole antibody detected in patients with clinical features of APS. There is some evidence that anti-β2GPI is more specific for APS.

## Recommended Reading

Amengual O, Atsumi T, Koike T. (2003). Specificities, properties, and clinical significance of antiprothrombin antibodies. *Arth Rheum* **48**, 886–895.

Bertolaccini ML, Khamashta MA. (2006). Laboratory diagnosis and management challenges in the antiphospholipid syndrome. *Lupus* **15**, 172–178.

De Groot PG, Derksen RHWM. (2004). Pathophysiology of the antiphospholipid syndrome. *J Thromb Haemost* **3**, 1854–1860.

Miyakis S, Lockshin MD, Atsumi T *et al.* (2006). International consensus statement on an update of the classification criteria for definite antiphospholipid syndrome (APS). *J Thromb Haemost* **4**, 295–306.

# CHAPTER 138

# Lipoprotein(a) Testing

Anne M. Winkler, MD and James C. Zimring, MD, PhD

Elevated lipoprotein(a) [Lp(a)] levels may be associated with an increased risk of atherosclerosis and thrombosis; however, current data do not provide a conclusive role for Lp(a) in this setting. Lp(a) is a complex of a apolipoprotein(a) [apo(a)] linked by a disulfide bond to a modified low density lipoprotein (LDL) that is complexed with apo B. The apo(a) component of Lp(a) is composed of multiple kringle-like domains with significant homology with plasminogen. As a result Lp(a) has the capacity to competitively interfere with the binding of plasminogen to fibrin or other cell surfaces, resulting in inhibition of fibrinolysis. These biochemical observations, along with early retrospective studies associating elevated Lp(a) with thrombotic disease, suggested a role for Lp(a) in the pathophysiology of atherosclerosis. However, subsequent prospective studies have provided conflicting results. Thus, there is little conclusive evidence demonstrating that Lp(a) can be used as a general screening test for atherosclerotic disease. Of clinical importance, cholesterol-lowering drugs and lifestyle changes do not alter Lp(a) levels. Thus, due to its potential lack of predictive power and the inability to therapeutically alter levels, monitoring Lp(a) in routine thrombosis evaluation is not currently recommended. However, in high-risk populations, or in the setting of a thrombotic patient for whom no alternate cause can be found, Lp(a) monitoring may have clinical utility.

**Laboratory Test:** Initial assays for identification of Lp(a) were based upon lipoprotein size and charge. Electrophoresis and ultracentrifugation can be used, but are time-consuming, expensive, and difficult to use for fine quantification. Newer immunochemical methods, such as enzyme-linked immunosorbent assays (ELISA) and immunoturbidimetric assays, are currently used for the direct measurement and quantification of Lp(a). Monoclonal antibodies initially used for ELISA demonstrated cross-reactivity with plasminogen; however, newer reagents utilizing capture antibodies against apo(a) and oxidized Lp(a) have minimized this interference, increasing the sensitivity and specificity of the assays. Immunoturbidimetric assays recognize the amount of Lp(a) by measuring turbidity as immune complexes form, and can quantify Lp(a).

**Sources of Error:** When Lp(a) levels contribute greater than 10 mg/dl of the LDL measurement, Lp(a) may migrate into an additional band. In this case it is important to perform quantitative immunoturbidimetric methods, and LDL-C measurements may need to be adjusted for Lp(a) levels by use of the Friedewald equation.

## Recommended Reading

Berglund L, Ramakrishnan R. (2004). Lipoprotein(a): an elusive cardiovascular risk factor. *Arterioscler Thromb Vasc Biol* **24**, 2219–2226.

Hilbert T, Lifshitz M. (2007). Lipids and dyslipoproteinemia. In: R McPherson, M Pincus (eds), *Clinical Diagnosis and Management by Laboratory Methods*, 21st edition. Philadelphia, AP: Saunders Elsevier, pp. 210–211.

# CHAPTER 139

# Laboratory Diagnosis of Factor Level Abnormalities Associated with Thrombosis

James C. Zimring, MD, PhD

Laboratory diagnosis of independent risk factors for thrombophilia has been an intense area of recent research. In recent decades, great progress has been made in identifying such risk factors – for example Factor V Leiden, the prothrombin gene mutation, lupus anticoagulant, etc. However, genetic analysis of familial thrombophilia indicates that additional risk factors remain to be identified. Moreover, acquired risk factors not strictly associated with heritable traits may also be important.

It has long been appreciated that natural deficiencies in anticoagulant proteins can lead to thrombophilic states. For example, defects in protein C, protein S or antithrombin can lead to thrombotic tendencies (see Chapters 132 and 133). It has thus been seen as logical that congenitally elevated levels of procoagulant human factors may be associated with increased risks of thrombosis; along these lines, levels of individual coagulation factors have been assessed as risk factors for thrombosis.

**Association of Factor Level Abnormalities with Thrombosis:** Based upon current data, elevation of certain clotting factors is indeed associated with thrombophilia, and in some cases constitutes an independent risk factor. Most notably, elevated plasma Factor VIII is an independent risk factor for venous thrombosis. Elevation of Factors IX and XI has likewise been associated with venous thrombosis. Furthermore, elevation of Factors V and VII, fibrinogen and VWF is associated with risk of arterial thrombosis. Finally, decreased levels of Factor XII are associated with thrombotic tendencies.

**Recommendations for Testing Individual Factors:** Of the above-named factors, some have been established as independent risk factors by randomized prospective trials, whereas associations of others are less well established and have relied upon retrospective studies. A comprehensive synopsis of these trials is outside the scope of the current text, but has been reviewed in the literature (Chander *et al.*, 2002). However, the current utility of these associations from the standpoint of screening for thrombophilia is tenuous; as described below, measuring these factors is not currently recommended as a part of routine screening for thrombophilia.

Each of the factors involved in coagulation is part of an intricate network of multiple proteins and factors cascading in a cooperative and cross-talking fashion. Thus, it is highly problematic to determine whether the association of an individual elevated risk factor with thrombosis is actually causal, or a correlation secondary to other interactions. Perhaps even more complicating is the problem that a number of coagulation proteins are acute phase reactants, the level of which can be significantly altered in the context of disease.

Factor VIII serves as the best example of the intrinsic problems in using isolated factors to screen for thrombophilia. Perhaps more than any other factor studied, baseline elevation of Factor VIII is associated as an independent risk factor for thrombosis.

However, accurately measuring the baseline Factor VIII in a given patient can be quite difficult. Baseline Factor VIII levels vary with blood type, are increased substantially due to inflammation and pregnancy, can be altered by physical activity and can change in response to therapy (e.g. in response to estrogen).

Additional complexities in using isolated factor levels to predict thrombosis lie in the nature of the testing methodologies currently employed. Current assays for measuring independent factor levels were designed to analyze patients with bleeding disorders, and have focused on detecting significant decreases. However, establishing increased factor levels, especially subtle levels, is not within the capability of the current generation of assays. Although substantially elevated factor levels can be demonstrated, the reproducibility of these systems is poor. Thus, in addition to the physiological problems of interpretation listed above, laboratory methodologies are currently insufficient for use of isolated factors in screening for thrombophilia.

**Current and Future Applications:** The above explanations are not intended to imply that there is no place in monitoring individual factors in the context of clinically confirmed thrombophilia. Whether correlative or not, substantially increased factor levels may nevertheless be meaningful. However, at the current time, such measures are not recommended for general screening for thrombophilia. Future improvements in the assays for accurately and precisely measuring increased levels may allow use in screening. Moreover, genetic traits that lead to increased levels may allow analysis at a different level. Indeed, the prothrombin gene mutation, which is an independent risk factor used in thrombophilia screening, is believed to increase risk of thrombosis by resulting in elevated levels of Factor II (thrombin). Future associations of polymorphisms with elevations of other factor levels may be discovered.

### Recommended Reading

Bertina RM. (2004). Elevated clotting factor levels in venous thrombosis. *Pathophysiol Haemost Thromb* **33**, 395–400.

Chandler WL, Rodgers GM, Sprouse JT, Thompson AR. (2002). Elevated hemostatic factor levels as potential risk factors for thrombosis. *Arch Path Lab Med* **126**, 1405–1414.

# CHAPTER 140

# Laboratory Management of DIC

Thomas C. Abshire, MD

Disseminated intravascular coagulopathy (DIC) is a syndrome characterized by the consumption and degradation of coagulation factors as a consequence of the unregulated and excessive generation of thrombin, and can result from a heterogeneous group of medical disorders, including sepsis, malignancy and tissue injury. DIC can lead to bleeding and/or clotting, but bleeding is the more common manifestation of DIC (75% of cases), with thrombosis as the primary clinical manifestation in the remaining 25%. As the tests utilized in assessing DIC as well as an overview of clinical DIC are discussed elsewhere (Chapter 110), this chapter will focus upon testing used to help manage the two major clinical problems associated with DIC: bleeding and thrombosis. First, however, the role of laboratory testing in the assessment and management of DIC will be addressed.

Laboratory Diagnosis of DIC: The following laboratory testing is essential in considering the diagnosis of DIC: prothombrin time (PT), partial thromboplastin time (PTT), fibrinogen, D-dimer, complete blood count with platelet count, and review of the peripheral blood smear. If the PT is inappropriately elevated greater than the PTT, then FVII and FV levels should be drawn. If the PTT is markedly elevated, a FVIII level should be drawn. If both the PT and PTT levels are elevated, a fibrinogen level should be performed, as severe DIC might cause markedly low fibrinogen levels. In most cases of DIC, it is not important to assay the contact pathway factors (FXII, prekallikrein, high molecular weight kininogen and FXI) due to their relative lack of effect on a bleeding diathesis.

As mentioned, it is important to draw a complete blood count and platelet count. It is unusual to have severe anemia from DIC, but thrombocytopenia is quite common. Additionally, mildly depressed platelet counts between 50,000/µl and 100,000/µl might also contribute to bleeding, not because of the platelet number but due to the impaired platelet function common with the increased fibrinolysis and elevated fibrin degradation products (FDP) and D-dimer. If there is clinical thrombosis, it is essential to perform a protein C and antithrombin level.

One of the more recognized laboratory features of DIC is microangiopathic hemolytic anemia, which occurs in about two-thirds of cases. Microangiopathic hemolytic anemia is detected by observing fragmented red blood cells (schistocytes) on peripheral smear; however, this is mostly a diagnostic finding, and the degree of hemolytic anemia is rarely sufficient to cause a clinical problem.

A special consideration in diagnosis is that most forms of DIC are intrinsically unstable states. It can be very difficult to confirm a diagnosis of DIC from a single specimen. Serial monitoring should demonstrate an increasing PT and PTT with a progressively decreasing platelet count. It could be argued that if the coagulation parameters and platelets in a patient are stable over 12–24 hours, then the patient is not actively in DIC (except for compensated DIC – see below). This is not to imply that it is necessary to wait for serial sampling before treating a patient with suspected DIC; however, serial samples should be drawn to assess and monitor the ongoing evolution of the consumptive coagulopathy.

## Monitoring of a Bleeding Patient in DIC:

### Prothrombin Time, Partial Thromboplastin Time, and Fibrinogen:
Bleeding is typically caused when disseminated coagulation activation has consumed clotting factors to levels no longer capable of maintaining hemostasis. The extent of factor consumption can be monitored by the screening tests, PT and PTT. An elevated PT is most sensitive to low FV levels, and an elevated PTT is sensitive to low FVIII levels as well as FV. Low fibrinogen levels do not usually affect the PT or PTT unless less than 50–75 mg/dl (depending on instrumentation and reagents); more moderate decreases in fibrinogen can be detected using a specific fibrinogen assay. The elevated PT and PTT may also be affected by other diminished clotting factors, such as FII and FVII (especially if there is excessive tissue factor release as the primary cause for DIC), or decreased contact factor activity such as low FXII and prekallikrein. Individual clotting factors can each be measured by standard one-stage clotting tests (see Chapter 126). However, as management of the bleeding aspects of DIC requires evaluation of the global level of clotting factors, the PT and PTT are typically utilized for monitoring in this setting.

### Fibrin Degradation Products and D-dimer:
Elevated FDP or D-dimer are seen in the majority of DIC cases, and are a measure of ongoing fibrinolysis. These tests are performed by agglutination based or ELISA assays (see Chapter 130). In addition to indicating fibrinolysis, FDP can also have a functional effect. They can inhibit platelet function as well as delaying conversion of fibrinogen and formation of fibrin monomer and fibrin. Elevation of FDP can also prolong the thrombin clotting time due to impaired fibrin formation, which may also complicate laboratory monitoring of fibrinogen and dysfibrinogenemia.

### Thrombocytopenia:
Similar to the previously mentioned consumption of coagulation factors, platelets are likewise consumed through activation and destruction in the abnormal microvasculature. Thrombocytopenia is one of the hallmark findings of DIC.

### Monitoring of a Thrombotic Patient in DIC:
The clinical manifestations of thrombosis include venous thromboembolism and arterial occlusion. These can manifest as end-organ dysfunction, such as renal insufficiency and hepatic dysfunction. Purpura fulminans can also occur, which clinically is visually similar to necrotic skin lesions and appears to be related to microvascular clotting within the skin. This latter condition is seen most commonly in DIC induced by infectious diseases such as: meningococcemia, varicella and Rocky Mountain Spotted Fever. It is now well known that there is inappropriate reduction in the natural inhibitor protein C in comparison to the other clotting factors. This is due to the relatively short half-life of protein C (similar to that of FVII). Occasionally there is decreased production of antithrombin (ATIII) or protein S, but these are less commonly encountered. When there is clinical or laboratory evidence of excessive thrombin activation, the replacement of these natural inhibitors of coagulation should be considered. However, such maneuvers are

only temporizing measures, and eliminating the initiating factor (e.g. treating the septicemia) is essential to breaking the DIC cycle.

## Unique Laboratory Findings Associated with Certain Conditions of DIC:

**Compensated DIC:** Compensated DIC consists of ongoing activation of the coagulation pathway with compensatory inactivation by the natural anticoagulant proteins. Thus, an unnatural equilibrium is achieved where clotting does not occur but thrombin activation is ongoing. This state, although physiologically stable for a time, is extremely dangerous, as loss of the compensatory mechanisms can lead to thrombosis and conversion into classic unstable DIC. Compensated DIC is extremely difficult to diagnose. This may occur in any clinical situation, but may be more common in children with hemangiomas and in adults with disseminated carcinoma. Ongoing thrombin activation can be monitored by measuring generation of prothrombin fragment 1.2, and compensation by antithrombin can be assessed by thrombin–antithrombin complexes. The net amount of active thrombin can be inferred by analyzing the generation of fibrin monomers, FDP and D-dimers. Occasionally, a slightly decreased FVII level is seen in conditions of DIC associated with excessive tissue factor release and, despite the low FVII, the PT may be normal.

**Early DIC with Sepsis and Hypoxia:** Occasionally, DIC induced by sepsis or overwhelming infections can initially present with isolated thrombocytopenia and only mildly depressed screening laboratory values; other classic DIC laboratory findings (see above) will follow within 1–2 days. In DIC associated with hypoxia, initially a markedly low fibrinogen level may be found, which will be followed in 1–2 days with a markedly prolonged PT and PTT as seen in classic DIC.

**Excessive Fibrinolysis:** Conditions of hyperfibrinolysis or primary fibrinolysis may be seen in certain malignancies (e.g. acute promyelocytic leukemia, metastatic carcinoma) or in some obstetric complications (e.g. amniotic fluid embolism). In these situations, it is essential to have the laboratory perform a test for FDP as opposed to D-dimer testing, since measurement of FDP is more indicative of excessive plasmin affect and fibrinolysis on each of the steps in conversion of fibrinogen to fibrin. The D-dimer can only be formed by the action of thrombin and FXIII forming cross-linked fibrin, which is then digested by plasmin. Accordingly, it is not the best indicator of excessive fibrinolysis. Additionally, the euglobulin lysis time (ELT) may be used. A short ELT is suggestive of hyperfibrinolysis (see Chapter 130).

## Recommended Reading

Asakura H, Wada H, Okamoto K *et al.* (2006). Evaluation of haemostatic molecular markers for diagnosis of disseminated intravascular coagulation in patients with infections. *Thromb Haemost* **95**, 282–287.

Emonts M, de Bruijne EL, Guimarães AH *et al.* (2008). Thrombin-activatable fibrinolysis inhibitor is associated with severity and outcome of severe meningococcal infection in children. *J Thromb Haemost* **6**, 268–276.

Levi M. (2007). Disseminated intravascular coagulation. *Crit Care Med* **35**, 2191–2195.

Taylor FB, Toh CH, Hoots WK *et al.* (2001). Towards a definition, clinical and laboratory criteria, and a scoring system for disseminated intravascular coagulation. *Thromb Haemost* **86**, 1327–1330.

Toh CH, Downey C *et al.* (2005). Back to the future: testing in disseminated intravascular coagulation. *Blood Coag Fibrinol* **16**, 535–542.

Voves C, Wuillemin WA, Zeerleder S. (2006). International Society on Thrombosis and Haemostasis score for overt disseminated intravascular coagulation predicts organ dysfunction and fatality in sepsis patients. *Blood Coag Fibrinol* **17**, 445–451.

# CHAPTER 141

# Laboratory Support for Heparin Monitoring

Anne M. Winkler, MD and James C. Zimring, MD, PhD

Heparin is one of the most widely utilized anticoagulants for thrombophrophy-laxis. Unfractionated heparin (UFH) exerts its anticoagulant effect by complexing with and potentiating the activity of the natural anticoagulant, antithrombin (AT). The anticoagulant activities of AT are mainly due to inhibition of activated thrombin (FIIa) and activated Factor Xa (FXa). Preparations of low molecular weight heparin (LMWH) maintain anti-FXa activity, but may have decreased relative amounts of anti-FIIa activity based upon the preparation method.

Both UFH and LMWH have relatively short half-lives of approximately 2 hours and 1 hour, respectively. Thus, frequent monitoring is required to maintain therapeutic doses whilst minimizing the risk of bleeding complications associated with supratherapeutic dosing. Two separate laboratory assays are routinely used to guide heparin therapy: activated partial thromboplastin time (PTT) and anti-Xa activity assay. The advantages and disadvantages of these approaches are discussed below.

**PTT:** The PTT measures the simultaneous function of multiple coagulation factors in the *intrinsic pathway*. The PTT is measured by adding a source of phospholipid and activation factor to recalcified plasma, and assessing coagulation by monitoring conversion of fibrinogen to fibrin (see Chapter 113).

**Interpretation:** Traditionally, the PTT has been used to monitor heparin therapy because of the sensitivity of the PTT to physiologic fluctuations in the activity of FII and FX. Prolongation of the PTT by 1.5–2.5 times the upper limit of the normal reference range is generally considered therapeutic. In addition, at the initiation of heparin therapy it is the goal to reach a PTT of 1.5 times the upper limit of the normal reference range within 24 hours.

**Special Considerations and Sources of Error in Heparin Monitoring by PTT:** Although the PTT is a routine test that is widely used for heparin monitoring, it has several distinct problems.

**Lack of Specificity for Heparin Effects:** First, the PTT measures the functional activity of multiple factors in the intrinsic pathway in addition to FII and FX; perturbations in any of these other factors can substantially alter the PTT. Deficiencies, factor inhibitors, and lupus anticoagulants can prolong the PTT, whereas acute phase levels of FVIII can shorten the PTT. Thus, there are multiple other independent variables that affect the PTT and that may change during the course of treatment. Determining the proper therapeutic value of a patient who presents with or acquires an altered baseline PTT is a clinical challenge.

**Difficulty in Standardization:** A variety of commercial assays exists to measure the PTT, many of which use different reagents and instrumentation. Unlike the prothrombin time (PT), there is no international normalized ratio (INR) for the PTT to allow standardization. To attempt to remedy this, it is recommended that clinical laboratories calibrate their therapeutic range using a minimum of 30–40 samples with values obtained by a protamine sulfate titration or anti-Xa activity assay. In addition, this standardization should be repeated with each new lot of PTT reagent, even from the same manufacturer.

Dynamic Range of the PTT: The PTT is sensitive over a range of heparin concentrations from 0.1 to 1.0 units/ml, but is not quantitative with concentrations >1 unit/ml. For this reason, the PTT is not useful in patients who require large doses of heparin, such as patients undergoing cardiac bypass procedures. In this case, a point-of-care activated clotting time (ACT) is often used.

## Chromogenic Anti-Xa Activity Assay:
The anti-Xa test quantifies the amount of heparin in a specimen by measuring heparin activity through the ability to inhibit FXa. In addition to monitoring heparin activity, the anti-Xa activity assay can be used to monitor fondaparinux (an inhibitor of FXa), but different calibration curves and reference ranges must be used.

Anti-Xa activity is measured by adding a known quantity of purified FXa to the patient plasma along with a chromogenic substrate for FXa. The presence of heparin in the specimen will inhibit the FXa and decrease generation of the signal from cleavage of the chromogenic substrate. Unlike the PTT, the assay does not depend upon other coagulation factors of the intrinsic pathway in the patient plasma. Thus, the number of uncontrolled variables is substantially decreased and there are fewer preanalytical and analytical variables affecting the result.

Since heparin requires AT to inhibit FXa, if the patient has an acquired or congenital AT deficiency, then the assay will underestimate the heparin levels in the patient. Some systems remedy this by adding AT into the FXa reagent; however, addition of AT is not included in all commercial systems.

Whether or not AT should be added to the assay is a matter of debate. Addition of AT will more accurately reflect the pharmacological levels of heparin in the patient, however, it may overestimate the heparin activity *in vivo* and lead to a false conclusion that a therapeutic level has been reached. This could lead to inadequate thromboprophylaxis. In contrast, not adding AT may underestimate the amount of heparin present. This is relevant if the patient has a transient AT deficiency. In such a case, resynthesis of AT by the patient's liver may shift a therapeutic dose to a supertherapeutic dose over time. This issue is not a matter of concern in patients unless they have decreased amounts of AT.

Interpretation: Because heparin is determined through inhibition of FXa activity, the amount of residual FXa activity is inversely proportional to the amount of heparin in the sample. A therapeutic range for thrombophrophylaxis is generally considered to be 0.3–0.7 units/ml.

## Comparison of the PTT and Anti-Xa Activity Assay for Monitoring Heparin Therapy:
Numerous studies have been performed comparing PTT and anti-Xa assays for monitoring heparin therapy. No statistically significant difference

between the two assays has been found when evaluating recurrent thromboembolism or bleeding complications. However, anti-Xa activity assays are superior in several regards, including time to first therapeutic level, quantity of heparin administered, number of dosing changes, and number of monitoring tests.

The anti-Xa activity assay has several additional advantages. First, the anticoagulant fondaparinux, which targets Factor Xa, can only be monitored by the anti-Xa assay and not the PTT (see above). Second, most preparations of LMWH have decreased anti-FIIa activity relative to anti-FXa activity. Thus, the anti-Xa assay is a better measure of LMWH activity.

**PTT and Anti-Xa to Monitor Heparin Resistance:** A subset of patients has been described who require an inordinately large dose of heparin to achieve a therapeutic range as measured by the PTT. Understanding of the causes and management of heparin-resistant patients is an evolving field. Currently, heparin resistance appears to occur as a result of multiple factors, such as a shortened PTT due to increased acute phase proteins (i.e. FVIII and fibrinogen), shortened heparin half-life due to plasma protein binding or increased clearance, deficiency of AT that removes heparin's pharmacological target, and use of medications such as aprotinin or nitroglycerin.

The proper management of such patients is an area of study, and may vary based upon the underlying cause of heparin resistance. However, currently, heparin is given and monitored by either a PTT or anti-Xa assay. A randomized controlled trial has shown that lower doses of heparin were ultimately given to patients being monitored with the anti-Xa activity assay; however, no difference was observed in clinical outcomes of recurrent venous thromboembolism or hemorrhage when using these two assays.

## Recommended Reading

Baglin T, Barrowcliffe TW, Cohen A, Greaves M. (2006). Guidelines on the use and monitoring of heparin. (2006) *Br J Haematol* **133**, 19–34 .

Eikelboom JW, Hirsh J. (2006). Monitoring unfractionated heparin with the aPTT: time for a fresh look. *Thromb Haemost* **96**, 547–552.

Hirsh J, Bauer KA, Donati MB, Gould M, Samama MM, Weitz JI. (2008). Parenteral anticoagulants: American College of Chest Physicians evidence-based clinical practice guidelines (8th edition). *Chest* **133**, 141–159.

Rosborough TK. (1999). Monitoring unfractionated haprin theraphy with antifactor Xa activity results in fewer monitoring test and dosage changes than monitoring with the activated partial thromboplastin time. *Pharmacotherapy* **19**, 760–766.

Spinler SA, Wittkowsky AK, Nutescu EA, Smythe MA. (2005). Anticoagulation monitoring Part 2: Unfractionated heparin and low-molecular-weight heparin. *Ann Pharmacother* **39**, 1275–1285.

Valenstein PA, Walsh MK, Meier F. (2004). Heparin monitoring and patient safety: a College of American Pathologists Q-probes study of 3431 patients at 140 institutions. *Arch Pathol Lab Med* **128**, 397–402.

Winkler AM, Sheppard CA, Fantz CR. (2007). Laboratory monitoring of heparin: challenges and opportunities. *Lab Med* **38**, 499–502.

# CHAPTER 142

# Laboratory Support for Warfarin Monitoring

Anne M. Winkler, MD

Warfarin has been the most widely utilized oral anticoagulant since its introduction over 50 years ago. Warfarin is a vitamin K antagonist, which exerts its anticoagulant effect by interfering with γ-carboxylation of glutamate residues of the N-terminus of coagulation Factors II (FII), VII (FVII), IX (FIX) and X (FX) through inhibition of enzymatic reduction of vitamin K, resulting in decreased coagulant activity. However, warfarin also inhibits carboxylation of the intrinsic anticoagulants, proteins C, S and Z, producing a less dominant, paradoxical procoagulant effect.

Warfarin is available as a racemic mixture of R and S enantiomers with a half-life of 36–42 hours, and its pharmacodynamics are strongly influenced by both environmental and genetic factors. As a result, frequent monitoring is required to maintain therapeutic dosing while minimizing the risk of supratherapeutic dosing. The prothrombin time (PT) modified in the form of the international normalized ratio (INR) is traditionally used to monitor warfarin treatment, and remains the mainstay of warfarin monitoring. Recently, genetic testing including analysis of mutations of the cytochrome P450 complex CYP2C9 and vitamin K epoxide reductase (VKOR) has lead to an explanation of the extremes of physiologic warfarin responses and a potential for more individualized treatment algorithms.

**PT:** The PT measures the function of coagulation factors in the "*extrinsic*" and "*common*" pathways, namely fibrinogen, FII (prothrombin), FV, FVII and FX. The PT is performed by adding thromboplastin (a source of tissue factor and phospholipid extracted from tissues such as brain, lung or placenta) to recalcified plasma and assessing coagulation (see Chapter 113).

Interpretation: Traditionally, the PT has been used to monitor warfarin therapy because of the sensitivity of the PT to the variability of the vitamin K dependent coagulation factors, FII, FVII and FX. Initially, the PT reflects the marked reduction of FVII by warfarin; however, with continued treatment, further prolongation results from reduction of FII and FX.

Special Considerations and Sources of Error and Warfarin Monitoring by PT: Although the PT is a routine test that is widely used for coumadin monitoring, reporting the PT alone has several distinct problems.

*Variability in Responsiveness to Thromboplastin Reagents:* Thromboplastin reagents vary in responsiveness to reduction of the vitamin K dependent coagulation factors, phospholipid content, and preparation. More specifically, responsive thromboplastin reagents produce more prolongation of the PT as compared to unresponsive reagents, and this variability can be measured by determining the international sensitivity index (ISI) for each thromboplastin reagent. This is accomplished by comparison of the PT of the working reagent of normal controls and patients who have received stable oral anticoagulation therapy for at least 6 weeks to results obtained using a standard reagent calibrated against the World Health Organization standard. In general, thromboplastin reagents with lower ISI values are more responsive and may result in a lower coefficient of variation (CV) as compared to a thromboplastin reagent with a higher ISI. As a result, the College of American Pathologists (CAP) recommends use of thromboplastin reagents with an ISI < 1.7 (moderately responsive) that have been validated for a specific reagent/instrument combination, of which over 150 are currently commercially available.

However, despite this recommendation, problems exist with incorrect manufacture-designated ISI values for a given reagent, and with additional variables such as the common use of automated clot detectors that can affect the reliability of the PT. As a result, local calibration using reagents of a certified PT value can be performed to improve INR results, and guidelines have been established by the Clinical Laboratory Standards Institute (CLSI).

*Standardization of PT Reporting:* Owing to the variability of the PT, adoption of the INR has improved oral anticoagulation monitoring and can be calculated using the following formula:

$$INR = (patient\ PT/mean\ normal\ PT)^{ISI}$$

Upon initiation of oral anticoagulation, INR monitoring should begin after two to three doses of warfarin; once stable, monitoring should occur at intervals no greater than 4 weeks. The American College of Chest Physicians (ACCP) has recently published evidence-based clinical practice guidelines for warfarin treatment over a range of clinical indications and additional information concerning the management of supratherapeutic INR values is published within the same supplement as recommended below.

The use of the INR to monitor patients with antiphospholipid antibody syndrome has been criticized, due to the varying effects of lupus anticoagulations on thromboplastin reagents. There is potential for spurious results without adequate reduction of vitamin K dependent coagulation factors, and therefore subtherapeutic INR results. In addition, use of the INR in initial monitoring and in patients with liver disease may be less reliable due to the variation in reduction of factors measured in the PT. Despite all these criticisms, the INR remains the gold-standard laboratory test for warfarin monitoring.

## Abnormal Warfarin Responsiveness:

Genetic Factors: As previously stated, warfarin is composed of R and S enantiomers, which are metabolized by different cytochrome P450 enzymes. The more potent S enantiomer is metabolized by the CYP2C9 enzyme, while the R enantiomer is metabolized by CYP1A2 and CYP3A4 enzymes. As a result, mutations in the genes encoding the cytochrome P450 2C9 enzymes produce impaired metabolism of the S enantiomer, resulting in increased half-life and therefore reduced dosing requirements. The

CYP2C9*2 and CYP2C9*3 have been the most characterized variants, and the specific demographics are displayed in Table 142.1.

| TABLE 142.1  CYP2C9 Allele Prevalence | | | |
|---|---|---|---|
| **CYP2C9 allele** | **CYP2C9*1** | **CYP2C9*2** | **CYP2C9*3** |
| Ethnic group | | | |
| White (%) | 79–86 | 8–19.1 | 6–10 |
| African American (%) | 98.5 | 3 | 6 |
| Asian (%) | 95–98.3 | 0 | 1.7–5 |

In addition, mutations in the gene encoding the VKOR protein and comprising the vitamin K oxide reductase complex 1 (VKORC1) have been identified as a cause of hereditary warfarin resistance, which results in increased warfarin dosing requirements. More specifically, three haplotypes have been identified, which have a reported prevalence of 58% in Europeans, 49% in Africans and 10% in Asians.

Lastly, an additional mutation affecting FIX propeptide has been shown to affect warfarin pharmacokinetics, causing reduction of FIX, which is not measuring using the PT. This mutation increases the risk of bleeding associated with warfarin, and has been identified in less than 1.5% of the population.

Recently, the FDA has approved assays for detection of 2C9 and VKORC1 mutants. Also, additional information concerning pharmacogenetic testing has been added to the warfarin-prescribing information. However, at the present time, and in the absence of randomized trials, there is no evidence to recommend universal screening of genetic mutations in patients beginning warfarin therapy, even though customized dosing algorithms exist.

**Environmental and Drug Interactions:** The pharmacokinetics of warfarin is heavily influenced by environmental factors and drug interaction, including prescribed medications, over-the-counter drugs, nutritional supplements, diet, and herbal products (see Table 142.2).

TABLE 142.2 Drug, Food and Dietary Supplement Interaction with Warfarin by Level of Supporting Evidence and Direction of Interaction (Section 1.1.2)*

| Level of Causation | Anti-infectives | Cardiovascular Drugs | Analgesics, Anti-inflammatories and Immunologics | Central Nervous System Drugs | Gastrointestinal Drugs and Food | Herbal Supplements | Other Drugs |
|---|---|---|---|---|---|---|---|
| | | | *Potentiation* | | | | |
| Highly probable | Ciprofloxacin Cotrimoxazole Erythromycin Fluconazole Isoniazid (600 mg/day) Metronidazole Miconazole oral gel Miconazole vaginal suppositories Voriconazole | Amiodarone Clofibrate Diltiazem Fenofibrate Propafenone Propanolol Sulfinpyrazone (biphasic with later inhibition) | Phenylbutazone Piroxicam | Alcohol (if concomitant liver disease) Citalopram Entacapone Sertraline | Cimetidine Fish oil Mango Omeprazole | Boldo-fenugreek Quilinggao | Anabolic steroids Zileuton |
| Probable | Amoxicillin/ clavulanate Azithromycin Clarithromycin Itraconazole Levofloxacin Ritonavir Tetracycline | Acetylsalicylic acid Fluvastatin Quinidine Ropinirole Simvastatin | Acetaminophen Acetylsalicylic acid Celecoxib Dextropropoxyphene Interferon Tramadol | Disulfiram Choral hydrate Fluvoxamine Phenytoin (biphasic with later inhibition) | Grapefruit juice | Danshen Dong quai Lycium barbarum L PC-SPES | Fluorouracil Gemcitabine Levamisole/ fluorouracil Paclitaxel Tamoxifen Tolterodine |
| Possible | Amoxicillin Amoxicillin/ tranexamic rinse Chloramphenicol | Amiodarone-induced toxicosis Disopyramide Gemfibrozil | Celecoxib Indomethacin Leflunomide Propoxyphene Rofecoxib | Felbamate | Cranberry juice Orlistat | Danshen/methyl salicylate | Acarbose CMF (cyclophosphamide/ methotrexate/ fluorouracil) |

| Causation | | | | | |
|---|---|---|---|---|---|
| Highly improbable | Gatifloxacin, Miconazole topical gel, Nalidixic acid, Norfloxacin, Ofloxacin, Saquinavir, Terbinafine | Metolazone | Sulindac, Tolmetin, Topical salicylates | | | Curbicin, Danazol, Ifosphamide, Trastuzumab |
| Highly probable | Cefamandole, Cefazolin, Sulfisoxazole | Bezafibrate, Heparin | Levamisole, Methylprednisolone, Nabumetone | Fluoxetine/diazepam, Quetiapine | | Etoposide/carboplatin, Levonorgestrel |
| *Inhibition* | | | | | |
| Probable | Griseofulvin, Nafcillin, Ribavirin, Rifampin | Cholestyramine | Mesalamine | Barbiturates, Carbamazepine | High vitamin K content foods/enteral feeds, Avocado (large amounts) | Mercaptopurine |
| Possible | Dicloxacillin, Ritonavir; Terbinafine | Bosentan; Telmisartan | Azathioprine; Sulfasalazine | Chlordiazepoxide | Soy milk, Sucralfate; Sushi containing seaweed; Ginseng | Cyclosporine, Etretinate, Ubidicarenone; Chelation therapy, Influenza vaccine, Multivitamin supplement, Raloxifene hydrochloride |
| Highly improbable | Cloxacillin, Nafcillin/dicloxacillin, Teicoplanin | Furosemide | Propofol | | Green tea | |

From Holbrook AM, Pereira JA, Labiris R et al. (2005). Systematic overview of warfarin and its drug and food interactions. *Arch Intern Med* **165**. 1095–1106.

## Recommended Reading

Ansel J, Hirsh J, Hylek E *et al.* (2008). Pharmacology and Management of the Vitamin K Antagonists: American College of Chest Physicians Evidence-Based Clinical Practice Guidelines (8th edition). *Chest* **133**, 160–198.

Flockhart DA, O'Kane D, Williams MS, Watson MS. (2008). Pharmacogenetic testing of CYP2C9 and VKORCI alleles for warfarin. *Genet Med* **10**, 139–150.

Gage BF, Lesko LJ. (2008). Pharmacogenetics of warfarin: regulatory, scientific, and clinical issues. *J Thromb Thrombol* **25**, 45–51.

Olson JD, Brandt JT, Chandler WL *et al.* (2007). Laboratory reporting of the International Normalized Ratio. *Arch Pathol Lab Med* **131**, 1641–1647.

# CHAPTER 143

# Prothrombin Complex Concentrates

Christine L. Kempton, MD

**Products:** Products are prothrombin complex concentrates (PCCs, also termed Factor IX complex concentrates; in the US these products include Bebulin VH and Profilnine SD) and activated prothrombin complex concentrates (aPCCs, also termed anti-inhibitor complex concentrates; in the US this product is supplied as FEIBA VH).

**Description:** Both PCCs and aPCCs are plasma derivates manufactured from pooled human plasma. Various purification (including ion exchange chromatography and DEAE cellulose adsorption) and viral inactivation (including vapor heat and solvent detergent) methods are employed in the manufacturing process. All products contain vitamin K dependent clotting factors, Factors II, VII, IX and X, though the relative proportions may vary. PCCs consist mainly of unactivated Factors II, IX and X, but also contain protein C and protein S and small amounts of activated factors. aPCCs also consist mainly of unactivated Factors II, IX and X but, in contrast to the PCCs, have higher amounts of the factors in the activated form (FVIIa, FXa). Prothrombin (Factor II) and Factor Xa are thought to be the critical components for the hemostatitic activity of aPCCs. Both PCCs and aPCCs vary in the amount of factors they contain; this variation is not only manufacture dependent, but also lot dependent.

**Indications:** These two products have different indications; neither is indicated for the treatment of hemophilia B (Factor IX deficiency) patients without inhibitors, as there are recombinant or purified Factor IX products for this indication.

PCCs have been used in congenital prothrombin or Factor X deficient patients. Although not FDA approved for use in hemorrhage associated with oral anticoagulants, off-label use of PCCs for reversal of warfarin anticoagulation is recommended by the 8th ACCP Consensus Conference on Antithrombotic Therapy for treatment of severe or life-threatening bleeding while on warfarin anticoagulation (see Table 29.2 for details). Alternate treatments in these patients include plasma or recombinant Factor VIIa, with product choice dependent on the clinical situation and product availability.

aPCCs are approved for the treatment of acute bleeds in patients with hemophilia A or B (Factor VIII or Factor IX deficiency) and inhibitors (see Chapters 96 and 97). They have also been successfully used in patients with acquired Factor VIII inhibitors (see Chapter 111). An alternate product used in the treatment of hemophilia patients with inhibitors is recombinant Factor VIIa.

**Storage and Stability:** Information regarding the purification, viral inactivation and storage of currently available products is presented in Table 143.1. After reconstitution, products can be stored at room temperature for up to 3 hours.

TABLE 143.1   Product information on currently available PCCs and aPCC

| | Purification method | Viral inactivation method | Prior to reconstitution |
|---|---|---|---|
| **PCC** | | | |
| Bebulin VH | Ion exchange chromatography | Vapor heat | 2–8°C refrigeration; do not freeze |
| Profilnine SD | DEAE cellulose adsorption | Solvent detergent | 2–8°C refrigeration; do not freeze; may be stored at room temperature ($<30$°C) up to 3 months |
| **aPCC** | | | |
| FEIBA VH | | Vapor heat | 2–8°C refrigeration; may be stored at room temperature ($<25$°C) up to 6 months |

**Dosage:** The activity of PCCs is labeled by the number of Factor IX IUs, upon which dosing is based. For warfarin reversal in the setting of an INR $<5$, a single 500-IU injection is thought to be optimal. Alternatively, a 25- to 50-IU/kg injection of PCCs can be used in the setting of more severely elevated INRs. Vitamin K should be used concomitantly for warfarin reversal and, since these doses are based on case reports and small clinical studies, careful monitoring of continued bleeding and INR level is necessary.

The potency units of aPCCs are designated as such that one unit is the amount that shortens the PTT of Factor VIII inhibitor-containing reference plasma by 50%. For treatment of Factor VIII and IX inhibitors, aPCC typical dosing is 50–100 U/kg every 8–12 hours. Daily dosing should not exceed 200 U/kg. aPCC dosing of 50–100 U/kg three times per week may be used to prevent bleeding in persons with hemophilia and an inhibitor.

**Adverse Reactions/Precautions:** Thrombosis and DIC have been reported to occur with both PCCs and aPCCs. Thrombosis is rare in persons with hemophilia receiving aPCC, occurring in $<1\%$ of patients. However, the risk of thrombosis increases when other thrombophilic risk factors are present (including the elderly or those with clinical indications for warfarin anticoagulation). Laboratory parameters of DIC should be monitored in those that are receiving treatment for multiple consecutive days (see Chapter 140). To decrease the risk of thrombosis, PCCs or aPCCs should not be given concomitantly with antifibrinolytics, or to patients in DIC.

Since PCCs and aPCCs are plasma proteins, allergic reactions can occur. Additionally, persons with hemophilia B and allergic reactions to Factor IX concentrates may also have an allergic reaction upon exposure to Factor IX contained in aPCC products.

Although the viral inactivation methods are able to reduce many viruses to undetectable levels, they are less effective at removal of parvovirus B19 and hepatitis A virus. Accordingly, infusion of these products carries a risk of infection with these viruses. Patients receiving these products may benefit from hepatitis A vaccination if not previously exposed. In addition, as prions are not removed by the treatments, there is a risk of transmitting spongiform encephalopathy. However, there have been no cases of prion transmission from PCCs or aPCCs to date.

Bebulin contains a small amount of heparin (<0.15 IU heparin per IU Factor IX). Therefore, this product should not be used in a patient with a history of heparin-induced thrombocytopenia.

## Recommended Reading

Aledort LM. (2004). Comparative thrombotic event incidence after infusion of recombinant factor VIIa versus factor VIII inhibitor bypass activity. *J Thromb Haemost* 2, 1700–1708.

Ansell J, Hirsh J, Hylek E *et al.* (2008). Pharmacology and management of the vitamin K antagonists: American College of Chest Physicians Evidence-Based Clinical Practice Guidelines (8th edition). *Chest* 133, 160S–198S.

Baglin P, Keeling DM, Watson HG *et al.* (2005). Guidelines on oral anticoagulation (warfarin): third edition – 2005 update. *Br J Haematol* 132, 277–285.

Baker RI, Coughlin PB, Gallus AS *et al.* (2004). Warfarin reversal: consensus guidelines, on behalf of the Australasian Society of Thrombosis and Haemostasis. *Med J Aust* 181, 492–497.

Leissinger CA, Blatt PM, Hoots WK, Ewenstein B. (2008). Role of prothrombin complex concentrates in reversing warfarin anticoagulation: a review of the literature. *Am J Hematol* 83, 137–143.

# CHAPTER 144

# von Willebrand Factor Concentrates

Thomas C. Abshire, MD

There are currently three Factor VIII (FVIII)/von Willebrand factor (VWF) containing concentrates on the US market, two of which have been approved by the FDA for treatment of von Willebrand disease (VWD) (Table 144.1). The amount of FVIII and functional von Willebrand Factor (VWF:RCo) contained in each product varies. Two of the products are labeled in FVIII units (Alphanate and Koate-DVI) and one is labeled according to VWF:RCo units (Humate-P). Since Alphanate was recently licensed by the FDA for treating VWD types 1 and 2 during surgery, the company has added VWF:RCo units on the vial as well.

All three products are derived from human plasma. The methods of viral inactivation are defined in detail in Table 144.1, and have been highly effective at effectively eliminating potential viral pathogens.

**TABLE 144.1 FVIII/VWF Plasma Derived Clotting Factor Concentrates Available for Treatment of VWD in the United States**

| Product | Manufacturer | Mechanics of viral inactivation | FDA approval for VWD | Ratio of VWF: FVIII |
|---|---|---|---|---|
| Alphanate | Grifols | 1. Affinity chromatography | Yes | 1 : 1 |
| | | 2. Solvent/detergent (TNBP/polysorbate 80) | Surgery; Type 1 and 2 NOT responsive to DDAVP | *HMWM not well retained |
| | | 3. Dry heat (80°C, 72 h) | | |
| Humate P | CSL Behring | 1. Pasteurization (60°C, 10 h) | Yes | 2 : 1 |
| | | | | HMWM well retained |
| Koate-DVI | Talecris | 1. Solvent/detergent (TNBP/polysorbate 80) | No | < 1 : 1 |
| | | 2. Dry heat (80°C, 72 h) | | HMWM not well retained |

*HMWM, high molecular weight multimers.

Humate P is the only product that is licensed for all types of VWD treatment. The ratio of VWF:RCo : FVIII is highest in Humate P (2 : 1), and this product also contains the largest proportion of high molecular weight multimers, which are thought to be most effective in platelet vessel adhesion. Alphanate has been recently approved by the FDA for treating types 1 and 2 VWD during surgery, in patients not responsive to DDAVP. Koate-DVI has been utilized to treat VWD, but is not licensed for this indication by the FDA. Compared to Humate P, both Alphanate and Koate-DVI have less well-retained high molecular weight multimers; the ratio of VWF:RCo : FVIII for Alphanate is 1 : 1 and for Koate-DVI is less than 1 : 1.

<u>Indications:</u> VWD results from a deficiency in quantity and/or function of the VWF, resulting in defective bridging of platelets to the vessel wall and/or destabilization of FVIII (see Chapter 95). The larger molecular weight multimers are best equipped to provide bridging of the platelet to the vessel wall, and therefore products that are deficient in these higher molecular weight multimers are not as effective in achieving this result.

The majority of VWD (75–80%) is type 1 (quantitative defect), and type 1 patients usually respond to DDAVP. However, in those who do not respond (e.g. type 1C or Vincenza type), replacement with a VWF containing product may be necessary. In type 2 VWD, which is characterized by a qualitative defect of VWF, DDAVP treatment may result in increased levels of VWF, which is often defective. This is particularly the case in VWD types 2A, 2B and 2M. The defect in type 2N is the result of improper binding of FVIII to the VWF. DDAVP response in type 2 is highly variable, and should be tested, especially in type 2A patients. DDAVP is often contraindicated in type 2B, and is ineffective in type 2N. Thus, for those patients who fail to respond to a trial of DDAVP, or who have type 2B or 2N VWD, VWF replacement is the preferred treatment. For type 3 VWD, there is no functional gene to synthesize VWF and FVIII is quickly degraded. Accordingly, treatment with a FVIII:VWF containing product for type 3 VWD is essential. When they become available, recombinant VWF or purified VWF (containing essentially no FVIII) will need FVIII products to be given simultaneously, at least initially, until the releasable pool of FVIII is mobilized to bind to the exogenously administered VWF. Finally, patients with platelet-type/pseudo-VWD have a defect in the GPIb binding site of platelets. The VWF is normal; therefore, these patients respond best to platelet infusions and not a VWF containing product.

When considering treatment for bleeding, invasive procedures or for surgery in patients needing a FVIII/VWF containing product, it is important to consider the function of both the VWF and FVIII when determining a treatment plan. Since the VWF initially allows for adhesion of platelets to the blood-vessel wall and subsequent aggregation, treatment of these patients is initially directed towards preventing mucosal bleeding and providing immediate surgical hemostasis. Subsequent treatment is focused on preventing the FVIII related bleeding symptoms (delayed bleeding). Accordingly, it is important to keep the FVIII level in a hemostatic range.

Treatment of VWD with FVIII/VWF containing products results in an increase in both VWF and FVIII. Additionally, both VWF and FVIII are released from endothelial storage sites (Weibel Palade bodies) and from megakaryocytes/platelets, similar to what occurs after administration of DDAVP. The usefulness of the VWF and FVIII released in this setting will be dependent upon the type of VWD being treated, and this fact must be taken into consideration when devising a treatment plan. Additionally, it is important to realize that the defect in VWD is in the VWF, and therefore the clinician cannot utilize a recombinant FVIII product, as this product will not contain the von Willebrand protein. Likewise, highly purified FVIII that does not contain VWF is not efficacious in this setting.

**Storage and Stability:** All three products are supplied as a lyophilized powder in single-use vials. They are reconstituted with Sterile Water for Injection, USP. Prior to reconstitution, the vials should be refrigerated (2–8°C) (with the exception of Humate P which can be stored at 25°C). Exposure to direct sunlight should be avoided. After reconstitution, the product may be refrigerated or stored at room temperature for up to 3 hours.

**Dosing:** The half-life of VWF is similar to that of FVIII, and dosing of a FVIII/VWF containing product is similar to the dosing of FVIII products. The half-life of VWF is approximately 10–12 hours, and 1 unit of VWF:RCo will raise the VWF activity by approximately 2%. As previously discussed, the hemostatic level of the VWF: RCo should be emphasized in the first 3 days after surgery or an invasive procedure, with a subsequent focus on maintaining a hemostatic level of FVIII for the 7–10 days after surgery or an invasive procedure. As mentioned, endogenous stores of FVIII and sometimes VWF (though it may be defective) are being released post-surgery, therefore, both VWF:RCo and FVIII levels should be monitored. It is important not only to keep these levels in a hemostatic range, but also to guard against the levels becoming too elevated – especially to a FVIII level greater than 200%, which might contribute to thrombosis. This potential concern is particularly the case in elderly patients.

The VWF:RCo should be maintained near 100% during the first 1–2 days after surgery, then kept above 50% for the first week after surgery and above 30% for the second week. These levels are initially maintained by 12-hourly infusions of an appropriate FVIII/VWF containing product, and then reduced to daily or every other day dosing. FVIII levels should also be maintained initially at 100% and then above 50% for the first week after surgery and above 30% for the second week or until healing is complete. Daily or every other day infusions of the FVIII:VWF containing product may be all that is needed to maintain adequate levels of both proteins.

Minor surgery usually only requires levels of VWF and FVIII above 30–50%. This is particularly the case with dental extractions, which may only need one dose of the product to keep the levels above 30%.

**Adverse Reactions of FVIII/VWF Containing Products:** A potential concern from a patient perspective is the possible risk of infectious exposure from a FVIII:VWF containing product, which is plasma derived. However, modern plasma manufacturing has essentially eliminated the risk of both lipid (e.g. HIV, hepatitis B and hepatitis C) and non-lipid enveloped viruses (e.g. hepatitis A, Epstein-Barr virus and cytomegalovirus) as a real concern. Additionally, donor selection and the plasma fractionation process remove any significant risk of other non-lipid evolved viruses or infectious prions such as Creutzfeldt-Jakob disease (CJD). Accordingly, patients can by most measures be assured of the safety of these plasma derived FVIII/VWF containing products. Note that at least one viral inactivation step is included in the manufacturing of all three products, and studies so far have found them to be safe against the previously mentioned viruses.

However, a more important potential concern is the risk of thrombosis with very high levels of FVIII – particularly in the elderly, who may have risk factors for stroke or cardiac disease. Other potential risks are antibody formation directed against the von Willebrand protein, seen occasionally in type 3 VWD. If this occurs, large doses of FVIII/ VWF containing products may overwhelm the antibody, but sometimes bypassing agents such as rFVIIa need to be utilized. Additionally, efforts may be directed against removing the antibody with intravenous immunoglobulin and immunosuppressive agents. Another concern is that of cost. It is important to keep in mind the ratio of the VWF:RCo:FVIII in the product of choice to best determine the overall cost to the patient. Utilizing a FVIII:VWF containing product with higher concentrations of VWF may be helpful in both treatment of VWD and reducing cost. Finally, fresh frozen plasma (FFP) or cryoprecipitate should not be utilized routinely in the treatment of VWD, but only in emergent situations when no suitable FVIII:VWF containing product is available. Plasma or cryoprecipitate are not virally inactivated and, in the case of FFP, too large a volume is needed to maintain hemostatic levels of VWF and FVIII.

## Recommended Reading

Berntorp E. (2006). Prophylaxis and treatment of bleeding complications in von Willebrand disease type 3. *Semin Thromb Hemost* **32**, 621–625.

Borel-Derlon A, Federici AB, Roussel-Robert V *et al.* (2007). Treatment of severe von Willebrand disease with a high-purity von Willebrand Factor concentrate (Wilfactin): a prospective study of 50 patients. *J Thromb Haemost* **5**, 1115–1124.

Budde U, Metzner HJ, Muller HG. (2006). Comparitive analysis and classification of von Willebrand factor/factor VIII concentrates: impact on treatment of patients with von Willebrand disease. *Semin Thromb Hemost* **32**, 626–635.

Federici AB. (2006). Management of inherited von Willebrand disease in 2006. *Semin Thromb Hemost* **32**, 616–620.

Federici AB, Mannucci PM. (2007). Management of inherited von Willebrand disease in 2007. *Ann Med* **39**, 346–358.

Federici AB, Castman G, Franchini M *et al.* (2007). Clinical use of Haemate P in inherited von Willebrand's disease: a cohort study on 100 Italian patients. *Haematologica* **92**, 944–951.

Lethagen S, Kyrle PA, Castman G *et al.* (2007). Von Willebrand Factor/Factor VIII concentrate (Humate P) dosing based on pharmacokinetics: a prospective multicenter trial in elective surgery. *J Thromb Haemost* **5**, 1420–1430.

# CHAPTER 145

# Factor VIII Concentrates

Amy L. Dunn, MD

Factor VIII (FVIII) concentrates are either manufactured from human plasma pools or genetically engineered. FVIII concentrates are indicated in the treatment of hemophilia A. Intermediate purity products that contain von Willebrand Factor (VWF), and which were previously used for the treatment of hemophilia A, are currently used primarily in the management of von Willebrand disease (see Chapter 144) now that more highly purified FVIII concentrates are available.

## Factor VII Products:

**Factor VIII Purified from Donors:** There is a variety of available Factor VIII products that are manufactured from donor pools of human plasma. Plasma concentrates are manufactured from either recovered plasma or source plasma from allogeneic donors who fulfill whole blood donation eligibility criteria and testing (see Chapters 5 and 11). This process is highly regulated through the FDA and the Plasma Protein Therapeutics Association (PPTA). Purification and potential pathogen-attenuation steps include cryoprecipitation, which concentrates the FVIII; ion exchange or gel permeation chromatography, which purifies the FVIII from the plasma; and monoclonal antibody immunoaffinity chromatography, which purifies the FVIII and removes contaminating proteins. Immunoaffinity chromatography has the added benefit of reducing non-lipid enveloped viruses such as hepatitis A and parvovirus B19.

Products that are purified utilizing murine monoclonal antibody chromatography have little residual VWF, and are referred to as high purity products. Those that do not go through this step have residual VWF of varying amounts and multimer pattern. These are commonly referred to as intermediate purity FVIII concentrates, and may also be useful in treatment of patients with VWD.

Transmission of pathogens is a major concern for products purified from human plasma. Pathogen contamination is minimized by utilizing a variety of blood donor screening techniques, including nucleic acid testing (NAT) and 60-day inventory hold periods. Additional viral attenuation steps are also utilized, which can include solvent detergent treatment that inactivates lipid-enveloped viruses such as HIV and hepatitis C, and heat treatment/pasteurization that inactivates both lipid enveloped and non-lipid enveloped viruses. These products are stabilized with human albumin. The final product is lyophilized; this has been shown to reduce the risk of hepatitis A contamination. A comparison of currently available products is presented in Table 145.1.

**Recombinant FVIII Products:** Expression of the cloned FVIII gene in recombinant expression systems has the promise of eliminating pathogen contamination from human donors. The process involves expression of the human FVIII gene by an animal cell line, purification, stabilization, lyophilization and final packaging.

TABLE 145.1 Comparison of Currently Available Plasma Derived Factor VIII Concentrates*

| Product | | Plasma source | Pathogen attenuation | Reconstituted volume per vial |
|---|---|---|---|---|
| Low VWF | Hemofil M | Pooled human plasma | MoAb IA, IE, SD | 10 ml |
| | Monoclate-P | Pooled human plasma | MoAb IA, pasteurization | 5–20 ml |
| High VWF | Alphanate | Pooled human plasma | IA, IE, SD | 5–10 ml |
| | Koate-DVI | Pooled human plasma | SD, heat, gel permeation chromatography | 5–10 ml |
| | Humate-P | Pooled human plasma | Cryoprecipitation, Al(OH)$_3$ adsorption, glycine precipitation, NaCl precipitation, pasteurization | 5–15 ml |

*Information on the above products is available in the package insert of each product.
VWF, von Willebrand Factor; MoAb IA, monoclonal antibody immunoaffinity chromatography; IA, immunoaffinity chromatography; IE, ion exchange; SD, solvent detergent.

Although the FVIII has not been purified from humans, some human or animal proteins can be required for proper culturing of the expression systems; thus, products are also subjected to pathogen inactivation. Three different generations of recombinant FVIII products have been produced. In general, the goal with successive generations of products is to reduce the use of human or animal proteins in the manufacturing process to decrease the chance of introducing pathogens.

*First Generation:* In the first-generation product, the FVIII gene is expressed by Chinese hamster ovary (CHO) cells. During fermentation the cells are grown in culture media containing human albumin, bovine albumin and insulin, to help the growth process and ensure production of a sufficient amount of FVIII. During purification, murine monoclonal antibody immunoaffinity chromatography is utilized to isolate the FVIII. Human albumin is added to stabilize the product. The product is then lyophilized and packaged into single-dose vials of varying potency (Table 145.2).

*Second Generation:* In the second-generation product, the FVIII gene is expressed by baby hamster kidney (BHK) or CHO cell lines. Human albumin, but no other animal protein, is used as an ingredient in the fermentation step. Instead, recombinant human insulin has replaced the bovine form, and no bovine albumin is used. The products are purified using murine monoclonal antibody immunoaffinity chromatography, solvent detergent, ion exchange chromatography, and ultrafiltration. In the stabilization process, no human albumin is added to stabilize the molecule. Instead, sugars, salts or amino acids are used as stabilizers. These products are then lyophilized and packaged into single-dose vials containing various amounts of FVIII protein (Table 145.2).

TABLE 145.2 Comparison of Currently Available Recombinant Factor VIII Concentrates*

| Product | | Host cell | Purification/ pathogen attenuation | Stabilizer | Reconstituted volume per vial |
|---|---|---|---|---|---|
| First generation | Recombinant | CHO | MoAb IA, IE | Albumin | 10 ml |
| Second generation | HelixateFS/ Kogenate FS | BHK CHO | MoAb IA, IE, SD,UF | Sucrose | 2.5–5 ml |
| Third generation | Advate | CHO | MoAb IA, IE, SD | Trehalose | 5 ml |
| | Xyntha | CHO | IA, SD, peptide affinity chromatography, nanofiltration | Sucrose | 4 ml |

*Information on the above products is available in the package insert of each product.
MoAB IA, monoclonal antibody immunoaffinity chromatography; IE, ion exchange; SD, solvent detergent; UF, ultrafiltration.

*Third Generation:* Third-generation recombinant FVIII is produced in CHO cells without the addition of human or animal proteins during fermentation, purification or final formulation. The final step for both products is lyophilization and unit-dose packaging (Table 145.2). Xyntha is a unique third-generation product which lacks the B domain of FVIII. The B domain is not necessary for coagulant activity and the resultant shorter molecule is secreted more efficiently in cell culture. In addition the mouse monoclonal antibody purification step, which is common to all other recombinant products, has been replaced by a synthetic peptide ligand purification step.

**Indications, Dosing and Administration:** Whether purified from human donors or from recombinant expression systems, FVIII is predominantly used to treat patients with a congenital deficiency in FVIII (hemophilia A). FVIII products can also be used in some cases of acquired Factor VIII deficiency; however, in the case of high-titer autoanti-FVIII inhibitors, bypassing treatments are often required (e.g. activated prothrombin complex concentrates or recombinant FVIIa).

There are several important considerations when caring for a person with hemophilia A. The first is that prevention of bleeding episodes should be a primary goal. The second involves treating bleeding episodes early and aggressively. Additionally, supportive and adjunctive measures for each bleeding episode in the context of a multidisciplinary team approach should be strongly considered. FVIII concentrates are licensed for use as intermittent infusions. These products have been utilized in continuous infusion, either diluted in IV bags or undiluted in syringe pumps, with good results. The continuous infusion products should be reconstituted in a sterile environment, and usually the container is changed every 12–24 hours; however, studies have found maintained activity for much longer periods of time.

**Processing and Storage:** Factor products are available as lyophilized powders that are reconstituted with diluents provided with the packaged factor prior to administration. It is important to handle the factor carefully, as extreme temperatures and agitation will diminish potency. Factor should be reconstituted immediately prior to dosing, to avoid loss of activity and minimize the risk of bacterial contamination. The products are available in single-dose vials containing differing amounts of FVIII protein.

**FVIII Dosing and Administration:** Upon reconstitution with the provided diluent, the product is delivered by slow intravenous push. In general, 1 unit/kg of infused FVIII product will raise the plasma level by 2%. The half-life of infused product is on average 12 hours, but may be shorter for individual patients – particularly young children. General suggested treatment guidelines are presented in Table 145.3, but should be individualized to the patient.

**TABLE 145.3 Suggested Approach to Treatment of Bleeding Episodes Based Upon Guidelines from Hemophilia of Georgia**

| Bleed site | Desired % activity | Length of therapy |
|---|---|---|
| Central nervous system | 100% | 7–14 days then consider prophylactic therapy for a minimum of 6 months |
| Persistent oral/mucosal | 30–60% | 3–7 days |
| Retropharyngeal | 80–100% | 7–14 days |
| Persistent nasal | 30–60% | 1–3 days |
| Gastrointestinal | 40–80% | 3–7 days |
| Persistent gross urinary | 40–60% | 1–3 days |
| Muscle | 40–80% | Every other day until pain-free movement |
| Iliopsoas | 80–100% | Until radiographic evidence of resolution |
| Joint | 40–80% | 1–2 days |
| Target joint | 80% day 1 | 3–4 days |
| | 40% days 2 & 4 | |

## Complications of FVIII Therapy:

**Inhibitor Development:** A common complication of treating hemophilia A with FVIII is development of inhibitory anti-FVIII antibodies. As the FVIII protein is altered or absent in hemophilia A patients, infused FVIII represents a foreign antigen to the immune system. These inhibitory antibodies occur in approximately 30% of patients with severe hemophilia A, and to a lesser extent in mild and moderate disease. They are more likely to occur in the setting of a positive family history of inhibitors, in association with large gene deletions such as introns 1 and 22 inversions, and in non-white patients. There is no clear evidence as to whether the rate of antibody formation is influenced by the type of product utilized; however, it has been suggested that plasma derived products that contain VWF may be less immunogenic.

Infection: Although extensive efforts are taken to screen out infectious donors, purify products and treat them with pathogen inactivation, the potential for pathogen transmission is not zero. In addition, the effectiveness of current pathogen-inactivation steps for removal of prions is not known; therefore transmission of spongiform encephalopathy is a theoretical risk.

Allergic Reactions: Trace contaminating proteins can induce allergic responses in hypersensitive individuals. These can be of human origin for human-derived products. In addition, as the recombinant expression systems involve cultured cells of hamster origin, residual contaminating proteins have the potential to cause a reaction in patients hypersensitized to rodents.

## Recommended Reading

Abshire TC, Brackmann HH, Scharrer I *et al.* (2000). Sucrose formulated recombinant human antihemophilic Factor VIII is safe and efficacious for treatment of hemophilia A in home therapy – International Kogenate – FS Study Group. *Thromb Haemost* **83**, 811–816.

Barrowcliffe TW, Raut S, Sands D, Hubbard AR. (2002). Coagulation and chromogenic assays of Factor VIII activity: general aspects, standardization, and recommendations. *Semin Thromb Hemost* **28**, 247–256.

Boedeker BG. (2001). Production processes of licensed recombinant Factor VIII preparations. *Semin Thromb Hemost* **27**, 385–394.

Di Paola J, Smith MP, Klamroth R *et al.* (2007). ReFacto and Advate: a single-dose, randomized, two-period crossover pharmacokinetics study in subjects with haemophilia A. *Haemophilia* **13**, 124–130.

Dunn AL, Abshire TC. (2006). Current issues in prophylactic therapy for persons with hemophilia. *Acta Haematologica* **115**, 162–171.

Hurst D, Zabor S, Malianni D, Miller D. (1998). Evaluation of recombinant factor VIII (Kogenate) stability for continuous infusion using a minipump infusion device. *Haemophilia* **4**, 785–789.

# CHAPTER 146

# Factor IX Concentrates

Amy L. Dunn, MD

Factor IX (FIX) concentrates are either manufactured from human plasma pools or genetically engineered without the use of human plasma or proteins. FIX concentrates are indicated in the treatment of hemophilia B.

## Factor IX Products:

**FIX Purified from Donor Plasma:** Two plasma derived FIX concentrates, AlphaNine SD and Mononine, are currently available. Plasma concentrates are manufactured from either recovered plasma or source plasma from allogeneic donors who fulfill whole blood donation eligibility criteria and testing (see Chapters 5 and 11). This process is highly regulated through the FDA and the Plasma Protein Therapeutics Association (PPTA). A comparison of AlphaNine SD and Mononine is presented in Table 146.1.

**TABLE 146.1 Comparison of Currently Available Plasma-Derived Factor IX Products***

| Product | Plasma source | Pathogen attenuation | Reconstituted volume per vial |
|---|---|---|---|
| AlphaNine SD | Pooled human plasma | DEAE chromatography, Dual affinity chromatography, NF, SD | 10 ml |
| Mononine | Pooled human plasma | MoAb IA, UF, sodium thiocynate incubation | 10–20 ml |

*Information obtained from package insert.
MoAb IA, monoclonal antibody immunoaffinity; IE, ion exchange; SD, solvent detergent; NF, nanofiltration; UF, ultrafiltration.

**Recombinant Factor IX:** BeneFIX is currently the only available recombinant FIX concentrate in the US, and uses no human or animal protein for stabilization or production – i.e. it is a third-generation recombinant product (see Chapter 145) (Table 146.2).

**TABLE 146.2 Recombinant Factor IX Product***

| Product | Host cell | Pathogen attenuation | Stabilizer | Reconstituted volume per vial |
|---|---|---|---|---|
| BeneFIX | CHO | MoAb IA, UF | Sucrose, amino acids | 5–10 ml |

*Information obtained from package insert.
CHO, Chinese hamster ovary; MoAb IA, monoclonal antibody immunoaffinity; IE, ion exchange; SD, solvent detergent; UF, ultrafiltration.

Prothrombin Complex Concentrates: In the past, prothrombin complex concentrates (PCCs) which contain varying amounts of FIX in addition to other factors (Factors II, VII and X) have been utilized as replacement products. As more highly purified FIX concentrates are now available, PCCs are not routinely used for this indication. While FIX concentrates are the preferred products, PCCs can be used under emergency situations when FIX concentrates are unavailable (see Chapter 143).

Indications, Dosing and Administration: Factor products are available as lyophilized powders that must be reconstituted with diluents provided with the packaged factor prior to administration. It is important to handle the product with care, as extreme temperatures and agitation will diminish potency. Factor should be reconstituted immediately prior to dosing, to avoid loss of activity and minimize the risk of bacterial contamination. Upon reconstitution with the provided diluent, the product is delivered by slow intravenous push. In general, 1 unit/kg of infused plasma derived FIX will raise the plasma level by 1%. The half-life of infused product is on average 24 hours, but varies between patients. Recombinant FIX should be dosed at 1.4 units/kg for young children and 1.2 units/kg for adults to achieve a plasma rise of 1%. General suggested treatment guidelines are presented in Table 146.3, but should be individualized on a case-by-case basis.

TABLE 146.3 Suggested Treatment Guidelines for Hemophilia Bleeding Episodes Based upon the Guidelines from Hemophilia of Georgia

| Bleed site | Desired % activity | Length of therapy |
| --- | --- | --- |
| Central nervous system | 100% | 7–14 days then consider prophylactic therapy for a minimum of 6 months |
| Persistent oral/mucosal | 30–60% | 3–7 days |
| Retropharyngeal | 80–100% | 7–14 days |
| Persistent nasal | 30–60% | 1–3 days |
| Gastrointestinal | 40–80% | 3–7 days |
| Persistent gross urinary | 40–60% | 1–3 days |
| Muscle | 40–80% | Every other day until pain-free movement |
| Iliopsoas | 80–100% | Until radiographic evidence of resolution |
| Joint | 40–80% | 1–2 days |
| Target joint | 80% day 1 | 3–4 days |
| | 40% day 3 | |

There are several important considerations when caring for a person with hemophilia. The first is that prevention of bleeding episodes should be a primary goal. The second involves treating bleeding episodes early and aggressively. Additionally, supportive and adjunctive measures for each bleeding episode in the context of a multidisciplinary team approach should be strongly considered. FIX concentrates are licensed for use as intermittent infusions. These products have been utilized in continuous infusion, either diluted in IV bags or undiluted in syringe pumps, with good results. The continuous infusion products should be reconstituted in a sterile environment and usually the container is changed every 12–24 hours; however, studies have found maintained activity for much longer periods of time.

## Complications of FIX Therapy:

**Inhibitor Development:** A complication of treatment with FIX products is development of inhibitory antibodies to the infused protein. In addition to limiting efficacy due to inhibiting the activity of the infused FIX, patients with these antibodies may also manifest life-threatening anaphylaxis and or nephrotic syndrome when challenged with these products. The rate of antibody formation appears to be similar between recombinant and plasma derived products.

**Infection:** Although extensive efforts are taken to screen out infectious donors, purify products and treat them with pathogen inactivation, the potential for pathogen transmission is not zero. In addition, the effectiveness of current pathogen inactivation steps for removal of prions is not known; therefore transmission of spongiform encephalopathy is a theoretical risk.

**Allergic Reactions:** Trace contaminating proteins can induce allergic responses in hypersensitive individuals. These can be of human origin for human-plasma derived products. In addition, as the recombinant expression systems involve cultured cells of hamster origin; residual contaminating proteins have the potential to cause a reaction in patients hypersensitized to rodents.

## Recommended Reading

Chowdary P, Dasani H, Jones JA et al. (2001). Recombinant factor IX (BeneFix) by adjusted continuous infusion: a study of stability, sterility and clinical experience. *Haemophilia* **7**, 140–145.

Dunn AL, Abshire TC. (2004). Recent advances in the management of the child who has hemophilia. *Hematol Oncol Clin N Am* **18**, 1249–1276.

Martinowitz UP, Schulman S. (1995). Continuous infusion of factor concentrates: review of use in hemophilia A and demonstration of safety and efficacy in hemophilia B. *Acta Haematologica* **94**, 35–42.

Warrier I, Ewenstein BM, Koerper MA et al. (1997). Factor IX inhibitors and anaphylaxis in hemophilia B. *J Ped Hematol/Oncol* **19**, 23–27.

# CHAPTER 147
# Factor VII Concentrates

Christine L. Kempton, MD

Recombinant Factor VIIa (rFVIIa; supplied as Novoseven in the US) is approved for use in patients with congenital hemophilia with inhibitor, acquired Factor VIII inhibitor and Factor VII (FVII) deficiency. In addition to its approved use, there is widespread use of rFVIIa for off-label indications, including trauma and surgery, and emergency reversal of warfarin, with variable success.

Although some clinical studies have demonstrated that rFVIIa reduces red blood cell transfusion requirements in some, but not all, surgical settings, improvements in more clinically relevant outcomes, such as reduced morbidity, mortality or length of hospital stay, have not been demonstrated. Moreover, there is a risk of inducing thrombosis. A recent review of 17 randomized clinical trials for off-label use of rFVIIa showed little evidence to support use of rFVIIa to treat massive bleeding; in addition, increased risk of thromboembolic events was suggested in patients with intracranial hemorrhage. In a phase III study, rFVIIa reduced hematoma growth, but did not improve survival or functional outcome. The off-label use and dosing of rFVIIa needs to be carefully evaluated on a case-by-case basis.

**Manufacturing:** rFVIIa is produced from the cloned human FVII gene that is expressed in baby hamster kidney (BHK) cells. The BHK cells produce the single-chain form of FVII. After secretion into the media, it is autocatalyzed to its active form. The culture media containing rFVIIa is then purified using a chromatographic purification process. Although no human serum or proteins are used in production, there are trace proteins from BHK cells in the final product.

**Pathophysiology:** At doses used to replace FVIIa to physiological levels, rFVIIa functions by binding to tissue factor; however, when doses that lead to superphysiological levels are used (e.g. treatment of hemophilia with inhibitors), rFVIIa also binds to the surface of activated platelets and can activate FX directly. Because rFVIIa directly activates X, with subsequent activation of thrombin and conversion of fibrinogen to fibrin, pathways requiring FVIII are bypassed. This allows rFVIIa to circumvent inhibitors directed against intrinsic pathway factors. For this reason, platelets and coagulation factors distal to FX are required for efficacy in promoting hemostasis.

**Storage and Stability:** rFVIIa is supplied as a lyophilized powder in single-use vials. Current vial sizes are 1 mg, 2 mg and 5 mg. The new formulation, as of August 2008, is stored at room temperature, and is reconstituted using the accompanying diluent (histidine diluent). After reconstitution, rFVIIa may be refrigerated or stored at room temperature for up to 3 hours.

**Dosage:** For a summary of dosage, see Table 147.1.

| TABLE 147.1 Recommended Dosing of rFVIIa | |
|---|---|
| Clinical indication | Dose |
| Congenital hemophilia with inhibitor | 90–120 μg/kg |
| Acquired Factor VIII inhibitor | 70–90 μg/kg |
| Congenital Factor VII deficiency | 15–30 μg/kg |

For treatment of congenital hemophilia with inhibitors, the recommended dose is 90–120 μg/kg infused over 2–5 minutes. Since the half-life is 2.3 hours in adults, and potentially shorter in children, repeat doses should be given every 2–3 hours, as needed.

Although 90–120 μg/kg is the recommended dose, neither the minimally effective nor the maximum tolerated dose has been determined. In clinical trials, 35 μg/kg and 70 μg/kg appeared to be equally effective for the treatment of hemarthrosis (61% effective); however, 90 μg/kg was superior to 35 μg/kg for management of perioperative hemostasis. Subsequently, 90 μg/kg given every 3 hours was tested for the home treatment of mild to moderate bleeds (joint, muscle and mucocutaneous bleeding) in patients with hemophilia and inhibitors, and was effective in 92% of subjects after a median of 2.2 injections. In another clinical study, a single injection of 270 μg/kg was compared with three injections of 90 μg/kg and was found to have similar efficacy and safety.

Use of a continuous infusion of rFVIIa for perioperative management has also been investigated. In an open-label randomized study, 24 subjects received either 90 μg/kg every 2 hours for 5 days followed by every 4 hours for an additional 5 days, or 50 μg/kg per hour by continuous infusion for 5 days followed by 25 μg/kg per hour for an additional 5 days. Both groups received a 90-μg/kg bolus prior to surgery. The hemostatic efficacy and safety of the two regimens was similar; however, the sample size is too small to conclude statistical equivalency.

For treatment of acquired FVIII inhibitors, lower doses (70–90 μg/kg) are recommended. This dosing range was determined from a compassionate use study and a registry supported by the Hemophilia and Thrombosis Research Society. The mean dose used was 90 μg/kg (range: 31–197 μg/kg). No dose response was seen when doses between 70 and 90 μg/kg were used. Doses should be repeated every 2–3 hours until hemostasis is achieved.

When treating FVII deficiency, the goal of treatment is to replace the missing clotting factor rather than bypass the tenase complex; therefore, lower doses (15–30 μg/kg) are effective.

No laboratory monitoring has been demonstrated to correlate with efficacy. The dosing interval and number of doses used should be based on the clinical assessment of hemostasis. In the setting of severe bleeding, or postoperatively, treatment should continue past the point that hemostasis has been achieved in order to assure maintenance of hemostasis.

**Adverse Reactions/Precautions:** rFVIIa represents a risk of allergic reactions in persons with known hypersensitivity to mouse, hamster or bovine proteins, as such proteins are introduced at trace levels during recombinant expression. Since rFVIIa will bind to tissue factor and initiate coagulation, it should be avoided in clinical situations in which tissue factor expression is increased, such as DIC, sepsis and crush injury.

Thrombosis is rare in patients with congenital hemophilia with inhibitor who receive rFVIIa (< 1%). Thrombotic complications have not been documented in case reports of rFVIIa used in congenital FVII deficiency.

Arterial and venous thromboembolic events have been reported in association with rFVIIa use in acquired Factor VIII inhibitors. Of 139 patients with acquired FVIII inhibitors who were treated with rFVIIa for 204 bleeding episodes, 10 patients had 12 thrombotic events. The increased rate of thromboembolic events in those with acquired FVIII inhibitors is most likely secondary to underlying vascular disease in an elderly population. Therefore, patients should be monitored for thromboembolic complications when using rFVIIa, particularly the elderly and those with other risk factors for thromboembolism (e.g. surgery, trauma, etc.). Furthermore, the majority of adverse events reported to the FDA occurred in those receiving rFVIIa for off-label indications.

## Recommended Reading

Johansson PI. (2008). Off-label use of recombinant Factor VIIa for treatment of haemorrhage: results from randomized clinical trials. *Vox Sanguinis* **95**, 1–7.

Key NS, Aledort LM, Beardsley D *et al.* (1998). Home treatment of mild to moderate bleeding episodes using recombinant Factor VIIa (Novoseven) in haemophiliacs with inhibitors. *Thromb Haemost* **80**, 912–918.

Mayer SA, Brun NC, Begtrup K *et al.* (2008). Efficacy and safety of recombinant activated Factor VII for acute intracerebral hemorrhage. *N Engl J Med* **358**, 2127–2137.

O'Connell KA, Wood JJ, Wise RP *et al.* (2006). Thromboembolic adverse events after use of recombinant human coagulation Factor VIIa. *J Am Med Assoc* **295**, 293–298.

Pruthi RK, Mathew P, Valentino LA *et al.* (2007). Haemostatic efficacy and safety of bolus and continuous infusion of recombinant Factor VIIa are comparable in haemophilia patients with inhibitors undergoing major surgery. Results from an open-label, randomized, multicenter trial. *Thromb Haemost* **98**, 726–732.

Roberts HR, Monroe DM, White GC. (2004). The use of recombinant Factor VIIa in the treatment of bleeding disorders. *Blood* **104**, 3858–3864.

Santagostino E, Mancuso ME, Rocino A *et al.* (2006). A prospective randomized trial of high and standard dosages of recombinant Factor VIIa for treatment of hemarthroses in hemophiliacs with inhibitors. *J Thromb Haemost* **4**, 367–371.

# CHAPTER 148

# Antithrombin Concentrates

Michael A. Briones, DO

Antithrombin III (antithrombin [AT]) is the major inhibitor of thrombin and Factor Xa, and therefore plays a critical role in regulation of the coagulation system. AT products are predominately administered to patients with hereditary AT deficiency for prophylaxis or treatment of thromboembolic complications. In addition, the use of AT has been investigated in patients with heparin resistance scheduled to undergo cardiac surgery necessitating cardiopulmonary bypass, and in patients with disseminated intravascular coagulopathy (DIC).

**Description:** AT is a plasma-derived, single-chain glycoprotein with a molecular weight of 58 kDa. It is a serine protease inhibitor (serpin), sharing about 30% homology in amino acid sequence with other serpins. AT has potent anticoagulant activity, which is significantly enhanced by heparin and vessel-wall associated glycosaminoglycans (GAGs). AT inhibits the activity of several activated clotting factors, predominantly thrombin and Factor Xa, but also others including Factors IXa and XIa. The presence of heparin dramatically increases factor inhibition by AT. The normal antithrombin activity level in adults is 80–120% (0.8–1.2 IU/ml), and in neonates is 40–60% (0.4–0.6 IU/ml).

**Indications:**

**Hereditary AT Deficiency:** Inherited AT deficiency is an autosomal dominant disorder with AT levels of 40–50% and a prevalence of 1 in 2000 adults. Approximately 50% of individuals with AT deficiency will have a thrombotic event. The diagnosis of hereditary AT deficiency is based on a clear family history of venous thrombosis as well as decreased plasma AT levels, and exclusion of acquired deficiency. Conventional treatment with heparin of acute venous thromboembolism (VTE) in patients with AT deficiency may be effective; however, if an additional thrombogenic factor, such as pregnancy, coexists there may be a variable response to heparin, and symptoms may progress despite adequate heparin treatment. In such situations, additional support with AT concentrate may be necessary. In addition, AT concentrates have been used in the perioperative period for surgical prophylaxis in patients with a known AT deficiency.

**Acquired AT Deficiency:** Acquired AT deficiency occurs more frequently than congenital deficiency. In the neonatal period, AT levels are reduced in premature infants, and in neonates with respiratory distress necrotizing enterocolitis, sepsis and DIC. Outside the neonatal period, acquired AT deficiency occurs in DIC, liver disease (decreased synthesis), nephrotic syndrome (increased urinary excretion), and with medications such as oral contraceptive and hormone replacement therapy, heparin and L-asparaginase. AT levels are not reduced during normal pregnancy, however, levels may be reduced in pre-eclampsia, eclampsia, and pregnancy-induced hypertension.

The use of AT in situations not associated with congenital deficiency is non-FDA approved, and is at the discretion of the prescribing physician. Currently, debate continues about the use of AT concentrates in patients with acquired AT deficiency. Debate is also ongoing regarding the use of AT in DIC (see below), sepsis, and those with relative heparin resistance during surgery and extracorporeal circulation. AT is not FDA approved for pediatric use but is used off-label, and has been utilized in pediatric patients such as in the setting of sepsis and DIC, in neonates with respiratory distress syndrome and in neonates with necrotizing entercolitis, with mixed results.

**Disseminated Intravascular Coagulopathy:** AT can be decreased in patients with DIC, where low AT levels are associated with a poor prognosis. Due to this association, AT replacement has been studied for the treatment of DIC. In the first clinical study, 51 randomized patients with shock and DIC were infused with either AT, heparin, or both AT and heparin. AT treatment was associated with significantly shorter duration of symptoms and more rapid normalization of coagulation parameters. In patients given AT and heparin, the addition of heparin showed no advantages, and resulted in increased bleeding but not increased mortality. Meta-analyses of several smaller randomized clinical trials with AT suggested significant reduction in mortality from severe DIC. It should be noted that the efficacy of AT in terms of reduced all-cause 28-day mortality has not been proven. Lacking a clear result regarding mortality, the documented improvement of DIC parameters and organ dysfunction, and of the reduction in length of stay, may justify the treatment with AT; however, current available evidence is not suited to sufficiently inform clinical practice.

**Heparin Resistance:** Heparin resistance is defined as requiring larger than typical doses of heparin to achieve a therapeutic dose based on laboratory monitoring, such as PTT or anti-Xa. Heparin resistance can be secondary to increased FVIII or fibrinogen levels, increased binding to plasma proteins, increased plasma clearance, decreased AT levels, aprotinin or nitroglycerin. Heparin resistance may occur in approximately 20% of patients undergoing cardiopulmonary bypass, and is associated with acquired AT deficiency. Therapeutic options include fresh frozen plasma (FFP) transfusion and AT infusion. Both have demonstrated efficacy in the treatment of heparin resistance; currently, no clinical trial has demonstrated a benefit of one versus another.

**Dosing/Administration and Stability:** AT concentrate is a plasma derived, sterile, non-pyrogenic, stable, lyophylized preparation of purified human AT. A recombinant form of AT is being developed and has been used in clinical trials of heparin resistance and hereditary AT deficiency, but is currently not FDA approved.

AT concentrate is supplied in 500 and 1000 IU vials. The potency assignment has been determined with a standard calibrated against a World Health Organization AT reference preparation. Dosage should be determined on an individual basis, based on the pre-therapy plasma AT level, in order to increase plasma AT levels to the level found in normal human plasma (100%). Dosage of AT can be calculated from the following formula:

$$\text{Units required} = \frac{[\text{Desired} - \text{Baseline AT level}] \times \text{Weight (kg)}}{1.4}$$

The above formula is based on an expected incremental *in vivo* recovery above baseline levels for AT of 1.4% per IU/kg administered. Thus, if a 70-kg individual has a baseline AT level of 57%, in order to increase plasma AT to 120%, the initial AT dose would be $[(120 - 57) \times 70]/1.4 = 3150\,IU$ total. A more practical dosing guideline is that 1 unit/kg will produce a rise of approximately 2%, with a half-life of approximately 60 hours.

Due to variability in AT recovery, initially levels should be drawn at baseline and 20 minutes post-infusion. Subsequent doses can be calculated based on the recovery of the first dose. It is recommended that, following an initial dose of AT, plasma levels of AT be initially monitored at least every 12 hours and before the next infusion of antithrombin to maintain plasma AT levels greater than 80%. In some situations – e.g. following surgery, hemorrhage or acute thrombosis, and during intravenous heparin administration – the half-life of AT has been reported to be shortened. In such conditions, plasma AT levels should be monitored more frequently, and antithrombin administered as necessary.

When an infusion of AT is indicated for a patient with hereditary deficiency to control an acute thrombotic episode or prevent thrombosis following surgical or obstetrical procedures, it is desirable to raise the AT level to normal and maintain this level for 2–8 days, depending on the indication for treatment, type and extent of surgery, patient's medical condition, past history and physician's judgment. Concomitant administration of heparin in each of these situations should be based on the medical judgment of the physician.

AT should be stored refrigerated (2–8°C). Freezing should be avoided, as breakage of the diluent bottle might occur. Reconstitution of the vial is with a suitable volume of Sterile Water for Injection, USP; usually a sterile double-ended transfer needle; and a sterile filter needle are provided with each vial. The rate of administration should be adapted to the response of the individual patient, but administration of the entire dose in 10–20 minutes is generally well tolerated.

**Adverse Reactions/Precautions:** It is important to note that the *anticoagulant* effect of *heparin* is enhanced by concurrent treatment with AT in patients with hereditary AT deficiency. Thus, in order to avoid bleeding, a reduced dosage of heparin is recommended during treatment with AT in this population.

In clinical studies involving AT, adverse reactions were reported in association with 17 of the 340 infusions during the clinical studies. Such reactions included dizziness, chest tightness, nausea, a foul taste in mouth, chills, cramps, shortness of breath, chest pain, film over eye, light-headedness, bowel fullness, hives, fever, and bleeding and hematoma formation. If adverse reactions are experienced the infusion rate should be decreased or, if indicated, interrupted until symptoms abate. Reinitiating infusion is at the discretion of the clinician.

## Recommended Reading

Abildgaard U. (2007). Antithrombin – early prophecies and present challenges. *Thromb Haemost* **98**, 97–104.

Avidan MS, Levy JH, Scholz A *et al.* (2005). Phase III, double-blind, placebo-controlled, multicenter study on the efficacy of recombinant human antithrombin in heparin-resistant patients scheduled to undergo cardiac surgery necessitating cardiopulmonary bypass. *Anesthesiology* **102**, 276–284.

Hoffmann JN, Wiedermann CJ, Juers M *et al.* (2006). KyberCept investigators. Benefit/risk profile of high-dose antithrombin in patients with severe sepsis treated with and without concomitant heparin. *Thromb Haemost* **95**, 850–856.

Jilma B. (2006). Antithrombin for severe sepsis? Try it again, but without heparin. *Thromb Haemost* **95**, 755.

Kienast J, Juers M, Wiedermann CJ *et al.* (2006). Kyber-Cept investigators. Treatment effects of high dose antithrombin without concomitant heparin in patients with severe sepsis with our without disseminated intravascular coagulation. *J Thromb Haemost* **4**, 90–97.

Spiess BD. (2008). Treating heparin resistance with antithrombin or fresh frozen plasma. *Ann Thorac Surg* **85**, 2153–2160.

Warren BL, Eid A, Singer P *et al.* (2001). Caring for the critically ill patient. High-dose antithrombin III in severe sepsis: a randomized controlled trial. *J Am Med Assoc* **286**, 1869–1878.

# CHAPTER 149

# Protein C Concentrates

Michael A. Briones, DO

Protein C is a natural anticoagulant. Upon activation, and in complex with its cofactor (protein S), activated protein C (APC) exerts its effect predominantly by inhibiting Factors Va and VIIIa (see Chapter 78). APC has also been shown to have profibrinolytic effects. Moreover, APC is now appreciated to have additional regulatory effects outside of the context of coagulation; in particular, APC also inhibits tumor necrosis factor production by monocytes, blocks leukocyte adhesion to selectins, and limits thrombin-induced microvascular damage.

Patients can have decreased protein C through genetic protein C deficiency, but in adults low protein C is more commonly an acquired state. Protein C is synthesized by the liver in a vitamin K dependent fashion. Thus, liver failure, vitamin K deficiency or administration of warfarin can all lead to low protein C levels. Finally, protein C can be depleted through consumptive coagulopathy.

Protein C replacement is a mainstay of treatment in congenital deficiency. In acquired deficiency, because protein C works in balance with multiple other coagulation parameters, the choice to replace protein C in a deficient patient must always be made in the context of the overall clinical picture.

There are currently two protein C products available: Ceprotin (used for replacement of protein C), and Xigris (used predominantly for treatment of sepsis). Ceprotin, a protein C concentrate (Human), is manufactured from human plasma and purified by a combination of filtration and chromatographic procedures, including a column of immobilized mouse monoclonal antibodies on gel beads. The manufacturing process includes steps designed to reduce the risk of viral transmission, including extensive donor screening and pathogen-inactivation steps (Polysorbate 80 [P80] treatment and vapor heating). Xigris (Drotrecogin alfa [activated]) is a recombinant form of human activated protein C (APC) that is manufactured as an inactive zymogen and is then enzymatically activated by cleavage with thrombin, followed by purification using monoclonal antibodies.

## Indications:

Ceprotin: Ceprotin is indicated as a replacement therapy for patients with severe congenital protein C deficiency, and for the prevention and treatment of venous thrombosis and purpura fulminans secondary to protein C deficiency. Ceprotin is approved for use in both pediatric and adult patients.

Xigris: Xigris is approved by the FDA for use in severe sepsis, based on the finding that it decreases mortality and the development of DIC in high-risk patients (defined by APACHE II scores > 25) by about 20%. Thus, clinical use of Xigris focuses on the non-coagulation based anti-inflammatory properties of APC. When used in sepsis, the clinician must always balance potential efficacy with risk of inducing hemorrhage. Use of Xigris may not be beneficial, and may be contraindicated in patients who do not meet APACHE II criteria, due to the risk of bleeding. Xigris is only approved for acquired deficiency in the presence of severe sepsis, and it is not approved for pediatric use. However, there are several documented instances of successful use in treating familial protein C deficiency.

## Dosage and Administration:

Ceprotin:  Ceprotin for acute episode/short-term prophylaxis is given with an initial dose of 100–120 IU/kg, followed with subsequent dosing of 60–80 IU/kg every 6 hours and maintenance dosing of 45–60 IU/kg every 6 or 12 hours.

Because packaging requirements and administration guidelines can be changed, dosing and administration should always be confirmed using the product insert provided by the manufacturer.

Xigris:  Xigris should be administered intravenously at a continuous infusion rate of 24 µg/kg per hour (based on actual body weight at start of infusion), for a total duration of infusion of 96 hours. The dose should not be adjusted during the infusion for changes in actual body weight. Each IV administration should be completed within 12 hours after the solution has been prepared. If the infusion is interrupted, Xigris should be restarted at the initial infusion rate and continued to finish the rest of the overall recommended duration of infusion. Should clinically significant bleeding occur, Xigris infusion should be stopped immediately. No dosage adjustment is required for age, hepatic impairment or renal impairment.

Because packaging requirements and administration guidelines can be changed, dosing and administration should always be confirmed using the product insert provided by the manufacturer.

## Storage and Stability:

Ceprotin:  Ceprotin is available in single-dose vials that contain nominally 500 (blue color bar) or 1000 (green color bar) International Units (IU) of human protein C. The 500 and 1000 IU vials are reconstituted with 5 ml and 10 ml of sterile water, respectively. This results in human protein C at a concentration of 100 IU/ml.

Ceprotin, packaged for sale, is stable for 3 years when stored refrigerated at 2–8°C. The product should not be frozen, so as to avoid damage to the diluent vial, and should be stored in the original carton and protected from light. The solution should be used within 3 hours. As with all products, Ceprotin should not be used after the expiration date on the vial.

Xigris:  Xigris is available in 5 mg and 20 mg single-use vials. Xigris should be stored in a refrigerator at 2–8°C, in its original carton. Freezing and exposure to light should be avoided. As with all products, Xigris should not be used after the expiration date on the vial.

## Adverse Reactions:

Ceprotin:  The most serious and common adverse reactions observed related to Ceprotin treatment are hypersensitivity or allergic reactions (itching and rash) and lightheadedness.

**Xigris:** Bleeding is the most common adverse reaction associated with Xigris.

## Recommended Reading

Bernard GR, Vincent J-L, Laterre P-F *et al.* (2001). Efficacy and safety of recombinant human activated protein C for severe sepsis. *N Engl J Med* **344**, 699–709.

De Backer D. (2007). Benefit-risk assessment of Drotrecogin Alfa (activated) in the treatment of sepsis. *Drug Safety* **30**, 995–1010.

# General Handbook References

(2008). *Standards for Blood Banks and Transfusion Services*, 25th edition, Bethesda: AABB Press.

Brecher ME (ed.). (2005). *AABB Technical Manual*, 15th edition, Bethesda: AABB Press.

Colman RW, Hirsh J, Marder VJ, Clowes AW, George JN (eds). (2001). *Hemostasis and Thrombosis: Basic Principles and Clinical Practice*, 4th edition, Philadelphia: Lippincott Williams & Wilkins.

Goodnight SH, Hathaway WE. (2001). *Disorders of Hemostasis and Thrombosis: A Clinical Guide*. United States: McGraw-Hill.

Harmening DM (ed.). (2005). *Modern Blood Banking & Transfusion Practices*, 5th edition, Philadelphia: F.A. Davis.

Hillyer CD, Hillyer KL, Strobl FJ, Jefferies LC, Silberstein LE (eds). (2001). *Handbook of Transfusion Medicine*. San Diego: Academic Press.

Hillyer CD, Silberstein LE, Ness PM, Anderson KC, Roback JD (eds). (2007). *Blood Banking and Transfusion Medicine: Basic Principles & Practice*, 2nd edition, Philadelphia: Elsevier.

Hillyer CD, Strauss RG, Luban NL (eds). (2004). *Handbook of Pediatric Transfusion Medicine*. San Diego: Elsevier Academic Press.

Hoffman R, Benz EJ, Shattil SJ, Furie B, Cohen HJ, Silberstein LE, McGlave P (eds). (2005). *Hematology: Basic Principles and Practice*, 4th edition, Philadelphia: Elsevier.

Kitchens CS, Alving BA, Kessler CM (eds). (2007). *Consultative Hemostasis and Thrombosis*, 2nd edition, Philadelphia: Saunders.

McCullough J. (2005). *Transfusion Medicine*, 2nd edition, Philadelphia: Elsevier.

McLeod BC, Price TH, Weinstein R (eds). (2003). *Apheresis: Principles and Practice*, 2nd edition, Bethesda: AABB Press.

Roback JR, Combs MR, Grossman BJ, Hillyer CD (eds). (2008). *Technical Manual*, 16th edition, Bethesda: AABB Press.

# Index